Man And The Land

A Cultural Geography

THIRD EDITION

GEORGE F. CARTER

TEXAS A & M UNIVERSITY

HOLT, RINEHART AND WINSTON, INC. New York Chicago San Francisco
Atlanta Dallas Montreal
Toronto London Sydney

Copyright © 1964, 1968, 1975 by Holt, Rinehart and Winston, Inc.
All rights reserved

Library of Congress Cataloging in Publication Data

Carter, George Francis, 1912-
    Man and the land.

    Includes bibliographies.
    1. Geography—Text-books—1945-   2. Anthropo-
geography.   I. Title.
G126.C274   1975   910'.03   74-14897
ISBN 0-03-003426-4

Printed in the United States of America
5   6   7   8   071   9   8   7   6   5   4   3   2   1

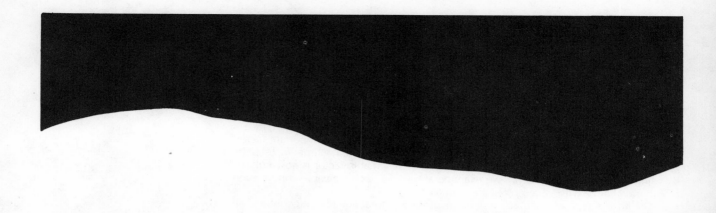

Preface

OUR great continuing interest in geography has been learning how people live in the different parts of the world. Ancient geographers believed that the places people live in actually make the people different; these geographers thought that climate is the strongest causative agent in the physical environment. This general view is best described as physical environmental determinism, and most people still accept it as an axiom.

I disagree with this view wholeheartedly and have written this book with the central theme that inquiry leads to the real causes of the differences in ways of life—a process that disproves the physical-environmental-determinist theory. Such inquiry serves to give the book a focus, a continuing thread that holds together a worldwide survey with excursions into great time depth. Too often texts become attempts to supply the vast factual background of a subject rather than develop the reader's ability to deal with a particular class of data. We could teach factual descriptive geography forever, but with this method the student is likely to know only the specifics he has learned. Yet when principles and processes are isolated and shown at work in real situations, they become the key to understanding the dynamic nature of the existing scene.

The difficulty here is that the role of principles and processes tends to be obscured by the numerous interactions between man and the land, for the varied races of man, with their differing cultural-historical backgrounds distributed over a diversified physical earth, create a most complex problem. The method used in this text reduces this complexity so that the principal factors—man, land, and culture—may be seen clearly at work. This makes it necessary to study one factor at a time, observing what happens when it varies while the other factors are held constant. As we begin, race is a thorny problem which, when examined, seems likely to be an unimportant factor. Land is then held constant by dealing with just one kind of land at a time: first the dry lands, then the humid tropical lands, and so forth. The variable factor is now culture. By using considerable time depth we examine almost every conceivable combination of cultural change with varied races and climates.

With this type of approach it is not necessary to treat equally all countries, all parts of the world, or all of time. Accordingly the extent to which some areas, climatic types, and nations are treated varies throughout the text. Our interest is less in nations and regions than in observing processes at work. For this purpose one area is best for exemplifying sequent occupance, another for showing the effects of isolation, and still another for illustrating the problem of have and have-not nations in terms of basic resources.

This book aims primarily to develop a way of *thinking* about man's relation to the land, and it is principle and process oriented. Although the main objective is to develop an understanding of how the present situation came about, sufficient earth description is included to give an overall view of the actual earth and the men on it. I hope that the student will gain from the book the ability to judge correctly in the weighting of race, physical environment, and cultural-historical factors in the development of the humanly occupied earth as we now see it and thus be better able to judge the probable direction and rate of change in differing areas with differing peoples and differing cultures.

This book, in use for 10 years, is now in its third edition. The second edition was extensively rewritten throughout. This edition has been brought up to date, especially in statistics, and in response to numerous requests I have added a section on the Soviet Union.

Man and the Land was written as an introduction to geography—all of geography. Many schools use it as their introductory text and find that students are motivated to want to know more about regional geography, soils, vegetation, and geomorphology. Once the student has seen how the different fields of knowledge help answer the fundamental questions, he will understand why he is asked to study particular subjects. He is *not* told "yours not to question why, yours but to learn or die." On the contrary, this text should leave him with an acute desire to know more. It has been our experience that students who have had this course look forward to specialized courses whereby they can deepen their understanding of the lands and people and processes of interaction developed here.

I am particularly grateful to those reviewers who provided suggestions for this edition. They are C. M. Dupier, Jr., Cumberland College; Elliot G. McIntire, California State University, Northridge; Douglas M. Orr, Jr., University of North Carolina, Charlotte; Charles A. Stansfield, Jr., Glassboro State College; Elizabeth Strasser, Trenton State College; and H. Christian Thorup, Cuesta College.

College Station, Texas   GEORGE F. CARTER
October 1974

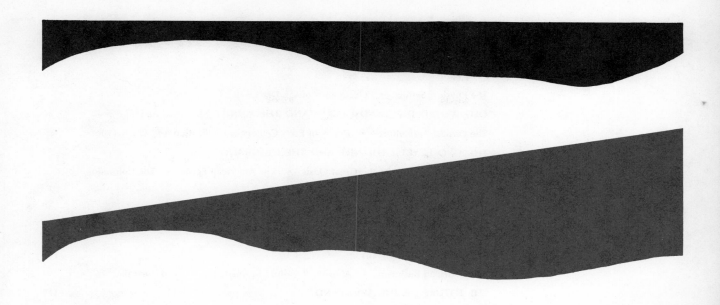

Contents

Maps

World Maps 514

Man And The Land

Introduction

The Nature of Geography

Geography holds in common with the other sciences both its content and its method. It distinguishes itself from other scientific disciplines by its approach—by what it asks the scientific methods to reveal about phenomena. Geography is a catalogue of questions, and the questions—not the phenomena, not the facts, not the method—are geographic.[1]

Geography is primarily concerned with the relation between men's ways of life and the places in which they live. This, of course, requires a familiarity with place location, for understanding and communication demand that we have certain basic types of knowledge. No one can be an adequate citizen in a late twentieth-century society without knowing the location of such places as Vietnam, Cuba, Ghana, China, and the Soviet Union, and this requires at least some familiarity with maps. Beyond that, there is in geography today a field of place-location theory that works on mathematical terms and deals with explanations for spatial distributions in rigorously scientific fashion. It is as far from grade school place-name learning as arithmetic is from calculus.

Furthermore, there are other highly developed fields in geography. Some people specialize in political geography, others in economic geography, and still

[1] F. Lukerman, "Geography as a Formal Intellectual Discipline and the Way in Which It Contributes to Human Knowledge," *Canadian Geographer*, vol. VIII, no. 4 (1964), pp. 167–172.

2

Chequeres in the Highlands of Peru near Pisac. *Man lives on a physical earth that varies naturally from place to place in climate, soils, and landform. Pursuing different ways of life in different areas, man modifies this earth as he brings to it such things as roads, farms, cities, and dams. Traditionally the geographer considered the distributions of man's ways of life on this earth and the nature of the relationship between the varying physical earth and mankind's varying ways of life. (Peruvian Transport Corporation Limited for World Bank)*

others in one aspect of physical geography (landforms, soils, plants, climates). While geography retains maps as a symbol of its interest in location, it goes far beyond simple names and places. Yet just as we cannot approach Milton or Shakespeare without having learned to read, so we cannot begin in geography without a knowledge of the earth and the location of the people and places on it.

Of course, maps reveal far more than simple place locations. A place can gain importance from its relationship to other places, and this is most easily seen on a map. Vietnam is close to China; Cuba is only 90 miles from the United States. Maps can show the distribution of the hot, wet climates of the earth and the distribution of underdeveloped nations. There is a startling correlation. All of tropical America, Africa, Southeast Asia, and Melanesia are underdeveloped. Who can stop here? Are we not compelled to ask: Is there a causal correlation between heat and humidity and lack of development? The distribution of the dark-skinned races of the world is centered in the tropics, and until late in the history of mankind, no light-skinned people were in the tropics. Why is this so? Is there a connection between climate and skin color? Climate and race? Climate, race, and accomplishment?

When we look at maps of food production in the United States, one area stands out for its production of corn and hogs. Is this because the soil and climate of the strip from Indiana to Iowa is particularly favorable to corn, and is corn the ideal food for hogs? Soil and climate maps reveal that there are some strong soil, climate, and corn correlations. But correlations are always suspect. A few years ago a series of maps was published showing an astonishingly high correlation in the United States between apple growing and voting Republican. The correlation was excellent, but there was no causal relationship. So for the corn belt we must ask: Is the correlation causal? Or is it a historical accident? Or is it some combination of history and physical geography?

We can ask other questions. Could other crops be grown there? Do people in other parts of the world grow corn and hogs in similar soil and climate situations? Or, for that matter, is there another situation like this? We will quickly find out by consulting maps. Curiously, the American corn-hog belt is unique. Similar climates are found elsewhere, but the American development is nowhere duplicated.

This finding leads to certain questions: How long has this peculiar concentration been there? Not long, the anthropologist replies, for the Indians had no hogs, and we developed the area only about a hundred years ago. Will it always remain there? Not necessarily, the economist and agronomist tell us. Thus a relatively

simple piece of geography can provoke a flood of questions the answers to which must come from an examination of the soil, climate, location factors, and history associated with the economics of the corn belt.

But what makes it geography? Corn can be studied botanically, agronomically, or economically. In fact, the economics of raising corn and hogs is so interlocked that there is a corn-hog ratio, used to express the relationship between the price of corn and the price of corn-fed pork, which tells the farmer whether it is more advantageous to sell corn as grain or feed it to a hog to sell as pork. Is this not economics? The economist is seldom interested, however, in the fact that the Indian way of life on this same land was utterly different and that in other lands with similar climates people follow different ways of life. But for the people known as geographers, these have always been intriguing problems. For as far back as our records run, there are men who have been trying to answer these questions, and such men have always been called geographers.

As we can see, geographers have an enormous task, for they are crossing the boundaries of numerous academic disciplines. Is geography then just an *omnium gatherum*, a bringing together of bits and pieces from everywhere? Is it the geographer's task to make a grand synthesis of knowledge about one area of the earth such as the corn belt and show how society has worked out a unique response to the physical earth at this particular place? There are some who think so, and surely the knowledge of various aspects of regions of the earth should be drawn together, even if there is some question about the feasibility or desirability of putting on paper all data about any area.

This type of study has been called regional geography, and bringing together various aspects of knowledge is called *holism*. A watch is more than gears, springs, and dials: the working assemblage is a watch. So too the corn belt is more than soil, climate, plants, animals, and man. It is the whole assemblage functioning there right now. It has a history, a present, and a future of some sort. We are almost tempted to treat it as a living thing—an organism.

Unfortunately, assembling all the data about an area, any area, tends to produce a catalogue, which is no more interesting than a list of place locations. Life drains out of the subject unless there is a unifying theme such as determinism, "landscape," or ecological relationship. Moreover, few geographers can handle all subject matter with equal expertise. If the geographer tries to cover everything, he must settle for knowing less in detail about all that he describes than the regional experts who supply the data. He tends to trade expertise for general knowledge, defending himself by the holistic argument. There *is* value in seeing whole systems. It is important to see economics at work on a landscape among a people who have a particular social structure and value system which they inherited from their past and adapted to their present solution. But there are also dangers. Notice this caution from one of the master geographers of this country:

I do not accept the idea that anyone can do the geography of a region or a comparative geography, when he knows less about anything he assembles than others do.—The ineptly named holistic doctrine leaves me unmoved: it has produced compilations where we have needed inquiry.[2]

We will accept Sauer's dictum and stress inquiry. Description we must have, for thinking cannot be done in a vacuum, but the real aim will be inquiry—exploration of the meaning of the facts.

In reading through the many discussions of the nature and history of geography, we are struck by several things. First, there has always been an interest in people and places. Man is interested in the rest of mankind. However, no one seems able to describe how other people live and what their lands are like without attempting to explain why they live in a different way in their different lands. Does variation in land (climate, soils, and so forth) make man physically different and also cause a variety of economic activities, beliefs, and achievements? These questions are the persistent theme of geography. Traditionally, people expect geographers to know where things are, which people live where, what they are like physically and culturally (their way of life), and ultimately why they live the way they do—an ancient geographical question which will be the focus of this book. The other focal points in geography have their fascination and value. No more is claimed for the study of geography presented here than the interest in man and his ways of life in different places—one of the ancient and continuing traditions. Because of this emphasis on man, it is often called human geography.

Human Geography:
Then and Now

From earliest times men have known that men living in other lands were different. Since the lands were also different, a natural conclusion was that the difference

[2] C. O. Sauer, "Education of a Geographer," *Annals of the Association of American Geographers*, vol. XXXXVI (1956), pp. 287–299.

in land made the difference in people. Among a people as learned as the Greeks we find such generalizations as the following: Men of favorable, gentle climates are easygoing and generous, while people of poorer environments are more active and penurius. Hippocrates thus characterized Europeans versus Asiatics. Aristotle called the north Europeans brave but thoughtless, unskilled, and lacking in political organization and hence unable to rule their neighbors.

Herodotus (born around 490 B.C.) gives us a glimpse of general knowledge in late Greek times. History and geography, sciences, and humanities were not then sharply subdivided, and Herodotus gathered all kinds of information about people and places and tried to account for what he learned. He thought the north Europeans lived in a land too cold for worthwhile mental activity. He concluded that the Africans south of Egypt had failed to advance because the climate was too hot. He heard that the Nile rose in high mountains with immense rainfall, but he doubted the story because everyone knew that the farther one went up the Nile in Egypt, the hotter and drier it got. He recorded and believed that the ants in India were as large as foxes and the natives gathered gold nuggets from ant heaps. He also recorded that these natives had a kind of wool that grew on trees, an obvious reference to cotton. Out of these compounds of fact, fiction, and preconceived notions the geographers spun their first explanations for the different ways of life on the earth. Their initial conclusion that the land (weather, climate, landforms, and vegetation) dominated man flourished for a long time.

In the sixteenth century, scholars believed that northerners were brutal, cruel, and enterprising; people of temperate lands were described as talented, energetic, and capable of command; tropical people were thought to be vengeful and cunning but gifted in judgment of what is true or false. Men were so convinced of these differences, which by now were attributed primarily to climate, that they advocated adjusting the laws of the various lands to fit the national characteristics. In the second half of the sixteenth century it was thought that a monarchy was the natural form of government for productive countries, a republic was ideal for poorer ones, and a democracy was suited for island societies. These generalizations came less frequently from geographers than from historians and political scientists.

In the nineteenth century, physical and human geography began to drift apart. The human geographers continued to stress man and his work. They described the land, the people, and their history. Some geographers stressed the history of the people and avoided drawing facile conclusions about climate and human achievement, but others attributed national character to the physical geography of the nation, just as the Greeks and medieval scholars had done. At this time Alexander von Humboldt[3] described both physical and cultural landscapes, carefully weighted the interrelationships, and generally avoided shallow conclusions. For example, he declined to accept clear desert air as a natural cause of Middle Eastern astronomy. If this were so, he asked, why is there not a like development of astronomy in all or at least some of the other dry lands that have clear skies? Here a relatively new element entered geographical theory. By now men knew enough about the whole earth to begin to compare similar lands to see if the same physical settings produced similar results. Von Humboldt, with his immense learning, was already beginning to glimpse both positive and negative climate and culture correlations. Not all deserts produced astronomers! He also saw that various physical environments in America produced Asiatic arts and governmental structures, and he attributed this to Asiatic influence in pre-Columbian America (before the time of Columbus, or before 1500), an interesting idea for a man to put forward in 1800.

In the mid-nineteenth century, we entered what has been called the age of Darwin. Evolution and the dominance of natural forces in shaping all living things overwhelmed our thinking. The field of ecology was born: the study of the interrelationships among the life forms in a given habitat, not only with one another but with their physical environments. In the insistence on the complexity of all life systems, with their almost limitless relationships between organism and environment, organism and organism, organically altered environment and other organisms, the ecologist was moving toward a holistic approach and coming closer to reality.

The ecological approach was brought to America and flourished, especially at the University of Chicago, early in the twentieth century. For a time, geographers, climatologists, botanists, soil scientists, and others worked together and emphasized the importance of the physical environment. Ellen Semple (1863–1932), an American who had studied geography in Europe, exemplifies the thinking of this time. Her view was that nature molded man:

Man is a product of the earth's surface. This means not merely that he is a child of the earth, dust of her dust; but that the earth has mothered him, fed him, set him tasks, directed his thoughts, confronted him with difficulties that have strengthened his body and sharpened his wits, given him problems of navigation or irrigation, and at the same time whispered hints for their solution.[4]

Miss Semple, who was a master of poetic writing, as this sample shows, ascribed to the earth a major role in shaping man and his thoughts. She extended this theory to such wide-ranging statements as: Deserts give birth to religious geniuses. Mountain dwellers are conservative. Climate makes the north Europeans energetic, provident, serious, and thoughtful rather than emotional; they are cautious rather than impulsive. People living in the genial climate of the Mediterranean, on the other hand, are easygoing, improvident, gay, emotional, and imaginative. In the still warmer climates of Africa the endearing qualities of the Mediterraneans "degenerate into grave racial faults." Attributing to the physical environment the major role in shaping man and his culture is called physical-environmental determinism, and it has a long history in geography.

No geographer has had more influence on American thinking about the relationship of the environment to man than Ellsworth Huntington, and few have more strongly advocated the dominance of the physical environment. In the stream of thought he is a product of the Darwinian influence, and his work exemplified the search for physical-environmental laws underlying human behavior. His numerous, highly readable books have influenced countless teachers, who in turn have communicated his ideas to multitudes of students. Huntington was a New Englander, and he spent his academic life at Yale. Early in his professional life he was enormously impressed by the ruins of great cities in Inner Asia which he saw as a member of the Pumpelly expedition. In keeping with the theory of his time, he attributed these ruins to climatic rather than political or economic changes, and he spent the rest of his very active life extending this theme to the study of most of the world.

He believed that there are optimum climates for physical work and for mental work. New England's climate (or its equivalent elsewhere) was considered the optimum for mental work, and on this basis he explained Great Britain's dominance in the world and New England's influence in America. Tropical climates

---

[3] In 1800 Von Humboldt traveled widely in America, making observations on geology, climatology, botany, archeology, economics, and politics. On his return to Europe he wrote extensively about his observations. His influence on both science and the humanities was enormous.

[4] Ellen Semple, *Influences of Geographic Environment* (New York: Henry Holt and Company, Inc., 1911).

he considered to be mentally deadening because of the monotonous temperatures, high heat, and humidity. Huntington thought of climate as the decisive influence on man. One of his themes was that the climatic changes in postglacial time (the last 10,000 years) were sufficient to account for some of the sweeping changes in history. Thus he saw the great Mongol invasions as a result of alternating cycles of arid and humid conditions in Inner Asia. He sought correlations between season of birth and individual achievement, between climate and national achievement, between the ozone content of the air on a given day and mental productivity, and so forth. In general he found a correlation between season of birth and achievement, between cool mid-latitude climate and national greatness, and between the higher ozone content of higher latitudes and the levels of civilization there as opposed to the lower levels in tropical lands.

Countercurrents have developed in the stream of geographical thought, especially in the twentieth century. In keeping with the growth of scientific and specialized knowledge, all subjects tend to be subdivided. Geography has been divided into human and physical. Human geography in turn has split into regional descriptions and analytical studies. The studies in physical geography have contributed increasingly to our understanding of the physical environment. Human geographers have struggled with the problem of how to assess the role of the physical environment in the development of ways of life. There was little doubt in earlier times: the way of life was determined by the physical geography, and man himself, including his mental ability, was formed by the physical environment. The Darwinian revolution reinforced this view, for it emphasized the lawfulness of nature. Although ecology tended to stress the complexity of interactions between man and the environment, it nonetheless supported the dominant role of the physical environment.

By the twentieth century doubts had arisen in various places. In France the possibilists stressed that man had free will and could choose in any environment, if not quite freely, then at least among a number of possibilities. In a jungle one can gather wild fruit or cultivate the trees or cut down the trees and plant rice or build a trading city such as Singapore. The emphasis was shifted from a man enclosed in a physical environment as confining as a strait jacket to a man free to pick and choose to some extent. This school of thought did not deny that some restrictions existed in the physical environment, but it shifted the center of interest and the decision-making process from the physical environment to man. Man was not viewed as a passive being remorselessly molded by the physical environ-

ment but free to choose between various possibilities and at times able to act directly on the physical environment to make it more to his liking. No idea is entirely new, of course. George Perkins Marsh wrote vigorously in the 1860s that man had changed the physical environment of the Mediterranean far more than the environment had changed man, but he was almost a century ahead of his time.

In mid-twentieth century Griffith Taylor, a geographer from Australia, where the arid lands place formidable limits on man's use of the land, proposed a stop-and-go determinism with nature operating the traffic lights. He believed that there is a natural best use for each region and the role of the geographer is to find this best use. The criticism of this view has been that best use is a matter of opinion and depends on the background of the people in a given physical environment. In other words, the proper place to begin to study the way of life of any people is with the people themselves, for it is only when we understand the outlook they have brought to their particular piece of physical geography that we can understand the decisions they have made. This has been the theme of the historical geographers, and it is the reason for their insistence on deep historical studies. They claim that people base their choices on ideas from the depths of their past. Possibilities exist in men's minds, and it is only as men perceive the possibilities that they actually exist. Today many human geographers are avidly pursuing the question of perception: how man sees his environment.

In this book we are concerned especially with human geography. There are two types of human, or cultural, geography that differ somewhat in the questions they ask. The proponents of the behavioral school ask about individual reactions. This approach allows them to deal with more detailed questions and satisfies the urge to know more precisely why certain things are done by specific people at definite times. The method is largely based on questionnaires directed at living people. It results in direct answers, but it is limited to the present, for we cannot ask the Egyptians of the first dynasty just why they did what they did. Another approach, the one used here, is that of the cultural historical geographer. Because we cannot determine just which individual performed a certain act, this approach deals with explanations based on facts not directly attached to individual persons. We know, for example, that agriculture was invented though we cannot say by whom. As this text will show, however, we still can learn a great deal about inventions and the circumstances under which they arise, the impact they have on the land, and the way in which people react to them. The differences between these two approaches are very like those between psychology and anthropology. Psychologists focus on individuals, anthropologists on whole cultures. The sociologist occupies a middle ground, paying most attention to groups and classes of peoples. The boundaries are arbitrary because one field blends into the other, but each field has a contribution to make. The approach taken here can well be called cultural historical anthropogeography since it tends to follow the approaches common to those fields. Anthropogeography is a term coined in the nineteenth century to recognize the closeness of these two fields. Unfortunately, it became identified with the physical-environmental-deterministic school of thought and subsequently went out of favor.

There are classic schools of thought in deterministic human geography which developed in the early twentieth century. Lothrop Stoddard, who wrote *The Passing of the Great Race,* is perhaps the outstanding advocate of the master race theory, a concept cynically and brutally applied by Adolf Hitler. The theory of Halford J. Mackinder, a British geographer, was an interesting opposite. He felt that whoever controlled the "World Island," the great Eurasian land mass, would inevitably rule the world. Great Britain's position, therefore, according to Mackinder's "geopolitics," should be to always make sure that no one power acquires control of Eurasia. What kind of land and what kind of people are on this World Island? Is mere space valuable? Is it significant that vast areas of Asia are arid or frozen and that the populous and productive areas are almost entirely on the fringes? In one sense, the answers did not matter. Mackinder's ideas permeated European political thought—British and Russian thought particularly. The Soviet Union's long-range strategic plan—wherein first Asia, then Africa and Europe are controlled, followed by Latin America and finally, in its isolated position, the United States—can also be said to follow this world view. When the Communists gained control of China, it looked as if the plan was progressing, but the Chinese decided to go their own way. The point is that geographic ideas have consequences, and most interestingly, they may play roles in the thinking of any political system, be it nineteenth-century capitalistic Great Britain or twentieth-century Communist Russia.

A differing view of the world was developed by Alfred T. Mahan, an American naval historian prominent in the late 1800s. His theme was that whoever controls the seas and their strategic crossroads controls the world. Such ideas as this motivated Great Britain to become the world's greatest sea power and seize control of the crossroads of the world seaways: islands down the length of the Mediterranean, Egypt and the Suez Canal, a line of holdings out into the Indian Ocean, and Singapore—all of these to defend the lifeline to India. How important are the seaways today? One must conclude that they are immensely important. No major industrial nation today could exist if the seas were closed. Is it necessary or desirable that one nation control the seas, as Great Britain did for nearly one hundred years? Would it be better if no one ruled the seas as is now the case? These are political geographical questions, and they will be given no more than passing mention here. For the present purpose, it is enough to call attention to the power and importance of geographic ideas—whether they are true or false—because they influence perceptions, decisions, and actions.

Still a third powerful set of notions involves air power. Alexander P. de Seversky's theory maintains that air power is the dominant control and the polar skies hold the key to world domination. If you were to look at a globe, you would see that the land of the world is arranged around the North Pole. Most of the land, most of the people, and a huge proportion of the industrial development of the world is concentrated in the Northern Hemisphere roughly between 30 and 60 degrees north latitude. With the advent of intercontinental ballistic missiles, however, particular location is no longer so important. De Seversky's insights into the importance of air warfare were nonetheless prophetic, and they influenced the geopolitical views of the great powers.

All of the ideas mentioned here have some factual basis. All of them are strongly based on some particular view of the physical geography of the earth. All of them have been influential in the grand strategies of nations. Should we say that it is physical geography that is decisive? Or should we say that it is man's *view* of his physical geography that is decisive? Notice the role that technology plays in these views. The theory of sea power and the concept of the World Island are more or less contemporaneous, and one is in part the check on the other. Air power introduced a whole new dimension, intercontinental missiles still another. And looming over all of this is the nuclear horror that makes us devoutly hope that we will not launch an atomic war. Any sound approach to geography, then, has to include various perspectives. It must view the actual earth, the earth as seen by man however erroneously, and finally come to deal with an incredibly complex set of notions—the actual, the perceived, and the possible future situation in its actual and perceptual combination that will be the basis for future actions. And it will not be even this simple, for there is no universal world view. Differing nations hold differing views at differing times and act accordingly.

We shall pay a great deal of attention to what

has happened in history. By seeing how and why men have acted in the past we will gain insights into what men are now doing and may do in the future. Are we momentarily hung up on a World Island view? How would you know if you did not know about the real earth and the real men making decisions about that earth? Our study will require a synthesis of anthropology and geography: hence, anthropogeography. No attempt to cover all modern lines of inquiry will be made here. References to thorough studies of the history of geography will be found at the end of the chapter. Most of these have been written by geographers for geographers, but what I have written here is meant for students who know little geography and need only an introduction. Most of them will view the world as being vaguely determined by the physical environment and will feel that this is *the* approach in geography. This discussion should indicate that the popularity of this ancient geographical theme is now at low ebb. Today the simplistic, physical-environmental determinism of Greek, Roman, medieval, and even nineteenth-century times is no longer an active school of thought. Echoes of it are still heard, but more often in primary and secondary schools than in universities, and there more often outside geography than in it. Only a few geographers today think that national character is formed primarily by climate or mountains or plains. The discussion today is much more concerned with how men make their choices. We realize now that this depends on the particular people's view of their situation and their perception of their environment, which can be thought of as an enormous complex of physical and cultural facts.

These, then, are the themes that run through human geography. They lead to the questions we will pursue: How dominant is the physical environment? How free is man in his choosing? What are the processes at work in man and in the physical world in which he lives? What is the nature of the action and interaction between these two sets of forces, the one humanistic and social, the other physical and mechanical?

Methodology

Historical models

Part of our problem is finding a way to learn the facts. We would like to know exactly how heat and cold, isolation and crowding, access to ideas and exposure to great minds affect individuals, groups, and nations. We would like, of course, only to approach these problems scientifically and experimentally, but there are limits to our ability to manipulate men and environment. We are not free to take, for example, 10,000 Irish men and women and place them in a hot, humid environment for 500 years to see what would happen. Ideally, we would take similar groups and put them in various physical environments (hot humid, hot dry, cold wet, cold dry, mountainous, plain, isolated, centrally located, and so forth). In this case we would be holding race constant while we varied the environment. It would be exciting to try this with different races:[5] Mongoloids, Negroids, Europoids. We would be following the techniques of the physical sciences, where the usual approach is to hold constant all factors but one to make it possible to study that particular one. Thus in the gas laws one learns that if all other things are held constant, temperature varies with pressure. In biology great effort is expended to hold all things equal while one factor is varied. From studies of this type we have learned that a millionth part of certain minerals, like copper, are vital to man's health. However, we cannot experiment freely with human beings, at least not in societies where the human being is considered to have inalienable rights. Another part of the problem is that we need knowledge now and cannot wait for centuries to see what the effects might be of moving men around the world.

On the other hand, we need not wait for the passage of centuries, for our records allow us to look back not only centuries but millennia. History can give us cases of people of every race, placed in nearly every climatic and landform type, in virtually every cultural situation from isolation to crossroads position. The further back we seek data the dimmer our view will be, and yet we can learn much about a 5000-year span,

[5] Race classification is a difficult problem. Race is a biological term and should be used to designate a group of genetically similar people. It should not be used for nationalities or religious groups. There is a Jewish religion, but not a Jewish race, for Negroes and Mongols may be members of the Jewish faith. Similarly, there is no German race, but a German nation that includes representatives of many races. How many races we name depends on how much detail we wish to include and what authority we wish to follow. In the broadest sense there are three main groups of men: the Negroid, the Mongoloid, and the Caucasoid. This grand division lumps many intermediate peoples into the overall classification—all the American Indians with the Mongoloids; all dark-skinned people of the southwest Pacific (Melanesians, Papuans, Australian aborigines) with the Negroids; all the white to dark-brown people of northwest Europe, the Middle East, and North Africa with the Caucasoids—but it has the great merit of simplicity. Race will be further discussed in Chapter 1.

something about a 50,000-year period, and a little about two million years ago. Man or man's immediate ancestors occupied much of the earth over this time span. To what extent did the earth shape him? To what extent did he shape the earth? At what period did the earth most shape man, and when did man most shape the earth? Is man to be dealt with like any other biological entity? Or is he so different that the usual biological-physical analyses must be greatly modified when considering his cultural growth and development? How do these questions about man affect our conclusions about geography as a type of human ecology? There are many questions. We must isolate some of the variables and hold them as constant as possible while we study the independent variable.

Climatic-regional comparisons

We can control the physical-environment variable by limiting our study to one type of environment at a time. Thus we can study all the hot wet regions of the world first. They will be alike in temperature and rainfall. Because these factors are dominant in controlling vegetation and through vegetation the soil, these regions can be physically very much alike. There will be variables, of course, in landforms (plains versus plateaus), in oceanic versus continental location, in proximity to civilization or isolation, to mention a few. The early geographers thought that one could generalize about tropical people compared with mid-latitude people. We can narrow this down by dealing separately with the wet tropics and the dry tropics, and we can examine first whether the peoples of the wet tropics are or are not alike. Alike in what? Biology (race), psychology, cultural achievement, economy?

After a discussion of the tropical regions, the method can be extended. Some climatic areas are most conveniently similar. On the west coasts of continents in higher mid-latitudes we find remarkably similar climate, landforms, vegetation, and soils. All these areas have fisheries off their coasts. All were glaciated in the recent geological past. Norway, British Columbia, the Panhandle of Alaska, southern Chile, the southern part of the west side of the south island of New Zealand are examples. If the physical environment is dominant in shaping mankind in mind, body, character, and achievement, should not the people in these areas be alike? If they are not, we must conclude that something other than physical environment is forming man and his way of life. This alternative does not exclude the physical environment as a causative agent in determining the ways of life of mankind on this earth, but it reduces the importance attributed to physical environ-

ment from the time of antiquity through the nineteenth century.

What other variables might there be? We can ask if the people in these areas are now physically alike. Northwest Europeans, American Indians, Mediterraneans, and Polynesians are or were in mid-latitude west coast regions, and these are biologically distinct peoples. Perhaps the cultural differences arising in these like physical environments are due to biological (racial) differences? If this is the answer, the similar physical environments seem not to have made man similar either biologically or culturally.

What is at work here then? Location? History? Some complex of land, man, and history?

This is the Gordian knot that we are going to try to cut. The aim is to develop an ability to cope with this problem, which has fascinated men from earliest times, as shown by the first written records, and still intrigues both student and scholar. It will be necessary to deal with physical earth science (climate, soils, landform, oceans, rivers), biological science (physical anthropology, aspects of biology and botany), and the humanities (history, anthropology, art). The material will be diverse yet unified because it is all focused on the relationship of man to the land.

The presentation in this book will feature case studies to illustrate the operation of similar process in differing lands and differing processes in similar lands. The use of specific cases will keep the discussion concrete, while the study of processes will allow the development of principles. If we can master the few principles of a subject, we can apply them to innumerable problems and situations. In this sense, geography is no different from mathematics: once the principle of division is learned, all numbers are divisible. If we can find the principles underlying the relationship of man to land, our understanding of the world we live in will be comparably advanced. Since the man-land system is vastly more complex than arithmetic systems, the task will not be easy. But if we consider one factor at a time, we can go far toward understanding the processes that underlie the man-land relationship.

The goal is the development of competence in thinking about the very complex interplay of physical, biological and humanistic knowledge. This broad, cross-discipline, nonspecialized approach to knowledge is at a low ebb today. Even the humanistic studies call themselves social sciences and often become narrowly specialized. At the present we are placing great emphasis on science and specialization. There are enormous values to be so gained, but liberal arts and a general education also have values. These values were beautifully summed up by Aristotle about 2000 years ago.

Note carefully what this great early philosopher has to say. The overall aim of general education is judgment. From this course you should emerge with a sound sense of how to judge good and bad reasoning in the exceedingly complex fields of physical environment, race, and culture. While facts are important, they are only a means to an end. They must be used correctly if we are to make sound judgments. Consider Aristotle:

There are, as it seems, two ways in which a person may be competent in respect to any study or investigation, whether it be a noble one or a humble one. He may have either what can rightly be called a scientific knowledge of the subject; or he may have what is roughly described as an educated person's competence, and therefore be able to judge correctly which parts of an exposition are satisfactory and which are not. That, in fact, is the sort of person we take the man of general education to be; his "education" consists in the ability to do this.[6]

[6] Aristotle, *Parts of Animals*, Bk. I, trans. A. L. Peck (Cambridge, Mass.: Harvard University Press, 1937), p. 53.

Review Questions

1. Are geographers the only people who use maps? Can geography be defined as the study of maps? Of places?
2. Why is it that a positive correlation between the distribution of two phenomena does not automatically prove a causal connection? Give examples of correlations between items that are causally connected; between items not causally connected.
3. What does the term anthropogeography imply? How would psychological and sociological geography differ from anthropogeography?
4. Why should geographers have been less deterministic than historians and political scientists even in the eighteenth and nineteenth centuries?
5. Describe three conflicting deterministic views of the political geography of Eurasia. Which one, if any, is correct? If they were all incorrect, would this mean that they were unimportant?
6. Since complex problems with several unknowns are hard to solve in mathematics, science, or social science, how can the extremely complex problem of the geographer be simplified? What solution is used in this book?

Suggested Readings

Broek, J. O. M. *Geography: Its Scope and Spirit*. Social Science Seminar Series. Columbus, Ohio: Charles E. Merrill Books, 1965.

Hartschorne, Richard. *Perspective on the Nature of Geography*. Association of American Geographers, Monography Series. Chicago: Rand McNally & Company, 1959.

James, Preston. *All Possible Worlds*. Indianapolis: Odyssey Press, 1972.

Taylor, T. Griffith, ed. *Geography in the Twentieth Century: A Study of Growth, Field Techniques, Aims and Trends*. New York: Philosophical Library, 1957.

Woolridge, Sidney W., and East, W. Gordon. *Spirit and Purpose of Geography*. London: Hutchinson University Library (Geography), 1951.

The Origin of Man
and the Problem of Race

CHAPTER ONE

THE study of man and the land deals with two complex variables. On the simplest level of comparison, there are various climates and distinctive races, and the races of men live in differing climates in different ways. Is there a cause-and-effect relationship between the type of man, the place he lives, and the way he lives? If so, is it a climatic cause or some other? To deal with this question adequately, we must take one part at a time, and we will begin by developing an understanding of race.

Mankind ranges in skin color from nearly black to nearly white, with the majority in mid-range—some shade of brown. Types of hair vary from the tight tufts of the South African Bushmen to the long straight hair of the Mongols of Inner Asia. Much of mankind falls in between with wavy to curly hair. The races of men vary in average size from less than 5 feet for males of some races to over 6 feet for others, a significant variation, and especially striking when individuals from the extremes are seen side by side.

It is not, however, body size or skin color or hair form that really intrigues us but such questions as: How did the variations in these characteristics arise, what do they tell us about human origins, and particularly, what do they tell us about qualities that are more than skin deep? Are the races of mankind different in mind as well as hair, in temperament as well as body? Some maintain that certain races are superior, while others insist that all are equal. How can we decide between these views? Is it important for human geography? We must conclude that it is extremely important whether the advocates of equality or inequality

Prehistoric Man. *The change in man over the past nearly one million years is shown here in reconstructions of three kinds of men. Pithecanthropus (left) is now placed in our genus and called* Homo erectus; *this type of man spread from Gibraltar to Java about the time of the early glacial period. Neanderthal man (center) dates back at least 50,000 years and is now placed in our species,* Homo sapiens. *Cro-Magnon man (right), best known from France where men of this type did wonderful cave paintings and fine work in bone and stone, is already racially a European. New finds are extending the length of time that sapiens type of men have been in existence, and it is likely that the Neanderthal and Cro-Magnon types, both now classified as* Homo sapiens, *lived contemporaneously and are best viewed as among the races of mankind. (Courtesy of The American Museum of Natural History)*

in ability are right. It is imperative that we are able to think correctly on this topic.

For a study of the relationship between man and the land, our work would be enormously simplified if we could deal with man as a unit rather than as a multivariant item—as if the races of mankind were relatively uniform in ability. We must be careful not to bend the data to suit either our convenience or humanitarian or religious notions—easier said than done. It is important to move as far as possible from the here and now, where our emotions are inevitably involved, to develop a line of reasoning based on the facts. We shall try to gain an understanding of the processes that have made men physically different and then see if a similar process analysis will lead to an expectation of mental difference.

ADAPTATION, PHYSIOLOGY, CLIMATIC IMPACTS

The Origin of Man

We can make several simple observations about man's origins. First, from man's physical makeup it is clear that he is not adapted to an aquatic life, and any relationship to that environment is so remote as to be unimportant for this discussion. Neither is he fur covered and thus adapted to cold. Nor is he adapted to aridity like the camel or the kangaroo mouse, who drinks not at all. He is clearly best suited to tropical and near-tropical humid conditions. It is quite significant that unclothed mankind relaxes in heat and humidity but perishes in even moderate cold. Man's present spread into cold climates was accomplished in spite of his physiology and was possible only because of his invention of such things as clothing, housing, and the use of fire. Stripped of cultural equipment, man proves to be suited to warmth; therefore we must look to the tropical parts of the world for his ultimate origins.

That man's physical being is clearly related to the animal world is easily demonstrated. In common with most of the animals of the world man uses iron as the basis of the oxygen exchange in his blood system. (Shellfish use copper.) There are scientific tests that show man's relationship to animals. We can prepare serums and compare the degree of reaction of the human body and the animal body to the same serum. When the serum is made so that the closer the relationship between man and animal the greater the reaction,

it has been proved that apes and monkeys react violently, horses only mildly, snakes hardly at all. Blood can tell much more. Blood types in man, in simplest form, are O, A, and B. These are also found in chimpanzees and gorillas of Africa, creatures that look a little like man, are anatomically much like him, and even sometimes behave as man does. This relationship can be overstressed. Men are not merely latter-day chimpanzees, for they share some details of musculature with certain monkeys of the Old World (Eurasia and Africa) rather than with chimpanzees or gorillas, which are in most measures man's nearest animal relatives. Further, man's mind is so organized that speech, tool using, and abstract thought are possible to a degree never seen in his biological relatives. Perhaps the best statement is that man branched off from some common stem fairly early in the Tertiary division of the Cenozoic era, and each line (monkey, ape, and man) has been following its own evolutionary course for some tens of millions of years.[1]

Within the tropical and subtropical parts of the world man quite clearly arose in the Old World and not the New World (North and South America), for there is an abundance of human ancestry in the Old World and none in the New World. It is true that there are monkeys in both the Old World and the New World, but the New World monkeys have long, prehensile tails and fur, while man is notable for his lack of an external tail and his relict hairiness. The Old World monkeys are often tailless and always hairy, so they are much closer to man in such details. Further, the Old World has many fossil forms linking living men to ancient subhuman and prehuman forms extending back tens of millions of years. In the New World no evidence of this kind is present. When we conclude that man evolved in the Old World and only spread to the New World after he was fully man, we have eliminated most of the earth as the possible home of man and need only examine the evidence from the warm tropical belt that extends from Africa to Asia.

In recent decades there has been increasing acceptance of the probability that man arose in Africa. Over thirty years ago when the late L. S. B. Leakey, a

---

[1] Only the physical origin of man is being discussed here. As a scientist I can point out the approximate time and place where man begins to behave in a manner quite different from all the other animals, but that is all I can do, and all that any scientist can legitimately do. As an individual, I can affirm that I believe in God and a spiritual part to man. Regrettably some scientists and some clergy go far beyond the limits of their field, often to the detriment of the subject discussed.

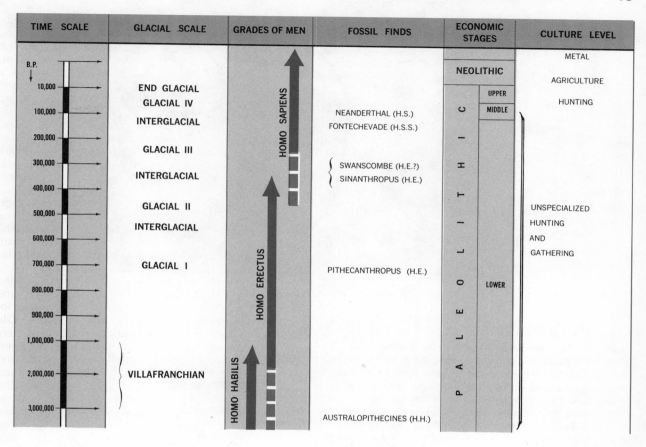

FIGURE 1-1   The Origin of Man

The general time scale for the Pleistocene and for the appearance of the various kinds of men
is shown here. (B.P. =before present.) Dating the Pleistocene and its subdivisions, placement of
the various kinds of men, and even classification of the kinds of men are subject to change.
This table is meant only to give a general perspective of the present view of the situation. The
Pleistocene is now thought of as having a duration of about three million years. The last half
was marked by the appearance and disappearance of glaciers. The Australopithecines are
Villafranchian and earlier. The Homo erectus stage of man extends from the Villafranchian
into the middle of the glacial Pleistocene. Homo sapiens appears somewhere in the middle
of the glacial Pleistocene. Some of the indefiniteness of these statements arises from
disagreements over classification. If Swanscombe man from England is Homo sapiens, as some
think, this type is the earliest sapiens man now known. Some experts, however, consider this
type to be a form of Homo erectus. More confusion is found in some current attempts to
differentiate between Neanderthal and later men as Homo sapiens and Homo sapiens sapiens,
but this seems to be unnecessary complication, for the later type of sapiens man is known to
precede as well as succeed the Neanderthal type of man. Despite the uncertainty over details,
the overall view indicates a progression in man and in his culture throughout the three-
million-year period, and it is the general picture that should be focused on.

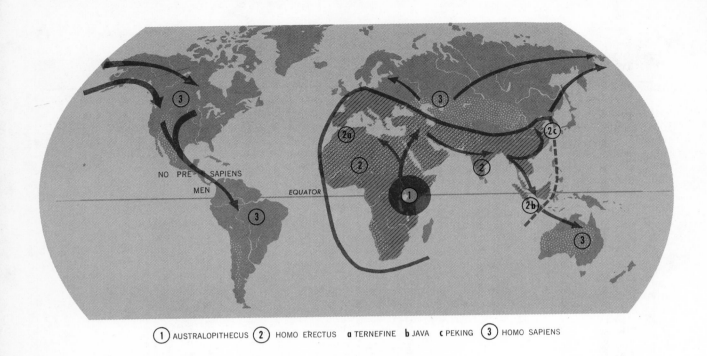

1 AUSTRALOPITHECUS 2 HOMO ERECTUS a TERNEFINE b JAVA c PEKING 3 HOMO SAPIENS

MAP 1–1   The Spread of Mankind over the World

Using a three-stage separation of mankind, the spread of man over the globe must have presented a pattern somewhat like the highly generalized one of this map. The first manlike beings were probably in Africa. From there they seem to have spread out about the beginning of the Middle Pleistocene, perhaps one million years ago. This early dispersion carried men across North Africa, into Southeast Asia, and northward into China, where man appears to have been living at Peking about 400,000 years ago. Extension into Indonesia was also possible during glacial periods when the lowered sea level permitted man to walk at least as far as Celebes. It is similarly possible that man at this time could reach the Philippines and Japan, but it is more likely that it was not till the appearance of sapiens men that these lands were reached. The New World and Australia were seemingly late in being reached. For the New World this may be explained by the rigors of the very cold climate that had to be endured at the Bering Strait gateway from the Old World to America. There were less serious barriers to man's reaching Australia, and early dates for man there would seem likely. The entry to America may have been plural: by shore-dwelling shellfish gatherers quite early and by skilled Arctic hunters somewhat later. This possibility is suggested by the two arrows. Note that the location of sapiens men in peripheral locations does not mean that they evolved there, but that only when sapiens men had appeared did mankind spread into these areas.

British anthropologist, advanced this notion, he was greeted with disbelief; yet today it is quite generally believed that man (*Homo*) appeared about two or three million years ago in East Africa. To put some of this in perspective: As recently as 1959 the theory that pre-man used bone tools was opposed with great vehe-mence by some scholars; an antiquity for man at Olduvai Gorge in East Africa was estimated by eminent scholars as about 400,000 years; the thought that stone toolmaking could be accomplished by pre-men was scoffed at. Today we know that men prior to *Homo sapiens* were making stone tools two to three million years ago. Recent studies show that wild chimpanzees make rudimentary tools, and we have known for more than thirty years that they do so in captivity.

On the basis of both protohuman skeletal material and selective use of animal bone in caves of South Africa, Raymond A. Dart had long pointed to evidence suggesting pre-*Homo* forms of toolmaking. The importance of his work has been included in the popular book *African Genesis* by Robert Ardrey. Recent studies suggest that extremely crude stone tools were also in use at this time.

Tool Using

Something had to give the creatures that were to be man an advantage. They were slow runners, poor climbers, worse swimmers and virtually fangless, clawless, and generally defenseless. How then did they ever come to dominate the earth? It seems to be that out of man's primate underpinning came numerous tendencies, patterns, drives, and aptitudes including family living, vocalizing, and toolmaking. Studies of the social life of chimpanzees and baboons suggest that social structure was extremely important for their survival. It probably was even more important for man, who lacks the fangs for fighting that all monkeys and apes have. If formidable fighters such as baboons and chimpanzees still must rely on band-structured society, how much more must early man have had to do so! Band structure leaves no physical traces, but stone tools last almost forever, and our view of early man is peculiarly biased by this fact. Nevertheless, since we do have the stone-tool evidence and little else for the early men, we must use what we have in order to try to understand these men and their ways of life.

The earliest known stone tools were found by L. S. B. Leakey near the base of the Olduvai Gorge in East Africa. At first these crude bits of broken rock were thought to be the work of the Australopithecine alongside whose bones they were found. Later, remains of a more advanced creature, now named *Homo habilis*, were found at the same and even earlier levels, and it is now considered likely that *Homo* was eating his cousin *Australo*. This is not very strange, for some tropical people today eat monkeys, man's cousins twice or thrice removed.

We find two things seemingly coinciding: an increased brain size and the making of stone tools. Did it require more intelligence to work stone? Or was the invention within the capability of man for a long time before it was actually made, just as the invention of

metallurgy was achieved by men who had not changed biologically for perhaps 100,000 years? Whether or not a nearly 50 percent increase in cranial capacity was required to devise the method of making stone tools, this was a revolutionary achievement, perhaps more important than the agricultural revolution. It increased man's ability to compete and probably marked the beginning of the extinction of man's nearest relatives, for competition is sharpest between closely related life forms due to their having similar habits, likings, and needs.

About one million years ago man of the *Homo erectus* stage spread with seeming rapidity from tropical Africa over most of tropical and subtropical Eurasia. Perhaps the spread was slower, and only our incomplete data lead us to see it as rapid. However, a rapid spread would be expected once man had tools that enabled him to compete successfully for new living areas and food supplies. What these new tools were and what they did for man we are only beginning to realize. In addition to axelike-edged cobbles and sharp flakes of stone, we now have suggestions from Leakey and others that man in these very early times used one of the most unlikely tools imaginable, the bola.

This particular tool is an interesting one to consider. It implies that men were using cords and stones, for a bola requires both. The easiest way to fasten a stone is to put it in a rawhide pocket and attach either a rawhide thong or a fiber cord. That men a million or more years ago had such skills and equipment is a startling suggestion, but if true it might explain why men could suddenly spread throughout the world. Man of one millenium was a naked, defenseless, timid creature who had to flee from most predators and could not hope to compete with chimpanzees or gorillas for food areas. In the following millennia man, armed with a stone in his fist, or even more wickedly with a stone on the end of a string, could strike beyond the length of his arm and give an enormous velocity to the stone-on-the-string simply by swinging it. From a nearly defenseless gatherer man suddenly became a ferocious predator. Whether the bola is as early as Leakey thinks is not as yet proved, but Leakey has a formidable record of being right.

Once man had tools, his perception of the earth must have changed radically. Most animals could no longer safely attack him, and few animals could dispute his feeding areas. Man was on his way to dominance of the world. One of the first measures of this progress may have been his rapid spread from tropical Africa to North Africa, India, and Southeast Asia. Soon, in one of the early periods of lowered sea level due to the

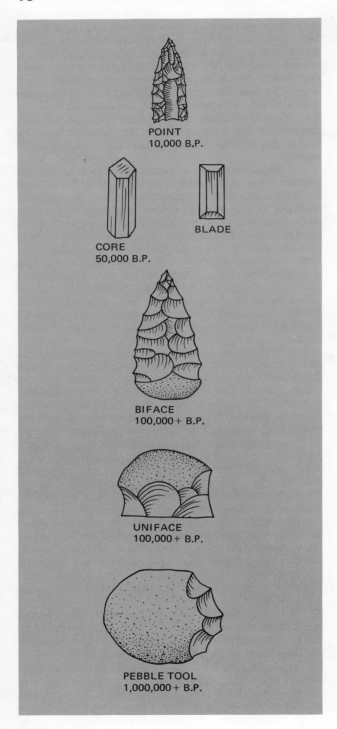

POINT
10,000 B.P.

CORE
50,000 B.P.

BLADE

BIFACE
100,000+ B.P.

UNIFACE
100,000+ B.P.

PEBBLE TOOL
1,000,000+ B.P.

lockup of sea water on the land[2] in the form of vast glacial ice sheets, he was able to walk out to Indonesia, where we find him on Java at least as early as 700,000 years ago.

Notice that for the question of man's origins, it is relatively unimportant that the physical world changed. The changing climates of the ice ages may during glacial times have made it easier for him to pass the desert barriers of North Africa and Arabia, but the thin green line of the Nile had always offered a corridor to the Mediterranean. Consequently, the way to Palestine and the rest of the world had always been available to him. Nevertheless, as far as our present evidence goes, it was not until man had tools, stone tools surely and probably some of skin and fiber and bone as well, that he passed through the long corridor and stepped out of Africa and into the rest of the world.

The Origins of Races

Once man had emerged from Africa south of the Sahara and spread from the Atlantic to the Pacific in the Old World, there were biological consequences. Any sexually reproducing population will vary. The

[2] The removal from the sea of water that formed ice sheets on the land was so great that the sea levels of the world were lowered more than 300 feet. Exposed shallow sea bottom then connected Java with Asia.

FIGURE 1-2   Progress in Tool Making
Man has been making stone tools for two or three million years. The earliest stone tools were pebbles broken to give a sharp edge. These are the pebble tools, which were first used in Africa and then spread throughout the world. This is a classic case of invention and diffusion. In Southeast Asia a pattern of making uniface tools developed and spread to Australia and America. In Eurafrica large, pear-shaped bifaces, called hand axes, appeared. Another way to make tools was to strike long, sharp slivers from a core. These slivers, called blades, were highly developed in Europe in the Upper Paleolithic period, but earlier, cruder forms have been found in Southeast Asia and America. Later finely flaked spear and arrow points were devised. The sequence shown here is partly chronological and partly geographic. It was not determined by the kind of stone available but by the ideas men had at differing times and places as to how a stone tool should be made. Time here is indicated in very round numbers.

variations will be somewhat random and hence different within each group of people. Unless a thorough mixing of genes through intermarriage occurs, the separated populations will gradually come to differ biologically. This process is aptly called *genetic drift*. All that is required for its action is imperfect gene mixing. This is greatly accentuated in small populations and primitive cultures with relatively limited mobility. Differentiation takes time, and we can gain perspective on man by looking at the animal world. It is estimated that it takes about eight million years for a genus of mammals to arise and about one million years for a species of large animal to originate. Within the last 300,000 years no new species have arisen among mammals of fox size or larger. Man is a large mammal, and separation of one million years could give rise to separation into species.

Species of animals do not naturally interbreed, even though species such as lions and tigers may do so under abnormal conditions, as in zoos. All of mankind interbreeds freely, and if we assume that divergence increases through time, early men should have been more like one another than present men are, and they too should have been able to interbreed freely. This seems borne out by the recent comment that at the same time level the early *Homo erectus* forms are extremely alike from Gibraltar to Java. What seems to have happened thereafter in man, given geographical isolation, restricted gene flow, and selection to fit differing climates, is that differences accumulated. There are obvious areas for such isolation: Africa south of the Sahara, western Eurasia, Inner Asia, Southeast Asia, India, and Australia. Seas, mountains, and deserts block off these areas from one another, not sufficiently

to prevent mixing but enough to restrict it. This situation is ideal for allowing divergence in detail while development runs parallel along the main line. In each of these areas we find variant groups of mankind: Negroes, Europeans, Mongols, Malaysians, Indians, and Australians. Some, such as the Australian aborigines, are nearly gone, and others, such as the Indians and Malaysians, are now blends of two or more races, but the general principle of differing men in isolated areas holds.

The noted physical anthropologist Carleton Coon interprets the record as follows: From an initial base on the *Homo erectus* stage of development, the men in the various parts of the Old World evolved into *Homo sapiens*, maintaining sufficient gene exchange to re-

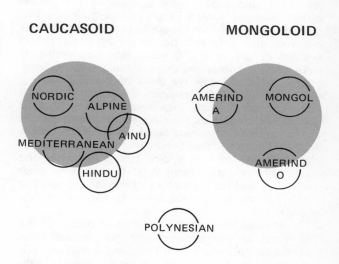

CAUCASOID **MONGOLOID**

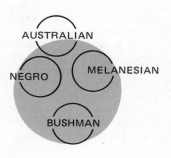

NEGROID

FIGURE 1-3 The Races of Man

The races of man presented here are focused around three centers. Placement within and between the centers is meant to suggest something of relationships. Thus the placement of the Polynesians virtually midway between the three major circles is intended to express the often stated opinion that they are a race with a strong mixture of Negroid, Mongoloid, and Caucasoid. Two types of Amerinds (American Indians) are shown here. The long-headed earliest men were in some way related to the Australoid-Negroid men. Later entrants are more like Europoids. They are designated by the symbol for their blood type. Most Indians south of the Mason-Dixon line in the United States were pure O in blood type. North of that line there was a mixture of O and A. B types, commonest in the Inner Asian homeland of the Mongols, are virtually absent among American Indians.

main a single species but diverging sufficiently to become distinct races. There are different explanations. Some scholars stress the isolation and speak of the modern races as having separate origins. Others stress the maintenance of sufficient gene flow to hold all of mankind in one species. The wide range of opinion can be seen by noting that articles by differing authorities in the field were once published side by side in the *American Journal of Physical Anthropology* in 1944. One expert advocated calling all fossil men including the Java man *Homo sapiens*. In present terms this would mean placing *Homo erectus* in *Homo sapiens*. The other expert wanted to place all living races in separate species: *Homo europus, Homo africanus, Homo mongolus*. The tendency in the last twenty-five years has been to classify more and more of the early men as *Homo sapiens*—for example, the Neanderthalers, formerly named *Homo neanderthalensis*. This race of mankind is best known in Europe where it was dominant during the early part of the last glacial period —about 70,000 to 30,000 years ago. Neanderthalers were formerly considered to be not fully erect-walking, brutish half-men. Now we think of them as walking fully erect and being clearly men. We even believe it probable that through interbreeding some of their physical characteristics were mixed into some groups of modern men. It should be obvious that the bones did not change in those twenty-five years but that our ideas of what they show did. This is typical of the whole field of racial thought; it is subject to great uncertainty and consequently to frequent change. The problem is to deal with the essentials and not get lost in the details.

If races arose in the continental blocks of the Old World, we should expect an African, an east Eurasian (Caucasian), and a west Eurasian (Mongolian) primary grouping. Within each of these categories subdivisions are indicated, but some groups almost defy classification. In ancient times India was peopled by dark-skinned, curly-haired, broad-nosed people. Into this group, especially from the northwest, there has been an intermittent flow of Caucasoid genes. The result is a mixed group with Caucasoid characteristics strongest in the northwest and many large groups of the ancient peoples still existing in the central and southern parts of India. Polynesians are another mixed and varied group. Australians are either an ancient race in their own right or a mixture of Negroid with some race that supplied genes for hairiness. As Table 1-1 shows, this would suggest a Caucasoid mixture, but no one knows how and when this could have occurred.

The racial characteristics listed in the table have been selected for their simplicity and visibility. Blood

Culture Change. *The world is in a period of accelerated cultural change, so the old and the new are often incongruously intermixed. Here in Ethiopia camels rest at the junction of an old camel trail and the new highway linking Wondo and Andola.* (Terrence Spencer for World Bank)

groupings and anatomical and physiological specialties have been omitted. It can be seen that the Caucasoids, Mongoloids, and Negroids are each marked by a group of similar characteristics. These are group characterizations—the description of populations rather than of individuals. In this table one of the oddities is the placement of the Ainu in the Caucasoid group. The majority of their traits strongly suggest that they belong there. Because of their location in Japan, however, they are sometimes placed in a mixed-race classification. It is an academic point today since there are few Ainus left, and most of those who remain are indeed mixed. In the past, however, they occupied much if not most of Japan, and this extension of a Caucasoid people into the Far East seemingly records some prehistoric spread. It is this type of riddle that suggests the need for great caution in discussion of race.

Other cautions arise from consideration of the Mongoloids. They appear here as a very uniform group; straight head hair, slight body hair, brown skin, round-headed, and short limbed. However, the people of Southeast Asia are sometimes dark brown, wavy haired, and slender bodied, characteristics that would suggest a Negroid mixture. This will be further discussed later. The American Indians are also complex in origin, for the early ones were longheaded, the opposite of the classic Mongol. While it is often thought that the American Indians are a simple example of a uniform Mongoloid people, this is clearly not true. There are indeed classic Mongols complete with the slant eyes resulting from an epicanthic fold. Others are small, dark people with curly to wavy hair, suggesting Asiatic Negrito mixtures. Still others are large-faced, heavy-browed people strikingly like the natives of Australia. Moreover, portraits of classic European, Negro, and Mongol types are to be found in the high cultures of Mexico and South America. The American Indians are, then, of mixed racial origins with quite distinctive differences to be found from one part of America to another. These differences are not associated with environmental situations and hence do not seem likely to have arisen in America. They are related to cultural patterns, thus strongly suggesting origins at differing times from varying places outside of America. Despite all these variations, there are broad similarities, and we need names to designate the various kinds of men. The names in Table 1-1 are the ones that will be used in our discussion.

Environmental Forces and Race

If the environment were to act on man, several obvious things would be true: the longer the action the more the effect; the less cultural equipment man interposed between the environment and himself the greater the effect. Race differentiation should increase through time but should have been most marked in early time when man had the least cultural equipment to interpose between himself and the environment.

Change can arise without selection. Variation goes on at a fairly constant rate, and some variations are neither advantageous nor disadvantageous. When they accumulate in a group that is genetically isolated, that group can be said to be drifting away from other

TABLE 1-1   Characteristics of the Races of Man

Race	Head Hair	Body and Face Hair	Skin Color	Nose	Other Characteristics
Caucasoid					
Nordic	⎫	⎫	Very fair	⎫	Longheaded, tall
Alpine	Wavy	Abundant	Fair	Narrow	Roundheaded
Mediterranean	⎬	⎬	Dark white	⎬	Longheaded, short
Ainu	⎭	⎭	Light brown	Medium	
Mongoloid					
Mongolian	⎫ Straight	⎫ Slight	Light brown	⎫ Medium	Roundheaded, short
Malaysian	⎭	⎭	Brown	⎭	limbed
Negroid					
Negro	⎫ Wooly	⎫	Dark brown	⎫	⎫ Long limbed
Melanesian	⎬	Slight	Dark	Broad	
Pygmy	⎭	⎬	Dark	⎬	Vary by groups
Bushman	Extremely wooly	⎭	Yellowish brown	⎭	Distinct body characteristics
Doubtful					
American Indian	Straight	Slight	Brown	Variable	A mixed race
Australoid	⎫	Abundant	Dark brown	Broad	A mixture or an ancient race(?)
Veddoid	Wavy	Moderate	⎫ Brown	Broad	A mixed race
Polynesian	⎬	Slight	⎬	⎫ Medium	A mixed race
Hindu	⎭	Moderate	⎭	⎭	A mixture becoming a race(?)

people in that characteristic. We know of the functioning of drift even today among religious sects that insist on marriage within the sect and thus come to reproduce the breeding conditions of earlier times. Traces of variation in nonadaptative genetic characters can then appear relatively quickly. This works most effectively in small isolated tribes and was probably most important during the earlier history of man. Later, as cultural equipment increased, both numbers of men and their ability to move about must have increased. This must have increased gene flow and decreased the rate of divergence. All in all one might expect racial divergence to be rapid early in the period of first dispersion, to slow down later, and to be reversing now. That is, given the high mobility and vast cultural equipment of modern man, physical selection should be minimal, and gene flow should be a greater force "averaging" mankind faster than the divergence factors can act. The condition of earliest man must have been the reverse, and hence race formation should have been most pronounced in the earliest times. A few examples can show how the process would work.

Skin color and climate

That dark skin color is advantageous for exposure to strong sunlight is clearest to those with fair skins. For them, exposure of large areas of the untanned skin to high sun intensities can be incapacitating and in extreme cases even fatal. Dark-skinned people, on the other hand, are naturally protected from harmful effects of solar exposure. Examination of a map of skin color in the Old World shows that there is a high correlation between sun intensity and skin color. The tropical and subtropical belt from Africa to Australia is characterized by dark-skinned people, and light-skinned people are concentrated in sunless northwest Europe. There are many modifications.[3] The darkest skin colors are not in the tropical rainforest areas where shade is normal, but in the low latitude, arid, and mountain areas where the insolation is intense and shade is at a minimum. The exceptions can usually be accounted for by movements in relatively late times: Europeans moved into Africa north of the Sahara in the Upper Paleolithic period and perhaps into Arabia about the same time; Europeans poured into northwest India about 1500 B.C. and probably also about 3000 B.C. Mongoloids have been moving into Southeast Asia in even more recent times and interbreeding with and displacing an original dark-skinned popula-

[3] Those especially interested should read papers by H. J. Fleure or W. F. Loomis listed at the end of the chapter.

tion. China south of the Yangtze is said by Chinese scholars to have been solidly Negroid as late as the Chou dynasty (approximately 1000 to 200 B.C.). This statement startles many scholars who do not follow Chinese archeology, but it has been repeatedly reported by Kwang-chi Chang.[4] There are remnants of small Negroids in the Andaman Islands, Malaya, the Philippines, and New Guinea, and the refuge location of these people suggests that they are the pure-blooded survivals of an original dark-skinned stratum. At some time in the past it would seem that all of the Old World south of the Mediterranean-Himalaya-Philippines line was Negroid. The correlation between tropical sun and dark skin must then have been nearly perfect.

The New World supplies a contrast. There is less striking correlation between skin color and climate among the Indians in America despite the fact that all the climates of the Old World are represented. The inferences are that the men who entered the New World were not originally very dark skinned and that they have been in the New World a relatively short time compared to the total of human time. Since many of the American Indians went virtually unclothed under high sun intensity, some of the people in Baja California, the Colorado Desert, and the Great Basin should have been black if selection and time can create black skins and if there had been enough time. Although there is disagreement on how long the Indians have been in America, no one argues that man has been here more than a small fraction of the time that he has lived in the Old World. If the first men to enter America were brown skinned, either because this was mankind's original color or because the brown-skinned races of mankind had evolved in central Asia before man entered America, there has not been sufficient time to redevelop black skins despite what would seem to be a strong selective pressure in that direction. The inference seems to be that development of the intensity of racial characteristics of the type illustrated by skin color must have been a slow process extending very far back into human history. It suggests that races are ancient. This is, of course, exactly what the process of biological divergence would lead us to expect. If man

[4] See, for instance, "A Working Hypothesis for the Early Cultural History of South China," *Bulletin of the Institute of Ethnology*, Academia Sinica, 1959, where he refers to the Mesolithic people of south China as "Oceanic Negroids (Melanesoid, Papuanoid, and so forth)" and concludes that "possibly some Negrito groups were the earliest inhabitants of Central and South China." Similar statements are to be found in his book, *Archeology of China*.

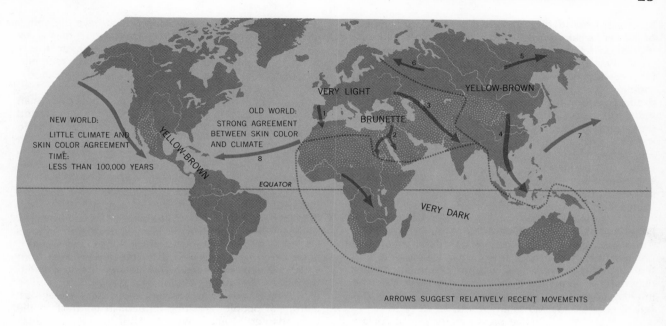

MAP 1-2   Skin Color and Climate
In the Old World there is a clear correlation between climate and the locations of the major divisions of skin color. The light-skinned peoples are in the northwest where the greatest amount of cloudiness throughout the year is found. The very dark skins are in the equatorial regions where the greatest intensity of sunshine is experienced. The yellow-brown skins are in the areas seasonally exposed to moderate intensity followed by low intensity of solar radiation. It was primarily the yellow-brown-skinned peoples that peopled America, and despite the passage of a few tens of thousands of years and a range of climate as great as that in the Old World, no comparable intensity of skin-color difference has developed. The pattern of skin-color intensity and insolation intensity is not perfect in the Old World, probably due to recent movements of mankind. Some of these movements are indicated by arrows. Arrows numbered 1, 2, and 3 show Europoids moving into Africa, Arabia, and India; arrows 4, 5, and 6 show Mongoloids moving outward from central Asia; arrows 7 and 8 show later movements by sea of Asiatics and Europoids to America. At some time in the remote past the distribution of skin color may have been much more closely correlated with the climatic zones than it is now. Very dark-skinned people probably originally occupied all of the equatorial zone of high sun intensity in the Old World.

spread out over the Old World at least a million years ago, racial divergence must date from that time. A mere 100,000 years in America normally would produce slight change if the rates were constant, and none would be expected in 10,000 years.

The reason for a relatively dark skin is obvious, but anyone who has suffered greatly from sunburn might wonder what possible function is served by light skin. One clue was gained when it was noticed in Lon-

don that rickets were more prevalent among dark-skinned slum children than among light-skinned children living under equally poor conditions of food, shelter, and clothing. Rickets is a vitamin-deficiency disease, the required vitamin being manufactured when the fats in the skin are irradiated by sunlight. In London, therefore, the coloring matter of the dark-skinned slum children screened out much of the penetrating solar energy. The light-skinned children, although also on a substandard diet, absorbed a maximum amount of the very weak sunlight of the London winter, enough irradiation to manufacture the needed vitamins.

Primitive man frequently experienced periods of famine when he ran short of food and thus of vitamin supplies. The critical period in western Europe was probably during the winter low-sun period when those individuals with a lighter skin had a survival factor in their favor. It seems significant that the earliest penetration by man into a cloudy, mid-latitude climatic zone was in northwest Europe, the region that is also the center of fair-skinned mankind. The only other apparent reason for this change in skin color could be genetic drift,[5] which seems less logical since the trait in question has obvious survival value.

Man was not able to reach the other cloudy mid-latitude (long-winter) climates until much later. The American regions representative of this climate were not reached by man until the Upper Pleistocene, and the Australian-Tasmanian occupation of similar climates may date to about the same time. Whatever the time interval, however, it was apparently not enough for this selection factor to achieve any result.

Today the selective pressures for skin color are almost nil. Better diets and knowledge of vitamins remove the need for light skins. Housing, clothing, and suntan lotions remove the need for dark skins. With natural environmental pressures removed, what will become of skin color? Presumably we will tend to lose the extremes, the partially albinized whites and the heavily melanized blacks, and revert to what may have been the original brown. Seemingly there will be little physical environmental pressure through selectivity in this trend, but a simple case of what is known as reversion toward the mean.

Heat, cold, and body form

Heat and cold place major stresses on men's bodies. Men are heat-producing machines. They burn food to operate body processes and to maintain an even body temperature. Departure from the normal 98.6° F. body temperature is accompanied by discomfort when the change is only a degree or two, by acute distress when it is even 5 degrees, and by possibly fatal results if it reaches 10 degrees.

Man's ability to produce heat is limited. If he loses heat faster than he can produce it, body temperature will fall, and this too can be fatal. For men living in cold climates, especially for ill-equipped men,

[5] Genetic drift is change due to chance variation and is not due to selection.

it would obviously be advantageous to have the ability to minimize heat loss. There are various means by which this can be accomplished: use of fur for insulation, circulation changes to minimize surface temperatures while keeping the deep body areas warm, body-shape adaptation to minimize heat loss.

The coldest climates early penetrated by man and the coldest winter area inhabited by man are found in the continental interior of Eurasia. The body build of the Mongoloid people, the race that originated in this area, has distinctive adaptations for cold, for they are short-limbed, thick-bodied people. This body form minimizes the area from which heat can be lost.

In a warm climate the problem is to dissipate the heat produced simply by existing as well as the heat produced by working, and unless this can be done, body temperature will rise. Heat is dissipated by perspiration, by evaporation from the lungs, and by radiation from the skin. The greater the surface area from which heat can be radiated and perspiration evaporated the better the cooling system will work.

It is probably no accident, then, that the Negroes, a tropical people, are long limbed, in contrast to the cold-adapted Mongols who are short limbed and thick trunked. The ability of the body to lose heat to its surroundings is aided by several factors: a sizable difference between the temperature of the body and that of the surrounding environment, dry air into which perspiration can be evaporated readily, and moving air, which removes the air saturated by evaporated perspiration and brings more absorbent or cooler air once again into contact with the body. In the warm, moist tropics, the air may be near body temperature and almost saturated with moisture. Heat loss is then difficult. In the dry tropics air temperature may rise above body temperatures, but the dry air then makes evaporative cooling very efficient. To achieve heat loss is still a problem, however. Not unexpectedly, tropical races often tend to be tall and slender, for this body form gives the maximum surface area to aid heat loss. In general the body forms of the races of men meet expectations for adjustment to heat and cold: blocky-built Mongols, slender Negroes, intermediate Caucasians.

Any generalization such as this is bound to have exceptions, especially when the racial classification we are using is so simple. American Indians vary considerably in body build, partly in agreement with the climatic situation but partly not; this variation probably reflects both diverse origins and varying lengths of time for adaptation. The Mongoloids vary from the blocky-bodied people of Inner Asia to the slender, small people of Southeast Asia. However, these latter people may

The Three Principal Races of Mankind. *The blonde Norwegian girl represents the fair-skinned, or Caucasoid, race that developed in northwest Europe. (Norwegian Information Service) The Negro boy, the racial opposite, represents the dark-skinned, or Negroid, peoples who developed in the tropical regions. (Smithsonian Institution) The Japanese couple represents the Mongoloid people who developed in Inner Asia. (World Bank) The differences among the races are clear, but much of the world is occupied by people of intermediate status between the extreme poles of differentiation.*

have some Negroid race mixture, which would account for the changes.

Closely related to body form in regard to heat and cold stress is physiological adaptation, which is closely related to cultural level, for the examples of physiological changes to meet environmental stress are found among primitive people. It is probable that they represent the situation of the ill-equipped men who early spread into nontropical environments. They are of particular interest for they represent a tendency in man to substitute bodily change for cultural change when an environmental challenge has to be met. Most animals change their bodies to meet such challenges, but man normally changes his culture. Seals gave up legs and developed flippers; we manufacture swim fins. Nevertheless, in a few cases we adjusted our bodies.

In their clothing the Eskimos represent one of the peaks of man's cultural adaptation to cold, yet the well-clad Eskimos still have to expose their faces and occasionally work barehanded. They are described as notably warm handed and relatively free from facial

frostbite, attributed to increased blood flow to face and hand areas. When individuals of different races were tested for resistance to chilling of their hands when immersed in ice water, Eskimos and Alaskan Indians had the warmest hands, Caucasians were intermediate, and the Negroes were the coldest—exactly as one would predict on an adaptation basis. The Indians of the Arctic regions have a somewhat higher basal metabolism than most of mankind. This accompanies a relatively poor cultural adaptation to the great cold of the northern forest regions of North America, for these people lived in flimsy shelters and were not nearly as warmly dressed as the Eskimos. The Indian people of Tierra del Fuego were far less culturally equipped for withstanding much lesser cold and met the problem by an increase in basal metabolism to 160 percent of our metabolism. In Australia, people with an equally primitive culture (virtually naked and unhoused) adapted to very moderate cold by developing the ability to drop the temperature of their limbs to as low as 54° F. while retaining normal body temperature in the vital trunk regions. It is notable that it is the culturally ill-equipped people who have made these biological changes. The well-clad and well-housed Norwegians show no such adaptations. The unclothed and unhoused Australian natives show extreme adaptation in a much milder climate. The culturally equipped Eskimos show less physiological adaptation in the high Arctic than the poorly equipped Indians of Tierra del Fuego living in a much milder climate.

In keeping with this line of reasoning, we find that of the major races the Mongoloids who seemingly originated in continental interior Asia, a land of great winter cold, are the most cold adapted. The Negroids are the most heat adapted, and the Caucasians with their predominantly mid-latitude location show intermediate adaptation to extremes of heat and cold.

Bodily adaptation to climate probably goes much further than we yet understand. The Mongoloid facial type is flat with a fatty layer under the skin. This affords a minimum of exposure of facial parts and is thus a defense against freezing. Mongoloids are also relatively beardless. Under the conditions of extreme cold in Inner Asia in winter, a beard catches moisture from the breath and soon becomes an ice mask. Europeans, however, have beards. They live in less cold environments, where ice is unlikely to accumulate in the beard, and the beard can serve as insulation.

If the long hair of the Mongols and the Europeans is useful against cold, is the short hair of the Negro useful against heat? It would seem so. Thick masses of hair around the neck and shoulders would be most unwelcome in heat and humidity, and women with long hair often adopt a short hairdo for the summer or pin their hair up. The hair of the Negro is permanently adapted to warm weather.

This list of adaptive traits can be extended. Noses with longer nasal passages are useful for warming and humidifying the bitterly cold, dry air of the continental interiors. But this is a good example of the problems and pitfalls in oversimplified explanations. A projecting nose is also easily frozen, and the typical Mongol nose is flat. The classic long Semitic nose does not seem to fit this rationalization either. However, the large, often-hooked Semitic nose did not originate in the eastern Mediterranean. It was brought into that region by the Armenoids, who came from some part of southwest Asia, an area of considerable winter cold. This trait could be a case of the migration of men in relatively recent times (the past few tens of thousands of years?) resulting in the scrambling of adaptive traits (fixed in the preceding few hundreds of thousands of years?). Although both genetic drift and selection probably influenced the development of races, most traits seem to have adaptative value. Racial differences seem to be due to a large extent to the millennia-long impact of the physical environment on the relatively unprotected bodies of our remote ancestors.

Environment and Accomplishment: Man in the Tropics

No matter whether man originated in Asia or Africa it seems clear that he is—as our cursory glance at his physical being would lead us to guess—a product of the warm and humid lands. If man originated in the tropics, why should that climate be the one for which he is most poorly fitted? If man evolved in a warm and humid climate, should not that climate be the one to which he is best suited, just as monkeys thrive in their tropical and subtropical homes and perish when they get too far out of them?

According to our previous reasoning, man should be more adapted to heat and humidity and to the relatively unchanging seasons of the tropics and subtropics than to any other type of climate. If his beginning is placed about 2,500,000 years ago in Africa, more than 2,000,000 years passed before he moved into the truly cold climates and remained there—and only a minority of mankind moved into the cold climate even then. The bulk of mankind quite sensibly stayed in the warm, comfortable humid climates. Our view is distorted by the fact that large groups of mankind now live in the higher mid-latitudes, but in terms of man's history this is a quite recent phenomenon. Since it was not in the "cool and stimulating climates" that man arose but in the warm and humid ones, this is where he should be physiologically at home, and one would think, mentally at home as well. Why, then, the claim that civilization flourishes only in mid-latitude climates?

Contrary to what is often claimed, there have been great civilizations both in the warm and humid as well as in the hot and dry climates (Indonesian, Indian, and Mayan in warm, humid climates; Mesopotamian, Egyptian, and Peruvian in hot, dry climates). We can well argue that more great ideas originated and more basic inventions were conceived in those regions than have yet come out of the "cool and stimulating climates"! The latter phrase itself is the result of a highly selective and momentary view of history. It reflects a view of the world as it was from 1850 to 1950, when the North Atlantic nations were probably at their peak of world dominance.

Physiology, climatology, psychology

If we try to measure the physiological impact of climate on man, we find both interesting data and difficult problems. It becomes surprisingly hard to determine just what we are measuring. For instance, it was once proposed to find out what the immediate effects of the physical environment were on groups of people at work. Two groups of people were selected: a group of women and a group of men, both doing a wiring-assembly job. For both groups systematic changes were made in the working conditions. Temperature, humidity, and air movement were varied. The two groups were subjected to all kinds of heat and cold. Until extreme ranges were reached, beyond anything that one would experience under natural conditions, the working groups simply ignored the changes.

For the women, a record was kept of their hours of sleep, their social engagements, their family affairs, and the like. It was thought that this might be impotrant to their work output. Instead it was found that the most important factor in the work situation was how the individual woman felt about the woman who worked nearest to her. The next most important factor was how the group was working as a whole. Introducing a new woman into the group was much more disturbing to their accomplishment than anything except the most outrageously extreme changes in the heat and the humidity. By and large, the women felt that they were participating in a very special study and felt secure in their jobs. As a result, they were content and willing and eager to work. This may be thought of as the role of culture. Perhaps we ought to change the old saw about "it isn't the heat, it's the humidity" to "it isn't the heat, it's the attitude."

The men were quite different. They did not work up to their capacity at all—neither in terms of their intelligence nor in terms of their dexterity. They did not even work to maximize their economic gain. Instead they acted as a faintly hostile group whose main function was to maintain group integrity, even if it cost them money individually. But again, there was little or no relation in all of this to the physical environment. The men maintained the same attitude whether their work environment was steamy as a tropical forest or as cold as the subarctic. Under severe environmental stress, therefore, the factors that determined achievement or lack of it were psychological-cultural, and physiological reactions appeared most strongly in terms of the differing psychology of male and female.

D. H. K. Lee[6] has reported a similar case. Under Arizona desert heat two sets of trainees were given identical training and tasks. Those in one group were uncooperative and were characterized as "tough guys." Those in the other were interested and cooperative. The first group showed high pulse rates and body temperatures that increased steadily with the air temperature. They were failing to meet the heat stress. The interested and cooperative group showed very little pulse and body temperature rise under the identical conditions. Here, then, mental attitude was controlling physiology to a measurable extent. This test was within one culture (standard American), it dealt with one race, and it involved comparable work in one environment. A minor attitude difference within this uniformity created the observed difference. What might the difference have been if any of the conditions were more varied?

In the Sahara it has been noted that Europeans and natives differ in their reaction to a vehicular breakdown in the searing heat in the midst of nowhere. The European displays great anxiety, rushes around trying to "do something," gets hotter, and uses up more of his bodily water resources. The native calmly lies in the shade of the vehicle and waits for something to turn up, thereby prolonging his survival time and increasing his chance of rescue. Americans, lacking experience with high heat, make the same mistake in Death Valley and occasionally pay with their lives. The Sahara nomads are Europoid in race, and so again the

[6] D. H. K. Lee, "Attitude Response to Environmental Stress," *Journal of Social Issues*, vol. XXII (October, 1966), p. 91.

important difference concerns racial and physical environmental constancy and cultural variability.

Physiological climatologists state that under warm, moist conditions of a moderately severe kind there can be expected some reduction of circulation efficiency, some discomfort due to sweating, sticky clothes, and a resultant reduction of patience, alertness, and judgment. But, they add, with a high degree of motivation a fairly high standard of performance can be maintained by selected individuals under conditions verging on those that could produce physical collapse. In other words, right up to the point of collapse, man can do anything if he really wants to do it. This is certainly what the experimental work with the women's group showed. Physiological literature on the subject contains descriptions of the true physiological failure of the body under extreme cases of heat and humidity. But it also indicates that under the conditions likely to be found in the shade of the tropics the capacity of the healthy, trained, and acclimatized workers is not likely to be much affected. Perhaps noteworthy is the fact that one-year-old children appear to stand the stress of heat much more readily than do adults. It is not clear just why this is so, but even the physiologists—who might interpret reactions as evidence of physiological strain more readily than psychologists might—agree that this equanimity may be a reflection of the child's power to ignore minor strains such as climatic discomfort.

It therefore becomes quite difficult to see just how climatic strains, particularly heat and humidity, could have the enervating effect claimed by those who see in the "deadly effect of the tropics" the reason for the absence of civilization there. And, if it is so difficult to make a convincing picture of physiological impact of the heat and humidity of the tropics, how much more difficult it is to make a similar argument of any convincing proportions for high altitudes, high latitudes, hills, or dales.

Since there are differences in race adaptations to climates—the Negro skin is better adapted to resisting the harmful effects of solar radiation, the Mongols are better adapted for great cold, and the Europeans are intermediate in most such measures—we might ask if one race should not accomplish more in the tropics than others. In fact, the adaptive traits are not simple in their action, they have both harmful and beneficial aspects, and they confer only small advantages. The Negro skin defends against ultraviolet (sunburn rays) but picks up more heat, as has been shown in tests on Negroes working unclothed in full sun under high heat alongside lighter-skinned people. However, wearing a white shirt that reflects much of the light would

approximately equalize the load. In a choice between disabling sunburn or lethal skin cancer, and a slightly higher heat load that can be averted by working in the cool of the day, in the shade, or at a slower pace, the selection was for protection against death and disability at a small cost in heat balance. Similarly, the selection for fair skins in north Europe traded some disability in sunburning for greater immunity against vitamin deficiency in childhood, which, when it created a malformed pelvis in girls, could lead to the loss of both mother and infant in childbirth.

These traits were far more important in man's past than in the present, for our increasing cultural equipment removes the need for physiological adjustment. The fair-skinned races can use clothing and suntan lotions to ward off excess radiation even in the tropics. Adequate clothing and housing allow the tropical man to live comfortably even in the Arctic. The success of races far outside the environment of their origin has been demonstrated by such extreme examples as these: The only man to reach the pole with Peary was a Negro; the first man to reach the top of Mt. Everest along with the acclimatized Nepalese was a low-country New Zealander. Other cases are so commonplace that we take them for granted; for example, the success of the Chinese, representatives of the cold-adapted Mongols, in such tropical places as Singapore. By means of cultural equipment, clothing, housing, heating, and cultural adaptation of the learned-behavior type (no rushing around in either high heat or great cold) most races can get along well in most climates. It is probable that only in very high altitudes (above 12,000 feet) are there conditions that relatively simple cultural adjustments cannot overcome.

Of even greater interest is the discrepancy between time of differential accomplishment and racial differentiation. Races were in existence some tens of millennia ago, and at that time there was much less cultural difference in terms of accomplishment than there is now, for at that time all of mankind led similarly simple lives. Today, with no discernible increase in race differentiation but, instead, a tendency toward greater uniformity within mankind (through extinction of some people like the Neanderthals, many American Indian tribes, most of the native Australians, all of the Tasmanians—and through mixing of races as in India, Brazil, and elsewhere), we find a widening gap in cultural accomplishment. If race differences, in part related to climatically impressed adaptive traits, played a principal role in success or nonsuccess in such climates as the tropics, the cultural differences should have been declining instead of intensifying during the past centuries. Since the contrary is true, it would seem logical that a more powerful force is at work. While this does not deny that there was any racial influence, it surely reduces the racial cause to a minor role. This is in keeping with the evidence reviewed here, which indicates that the racial adaptations that produce a few advantages are more readily offset by cultural equipment that is more and more available.

Race and Mentality

Analysis of bodily differences has shown that by taking the function of the differing external physical elements that mark us off into races, we can see the processes that would quite naturally make us different in skin color, hair form, and body form. The question remains: Does this apply to the brain? Will the same kind of process analysis lead to an expectation of differing mental makeups to fit the varied climates? As we have seen, the early geographers answered this in the affirmative. We will test this idea here by taking the same process-and-function-oriented approach that led to an understanding of the racial differences as a natural product of hundreds of millennia of separation in differing climates.

Although the body is a heat-producing machine that must function within narrow temperature limits, the brain is not limited this way. It is dependent, of course, upon the bodily adjustment to the surrounding environment, but its function is not the parrying of direct insolation or great cold, though it may direct some of the bodily mechanisms that are involved in these tasks. The brain is best described as an enormously complex computer. Its function is problem solving. Our task is to examine the nature of the problems that it solves to see if they vary significantly with climatic or other physical factors in accordance with the processes that led to racial differences. If not, it will be difficult to make a case for parallel changes in body types (race) and mental types. That is, if selection is a process that goes on through time by adapting an organ or organism to its task, and if the tasks are different (as are the bodily features we have just discussed), the organ must become different. However, should we find that minds always have similar tasks, we should not expect differences.

The brain regulates numerous processes, such as our automatic functions (breathing and heart operation, for example). There is some room for selection even in such vital functions as heart rates because they vary over a narrow range and the variations follow climatic zones. The brain also controls our conscious

muscular activities, but throwing a spear, driving a car, or operating a typewriter are the same tasks in all climates. It is difficult to suggest a type of work that requires significantly different mental development in contrasting physical environments, such as mountains versus plains or forests versus grasslands. Mountaineers may climb more slopes than plainsmen, and the plainsman may habitually have broader vistas than the forest-dweller, but they all must coordinate hand and eye in tool using, which was crucial for most of man's history. There is no physical environment that rewards, and thus selects, a lack of coordination. On the contrary, there is positive selection for ability to utilize the equipment of one's culture.

The plainsman's vaunted distant vision can have a simple explanation as epitomized in stories. When a plainsman said to a city dweller, "Wonder what scared those horses?" all the city dweller could see was a distant cloud of dust. The plainsman explained that the cloud of dust had to be made by horses because it moved too fast for cattle, and he also knew there were no cattle in that pasture. His eye had not seen any more than the city-man's eye, but his brain was stocked with data that allowed him to read more from the facts available. Another story tells of the city man and the woodsman walking down a city street. The woodsman amazed the city fellow by hearing a cricket, even over all the traffic noise. A few minutes later the woodsman walked on unnoticing when the city dweller came to a sharp alert and exclaimed, "Somebody dropped a dime!" What the ear hears or the eye sees is to an astonishing degree culturally conditioned, and cultural conditioning goes on in the mind. We have too often taken these culturally patterned responses as evidence of sense differences: greater visual acuity for plainsmen or better hearing for woodsmen. We should wonder about attributing independence to mountaineers and subservience to plainsmen, but these are topics that will be dealt with in the appropriate sections to follow.

If we return to consideration of the mind as a problem-solving machine, we must ask what are the great problems that men face. Is one major problem avoiding danger, as hunters fear man-eating tigers and people who make their living gathering wild plants watch for poisonous snakes and those who live in modern cities worry about criminals in the streets? Most of us rarely encounter criminals, so selection for minds able to cope with that problem must be slight. This is true for most of the extreme problems in other ways of life. They are exceptional, and to the extent that they operate, they tend to select similarly. The better mind avoids tigers, snakes, and criminals, or it more quickly registers the lurking stripes, or coils, or loiterers. There is always selection against the stupid, but this acts constantly in all environments and in all ways of life. A stupid mistake in the Arctic may bring death by freezing, in the desert by thirst, in the wet tropics by poisonous snakes. The brighter man in the Arctic gets more seals or caribou, in the desert finds hidden water and scarce game, in the jungle avoids the tiger and traps the elephant. In all of nature there is constant selection for better minds, but it is hard to see where there is a significant difference from one physical environmental situation to another. The problems differ in detail, but the function of the mind is to solve all kinds of problems. The ones we have been discussing can be called physical problems. There are also what in contrast we may call social problems.

Social problems are even more alike for mankind than the physical problems and much more important. Administrators in all endeavors agree that their greatest difficulties are in the area of human relations. Able men defeat themselves because they cannot get along with others. Men succeed or fail in politics depending on their ability to get along with, organize, and impress others. As we have seen in the experiments with groups of men and women at work, group attitudes were the most important factor governing their productivity. For every individual, his relationship to those around him is his greatest single problem, and it accompanies him from birth to death: child to parent, parent to child, sibling to sibling, schoolmate to schoolmate, scholar to teacher, husband to wife, employee to employee, and employee to employer. Each individual's success derives from his ability to solve these situations, and problems of heat and cold, wet and dry, pale into insignificance by comparison.

Most of these problems are the same for all mankind. Men must be born, they all grow and learn, most of them marry and earn, and all eventually die; then the ongoing generation universally faces the problems of death and its aftermath. From people to people these situations vary in form but not in essence. Therefore all mankind has been subjected to selection for a mentality suited to solving these problems. One can, of course, overgeneralize the processes. Animals all get born, have siblings and parents, mate, and die. Man, however, has a distinctive pattern for his life.

Men live in groups, and we assume that they have done so for a very long time since this is the pattern of living for gorillas, chimpanzees, baboons, many monkeys, and all men in recorded time. Living in bands has great protective value, as has been shown by studies of groups like the baboons. The old male baboons form a loosely organized defense perimeter

for the females and young, and the band fears no predator smaller than a lion, not even the leopard. While a leopard can quickly kill any single baboon, a group of males can kill any leopard, and one case has been recorded where two male baboons attacked a leopard that was blocking passage to the band's night haven. Although the two baboons lost their lives, they killed the leopard, and the band was saved. The experience accumulated from such encounters has taught these predators to avoid baboon bands and limit their forays to picking off stragglers.

Selective pressures are visible in this pattern of survival for males who cooperate in defense of the band, for social coherence rather than independence where individual baboons go off on their own, for ability to live together without quarrels that would destroy the protective cohesiveness of the band. The minds of all baboons have been so selected. But so have the minds of chimpanzees and gorillas and men. Survival has always depended on such traits, and these traits are not climatically, physiographically, or vegetatively influenced.

In the chimpanzees and gorillas the bands are somewhat smaller, and the behavior patterns even more like man's. Children stay with their mothers for years instead of months. The bands form stable units of males, females, adolescents, and infants, and the strength of the unit is in its coherence. The adult males form a strong defense. The members act as a loose but vital unit; they sound alarms and move, or stand and fight, with some cohesion. Group action underlies much of the survival of man; indeed since man is less equipped with fangs than his anthropoid cousins, he may have had (must have had?) an even stronger band structure. If so, he must have been even more strongly selected for social life.

The point is that this selection is for a mental type—a selection for minds capable of dealing with the problems of group living. These problems would be similar in any climate where early men lived. Indeed, one can ask if they have changed fundamentally even now. By a process type of analysis, we are first led to expect bodily *differences* to arise in groups of men isolated from each other in differing climates. By the same type of analysis, we are led to expect that by selection men's minds would be held to *similar* development, because the mental problems remain much the same.

We may profitably expand this a bit. Primitive man probably inherited a band-structured social organization. Australopithecines probably lived in bands, and *Homo erectus* quite surely did. The band structure had great survival value, and therefore selection for minds able to solve the interpersonal relations of living together was strongly rewarded, and those who were unable to deal with these problems were eliminated. Interestingly, sex as a basis of the social structure recedes in importance. Its importance is for family formation, and this seemingly is later than band-structured society.

Even early societies show sociocultural value systems. For example, of the *Homo erectus* men who lived at Peking, we have parts of 25 skeletons. Two of these are from individuals over 50 years old. This is great old age for primitive people, and in that time (between 300,000 and 400,000 years ago) these individuals must have been relatively nonproductive. A social structure and value system must already have existed that permitted their maintenance instead of their elimination. This suggests that even at this early date very human mental attitudes existed, and the implications are that the fundamental problems of mankind were being solved by minds already much like ours. Thus it appears that both bodily differences and mental similarities appear early. Seemingly, as men diverged bodily into races, they held together in mind, for while their bodies were responding to differences, their minds were responding to similarities. There are clearly understood reasons for bodily divergence into the races of mankind that we see today, but there are no discernible reasons for mental divergence. Indeed, the contrary is indicated; men's minds should be alike. The point should be amply made by now that selection in man has been two-pronged: toward differences in bodies to meet physical environmental challenges, especially early in human history when man could interpose little between his body and the environment, and toward similarities in minds, since the major selection was toward common social values.

These processes operate through time. It might be argued that man, to a considerable extent, makes his own environment, and that some men have been living in significantly different man-made environments for so long that they could now be distinctively different mentally: there are differing, genetically adapted mental types selected for hunting, for farming, and for city dwelling. There certainly are great differences in the amount of time that various groups of people have been exposed to these differing ways of life. There are a few people left on earth who are living in the Lower Paleolithic, unspecialized, hunting-and-gathering stage and perhaps an equal number living in the Upper Paleolithic, specialized hunting stage. A great many people in the world are still living on a Neolithic level: simple villagers making their living from primitive agriculture. Only a small fraction of the total of mankind lives an urban life, and for most of them it is a phenomenon of the last few centuries. Unless the selective process operates very rapidly, few people indeed can be expected to show biological adaptations to urban life.

While there is considerable argument about just when "men" becomes "man," almost everyone would agree that by 500,000 years ago men very much like modern man were on the scene. Their way of life was that of unspecialized hunters and gatherers, somewhat like the people of Australia or the Great Basin in the United States until a century ago. A few people became specialized hunters about 50,000 years ago, still fewer people became farmers 10,000 years ago, and even fewer people became urban people about 5000 years ago. It is difficult to say just what effect this had on human minds.

One might say that on a 500,000-year time scale, mentally we should be 87 percent hunters and gatherers, 10 percent hunters, 2 percent farmers, and 1 percent urbanites, but this would be true only for the people in the Near East, where these shifts seem to have appeared earliest, especially for the agricultural and urban developments. The rest of mankind would have even shorter time spans in which to be mentally changed to fit the new ways of life. This proposition assumes mental adaptation.

The history of the past 5000 years is filled with examples of people who jumped in very brief spans from one level to another of this sequence, with no measurable genetic mental changes. Apparently they had the inherent ability to live a fully civilized life even though they had until a few centuries ago lagged in a Paleolithic way of life.

The fact that so many races of men have adapted so quickly and so successfully to the requirements of modern urban commercial life, even though they started from entirely differing ways of life, suggests that the fundamental requirements for each way of life are much the same. Life's problems are social and interpersonal and not basically different for the men of Paleolithic, Neolithic, or urban way of life. If, then, some races have had representatives in the urban way of life longer than others, there is little evidence that they have gained a genetically based—that is, racial—advantage.

It has been argued that we are actually deeply imprinted with a territorial imperative, to use Robert Ardrey's phrase, which derives from near the beginnings of our origins. It is certainly true that the simple hunters and gatherers defended their territorial rights jealously, for the very good reason that their lives depended on the food that the territory provided, but hunters defend their hunting territories, farmers their land, and modern nations their sovereignty. What is striking is how much alike we are, whether we look at race or cultural level or place.

Twist and turn this as we will, then, the argument from the nature of the mind as determined by its problem-solving function brings us back again and again to expect considerable uniformity in the minds of men and provides us with no clear basis for expecting significant mental differences in minds developed under varying physical or even economic environments. This is not to say that there are no such differences, but to suggest that both in terms of function and of time these differences are seemingly minor, if they exist at all.

Tests of racial differences in abilities

Tests of race variation in general intelligence and in special abilities fail thus far to support any clearcut racial distinctions. Each race seems to have a normal distribution curve of ability: a few people of very high ability, some of good ability, a mass of average ability, some of poor ability, and a few people of very limited ability. In terms of the similarity of man both physically and socially, as we have seen, this is about what we might expect.

Intelligence tests have shown differences associated with race. In America, Negroes on the average test lower than whites; however, southern Negroes test lower than northern Negroes, and educated northern Negroes test higher than semiliterate southern whites. These averages could be misinterpreted as being the result of the stimulating climate of the temperate North versus the deadening climate of the tropical South. It is very probable that educated southern Negroes could be found who would test higher than a white group from Chicago, New York, or New England. The meaningfulness of intelligence tests has been greatly improved, but they still cannot avoid the effects of varying opportunity, motivation, and the deep psychological impacts of varied social positions in society. Further, averages are one thing and individuals another. The brighter individuals of any race are superior to the average of any other race. When dealing with individuals, and we usually do, we should keep that in mind.

An example of the difficulty of making such measurements and of how hard it is to eliminate the social factor is found in Japan. One group of people there is treated almost exactly as the blacks are treated in America. These Japanese, however, have no external racial mark. Those discriminated against are recognizable only by family names and their neighborhood and social associations. Nevertheless, they fall into the last-hired, first-fired syndrome with all of the associated penalties—financial, social, and psychological. These people, physically indistinguishable from the other Japanese, score about 10 percent below the national norms—almost exactly as the American blacks do. So, if one type of discrimination is not racially caused, how can we be sure that the other is? Perhaps social pressures suffice.

Race and Achievement

We can try another approach. If we look at the races of the world in terms of the locations of centers of great cultural achievement, we find the following picture. The enormous cultural achievements centered in Mesopotamia, Palestine, Egypt, Greece, and Rome are the work of the Mediterranean race, the basic stratum of the so-called European race. The achievements of India must be credited to a mixed race: Europoids entering northwest India about 3000 B.C., carrying agricultural ideas with them, and a further flow of men and ideas from the Mesopotamian region about 2000 B.C., combined with the aboriginal dark-skinned peoples of India. The Chinese accomplishment is in part the achievement of the Mongoloid people, stimulated to some extent by ideas streaming eastward from the Middle East. Accomplishments in Southeast Asia (Burma, Thailand, Indonesia, Vietnam) may be the results of ideas coming in from both India and China, which were utilized by the slender, brown-skinned races of that area. The most recent work in Southeast Asia is producing increasing evidence that the truly great, astonishingly early achievements in that area are the work of the small, dark people of Thailand, Vietnam, and south China. Should these findings become established facts, we will have an example of an early, highly original center of cultural growth attributable to a dark-skinned race. In America, culture first reached the level of civilization in the Middle-American area (southern Mexico to northern Chile). This accomplishment was the work of the American Indians, people of Asiatic origin, and may have resulted from the stimulus of ideas reaching them within the last few thousand years via the Pacific.

This review conspicuously omits such regions as Australia and Tasmania, Melanesia, Africa south of the Sahara, the subarctic areas of both the Old and New Worlds, and most of the subtropics of the New World as well. These areas include representatives of all the major races. Hence all races have examples of cultural laggards. Another assessment of the facts just reviewed would stress the fact that every civilization mentioned has had Negroid influences. Negroes were in north Italy at Grimaldi in the Upper Paleolithic period, and traces of Negroids are prominent in the Middle East, Egypt, and India. Even China was Negroid at least in its southern half during the first millennia B.C., and there is growing acknowledgment of the important contributions that south China made to the cultural complex we call Chinese civilization. As pointed out earlier, there is a strong Negroid element in the American Indians, and in some of the earliest civilizations in America (the great stone heads of the Olmec and the Danzantes of Monte Albán in Mexico) Negroids are prominently portrayed.

Much evidence now points to the Middle East as *the* center of the origin of civilization. Egypt early received impulses from this center. Later the stream of ideas flowed westward up the Mediterranean and northwestward into Europe. North Africa, like South Africa, made abortive starts toward civilization. Carthage is an example of an area where the cultural loss was neither racially nor climatically caused. Ideas also flowed northeastward to China and eastward to India. Trade surely was a major factor in this cultural flow, for we have evidence of sea trade going from Mesopotamia to the Indus Valley of India via the Bahrain Islands in the Persian Gulf as early as 2000 B.C. From India and China, developments in Southeast Asia and Indonesia were stimulated. With due allowance for deserts, mountains, and distances (and even greater allowances for the accidents of history), the effect is somewhat as though a rock were thrown into a somewhat irregularly shaped puddle: the ripples agitate the nearby waters greatly, the distant waters only later and slightly, and the sheltered distant coves hardly at all. It is a peculiarity of man on earth that the Old World race farthest from this center of activity is the Negro race. Melanesia, Australia, and Tasmania are remote lands; Africa south of the Sahara is also. This explanation is not the only one possible, but it will be left here for the moment and the details dealt with in later chapters.

Distance alone is an insufficient answer to differences in development, as is illustrated by the happenings in America. A massive flow of ideas was carried by peoples across the immense Pacific to set off the civilizational growth in Indian America.[7] Note that this

[7] This topic will be dealt with in some detail in Chapter 2, and evidence of trans-Atlantic flow of ideas will also be presented.

flow of people and ideas bypassed Melanesia. Was this because Melanesia is a great Negro area or because there was nothing in Melanesia to attract these people? For example, after 1500 the Spanish explored the Pacific and established an outpost at Manila. They ignored black Melanesia, but they also ignored brown Polynesia. They were not discriminating against race but making an economic choice. Neither Polynesia nor Melanesia was economically developed. Nor were they well located for trade with China and Southeast Asia, areas of spice, silk, porcelain, and other desirable goods that were the goals. Race had little if anything to do with the decision. However, the decision had far-reaching results for the races. The black-skinned Melanesians were left to their Neolithic way of life, and only late in the twentieth century have some of them finally seen white men, firearms, new crops, and domestic animals and met the worlds of new ideas that the Filipinos had thrust upon them four hundred years ago.

When Asiatics contacted America and their ideas took root in Middle America and began to grow, the spread of culture was again roughly symmetrical. Since the racial base of pre-Columbian America can be taken as uniform, and within this one race every cultural stage from Lower Paleolithic to the Bronze Age has been found, we see that cultural growth closely parallels proximity and opportunity to receive ideas. When the Europeans entered America, they made economic decisions about where to settle and what land to develop. The Spanish elected to ignore California and develop Mexico. The Indians of Mexico, after an initial setback, received an immense cultural boost while their racial cousins in California lagged for centuries. The original differences, as will be brought out more fully below, were due to the accidents of pre-Columbian cultural history. This set the stage for the Spanish decision, and these decisions accentuated the differences. This reasoning is an example of what can be deduced by using time and space to test ideas—as if in a laboratory experiment. We can conclude that racial arguments based on past performance therefore are inconclusive.

Finally, the European view of cultural history can be viewed from a number of vantage points. If one were to take a learned Egyptian's view of the world in 2000 B.C., it might be as follows: The perfect climate for mental achievement is the hot, desert climate. Here a brown-skinned, dark-eyed, slender, superior race has developed to the peak of human achievement. Look at the Egyptian-Mesopotamian-Palestinian achievements. Note also that farther south it is too hot; men are burned black and accomplish little. And farther north, on the other side of the Mediterranean, men have to endure cold winters, are lighter skinned, and accomplish little.

Yet 1500 years later it was the Greeks, men "on the other side of the Mediterranean," who would be the most accomplished scientists, architects, and dramatists of the world. Their view was that *they* had the optimum climate. Egypt was too hot and equatorial Africa more impossible. The lands to the north of them were even worse. Winters were horribly long and cold and summers short. The bleached-out people of that land not only had never accomplished anything (good evidence therefore of their racial inferiority), but they obviously were never going to accomplish anything because of their terrible climate.

Two thousand years later the racist theory held that these northern freakish blonds constituted *the* superior race, and the physical-environment determinists claimed that these cold northern lands were mentally the most stimulating in the world.

It should be clear now that neither race nor physical environment (especially climate) can be *the* answer. It seems doubtful that even in combination they yield a satisfactory answer; we must look much more deeply into the problem. The final answer does seem to be the presence of achievement when the opportunity is also available. The Negro race now has that opportunity, though it has had it only briefly. Men in New Guinea and Africa south of the Sahara are somewhat in the position of the northwest Europeans who, brought suddenly and forcibly into contact with the advanced cultures of the Mediterranean by the 500-year-long Roman conquest of their region, were jolted out of their primitive, tribal, agricultural way of life, and emerged, 1000 years after Rome collapsed, as the world's dominant people. Had men from Mars viewed the barbaric state of Europe after the fall of Rome, would they have been justified in deciding that the blondish race was mentally inferior because after 500 years of opportunity they were still barbarians? The difficulty with too many views of modern problems is the lack of such perspective.

Today the emergent people seem to expect instant civilization yet cannot in the nature of things accomplish this miracle. Their critics too easily attribute their failure to their tropical habitat or to their race, depending on the circumstances of the emergent people and the bias of the critic. No people are likely to accomplish in a decade, or even a century, what other peoples have required millenia to accomplish. Even given modern rapid communications, improved standards of living, the free education available to so many today—and unheard of in the past—it seems unrealistic to expect a mass of mankind of whatever race suddenly

to blossom. Individuals may make rapid changes, but masses of men change slowly. This theme will be illustrated here again and again for differing races in differing physical and social environments. It should provide a saner base for viewing potential achievements of all people of whatever race anywhere.

HYPOTHETICAL MAN ON AN UNREAL EARTH

One way to deal with complex problems is to simplify them. Much of the confusion over the relationship between men and climates and locations is that there are so many combinations of man and land. Our earth is almost infinitely complex. In a sense, no two parts of it are alike. Climate, soils, geology, vegetation, and drainage combine in a myriad of forms. Man seems equally complex. He is divided into numerous races, is a product of highly varied cultures, and is scattered over this physical earth like a mixed-up jigsaw puzzle. The easiest way to simplify this problem so that we begin to see the nature of it is to imagine an unreal earth populated by a hypothetical people.

A Uniform Earth

Instead of the complex physical earth of reality, we can imagine an earth that is a simple plain. Let this earth have no oceans dividing the continents. Let the climate be uniform: no tropics, no polar climates; instead, a mild subtropical climate everywhere. Let the vegetation be a parklike landscape with more grass than trees. Soils will be uniform too. We will assume a uniformly random distribution of animals: many small animals and some big game. We can also assume similar geology everywhere, but it need not be all one great limestone plain. It can be somewhat like the present crust with basalts, limestones, coal measures, and granites, but with these differing materials so distributed that any area 100-miles square has approximately the same array of minerals. Our physical earth is then uniform from pole to pole; there is no advantage to any location in respect to oceans, climates, preferred soils, or other resources. There would be no basis whatever for a difference among men caused by the physical environment, for the physical environment would exert the same influences on all men.

The population of this earth will be made up

of one kind of man, uniformly distributed over the earth. There will be no question of whether one group will be different in innate capacity from any other group. There will be individual variation, of course, but the percentage distribution of morons, geniuses, and just average people will be the same everywhere. All these people will speak the same language, and they will all have the same way of life: the simplest of human economies. They will be hunters and gatherers with no specialization either toward hunting or gathering. This stage is probably where man did begin in his cultural climb, so it makes an ideal starting point for discussion.

We now have a most unreal world, but a most interesting one from the point of view of considering the origin of different ways of life on the earth. There are no physical-environmental causes that could lead to differences—we have simply eliminated them. We have done the same for any possible racial differences. What will happen? Will things remain unchanged? If change occurs, will it be the same everywhere?

If we were to find that change does occur, then we would have learned something important. It would suggest that neither differences in physical environment, nor in race, nor in way of life are necessary for change. This would not mean that race and environment would have nothing to do with cultural change, but it would certainly change our perspective on the causative factors of change.

Invention as the Cause of Change

Suppose that in this uniform world a group of people are gathered around a fire for their evening meal. A stew of meat and roots and wild seeds is cooking in a vessel on the fire. This assumes that these people possess pottery—an unusual thing among hunters and gatherers, but it does occur among such people as the Diegueño of southern California. Everyone in the family is dipping into the kettle to fish out some food. The food is hot, and the efficiency of their eating is being hindered by the scorching of their fingers. One of the young men of the family in desperation takes a stick and spears a choice piece of meat, removing it from the kettle without scorching himself.

This is an invention: a special way of doing something has been created. The family would undoubtedly be astonished. They now have important decisions to make. It is easy to see that a natural reaction would be anger. The young man is getting away with the best food and not even burning his fingers. His father might deal him a clout on the side of the head, and that would be the end of that invention. Or the family might look on in amazement and not copy him at all, shrugging it off as just another of the eldest son's idiosyncracies. Or, finally, they might all seize sticks and learn to spear pieces of meat without burning themselves. In this last case, cultural change has occurred; this family is now different in one trait from all the other families in the world. Note that neither race nor physical-environment differences have played any part in this process.

The process of invention is difficult to understand fully and deserves much more discussion than can be given it here. Do inventions occur because of "need"? This one was certainly "needed" because without it there was always the possibility of burning one's fingers. But generations had come and gone without anyone else feeling this need. The need was always there, but apparently here was the first person to realize it. Once the invention was made, of course, the "need" became apparent. This, in fact, seems to be the usual rule, and it greatly diminishes the force of the old saying "necessity is the mother of invention."

Need and invention are firmly fixed in most people's minds as following logical sequence, but a number of cases of inventions that preceded the need for them will be mentioned here. A few of them will reappear later. For example, the six-shooter, the windmill, the steel plow, deep-well drilling—all of them inventions needed for the American settlement of the grasslands—were made elsewhere for other purposes. Investigation of each of them would show remarkably little "need." The six-shooter was invented by a New Englander who had little need for a multishot side arm that turned out to be exactly what the Texas Rangers and other Westerners needed. Other, and perhaps better examples, are to be found in the field of animal domesticates. Cows were not domesticated for milk, nor chickens for flesh or eggs, nor sheep for wool, nor horses for riding. Chickens were first domesticated for magic and ritual, seemingly filling a psychological need of the people. Eating chickens and chicken eggs came thousands of years later. Airplanes, without which we would have great transportation difficulties, were invented by two bicycle repairmen who were simply playing with the idea of flying and invented a worthless machine (by our present standards) that has become a necessity. Nitrous oxide was not invented as a pain killer, nor were its first users—who got their kicks from sniffing it—trying to escape pain. Its anesthetic properties were discovered almost by accident.

Now we need anesthetics. In one sense, we always did, but we didn't invent them out of need; we invented them and then realized we had always needed them. It is in this sense that we say that need follows upon invention, and that necessity as the mother of invention is a cliché that more often misleads than aids us. The great question is why one individual suddenly breaks through the accepted pattern to create something utterly different.

Sometimes we can get glimpses of the process. In the case of the meat spearer, there were prior conditions that set the stage for the invention. Clearly, if these people had not cooked their food, the need to avoid handling hot foods would not have arisen. Probably these people had some tools that they used in their hunting and gathering. If we study primitive people living on this cultural level, we find that they often used pointed sticks or twigs—to pry lizards out of cracks in rocks or hold spiny fruits such as cactus apples. If we assume that in our example such things were in use for these purposes, we see them as antecedents, as the prior conditions that set the stage for the invention. In all of human history there has probably never been an invention without antecedents, for the rather simple reason that our prehuman ancestors were already tool users and, as the current studies of monkeys and apes show, were already making inventions. Emerging men were already culture bearers and made their inventions in terms of the antecedent knowledge they already possessed. This has not changed through time.

Ultimately, we are looking at the problem of how the human mind works—how problems are solved. The process of invention occurs when certain individuals are able to hold more ideas in view at one time and see them in more new combinations than most other people can. These are the creative people in all societies; so far as we know, they appear at random in all races. Perhaps it is not random chance but that—like musical and artistic aptitudes that are more easily recognized—they develop in certain family lines with greater frequency than in others. However, such considerations need not bother us here, for we have assumed a uniform race situation, with family lines of varied ability uniformly distributed over the earth.

Frequency of invention

Under these circumstances, the question arises whether an invention might be made more than once by separate individuals independently. There are numerous cases of simultaneous inventions—the telephone, the use of anesthetics, the application of cal-

culus. All these inventions came late in time when the inventors were part of a group working on common problems from a common background, directly or indirectly in touch with one another's work. When we can put our finger firmly on the history of one of these inventions, perhaps an innovation in thought such as Darwin's theory of evolution, we find that the idea was "in the air" and more or less clearly stated by a number of men for about a century. Thinking along these lines can be found even as early as the Greeks. In this light the simultaneous "invention" of evolution by Darwin and Wallace seems much less mysterious.

To take another example, all of the alphabetic writing in the world can be traced back to a single point of origin. The same is true of the wheel, the true arch, and a great number of other inventions or innovations. The list of inventions for which there is much evidence of separate origin tends to shrink as our detailed knowledge grows. The requirements for the creation of a new idea, getting it accepted locally, and spreading it afar are apparently so rigorous that the independent appearance of inventions is kept to a minimum.

Diffusion

It is not enough that the invention must occur, even though this itself is difficult. Once made, the invention must be adopted by the immediately surrounding group; if the invention is really to survive, it must spread to others—there must be diffusion.

This step, of course, is easy to put into our hypothesis of the uniform earth. Suppose that the young man who has made this invention has been praised rather than beaten, and that the whole family is now using these clever little skewers to snatch things from the stew pot. Let us also suppose that this young man visits a neighboring family group in the next valley, where there is a daughter about his age. If this is a clever young man, he is likely to arrive at this neighbor's camp with some game, perhaps a pair of rabbits, and get himself invited to stay to dinner. Finally the critical moment arrives—the food is ready, but hot.

Should the young man use his meat spearer now or wait until he is safely married into the family? If he uses it now, should he eat the piece of meat himself, hand it to the girl, or offer it to her father? This is obviously touchy business. Maybe the mother is the real head of the family; or maybe they will think him an apple-polishing sissy for not having the courage to go ahead and eat it himself. Or, they may, no matter what he does, think his act a foreign affectation and run him out of camp. Although this account has almost reached the comic-strip level of discussion, it makes simple and clear the type of situation that is easily documented in the history of invention. These processes of invention, acceptance, and diffusion are real, and they have acted through time to limit severely the appearance of new ideas.

We have here another example of the uniformity of the problems that face all of mankind. Most of mankind goes courting. In societies where marriages are arranged by the elders, the problems of impressing the other family by proper behavior are then simply transferred. It doesn't matter whether the approach is made by the young man, his maiden aunt, or a professional go-between. Success depends on skill in dealing with the other person or persons and requires at least a minimum of judgment of situation and character as well as good timing and correct knowledge of the local pecking order. Those who succeed get married, and to the extent that their success depended on genetic superiority, they pass on that superiority to their offspring.

We can suppose that the diplomacy of the young man is up to the occasion this time, and that soon another family joins the magic group that eats with the aid of wooden skewers rather than with their unaided fingers. This is a momentous occurrence: without race or environment entering the picture, the world has become differentiated culturally into those who do and those who do not eat with their fingers.

One can ask why this might not occur again and again, or at least several times over the whole earth. The reply is that no invention is obvious until made, and that in all cases for which we really know the history (true arch, alphabet, wheel, and so on), there is only one origin. As our knowledge of cultural history grows, the importance of diffusion increases, and the role of independent invention diminishes almost to the vanishing point. There is no reason to think that man has ever been different.

There are other possibilities to account for the spread of an idea. It might be that some family living quite far away from the area where the meat-spearing-stick had been invented would hear rumors that something of this sort had occurred. They might not hear about the pointed stick, but they might learn that something besides fingers could be used to get meat out of a boiling pot and be stimulated to try to do something like this themselves. They might try a split stick and attempt to jam the meat into the split. Or they might try a split stick and use it like a fork. Or instead of a stick they might use a splinter of bone. It would also be possible to use a stick with a sharp angle to it and to use this like a hook. Perhaps there are other possi-

bilities, but these should suffice. First, we should note that no one had tried any of these ideas before. It was only after one group had been stimulated by the notion that this kind of thing could be done that they were moved to try it. Second, notice that some of these solutions—the hook, for example—are sufficiently different from the first one. If we did not know that the stimulus had occurred, we would have difficulty in deducing it from the evidence of the implements. Stimulus diffusion is probably more important in human history than we realize, simply because the relationships are often not obvious.

In the natural course of things, this idea would be bound to spread. One might assume that it would spread uniformly in all directions over this uniform earth, but this may not happen. Poor personal relations between one group of families and another could cause the rejection of the idea by a block of people in some area. The people beyond these curmudgeons might have welcomed the idea if it had ever reached them, but it cannot. The spread of the idea might be very uneven, and the differentiation of the world in terms of this idea is not likely to be a simple concentric pattern outward from the original center.

Change and Its Consequences

We can assume that changes would go on at some very slow rate all over the earth. We can document the fact that for mankind the rate of change through most of history has been one of incredible slowness, and that for all branches of mankind out of touch with the explosive growth of ideas in one or two centers, there has been virtually no advance. Certain changes did occur on this natural earth of ours, however. Some men became experts at hunting and specialized in that. Others became increasingly interested in plants. These differences we can assume would arise on our uniform earth also.

Specialized hunters

Hunting is a man's game. It suits his personality. It features great physical coordination, cunning and planning, and a big payoff. It is a gambler's game. After watching his prey for days and planning when to hide at the point that the buffalo is bound to pass on his way to the water, the hunter must take split-second action. There is a half-ton of meat in the pot—or nothing. The gathering of roots and seeds and berries was the woman's work. Her time was occupied with child-bearing and child-caring; she was not free to go the long distances and do the strenuous things that the man did. She may not make a killing at seed gathering, but there will surely be a little gruel for everyone. Women concentrated on the plant foods.

Some men are better athletes than others, and hunting is an athlete's game. In a family where the father is a great success with the bow and arrow, there will be more game brought to camp. The sons will be trained by their father and will also bring in plenty of game. Mother and the sisters will be kept busy dressing hides and drying meat. There will be less and less time for them to get out and gather seeds and roots and berries. It does not matter, of course, for there is plenty of nutritious meat in camp. Pursuit of game normally concentrates on the biggest game; one arrow can kill one rabbit or one deer. Big game is limited in number. If one comes to pay more and more attention to it, it becomes necessary to move more frequently to find areas where there are abundant supplies of the desired animal. It is easy to see how a family that specialized in hunting would become highly mobile, would have more sedentary women, and would perhaps drift toward male dominance since the men were the important breadwinners.

The changes started in this one family will have a chance to spread. They may not, of course. The sons may not like hunting, or the neighbors may frown upon undue emphasis on it. In this case the idea is rejected, and the start at cultural differentiation dies. However, let us suppose that the idea catches on. The daughters-in-law and sons-in-law bring close ties with other families in the region. They tell of this way of life with its abundance of rich meat and warm skins. It sounds good, and presently the people for a considerable area around are paying more attention to hunting and less attention to other pursuits. Another culture area has now arisen: specialized hunters have appeared. They could, of course, appear more than once, in more than one part of the earth. There could, of course, be several kinds of hunting. One group might hunt elephants and be content with making a kill every two weeks. Another might specialize in deer and aim to make a kill twice a week.

The consequences of these differences would be slight at first, but they would be cumulative. In the end they would lead to very great differences in all aspects of life. Emphasis on hunting would probably lead to improvement of hunting tools. Presently there might even be the invention of totally new hunting equipment; new ways of doing things would certainly be invented. These new activities and the thoughts associated with them would require appropriate words to express them. Vocabulary differences would then arise. Languages differ for less cause than this. Compare British English and American English. These two languages are drifting apart despite the shortness of time that we have been separated and the closeness of the ties that we have maintained. On the primitive level, there would be no contact between the peoples on one side of this uniform world we are discussing and the other. Although there would be no barriers other than distance and people, these would be quite enough.

We can also consider the effects that hunting might have on man physically. Hunting places a premium on certain qualities. The more successful hunter will often have more of these qualities than do others. In times of famine, he and his family will be better off, more of them will survive and have children, and thus more of the good hunter's favorable traits will be passed on to the descendants. The traits desirable in a hunter are not necessarily the same as those that make a good farmer or fisherman. However, this kind of speculation is in a difficult area. We cannot definitely know what makes a good hunter or farmer, and most of what we know suggests that cultural tradition is often more important than biological inheritance. The possibility remains, nevertheless, that differing ways of life might conceivably lead to differing physical make-up.

A still further change would presently make itself apparent. An emphasis on hunting and a lessening of interest in plants would slowly but inexorably shift the whole aspect of the landscape. Plants harvested by man grow differently from those that are not—but more of this later. Animals also respond to hunting, for hunting tends to strike a balance and keep the level of game fixed. Hunting tends to influence not only the animals hunted but also all the animals even remotely concerned with the primary game animal. And through them, a whole range of other animals are affected, as well as the plant world on which they feed.

For example, if much large game is regularly killed, there may be plenty of offal around, and scavengers such as the jackal, wolves, buzzards, and all similar animals will increase in numbers. This in turn may have extensive repercussions. The existence of more wolves, for example, may mean that more dens are dug that will eventually supply homes for still other animals. More wolves may mean that the number of rabbits is kept down, and this trend will allow more grass to grow, which in turn may favor the livelihood of still another animal. The chains of cause and effect that exist in nature are astonishingly intricate and far-reaching; a change in any link in one of these chains may be felt in distant and seemingly unrelated ways and places. The effect of so simple a thing as shifting major attention toward hunting most certainly would then have social, cultural, and physical-environmental effects, and probably would react on physical man also.

Fire

There is another thing that hunters over much of the world have learned to do that has a tremendous impact on their physical environment. They have learned that fire can be used to drive game in desired directions. Sometimes the purpose is to drive game past the hunters, at other times to drive it into traps or to stampede it over cliffs. Fire is highly effective: animals will flee before it. Man can judge wind and distance and, knowing the habits of the animals, can drive the game where and when he wishes. Of all the deadly tricks that the hunter has mastered, this may be the most extensive in its effects, for fire can almost totally remake the land.

Fire is the enemy of perennial vegetation such as trees; frequent fire is very likely to create a grassland. When the land is characterized by open forests, with grass growing through them and with a pronounced dry season, frequent fire is certain to limit tree growth and give the grasses a competitive advantage. To thrive, grass needs sunlight; therefore trees can shade grass out. In rainy lands, grass predominates in the open places where rocks outcrop, along the frequently disturbed stream banks, and in similar spots where perennial vegetation cannot maintain a foothold. In more arid regions, trees may not be able to achieve sufficient density of growth to maintain a dense ground shade. The roots of the trees must be able to spread out to absorb sufficient water. Grasses will then appear between the trees. During the wet season, there is water enough for all. The grasses hurry through their life cycle and get through the following dry season by dropping their seeds in the ground to await the next favorable growing season. The tree survives by a deeper root system and often by such protective devices as shedding its leaves.

But fire comes with the dry season. The grass is consumed, but it does not matter. Only the dry straw is there to burn; the grass has done its work. Although some of the seeds may be burned, many of them are lodged in the ground and a future generation of grass is assured. The tree, however, is vulnerable, especially when it is young. The seedling's stem may be heated through and killed. If the tree is a little larger, its thin bark may be roasted and the living tissue beneath the bark killed. Large trees, of course, often have bark thick

enough to withstand considerable fire, but if the fires come with such frequency that the young trees cannot survive and the old trees succumb one by one, only the grasses will remain. They will flourish on the increased sunlight and the decreased competition of the trees for the precious water. With the trees will go the animals dependent on trees: the tree squirrels, the arboreal birds, and all the rest. With the increase in grasses a different group of animals will increase, among them the large grazing animals man is interested in.

Although these effects are most visible on the edge of the dry lands of our real earth—as opposed to the hypothetical earth that we have been discussing—they are equally true for all of the earth. Even in tropical rain forests, man creates grasslands by his persistent setting of fires. We will continue this discussion later.

When vegetation alters, the soil is also changed. Grasses have deep, fibrous roots. Their decay adds organic matter through a thick layer of the earth. This is the opposite of the action of trees: they tend to live for a long time because their roots are more permanent. By dropping leaves, as well as fruit and branches, trees add organic matter to the surface of the earth, rather than to a thick layer. Grassland soils are therefore blacker, richer in humus, and generally very different from the soils formed under forests. One must add that this is an enormously complex matter. Some grasses multiply from roots, and when their tops are burned off, they simply sprout again. Trees vary greatly in their reaction to fire—some need fire to give them a start. However, fire generally favors grass over trees.

Our uniform physical earth therefore develops variety in its soils and vegetation and its associated animal life through the divergence of activities that we can expect from man. Instead of a picture of the physical environment forming man, we see that man, in this case a specialized hunter, can shape his physical environment.

Harvesting

In other areas of our uniform earth the way of life might change little, or a different type of change might occur: the women might shoulder more and more of the burden of supplying the food. We know that among some primitive people the women claimed the rights for gathering in certain areas. Here they knew virtually every kind of plant and its uses: they knew which grass bore seeds, where it grew best, and when the time was ripe to harvest it. They also knew the plants that had edible tubers. Australian women knew that if they cut off the vine from the tuber and stuck it back in the disturbed ground, the vine would grow and in a year or two another tuber would appear. Such knowledge verges on the beginning of agriculture. This "invention," or recognition of cause and effect, seems to have been one that did not lead directly to agriculture, however, though it set the stage for it. It is of interest to note that this plant knowledge belonged to a people who were Paleolithic in cultural level (that is, they did not have the bow and arrow, pottery, ground stone, agriculture, or any domestic animal except the dog).

The Australian situation suggests that agriculture had an exceedingly long prehistory, that domestication of plants was the work of women, and that they drifted into it perhaps quite unconsciously. If agriculture did develop in this way, it would help explain why basic domestic plants were already fully established when we first find firm evidence of the existence of real agriculture about 10,000 years ago. In general, our domestic plants—such as wheat, barley, corn, and rice —are so different from their wild ancestors that we are unable to point to specific ancestral forms, only to the general plant family from which they are derived.

Other evidence suggests how women's work could change plants. Among certain California Indians, the women gathered wild grass seeds by using a seed beater, an implement that looks a bit like a small tennis racket. The women walked through the grass, striking the seed heads with the seed beater so that the seeds were thrown toward a large basket tray held in the other hand; the seeds caught on the tray were then dumped into a basket. Large quantities of seed could be gathered in a day in this way, depending on certain conditions: the grass had to ripen uniformly, and it had

to hold its seed firmly and not shed them when any small breeze shook the plant. These conditions set the scene for the Indian women to make an efficient harvest. All this was fine for the Indian women, but just why should the grass have cooperated? If the Indian women got all the seeds, it meant extinction for the plant. Most grasses have loose seed heads, and they often have hair devices that make the seed cling to passing animals. In addition, plants often ripen seed over a considerable span of time, adding to the survival chances of seed. Selection in nature encourages such survival traits. How, then, did domestic grasses that hold their seeds tightly, ripen uniformly, and have no devices for being spread about—wheat, barley, rye, oats, rice, and corn, for example—come into existence? Did the Indian women select and encourage grasses for this quality?

Probably not. They simply gathered wild seeds; the method was vigorous but inefficient. Although they got a lot of seed, they scattered a lot also. The most that a heavy-seeded grass could hope to gain from a passing breeze was an inch or two of seed dispersal, while a brawny Indian woman striking the seed head a smart blow to break the seeds loose and send them flying toward her basket might well give the seeds a

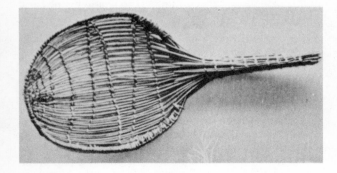

Seed Beater of the Type Used by the California Indians. (Smithsonian Institution)

(Top left) **Hopi Indian with Hoe, 1902.** *Hopi agriculture is better described as taking place in spite of the physical environment rather than in response to it. Maize is being grown here in a near-desert environment by enormous exertion and with special varieties of maize.* (History Division, Natural History Museum of Los Angeles County) (Top right) **Hopi Basketmaker, 1900.** *Weaving baskets is traditionally woman's work.* (History Division, Natural History Museum of Los Angeles County) (Bottom) **Zuñi Pueblo, 1897.** *The Hopi pueblo, which virtually duplicates the Zuñi pueblo, is a fortified mud and stone structure that resembles an apartment house in that it is inhabited by numerous families. The Hopi exemplify one way of life in an arid environment. Despite near-desert conditions, they are farmers. They build in the mud and stone that is natural to the region. Women weave baskets with material derived from the native plants. All of this may seem natural and suited to the environment, but not all arid lands are so used. Few lands as arid as the region of the Hopi are farmed without irrigation. Many arid lands do not use mud and stone to build structures such as pueblos. While basketry is extremely common and native plant materials are usually employed in basketmaking, the details of design and shape vary greatly from one arid region to another. Even in uniform environments there is great cultural diversity.* (History Division, National History Museum of Los Angeles County)

dispersal of a foot or two or three. If Indian women for thousands of years gathered seeds annually, there would automatically be a steady selection applied: the grasses that held their seeds firmly until the women came to disperse them might pay a price in some seed being caught by the women, but the rest of the seed would get maximum dispersal. Thus without any plan the plant and the woman would become increasingly dependent on each other. The plant would be modified in a direction that seemingly was not to its best survival interests in nature, but which actually was when women with seed beaters were part of that nature. Not all women gathered seeds in this way; therefore not all grasses would be so affected. In our uniform world, only in that area where the women worked out such a system would such grasses arise, and only there would parts of the physical environment be characterized by increasingly large areas of one kind of grass with these peculiar characteristics. Again, the activity of mankind would be altering the physical environment.

In California, where the women gathered seeds from the wild grasses, a striking thing happened. When the Indian women disappeared, the native California grasses nearly disappeared too. More than the disappearance of the Indian women was at stake, of course. The Spanish appeared and later the Americans. They introduced cattle and subjected the native grasslands to heavy grazing, and they also introduced whole galaxies of weedy competitors for the soil and sun of California. Nevertheless, one of the factors of change was that the Indian women no longer annually gathered the wild grass seeds, striking the seed heads with their seed beaters and giving the seeds a maximum dispersal.

Other Examples of the Effects of Cultural Change

There is an even clearer case of the effects of cultural change. In one part of Indonesia the matrilineal tribes live in forested areas, and the patrilineal tribes live in grasslands. There is no appreciable climate difference between the two areas. According to the climate, both should have been covered with tropical forest. At first sight it may seem strange that counting descent from your father should discourage tree growth, while counting it from your mother should encourage the trees. The answer is found to lie in the way in which these two groups make their living. The matrilineal society is a farming society. The forest is their long-term rotation crop and is valued as such. The patrilineal people are cattle raisers. The forest supplies no sustenance to their animals, who want grasses particularly young, new grass. At every opportunity they burn the land, and the fierce fires running through stands of grass 6 feet high kill trees that stand in the grassland or on its border. Slowly the forest edge is driven back, and the wet tropical forest gives way to grassland. Hence, even in Indonesia, in areas of uniformly tropical, rainy climate, farmers retain the forest while the cattlemen change the landscape to a grassland.

There is almost no end to the process. The attention of farmers is on rain and sun. Accordingly, their religious rites were different from those of the hunters. The hunter had special words for the implements he used and the animals he hunted. The farmer too had his own specialized vocabulary, for *clod* and *weed* are farmers' words, just as *scent* and *spoor* are hunters' words.

That changing activities can change language is amusingly clear in our own society. Few urbanites can now define *hame, breeching stay, lazy strap, trace, crupper.* Perhaps many would not even be sure where these terms come from. In fact, these are terms for parts of a horse's harness. Fifty years ago anyone would have recognized them instantly. We have also lost a rich and salty vocabulary filled with *royals, dolphin strikers, jiggers, top gallants,* and so forth. Of course, it has not been one-way traffic; our vocabulary is now rich with *carburetors, accelerators, distributors, spark plugs,* and a dictionary of other terms that were unknown to our nonautomotive great grandfathers. Divergence in way of life clearly changes vocabulary, and continued separation and change slowly penetrate the structure of language. Changes could be far-reaching and thorough even on a uniform earth, and a vast diversity of languages could be expected.

Agriculture and population growth

The greatest changes, however, would come when the magical threshold of purposeful agriculture was finally crossed. Most of man's early activities led to economically limited ends. The hunting and gathering way of life cannot maintain a large number of people on the earth. Gathering requires wide areas over which man can wander to gather first this wild harvest and then another. Hunting also requires large areas, for even under the best of conditions there are only so many animals per acre of land. It takes not acres but square miles of hunting territory to support people on this cultural level. Domestication and herding animals by pastoral nomads can support only a few more people per square mile than hunting does. The animals require food and therefore extensive areas for grazing. The purposeful raising of plants changes all of this. With agriculture, permanent dense settlements become possible. All of the earth-modifying effects of man were multiplied as the number of men increased.

Summary

The point of our hypothesis should be clear by now. On a uniform earth, with neither human biology nor physical earth-differences to serve as compelling forces to bring about change, change would still occur. Various changes in way of life are possible. Some would alter the landscape in one direction and others in quite a different way. Some would not alter the population density and would therefore be self-limiting. One of these ways of life, the agricultural basis of existence—springing from the lowly plant gatherers—would potentially lead to dense populations, the specialization of human activity, and the basis of civilization as we in fact know it.

We do not, then, need a diversified earth to create variety in the ways of mankind. This does not mean that our own diversified earth had no effect on the development of the many different ways of life that have actually come to exist. The isolation of some areas, the dryness of others, and the great cold to be dealt with in still others all have had their effects. The purpose of the uniform-earth hypothesis is not to deny these effects but to put them into perspective.

If a change is probable on a uniform earth occupied by uniform man, we are cautioned against simple conclusions about either race or physical environment as driving forces in the development of the ways of life or accomplishment on the actual earth. Since man contains the potential for change irrespective of race or physical environment, every case of action or lack of action resulting in great achievement or little, complex culture or simple, has to be most carefully evaluated to determine the real causes.

The plan of this book is to reduce complexity a little. As the first step, we tried to clarify the problem of race by indicating that we cannot expect mental differences to parallel the physical differences that mark the races. Mankind is probably closer to uniformity mentally than physically (in traits such as skin color). Since it is mental ability that is critical, the man we imagined populating the uniform earth is nearly approached on our actual earth. If now we could deal with the physical earth on some basis that reduces the complexity of the major variables of physical man, physical earth, and cultural man (or race, environment, and way of life), we could simplify the enormously complex problem of the interaction among these variables. Then we should be able to see at work the processes that have created the culturally diversified earth that actually exists.

We can assume that within a climatic zone, such as the arid lands, a certain amount of physical-environmental uniformity will be found. When there is a lack of moisture, drought-resistant plants, particular kinds of soils, landforms, and animals that have adjusted to these physical limits create a basic uniformity. There is still variation in detail, of course, but the overall uniformity approaches a control over this physical-environmental aspect of our problem. If race differences are reduced to minimal levels,[8] and if physical-environmental cause is limited in its effect by consideration of one major environment at a time, the variations in way of life, including achievement or lack of it, cannot lightly be attributed to race or environment. By use of the history of cultural growth in differing parts of the world with similar climates, the role of cultural history can then be clearly seen.

Now it is time to begin examination of the actual ways of life of the real men on this planet we call earth. By the accident of its origin, its geological history, and its setting in our solar system, the earth has a surface divided between land and water; the land is diversified by mountains, plains, and plateaus; differing climates give some uniformity to the minor landforms. The vegetation, plants, animals, and the soils of each of these climatic zones must therefore be examined to see what mankind has done with each kind of landscape.

[8] More evidence of the inadequacy of the race argument will be presented in appropriate sections that follow.

Review Questions

1. Why should a geographer be concerned with the problem of racial differences?
2. Under what conditions would the physical environment have the maximum effect in the formation of races? What does this suggest about the time of race divergence?
3. What is wrong with the argument that man does his best work in higher mid-latitudes because he is physiologically most at home in such areas?
4. Since separation and adaptation combined with genetic drift have led to clearly marked bodily differences, is it to be expected that there would be an equal degree of mental differences? Explain.
5. What is the effect of changing the time of viewing human accomplishment from millennia to millennia—shifting from the twentieth century, to the first century, to 1000 B.C.?
6. Discuss whether "necessity is the mother of invention."
7. What is an antecedent? What is the role of antecedents in invention?
8. What is diffusion? Give some examples from the present.
9. What does the relation of man to plants and the origins of agriculture suggest about man's inventiveness?
10. What is the significance of the evidence for Negritos in China?

Suggested Readings

Carter, G. F. "Pre-Columbian Chickens in America." In *Man Across the Sea.* Edited by Carroll L. Riley, J. Charles Kelley, Campbell W. Pennington, and Robert Rands. Austin: University of Texas Press, 1971.

Coon, Carleton S. *The Origin of Races.* New York: Alfred A. Knopf, Inc., 1962.

Fleure, H. J. "The Distribution of Types of Skin Color." *The Geographical Review,* vol. XXXV, no. 4 (1945), pp. 580–595.

Kates, R. W., and Wohwill, J. F., eds. "Man's Response to the Physical Environment." In *The Journal of Social Issues,* vol. XXII, no. 4 (1966). See especially D. H. K. Lee's article, "The Role of Attitude in Response to Environmental Stress."

Loomis, W. F. "Skin Pigment Regulation of Vitamin D Biosynthesis in Man." *Science,* vol. 157, no. 3788 (1967), pp. 501–506.

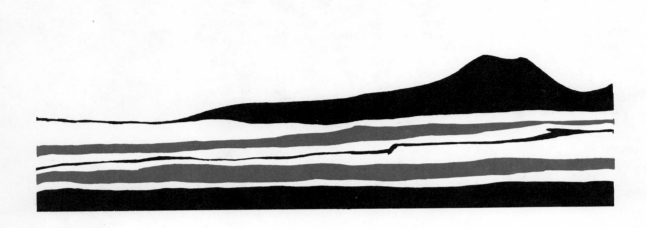

Arid Lands

CHAPTER TWO

ARID lands on each of the continents occupied by representatives of all the races present examples of similar environments inhabited by different varieties of mankind. There are Negroid people in the arid lands of South-West Africa, Negroid and Caucasoid people in the Sahara, Mongoloid and Europoid people in the interior arid lands of Asia, American Indians followed by Caucasoid people in the arid lands of America, and Australoids in the interior of Australia followed by Caucasoid people. With these examples we can compare different races in the same environment. We can also compare cultural accomplishment. Before 1500 there were people living in exceedingly primitive conditions in the Australian desert and in part of the North American desert. Pastoral nomads lived in some of the Old World deserts but nowhere else in the world, and a high civilization existed on the desert coast of Peru and on the riverine oases of North Africa and the Middle East.

Although these environments are not identical, they have strong similarities, especially if subdivided into warm deserts and cold deserts. The dry lands of Inner Asia are marked by great winter cold, similar more to the Great Basin country in North America than to the desert lands of the southwestern United States, where the winters are mild. The warm Sahara is more similar to the Australian desert than to Inner Asia or the Great Basin country in America. But, by limiting our studies to areas of greatest similarity, we can hold the environmental factor fairly constant.

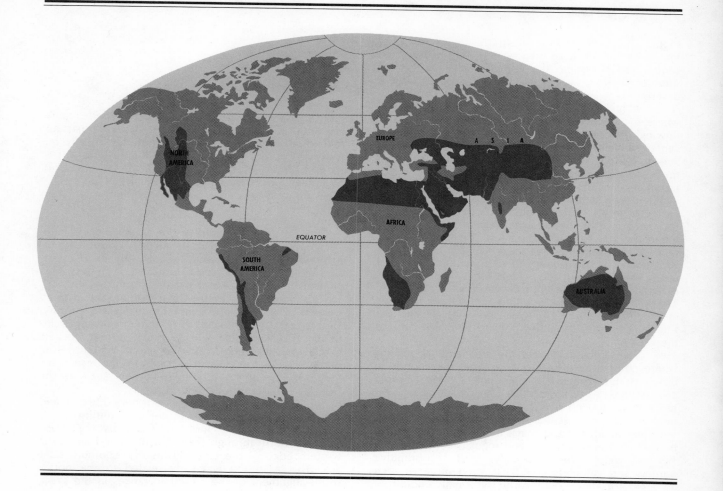

When we then find different races doing different things, we must return to the question of race as a possible causative factor in cultural response. The race question will be discussed throughout the book because it is too important and too difficult to be set aside with the brief treatment so far given it. In this and the other climatic sections race, environment, history of exposure to ideas, and other factors will be twisted and turned to examine the maximum number of combinations of causes of cultural response, growth or stagnation. We will consider changes through time with the race factor remaining constant and sometimes with it changing.

We have reduced but not eliminated race as a cause of cultural differentiation. We will now limit our consideration of physical environment to one type, the arid lands, or to a subvariety of this type to study the cultural differences that are found despite the relatively uniform conditions.

Arid lands are defined by their lack of rainfall. The exact definition depends upon who is measuring. The climatologist would say that in these areas the amount of precipitation is exceeded by the amount of evaporation. From this point of view, there are many far-reaching results. For instance, there is no permanent water table at a small distance underground; streams do not run regularly; springs are rare. There is no fixed amount of rainfall. San Diego, California, with about 10 inches of rainfall in a year concentrated in the cool winter months, is not quite a desert, although it is certainly arid. On the other hand, 10 inches of rainfall in an area on the south side of the central Sahara, where the rainfall comes in the hot summer months, would be characteristic of a true desert.

A hydrologist thinks of the desert as an area where the depressions in the land fail to fill to overflowing with water. Drainage is therefore toward the center of these basins, and there are no permanent streams—except for rare rivers such as the Nile and the Colorado that rise in high mountains or in adjacent moist areas and flow through a desert region on their way to the ocean.

A geologist thinks of the desert with pleasure, for there is little bothersome vegetation to obscure the rocky structure of the region. The bedrock outcrops throughout the area except in the depressions, which serve as catch basins for the surrounding higher areas. The barren surface is ideal for seeing the effects of wind and water sculpturing the face of the earth. The soil scientist sees the aridity of the land reflected in the concentration of unleached minerals and the thin and slightly developed soil mantle over most of the area.

The agriculturist sees the arid lands as regions where crops cannot be grown without irrigation. The desert is not attractive to him unless there is some exotic source of water, a large flowing spring or, better, a large flowing river; otherwise, for him occupation of the desert seems limited. The pastoralist, on the other hand, can use much more of the desert, for he does not need as large and regular a water supply but only water holes a few miles apart. The primitive hunters and gatherers could use even more of such an area, for they could carry the small amount of water they needed for a day or so and hence could range relatively far from water holes. This variable view of the usefulness and desirability of this one type of landscape is the principal theme that will concern us here.

Does the desert dictate a particular type of house? Since mud and stone are common in the desert and wood is scarce, will we find the mud house to be the "natural" response? Since most of the land is too dry for agriculture, is the pastoral-nomadic life to be expected in most if not all arid lands? Wherever great rivers flow through deserts, providing a year-round water supply and often, in the low-latitude deserts, nearly year-round crop-growing seasons, can we expect an Egypt to arise? There are no great world powers in the deserts today. Has this always been so? Will it always be so? And, no matter what the answer is to either question, just how, if at all, does the physical setting of the desert affect the answer?

The Physical Setting

The deserts of the world are found in certain regular positions on the earth. They are neither polar nor equatorial. They appear in mid-latitudes, mostly between 20 and 30 degrees. When they extend out of these latitudes, it is due to the shape of the land. Mountain ranges may cause extensions of deserts by blocking moist air. The strong north-south orientation and extension of the American deserts are due to this factor.

The greatest desert in the world is the Sahara. To some extent the adjacent Arabian desert and Indian desert (the Thar) may be considered an extension of it. These are typical low-latitude deserts. This general desert region also extends northeastward into the southern region of the Soviet Union and over to western China, typical mid-latitude desert country. Desert extends northward here through the deep interior of Asia where the moisture-bearing oceanic winds can penetrate only with great difficulty. The distance limits the amount of moisture that can reach this area, and great mountain barriers such as the Himalayas block off the moisture.

In North America, arid northern Mexico and the adjacent desert regions in our own southwest are in mid-latitudes. The North American dry region extends far north through the Great Basin and along the western edge of the Great Plains into southern Canada. This extension is clearly related to the north-south alignment of the mountain ranges in the West. The mountains block off the major air currents, which in these latitudes are flowing off the moist Pacific toward the land. In Washington and Oregon extremely high rainfall occurs on the western side of the mountains, and extremely dry conditions prevail not many miles away on the eastern side of the mountains because of this physical barrier. The important extent of arid lands in the United States appears clearly on the map, for nearly one-third of the country is arid.

In South America the east-west extent of arid lands is not great because little breadth of land exists there between 20 to 30 degrees latitude. The strong north-south line of the arid land is again due to the arrangement of the mountains in relation to the winds. On the Peruvian coast, desert conditions prevail because the winds in this area do not blow onshore strongly and because along that shoreline there is a major cold ocean current. The air moving toward the

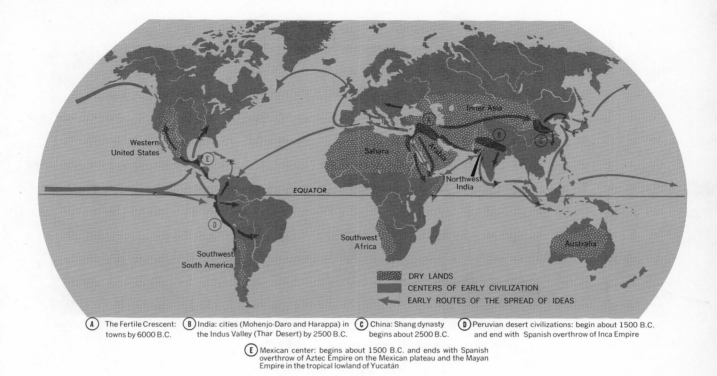

(A) The Fertile Crescent: towns by 6000 B.C.

(B) India: cities (Mohenjo-Daro and Harappa) in the Indus Valley (Thar Desert) by 2500 B.C.

(C) China: Shang dynasty begins about 2500 B.C.

(D) Peruvian desert civilizations: begin about 1500 B.C. and end with Spanish overthrow of Inca Empire

(E) Mexican center: begins about 1500 B.C. and ends with Spanish overthrow of Aztec Empire on the Mexican plateau and the Mayan Empire in the tropical lowland of Yucatán

MAP 2-1   The Arid Lands in Relation to the Early Centers of Civilization and the Spread of Ideas
The sea routes reaching the length of the Mediterranean and up to Great Britain were functioning by at least 2500 B.C. About this time overland contacts in the vicinity of the great bend of the Yellow River had begun to develop what was to culminate in Chinese culture. Sea routes led from Egypt down to the Red Sea, to the Indian Ocean, and thence to East Africa or India. Indian contacts reached into Indonesia, where they met Sinitic influences. We are just beginning to glimpse the importance of a trans-Pacific contact in stimulating cultural growth in native America.

In the arid lands there is a direct relationship between cultural status and location in relation to these centers of early cultural growth and the routes over which cultural ideas spread. Note that the arid land of Africa not only lies at the opposite end of the continent from early-developing Egypt but also is off the main routes of early contacts. The arid lands of North America were relatively distant from the cultural center in Middle America and were accordingly culturally retarded. In South America a center of development lay in the coastal desert, while the culturally retarded areas were to be found in the adjacent humid regions.

continent is cooled in its lower layers by this cold current and becomes very stable. Normally, to get moisture out of air, it is necessary to make it rise, expand, and cool. When the air has its heavy layers at the bottom, it is like a toy with a weighted base: it cannot be overturned easily. The air cannot be made to rise in violent uprushes, as occurs in thunderstorms. Such stable air is not a good source of moisture; therefore deserts come right down to the edge of the sea where the atmospheric circulation favors drought

and a cold current offshore reinforces the effect. This is also the reason the desert comes to the shore in Lower California, in North and South Africa, and in western Australia.

The desert crosses over the Andes in southernmost South America. The winds shift in these latitudes; instead of being from the east, as they are in latitudes 10 to 30 degrees, they are from the west. The shift reverses the wet and dry sides of the mountains, and the effect of a mountain mass in blocking

The Rocky Desert. *Rock-strewn desert surfaces are extremely prevalent, and they are probably the most common landscape in the arid lands. This scene is in an area of northern Australia where weathering of a rock surface has produced these rounded forms.* (Australian News and Information Bureau)

the prevailing winds is then easily seen. The area downwind of the mountain mass is the dry area and is said to be in the "rain shadow" of the mountain.

In South Africa, in the mid-latitudes on the west side of the continent—and coming to the margin of the ocean, which here also is cold for its latitude—the arid lands are extensive. Australia, its continental mass coinciding with the center of the desert latitudes, is largely arid. The vast arid interior is often spoken of as the dry heart of Australia. Very clearly, only the margins that have at least seasonally inblowing winds have much moisture.

Climate

The climatic characteristic of the deserts is that the amount of evaporation exceeds the amount of precipitation. For example, it was calculated that the total annual precipitation at Chacance, Chile, in 1918 was 0.04 inch. The potential evaporation was calculated to have been 182.5 inches. Needless to say, this is extreme desert. Deserts are also marked by great variability in precipitation. It is too often a feast or famine situation. Doorjabi in the Thar desert of northwestern India has an average annual precipitation of 2 inches; one year 34 inches of rain fell in two days.

Deserts have great ranges of temperature. On the basis of their temperatures they can be divided into two types: the low-latitude deserts and the mid-latitude deserts. In both types the change from day to night can be extreme, varying on a very hot day from temperatures in the 90°s to a night with a temperature near freezing. This is due to the very clear, cloudless air, which not only lets the solar radiation come in freely during the day but also lets the earth radiation escape almost as freely at night.

The amount of rainfall that sets the boundary between the arid lands and the humid lands varies with the temperature and the season when the rainfall occurs. A little rainfall in the cool season wets the ground more effectively than the same amount in the hot season.

It is the low-latitude deserts that deserve the description "hot deserts." Even though such deserts can be cold on winter nights, during the period of high sun the weather can only be called continuously hot. At Yuma, Arizona, during the middle of summer the average temperature is about 85° to 95° F. In one hot spell the maximum temperature exceeded 100° F. for all but one of 80 consecutive days. Death

Desert Mountains. *Looking toward the Dead Sea in the Negev, Israel.* (Embassy of Israel)

Valley has recorded maximum high temperatures in the low 130°s. Azizia, Algeria, has a record of 136.4° F. Such heat must be experienced to be appreciated. It is possible to blister the skin by touching metal objects in the direct sun. Even objects in the shade are hot to the touch. Since the air is much hotter than the 98.6° F. of normal body temperature, the more air that contacts your body, the warmer you will be. Under these conditions it is best to wear clothing for insulation against the heat. Yet even in these areas in the low-sun period the temperature can drop surprisingly low. Campers in the bottom of Death Valley on winter nights may wake to find their sleeping bags covered with frost. Winter days may be pleasantly warm but the nights can drop to the cool 40°s. High annual and daily ranges of temperatures in the desert are the rule. At Bil Milrha, in the Sahara, a low of 30° F. and a maximum of 99° F. were recorded within 24 hours.

Rainfall in the low-latitude deserts is meager. Cairo has 1.2 inches; Lima, Peru, 2 inches; Yuma, 3.3 inches. Such figures are not very helpful, however. A few days may have heavy rain, only to be followed by many months of virtually no rain. For example, Helwan, Egypt, once received 30 inches of rainfall, one-fourth of which came in seven storms. These ir-regular heavy downpours are characteristic of the desert regions of our own Southwest, where they come in late summer and often send great rushes of water down the dry washes.

Dry washes with their smooth sand floors and meager vegetation often attract campers from the rocky and nearly vegetation-free desert surfaces. However, a sudden storm on a mountain range miles away may throw a flood of water into the head of the dry wash, and this flood can run for miles and strike the campers with sudden deadly force. This potential danger led the French Army at one time to forbid its troops to camp in the *wadis,* the dry washes of the Sahara.

The rainfall pattern of these desert regions is, of course, poor for agriculture: it is erratic in distribution over time, local in nature, and irregular in amount. Does this prevent the agricultural use of the deserts? We will see that it does not.

Relative humidity in arid lands is usually low. This varies somewhat, however. Heavy dew can be experienced in the Colorado Desert, and the Yuma area can be very humid when air masses from the Gulf of Lower California move in. Generally, however, the high temperatures are accompanied by low relative humidity: evaporation goes on at a high rate,

Dunes in the Desert. *Sand dunes such as these in the Rub' al-Khali area of Saudi Arabia are more common in Arabia and North Africa than in the other deserts of the world. In no desert do they make up a major part of the landscape, and in most deserts they are relatively rare.* (Arabian American Oil Company)

and perspiration evaporates very rapidly, thereby cooling the body, with the result that 90° F. temperature is not felt to be excessively hot. Daily weather in the dry lands is, of course, sunny. There are exceptions along coasts, for example, in Peru and along the coast of Lower California—where the offshore cold current creates a persistent coastal fog condition, causing a cool, foggy desert! But most low-latitude deserts are more like Yuma, where in the month of June 97 percent of the possible sunshine is actually received.

The mid-latitude deserts are found in such areas as Inner Asia, the Great Basin area and adjacent regions in the United States, and Patagonia in South America. Here the summers are better described as warm rather than excessively hot. The winters can be bitterly cold both because of the latitudes (40 to 50 degrees) and because of the continental interior position that places the lands out of reach of the ameliorating effects of the sea. Temperatures as low as 32° F. for the month of January and 80° F. for the month of July are experienced at interior locations in these deserts. As in the low-latitude deserts, both the daily and the seasonal ranges of temperatures are very high.

There may even be some snow. Rainfall may occur in the winter, but in areas such as our Great Basin there is a tendency for a spring and fall concentration of the meager rainfall. With lower temperatures, there is a somewhat lower evaporation than in the low-latitude deserts. Humidities are low, but wind velocity is often high and the frequency of wind above average.

The cool coastal deserts deserve a brief description because they are so different from what is generally thought of as desert. Examples are found on the coast of Peru, in the Atacama Desert in Chile, in Lower California, and in the Namib Desert of South-West Africa. Here the land margins are much cooler than the interiors. Temperature gradients of 1 degree per mile are not unusual: temperatures of 70° F. on the beach, and in the 90°s 20 miles inland. This odd desert is a product of the movement of the great air and sea currents of the world. In the dry belt, the movement of the air is nearly parallel to the west coast. The offshore water is moving equatorward and is very cold for its latitude. The water a few hundred miles offshore is warm. The air over this warm water is also warm and has a high moisture content. When

Typical Rocky and Thinly Vegetated Landscape on the Edge of the Colorado Desert. *Such vegetation cover as is seen here is more typical of steppe conditions than of extreme desert. Here a riding party meets for breakfast in a landscape Americans avoided until recently.* (Palm Springs Chamber of Commerce)

it moves across the cold water, it is chilled in its lower layers, and condensation of the moisture begins, causing a fog to develop. The fog usually moves in over the edge of the land during the night and only slowly retreats during the day. Gray, overcast weather predominates on the coast, and it looks as if rain were certain within the hour. Instead, during the summer when this condition is most common, it never rains. The land is extremely dry, even though there usually is a heavy dew each night.

Landforms

It is difficult to say anything simple about desert landforms. Mountain masses are stripped of their weathered products; the rocky structures stand out; and the network of streams is a relatively coarse one, the streams themselves being dry except immediately after a rainfall. The gradients of the streams are very steep, tending to change abruptly upon coming out of the mountain mass. The weathered material brought down, often in violent floods resulting from the infrequent dumping of moderate amounts of water on the nearly bare rocks, is then dropped in vast quantities at the foot of the mountains.

The water rushing out of the mountain drops its burdens of boulders, gravel, sand, silt, and clay in that order. It piles these materials up in great fan- or cone-shaped masses at the mouths of the canyons, with the coarse materials at the apex of the fan. The clays are carried to the nearest flat area where the diminishing flow of water sinks into the soil. Everything is soon dry again. The wind then goes to work. The finer materials are moved around. The clay and silt-sized particles are small enough so that the wind can pick them up and carry them for great distances. The sand-sized particles the wind can lift only a few inches, or more often, can only bounce along the ground. The wind can blow out specific areas, leaving scooped-out hollows called "blow outs." The sand is sometimes piled up in dunes, but dunes are actually hard to find in most of the deserts of the world. Only the African-Arabian areas have really large areas of them.

Vegetation

One reason there are fewer sand dunes than the Hollywood version of desert would lead us to expect is due to the surprising amount of vegetative cover found in the desert. Truly barren land is very rare. On the coast of Peru there are sizable areas that are free of vegetation, and the Sahara has large barren tracts. The deserts of the United States, however, have some plant cover everywhere: creosote bush in the Colorado Desert and sagebrush in the Great Basin. The plants in desert areas are widely spaced and thinly leaved, but there are enough of them to make a sparse cover.

Two general kinds of plants must be distinguished, the annuals and the perennials. The perennials face the difficult task of living for long periods of time on very little water. They must have extensive root systems and small leaf systems, for roots gather water and leaves expend it. Some plants have no leaves, but specialized stems that take over the function of the leaves. Others have leaves at one time of the year, but none during the long dry periods. There are innumerable adaptations: corky bark, thick stems with hard exteriors to protect against sand blasting, resinous coverings to cut down on evaporation, pulpy interiors to store water. With all of these adaptations, plants must still grow widely spaced so as not to compete too strongly for the little water available.

One might expect that desert plants, even the perennials, would be short-lived. In fact, they often prove to be relatively long-lived. Such things are learned in rather odd ways. The ironwood trees growing along the dry washes of the lower Colorado River were largely cut down toward the end of the nineteenth century to supply firewood for the mines then operating in the area. The large stumps are still there. The trees that have grown up in the past 70 years are much smaller; therefore, the stumps must have been made from quite old trees. At a famous paleontological-archeological site, Tule Springs in Nevada near Las Vegas, pictures taken 25 years apart have shown

that even the little shrubby plants live for long periods of time, for nearly every bush around the site had lived through that period of time. A study of the growth rings of such plants has shown that some have lived for a hundred years. It is because the desert perennials are such highly successful adaptations that most of the desert areas of the world are better covered with vegetation than most people expect.

The annuals solve the feast-and-famine water situation of the desert in a totally different way. The cycle from seed through plant, flower, and back to seed is accomplished in one brief, almost fiercely rapid burst of growth. The seed stage is the dormant period. It matters little how long the environment is hot and dry, for the seeds are lodged in cracks in the ground, buried in the drifting sand, or hidden in cracks or under the edges of rocks. When rain does occur, the seeds will quickly germinate and repeat the brief life cycle. This burst of growth of short-lived annuals, often so short-lived as to deserve the term ephemerals, characterizes most of the deserts of the world. In the American Southwest, large areas may suddenly be carpeted with brilliantly colored blooms. The bloom is short-lived, however, and the areas of greatest color change rapidly. How can man use a plant resource as short-lived as this?

Animals

There are also numerous desert animals, many of which were shown in Walt Disney's documentary *The Living Desert*. The film has given a misleading picture of the animal life there, however. It might be many years indeed before you would see a fraction of what is shown in that motion picture. There is life in the desert, but it is not as rich as that found beyond the desert. It is easy to see why this must be so. Animal life depends on the plant world for its food, and while the desert is not bare, it is sparsely covered with plants. There simply is less leaf, seed, and root material for animals to feed on. Accordingly, there must be less life.[1]

[1] In many years of work in the deserts of America, I saw only two rattlesnakes, five coyotes, and one fox. Further, these were seen in areas so wild that two coyotes and the fox walked right into the camp looking for food. The coyotes came into our camp in Lower California. The fox came to call in the Death Valley region and stayed to take food from our hands. If animals in the mid-latitude forests were as tame, the situation would be similar to that of the bears in Yellowstone Park.

The desert animals, of course, are more numerous than the casual observer ever sees, especially while whizzing along in an automobile. There is evidence of their existence in the carcasses along the highway and the attendant ravens who harvest the slaughter. One of the oddest records of how many small animals there are in the desert is made by counting the number trapped overnight on freshly tarred roads. An impressive number of mice are so caught. As Disney correctly showed, these little harvesters of the seeds and leaves and roots of the desert are the base of the food chain. They feed the foxes and coyotes and the hawks—and, of course, could serve as food for man. In times past they actually did.

American deserts once had camels, horses, deer, antelope, and ground sloths. Indians hunted and ate all the horses, sloths, and camels, except in South America where the native camels survived, and modern Americans have virtually exterminated the herds of antelope that existed in the early twentieth century in the Southwest. Deer and mountain sheep survive in small numbers in remote desert mountains but elsewhere have been exterminated. This in itself is an interesting measure of the relatively low productivity of the dry lands. In the densely populated, humid eastern United States deer flourish, and game-management departments often worry because not enough of them are killed annually during the hunting season to ensure that the remaining deer will not starve to death due to overpopulation of the range. In the dry lands far fewer hunters killing just a few animals per year have almost wiped out game-animal populations.

The reason for the disappearance of many large animals in America is hotly debated. Many factors such as fire, climatic change, and disease played roles, but man was probably the decisive agent, for he was the only new factor on the scene. All the others had plagued the animals for millions of years. There is no convincing reason for the disappearance of the ground sloths from the American deserts except that man

(Upper) **Steppe Landscape, South-West Africa.** (Gordon Gibson for Smithsonian Institution) (Middle) **Grazing Comparison.** *Overgrazing can lead to erosion and steady worsening of the conditions for plant growth. The wind may begin to blow the soil away, or great gullies may be ripped through the land, creating an aridity that is out of keeping with the real climatic condition.* (USDA Photo) (Lower) **Pachypodium Lealii.** *This plant, found in parts of Africa, can grow 15 feet tall.* (Smithsonian Institution)

killed them all. This was an irreplaceable loss. The ground sloths ate the creosote bush along with other desert vegetation; nothing else eats the oily creosote bush. When early man in America extinguished this slow-moving, bear-sized herbivore, the world lost the only animal capable of grazing on this otherwise useless vegetation that covers vast areas of our arid West.

It is sometimes said that primitive men are great conservationists, but there is little support for such a view. Eskimos kill ducks and geese, pile them in mounds, and then eat only part of the pile. The Eskimos, once they got firearms, virtually exterminated the muskoxen. The Africans today are slaughtering their big game at a rapid rate. In *The Hunters,* a film about the Pygmies, one sees men robbing birds' nests of the baby birds.

It seems very probable, then, that the extinction of the American horses, camels, giant ground sloths, mammoths, and mastodons in America was the work of American Indians who rather suddenly obtained by invention or diffusion the weaponry and skills to kill big game. They wielded their weapons so efficiently that by the time the sixteenth-century Europeans arrived, all memory of the vast numbers of big game was long gone, with the possible exception of a folk memory of the mammoth in the eastern United States. The horse, a native of America, was a terrifying new animal to the descendants of the men who, millennia before, had eaten the last of the American horses.

Domestic animals

The early men of the Old World domesticated the camel instead of exterminating it. It is an amazing animal whose abilities are somewhat different from what is usually believed. For instance, while it is true that camels can go on for a long time without water, they have to be trained to endure this deprivation. Their humps are fat storage areas (incidentally, the American camels have no humps). Going without water for long periods of time under conditions of high heat, the camel can withstand a degree of dehydration that would kill a man. Then, given access to water, the camel in a matter of minutes can drink up to 25 percent of his body weight, almost immediately returning back to his normal state. Man cannot stand such dehydration, and any such water intake after heavy water loss would be deadly. As we mentioned earlier, such characteristics indicate that man is not a desert animal in his origin. The camel can drink the alkaline waters of the desert, eat its salty and oily shrubs, and walk over the slick rock surfaces or its soft sand surfaces on its tough, pillow-padded

feet. It gives milk, has a fine wool, a useful hide, and edible flesh.

There is a decided difference in the vegetative cover of the Old World desert area of Sahara-Arabia and that of the rest of the deserts of the world. The Sahara-Arabia region is much more barren. Whether this is a climatic effect or in some way the work of man is problematical. The Sahara-Arabia desert is drier, and at first glance this seems to be sufficient cause to explain the difference. A look at the American desert, however, makes one wonder.

The effects of the introduction of domestic animals into the arid lands of our own Southwest show what the grazing animals can do to a landscape. The first written accounts of the Southwest do not describe it as a desert. There were good stands of grass along waterways; most of the valley floors were broad and flat, and water tended to spread over them after summer thundershowers. This natural irrigation of these valley bottoms supported the grassy cover. After the pacification of the Indians and the introduction of domestic animals, there was a great increase in the number of horses, sheep, goats, and cattle. Shortly thereafter, huge gullies appeared throughout the country. The broad, flat bottom lands became trenched with deep, steep-sided canyons called *arroyos.* The water could no longer spread out over the valley bottom, but rushed down these narrow slots.

There has been bitter debate as to the cause of this condition. In the past, long before the Europeans erupted onto the scene, such arroyos existed. They had formed and healed a number of times. One great local storm can start such a gully. A long dry spell, by starving the vegetation cover and thus denuding the surface of the land, may start gullies over a whole area. What, then, is the cause of gullies: man or nature? This is probably not an either-or proposi-

Animals of the Dry Lands. *Three examples are shown here. The camel (top) is the true desert animal. He can eat the harsh and spiny vegetation, drink strongly alkaline water, and go for long periods without water. (MIFERMA for World Bank) In the American Southwest the peccary (lower left) is found in the cactus belt, on the border of the extreme desert. (Texas Parks and Wildlife) The antelope (lower right) is found now in the grassy semidesert or steppe country but formerly ranged widely into the desert. (Texas Parks and Wildlife) Similar animals are, or were, found in the African and Inner Asian deserts.*

tion, for either one can cause gullies. Long periods of good rainfall of low intensity allow much of the water to sink into the land. This type of rain supports vegetation, the runoff is slowed, and deposition (the collection of sediment) then occurs in the gullies. The process can lead to the healing of the land. A period of little rainfall with much of it concentrated in strong downpours, on the other hand, can lead to gullying, with or without incursions by domestic animals.

To overload the land with domestic animals, however, insures that gullying will result no matter what the short-range climate does. The explosive growth in the number of cattle on the land is closely correlated in time with the appearance of extensive gullying, and since there seems to be an obvious causal connection, this is a clear example of the effect of man on the land.

This phenomenon may in part explain the very barren appearance of the Sahara and the Arabian deserts, with their much greater expanses of dune areas. Man has been using domestic animals in the American Southwest for a mere 100 years or so, and the landscape already looks tattered and torn. In the Old World the figure may be as high as 10,000 years and is certainly as high as 5000 years. If men with domestic animals can create spots of desert in good arid lands such as our Southwest in 100 years, how much could they do in 5000 years?

It is difficult to envision the impact of grazing animals on arid or marginally arid lands. Palestine, an area for which we have relatively good records, ranges from desert to very well-watered lands: compare Jerusalem with 26 inches of rainfall to San Diego, California, with 10 inches. The climate has not changed since Biblical times when the area was described as a land of milk and honey. Today it looks like a desert, but close studies show that the same plants described in the Bible still thrive in protected places and that the springs described there still flow. The desertlike appearance of the land is the work of man and his grazing animals. What the sheep won't eat the goats will, and what the goats reject the camels eat with relish; it is the thousands of years of overgrazing that has made Palestine into what appears to be a desert. These lands and the adjacent deserts once had wild elephants, gazelles, ostriches, lions, camels, and other game animals. These findings and much parallel data indicate that the Old World deserts were probably much like the New World deserts in vegetation cover and wildlife prior to the coming of domestic animals. In comparing the Colorado and Nile river valleys (in a later section) it should be remembered that these lands were probably even more alike in the past than they are now.

The drying up of the Sahara is represented here by drawings found on the rock walls of caves and

The Desert. *The endless sweep of empty land so characteristic of the desert is exemplified by this gravel road in Iran. The man on the donkey only increases the sense of empty space.*
(Pierre Streit, Black Star, for World Bank)

canyons at Tassili, one of the plateaulike outliers of the Ahaggar Mountains in southeastern Algeria in the central Sahara. In Neolithic times the men of that region depicted the hippopotamus, an animal never found away from a living water supply. Some thousands of years later the hippopotamuses are gone, but cattlemen found pasturage where today there is extreme desert. As late as 3500 years ago the drawings depict bird-headed gods wearing Egyptian-style skirts and loincloths of Eighteenth Dynasty style, associated with a cup of pre-Dynastic style. Egyptian influences extended far indeed. Today the area is extreme desert. Undoubtedly the change from a land suitable to hippopotamuses to that fit for cattle is a measure of the glacial to modern climatic shift. The change after 4000 B.C. to the present condition may in part be due to further climatic change, but it may also be due to overgrazing and erosion, which perhaps tipped the balance against the vegetative cover that had clung to this region into the drier postglacial time. If cattle and sheep can so greatly deteriorate the American Southwest in a century, what should we expect in millennia? A Sahara, perhaps?

Using domestic animals for harvesting vegetation—much of which is inedible for man—and then either eating the animal or using its products, such as wool, milk, and butter, is just another way of using the land. This method allowed man to work more land more intensively than he could by the simple gathering way of life. In turn it altered the land, and it seems probable that the alteration is for the worse, ultimately limiting the usefulness of the land for man.

Domestic animals existed everywhere in the Old World dry lands before 1500, except in Australia and some parts of the South African area. In the New World in pre-Columbian times, domestic animals were either absent or unimportant (as in Peru). This leaves the following questions: Why did the man of the Old World domesticate animals? When and where did domestication develop? And

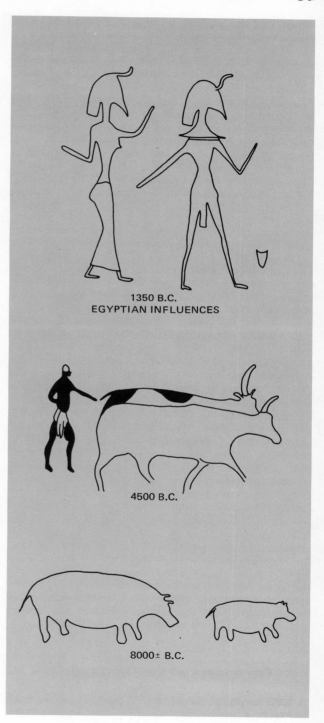

1350 B.C.
EGYPTIAN INFLUENCES

4500 B.C.

8000± B.C.

FIGURE 2-1   The Sahara: Rock Paintings from Tassili in the Ahaggar Mountains in the Central Sahara

Environmental and cultural change in the Sahara is depicted in paintings on rocks in the Sahara. Here we find water-loving hippopotamuses toward the end of the glacial period, grassland-grazing cattle about 6500 years ago, and Egyptian influences, as shown by bird-headed human figures, about 3300 years ago. Today the landscape is at times as austere as the moon. Climatic change is surely implied by the hippopotamus, but the change from cattle range to moonscape is more likely the work of man.

which animals were first used? Actually we know little about this subject with any certainty.

We know that the domestication of animals was carried out only in certain areas. We also know that this areal limitation had almost nothing to do with the availability of animals. The folklore that the American buffalo could not be domesticated or that the African elephant could not be tamed or that the American caribou could not have been used just as easily as the Asiatic reindeer has no basis in fact. However, there seems to have been only one area in which the domestication of animals occurred early. This was in or near the great center of origin of ideas in the Middle East. Animal domestication therefore would seem to be another idea we owe to this time, place, and people. Here the cow, pig, donkey, sheep, goat, and camel were probably domesticated; marginal to this area, horses, reindeer, elephant, chicken, geese, and ducks were domesticated. The only other area that had domestic animals was Middle America, and even this may have been due to ideas that originated in the Middle East and reached America from across the Pacific.

Once the waves of pastoral nomads began their endless sweeps across the dry lands of the Old World, the total ecology of the land must have begun to change. Animals graze selectively, and the plants they prefer are therefore selected against, while the "weeds" receive a chance to flourish. The land thus stripped of vegetation can blow or wash, creating a differing ecological condition, one in which plants preferring disturbed ground increase at the expense of plants preferring stabilized ground. The animals from rodents to horses that depend on the vegetation must also change or perish. With the changing rates of runoff, the ground water table changes, and springs come or go depending on these shifts. Man, dependent on plants, animals, and water, must adjust his usage of the land to these shifts. As the land deteriorates, man's pressure on the remaining vegetation increases, for he needs wood for tools and housing, for firewood to cook food and heat his home. With both man and beast using the limited vegetation of the arid lands, the plant cover can be severely damaged or even totally destroyed. It is important never to lose sight of the fact that man himself has created in part the conditions to which he must then adjust.

Occupancy of the Dry Lands

A brief survey of the history of man in the dry lands produces a feeling that anything can happen there.

In the Old World the great early centers of civilization were near the low-latitude hot deserts. Egypt, Mesopotamia, and the Indus Valley are examples of riverine oases. They are near the center of the growth of civilization, possibly the origin point of civilization for the entire world. At the same time, Australia and South Africa had primitive ways of life. The Australian natives were simple, unspecialized hunters and gatherers. The Pygmy people of the South African deserts were also hunters and gatherers but placed a little more emphasis on hunting. In North America the southern desert area and the Colorado Plateau had agriculturists, while in the Great Basin the people were simple hunters and gatherers. In South America the coastal desert of Peru had a series of city-states based on the irrigated valleys: this was an irrigation civilization. In the Old World there was a vast belt of pastoral nomadic peoples in the dry lands of Africa, Arabia, and Inner Asia. But such a way of life was totally lacking in Australia and late to appear in the South African dry region and the American dry lands. It is difficult to generalize these findings in terms of the compelling effect of the physical environment. If the environments are similar and if there are no great differences in the inherent ability of mankind, why then are there such great differences in the way in which the land is used?

OLD WORLD DRY LANDS: EGYPT AND THE MIDDLE EAST

Can we come to any conclusions from the fact that the birth of civilization seems to have occurred in the great riverine oases of the Old World? Egypt is often described in early geographical works as a perfect example of the force of physical environment. Here was a land designed to challenge man and encourage the growth of civilization. The desert climate provided a year-round growing season, and the annual floods of the river renewed the soils so that there were no problems of soil exhaustion.

The floods were a natural means of irrigation. What was more natural, runs the theory, than to broaden the area touched by the flood waters, begin to manage the waters, and drift into irrigation agriculture? With an agricultural base thus assured and the desert expanses on either side preventing easy access by enemies, it seems logical that the number of men increased, that the good food supply freed some men to pursue learning, and that among the

Near Cairo, Egypt. *The contrast between barren desert and the irrigated area is striking and typical of oases great and small.* (American Geographical Society)

subjects studied and developed were the heavens (to establish a calendar for predicting the time of the flooding of the river) and geometry (to be able to reestablish boundary marks when the flood subsided). Additional knowledge assured more food, more people, and more leisure, leading to the development of various classes among the people—priests, kings, scholars, workers. And in the vast isolation guaranteed by the desert the distinctive Egyptian civilization developed. These are the old familiar explanations, but do they really explain the birth of civilization?

The Earliest Agriculture

It was once thought that Egypt was the earliest civilization and that all the other areas drew from it. Today this view is being reversed. The earliest agriculture, the earliest villages and towns are being found in the Fertile Crescent—that sickle-shaped area reaching from northern Palestine through modern-day Syria and Lebanon into northern Iraq and southern Turkey. This is not a desert region; most of the early

agricultural sites found are in the 8-to-24-inch rainfall zone. Since these are areas of winter rainfall concentrations, the climatically comparable areas in America would be from San Diego to San Francisco. It is only semidesert at worst, and in these areas the first known agriculture was found. At present the earliest dates are about 10,000 years ago, but the well-established agricultural villages of that date suggest still earlier beginning dates. This dating takes us back to the end of the glacial period. We know that at that time the climate was quite different in these regions: they were better watered than they are now.

The impulse to invoke some physical-environmental cause for human behavior seems nearly irrepressible. Thus today one finds much speculation about the change from the moister time of the glacial period to the warmer, drier times of the postglacial (modern) climates as having in some way triggered the invention of agriculture. The thought here is that the drying up of the land led men, through necessity, to propagate the wild grasses that they had been harvesting but which were beginning to fail as the postglacial warmth and drought took over. Most recently it has been discovered that the glacial period in the

Egypt. *The sphinx figure is a record of Egypt's ancient past. The palms and the plots of land enclosed by the dikes on which the people are walking record the immemorial land use. The palm trees are, or course, date palms. The dress of the people and the carrying of burdens on the head are also holdovers from past times. Much of Egypt is little changed, while Cairo and some few other cities are far advanced into the twentieth century. (TWA Airlines)*

Mediterranean was a time of cold, dry climates. Post-glacial time has been marked by greater warmth and *more* moisture. Conditions were improving, not worsening. Undaunted, those who seek their answers in physical-environmental causation have shifted to "the improving environment encouraged them" from "the worsening environment drove them" to explain the development of agriculture. One should be suspicious of explanations that work as well backward as forward.

There are many ways of testing these hypotheses. If these are causes, they should have an effect wherever there were similar conditions. Very clearly this did not happen. Only in Southwest Asia can we

say with any certainty that man initiated plant and animal domestication at such a time and in such a place. In all the arid-humid border regions of the world where similar climatic shifts were occurring at the same time, no such event took place. All of mankind, as we have discussed earlier, had long been dependent on plant foods. Indeed, man had been dependent on plant foods so long that we can be sure that plant-dependent mankind had gone through periods of shifting climates again and again. But only with *this* climatic shift, and only at *this* place did man suddenly (as nearly as we can see) begin to take plants and animals in hand and manage them. The initiation of domestication looks much more like a unique

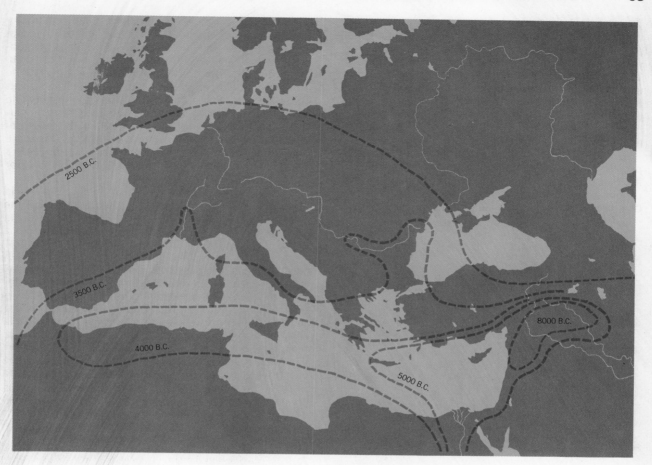

MAP 2-2   The Spread of Agriculture into Europe
The beginning of agriculture is in the Middle East, where settled agricultural villages date to about 8000 B.C. The initial spreading out was very slow: note the relatively small area reached during the first 3000 years. Thereafter the spread of agriculture was quite rapid, and there is a strong suggestion that it was greatly facilitated by sea routes: note the extension down the Mediterranean. Finally, agriculture spread throughout the European peninsula, reaching the English channel about 2500 B.C. The distance is about 3000 miles. Most of the spread occurred between 5000 and 2500, and a rate of about 1 mile per year seems indicated for this period.

event, comparable to an invention such as the wheel, than an environmentally caused, or stimulated, or conditioned response.

The origin of agriculture is of utmost importance to man's history. Without it the growth of our present civilizations would have been impossible. Yet this great invention, quite properly called the agricultural revolution, is very poorly understood as to its

origin or origins. Since it is of such importance and illustrates so well the kinds of questions that are involved in the origins of inventions and diffusions, it is worth extended discussion.

Is agriculture a process that mankind almost unknowingly develops through his use of plants? This is perhaps the dominant thought today. But if this is so, why has agriculture not sprung up innumerable times? And if this process just naturally leads to plant domestication, and mankind has been dependent on plants to some extent for more than two million years, why does this process mature so belatedly: a mere 10,000 years ago. But if agriculture is an invention, why are there allegedly eight or so centers of origin instead of the one or at most two that we might expect? The eight centers suggest more of a process-type development in which mankind repeatedly arrived at the idea of agriculture.

In his book *Agricultural Origins and Plant Dispersals*, geographer C. O. Sauer came close to suggesting that there were only four centers of plant domestication: one for root crops and one for grains in the Old World and a similar situation in the New World.

Others suggest that there may have been but five or three centers of agricultural origins. Perhaps there was only one! It is possible to point to evidence suggesting that the New World centers are the result of stimulus diffusion from the Old World. The New World centers are younger, and although the domestic plants are different, they form curious pairs with the Old World plants. Maize is a hard-seeded grass comparable to wheat and rice. The studies on the origin of maize show that wild maize was not an attractive source of grain but has been made so by man's manipulation. Why, then, did men single it out for improvement?

The pairs include the beans. In the Old World various legumes were domesticated: the pea, the broad bean (*Vicia faba*), the black-eye bean (*Vigna sinensis*), and in China even species of phaseolus beans. In America an array of beans was also domesticated but mostly phaseolus beans. These included the lima bean (*Phaseolus lunatus*), which contains cyanide, at times to deadly levels. We might wonder why this was domesticated. The wild tepary bean (*Phaseolus acutifolius*) has seeds that are so minute that it is difficult to understand just why this plant was selected as a seed source. One possible answer is that men arriving in America had not only the idea of agriculture but also rather specific notions of what plants were possible agricultural plants, and that beans were among them. In the array of possible useful plants they would find in this New World would be

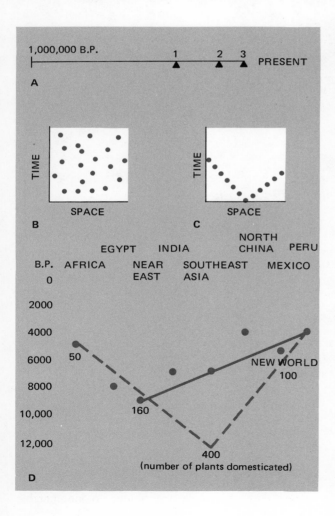

FIGURE 2-2 Agricultural Origins
A. *This line represents the last million years of man's dependence on plants. Number 1 indicates the appearance of modern types of men, number 2 the appearance of such grain-grinding equipment as the mano and metate, and number 3 the time at 10,000 ± 2000 years ago when all the primary centers of agricultural origins appear. Since need, opportunity, and numerous pressures (population, climatic change, logistical problems) had arisen in innumerable times over the million-year span, why do we have this concentration of invention?* **B.** *If the invention of agriculture was a process resulting from the use of plants, the need for better food supply, and varied opportunities and pressures operating on all mankind everywhere, then agriculture should have arisen randomly through time and space, as shown in B.* **C.** *If agriculture arose at a given point in time and space, the situation should be as shown at C.* **D.** *The actual situation as we now believe it to be is shown in D. The earliest agriculture appears in Southeast Asia. All other agricultural beginnings are later. The inference is that the idea, though not necessarily specific plants, spread throughout these parts of the world at these times.*

beans, recognizably like the beans they were accustomed to grow in their homeland.

The list of plant pairs is intriguing. In America the solanum was developed for its fruit, the tomato, and for its root, the potato. In Asia the solanum was developed for its fruit, the eggplant, virtually a tomato that requires cooking. Cotton was domesticated in both the Old World and New World, despite the fact that the wild cottons are very poor sources of fiber. We have the curious situation that plant pairs were repeatedly domesticated and that those of the New World were not at all obvious sources of food and fiber. A list of plant pairs is given in Table 2–1. The aim here is not to determine the origin of agriculture, but to show how questions arise and something of the line of attack that can be taken in solving such problems.

TABLE 2-1   Domestic Plant Pairs*

Plant Genus	Use in Old World	Use in New World
Amaranthus	greens	grain
Canavalia	sword bean	jack bean
Cucurbita	melons	pumpkins and squashes
Diospyros	persimmon	sapote
Gossypium	cotton	cotton
Hibiscus	fiber	fiber
Ipomoea	aquatica	sweet potato
Lepidium	garden cress	garden cress
Lupinus	legume	legume, Peru
Oxalis	vegetable, China	oca, Peru
Phaseolus	Southeast Asian beans	beans, Mexico and Peru
Phalaris canariensis	canary grass	canary grass, Kentucky
Setaria	grain, China	grain, Mexico
Solanum	eggplant, China	tomato, Mexico and Peru

* With rare exceptions these pairs link Southeast Asia (China to India) with Mexico and Peru. An outstanding exception is *Phalaris,* which is shared by the Mediterranean and an abortive agriculture in the southeastern United States.

It is, of course, necessary to deal with the related problems of time and of time and voyaging capabilities. The earliest American agriculture now known is in Mexico and dates to about 5000 B.C. This is 3000 years later than the earliest Old World agricultural beginnings. Only a few years ago it was adamantly

maintained that early men could not have crossed the world oceans, and there was severe criticism of those who suggested that since Shang dynasty (approximately 1600 to 1000 B.C.) art was present in America, men must have been crossing the world's greatest ocean in sufficient strength to have a cultural impact on America. Now, however, we have hard data to show that pottery making was introduced on the coast of South America sometime between 3000 and 4000 B.C. by people coming from the south islands of Japan where pottery had been made for at least 3000 years. An equal step back in time would suffice to bring to America men already acquainted with the idea of domestication of plants but carrying none of their native plants with them. This step has not been made, but in my view it is expectable. It is suggested here that these men would have quite naturally looked around in America for plants that resembled the domestic plants they had at home and would have begun their domestication efforts with these familiar plants—hence the remarkable number of plant pairs found in America and the Old World, especially Asia.

Later, of course, there were exchanges of plants. The coconut was brought to America, and the sweet potato was carried back. The Chinese have repeatedly reported the presence of peanuts (strictly American in origin) in Chinese archeology as early as the Lung-shanoid period. This falls about 3000 B.C., the time when Asiatic potters reached America and taught the American Indians that art. As with the repeated reporting of Negroid people in China, the reporting by Professor Kwang-chi Chang of the peanuts found in Chinese archeology has been largely overlooked—in its way, an example of the failure of ideas to spread.

As for the origin of agriculture, it should be said that there are more things unknown than known about the problem and that we cannot at present determine with any degree of certainty the meaning of the eight (or is it five or three or one?) centers of origin for agriculture. It is now widely admitted that there were pre-Columbian contacts across the world oceans at what were once considered unbelievable times. The considerations advanced here suggest that we have not yet grasped the full extent of these contacts nor examined all the ramifications of the time factor. From this lengthy digression, we can now return to the Middle Eastern agricultural origin where we will be on firmer ground.

The spread of agriculture

Agriculture seems to have started in the well-watered mid-latitudes rather than in the desert; instead

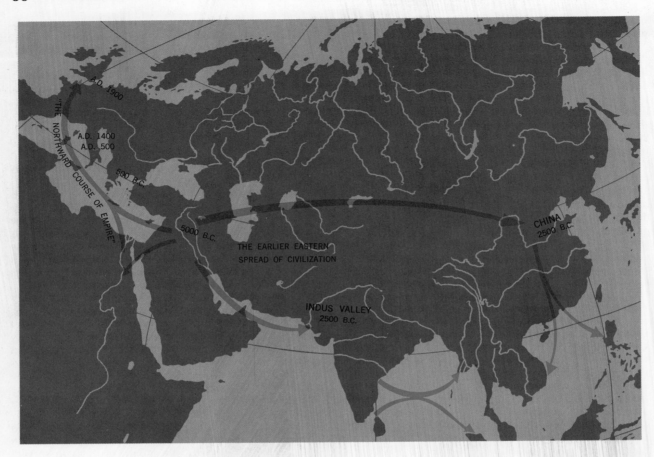

MAP 2-3   The Spread of Civilization from the Middle East
The beginning of agriculture in the Middle East gave that region an enormous lead over the rest of the world. The later spread of ideas from this center has often been described as the "northward course of empire." This is indicated here in round numbers: 5000 B.C. for the beginning of cities and states in Mesopotamia, 500 B.C. for the flowering of Greece, A.D. 500 for the end of Rome, A.D. 1400 to mark the Renaissance in Italy, and A.D. 1900 for the period of greatest dominance of the North Sea nations. Ethnocentric westerners have written of "the northward course of civilization." This overlooks the simultaneous eastern and southern expansions.

of starting with irrigation, agriculture probably started in the rainy, hilly areas. If wheat and barley are the first crops of this part of the world, agriculture did not start in the flood plains, for they would produce swamp plants such as rice. It now seems much more likely that agriculture started in the humid uplands and that the first agricultural villages were there. It was probably much later that well-established agricul-

tural peoples gradually and slowly moved down into the valleys of the great rivers that flowed through the low deserts.

A glimpse of how slow this process may have been is shown by the dates we now have. In the Middle East there were agricultural villages about 10,000 years ago, but it was not until 4000 B.C. that the "proto literate," or beginning-civilization stage, was reached. Egypt and China probably were influenced by the appearance of domestication in the Middle East. Egypt, the nearer, started about 3500 B.C. and China about 2500 B.C. The origins of agriculture in China involve the north versus the south, a complex question that will be avoided here.

When we consider that the date for the earliest agriculture may be moved further back in time, while the dates of beginnings of the cities and states and dynasties are less likely to be changed, we realize that thousands of years were required for the spread of agriculture and the building of populations in this area. This is also portrayed in Map 2–2, where the slow spread of agriculture in Europe is shown. The equivalent distance in the United States would be from Georgia to northern California.

If agriculture began in the foothills of the northern part of the Fertile Crescent, it spread east and west through subtropical humid lands and eventually northward into mid-latitude humid lands. While it extended early into the adjacent oases of Mesopotamia, Egypt, and northwestern India, the spread into Africa south of the Sahara was slow. In the arid lands of South Africa, even though there were possibilities for irrigation, men did not take the first steps toward agriculture and civilization.

The domestication of animals began with sheep about 11,000 years ago. The pig was domesticated about 8500 years ago, and domestic cattle appeared about 7000 years ago. The domestication of animals seems to have centered in the humid area to the north of the arid parts of Mesopotamia. Man's use of arid lands by moving herds of animals about to harvest the sparse vegetation and then living on the animals must have been developed in the Old World sometime thereafter. The idea spread over North Africa and Arabia and throughout the arid interior of Asia from the Black Sea to China. The idea of cattle keeping spread slowly into South Africa where the arid lands belatedly came to be used by pastoralists. No other arid land had a pastoral-nomadic way of life. The requirements were there: the difficult dry environment, the potential domestic animals, and the human need for a more controlled food supply. Nevertheless, only in areas adjacent to the Middle East did this type of adjustment to the land occur. Seemingly, the idea arose only once. It is the repeated findings of suggestions of single origins despite similar environmental challenges (here, dry lands) and natural opportunities (here, domesticable animals) that suggest the rarity of independent inventions and the dominance of the spread of ideas (diffusion).

The potential of this way of life is best shown by what happened in Inner Asia. Here the horse, camel, goat, and sheep provided a nomadic way of life that supported large numbers of people in comparative luxury. Their numbers and mobility, when organized under great leaders such as the khans, allowed them to use their central location to win control of central Eurasia. At times their political control extended into central Europe or into India from a base in China. This use of the arid to semiarid lands is a potentially productive one. The failure of mankind to duplicate it anywhere in the world beyond the Middle East seems all the more remarkable.

The arid lands of the Old World close to the Middle East for thousands of years produced more new ideas than any land in the world. Hot, low Mesopotamia and hot, low Egypt were long the great centers of civilization. From them ideas streamed out to ignite the fires of civilization in the eastern Mediterranean and in China and to greatly affect the growth of civilization in India. The low-latitude deserts have the highest summer temperatures in the world. If heat can inhibit human activity, especially mental activity, how can it be that these areas produced the wheel, the arch, mathematics, writing, astronomy, calendars, and a host of other inventions?

Decay of Early Centers of Civilization

Later these areas lost their intellectual and cultural lead. After Greco-Roman times Egypt and Mesopotamia recede to minor significance both politically and as sources for ideas. There is no great climatic change about that time; the land and sea and even the people remain as they were. It is the cultural world that changes. The centers of creativity move to Greece and Rome and Constantinople (now Istanbul), and later, in one direction, to northwestern Europe. We tend to overlook the fact that there was also an eastward movement: China and India were functioning as major centers of creativity at this time, and civilizations were flourishing in America too.

The shift away from the ancient centers in the Middle East is not easily explained. There is no obvious physical, geographical-environmental cause. The crossroads position of the Middle East continued and still continues. The Arabs pursued their trade between the Far East and the Mediterranean. The building of the Suez Canal emphasized the importance of the area for the transfer of goods, men, and ideas in the late nineteenth and the twentieth century. Today the major airlines of Eurasia funnel through it.

The reasons for the long decadence of the Middle East and Egypt are far more subtle than mere shifts in climate or even a shifting economic position. Even to say that the causes lie in history begs the question. The real causes lie in the spirit of men and nations, and this question has hardly been touched. What happened to the outlook of these people who once dreamed the mightiest dreams, conceived the finest art and architecture, and laid the basis for our modern science in the midst of the desert furnace heat in the semiarid foothills and the riverine oases that still dominate the land?

The geographer can state that the land is still there with the same climate, landforms, and soil. There has been some extinction of animals, considerable destruction of vegetation, and some soil erosion. But these are minor items. The same races of men are present that produced the inventors of writing, the codifiers of laws, the architects of pyramids, and writers of the lofty prose of the Old and New Testaments. The land remains in one of the crossroads of our world. All that is required to make the great oases of Mesopotamia, Egypt, the Indus Valley, and the series of riverine oases between the Black Sea and Lake Balkash bloom again is the will to do so and the political and economic institutions to allow those who would to do so.

In the past, human energies in these areas were harnessed under some form of absolute rule, often combined with the idea of a divine priest-king, and the major economic base of society was agriculture. The early irrigation agricultures were efficient for their time, for they could produce a surplus of food to support a small percentage of the population in non-agricultural pursuits. Under the authoritarian governments of the day, this surplus went to the aristocracy and the priesthood, if different, and to the artists, architects, and scientists selected by the ruling group.

The group freed from the grinding daily toil was minute compared to the total population. Nonetheless, 5000 years ago the move was a great advance. At least a few men were freed for the pursuit of art, science, and philosophy. In the course of history there are many reasons for the decay of an empire. Political shifts of power, shifting trade patterns, colonial status, and the failure to participate in the industrial and social revolutions that were transforming the Western world all contributed to the downfall of Egypt. The former seats of invention and learning stagnated while much change was taking place in Europe: the freeing of larger percentages of people for learned pursuits,

the growth of widespread education, and the potential extension to the mass of the people of the right to choose their own way of life and to produce to their limit of ability and energy. In Egypt a farmer remained a farmer, and opportunity existed only for a few. Productivity of the land and of the people remained at relatively low levels, while in the Western world areas formerly unproductive became immensely productive. The physical geographies have not changed; the set of men's minds have changed, and the physical earth has produced accordingly.

Ruins in the desert

Nevertheless, there have been real, not just relative, declines in the productivity of the Old World dry lands. Inner Asia has a great series of ruined cities. Most of these lie along the rivers that come down from the high mountains and head out to the desert. The discovery of these abandoned cities led to theories of climatic change as the cause of human accomplishment and failure. Careful study of the evidence of climatic change has not supported these theories. Instead, the growth and destruction of these ancient cities prove to be closely correlated with the political and economic happenings in the region. It has been the stability of governments and the flourishing of trade or the dominance of nomads that have determined the success or failure of the riverine oases people.

In southern Palestine there have at times been cities in the middle of the desert. No stream is there now to support them. Seemingly they could not exist in the present climate; and there does not now seem to be any reason for their existence. Since they have existed, it is easy to conclude that there must have been a climatic change. A further look at the matter, however, indicates that this is not so. Study of the houses shows that every drop of water that fell was channeled into deep cisterns where it was kept safe against evaporation and undoubtedly used most sparingly. The climate must have been as arid as it is now. Trade was the reason for the existence of these cities. At that particular time in history the Gulf of Suez was blocked for political reasons, and trade from the Indian Ocean to the Mediterranean was diverted into the Gulf of Aqaba, where the goods were sent overland. It is one of the curiosities of history that current Egyptian-Israeli relations re-created this situation and that the Israelis reestablished this ancient trade route. Further, they have studied the ancient land-use solutions, such as building diversion dams and concentrating the slope wash onto the areas of good soil, and

Ruins of an Ancient Church in the Negev, Israel. *Such ruins are common in dry lands and are the usual source for the idea that there has been a climatic change. The Israeli redevelopment of the Negev illustrates the role of politico-economic forces.* (Embassy of Israel)

from the shores of the Indian Ocean: African gold and ivory, spices from India, incense from south Arabia. First the donkey caravans moved goods across the desert of southern Arabia, and then, after 2000 B.C., the camel came into use and the trade multiplied.

These people also developed an irrigation system adapted to the environment. They built numerous small dams to spread the flash floods that dominate arid regions. The spread-out and slowed-down waters dropped their silt load, and the water percolated down into the alluvium to form a storage reservoir instead of rushing off the land. This made agriculture possible. The silt accumulated to depths of 50 and 60 feet in the valleys, forming broad, flat floors with high water-storage potential. Man was again making over the landscape to suit his needs. When the trade monopoly ended, partly because the Europeans discovered the route around Africa to the Indian Ocean, the wealth faded, and the dams were neglected. Soon gullies were ripping through the deep silt deposits, disrupting the underground storage areas, ripping the fields to pieces, and ruining the system, so the land reverted to nomadic herding. It went from one state to another without benefit of climatic change. The changes were largely external and cultural: ocean routes required ports, overland routes called for cities, and the breakdown of trade led to the collapse of the cities. Scholars without historical insight concluded that the climate had changed. Not only is so simple an explanation inadequate, but we must also realize that people in similar environments—South-West Africa, Australia, or Baja California—did not take advantage of opportunities for irrigation. In still other areas, the American Southwest, for example, these practices appeared again. The reasons for the parallel case of the Southwest will be explored later.

Similar dramatic changes are found in North Africa where before Roman times there were no camels at all. Along the northern fringe of the Sahara in what is really the arid end of the Mediterranean, there were large, flourishing agricultural colonies. Carthage was one of these. Later, Rome greatly developed this land. It became a granary, a source of olive oil and wine, a rich colonial area. Thereafter it declined. Cities dwindled or were abandoned; areas that had been in olive groves or vineyards or wheat fields became poor pastorage. The explanation often given was climatic change. Certainly when land goes out of agricultural use and becomes a barren pasture, it *looks* as if the climate has changed.

During Roman times the land was governed strictly; Romans enforced the peace, and Roman law

by using the same methods on the same land, they are again bringing it into cultivation. Obviously the change is motivational, political, economic, and social—not climatic.

Southwestern Arabia is a classic case for testing the climatic-change theme of the environmental determinists. In Roman times it was an area of great wealth, and in the second century B.C. Agatharchides wrote, "For no nation seems to be wealthier than the Sabaeans and Gerrhaeans, who are the agents for everything that falls under the name of transport from Asia and Europe." Trade was the key to their wealth, and the trade they controlled funneled into their land

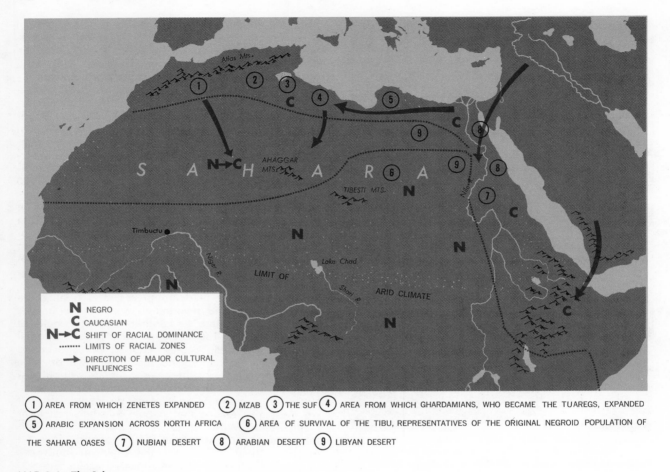

MAP 2-4   The Sahara
*The Sahara is a vast dry area approximately the size of the United States. In the main it is low,
but it does contain two large volcanic masses, the Ahaggar and Tibesti mountains. From these
mountains minor amounts of water flow down into the surrounding basins and create oases.
Larger oases are created by the streams flowing southward out of the much better watered
Atlas Mountains. Such streams, even when they sink into the sand, often flow immense
distances. Water also enters the Sahara in important amounts by the Nile, which flows
through to the sea, the Niger, which enters the desert but turns back near Timbuctu, and the
Shari, which ends in Lake Chad.*

prevailed. Since the land was meant to be productive, disturbances that would reduce productivity were put down. Nevertheless, in later Roman times the land began going out of production, men were leaving the land, and the area was in an obvious decline. Was this the result of climatic change? Certainly not. Nomadic raids? Perhaps, but only secondarily so. A very good argument has suggested the cause was excessive taxation.

In early Roman times the tax was the *decumanae,* the "10 percent." Later it began to double, until it rose to about 40 percent. The tax was a combined land and poll tax. The people with the largest families paid more. If a farmer went bankrupt and fled the land, the payment of his taxes was divided among his neighbors. This meant that a farmer who was barely getting by when there were 10 farmers in his district might face a 10 percent tax rise when one of his neighbors went out of production. This might be ruinous to him. But when he too went out of production, the remaining farmers had to add his tax burdens to theirs as well. This pyramiding of taxes seems guaranteed to drive men off the land or out of production of any kind, and this is what happened toward the end of the Roman Empire. The abandonment of the land was general. Not just the edge-of-the-desert lands that were exposed to climatic risks and to barbarian raids were abandoned; all over the Empire, in the well-watered lands and in the heart of the Empire where nomads could not as yet raid, land was being abandoned.

Another source of change: camels, nomads, and farmers

The appearance of the camel gave great mobility to the desert dwellers and thereby altered their life in the Sahara. This shift came surprisingly late. Camels were first introduced into Egypt about 500 B.C., but it was not until late Roman times that they came into wide use in the western Sahara. Prior to that time oxen and the little native elephants of North Africa were the pack animals used by the caravans. Movement through the immense dry wastes of the inner Sahara was greatly limited, for neither oxen nor elephants are desert-adapted beasts. Prior to the time of the camel, the oases people had been dominant. Now men could wander farther over the desert, and greater numbers of men could live away from the oases. Furthermore, their great mobility enabled these men to concentrate sizable numbers in an attack on any single oasis. The balance of power shifted toward the pastoral nomad.

Settled people suffer from nomad pressure even without actual warfare. In the arid lands, olives and vineyards can be established in good spots by careful selection of the plots of land where water is concentrated during the occasional rains. It takes years to get vineyards and olives established and bearing. The trees and vines are scattered over a wide area with some tucked in this canyon and a few at the mouth of the next, and some more around the corner of the hill where water is shed off a rocky expanse and concentrated onto a patch of good soil. The work of years can be undone in one night by a herd of camels. Camels browse; they have tremendously powerful jaws and teeth that can shear off a small limb. With the breakdown of law and order the pastoral nomad could wander into the edge of the settlements and cause damage with his herds before the settled people could muster the number of men required to drive him off.

If the settled people lost the struggle against destructive taxation and destructive nomads, the land that they abandoned would cease to produce wheat

(Top) **Camel Caravan in Algeria.** *The barren desert appearance is probably greatly increased by the presence of camels, goats, and other herd animals for thousands of years.*

(Bottom) **Biskra, Algeria.** *An oasis town on the south side of the desert mountains, its location in relation to the Mediterranean climate, the sea, and the desert is much like that of Palm Springs, California.*

and wine and olive oil. It would become pasture land occupied by nomad herders living in tents. The aspect of the land would change in such a way as to suggest a climatic change. The causes, however, would be cultural. These particular changes occurred in the Sahara after the Roman period, and the result was the extension of arid-appearing conditions far into the zones that are climatically humid. Much of the North African margin is humid land that could well support a forest. Goats, camels, and man have removed most of the forest and made the land appear far drier than it really is. The "natural" vegetation can be seen today only in protected areas.

The shift of dominant peoples in the Sahara was complex. In the eastern Sahara, Egypt is an immense, populous oasis with easy access to the narrow desert between it and the Red Sea. Because large settled populations had ready access to this desert strip, the nomads were controlled. To the west of Egypt lies the immensely dry Libyan desert, but the few nomads there could not muster large enough forces to upset the dense settlements along the Nile. Hence in the eastern Sahara the farmers dominated the nomads.

In the western Sahara the conditions are reversed. No large oases exist in the desert, which is vast and somewhat less formidable than the Libyan desert. With the coming of the camel, the balance of power shifted to the nomad. Note that the natural environment remained the same and the change was dependent on the cultural shift.

The reversal of power changed the racial makeup of the Sahara. Prior to the introduction of the camel only the northern fringe of Africa was held by racially Mediterranean peoples such as the Berbers. When the Romans introduced the camel into their territory, these people became the first camel nomads of the western Sahara. It is one of the oddities of history that a Jewish tribe, the Zenetes, not mentioned earlier in the literature of the area, seem to be the first camel nomads of that region. They appeared in about the sixth century A.D. and spread into the desert area founding a string of oasis towns. Another group that obtained camels and spread into the desert were the Garmantians, a people identified later as Tuaregs.

The spread of these mounted, nomadic, warlike peoples was at the expense of the previous occupants of the oases. These were a Negroid people who had made their living by growing grain. Their entire way of life differed from that of the invaders. The Negroes grew the grain, crushed it in characteristic grindstones that were totally different from the rotary querns of the invaders, and used the bow and arrow for hunting and war. Grain growing and flour were replaced by date growing and meat and dairy products. The rotary quern replaced the ancient grinding stones, and the spear replaced the bow and arrow. The change took a long time and in fact is still continuing. In part, however, the region has come to a semiequilibrium.

The camel nomads dominate the area. They own the oases, and the oases people usually work the land for the nomads. The town-dwelling oases people are largely Negroid. This reflects two facts. First, the nomad considers the town-dwelling–farming way of life to be beneath him. Secondly, the oases are centers of malaria, and the Negro people as a race have more resistance to malaria than the white race. As a result there is a steady elimination of the whites that do settle in the oases. A complex relationship has been established in which the nomads own and protect certain oases and the oases dwellers furnish dates, grain, and other materials for the nomads. In return they get not only protection, but meat, wool, hides, and milk products. Only in the most remote part of the Sahara, near the Tibesti Mountains, for example, do the Negro people still hold out in some purity of race and freedom.

This tremendous racial and cultural shift started, then, in later Roman times because of the highly mobile way of life that the camel made possible. Since the Mediterranean peoples took up this way of life, they were the ones to gain by it. The great push into the Sahara by the Mediterranean peoples was accelerated by the expansion of the Muslims, and its continuing spread was fanned by the religious drive of Islam. It is typical of mankind that the Negro people clung to their grain-farming, sedentary, bow-and-arrow way of life and lost control of their ancient homeland. In the American desert the same thing happened to the settled Pueblo farming people when the warlike nomadic Athabascans invaded their territory. In the latter case there was no major racial difference, and the comparative cultural method thus cautions us against assuming that there was some racial causation in the failure of the Negroids to shift to camel nomadism.

Oasis Types

There are many types of oases. South of the Atlas Mountains in Algeria there is a great sheet of exposed, barren limestone. This is the Shebka. The city of Ghardaia sits in the midst of this desert, and in the surrounding deep-walled canyons live the Mozabites, the inhabitants of the Mzab, the area about Ghardaia. Mile upon mile of the desert land is a white, harsh, bare rock surface that has been cut into a maze of badlands. This surface is 2000 to 2300 feet above sea level. The limestone is underlain by water-bearing strata, and springs appear where valleys cut through this material. Where the aquifer is not reached by the deep valley cuts, wells are dug in the bottom of the valleys to reach the water-bearing strata. The depths of these wells vary from 28 to 180 feet, and there are 3300 of them. Since the water does not rise in these wells, it must be raised by buckets. Men do most of this work, but animals such as the donkey are also used. Since the water irrigates the gardens, a vast amount of labor must be expended for enough water to keep the gardens alive.

Little rain is received. A rainy year may have two or three showers, while a dry year has none. Dams are built on the large, medium, small, and even tiny valleys to store the water that comes so infrequently. The crops are grown along the narrow valley bottoms in small plots of land that are more like gardens than fields, for the use of the land is intensive, and fruit trees have vegetables or grains growing under them. Dates, figs, pomegranates, apricots, peaches, barley, beans, and grapes are all intimately intermixed. If a man works hard, he can win a modest subsistence and no more, yet the people cling to this way of life. Today many of the men spend their middle years outside the area, usually in Algeria, working as merchants or bankers and forming a separate, exclusive Islamic sect. These outside earnings make possible the maintenance of the oasis: Life outside is only a means to gain the things that the oases cannot supply. The social life, the sense of belonging, the nonmaterial things that these people value in life are found in their desert oases—lands that to us would be unthinkably isolated, barren, and laborious.

Everyone in this type of community plans to have two houses. One house is located near the gardens and may be far from the village; the second house is in the village or town, for social life is prized. The houses are made of limestone, and a mortar made of burnt limestone and gypsum is used to cement the rocks together. It is too facile a conclusion to say, "Of course, houses of rock in a rocky desert." A great deal of fuel must be used to burn limestone to cement; it could as well have been used to construct a wooden house. Indeed, why is so much emphasis placed on a house in a land where it rarely rains? Why not hollow out the limestone valley walls? If a tent is sufficient for the nomads, why do oases people need so permanent a house? And if a stone house is desired, why must it be cemented with mortar? Stone walls can be set up without mortar, and in dry lands mud can be used for mortar that will stand for centuries. The people of the Mzab feel that life is best lived in a stone house with real mortar between its rocks and located in the bottom of a dry valley cut into an utterly barren rock mass in the midst of the world's greatest desert.

Near the Mzab are the people of the Suf, a great sea of sand dunes. Property here is not counted in land, for land is unlimited. It is not even measured in water, for water underlies vast areas of these dunes. Property is considered in terms of date palms. The date palm flourishes if it has its feet in the water and its head in the furnace. The only practical limit on how many date palms a man may have in the Suf is the limit of a man's strength. Each palm is the result of digging down to the shallow water table, planting a palm, and then keeping any excess of sand from accumulating and choking the tree. Successful growth requires the constant digging away of the sand, which places a limit on how many trees one man can maintain. To most people the area would look utterly useless; certainly without the date palm it would be much less useful. As it is, the people live in this great sand sheet following a way of life where half of their effort goes into nomadic herding and half into tending the palm trees. The two activities are, of course, supplementary. Man cannot live on dates alone, but dates are an excellent supplement to the milk and cheese of the herdsman.

Given a vast sand sheet and a way of life based on tending trees widely scattered over this dune-swept expanse, what type of dwelling would seem most natural? Tents—as used by the pastoral fraction of the population—or palm-thatched houses with stout palm log frames? A thatched house would give excellent insulation and ventilation and, in a dry area, would last indefinitely. It would also be cheap and easy to make using the material available. Yet the Sufis do not build that kind of a house. Instead, they mine calcareous cemented sand sheets from the sand dunes. Houses are built on a square floor plan, one square room at a time, each room being roofed with a dome of stone. Villages are large and may have up to a thousand of these square-shaped, dome-topped rooms. Here in the midst of the desert, then, one finds a fixed house utilizing one of the most advanced of architectural devices for the housing of a very simple society that is at least half nomadic.

The dominant house type in the oases of the arid lands of North Africa, Arabia, and Inner Asia is made of mud. Wet earth of sufficient clay content to dry in the sun to a bricklike hardness can be used in several ways. It may be made up into bricks and built into walls with more mud used as a mortar. This is the adobe brick construction spread widely through the Americas by the Spaniards. Another method is to construct a flimsy post-and-pole structure and to plaster this with mud. This is the equivalent of our lath and plaster or stucco construction. Another possibility is to use mud mortar to build a stone wall. Still another is to use the mud like poured concrete, pouring it in forms and thus constructing a solid mud wall without bothering with brickmaking.

In dry lands clay houses are useful, for the rainfall is slight and the unfired clay will stand as long as it is kept dry. This fact has led to the incautious statement that building clay homes is the natural way to do things. Inspection of the distribution of clay buildings should raise some questions. As we have just seen, there are various house types in use in the Sahara side by side: tents, two kinds of stone houses, and in the Arab villages square-built, flat-roofed, mud-walled houses. The tradition of building with mud extends on across the dry lands to China. However, in the dry lands of South Africa there were no clay houses, but only grass shelters. In the dry lands of Australia there were no houses. In the American dry lands mud and stone laid in mud mortar were commonly used in those areas where influences from the Old World are suspected (northwest Argentina to southwestern United States); where these influences are unknown, this type of building does not appear. Thus the Pueblo people built mud and stone houses, but their neighbors in the adjacent, colder, arid interior land of the Great Basin did not. Building in mud, then, seems to be an invention possibly made only once or twice, and despite the environmental invitation, or need, or stimulus, it was not duplicated in other areas that were physically similar.

Building in mud can be a trap. In northern Peru, for instance, it rarely rains. However, when air flow along the Ecuadorian coast reverses, as it does every 20 years or so, heavy rains come, and the adobe houses equipped with sunproof but not waterproof roofs revert to mud and collapse. However, if adobe bricks are set on any foundation that keeps ground water away from the base of the walls, and are defended by a good roof with broad eaves that keep water off the walls, they will stand indefinitely in any climate. Adobe buildings built by the Spanish and Mexicans in California still stand despite the rains and floods of two centuries. A thick adobe wall gives wonderful insulation against heat, cold, and noise. It is good in any climate if properly built and protected. We are culture bound or we would build more of them everywhere. But at least it should be clear that mud dwellings are not simply a response to dry environments.

We may ask, if desert dwellers preferred a house of stone, why was the roof a dome? Surely palm logs would supply roof timbers for an ordinary house type. Domed mud houses are not uncommon in the Middle East, so this may be a transfer of ways of doing things (diffusion) comparable to the spread of the Danish log cabin from northwest Europe to America. Actually a thick-walled, high-domed house is cool and airy and well suited to hot climates. Why, then, have more people not accepted this obvious environmental invitation to build a desirable house type?

In the belt sweeping from the Sahara through Arabia to China the land is occupied by a combination of oasis agriculturalists and pastoral nomads. Since European cultural roots are so deeply embedded in these regions, we tend to think this is the natural way to use arid lands. This conclusion overlooks the fact that even in the Old World not all arid lands were always so used. Arid South Africa belatedly became partially pastoral but had no oasis agricultural development. Australia had neither pastoral nor agricultural development. There is considerable doubt about whether oasis agriculture plus pastoralism is the natural or obvious way to use dry land.

Detailed studies of the variations in ways of using the oases for agriculture or the desert for pastoralism show that there is less similarity in detail even within these systems than one might expect under a severely limiting environment. The delayed spread of the camel and the date palm, the variety of house types, including the nonuse of potential types (palm log and thatch, for example) suggest that man chose rather narrowly among a wide range of possibilities of land use. This can be made even clearer by taking two arid lands of very similar physical geography and studying men's different perceptions of the possibilities in these similar physical environments.

THE TWO EGYPTS: THE NILE AND THE COLORADO

In the great oases, such as the Nile, the Tigris-Euphrates, the Indus, and the Colorado-Imperial valleys, the water supplied by rivers that rise in high mountains outside the desert can provide for a large population using large-scale production. Such an area may develop into a center of political power and cultural creativity. At the other extreme are springs in the desert that can only support a few families of gatherers or seasonally a tribe of pastoralists. In between are medium-sized streams flowing into the desert and odd situations of special water conditions such as the Suf and the Mzab.

One must note that large streams flowing through a desert do not guarantee a growth of civilization. In Egypt, Mesopotamia, and the Indus, yes. In the Colorado of America, the Orange of South Africa, the Murray-Darling of Australia, no. The reasons for this and the nature of some of these oases will be explored here.

Egypt as an Example of the Great Oasis

Egypt does not owe its start to the annual flood of the Nile that enticed man into irrigation agriculture and into becoming the first great center of civilization. That theory was popular at the opening of the century when our knowledge of cultural history was relatively dim. Recent discoveries suggest that plant and animal domestication went hand in hand on the rainy mid-slopes of the mountains surrounding the Tigris-Euphrates (Mesopotamian) plain. The picture now emerging is that plant and animal domestication began earlier than 10,000 years ago, spread slowly into adjacent regions, and led eventually to the development of irrigation in the foothills, extending into the great river valleys where civilization developed. This was a long, slow process, for civilization (urban life with all it implies) does not appear in Mesopotamia until about 3500 B.C. By this time large towns, great temples, city-states, pictographic writing, and metallurgy were in existence. The accomplishment was great, but so was the time interval: 4500 years.

Egyptian development must be understood in this setting, where basic inventions were made nearby (plant and animal domestication) and spread to their area along with the associated cultural growths: irrigation, cities, writing, metallurgy. Herodotus (fifth century B.C.) says that the Egyptians themselves asserted that their culture owed its beginning to influences from the north, but so overwhelming was Egyptian cultural dominance in the fifth century that the learned Greeks did not believe it. Nevertheless, the evidence to date suggests that the history of Egypt begins in the neighborhood of Thebes (near modern Luxor) distinctly later than the history of Mesopotamia. Further, the beginning at Luxor suggests an arrival of influences via the Red Sea and the Wadi-el-Hammamet. This suggests early maritime activity.

The initial introductions apparently preceded a developed civilization in Mesopotamia and started a simple agricultural village growth in Egypt. Dates for

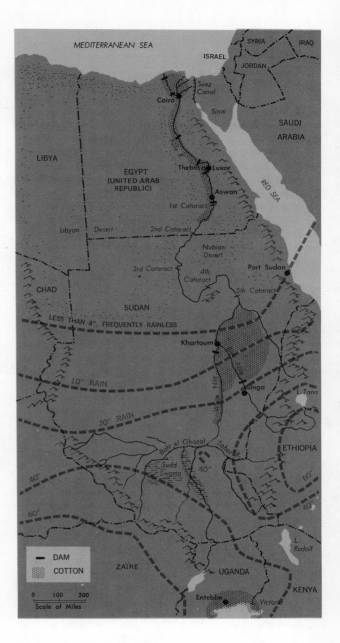

MAP 2-5 Egypt and the Nile Valley

Egypt is the typical great riverine oasis. The entire country has less than 4 inches of rainfall per year. The agricultural development is therefore concentrated along the narrow Nile Valley. Note the high rainfall at the headwaters of the Nile and the frequency of cataracts in the river. The later greatly inhibit the use of the river for transportation. (Reprinted from Focus, American Geographical Society)

75

early periods are difficult to determine, but 4000 B.C. for late Nakada I and approximately 5000 B.C. for agricultural introductions give an idea of the time level. The sequence began with agricultural villages of wattle and daub huts and the use of plain pottery. It developed into larger villages, with pottery becoming complex and copper in use. Progress continued with rectangular brick houses, basin irrigation in widespread use, large temples, and city-states.

Egypt and Mesopotamia showed parallel developments, but the earliest appearances were always in Mesopotamia. The same names were used for the same kinds of gods: polytheism, creator gods, dying sons of gods, fertility gods. Similar evidence is found in cylinder seals, art motifs, and architectural beginnings in brick with great variety of details like the earlier Sumerian structures. Writing in Egypt begins with pictographs, and the earliest of these contain Sumerian signs from the period about 3500 B.C.

The Egyptian growth, then, seems to be clearly a case of cultural transfer and stimulus from Mesopotamia. The process went on over a long period, beginning at least as early as 5000 B.C. with the spread of simple village agriculture. Under a steady bombardment of new ideas from the explosive cultural growth in nearby Mesopotamia, 600 miles from Cairo, it took about 1500 years to develop a civilization. Any change in this picture is likely to lengthen the time required, for the agricultural beginnings in Egypt may prove to be even earlier than 5000 B.C. The lesson of the slow spread of ideas and cultural growth is an important one. So too is the theme that even the greatest cultural growth owes much to borrowing from neighboring people; conversely, those who have no bright neighbors seldom shine. This does not mean that borrowers are not also creative.

The ideas brought to the Nile valley were used in Egyptian fashion, not in Mesopotamian fashion. Instead of canal irrigation the Egyptians developed a system of basins fed by canals and protected by dikes. Canals led water from rivers to basins that connected from one to another, usually in sets of four or five, with the outermost basin at the edge of the desert, often 9 miles away from the river. The complex network of basins interconnected north-south and east-west made land travel nearly impossible on the flood plain, so the Nile became the highway. Naturally? No, artificially, at least in part, for without the basins and canals the levees along the Nile and the other routes along and across the flood plain would have been the "natural" routes.

With the basin system the land was flooded once a year by the Nile. With any system designed to move water onto fields in other seasons, not one, but three or four crops per year could have been raised. The water was allowed to stand for 40 to 60 days, and then with the receding flood it drained off. The deeply soaked land was planted, and the crop grew on the water stored in the soil. The crops planted included barley, wheat, flax, fava beans, lentils, and onions. Figs and dates came early, and vineyards were added later.

All of this must be seen as part of a particular physical and cultural geography. The Nile rises near the equator and receives steady year-round rainfall from that region. To this is added a large increase of rain in the summer from Ethiopia. These rains cause the Nile to flood Upper Egypt in July and August and Lower Egypt in August and September. The Nile carries a heavy load of silt, which has played a role in maintaining fertility and in filling the valley. The rise due to silt being added to the valley floor is about 4.5 inches per century. In a country where monumental buildings on the flood plain may be 4000 years old, this is a noticeable fact, and for archeologists seeking really early material (the earliest agriculture, for instance), it is a positive embarrassment. The evidence on the flood plain of the first agriculturists, if they date back to 5000 B.C., would be buried about 25 feet.

The basin-flooding system was developed to take advantage of the 25-foot rise of the river by soaking as much land as possible, thereby storing water for the fall and winter growth of crops. Wheat and barley are crops that originated in areas of winter rain and summer drought. They must originally have grown best during the cool rainy season, ripening as the summer drought came on. Planting in wet ground in the fall, growing during the cool season, and harvesting from the dried out basins closely follows the natural growth conditions for these crops.

The basin system fitted the particular crop needs well. But what would have happened if the Egyptians had had maize, a plant that grows best during the warm season? Would its presence have led them to develop a canal irrigation system designed to use the summer flood waters for immediate planting? Would this have led to a year-round irrigation system, to a much larger population, and an entirely different history of the Middle East? No one can answer such hypothetical questions, and they are introduced here only to emphasize the role of culture. The basic crops of Egypt came from a Mediterranean climate marked by cool-season rain and warm-season drought. The Egyptians took these foreign plants and adjusted the Nile floods by holding the water in basins instead of using it at once, so that the plants were grown in the appropriate season. The ideas, the agriculture, the particular plants, and the arts were first introduced, then the Nile was adjusted to them all. This is quite different from the old view that the Nile beguiled men into beginning agriculture because of its annual flood.

The agriculture was a simple type that kept the land in production only one-third of the year. The basins tended to become alkaline; the canals silted rapidly because wind often blew great masses of sand or silt into them. Usually the floods were dependable, sometimes they were excellent, but sometimes they failed, or the flood was too great (as portrayed in Biblical accounts of the seven fat years and seven lean years). The Egyptians studied the Nile floods, installed gauges to measure river rises, predicted flood heights from initial readings upstream, and took protective measures accordingly in the downstream areas. Eventually, from accumulated records, the officials could predict the annual crop from the initial rise of river level in Upper Egypt.

With their industrious working of the land (about 60 million acres were irrigated) and careful control of floods, the Egyptians built a large and stable population that supported a governmental, educational, and artistic class whose products are so justly famed. They engaged in trade, bringing wood from Lebanon, copper from Sinai, gold, ivory, Pygmies from East Africa, and spices from southern Arabia and beyond.

The government was the key to Egyptian success, for without a stable government canals became clogged, basins salinized, floods unpredicted and uncontrolled. Invaders—Hyksos, Ethiopians, Nubians, Assyrians, Chaldeans, Persians, and Greeks under Alexander the Great—all were absorbed by the Egyptians, and the governmental structure was maintained. Julius Caesar and Napoleon, two experienced judges, commented on the Egyptian success in exacting absolute obedience to a government that needed such obedience to operate its system. Napoleon's comment about the Egyptian government was:

In no other country does the administration exert so much influence on public prosperity. If the administration is good, the canals are well dug and kept up, the regulations of irrigation are carried out justly and the inundation is far reaching. If the administration is bad or weak or corrupt, the canals are blocked with mud, the dikes are in disrepair, the regulations on irrigation are disobeyed and the systems of inundation are thwarted by the sedition of individuals or localities. The government has no influence over the rain or snow which starts the flood, but in Egypt the government has direct influence on the extent of inundation which takes place. This is what makes the difference between Egypt's government before the Ptolemies (Alexander's successors)

which was already decaying under the Romans, and which was ruined by the Turks.[2]

One must conclude from this that Egypt was not the "gift of the Nile," an inevitable response to the flooding habit of the river. Instead the Egypt we know was the product of ideas imported from outside that were given a special form in Egypt. This borrowing of ideas and adapting them to particular needs and tastes is a normal occurrence in mankind. It might be better to say that Egypt was a product of the Egyptian peasant, directed by an autocratic bureaucracy, in the application of Mesopotamian ideas. How different things might have been can be seen by considering the "American Egypt."

The American Egypt

America has an area that is, physically, remarkably like Egypt. However, there is no resemblance in the history of the two places. If the environment is a powerful force in determining development, should we not expect two similar environments to provoke, stimulate, encourage, even demand similar responses?

The lower valley of the Colorado River is, like Egypt, a narrow ribbon of green stretching through formidable deserts. It is fed by a stream rising in high mountains a great distance away. Below the cataracts of the Grand Canyon the alluvial flood plain stretches to the sea. Like Egypt, there is a depression off to one side into which water sometimes flows naturally and which can be irrigated by controlling this natural overflow. In Egypt such an area is the Fayum; in America the comparable region is the Imperial Valley.

Like the Nile, the Colorado River rises far from its lower valley in high mountains. While the Nile rises near the equator and flows north, the Colorado rises in Wyoming and flows south, but the mouths of the two rivers are in the same latitude. The major floods of the Colorado are fed by the spring and summer snow that melts and flows into the area below the Grand Canyon during the early summer. Like the Nile, the Colorado carries a heavy burden of rich silt that has been filling the lower valley at a geologically rapid rate. Both the Nile and the Colorado have a series of cataracts. Below the Grand Canyon the Colorado passes through a series of broad basins such as the Blythe and Palo Verde. Below Yuma the Colorado

[2] A. Moret, *The Nile and Egyptian Civilization* (New York: Alfred A. Knopf, Inc., 1927), p. 34.

has built an immense delta into the Gulf of Lower California just as the Nile has built a delta into the Mediterranean. Both river valleys are marked by very high summer heat (July average about 90° F. and absolute maximum above 120°), very long, nearly frost-free growing seasons, and sunny, cool winters (January average about 55° F.).

What happened in the American environment? At first glance the answer seems to be *nothing*. The lower Colorado River valley was occupied by a group of people known as the Yumans. Actually, they included many groups of people with a similar language who were strung out along the river. In the delta were the Cocopa. The Yuma lived near the city of that name, and the Mohave occupied the area near Needles. There were other tribes, but they were exterminated or driven out by these major groups shortly after their first Spanish contact in the sixteenth century. In the description below the Yuma will be used as a representative tribe.

Yuma agriculture

The Yuma lived a leisurely and simple life; they developed some agriculture, but they did not work hard at it. Before the annual flood of the river they cleared small plots of land in the river bottom. A good growth of arrowweed was a sign of good land. They cut down the arrowweed, piling it in stacks to dry, and just before the arrival of the floodwaters they set the piles afire. It was considered good to time the burning so that the ground was still hot under the brush piles

when the floodwaters finally reached the field areas.

When the floodwaters receded, the Yuma waded out in the mud and planted their crops. They had a small variety of maize, a specialized type of bean, and an unusual kind of pumpkin or squash. All of these plants were able to utilize the moisture in the wet ground to make a rapid growth under furnacelike heat, producing a crop very quickly. The corn plants only grew to heights of 2 or 3 feet, but they produced ears of corn about 5 inches long. By comparison, our modern 10- to 12-foot-tall corn would yield ears 2 feet long. The beans also had the ability to absorb water very quickly and sprout in half the time that our modern beans require. They grew very rapidly despite the high heat, while modern beans drop their flowers and set no fruit when the heat is as high as the norm for the lower Colorado River valley in summer. The pumpkins also grew quickly and provided a quick harvest.

These plants were not domesticated by the Yuma Indians. They borrowed them from their neighbors, the Hohokam people, who formerly lived in southern Arizona along the Gila and Salt rivers. They in turn had received their plants from neighbors to the south of them, in adjacent Mexico. Probably the whole set of plants they used were domesticated outside their desert areas, but the situation is not very different from that in Egypt. It is doubtful that the Egyptians domesticated any of their plants either. The homeland of wheat cannot be determined definitely as yet, but it is more likely to be in Southwest Asia than in Egypt. The Egyptians were also borrowers.

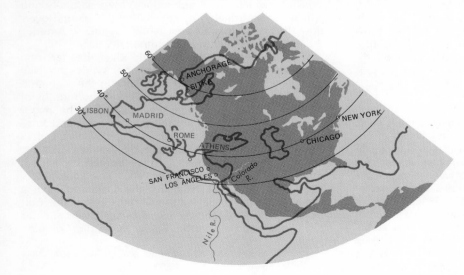

MAP 2-6   Eurafrica Superimposed on North America

As the map shows, the mouths of the Colorado and the Nile lie in the same latitude. Both rivers rise in mountains and flow through deserts to the sea. Both have narrow, flat-floored valleys that are subject to annual overflow by silt-bearing floodwater.

This map is also useful when considering other lands. Los Angeles, for example, is seen to be in the latitude of North Africa; San Francisco, in the latitude of Greece, Sicily, and Lisbon. Notice that Great Britain and Scandinavia are located in the latitude of the Panhandle of Alaska and the Aleutian Islands. Major cities have been added to facilitate comparisons.

Potential resources

The Yuma, however, never tried large-scale production of food. A small clearing and a small crop were sufficient for them. If this single harvest meant that during part of the year they would go hungry—so regularly that one season was known as "the starving time"—this recurrent deprivation nevertheless did not make them get busy and clear more land. They did have other resources: each summer there was an abundant crop of mesquite beans. These nutritious beans, closely resembling large string beans, grow in clusters on the mesquite trees that form immense thickets along the Colorado River. It was only necessary to be inured to being pricked and scratched by the claws of the mesquite tree in order to gather a huge harvest. The fleshy pods are stringy but full of starch and sugar, while the seeds are very hard and must be reduced to meal to be utilized. The major way of using these beans was to put them down in a pit, well moistened, to let them ferment slightly, thus softening the fibers in the pods and converting more of the starches to sugar. The resultant material was then chewed, and the remnant of fiber formed a quid that, when well chewed, was spit out.

Apparently there was no attempt whatever to domesticate the mesquite trees. Why bother when they were so abundant? Yet why did *some* people in the world decide to domesticate trees? There was no selection for larger, sweeter, and less stringy beans among the Yumans, and the fermenting of the beans never led to making a fermented drink. Would it not have been easy for some Yuman to put a few of the fermenting beans in a pot of water, forgetting them for a day or two, and then find that the frothy water when drunk gave him a "heady" feeling?

Similarly, palm trees grow wild in the Sonora Desert. They are found in little clusters in the canyons in the desert mountains. They have fan-shaped leaves and bear a round, fleshy fruit with an oily kernel. Elsewhere in the world, palms have been domesticated—for their sweet flesh in the North African–Arabian area, for their oil-rich meat in the Pacific area. However, these American palms were treated like mesquite trees. Even more notable, Arabian date palms were introduced into Lower California in the seventeenth century and flourished there. However, their cultivation did not spread, and not until the development of the Imperial Valley in the twentieth century did date growing finally take hold. Today, in the Coachella region of the Imperial Valley, ironically, near Palm Springs—which was named for the native palms growing there—there is now a sizable date industry. Palm trees were available there for thousands of years. Edible domestic

Date Palms in the Palm Springs Area of Southern California.
The orderly planting marks the American farming treatment of this tree as contrasted with the irregular pattern in the Old World oases. These trees are the result of a U.S. Department of Agriculture expedition to Arabia and North Africa to get young palms that were the offspring of the finest date-producing trees. Today the annual yield from this American oasis is over 20,000,000 pounds of dates of the highest quality. The opportunity for this productivity existed from the moment the Spanish discovered the American desert, but it was not realized until about 1900. (Palm Springs Chamber of Commerce)

date palms were introduced in adjacent California in the eighteenth century, but it was 200 years later as an American enterprise, rather than Mexican or Spanish or Indian, that palm trees of entirely new introduction finally were used in the environment for which they are so well adapted.

Available grasses were not domesticated, and tree fruits were not cultivated—neither the unlikely palm nor the obviously rich mesquite. These things did not happen in the world of the Yuma Indians, while the advantages of nature were tapped and made to supply a base for further achievements in the Egyptian world. Why the difference? The climates were equally hot; blisteringly hot summers did not stop the growth of culture in Egypt. The ideal winter weather, bright, sunny, warm days followed by cool, pleasant nights, did not stimulate the Yuma. The landscape, the soils, the rocks, almost anything that one can think of in the physical environment were alike in both places.

Just as there was copper in the Sinai Peninsula adjacent to Egypt, so there was copper in adjacent southern Arizona. There was abundant granite and marble and limestone for monumental buildings and works of art. Asphalt was available in southern California. No raw material of importance to Egypt was lacking in the areas immediately adjacent to the lower Colorado River valley. Why did a wealth of ideas take root in one area and not the other? Obviously, it was not the direct physical environment that was the causative agent.

Access to ideas

Perhaps there was an indirect environmental factor such as the ease of access to ideas. After all, no people can invent more than a fraction of their culture. Civilizations have largely been pieced together from the inventions and achievements of other peoples through time and over a vast geographical spread. For instance, the paper in our books is a Chinese invention, and movable-type printing is also said to be a Chinese invention. Our writing is an adaptation made by a people on the shores of the eastern Mediterranean from a basis of Egyptian writing. Our breakfast coffee is of African origin. The tomato was first developed by the American Indians. Our use of the 360-degree circle is derived from the Sumerians and so on. Perhaps the Yuma had no one nearby to borrow from.

However, this is not true. Eastward up the Gila and Salt rivers they had neighbors who had smaller streams to work with and used them to irrigate large areas. The Hohokam people raised not only corn and beans and squash, but also grew cotton, which they

wove into fabrics. They traded with the Mexican Indians for copper bells and parrots and did beautiful inlay work on shells they received in trade with tribes along the Gulf of Lower California. They built great adobe compounds around very large, multistoried structures

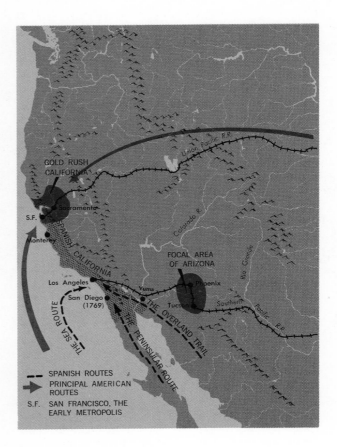

MAP 2-7   Development of the Low Desert in the Post-Columbian Period

The Spanish reached California both by land and by sea. To them the Colorado Desert was a formidable barrier to be crossed hastily. They attempted to settle in Yuma, where they wished to hold the strategic river crossing, and in El Paso and Tucson, frontier outposts guarding the route to Santa Fe. American interest early centered on the San Francisco–Sacramento area, the gateway to the all-important gold fields. Only after a century of contact has attention begun to shift toward the enormously productive and scenically attractive desert regions of California.

that had windows so located that they served for astronomical sighting. Hohokam pottery was well made and tastefully decorated. The Hohokam people even knew how to use acids to etch designs on shells.

All these ideas and many more were available to the Yuma, for they were great travelers, visiting the Hohokam and learning of their accomplishments. Indeed, they had obviously borrowed their corn, beans, and squash from that people, for the varieties and species that the Yuma used were exactly the same as those of the Hohokam. Their pottery is even a poor imitation of the finer product of the peoples on the Gila and Salt rivers. The Yuma traveled to the adjacent Pueblo lands, where they saw people living in snug stone houses with rooms set aside for granaries where a whole year's supply of corn was usually held in reserve for time of famine.

The Yuma had access in their cultural environment to the equivalent of the Middle Eastern architectural ideas: great adobe structures, adobe brick making, dry-wall stone masonry, and stone and adobe wall construction. All of these could have been used and have at times been assumed to be the natural response to arid lands. The Yuma, however, had inherited a system of building by erecting four posts with poles laid from corner post to corner post. For a summer residence they roofed this with a thick layer of brush, put up no walls, and lived in the shade. A winter house had walls made by leaning poles against the roof, covering these poles with a brush thatching, and then covering the whole structure with earth. The result was a dark, airless but warm structure useful in the nearly rainless land. It could equally well have been a stone, mud-brick, fired-brick, wattle-and-daub, or even a wood structure, for mesquite, cottonwood, and willow were available in abundance.

Yuma personality

Nevertheless, the Yuma simply did not feel moved to work hard, dig canals, and irrigate large areas to be well fed. They did not feel any urge to build either great adobe compounds or even snug little adobe houses, much less the strong and secure stone houses. The physical environment contained the raw materials for doing all of these things, and the ideas were available in adjacent areas. The Yuma knew those adjacent peoples and their way of life. They simply did not choose to live a kind of life other than the one that they had. This choice meant that food production was on a practically day-to-day level and everyone had to participate in harvesting it. They did not have a dependable surplus or a political organiza-

tion to channel a surplus to a creative few. The Yuma simply went along for thousands of years little changed in way of life. Perhaps, then, the Yuma were inferior genetically, mentally limited in some way. No one who has worked with these people in modern times has ever thought so. They are a lively, alert, often humorous, and intellectually curious people.

This leaves us with no apparent answer to the discrepancy between the development in Egypt and in its American counterpart. The physical environment in each place is very similar, and able people occupied both areas for long periods of time. They were exposed to earlier climates more favorable than the present one, and they experienced the stimulus of the great climatic change at the end of the glacial age—with accompanying rising sea levels that led to the filling of river valleys with silt, which created broader and broader flood plains subject to annual overflow. In one area, the idea of agriculture was borrowed from adjacent people, and on this agricultural base a great civilization was built. In the other, the idea of agriculture was borrowed, but very feebly developed, and very little progress was made toward civilization.

The element of time

Could it be that the development of the idea of agriculture takes a great deal of time? After all, Egypt did not spring into greatness overnight, and neither did any of the Mesopotamian centers. In the Middle East the beginnings of agriculture go back to about 8000 B.C. Agriculture-village ways of life existed for nearly 5000 years before the appearance of cities, states, and empires, and the development of arts and science and all the other marks of the growth of ideas and populations. Perhaps, then, in the American Egypt there simply had not been sufficient time for the ideas to take root and develop.

The appearance of agriculture in the southwestern part of the United States is dated now about 3000 B.C. This is the earliest find now known, and it is in southern New Mexico, a considerable distance from the Colorado River. It has been suggested that for 3000 years men continued to live an unsettled life, planting small plots, leaving, and returning only to harvest them. Probably it was much later that irrigation agriculture became established in the Gila and Salt river valleys adjacent to the Colorado River area. An informed guess would place this development about the time of Christ. Some time thereafter the Yuma must have obtained their agriculture from the Hohokam people. The time lags may have been on this order: 3000 B.C. for the appearance of agriculture in New Mexico, 500 B.C.

Agriculture: Bonanza or Dead End? *Agriculture has not always been a success story; instead, much of mankind found it a trap. It multiplied population, eroded the soils, and finally led to a desperate struggle to maintain life. Studies of the simple gathering way of life show that men living by this means had a great deal of leisure. Why, then, did men take up agriculture, and why has it swept nearly the entire world?* (Paul Conklin for World Bank)

for the appearance of irrigation agriculture in the Ho-hokam area, and perhaps A.D. 500 for the appearance of agriculture on the Colorado. The Yuma may have taken up pottery making and simple farming about this time. They are unlikely to have had more than 1000 years to develop the idea of agriculture before the Spanish first saw them about 1540, and the time may well have been shorter. It takes time for ideas to spread, to take root and grow, and the number of years involved here does not seem to have been enough.

It is easy to find in history abundant testimony that ideas spread slowly. Japan, for example, as late as A.D. 500 lacked writing, despite its proximity to China where civilization had been flourishing for 2000 years. Japan is also a good example of the rapidity of cultural change in a given geographical setting when human minds in the area alter from resistant to change to avid for change. The problem for the American Egypt finally comes down to the question, what was the attitude of the Yuma mind?

The Yuma value system
The Yuma people had a peculiar turn of mind from our point of view: they did not value the material

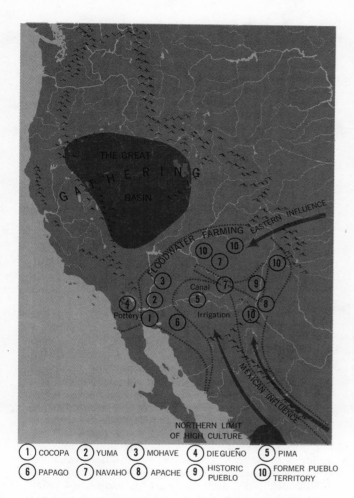

COCOPA (1) YUMA (2) MOHAVE (3) DIEGUEÑO (4) PIMA (5)
PAPAGO (6) NAVAHO (7) APACHE (8) HISTORIC PUEBLO (9) FORMER PUEBLO TERRITORY (10)

MAP 2-8 Cultural Influences on the American West before 1500

Some of the cultural influences on the American West in pre-Columbian times are indicated here. Influences from the centers of high culture in Middle America were the most significant, reaching the Southwest in several waves and by several routes — and at differing times. Early introductions brought in agriculture that was dependent on rainfall, or on the natural concentration of floodwater runoff in certain favored spots. Some influences seem to have reached the Southwest by a route from Mexico to the Mississippi valley and then crossed the Great Plains to reach the Southwest. Finally, rather advanced ideas, including canal irrigation, probably reached the Southwest via a route on the west side of Mexico that had previously functioned to transmit such ideas as pottery and primitive agriculture.

Those areas far from the sources of ideas received the least stimulus. Thus canal irrigation stopped in Arizona, though it was fully applicable in adjacent California and Nevada. Floodwater farming spread a bit farther than canal irrigation, and pottery spread beyond the limits of agriculture. The Great Basin, which had the needed raw materials for pottery, possessed also the human needs for pottery and agriculture and a physical environment containing many oases — including rivers such as the Truckee that rise outside the desert and flow into it. But the Great Basin was too far from the source areas for their cultural ideas to reach there early enough to take root and grow in pre-Columbian time. Given more time, such influences almost surely would have penetrated these areas, but the needed time might well have been measured in millennia.

comforts of life. As we have noted, they were courageous, curious, and humorous, but they did not seem to be at all avaricious. Since in the Yuma view of the world material wealth did not count for much, men were not admired for the size of their houses, the extent of their fields, or for the accumulation of any kind of wealth. Respect was paid to men of prowess in battle, but the greatest distinction—the thing most desired—was to have good dreams! The powerful dreamer was the fortunate man. It is difficult for us to understand their psychology, but their descendants have been equally at a loss to appreciate the modern Western obsession with material things. Part of the Yuma "lack of progress" in the prewhite period, as well as in the postcontact period, can be understood as their failure to fit into our *definition* of progress.

Progress necessitates hard work to acquire more food and goods to support more people, who in turn produce and require more goods and produce more people. Perhaps the Yuma looked at the consequences of stepping on the escalator that leads to civilization and decided that it looked like a treadmill.

There were features in their way of life that reinforced a negative attitude toward material things. At an individual's death, all of his property was destroyed with him. Since the Yuma practiced cremation, a man's belongings were burned with the body—nothing was kept to remind the family of the deceased. Why accumulate property if it will be destroyed? In recent times, fine horses have been killed, and expensive saddles bought with half a lifetime's savings have gone up in smoke.

There was no reason to work hard at preparing large fields either. Property was largely communal. If you had a very fine crop of corn, your relatives and friends and neighbors would drop in to admire it and help you eat it up. The rules of hospitality demanded that you cheerfully supply all comers so long as the crop lasted. Since there would be no surplus for your family no matter how hard you worked, what was the use of working hard?

The Yuma therefore went on living the same, simple economic way that their forefathers had lived for generations uncounted, gathering the wild roots, seeds, and game that nature provided. Even though food was freely shared, there were very few people to share it. Sharing with the one or two families that formed your branch of the tribe was a small burden, whose compensation was the knowledge that when you were low in food and other families had good luck, you would be cared for automatically.

An innovation in agriculture would increase the number of people due to the increase in food. Increase in numbers would make sharing more burdensome, but better agriculture would also reduce the need to share. Nonetheless, the carry-over of old ways of life, combined with an unworldly, nonmaterial outlook, slowed the rate of change. The cataclysmic arrival of the white man ended slow change and substituted the white man's ways. Had there been the time, it is possible that the lower Colorado would have seen the steady growth of population density, increased clearing of land, the addition of irrigation, the growth of communal effort in digging canals, the accompanying growth of political and social institutions, the appearance of genuine political leaders, and the slow growth from villages to towns to cities and possibly to empire.

Would there then have been a duplication of Egypt in America had there been a period of perhaps 5000 years when the Yuma were left undisturbed to work out their own development of the application of agriculture in a river valley oasis? We can probably assume that there would have been growth of population and the appearance of cities and arts. We surely could not expect an Egyptian cast to the development, but rather something distinctly Yuma. Just what the social, political, and economic structures might have been and what kind of development of arts would have occurred, we can only imagine. This illustration shows us that similar environments do not produce similar things. We must consider human, cultural, and historical factors as well as the physical; moreover, the time element is enormously important. In the Yuma case, time was short, and their degree of contact was slight. Perhaps their liking for warfare prevented ad-

vanced people from invading their territory and forcibly introducing new ideas. History would probably have been different if the Aztecs with their organized military had lived near Tucson and Phoenix. But that is the point. The determining causes were not the physical environment but the ideas that existed in and adjacent to it.

The Colorado River Valley Today

What, then, has happened in the lower Colorado River valley, now that we have taken it over? Has it become another Egypt, a center of culture and population growth, invention and the arts? Where are the universities, the great monumental art works, the museums, the libraries? There are none in the lower Colorado River valley and only some in adjacent Arizona.

The southern California desert

The American development of California centered around San Francisco and Sacramento, drawn there by gold. That area became the early center of population where the capital, the state university, and the first major city were located. It was the center of culture, learning, and finance. Until the twentieth century southern California remained a sleepy sun-soaked backwater of the economic and social currents flowing into California. The desert land of the Imperial Valley and the Colorado River was the backwater beyond the backwater. Central California boomed in the 1850s, southern California began its explosive growth after 1900, and the desert land of California did not begin to boom until the 1950s.

The development of the land has been relatively slow. The base of development has been agricultural, with irrigation dependent on water from the Colorado River. Large stretches of the Colorado River valley were set aside as reservations for the Indians. The largest section of open land lay in the Imperial Valley, that great depression 50 miles by 100 miles with its floor 250 feet below sea level. When the white man first saw this area, the valley was dry, and the low spot in the valley was covered with a deposit of salt.

The first Europeans considered this area a dangerous desert that was extremely difficult to cross. For the Spanish and the Americans it was an obstacle to the overland route to California. Later the American railroads pushed across it, following a route around

Aerial View of Palm Springs, California. *Palm Springs bills itself as the winter golf and swimming-pool capital of the world. Its development has been that of a boom town since the warm, sunny climate and stark scenery of the desert gained in public appreciation in the mid-twentieth century.* (Palm Springs Chamber of Commerce)

the salt flats at an elevation at times below sea level. The appearance of modern transportation changed the utility of the area; suddenly its products could be marketed. The first product was the salt itself, lying on the floor of the ancient lake, available for the scooping up and exportable by loading on the adjacent railroad.

At times of high water there was a weak flow from the Colorado River toward the Imperial Valley. This was just sufficient to fill the ancient channel of the river that looped south into Mexico and then turned northwest toward the valley. With the coming of the railroad and the growth of population in California at the end of the nineteenth century, and with a developing market in the East, men looked more appraisingly at the vast expanse of gently sloping alluvial land of the Imperial Valley and the almost frost-free climate. They saw it with the eyes of agriculturally and commercially minded people, for they now had the technical means to develop it and to move the produce to distant markets.

The natural path of the waterway was opened up, canals were dug, and header gates installed. Thousands of acres of land were soon producing crops under the desert sun. The intensity of solar energy, available almost every day in the year because of a nearly 365-day frost-free growing season, made it possible to produce many crops in great abundance. Winter crops could be followed by summer crops. Sometimes the land could be made to yield three crops in a year. When the soil was put into alfalfa, it was found that 10 crops per year could be taken off the land. To the men of our modern Western culture, this looked like a plant-producing situation of almost unlimited productivity requiring only transportation adequate to link it to the consuming areas. Once the productive areas of the East could afford the luxury of fresh vegetables in winter and exotic fruits all year round, the Imperial Valley became a virtual bonanza.

The productivity of the Imperial Valley is shown in Table 2–2. A return of over $488,800,000 from less than half a million acres under irrigation indicates an average yield of about $1000 per acre. The pattern of land ownership shows the startling fact that the average farm is over 180 acres. Such size is in keeping with the progressive use of heavy equipment and the mechanization of the whole agricultural process, for large capital outlays and large units are required to justify such expenditures. Comparison with figures from 1967 shows that in a 5-year period there have been significant shifts in production. Both the crops and the shifts in crops are of great interest. The vegetable production alone is a huge business. Seventy-three million dollars worth of lettuce is hard to visualize. Alfalfa production is worth $62 million. Much of this alfalfa goes to feed livestock, which in recent years has moved from third to first place in value. Feed lots are so numerous that whole areas of the valley give off an unpleasant odor. When a visitor comments on the smell, the natives reply, "It smells like green money to us." The relatively rapid shifts from one crop to another, depending on market demand, are typical of large-scale, highly capitalized agricultural enterprise. The immense growth in the alfalfa and cattle industry in so short a period is an excellent example. It contrasts sharply with traditional farming methods whereby men may, for centuries, persist in producing the same crops almost irrespective of markets and profits. Much of the enor-

mous alfalfa production goes to feeder lots in the valley to supply the cattle moved to the great urban centers of southern California. The valley is strongly concentrated today in cotton, cattle, sugar, and vegetables.

The wealth thus produced on the land did not stay in the valley, however; most of it flowed out. The development was one of large-scale properties, usually owned by men with considerable capital who employed others to manage their farms, which were run with a largely shifting labor population that was often foreign. There was little improvement on the land to show for the wealth that the land was producing. The fine homes were "on the coast" or "in the East." The valley was extremely hot in summer, dusty at all times, and much of it was a dreary desert landscape that did not appeal to the recent emigrants from the beautiful, green, tree-filled land of the East. Consequently, no one felt an interest in promoting great universities, beautiful cities, or other evidences of flourishing civilization in the Imperial Valley.

Today the desert has been "discovered." On the west side of the Imperial Valley lie such resorts as Palm Springs and Borrego Valley. Yuma, once considered unlivable, is now a fashionable winter resort. The bright, warm, sunny days and cool, pleasant nights make it a most desirable area for winter resorts. The desert scenery, instead of being considered dreary, is now captured on canvas. The desert is eagerly sought after for winter homes, while air conditioning has become ever more prevalent and comfort in the area is no longer limited to the winter time. As men's ideas about the desert continue to change in the future, the vast oasis we call the Imperial Valley will change too.

The future of the Imperial Valley will not, of course, be Egyptian, for the cultural background is American, featuring an industrial-type farm management that makes a maximum use of machinery. In Egypt there is an immense base of poor and uneducated, and hence unskilled, men on the land; therefore labor tends to be accomplished by unaided manpower. Further, Egypt has not yet become a highly productive nation that has a market for specialty crops. The lower Colorado River region, on the other hand, with an educated people who have a high standard of living—and with mechanization removing the drudgery from farming—is likely to attract a permanent settlement of families with good incomes who will demand schools and all that is needed to make this a cultural center.

There are other advantages and disadvantages. The entire development is dependent on Colorado River water. Most recently, attention has focused on

TABLE 2-2   The Productivity of a Great Oasis: The Imperial Valley

Crops	1967	1972	1973
Field	$84.7	$104.7	$168.5
Vegetable	72.5	93.1	119.1
Livestock	70.5	119.7	190.6
Fruit and nut	.8	1.4	2.7
Seed	2.6	2.9	6.4
Apiary	.4	.7	1.5
	$231.5	$322.5	$488.8 (millions)

Million-Dollar Crops: 1973 (in millions)

Lettuce	73.0	Asparagus	5.0
Total alfalfa	62.2	Cottonseed	4.6
Sugar beets	32.0	Tomatoes	4.3
Cotton	30.0	Rye grass	3.7
Wheat	26.0	Watermelons	3.0
Cantaloupes	12.5	Barley	2.5
Sorghum	8.0	Alfalfa seed	2.4
Carrots	7.0	Pasture alfalfa	2.0
Onions	7.0	Sugar beet top	1.0

Source: Imperial Valley Weekly (March 7, 1974), El Centro, California.

the geothermal energy potential of the valley. The earth's crust is splitting up the axes of the valley, and enormously high temperatures come relatively close to the surface of the earth. Plans have advanced to tap this heat reservoir and use it to supply power to factories and generate electricity. Should such plans materialize, the valley could become an industrial center. If this happens, the valley will have changed from a desert with primitive people to an oasis of great agricultural wealth to an industrial center—all within 100 years.

The southern Arizona desert

The low desert, as found in the lower Colorado and Imperial Valley area, also extends eastward up the Gila and the Salt rivers of southern Arizona. These valleys were the home of the Hohokam people who used great canals to irrigate miles of the desert and were clearly many steps ahead of the Yuma. It is significant that they are closer to Mexico, where one of the great centers of civilization in pre-Columbian America began in the south. Proximity to ideas, possession of the ideas for a longer time, and greater receptivity to them—perhaps even colonization by a more advanced people—placed these people much further on the road toward civilization. Probably the most important factor was that they lay at the end of the Sonoran Desert closest to the center of cultural growth and stimulus in the Mexican area. Some catastrophe, however, greatly reduced the Hohokam just before the arrival of the Spaniards. It seems likely that warfare with the raiding nomadic peoples, such as the Apache and the Navaho, nearly destroyed them. Settled farming people always have difficulty with the nomads. One raid can destroy a year's work in the corn fields, and the nomads can melt into the surrounding desert, leaving their victims faced with virtual starvation. Compare the conquest of the Sahara by the camel nomads. But some other cause may also have intervened; for example, a new disease introduced into an unprepared population of the Hohokam area could well have been original size. Had this happened sometime after 1520, the time of the Spanish arrival in Mexico, the dense population of the Hohokam area could well have been reduced to the situation that the first explorers recorded.

The Spanish evaluation of these same southwestern lands is significant, for they did little with any part of them. They sent land and sea expeditions north from Mexico in 1540. The land expedition reached the Colorado River, but the land was not reported as offering great opportunity. It was a marginal land, far from the Spanish centers of interest in Mexico and Peru where there were large populations of civilized natives and great sources of gold and silver. The low, hot desert lands were only something to get across on the way to the frontier outposts of Santa Fe and Monterey, the mission frontier in California. Missionary zeal was the major driving force that led to settlement. The land in what became southern Arizona and adjacent New Mexico and on over to El Paso was not viewed as an asset. It was too dry, too hot, and too far away; it remained a poor missionary and political frontier.

The early American view of the situation was similar at first, occasional reports stressing the desert character of the land. Only a few of the people on the way to California during the gold-rush days of 1849 took the southern route that led to Santa Fe and then through southern Arizona and across the desert from Yuma to California. The surveys for the railroads in the 1860s, however, found that there was a feasible low-level route through this southern country. A railroad was pushed through, linking this area with the East; that it also linked this southern Arizona desert country with the West was considered relatively unimportant. The West was not populous, and it did not need to import agricultural produce or raw materials.

The mile-high Colorado plateau of northern Arizona endured not only the cold winters but the fierce Navaho and Apache warriors as well. They were as great a block to settlement for a time as were the Iroquois in the East in an earlier time. This temporary barrier left the southern part of Arizona as the major open land. Here was a low-latitude desert, thinly settled by a peaceful people and marked by the lines of a great irrigation system, some of whose abandoned canals could still be used. The arrival of the railroad facilitated the shipment of the area's produce to the rapidly industrializing East. The productivity of the East allowed the purchasing of the type of crop that could be grown in this immense oasis, watered by the Gila and the Salt rivers. The physical geography had not changed a bit, but southern Arizona was no longer too far, too hot, and too dry to be an active and productive part of the nation.

For these reasons—the location of hostile Indians principally in the north and the arrival of the railroad on the irrigable plains located in the south—it was natural that the southern part of Arizona developed first. Arizona became a state in 1912. Its capital was located at Phoenix in the center of the irrigated area, and the state university was established at Tucson. It was southern Arizona that had become the center of population, government, productivity, and culture. There was less absentee ownership and distant managership of the land than in the Imperial Valley. Southern Arizona with its populous and cultured oases on the Gila and Salt rivers, then, has an entirely different aspect to it than the oases of the Imperial Valley and the Colorado River valley, which are both far from the center of early growth and development in California.

What might have happened had the state boundaries been laid out differently? Suppose the southern California desert had been detached from the rest of that state and included within Arizona's borders. Suppose the western boundary of that hypothetical state had been along the ridge of the peninsular range just west of the Imperial Valley. Would there have been a Tucson at Yuma and a Phoenix in the Imperial Valley? Would Yuma, as a location midway between the two great oases, have been chosen as the state capital? Would it early have assumed an important social, political, and cultural role in the Southwest? This seems possible. Clearly, we cannot now know in detail what would have happened, but we can be sure that things would be different. The human assessment of the land and the type of growth and development there would have varied, depending on how even the minor subdivisions within the national state were drawn.

So much then for the element of value judgment and the use of the low desert countries of the Southwest. The setting is not greatly different from that of the Fertile Crescent of the Old World. An Egyptian environment (the lower Colorado) lay at one end of the crescent. At the other lay a Mesopotamia with the Gila and the Salt taking the place of the Tigris and Euphrates. To the north of this "Mesopotamia" lay highlands, just as in the Old World. The physical setting, though on a smaller scale, is sufficiently similar to have provided the physical challenges that are found in the seat of the origin of civilization in the Old World. The differences, therefore, were not determined by the physical geography.

The Arid West

If a larger view is taken of the arid West in America, a similar picture comes into focus. This is a vast region. It extends from the Sierra Nevadas to the Rocky Mountains and almost from the present-day Canadian border to far below the Mexican border. This is the American equivalent of arid Inner Asia. Within the United States the northern part is called the Great Basin. It centers on the state of Utah but includes most of Nevada and parts of Oregon, Washington, and Idaho. This is an area of irregularly distributed, short mountain ranges, and the level of the land is high. South of the Great Basin lies the Colorado plateau country. This again is high land, less mountainous but characterized by such immense features as the Grand Canyon, making this a rugged land of great relief. The region's interior location, shut off from moist winds by the Sierra Nevadas to the west and the Rockies to the east, contributes to the aridity. Yet the differences of elevation are so great that it is a land of contrast over short distances. It is possible to start in one of the broad basins in a condition of extreme aridity and then ascend one of the adjacent mountain ranges to pass from extremely sparse sagebrush-covered land into a zone of scrubby piñon pine, then on into a zone of large Western yellow pine, and finally into an alpine meadowland with running streams, aspen, and rolling green meadows. Certainly not a desert situation! Sometimes sizable streams come down from these high mountains and create oases in the desert.

Pre-Columbian life

In the Great Basin before 1500 the way of life was that of hunting and gathering. Small bands moved intermittently about the land, gathering roots and seeds and catching small game. These people have been variously designated as seed gatherers, unspecialized hunters and gatherers, and the Desert Culture people. All these terms describe much the same thing. The people who occupied most of this territory were living a very simple life, wandering in small bands, gathering wild seeds and roots and nuts, and doing some hunting. But their way of life was as unspecialized as possible. If they had any specialty, it was in the gathering of plant foods, hence the designation seed gatherers. They were very close to our theoretical man on the hypothetical earth. It is probable that their way of life was very close to that of all mankind in the early stages of man's development. We know that they had been living like this for tens of thousands of years. The people of the plateau region followed this same way of life until about the time of Christ. They then rapidly changed over to the agricultural town-dwelling way of life. The people of the adjacent Great Basin did not change.

Why in this vast area did man cling so long to so simple a way of life? Was it the difficult aridity, the high elevation with its numbing winter cold, and the summers with their hot days? But what about the stimulation of the greatly varied environments to be found from the desert basin floors to the alpine mountain environments? Why did people on the Colorado plateau suddenly change? Why did the wonderful oasis at Salt Lake City not seem as useful to these people as it did to the Mormons? And, after all, this area is not too different from the lands to the north of the mountains that divide Inner Asia from southern Asia. Along the oases created by streams coming down from the high mountains and flowing toward the Caspian and the Aral seas and down into the desert heart of the Tarim basin, great cities flourished for thousands of years. Again the physical environment, forbidding as the desert stretches are, is an insufficient answer. If we brush aside the question of the potentiality of the human stock as unanswerable, what reason can we find? What could be done in such an area?

If we look at the rest of the world, we see considerable range of possibility. In Australia in the period prior to 1500 a comparable area was occupied by people with a similar level of culture. In South Africa the same situation prevailed, for the cattle herding was clearly a relatively late introduction. In North Africa and Arabia and Inner Asia settled people practiced agriculture wherever water allowed, and pastoral nomads moved herds about over the rest of the land. Why was there none of this in Australia and in North America and little of it in South America and South Africa?

It is notable that in all of these cases the distance from centers of ideas is of significance. South Africa was far from the active axis of civilization that stretched from Egypt to China. The Inner Asian area lay on the routes that led from one of these areas to another. Australia lay completely out of touch with all of the activities of the developing civilizations of the world, though as we will see, mere distance is an insufficient answer even there.

In America the centers of growth lay in Middle America. Although northward extensions of this American civilization strongly influenced the Gila-Salt valleys and the Colorado plateau, they barely reached the Great Basin. The Great Basin people lay beyond the zone of effective reach of these ideas. Agricultural beginnings spread into the edge of the Great Basin, but about A.D. 1000 raiding peoples appeared, and the farming folk withdrew. The people of the Great Basin remained hunters and gatherers because they were too far from any of the currents of ideas and change to benefit. Not that ideas never reached them, but the contact was not long sustained and meaningful enough for them to change their way of life, a long and slow business.

Moreover, mere environmental stimulus will not do. The people of the Great Basin faced extraordinarily stimulating conditions. During glacial times the Great Basin contained many lakes. Fish, waterbirds, and a host of aquatic plants were available to these people,

MAP 2-9 Pluvial Lakes and Streams of Nevada
The magnitude of the changes that resulted from glacial-rainy periods in the arid lands of the world is demonstrated by this map of the lakes and streams of Nevada during such a period. The area is now one of America's driest. The Great Salt Lake in Utah, Pyramid Lake in Nevada, and Lake Tahoe on the Nevada-California border are the principal remnants of the water bodies shown here. The physical environment of this interior land must have been vastly different prior to the great climatic change that occurred about 10,000 years ago. (Adapted from compilation by Hubbs and Miller.)

and they used them. With the end of the glacial period the lakes dried up, and the whole environment changed drastically. If necessity could compel a people to take up agriculture or change in some way to meet a challenge, these people surely should have changed. The record, however, is one of remarkably little change. Nor were these people subjected to this physical stimulus only once. There were several climatic fluctuations sufficient to change this lake environment greatly. Men were alternately subjected to good and poor environments. However, they ate up the potential domestic animals, such as horses and camels; they did not culti-

vate the nutritious local plants; nor later in time did they take up the already domesticated plants that their neighbors had. Yet there were numerous oases, and some of them were large enough to have supported towns or cities, even city-states. (Outstanding examples today are Salt Lake City and its adjacent cities in Utah; Reno and Carson City in Nevada.) Neither the agricultural nor the pastoral opportunities were perceived by the seed gatherers of the Great Basin.

(Left) **Paiute Woman Seated outside Her Brush Shelter Making a Basket, 1873.** *The racketlike implement on the shelter is a seed beater, and the large flat basket was used to catch the seeds. The small-mouthed basket is pitch-covered, and it served as a water bottle. This crude brush shelter was in use in the same area as the Pueblo stone houses and Navaho earth-covered homes.* (Smithsonian Institution, Bureau of American Ethnology)

(Below) **Navaho Woman at Her Loom, about 1880.** *The earth-covered log house of the Navaho is one of several house types in simultaneous use among the Indians of the Southwest. The Spanish and the Americans introduced still other types of houses in the same geographical setting. The Navaho learned weaving from the Pueblo, an example of diffusion.* (Smithsonian Institution, Bureau of American Ethnology)

Agricultural villages on the Colorado plateau

South of the Great Basin, on the Colorado plateau and along the Rio Grande River, there are farming peoples. They live in towns and are called the Pueblo people, a name that is the Spanish term for town. The Pueblo people were originally seed gatherers or unspecialized hunters and gatherers wandering over the vast arid Colorado plateau in much the same manner as the people of the Great Basin in historic times. Ideas coming up from the south where agriculture had long been established reached these people sufficiently early and with sufficient impact for them to change their way of life.

Earliest agriculture in the Southwest, unquestionably moving northward from Mexico, appears in the rugged mountainous fringes of the Colorado plateau, where increased elevation decreased the temperature and increased the rainfall and aided the possibility of agriculture. Corn cobs from a cave in New Mexico have been dated by radioactive carbon (C–14) as about 3000 years old. By careful site selection such an agriculture could have been spread almost indefinitely into the plateau and Great Basin regions to the north. Eventually this was done to a limited extent. It is noteworthy that the earliest introduction of agriculture was not in the low, hot lands that required the relatively advanced skill of irrigation with canal digging and all the population density and political organization that this monumental application of labor implies. Instead the beginning was made in the foothill regions where the plots were small and little more than the natural rainwater was needed, enhanced in later time by the selection of field sites that natural runoff tended to flood.

With the introduction of the idea of agriculture, a great part of the arid western United States became a potential agricultural region. This new idea could have changed the human geography. With agriculture all the other ways of life could go on with little change. Only a more settled way of life for part of the people during the crop season was required. The return would be large: a relatively certain and large addition of food.

Yet men did not leap at the opportunity, for agriculture did not become established on the plateau region until after the time of Christ—there was a 2000-year lag. This situation resembles that of the Yuma. There it is clear that a human decision was made—the environment certainly did not make it. Perhaps the weakest part of the discussion of the Yuma was the attempt to show that they are not inherently (genetically) inferior people. It could be said that since they never *had* accomplished anything it was evident that they *could* not. With the plateau people, however, we can test this reasoning. For many thousands of years

Walpi Pueblo, Arizona, 1879. *This photograph shows the multistoried building of the Pueblo, with its ladders for access, its old-style doorways, and a rabbit-fur robe hanging near the top of the ladder. The Pueblo people began using stone for building shortly after the time of Christ, a change made perhaps in response to an inflow of ideas from Mexico rather than as a pure response to environmental need.* (Smithsonian Institution, Bureau of American Ethnology)

they had lived as hunters and gatherers; they were among the cultural laggards of the world. About 4000 years ago ideas began to reach the area to the south of them, and for 2000 years the plateau people took no advantage of the new opportunity.

For 2000 years, therefore, their refusal could have been blamed on something genetically lacking. But then they took up these ideas with such a spurt of energy that they shortly developed their own distinctive culture, rapidly changed in numerous ways, and went from one of the simplest cultural people in the Southwest in pre-Columbian times to the culturally advanced Indian group known to us today as the Pueblo people.

Had our assessment of the geography and the people been made before 1000 B.C., we might have said that the environment of these arid high plateaus prevented the development of culture and the growth of populations, and that the one thing that would *not* develop here would be a town-dwelling society. Another person looking at the lagging status of these people might have suggested that they were simply inferior human types. Yet at A.D. 1000 town-dwelling societies with highly developed skills in agriculture and pottery and with a rich ceremonial life were flourishing in these very regions and among these very same people.

Parallels to this situation are numerous. The well-known case of the Japanese spurt in modern technological development in the late nineteenth century— a conscious decision to end their isolation from the advances of the West—is a good case in modern time of the role of human decision in accepting or rejecting ideas. These changes are usually slower than they appear, for beneath the surface pressures are built up leading toward the sudden spurt. Thus the Japanese case had antecedents that make the abruptness of the shift less miraculous than it sometimes seems.

The Europeans have also been cultural laggards at various times, and, like the Japanese, at various times one group of Europeans or another, after a period of preliminary absorption of ideas, have burst forth in a cultural or political growth. The Greeks lagged behind the Egyptians, the Romans lagged behind the Greeks, and the Britons lagged behind the Romans. Yet each of these people learned from, and eventually exceeded, the people who had once been far ahead of them in cultural accomplishment. Even the 2000-year lag in the spread of agriculture in our Southwest from the area to the south of the Colorado plateau onto the plateau itself can easily be matched. In Europe cities were flourishing south of the Alps by 1000 B.C., but there were none north of the Rhine until almost A.D. 1000. Bronze, which appears in the Middle East about 2500 B.C., took 1000 years to spread into northwest Europe.

Strong influences suggested by great change over a brief period

The change on the plateau was dramatically sudden and far-reaching. Up to the time of Christ the people later known as the Pueblo Indians were living in small bands, hunting and gathering in their ancient way. When agricultural beginnings were made, pottery appeared, and house types began to change. The old crude shelters of grass and brush and the ancient semi-subterranean houses were replaced by surface houses made of stone laid in mud for mortar. This is a larger change than is usually admitted.

Houses on the Colorado plateau could be either semisubterranean or surface types, made of log, stone, or grass. The environment apparently did not demand one particular type. In an arid region marked by bitter cold in winter, the semisubterranean house would have had great advantages. It would be snug and warm in winter, and if there were not too much moisture, it would stay dry. The log house, on the surface and well insulated with a thick layer of earth, would also be snug, dry, and warm. The least desirable in such a cold winter climate would be the grass-thatched flimsy shelter that probably was used throughout most of the arid West for tens of thousands of years.

The stone house that seems so "natural" to the arid regions was late to appear; it developed rapidly and included some odd features. The first stone houses are thought to have been granaries or storerooms built on the surface by some of the early agriculturalists. Soon they were being built in groups, and shortly thereafter they were being lived in. Little groups of small rooms linked together into "L" and rectangular shapes soon dotted the plateau region and were spread far north, well into Utah. There were no stone houses among the Indian peoples to the west, north, or east of the Southwest, but stone houses *are* characteristic of Mexico. This suggests that the idea may well have been introduced from that direction rather than developed locally.

The plant evidence suggests early introductions both from the south and from the east. Later it is clear there were large cultural introductions from the south. These included such things as the cotton plant, ceremonial ball courts, and such direct imports as copper bells manufactured in Mexico. It appears that for a long period ideas did not move into the plateau region at all, but then a critical point was reached, and ideas flooded in. Once this initial change toward a settled way of life was made, the path was open to the acceptance of many ideas that had long been available on the southern edge of the plateau. The result: Pueblo culture.

The Pueblos include several people who are physically and linguistically distinct but who share a common culture. They can be thought of as resembling the German-French-Italian composition of Switzerland, except that the Pueblos did not develop a national unity. The Hopi are a very small people physically and are Uto Aztecan in speech. Their relatives are in the Great Basin. The other Pueblo people are taller, and some of the Rio Grande people are related in speech to the Great Plains' tribes. Like the plant evidence, this suggests that men and ideas from the plains and from the low deserts met on the Colorado plateau and sparked the flame of culture change.

We know that ideas were coming into the Southwest from the south where building in stone was a well-developed characteristic. Along with the ideas of agriculture, pottery, cotton, and weaving that reached the plateau people by various routes, the idea of building in stone was probably imported. Then, when the raiding peoples appeared—the Navaho and Apache —the Pueblo people reacted by building larger and more compactly grouped stone dwellings until the classic Pueblo fortress, apartment-house type of dwelling evolved. The conclusion to be drawn, therefore, is not that a direct reaction to a dry environment made it "natural" to build in stone. A different interpretation suggests that the idea itself was the important determinant, and that secondarily the politico-economic situation was significant. The fact that the physical environment could supply large quantities of layered sandstone easily broken into bricklike shapes—and hence readily used for construction purposes—was only of minimal importance.

Just how these arid-land seed gatherers were jolted out of their millenia-long wandering way of life, induced to take up agriculture in this difficult region, taught to build stone houses, and soon brought to live in towns has never been satisfactorily answered, though recently suggestions have been made that after a long period of indirect contacts, more direct contacts were initiated. From the evidence of slowness of change elsewhere (for example, among the Yuma and the Great Basin tribes) we may suspect extremely strong influences from outside. Since the amount and rate of change are comparable to the changes that occurred in early Egypt under Mesopotamian influence, comparable influences seem likely to have been operative in the Southwest. And since the center of American cultural growth was in Middle America, these influences must have come from there by various routes.

The results were spectacular. Agriculture began about the first century A.D. Stone houses appeared shortly thereafter, and the houses almost at once were built in groups. By A.D. 1000 the plateau country had agricultural settlements scattered all over it; then the area of settlement shrank southward, and the settlements became larger, while the buildings were unified and fortified. Some of the best agricultural land in the Southwest was abandoned, and some of the remnants of the settled people, the Hopi, for instance, ended up on very marginal land.

The great drought

This shrinkage, with migrations of tribes from place to place, is often attributed to the great drought that occurred at the end of the thirteenth century. This is a typically environmental-deterministic explanation and deserves examination.

There was a drought period at the end of the thirteenth century. We know both the time and duration of this drought from the study of tree rings. Trees depend on the weather for their growth. For Arctic trees summer warmth is critical, for arid-land trees moisture is critical, and, as it turns out, winter moisture is the most important kind for the coniferous trees in our Southwest. Good years are marked by wide rings, poor years by thin rings. Within an area experiencing similar weather, trees growing under similar conditions—preferably on dry hillsides where even small variations would affect the tree's growth—should show similar sequences of good and bad years by similar sequences of thick and thin rings. This analysis has been utilized to build a calendar for the Southwest that is also a record of past weather. Living trees provide a starting point, for their last ring formed the year they were cut. The rings of a 200-year-old tree can be matched at its early end with a log from an early building. Its rings can be overlapped onto a log from a pre-historic building and so on.

Such a tree-ring calendar has been assembled that reaches back almost 2000 years. It shows many periods of drought, a notable one occurring during the last quarter of the thirteenth century. This was also a time of great unrest among the Pueblo peoples. Great pueblos were abandoned, and whole areas such as the Mesa Verde vacated. The conclusion is usually drawn that the great drought was the cause. Is this really an accurate explanation?

First, it should be noted that the drought at the end of the thirteenth century was broken by a few good years. Second, there were other periods of severe drought in the past. Finally, the weather that is critical for tree growth in the Southwest is winter weather. Moisture at that time supplies the ground water for tree growth in spring. For the Southwestern Indian agriculture, the winter rains supply the moisture the plants need for their spring growth, and the late summer rains are critical for completing the crop. These two periods of rain in the Southwest come from quite separate weather systems. The winter rains come from cyclonic storms along the polar front. The summer rains are thunderstorms which result from in-drafts of air from the Gulf of Mexico.

These two air systems do not necessarily correlate: wet winters need not be followed by wet summers. Hence a wet winter that produced a wide ring on a tree might tell little or nothing about the summer rainfall that would be important to the success of the

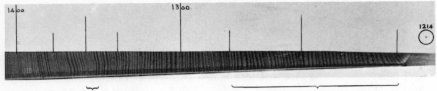

Years of extremely slow growth—little rainfall

Years of rapid growth—plenty of rainfall

(Upper) **Tree Rings from a Douglas Fir from New Mexico.** *The great variation in ring width is related to growth conditions. In the Southwest, the critical factor is usually moisture. There was a period of great drought from 1276 to 1299, and a much greater drought between 1338 and 1352.* (H. S. Gladwin)

(Lower) **Specimen of a Tree (charcoal) from Gila Pueblo, Arizona, Dated by Its Ring Pattern at A.D. 1285.** (H. S. Gladwin)

Indians' corn crop. Considering all these factors, why do we cling to the idea that the Great Drought caused the abandonment of much of the Southwest?

Man always seeks an answer to problems. He seizes quickly any apparent correlation as an answer. Once an answer has been advanced, it tends to become *the* answer. New evidence and more critical treatment of the old evidence then has an uphill fight, for it must displace an idea that already satisfies the need for an explanation. Mankind resists such change. This kind of resistance to change explains the long periods of time during which most people have clung to their old ways of life.

It is established as fact that the idea of an agriculturally based, settled way of life appeared in the semiarid plateau region and spread for nearly a thousand years. Thereafter, an immense shrinkage occurred. Vast areas that men had used for a farming way of life were abandoned. If it was not a climatic change, what then was the cause?

Nomadic raiders

There are other things besides climatic change that can cause farmers to abandon areas of good land.

Nomadic invasions that are beyond the farmers' organized ability to withstand would be one possible cause. The Pueblo people had no organization that welded them together. Each town stood by itself. They were easy marks for an invading people. Any warlike nomadic people entering this environment would have found poorly defended, small settlements sprinkled over the landscape, each producing abundant food in the widely scattered good patches of naturally watered land. Raiding these little villages would have provided food for the nomads with little more risk than the raiding of beehives.

There are warlike, raiding, nomadic people in the Southwest, the Navaho and the Apache. Linguistically these people are Athabascans, and their origin is in western Canada. The degree of differentiation of their speech from that of their Canadian relatives is not great. Therefore the amount of time that they have been separated from their northern relatives is probably not great.

The Athabascans are hunters. About a thousand years ago, hunting or wandering into the Great Basin or Colorado plateau country, they would have entered a land sprinkled with tiny villages of farming people. If these hunters raided the farmers, the settled people would have been placed under terrible pressure. The unorganized villagers intent on tending their tiny, widely scattered fields could easily be ambushed on their way to or from work. Or a raiding band could readily strip a field at harvest time and disappear into the surrounding vast area. Settled people are at the mercy of the hit-and-run tactics of nomadic people. Settlement must be so dense that the nomadic people face superior numbers, or they will pick off the isolated groups one at a time. As we have seen, in Africa, Egyptian numbers were too great for the nomads, and the settled people dominated. In the western Sahara, on the other hand, the small oases were dominated by the nomads.

In this situation there is a type of environmental causation at work. In the arid lands agriculture can be practiced only on the scattered good spots. This was true in the Negev of Palestine and in the American Southwest, both areas once farmed and then abandoned. In America, land use was based on careful selection of those bits of land where water was naturally concentrated in rocky canyons and spread out over the alluvium at the foot of the hill or in sand patches that allowed little or no water to run off and little water to evaporate. Since favorable spots would be scattered, an agriculture based on this land use must have widely dispersed fields, and the farmers must scatter widely to work their land. In a sense, the environment determines this land use, hence this exposure to ambush. In fact, the land could be used otherwise; for example, irrigation works could be installed. Even minor ones would allow larger areas to be farmed, keeping larger groups of men together and thus lessening their vulnerability to raids. Or if the people had never taken up agriculture, they would not be "sitting ducks" for raiders. The choice lay with man, and the choice had consequences. The physical-environmental "influence" was dependent on man-made choice.

The Pueblo people reacted to the nomadic raids by concentrating into larger groups and so arranging their houses that they formed a natural fort. Lower floors had no doorways or windows, and entry was by means of ladders that could be pulled up, leaving no easy entry for raiders. But much good land had to be abandoned, and even the biggest fortified apartments were vulnerable, for though the fortress might not be taken, the fields could still be raided. What use is a fort if the food supply cannot be protected? It could not have been drought that drove the Pueblo people out of the lands, for they abandoned some of the best lands. The town-dwelling Hopi were driven to settle on some of the poorest agricultural land in northern Arizona. If drought were the cause of abandonment of land, therefore, the Hopi would never be found

where they are, and such well-watered lands as the Mesa Verde with its many great pueblos would never have been left to the raiding peoples.

Some conclusions

In pre-Columbian times the arid West can be seen as an area where the degree of development of culture was directly related to the distance from the major centers where ideas originated in Middle America. The southernmost border of what became the United States had the most advanced culture, the adjacent peoples had lesser degrees of cultural development, and those beyond them were almost untouched by the stream of new ideas and remained simple gatherers. They all occupied arid lands. It was apparently not the aridity but the access to ideas that was important. The geography of position was more important than the geography of the environment.

Modern-day occupation of the land has reversed things to some extent. The physical environment has not *changed,* but the focal points of cultural growth and stimulus have changed. The available technologies and the views on land use have also altered tremendously. Most of the land has gone from a local subsistence economy to a commercial economy based on specialized goods produced for the great consuming centers in the East. Ideas no longer come in large part from Mexico but from the East and lately from the West. The arid West is no longer as marginal as it once was, but it is still a land of limited use. Is this situation physically determined? Within limits the answer must be yes. The shortage of water is decisive now—but it need not always be.

We are in the midst of a technological revolution. Chemistry is struggling with the problem of desalting the sea water at a cost that will make this inexhaustible source available, first for urban uses and later for agricultural use. Simultaneously, the physicists dream of a solution to the release of atomic energy that may make this water available most anywhere. Should the physicists' dream materialize, the finest plant-factory areas of the world will prove to be the low-latitude deserts where the growing season is almost 365 days long, the soils are loaded with nutrients, and the control of water through irrigation makes possible the maximum yields of plant materials that can supply the food and technological requirements of the world. The physical environment will not have changed, but our ability to use it will have changed.

The lessons from the American Southwest are these: Ideas are the primary cause of change—our assessment of the world changes at least potentially with each new idea—and determine man's use of the physical world. It is not the physical environment that determines what man does. No one people creates many ideas, hence access to accumulations of knowledge becomes the key to any people's perception of the possibilities in a given physical environment.

SOUTH AMERICA

The history of the arid lands of South America presents both similarities and contrasts with those of North America. The distribution of the dry lands in South America is normal as to latitude, side of continent, and location in the rain shadow of mountain ranges. The arid area extends over the mountains and into northwest Argentina, where there is a rain shadow, dry land that resembles in many ways the basin and range country of the North American arid region. The shape of the land, however, has caused unusual distribution of these dry lands. The Peruvian-Chilean coastal desert, for example, is a long narrow ribbon running parallel to the coast and sharply limited inland by the presence of the high Andes Mountains. As with the mid-latitude deserts of North America and Africa, the desert comes right to the coast, and there is a cold current offshore. In South America the cold current is very marked in temperature, salinity, and rate of flow, creating a special bioclimatic setting.

In terms of physical geography, the coast of Peru is more like the coast of Lower California and adjacent southern California than any other part of the earth. Points of similarity include a cool current offshore, a strong inversion formed over the cool current and blown onshore to give a high degree of cloud cover and coastal fog, desert or semidesert conditions right to the coast, increasing sunshine and rising temperatures inland, and better watered and cooler highlands farther inland. In South America almost all of these factors are heightened in intensity: the water is colder, the mountains higher, the desert lands barer. Nevertheless, the parallel is a close one. Comparisons between the northwest Argentine and the American Southwest are similarly close. Both areas are interior dry lands with basins and ranges, hot summers, and cool winters. The question then arises: What has man done with these similar environments? In pre-Columbian America we are presumably dealing with one relatively restricted type of mankind in comparable environments, and we are therefore able to hold these variables constant while we try to understand the role of each.

In contrast to the North American dry lands

where the developments can be viewed as marginal seepage from the Middle American centers of high culture, the South American dry lands of the Peruvian coast and highlands and adjacent Bolivia had an advanced civilization. It was a civilization with a long history and many brilliant accomplishments in engineering, art, agriculture, and political and economic organization. The region in some ways is reminiscent of Mesopotamia: a dry land center of cultural and political ferment and growth.

Agriculture was developed in oasislike valleys watered by streams flowing down from the melting snow of the Andes. These streams were most skillfully utilized. The water was spread out over the valley bottoms through the use of canals, and when there was not sufficient land for the available water supply, canals were constructed that carried the water to areas where it could be used. In addition, water that flowed underground was tapped by digging tunnels into the slope, sometimes called horizontal wells, or guanats. The building of long tunnels with frequent vertical shafts—to remove the dirt and supply ventilation for the workers—is a characteristic idea these people shared with the agriculturalists of the Middle East.

In these valley oases, city-states developed, separated from each other by desert, and at various times one or another came to dominate considerable areas. Culture was also developing on the adjacent highlands, and city-states there grew into empires. One early empire sprang from the city of Tiahuanaco, whose art styles influenced much of the Peruvian and Bolivian area. The highlands and lowlands, the city-states and empires, the irrigation agriculture and mud-walled cities are all reminiscent of the Middle East. Does this resemblance indicate the determining influence of the physical environment?

One school of thought holds that this type of environment, along with the idea of irrigation agriculture, will inevitably create city-states with their structure of highly centralized political control. It is argued that the building of canals and the control of water require absolute political control. We will return to the possible causes of growth of political structure presently. For the moment, note only that in Peru and Bolivia the advanced ideas that lead to cities, states, monumental buildings, and irrigation were present.

In the similar physical geography of North America none of these things existed. In the coastal desert of Lower California, with its mountainous relatively well-watered central spine from which streams descended toward the coast, a related race of men remained in a primitive hunting and gathering way of life, having virtually no clothes, no houses, no government. The contrast with the coastal desert of South America could hardly be more extreme.

Origin of Pre-Columbian South American Civilization

Why was this part of arid South America a center of great civilization when its corresponding area in North America was virtually a cultural void? Why is this the only Southern Hemisphere dry land to achieve the level of civilization in pre-Columbian time? Since the physical environments are not different, they cannot supply an answer, so we must study the culture of these people to compare their ways of doing things with those of peoples in other parts of the world.

To begin with an obvious comparison, the Incas kept their tax rolls, census rolls, and other records by tying knots in strings. It was a complex system, with the position of the knot determining the value of the unit. Thus a knot in one position represented the unit 1; in another position, 10; and in another, 100. Both a decimal system and a place-value system were used, just as in our own system where the value of 1 is determined by how far it is from the decimal point. The zeros, of course, are just spacers to fix position.

Such a mathematical system is rare in the world. It appears in the Old World centers of civilization, including India, and decimal systems are found in early China. China very early received from India the basis for the abacus, a system of beads on strings in a frame on which computations in a decimal system can be carried out rapidly. The Chinese annals state that before they had writing, they kept their records by tying knots in strings. In the adjacent Pacific, Polynesian peoples kept their records by tying knots in strings, and when the records were mathematical, they used a decimal system. The Polynesian and Peruvian records state that the Polynesians, at least occasionally, came to the coast of Peru. The probability that the Incas did sail out into the Pacific, as they say they did, has been greatly increased by Thor Heyerdahl's demonstration that the sailing rafts they had were quite capable of the job. (This judgment has been further supported by Clinton Edwards's and E. D. Doran's study of these rafts and other boats of the Pacific.)

Evidence of contacts

The evidence of contacts can be greatly extended. Only a few examples will be discussed here with a brief listing of the items that support the notion that very early, extensive voyages were made that could have brought civilization to the dry-land west coast of South America from Asia and elsewhere. With this evidence we can no longer use the South American Indian civilizations as examples of the uniformity of man's perception of arid land environment. We should no longer accept the concept that he just naturally responded by beginning irrigation and proceeding to civilization.

The earliest pottery now known in America was found on the coast of Ecuador. In manufacture and art style it duplicates material from the south islands of Japan, and radio-carbon dating of the material shows that the dates are the same in both areas. This indicates that people from the islands of Japan had in some way reached the coast of Ecuador during their middle Jomon period dating about 3000 B.C. The Japanese prehistorian Nishimura pointed long ago to evidence of the priority of rafts in Japanese navigation and concluded that the South American Indian rafts were derived from Asiatic influences, but little attention was paid to his claim. More recently it has been pointed out that the name for this sailing raft, *balsa,* is of Chinese origin. In addition to this very early pottery, there is evidence of extensive Asiatic influence in Ecuador a few centuries before Christ. Students of art history point to Shang Dynasty parallels in Peruvian art, and the evidence thus suggests contacts over a long period of time.

Much of the evidence has been dismissed as interesting but insignificant because man can obviously invent things, and ideas may well have arisen in separate areas independently. Who can say that the idea of the zero is not a "natural" one that will arise whenever men get to a certain stage of development where they must use figures to deal with censuses and tax rolls? If pyramids are found in two different areas, how can this prove contact when a pyramid is one of the "natural" shapes for a monumental pile? If men draw similar-appearing mythical animals, how can this prove contact, since all men are basically alike and are apt to have similar hallucinations? Yet it is possible to turn to something that man cannot invent and that is found on both sides of the Pacific. Games, for instance, require a complex idea, one with many parts that are not necessarily—that is, functionally—related.

A domestic plant is something that man cannot invent. Man domesticates plants; he does not create them out of nothing. Of the domestic plants that were common to both the Old World and the New World, the most interesting is the sweet potato. This plant was widespread in the Middle American areas in pre-Co-

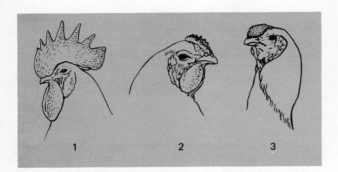

FIGURE 2-3 Races of Chickens
The races of chickens are as distinctive as the races of men and can be easily traced around the world. The chicken types shown here are the Mediterranean cock, the Asiatic cock, and the Malay cock, respectively. (Adapted from G. F. Carter, "Pre-Columbian Chickens in America," Man Across the Sea, ed. Carroll L. Riley et al, University of Texas Press, 1971)

lumbian times and had a considerable distribution in Polynesia also. It is an American plant. It is the same plant in America and in Polynesia. Further, in part of the Inca empire, the name for the plant is the same as in Polynesia. In the northern part of the Inca empire, the name is *kumar, umar,* or variations on them. In adjacent Polynesia, the names are *kumara, umara, umala, uwala.* The same plant, used in the same ways, known by the same name by peoples who each had traditions of contacts with the other, seems to provide absolute proof of contact.

Further information can be derived from this evidence. We have seen that people do not necessarily take up ideas quickly. In fact, there may be a lag of 1000 years or more before obviously useful ideas are adopted. We cannot assume, then, that a single, fleet-

ing contact led someone to adopt a strange food plant. There must have been contact over a fairly long time so that the receiving group could become used to the new plant, could come to like the taste of it, and so become interested in how it was grown and harvested, stored and cooked.

We can now use the closeness of this parallel to extract still more meaning from this situation. The introduction of the word "corn" into America was the result of the entry of a colonizing people who came to dominate the areas where this word is used today. This suggests that in northwest Peru and adjacent Ecuador we might expect to find evidence of such contacts. Polynesian adzes are found along this coast, and they are called by the Polynesian name, *toki.* The Spanish recorded several graphic descriptions of the coming of

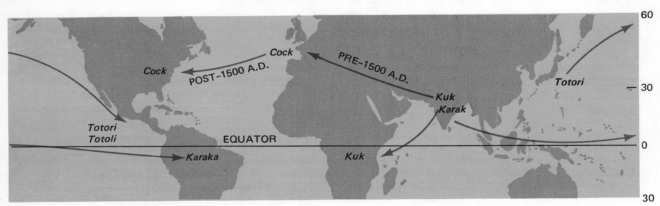

FIGURE 2-4 Names for Chickens around the World
The names for things which move from one place to another are often carried with them. Here we see the spread of the names for "chicken." The word "cock" traces back to "kuk" in India. The same word spread to Africa. Another Indian word "karaknath" spread to South America. A Japanese word "totori" spread to northwest Mexico. (Adapted from G. F. Carter, "Pre-Columbian Chickens in America," Man Across the Sea, ed. Carroll L. Riley et al, University of Texas Press, 1971)

people from overseas to this area. It is here that the Asiatic sailing raft was found. It is in this area that pottery from Asia first appears about 3000 B.C. and where there is archeological evidence of another full-scale colonization about 200 B.C.

Recently my own research has focused on the chicken in America. The homeland of the chicken is in Southeast Asia. Among chickens there are sharply defined races just as among man. To cite just a few traits, Mediterranean chickens lay white-shelled eggs, and Asiatic chickens lay brown-shelled eggs. Fluffy-feathered and feather-footed chickens are Chinese, and seminaked chickens are Malays. Uses of chickens are also quite specific. The ancient pattern of chicken use was for divination and sacrifice and not for eating flesh or eggs. In parts of Asia this pattern still persists. In South America many students have commented on the dominance of Asiatic chickens. Almost 300 years ago Father José de Acosta commented on the fact that the names for chickens in America (he referred specifically to Peru) were not Spanish. The literature shows that many South American Indians even today view eating chickens or eggs with horror. And, finally, among the Arawak Indian tribes who occupy large areas of the Amazon basin, the name for the chicken duplicates a Hindu word for chicken: *karaknath* in India and *karaka* in the Amazon basin. In India this name is specific for the melanotic silky chicken, a very strange breed, and this kind of chicken is widespread in South America. This kind of evidence made it possible to predict that archeological chicken bones would be found in an American site, and, indeed, such finds have now been made. Occurring as they do, far from the probable point of entry of the chicken into America, they by no means date the earliest chickens.

Evidence for the very early massive transfers of ideas to America across the world oceans can even be extended. The chicken is a particularly useful example, for shipwrecked sailors do not carefully feed and water chickens; chickens do, however, have to be fed and watered and carried in breeding pairs if they are to be introduced into a new area. The evidence is quite clear that breeding pairs of many kinds of chickens were purposefully carried to America, apparently solely for divination, ritual, and curing. It gives an interesting perspective on the nature of the voyages to America: well-equipped mariners, not shipwrecked, starving fishermen, made deliberate and relatively swift voyages to America.

These data can be pressed even further to reveal still more evidence of what went on and who was involved. Over 80 years ago it was noted that *kumar* is an ancient word in Sanskrit, the Indo-European language introduced into large parts of India about 1500 B.C. The word has very wide meaning in India and may refer to many plant materials, to young growing things, and so forth. This is a sign of antiquity of a word. In America the word *kumar* (and its variants) is restricted to a small area in northwest South America and has a restricted meaning: sweet potato. This suggests a recent development. It is always helpful to look at parallels, and there is a rather close one in *Zea mays,* or Indian corn. The Spanish took up the Caribbean Indian name, and from this came the word "maize," the name by which this American plant is known over much of the world. When the British came to America, however, they chose not to use one of the local names for this new grain but one of their own words. They could not use "wheat" or "barley," for these had precise meanings. They did have an old word, however, that had wide meanings but included the sense of a grain, and that was "corn." One finds the "grain" meaning in the English corn laws; we would say grain laws. But "corn" in English has more meanings. It originally meant any small hard object, hence a grain. But we also use it to designate pepper berries, which we call peppercorns. We even call a hard lump on our foot a corn. The parallel is a patricularly neat one, involving as it does the transfer of an old word with wide meaning, including some plant usages to a specific plant in a new world.

We are unlikely to know just how the first trip to America was made. It was probably accidental. Japanese fishing boats, disabled on the fishing grounds off Asia, drift to America automatically. In the nineteenth century they arrived with considerable regularity. When the Europeans first reached the Indians of the northwest coast of America, they had Japanese fishermen as slaves. The sweep of the currents and winds is such that boats off the coast of America would have been carried down the coast to California and beyond. The people in America would learn from accidental landings that there was land to the west of them. Any castaway who won his freedom might well organize a return trip and be able to report the existence of a great land to the east of Africa. A South Pacific drift would also lead to America, just as any drift voyage from Middle America would lead to Polynesia and Asia.

Although we will never know the details of the first contacts, the biological evidence provides absolute proof that voyages were made. Sweet potatoes, cotton, and chickens are not naturally dispersed across seas; man is required to transport them. But if the contacts between the peoples on the two sides of the Pacific were enough so that they came to know and value some

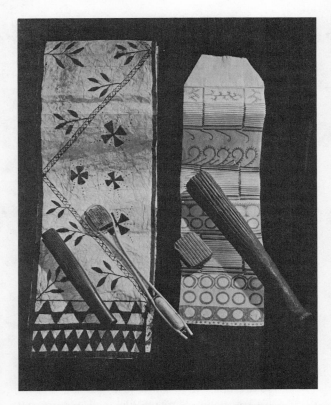

(Opposite, far left) **Bark Cloth and Bark Beaters from Oceania** (left) **and America** (right). *Not only was bark cloth, or tapa cloth, prepared on both sides of the Pacific but the implements used in beating the soft inner bark into a feltlike material were identical.* (Courtesy of The American Museum of Natural History)

(Opposite, left) **Pan Pipes from the Solomon Islands** (left) **and from Bolivia** (right). *The degree of similarity in these instruments on the two sides of the Pacific extends to the method of binding them, the scales used, and the religious-ceremonial notions associated with them.* (Courtesy of The American Museum of Natural History)

(Opposite, below) **War Clubs of Melanesia Compared with Those of Peru and Mexico.** *Such similarity of form and function is virtually conclusive proof of the spread of ideas from one region to the other.* (Courtesy of The American Museum of Natural History)

of each other's economic plants, they must have exchanged other ideas also. There is abundant evidence that they did so.

The game of parchisi was known in Southwest Asia and in Middle America in pre-Columbian times. The shape of the board and all the rules of the game are the same. Mathematically the chances of independently duplicating such a game can be calculated to be between 100,000 and 1,000,000 to 1, depending on how many independent variables the game is assumed to have.

For those who are not mathematically minded, this can be put into familiar terms. If you were in a distant jungle land and saw a group of natives moaning, chanting, and swaying as they crouched in a circle on the ground, you would undoubtedly think that you had stumbled upon a primitive ceremony. When you came close, however, you would find that goods were changing hands. Gambling! "Ah, well, natural to mankind," you might say. But then you would see that they were casting lots. "Naturally," you might think, "they have to invoke the laws of chance somehow." But the device they are using is a pair of cubes. Puzzling—it could have been a device for heads or tails or something else. However, cubes are a natural shape. Then you notice the numbering of the faces on the cubes with dots from one to six. Now why dots? Among a jungle people, why shouldn't the six sides be marked by symbols for tree, monkey, parrot, snake, fish, or man?

By this time your suspicions are thoroughly aroused. Listening closely you find that you can understand part of the chant. As one of the men winds up to throw the dice he chants: "Come on, seven." Now, if you know anything about the American dice game, you will know that American soldiers have been in this jungle before you. There are entirely too many improbable occurrences of nonfunctional elements combined into an arbitrary game to have just happened by chance. This is what the earlier figures expressed in the precise language of the mathematician.

As an interesting aside, why do we mark our dice with dots instead of numbers? Is it significant that dice exactly like those that we use today were used in Mesopotamia as early as 3000 B.C., and that in Mesopotamia the circle is divided into 360 degrees, just as we divide our circle today, and that much of our culture is derived from the eastern Mediterranean? This kind of evidence is proof that the roots of our Western European culture lie in the Middle East. Similar evidence linking Middle America and Southeast Asia should be given a similar interpretation.

But if plants and games, what else? The list is long. Pan pipes are blown on both sides of the Pacific and on both sides of the Atlantic as well. However, the pan pipes in America blow the same notes in the same scales as those used on the Asiatic side of the Pacific and have such improbable ideas associated with them as that pairs of pan pipes represent male and female. There are resemblances also in metal objects. A list of Asiatic cultural elements found in Middle America, indicating something of the extent of the parallels in ways of thinking and of doing things on the two sides of the Pacific, is given here.

Mathematics and Calendars
Time counting by permutations
The zero: early appearance in India and Mexico
Place numerical systems: India, Mexico, and Peru
Knot-in-string records (before script): Peru, Polynesia, China
Zodiac: Asia and Mexico
Music
Pan pipe: identical notes, similar scales, ceremonial ideas, including male-female concept for paired pipes and alternate notes
Nose flute
Conch-shell trumpet, with similar names and blown in similar ways
Hollowed-log drum with slit opening used for signaling in Middle America, Polynesia, Southeast Asia, and Africa
Art
Jade: highly valued in Mexico, especially by Mayan culture, and in China—similar concepts associated with jade in both areas

Chinese Chou dynasty art motifs duplicated in Maya area

Indian art motifs in Mexico: trefoil arch, sacred tree, tiger thrones, lotus staff, lotus panels with fish-vine-man-vine-fish elements, serpent-worship elements

Technology

Alcoholic and narcotic beverages prepared in similar ways and used in similar ceremonies

Metallurgy with similar alloys, weapon forms, and casting methods

The wheel: principle known and used on toys in America

Fore- and aft-rig sails used on centerboard rafts: Asia, Polynesia, Peru

Race

Vivid portraits in pottery and stone of various races: Mongoloid, Caucasoid, and Negroid

Legends giving detailed descriptions of robed and bearded learned men visitors who came and went by sea

Chineselike palm prints among the Mayans

Many agricultural tribes having the Diego blood antigen that is also found in parts of Southeast Asia

Legend

Peruvians claiming to have crossed the Pacific

Chinese fifth-century document probably describing a missionary effort in Middle America

Polynesians claiming voyages to and from America

Numerous myths alike on the two sides of the Pacific

Agriculture

Sweet potato, an American plant, carried with its name into Polynesia and probably carried by people from Borneo clear on to Africa well before A.D. 1500

Numerous other plants probably carried between areas: coconut probably carried to America, cotton probably carried to America, and later new forms from America carried into the Pacific

Similar types of systems for both agricultural terracing and irrigation

Distributions of traits should always show the worldwide distribution, making clear that the things mapped are actually the same things wherever found. In this list, time counting by permutation is highly specific and strongly developed in Asia and parts of Indian America but seems alien to the Mediterranean civilizations. The zero, on the other hand, probably was first used in the Middle East where its development can be traced through nearly 2000 years before its appearance in developed form. It could theoretically either have spread to America via the Atlantic or the Pacific or have been independently re-invented in America. Some items, jade for example, were anciently prized in the Mediterranean world, but the ideas associated with jade in the New World are specifically Chinese, so it seems

that the idea came to the New World by way of the Pacific.

Agricultural terracing of mountain sides is another trait shared by Asia, the Mediterranean, and the Andean regions. Are these independent inventions? How many times was terracing invented? The very extensive terracing of the Mediterranean duplicates the Andean terracing by its rock walling and its tendency to rise from the foot slopes to the tops of the mountains. In a worldwide survey of terracing, geographers J. E. Spencer and G. A. Hale noted the similarity between the Mediterranean and the Andean region, and they pointed out that since terracing was present on the Atlantic islands at an unknown early date, it was at least as feasible that the idea came to America across the Atlantic as across the Pacific. There are, then, several possibilities: The idea of terraces was locally invented in response to the environmental challenge, it was introduced via the Atlantic, it was introduced via the Pacific, or some combination of these things occurred. It seems quite clear that at the moment we do not know just what did happen. The environmental response theory, the notion that psychically similar mankind, when faced with a steep land, just naturally terraces it is weakened by two things. First, many people did not. Second, in the Peruvian case, data indicate the pre-Columbian arrival of ideas from Asia and Europe precisely where terraces are prominent in the landscape. The presence of agricultural traits such as quanats and siphons, both found in the eastern Mediterranean and not in the Orient, seems to tip the scale toward a possible Mediterranean introduction of the idea of terraces. Again we are looking at a man's perception of the correct solution of a landscape problem (how to grow crops on steep slopes) as a learned process rather than a spontaneously, locally invented one. This is so important a question that we should examine it more thoroughly.

John Howland Rowe has noted that we can compile a list of 60 traits that are found in the Mediterranean and the Andean region of pre-Columbian America. These include the following items: lateral wells, entasis in architecture (making columns sufficiently out of plumb that they will appear straight despite the effects of perspective), a highly specific axe type known only in Egypt and the Andean region, sandals and women's clothing similar in detail to Greek dress, and a whole Greco-Roman ceremonial complex—oracles, vestal virgins, new fire made with concave mirrors, the use of sacrificial animals that must be domestic and must have a specific color for the specific ceremony. His list could easily be lengthened to 100 or more items, partly by transferring some from the Pacific to the Atlantic list and partly by adding items that he did not mention. Rowe claimed that these lists proved nothing. He called for worldwide, highly specific studies of individual traits.

D. Randall Beirne supplies such a study of axe types. He first clarified the names applied to axe parts, then classified the axe types of the world by their method of hafting, and finally surveyed the world distribution of axes and their relatives, the adzes. He found numerous Mediterranean axe types in America, especially in the Caribbean and northern Andean region, with evidence that from the latter area they spread down the Andes and into Peru and Bolivia. Several types of adzes spread from Asia to America across the Pacific, where they are found on the coast of Ecuador. Seemingly, then, there were two streams of ideas reaching America, meeting in the tropical part of America. The Asiatic influence is now getting scholarly attention, and the study of Atlantic influence is just being revived after a half century of neglect.

It has long intrigued men that the Aztecs and Incas welcomed the Spaniards as returning lords to whom they owed allegiance. This is the blond god myth. It has been explained away in the past but once again is becoming increasingly difficult to ignore, for there are portraits of European men in pre-Columbian America. There also are traces of fifth-century Christianity, for example, the belief that one could confess and be forgiven only once, as well as Rowe's list of 60 Mediterranean traits. We may ask for plant proof parallel to the Pacific case: it is there. A pineapple is portrayed on the walls of Pompeii. A Roman pottery head stylistically belonging to the second century has been found in Mexican archeology. Why are there no written references to America? Actually the classical records contain many references to lands beyond the Pillars of Hercules, the Straits of Gibraltar, but since they are called islands, they have been neglected. Some of the accounts described navigable rivers on these "islands," although none are found there and many exist on the continents of America.

We face the probability that a number of discoveries of America were made and lost after a period of active contact. The Norse discovery of America about A.D. 900 maintained active contact via Greenland for 500 years before politics, plague, and piracy wiped out the colony. The loss of memory about these recent events should caution us against assuming that similar discoveries could not have happened in the more distant past. Which traits went which way and just how much influence came to America via the Pacific and how much via the Atlantic are under active investigation. At least it is definite that ideas reached America via both ocean routes, despite the fact that some people are still obsessed with the notion that the oceanic barriers were only breached in 1492.

The problem of spanning the oceans

The Pacific has been thought of as a great barrier. This may be because the people looking at it were landsmen and knew little about the history of shipping. It is difficult to pin down the history of ships in the period before writing, for wooden ships and fabric sails have seldom been preserved. There has also been a tendency to treat the first recorded ship as *the first ship*. This is a fundamental error that permeates thinking on almost all such problems: the first man found is the earliest man, the first stone implement found is the earliest implement. All this is probably wrong, for the chances of finding the earliest anything is extremely remote. The earliest ship known is found in an Egyptian mural, but it surely is not the earliest ship. As an example of how knowledge grows, consider a recent finding resulting from study of the source of obsidian used in Mesolithic Greece about 9000 years ago. It was found that this particular obsidian was being imported by sea from the island of Delos. Sea passages of up to 20 miles were necessary, and moderately sizable boats and skilled mariners were required. This single finding added a thousand years to the positive evidence for shipping.

Shipping was not even dependent on boats, as we too often assume, and certainly not on wooden boats. The Eskimos and the Irish, to name only two, used skin boats and made notable voyages. The Eskimos repeatedly reached Great Britain in their kayaks, and the Irish traveled great distances in the curraghs, small ships capable of carrying up to 60 men. That the Eskimos had such boats—their umiak is the equivalent of the Irish curragh—suggests that such boats were within the technological ability of the Upper Paleolithic people. But even this possibility does not tell the story.

Any buoyant material may be bundled up and bound together to create a passable craft. Craft made of reeds, bark, bamboo, and similar material are found extremely early in history, widely distributed over the world, frequently in the hands of marginal peoples. People of Lower Paleolithic culture level, such as the Tasmanians, used primitive craft to reach Tasmania, and their ancestors probably used similar craft to reach Australia. Evidently man was moving about the world across the broad oceans much earlier than we have thought. Lower Paleolithic people, such as the Tasmanians and Lower Californians, used bundle rafts.

Upper Paleolithic people, such as the Eskimos, used skin boats. Plank boats may have been used as early as Neolithic times.

The ultimate age of watercraft will surely be determined in Australia, for that continent has been cut off from the rest of the world since the time when only primitive pouched mammals existed. Man and his dog are the only terrestrial mammals (bats excluded) who reached Australia, and they had to have means to cross ocean barriers that stopped all other land mammals. At whatever time we date man in Australia, we shall also date the existence of watercraft. Currently man's arrival in Australia is placed at something more than 32,000 years ago, but the date is likely to move to earlier times.

In Southeast Asia the brilliant solution of the sailing craft with the centerboard was invented early, but we have no means of knowing just when. The early mythological history of Japan does not mention boats, but rafts. The sailing raft that was able to sail windward, tack, and perform better than a square-rigged ship made it possible for the Asiatics to move freely about the seas beginning in very early times. A number of students have pointed out that the sailing log raft used off the coast of Peru is an isolated phenomenon in America that in many details—kind of mast, sail, use of centerboards—resembles Asiatic rafts. In Asia these log rafts are numerous and widespread, suggesting certainly that Asia is the homeland and America the area to which they spread. If the raft went, men went on it, probably carrying ideas and plants. In one of the accidents of history the raft was later replaced by canoes in Polynesia, so we lost sight of the connecting link between America and the probable homeland of this ancient craft. We conclude therefore that the sailing raft, because it appears in America in one of the areas where we suspect Asiatic contacts, may have a great antiquity on both sides of the Pacific, but its probable home is Asia.

The Europeans were relatively poor boatmen. They were oarsmen who built good, sound hulls capable of weathering severe storms; the Phoenicians were able to circumnavigate Africa when they were sent out by the Pharoah Necho in 600 B.C.

The field of scholarship is filled with hidden evidence. Ninety years ago an inscription was found in Brazil. It seemed to be Canaanite writing, but it contains flaws which led the experts to decide it was a forgery. The stone was then lost, and our knowledge of it came from poor copies. Recently a good copy came into the hands of Cyrus Gordon, an expert on Middle Eastern writing. When he examined the inscription, he found no flaws. At the time that the stone was rejected as a forgery, Canaanite was poorly known. Since then, all of the grammatical constructions thought not to belong in Canaanite had been found to be authentic to the writing. This validates the inscriptions, for who could have invented constructions that were not yet known? The stone records the landing on the coast of Brazil of a boat that had set out from Sidon, in Lebanon, and along with a fleet of vessels had circumnavigated Africa. This boat had become separated from the rest and touched this strange shore. They inscribed the stone to commemorate the event and claim the land for King Hiram of Tyre. The time was about 500 B.C. As another interesting aside, they recorded that women were on the voyage. We have long suspected that women were taken on these voyages, but this is our first evidence. It helps to account for the appearance of cultural traits associated with women, such as the Greek woman's dress in Peru or pottery making in Ecuador.

The Atlantic is a small ocean compared with the Pacific, and the spate of voyages in the stormy North Atlantic in the late 1960s proved that even tiny boats were capable of crossing it. In the tropical Atlantic the crossing would be even easier, for the winds and currents blow constantly toward America and a disabled vessel in the latitude of the Mediterranan is automatically delivered to the Caribbean. In the mid-1900s a Scandinavian made this crossing in a fragile, canvas canoe. The probability seems very high that once Europeans began voyaging outside the Straits of Gibraltar, some of them were disabled and drifted to America. We know that Mediterraneans were reaching England by sea at least as early as 1500 B.C., and coastal voyaging along the Portuguese–North African coast was probably much earlier. The arrival of European ships and men and ideas in America beginning at least by 1500 B.C. seems feasible.

All of these considerations show that we cannot safely say that developments in America were the result of men (assuming all mankind to be similar) coming to the same conclusions when faced with the same problems in similar physical environments. Quite the contrary, it appears as if the similarities may be due to the wholesale importation of ideas.

Instead of thinking that the culture of dry lands of the Pacific coast of America was developed in isolation from the Old World, as an example of an environment in which man reacts to the desert, we realize that Old World cultures of the time level of 2500 B.C. and later—for the contacts were continued over a long period of time—may have reacted to the dry-land environments of America. The cities in the desert with their adobe walls and pyramids, their irrigation canals

Brazilian Jangada. *This vessel is used by fishermen in South America, including northeastern Brazil, Peru, and Ecuador, and it is typical of fishing vessels in southern India today. The identical sail is still in use also in the Caroline Islands. (Brazilian Embassy)*

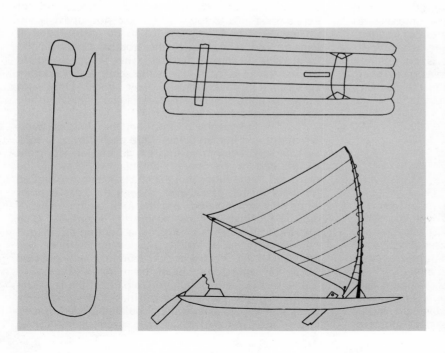

(Far left) **Prehistoric Centerboard from Peru, pre-Inca.** *The centerboard is a movable keel and enables a fore-and-aft-rigged ship (opposite of a square rigged ship) to sail upwind. Both the fore-and-aft-rigged sail and the centerboard employ the same principle of aerodynamics as the wing of an airplane. Furthermore, multiple centerboards can be adjusted to balance the forces of wind, sail, waves, and currents as desired and thus play the role of the rudder. This sophisticated piece of sailing equipment is ancient in the Pacific. The proportions of this ancient Peruvian centerboard are like those of centerboards used on modern racing sail boats. In Peru they are more than 1000 years old.*

(Left) **The Sailing Raft of Brazil in Plan and Profile.** *In the Pacific and in America rafts had Asiatic names: Brazil, jangada; Japan, ikada; old China, palsa; Ecuador, balsa; modern China, pa; Polynesia, pae pae.*

105

and lateral wells, their divine priest-kingships—complete with trappings such as litters and umbrellas as insignia of rank—all may be seen as a direct importation of ideas. Scholars must determine which ideas came from which countries, but this is no different philosophically from sorting out the roles of Egyptian, Cretan, and Middle Eastern influences in the development of Greece.

The Spanish Evaluation of the Desert

The Spanish found this area to be a vast, organized empire, containing great cities, linked with footpaths, equipped with granaries, and supported by terraced mountain sides and irrigated desert oases. There were also great temples roofed, lined, and furnished with gold and silver extracted from rich local mines.

The Spanish looted the temples, shipping back to Spain the vast wealth of gold assembled through centuries. Thereafter the Spanish took over the mines, and the precious metal became Spain's great source of wealth. What else could they take? Potatoes, one of Peru's great gifts to the world, could not survive a trip up the coast of South America to be transshipped at Panama and sent through the tropical Caribbean on to markets in Europe. Besides, what European would eat so strange a food? It was the same with the other agricultural products of the New World. Only such things as gold, silver, gems, and sugar could last and pay the costs of the long trip.

The geography of location, the state of transportation, and human values were each playing a role. Such factors determined that the general development of this part of the Spanish colonial world would be sacrificed to the support of the mines. Men were drafted for work in the mines, even though the coastal oases were thus depopulated, and the men of the lowlands died in the high altitudes where the mines lay. Even the distant grasslands of Argentina became an adjunct to the mines, supplying the needed mules, beef, and mutton. Under the impact of economic and political disruption, enforced labor in the mines, and unaccustomed diseases, the population in parts of the land was greatly reduced. Where great cities had stood there were now a few great, thinly populated landed estates. Where the land had once been used intensively to

(Opposite page) **History in Architecture in Peru.** *The scene through the window made of huge shaped stones is of Machu Pichu in the Andes of Peru. Two styles of stonework can be seen: enormous stone blocks cut and fit together, overlaid by smaller, poorly shaped blocks.* (Panagra)

(Left) *The Church of San Marcello in Lima was built by the Spanish in 1584. It records the transplantation of the Spanish rococo style to the New World.* (Pan American World Airways)

(Right) *Modern buildings in Lima portray the impact of the twentieth century on Peru.*

support many people, it was now used extensively to support a few. The pattern thus set has persisted to the present.

The dry lands of South America cross the Andes just south of 30 degrees latitude. Southward in the rain shadow of the Andes there is a long strip of dry land, lying principally in Argentina. In northwest Argentina the land resembles somewhat the Great Basin in North America, where streams descend from isolated, short mountain ranges into the dry basins between the mountains.

In pre-Columbian times the use of this land greatly resembled that of the American Southwest. Town-dwelling people utilized these streams and other areas of water concentration for agriculture. They made pottery and built stone houses grouped together much like those of the southwestern Pueblos. Is this kind of geographical adaptation environmentally determined? At first glance it might seem so.

However, both areas are marginal to centers of culturally advanced areas: northwest Argentina to Peru, the American Southwest to Mexico. Location in relation to civilization is not raw physical environmental determinism; the first cause is culture. If northwest Argentina and the American Southwest were indeed alike in this early period, we have the suggestion that the Mexican and South American cultures that influenced them shared certain basic values that gave rise to similar appearing cultures in arid peripheral areas. The basic relationships of the Middle American cultures is today

107

increasingly recognized. When we explore details, we find such telltale similarities as T-shaped doorways associated with stone-house construction in the Southwest, Mexico, and the Andes. Such likeness seems indeed to stem primarily from a sharing in a basic stock of ideas and only secondarily from the sharing of a similar physical environment in which these ideas were worked out. In both areas cultural level declined with distance from the cultural centers. Beyond the farmers of the difficult arid lands were simple hunters and gatherers. It did not matter whether "back of the beyond" was the rich land of the Argentine pampas or the fertile valleys of California. If ideas did not penetrate effectively, these better lands lagged behind the relatively poor physical environments closer to stimulating centers.

In the American Southwest the cultural sequence after 1500 was from that of a marginal area to integration into one of the major economically, politically, and culturally productive areas of the world. In Peru the change went from a center of political power and cultural growth to the status of a colonial dependency of a distant power to ultimate independence, but as a small nation distant from the center of world power and not integrated into the adjacent region. The crude physical environment remained unchanged; the geographical position in relation to cultural forces had changed. The achievements of the men on the land paralleled those changes. While race changed considerably in the Southwest and culture advanced, race remained little changed in Peru and culture declined. It seems clear that the causes are not racial, however, for both the Peruvian natives and the Spanish have records of great achievements in former times. But so do the Indians of the American Southwest. Both race and physical environment are therefore inadequate as explanation, especially for the arid lands of South America.

The arid coastal lands of Peru and northern Chile are not poor. The irrigated land is a rich potential; the problems are those of finding and marketing material desired in the world market. At present, production centers on cotton and sugar cane. Development of the adjacent highland areas, perhaps as industrial centers, could well lead to a shift toward the production of locally desired food, for the coastal desert oases of Peru and Chile could in a productive, industrialized society play the same role as the Imperial, Colorado, and Gila-Salt oases do in the American economy.

The cold offshore ocean current of Peru creates one of the world's richest fisheries. Upwelling of deep water brings oxygen and nutrient-rich waters into the zone of sunlight, and the result is an explosive growth of plankton that supports a huge population of fish of sardine size, known in Peru as *anchoveta*. The tuna and larger fish feed on these anchovies. The Incaic people used this fishery to some extent, but lack of transportation and refrigeration makes it difficult to use a fishery to feed any but the people in the immediate vicinity. The fish population is so huge that it supports millions of birds, cormorants in particular, which roost on rocky islands off the desert shore. In this arid climate their droppings pile up tens of feet thick. The Incas appreciated the fertilizer value of this material and hauled it ashore to manure the irrigated fields of their coastal oases. Later this material, called *guano*, was hauled by boatloads to Europe and America, for it was recognized as being extremely rich. It not only had all the major plant nutrients, but it contained most of the micronutrients too. Even though our farmers did not then know of the role of the micronutrients, the almost magical rejuvenation of old worn-out lands, when treated with guano, made this a most desirable product. The Peruvians eventually declared it a natural resource and forbade its export.

Very recently Peru has turned to the direct harvesting of this immensely rich fishery. It is difficult to put this in perspective. The fishery has always been there. Its presence is obvious enough because of the millions of birds fishing and the incredible number of tons of guano being deposited on the offshore islands. The Incas had fished. The Spanish have an ancient tradition of deep-sea fishing. Why then did it take until the latter third of the twentieth century for the Peruvians to take notice of the immense wealth off their shores? One could make several observations. First, it is not unusual for people to fail to see a resource for what it is. Second, the Spanish heritage was such that the initial focus was very heavily on gold and silver and thereafter on great haciendas that provided a rich life for men oriented toward the land. Third, in the colonial period there was no easily accessible market for fish. Thus with no tradition of fishing there would be an understandably delayed realization of the great wealth beckoning from the marginal sea.

The appreciation of the wealth of their fishery was in a sense thrust upon the Peruvians. American tuna fishermen began operating out of San Pedro and San Diego. At first they limited themselves to the local waters, but the lucrative business led them to build larger boats and begin fishing on the high seas and in Mexican waters. The tuna boats grew bigger, and soon the American fishermen appeared off Peru's shore. The Peruvians reacted as if pirates had hove over the horizon, but with the prevailing three-mile limit to territorial seas, they seemingly could do little. President Truman, however, opened the door for them by proclaiming American rights to minerals on the continental shelf of the United States. We are blessed with very broad shelves especially on our east and Gulf coasts, and our claims extended to as much as 100 miles at sea. Our move was largely triggered by the developing offshore oil fields. When the Peruvians looked to see what their continental shelf situation was, they found that they had none. The Andes plunge from their immense heights straight into the sea and go right on down to depths commensurate with their heights. This is due to the movement of the floor of the Pacific toward the west side of South America. At the continental margin the sea floor dives beneath the continent, and it is due to the titanic thrust of this movement that the edge of South America is crumpled up and the Andes thrust into the skies. This geological formation also creates the volcanic activity and the destructive earthquakes of the region. Rather incidentally it left the Peruvians with no continental shelf to claim. They reacted by claiming control of the sea off their shore for 200 miles and began vigorously to protect their fishery. Other countries have been reluctant to accept this unilateral changing of international law, and the United States has continued to protest the action. More and more countries are taking similar action, however.

Belatedly the Peruvians also began to do something on their own about harvesting this resource. They started by going into the fish-meal business. The small fish were netted in quantities and hauled to shore factories where rather simple processing extracted some of the fish oil and prepared the rest of the material as a fish meal. This has a ready market in the world as a protein supplement for chicken and cattle feeds. Because the world price was high and the demand huge, fortunes were quickly made in the business. This, of course, led to rapid expansion, with the usual dislocations including a temporary glut of fish meal. The world market, however, is virtually insatiable for raw materials, and after a temporary setback the business has gone ahead.

The effects in Peru have been far-reaching. Little coastal villages have exploded into towns. The tiny boat-building business has become a major industry. Whole sets of auxiliary industries from engine manufacturing to net making have sprung into being. Wages in some of these endeavors are very high for an underdeveloped country, and illiterate Indians are serving as ship captains and making large salaries. The Peruvian economy has received a boost, and the income is filtering out into the people at an unusual rate. The measure of the magnitude of this change is to be found in the

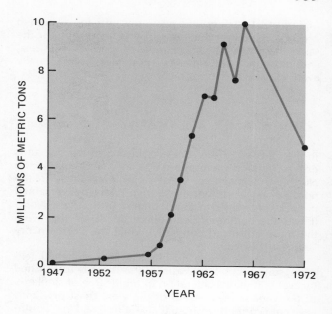

FIGURE 2-5 Growth of the Peruvian Anchoveta Fishery
A growth from less than half a million tons to over 10 million tons occurred in a 20-year period. Most of the growth took place in the five-year period between 1957 and 1962. Overfishing led to such precipitous decline that in 1973 fishing was restricted to very low levels, and the world fish-meal market was disturbed. No resource is limitless, as this example illustrates.

fact that Peru has vaulted from an insignificant place in the world fishery to the number one nation in tonnage of catch.

Despite this achievement fishery ecologists shake their heads. They point out that the strip of upwelling water is a relatively restricted part of the ocean. There is a limit to how much fish can be taken from it without damaging the safe yield. Already the sea birds are suffering as man harvests the fish on which they have been so long dependent. This will inevitably lower the yield of guano, but the lower yield is more than offset by the enormous return achieved by harvesting the fish directly. By the early 1970s the fishery went into a steep decline that compelled the Peruvians first to close the fishery entirely and then to reopen it under strict control to prevent its destruction.

The fishery is but one of the resources available in Peru. The Andes are mineralized and rich in copper, gold, silver, and other metals. The limited development of arid South America is not due to environmental poverty; rather, it is a measure of the continuation of a preindustrial society and economy into the twentieth century. There is nothing remarkable about this, for the same thing can be said of most of Asia, all of Africa, and most of Latin America—examples from virtually every climate and every race.

AFRICA AND AUSTRALIA

In the Old World the course of history in the dry lands of the Southern Hemisphere was not the same as in the Northern Hemisphere. Great civilizations arose in the northern dry lands, but the southern dry lands remained in backward stages of development. In the New World it was the Southern Hemisphere dry lands that were the more advanced culturally, while the northern dry lands languished.

That there was no physical environmental (climate, soils, vegetation) basis for this is easily established by even a brief consideration of the degree of similarity of the various regions. All of the west coast dry lands have cold currents offshore (Baja California, Peru, South-West Africa, northwest Africa, and west Australia). These currents cause extreme aridity along the shore but also create a potentially rich fishery offshore, for in the cold waters on these coasts is a phenomenon known as "upwelling," which brings up nutrients from the oceanic deeps. The fishery was only potential. The fish were there, but in most cases the people were not equipped to harvest them. Each major arid land has at least one river flowing through it: the Colorado, the Nile, the Indus, the Orange (South Africa), and the Murray-Darling system (Australia). Had men understood the idea of agriculture in South Africa, the Orange would still have been difficult to utilize extensively, for it is deeply entrenched, but it was not used for irrigation at all. Similarly, this resource in Australia was left untapped. Animal domestication was also unknown; hunting, not herding, was the only opportunity that men saw.

South African Bushmen

In the South African dry lands there were two ways of life: cattle raising in the steppe lands, and hunting and gathering in the desert. Hunting and gathering, the earliest, is still the characteristic way of life of the Bushmen, who shift their habitations from one water hole to another depending on the location of game and the seasonal abundance of vegetable crops. The women are the gatherers and the men the hunters. The woman's characteristic tool is her digging stick, the man's his bow. There is no pottery; ostrich egg shells serve as cups. Except for the use of ostrich shells the way of life is the same as that of the gatherers of the Great Basin—and omitting the bow, it would be the same as that of the Australian natives.

The Bushmen are one of the most distinctive races of mankind. They are dwarf in stature, have a strongly triangular face with Mongoloid, slanted eyes, yellowish skin color, and extremely kinky hair. The women have a tendency toward extreme fattiness of the buttocks combined with slenderness elsewhere, and both men and women have other anatomical peculiarities. Not only is art of the type still produced by these people found all the way into Spain, but Negroid skeletons and statuettes portraying the anatomical peculiarities of the Bushmen have been found in northern Italy in a cave at Grimaldi. It appears that these small people once extended all the way to the Mediterranean, and that until perhaps 1000 years ago they held most of East Africa. The traces of culture higher than is presently found in that area seem attributable to them. This tends to show that the Bushmen are not genetically inferior, since they have at times been bearers of higher culture. They were largely displaced in relatively late times by the expanding large Negro people, frequently with a lowering of culture level. A parallel situation occurred with some of the American Indians. For example, a highly developed Indian civilization was destroyed in the Maya country of Yucatán, while other Indians continued to live as simple hunters and gatherers in the Great Basin. Had we no knowledge of the accomplishment of the Maya, we might be tempted to judge the ability of the American Indian by the poor achievement of the Great Basin people. The tendency to judge the Bushmen by the accomplishments of a small remnant is probably equally faulty.

The origin of so distinctive a physical type has long intrigued scholars. Because we now know these people from the desert areas of southwest Africa, it has been thought that their distinctive characters were adaptations to the desert. Yellow skin and kinky hair are well suited to the high humidity and moderate heat of the tropical forest. Small body size is useful in any hot area. Fat storage will aid any people who lack an assured food supply and is especially helpful for mothers who may have to nurse infants through starving times. But much of mankind faced similar needs without developing equally marked racial traits. Isolation and genetic drift, as well as selection, must have been at work. This conclusion is strengthened by our increasing knowledge of the former, very much wider distri-

Native Housing on the Outskirts of a Town in South-West Africa. *Except for the children, this scene could be duplicated in numerous places in Latin America, North Africa, Arabia, India, and elsewhere. It records a cultural stage and an economic condition rather than an environmentally or a racially determined status. Compare Talara or Palm Springs, both much more arid than South-West Africa.* (Gordon Gibson for Smithsonian Institution)

bution of the Bushmen, including their occupation of part of the Mediterranean basin about 20,000 years ago.

From the worldwide distribution view, the Negritos appear to be the early type of Negro, for they are found in the refuge areas of the world. In Borneo a Tasmanoid (Negrito) skull was found at the 40,000-year level in a cave deposit. This would predate any known large Negroid skeletal material. Several scholars have come to believe it possible that the large Negroes of the world appeared relatively late and are the result of admixture with either Mongols or Europeans. The Bushmen's racial peculiarities may be an inheritance from the very ancient period of racial divergence that was hypothesized earlier. However, the Bushmen of South Africa, whatever the environment that molded them, are quite different from other Negritos, and if the Negritos are a basic stratum in mankind, it is a very ancient one. An alternative hypothesis would be that dwarfing occurred repeatedly in the Negro race, and distinctive differences among the Negritos of the world have been cited as evidence of lack of relationship between them. The differences could be cited equally well as evidence of great antiquity for the Negritos, with the vast amount of time allowing for the observed degree of divergence.

On the margins of the South African dry lands, the cattle-using Hottentot people have been moving in on the Bushmen, introducing another way of life. In Africa the use of cattle for pastoral life is clearly an idea that has been working south from origins in the Egypt-Ethiopian civilizations. The farther from the center of origin of ideas, the more primitive the culture. Moving south from an Egyptian center, there is a drop in level of civilization; the most primitive use of the land has persisted in the far south. Had the Bushmen chosen to do so, they could have practiced a pastoral, nomadic way of life, but perception of this use of their environment was too strange for them to grasp readily.

South-West Africa, the homeland of the Bushmen, today is a sparsely settled country of about 600,-000 people distributed over 300,000 square miles. One-fourth of all the Caucasians in the country, mostly Afrikaans-speaking, live in the capital city of Windhoek. The major use of the dry upland is pastoral with cattle and sheep grazing on large holdings. There are also extensive game preserves and native reserves. From the uplands there is a steep drop to the dry coastal lowlands. The ancient gravelly beaches of this zone produce diamonds with an annual value of tens of millions of dollars. Offshore the cold, upwelling water supports a valuable fishing industry based on pilchards and rock lobsters. An oddity is the raising of karakul sheep for the curly black fleece. The lambs must be killed and skinned within 36 hours of birth. This fits the sparse

111

browse of this region well, for the lambs need not be fattened for the market, and the ewes need not provide for more than a small proportion of their young. The extensiveness of the land can be seen by the size of some of the karakul farms, which range from 20,000 to 60,000 acres. It is interesting to note that the fishery potential has always been there, but that men equipped to harvest it have only now arrived. Similarly, diamonds were just gravel to the Bushmen. The pastoral potential has also been latent, but even if the Europeans had not arrived on the scene, the idea of cattle keeping probably would have been introduced from the northeast. The karakul lamb business is a spectacular case of the interrelatedness of the world today, since luxury products produced in so strange a place can be sent into the world market. South-West Africa's future is tied to water, just as is much of that of the American West. The only major source seems to be the Orange River, and major engineering works would be necessary to make its water available to the dry uplands.

Aboriginal Culture in Australia

Finally there is Australia with its desert heart, for half of the interior land is desert or semidesert. Australia is of extreme interest to the human geographer. In pre-Columbian times, no matter what the climate was, the culture level remained much the same. The entire continent including Tasmania was in the hands of people who lived on a hunting-and-gathering level of existence. They were a short-statured, bearded, prognathous, dark-skinned, beetle-browed type of man who, in many characteristics, falls between the Negroid and the white race. These people are probably best considered as a separate variant of mankind that arose in Australia through very long isolation. They had no agriculture, no pottery, no ground stone tools, and no domestic animals except the dog. By these terms the aboriginal Australians were living on a pre-Neolithic level. They were survivors of the ancient Paleolithic way of life that once characterized all of mankind for the longer part of his existence.

In Australia there are many climates in which this way of life was practiced. In the north there is a season of great rains in the summer, followed by a very long dry season in the winter. Winter in northern Australia, of course, is not to be thought of as a time of cold. The east coast of Australia ranges from a wet tropical northern coast to a subtropical rainy central coast, until in its southern corner the land has a distinct winter, and the mountains have a heavy snow

Gibber Plain near Woomera, South Australia. *This scene could as easily be in the American, North African, or Arabian deserts, for such rock-paved areas characterize parts of all deserts. While the scene emphasizes the stark harshness of the desert, it must be remembered that after a moderate rain, areas of this type are quickly covered with short-lived plants and can be carpeted with greenery and flowers for a few weeks. (Australian News and Information Service)*

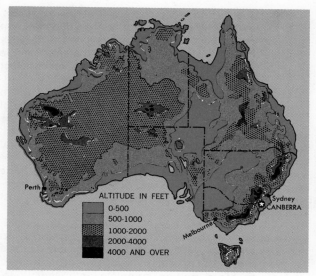

LANDFORM

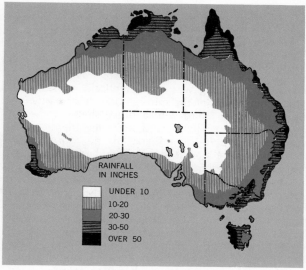

AVERAGE ANNUAL RAINFALL

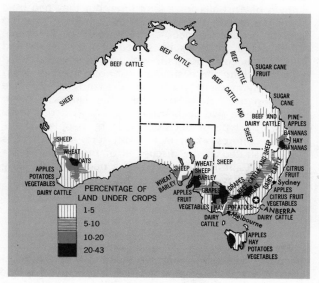

LAND USE

**MAP 2-10   Australia and Tasmania:
Landform, Rainfall, and Land Use**

*The general low altitude of Australia is clearly indicated by the
very small areas over 2000 feet and the very extensive areas
under 1000 feet. It is a land of extensive plains and low rolling
uplands with small areas of moderate-size mountains.*

*The range of rainfall in Australia is very high: from more
than 50 inches in some regions to less than 10 in others. The
values of these rainfalls are also variable. Fifty inches on the
tropical north coast coming during a period of heavy rainfall
gives rise to flood conditions that in the succeeding period of
little rainfall under tropical sun soon give way to severe drought.
Fifty inches of rainfall in south Australia creates a much wetter
situation due to less evaporation. The 20-inch rainfall line
approximately marks the zone of transition from agriculture to
cattle and sheep raising, from farming to ranching. The area
with less than 10 inches of rainfall is desert of very low
productivity.*

*The importance of rainfall in the utilization of Australia
becomes clear when the rainfall map is compared with the
land-use map. Very little land with less than 20 inches of
rainfall is used for crops, except in the area of winter rainfall
in the south. Equally interesting, however, is the extent of
untilled land in the areas of high rainfall in the north. The
land-use map shows how small is the area under cultivation. The
large, almost unused arid interior stands out strongly in contrast.
The nonintensive level of land use is well indicated by the
broad areas of cattle and sheep ranges, which are almost all
on natural, unimproved pasture land. (Reprinted from Focus,
The American Geographical Society)*

cover. Southwest Australia has a Mediterranean type
of climate, as is also found at the tip of South Africa,
in California, and in central Chile. Tasmania, inhabited
by a people related to the Australians and having an
even simpler culture, has a cool marine climate like
that of Great Britain and the Pacific Northwest coast
of America. As elsewhere in the world where this way

of life is found, the digging stick was the woman's chief tool, and her duty was the gathering of roots, seeds, and berries. The men hunted, and the throwing spear was their chief implement.

Housing was virtually nonexistent. A shelter that cut the wind but did not cover them was as much as they built. Clothing was also practically nonexistent. In the cold nights of the mid-latitude deserts a family curled up about a fire with a brush shelter to windward and slept on the ground while the thermometer dropped to the freezing mark. In the morning the natives would be almost torpid. They had the remarkable ability to adjust to such rigorous conditions by lowering their body temperature, especially in their limbs, retaining vital heat in their body cavity. When the sun came up and they resumed exercise, they regained their normal body temperature and went about the daily business of hunting and gathering. If need could drive men to clothing and housing, surely these people should have felt the need. One would expect these unclothed and unhoused natives to have been stimulated, encouraged, or driven by the physical environment to do something about their "needs." But they apparently felt the cold without "feeling the need" to do something about it. Until 1500, when the Europeans began to spill across the whole world, Australia lay outside the orbit of the world of ideas, and these simple folk were left alone to follow their own bent—which did not include "progress" in the sense that we think of it.

It is easy to conclude that there must have been something wrong with the Australians as human beings, but there is no evidence of it. Were they mentally deficient? No one can say that they were, or are. They simply did not take a great interest in the material things that so concern us. On the other hand, they had interests that we largely ignore; in particular, they kept close track of family relationships. It is a peculiarity of our society that we do not have names to describe the types of family relationships that exist. We lack any term to distinguish between our aunts and uncles who are our father's brothers and sisters and those on our mother's side. We have the single term "cousin" for the large number of relationships in several generation levels. We feel the need of kinship terms and so extend the titles "aunt," "uncle," and "cousin" to people who are only distantly related. This would all seem ridiculous to a native Australian. He would think we were wrong to be so concerned with houses and clothing while not keeping track of our family relationships.

The Australian lavished attention and inventiveness on his kinship system. He also had a vast system of secret societies, organizations resembling our modern fraternal orders with their long series of grades. Advancement from one level to the next depended on age and on learning the extensive ritual and mythology that belonged to each. In this way an illiterate people had an extensive mythology that seems inconsistent with their low economic stage. Like the Yuma, their interests in life were not the material things that are so important in our society today but the immaterial things such as the importance of relationships and the achievement of status in the secret societies, which maintained the traditions of their people. The physical environment was nothing more than the stage on which this play was acted. It did not matter to these players whether the stage was tropical woodland, temperate forest, or desert scrub. It was the plot that counted.

Why should Australia have lagged so long at this cultural stage? Was it isolation? This seems an obvious answer at first glance. However, New Zealand is even more distant, yet it had the more advanced Polynesian Maori culture complete with agriculture. Northern Australia is close to the outlying Indian civilization that centered on Java. The Asiatic people had known of Australia for a long time, but they had little incentive to take over the country, for there were no settled, productive people there. How could simple hunters and gatherers be seized, made subject, and taxed? They could not trade with natives who produced almost nothing and desired almost nothing. They could not simply move in and occupy the land, for primitive as the natives were, they held the land and were fiercely possessive of their hunting and gathering territories. The type of decision made by peoples of advanced cultures when faced with such a prospect is well illustrated in the discussion of the Spanish reaction of a California occupied by a people of similarly low culture level (see Chapter 4).

Arid Australia

Modern Australia is one of the latest of the European colonial developments. The British found a land occupied by a people at a much lower level of development than their own. In Australia there was even more marked a contrast than in the United States, for in the eastern United States the Indians were mostly farmers, and in our South many of them were town dwellers. The British "solution" in both America and Australia was the removal of the native population. The causes and reasons given for this policy and the history of attempts to get along with the natives are a complex matter. The role of outright warfare in reducing the numbers of aboriginal Australians has been overemphasized, and indirect destruction through the introduction of European diseases to which the native population had no resistance underestimated. The effects of such diseases can be gauged by examining episodes in the extermination of some American Indian tribes. The Mandan went from 2000 to 200 people in one winter simply as a result of the introduction of smallpox. The remnant joined adjacent tribes, and the Mandan as such ceased to exist.

In Australia, after their initial settlement in the humid lands of the east coast, the British within 25 years penetrated the dry interior of the continent. Rapid exploration showed the vast extent and the extreme aridity of much of this land. The colony needed products to export to the mother country that could stand the rigors of a long ocean voyage, both in terms of cost and in resistance to deterioration. Although the hot, dry lands of Australia could produce watermelons and lettuce for the British market, in the nineteenth century there was no way to get perishables across the oceans. Even today the cost of jet transportation is prohibitive, so Australia cannot play the role of an Imperial Valley to supply Great Britain's need for vegetables in winter.

Products from the arid lands had to be nonperishable, in demand in the home country, and of sufficient value to stand the transportation costs. Wheat, the staff of life for the Europeans, can be grown on the margin of the arid lands. It is a relatively nonperishable product and has high value for its bulk. Sheep can be grazed in the arid lands and moved about to take advantage of the uneven distribution of rainfall. Wool was in high demand in the mother country, could last during the long trip, and therefore was profitable to send through the tropics and nearly half way around the world to the market.

Both wheat and sheep are traditional products of the land in the British Isles, but British wheat and sheep are not necessarily the best kinds to develop in Australia. After all, Great Britain is a cool, moist country of nearly perpetual green. Arid Australia is a tawny yellow country of great heat and frequent drought. Although the idea of growing wheat and raising sheep for wool was natural to the British colonists, the transfer of these ideas to a new and totally different physical environment was not a "natural response to the physical environment." Instead, the potential market led these men to develop such land-use possibilities almost in spite of the physical environment.

Arid Australia remains a land of thin settlement, with extensive rather than intensive use of the land,

and it is constantly struggling with the problem of water. Even at the present stage of technological development, the use of this great dry heart of Australia is severely limited by the lack of water.

The soil in much of arid Australia is poor, largely due to leaching during long periods of weathering under tropical wet climates of the past. During the great climatic shifts that marked the time of the glaciers, Australia, like North Africa, shifted from wet to dry, and alternated in 100,000-year periods between tropical rainy weathering and arid weathering. Much of arid Australia now has the leached and impoverished soils typical of a tropical rain forest situation. Today we can add the needed tons of phosphates, calcium, and nitrogen to poor soils when economic conditions warrant. In many of the most productive farming lands of the world the soil owes its fertility to these additives and not to the natural condition of the soil. The vast expanse of gently rolling land in arid Australia could have the added nutrients if demand for food production warranted the expenditure and if there was water. Only water is still beyond our ability to supply to these arid regions.

THE FUTURE OF THE ARID LANDS

If one tries to assess the long-range future of the arid lands, one immediately faces the problem of water. Our major use of the land is for agriculture, and the primary need of agriculture is water. But this is the great shortage in the arid lands, and this suggests that, barring some major breakthrough in technology, the future of these lands is very limited.

We are, however, on the verge of a major technological breakthrough, or rather of a series of them. One of these is the desalinization of waters including sea water. In 1952 the cost of desalinizing 1000 gallons of sea water was $4.00; in 1956, $1.75; in 1965, $1.00. The goal is to reach a cost between $.10 and $.20. To reach these last figures huge plants at the seashore, with large savings due to the scale of the plants, will be required. If these figures can be realized, the water will be cheap enough to be used for agricultural purposes.

If, in addition, the projected developments in atomic energy come through on schedule, we will have virtually unlimited power for desalinization and for pumping water long distances. The technology of pumping economic materials for long distances is already highly developed for gas and oil, and if power costs were to change radically, water could also be moved long distances. In California it already is being so moved (see Chapter 4). We must then take a look at the atomic power situation, but it is worthwhile now to review briefly man's use of energy.

Man was first limited to his own muscular output. With the domestication of animals he began to utilize their power, though only slowly. Water mills and windmills were developed in the early centuries of the Christian era. The steam engine marked the beginning of the use of fossil fuels—coal and petroleum—and resulted in a quantitative jump in the power available to man; the internal combustion engine simply extended this availability. The fossil fuels, however, are not of infinite supply, and with the exploding world demand for power, the future looked pretty grim until atomic power was developed. The first atomic power plants were relatively inefficient and had the further disadvantage of producing very long-lived radioactive by-products that have posed terrifying problems of disposal. The next step was the development of the neutron-rich fast-breeder fission reactors. The current development is in fusion energy, which is described as the basic energy process of the sun. This prospect has moved from the stage of discussion of its scientific feasibility to a consideration of the technologic, economic, and social problems of the system. The sources of desirable materials such as deuterium or tritium from the sea are considered sufficient to meet projected future needs for millions to billions of years. Among the desirable features of a fusion reactor using a mixture of deuterium and tritium as fuel would be that it could be used to "burn" various fusion products and thus largely solve the problem of disposing of radioactive wastes. The system also is advantageous in that it does not burn hydrogen and carbon and hence does not release carbon dioxide to the atmosphere. Our burning of fossil fuels, on the other hand, has released so much carbon dioxide that we are in danger of changing the climate of the earth. There would be no release of dangerous radioactive wastes, and there would be no possibility of a runaway reaction.

This very brief description of an exceedingly complicated matter is intended only to indicate that there are energy sources on the horizon at a time distance of between 10 and 50 years, depending largely on the amount of funding supplied for development, that will provide power safely, cheaply, and in virtually unlimited amounts for the duration of the earth. The projected date for demonstration of desalinization plants capable of producing water economically for agriculture is about 1980, although 2000 may be more realistic. These two technologies are already foreseen as making possible what is called the agri-industrial complex. Such a complex would start with a huge nuclear-powered plant located on a seacoast at a site selected for proximity to good land with a suitable climate. The desert lands with their long growing season and very high percentages of sunshine are prime areas for such locations. By utilizing the economics of scale, location, and combined agricultural operations with food processing, and with foreseeable improvements in yields through plant breeding, fertilizing, new methods of irrigation, and so forth, it is already demonstrable that food can be produced cheaply and in large quantities with desalinized sea water.

The lands that will benefit most will be the arid lands near the sea and of low elevation. Australia is one of the areas that could best use such a development, but the Sahara, the Arabian peninsula, coastal Peru, and Baja California would all benefit. Areas like Inner Asia and the arid plateau of northern Mexico would present various difficult problems, but the major cost would be for the energy to lift the water to such heights or for such distances. For answers to that problem we must wait to see just how the costs of the looming fusion energy process work out.

The basic point of this discussion is, however, that technology is now moving so fast that it is difficult to assess the future. We clearly must get through a difficult present without poisoning or blowing ourselves up. If we can manage this, we may see a nearly total reassessment of the world as to its productivity and habitability. With unlimited energy we can air condition desert residences even more extensively and cheaply than we now do, and we can extend this amenity to the rest of the world. Simultaneously, we may be able to make the desert blossom like a rose. A glance at the map of arid lands at the chapter opening will show that immense areas of highly useful land would be made more meaningfully a part of the habitable world if these developments are realized.

Summary

The arid lands are limited by the shortage of water, but they have a great potential in their abundance of sunshine, the long growing season in the low-latitude dry lands, and the chemical wealth of their unleached soils. Where water from adjacent humid regions is concentrated in these areas, there are rich oases in the desert. These may vary from tiny oases around small springs to extensive ones such as the Nile and the Colorado river valleys to other potential Egypts such as the Orange River of South Africa or the Murray-Darling river system of Australia.

In each of these dry lands of the world what happened was entirely different: great civilization and empire developed in Egypt and Peru; the simplest hunters and gatherers lived in the American Southwest, in Australia, and in South Africa. In the Old World there was a development of pastoral nomadism that had no counterpart in the New World or in Australia. The stage of development of culture in the Old World can be seen to be rather directly related to access to ideas stemming from the great cultural hearth in the Middle East. In the New World the spread of ideas radiated from a Middle American center, and there was some symmetry to the spread of the ideas from this center and in the effects that these ideas produced in the arid lands they reached. For example, on the margins of the civilized center of Mexico lay the simple town dwellers of our Plateau region (the Pueblo people), and on the margins of the civilized center of Peru a similar culture existed in northwest Argentina. Beyond these regions the simple hunters and gatherers remained untouched by advanced ideas.

In post-Columbian times the use of the arid lands has had a varied history. The greatest changes occurred where the land was in the hands of the least culturally advanced peoples, who were largely dispossessed by the Europeans. The change in Egypt is slight compared to the change in Australia. The spread of domestic animals as a useful means of harvesting the arid lands, replacing the simple gathering folk with pastoral or ranching people, has been nearly worldwide. There is no great civilization or empire now based on the dry lands or on any major oases. The bases of power that once lay in Mesopotamia, northwest India, Egypt, and Peru now lie in utterly different climates. These changes have all occurred in a time of relatively constant conditions of the physical world. The causes cannot, then, have been the physical environments.

In the not distant future, revolutionary changes in availability of water and of unlimited cheap power promise to change almost totally the potentialities of these lands. Vast areas of presently barren, unproductive land may change to the high level of productivity now seen in the Imperial Valley of California. The impact of such changes are difficult to imagine, but the potentials are enormous. Australia's dry heart could become productive and populous; the Sahara could become a prized land for its potentially huge productivity near one of the great centers of human population concentration. The pessimists today are underselling technology, though they may be doing us all a service by heading off potentially long-term harmful actions in the immediate present.

Review Questions

1. Why are there not more animals in the deserts than in humid regions, since man scarcely disturbs the desert wildlife?
2. Discuss climatic change and land-use change as factors in causing such landscape changes as denudation and gullying in arid lands. Cite examples.
3. Why were there no pastoral nomads in pre-Columbian America? Were the causes climate? Vegetation? Animals?
4. How many possible house types are there in a desert situation? Give examples of known types. Can you think of any types *not* used that were within the technological ability of pre-twentieth-century man?
5. How did the idea arise that civilization originated in Egypt? What were some of the results of this explanation?
6. What is the size of the average farm in the Imperial Valley? Is this good or bad? Do your reasons apply to other countries?
7. "The climate of California is different from that of the Southwest. Agriculture spread into the Southwest but not into California. This is because the climate of Southern California is different from that of the Southwest." Discuss these statements and the conclusion as to logic and fact.
8. Compare the inferred effect of the Navaho-Apache raiders in the Southwest with the camel nomads in the Sahara. What are the similarities? The differences?
9. Since the Spanish settled the desert lands of South America, why did they not do likewise in North America?
10. Why were the arid lands of South Africa not made the basis of an irrigation civilization? Was this due to race? Physical geography? Location? How do you know? If you don't, how would you propose to find out?

Suggested Readings

Aschmann, H. H. *The Central Desert of Baja California: Demography and Ecology.* Riverside, Calif.: Manessier, 1967. Originally published in Ibero Americana, no. 42. Berkeley: University of California Press, 1959.

Bowman, Isaiah. *Desert Trails of the Atacama.* American Geographical Society Special Publications, no. 5. New York: American Geographical Society, 1924.

Braidwood, Robert J. "The Agricultural Revolution." In *Man and the Ecosphere.* Readings from *Scientific American.* San Francisco: Freeman and Company, 1971.

Cameron, Roderich. *Australia: History and Horizons.* London: Weidenfeld and Nicolson, 1971.

Carter, G. F. "Evidence for Pre-Columbian Chickens in America." *Man Across the Sea.* Edited by Carroll L. Riley, J. Charles Kelley, Campbell W. Pennington, and Robert Rands. Austin: University of Texas Press, 1971. In this book will be found a series of pertinent articles on plants, animals, and boats, a lengthy review article by Jett, together with critical review articles for each major section of the book.

Daumas, General E. *The Ways of the Desert.* 9th ed. Translated by Sheila M. Ohlendorf. Austin: University of Texas Press, 1971.

Doran, E. D. "The Sailing Raft as a Great Tradition." *Man Across the Sea.* Edited by Carroll L. Riley, J. Charles Kelley, Campbell W. Pennington, and Robert Rands. Austin: University of Texas Press, 1971.

Dunbier, Roger. *The Sonoran Desert: Its Geography, Economy and People.* Tucson: University of Arizona Press, 1968.

Edwards, Clinton. *Aboriginal Wateraft On the Pacific Coast of South America.* Ibero Americana, no. 47. Berkeley and Los Angeles: University of California Press, 1965.

Flannery, K. V. "The Ecology of Early Food Production in Mesopotamia." *Science,* vol. 147, no. 3663 (1965), pp. 1247–1255.

Ford, Thomas R. *Man and Land in Peru.* Gainesville: University of Florida Press, 1962.

Huzayyin, S. "Changes in Climate, Vegetation, and Human Adjustment in the Sahara-Arabian Belt." In W. L. Thomas, Jr., ed., *Man's Role in Changing the Face of the Earth.* Chicago: University of Chicago Press, 1956.

Jett, Stephen. "Pueblo Indian Migrations: An Evaluation of the Possible Physical and Cultural Determinants." *American Antiquity,* vol. 29, no. 3 (January 1964), pp. 281–300.

Stamp, Sir L. Dudley. *History of Land Use in Arid Regions.* Arid Zone Research, no. 17. Paris: UNESCO, 1961.

White, Gilbert F. *Science and the Future of Arid Lands.* Paris: UNESCO, 1960.

FOCUS (American Geographical Society leaflets)

Algeria, Egypt, Libya, Morocco, North African Arid Realm, South-West Africa, Iraq, Pakistan, Australia, Peru, United States Arid Realm

The Wet Tropics

CHAPTER THREE

No other climatic region has so much folklore associated with it as the wet tropics. This land, straddling the equator in an irregular belt of 20 to 40 degrees in width, is distinguished from all the other regions of the earth by the lack of substantial seasonal variations, particularly by the lack of winter cold. In the wet tropics the difference between day and night provides variation in climate that is greater than between winter and summer. The coolest months average about 65° F., and palm trees mark the landscape. These lands occupy 20 percent of the land surface of the earth. Because these are well watered lands not handicapped by short seasons or great cold, they should offer mankind an unusual opportunity not found in some other areas of the world. And, as we will see, progress has indeed varied in these lands.

The Physical Setting

Tropical lands need subdivision. The tropical rain forest has no dry season, the savanna has a distinct dry season, and drought lays a firm hand on the land. Although the monsoon climate has a dry season, the rainfall during the wet season is so great that the effect of the drought is lessened. Mountains interrupt the tropical belts, and above 3500 feet the temperatures fall below those typical of the tropics.

Tropical rain forest

The tropical rain forest is characterized by uniformly high temperatures with heavy rainfall well dis-

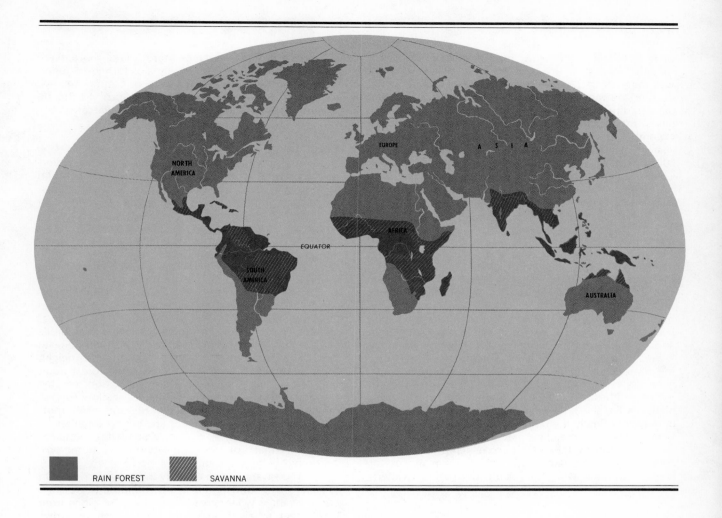

| RAIN FOREST | SAVANNA |

tributed through the year. It is typically located astride the equator and extends poleward approximately 5 to 10 degrees of latitude. The length of day does not change on the equator, and hardly varies near it. The noonday sun is always nearly vertical in the sky, so that the amount of sunlight or insolation is at a near maximum throughout the year, varying little. The results are annual average temperatures of about 77° to 80° F.

Many cities in the United States have summermonth temperatures higher than this, but in the tropics these temperatures vary little through the year. The annual range from the hottest month to the coldest month is less than 5 degrees. For instance, Belém in Brazil and Singapore in Malaya have an annual range of 3 degrees, although the daily ranges are much greater than this and may run from 10 to 25 degrees.

Bolobo in Zaïre (formerly the Congo) has a daily temperature range of 16 degrees and annual temperature range of only 2 degrees. Even the absolute ranges are not great. Santarém in Brazil, for example, has had a maximum of 96° F. and a minimum of 65° F. Compare Baltimore's 107° F. maximum and −7° F. minimum, and some of the imagined horrors of the tropics begin to disappear.

However, the wet tropics are humid, and humidity increases the body's reaction to heat. The nights too are humid and cool off relatively little. The tropics are not cool, but the question of discomfort must be kept in perspective. The heat load is considerably greater during a heat wave in Chicago, St. Louis, or Washington than it ever is in the tropical rain forest.

Rainfall is heavy and regular, and a great deal

121

of it comes in the form of hard thundershowers. Although there is no dry period, some months are less wet than others. At Belém it rains 90 percent of the days in the rainiest season but only one-third of the days in the "dry" season. The amount of rainfall is heavy but often varies considerably. For instance, at Colombo in Sri Lanka (formerly Ceylon) the wettest year on record had 139 inches of rain and the driest, 50 inches. The wet tropics also have spots such as Cherrapunji in India that compete for the wettest place on earth, with such rainfalls as 400 inches per year, 900 inches being recorded in 1861.

Monsoon regions

The monsoon regions, most typically developed in India and Burma, are intermediate between the rain forest and the savanna climates. They have the heavy rainfall of the former and the dry season of the latter. They are usually found along tropical coasts backed by mountains, where the seasonally onshore winds spill heavy loads of moisture. Total rainfall is often high and markedly seasonal. The range of temperatures is greater than in the rain forest climates, annually varying some 12 to 14 degrees. The highest temperatures generally come just before the rainy (summer) season begins. With the approach of summer the sun mounts higher in the sky, and the parched land gets hotter and hotter, until finally the rains come. Then the moistened ground and the spreading green under the increasing cloud cover bring the slightly lower temperatures that continue until the rains cease and the sun moves to lower elevations, when the warm, dry period (winter) begins.

Savanna climates

In the savanna climates there is more seasonality of rainfall and less rainfall than in the monsoon climates. The forest that characterizes both the rain forest and the monsoon regions gives way to a tree-studded grassland that merges toward the poles with the low-latitude dry lands. This is the transition between the desert climates and the equatorial wet climates. Tropical rainy summers alternate with desert dry winters. Rainfall is strongly concentrated in the summers but usually amounts to only 40 to 60 inches per year. In these hot regions this is not a great deal of precipitation, for rain falling on hot earth and under an intense sun evaporates quickly. In the cooler mid-latitudes this would be enough rainfall for a forest cover!

Where the onshore winds on the tropical east coast of continents bring both evenly distributed rainfall and stable, oceanic temperatures to the shore, there are extensions of the tropical climates to mid-latitudes. Such areas extend up the east coast of Central America to Yucatán, down the east coast of Brazil to Rio de Janeiro, and down the east side of Madagascar.

Vegetation

Each of these climatic regions has its characteristic vegetation. The rain forest has trees 100 feet and more in height and is very exacting in its requirements: There can be no prolonged dry season, and there must be at least 60 to 80 inches of rainfall per year. Elevation brings cooler temperatures, and the tropical rain forest generally does not go up to more than 2000 feet elevation. Few areas meet such conditions; the Amazon basin, the Congo basin, and the equatorial inlands from Sumatra to New Guinea are the principal rain forests.

The tropical rain forest is composed of layers of trees. The forest giants reach up 130 to 160 feet, and their tops protrude from a green sea of foliage. Trees of 100 to 130 feet in height form a continuous umbrella of green and intercept much of the insolation. Beneath them is an understory of trees 50 to 80 feet in height. These layers leave little light to reach the forest floor, and consequently there is little vegetation on the ground. The trees are straight trunked, shallow rooted, and stand on mats of surface roots and prop roots that form buttresses. The trees are bound together by strongly growing vines, or lianas. Epiphytes, nonparasitic plants that use trees for support, are numerous, the orchid being an example.

The savanna varies greatly in its plant cover. On its equatorial boundary it is nearly semidesert; these are grassy savannas where greatly varied grasses range from 1 to 12 feet in height—elephant grass sometimes grows 16 feet high. There are also wooded savannas where the trees are moderately abundant; such an area is sometimes called a "park forest." Trees in savannas may reach a height of 50 feet and often have thick bark. There is considerable controversy over the origin of the grass savannas—as there is with the grasslands (see Chapter 7). Are they natural, or did man help to create them by setting frequent fires? That a woody cover of the land is a "normal" situation rather than a grass savanna is suggested by the dry deciduous forest that often forms the transition between the rain forest and the dry lands. The dry deciduous forests are stunted-tree and thorny-bush areas. Rainfall ranges in the wooded savannas from 15 to 45 inches per year, but the dry season is very long, sometimes lasting up to nine months. Again, the rain comes in the summer when evaporation is high.

(Left) **Banks of the Amazon.** *Typical selva vegetation appears here except that the river forms an opening that allows light to reach the forest floor. Hence there is a growth of vegetation near the ground that would not be found 100 yards in from the bank. The cultural poverty of the Amazon is also illustrated by the poor housing clinging to the bank of the stream. Yet this is one of the world's most navigable rivers running through one of the greatest rain forests there is.* (American Geographical Society) (Right) **Daule River, Ecuador.** *Savanna-type landscape in the seasonally wet-and-dry tropics is shown here. The steep bands and low position of the landing stages indicate that this is the dry period.* (World Bank)

Ethiopian Scene. *Much of the tropics is dry land from brush to grass. Some of the apparent aridity is due to long usage by man, with erosion and impoverishment of the land and the vegetation due to his agricultural practices. This is the arid margin of the wet tropics in Ethiopia.* (Kay Muldoon for World Bank)

Such woody plant cover is common in Queensland, Australia, and is found on the Pacific slope of Mexico south of Lower California down to Panama. There are large areas of this type of vegetation in Africa, especially in the Sudan and East Africa. In Brazil this type of savanna is very extensive and has the special name of *caatinga*. It covers 300,000 square miles of northeastern Brazil and is present over large areas to the south. This vast, scrubby tree forest is deciduous: in the dry season, it is a barren, gray area looking like a mid-latitude beech forest in winter; with the onset of the summer rains, it puts out a profusion of colorful blossoms and rapidly becomes green.

Tropical soils

The soils of the tropics are one of the limiting factors in the agricultural use of the land. The process of weathering of rock that produces soil is a complex one and goes on through time. The weathering is directional; that is, there is a continuing breakdown of the rock material into simpler and simpler compounds. In humid regions the soluble materials are constantly being removed by the percolating waters. The chemical action of the rain water is greatly increased by the organic compounds resulting from the decay of the vegetable material derived from the forest cover of the land. In the rain forest regions the immense forests produce a huge amount of vegetable material, all of which eventually falls to the ground and decays, providing humic acids and other organic compounds that aid in the chemical breakup first of the solid rock, then of the rock fragments, and, last, of the finer particles. The end of the process can be an accumulation of the insoluble parts of the rock, with most of this material being quartz. Such material is of no use for growing plants. Since this weathering process goes on through time, it is clear that the longer any land area has been left undisturbed subject to humid tropical weathering, the poorer the soils on that area will be. This tends to be true in all humid areas, but since the speed of chemical reactions increases with temperature, impoverished soils are created more rapidly in the hot, wet tropics than elsewhere.

Tropical soils therefore tend with age to become poor. The old flat lands of the tropics tend to have ancient, deeply weathered covers of mineral matter that has been severely leached and hence have little remaining soluable materials for plant food. These are poor soils. Where young volcanic materials cover the landscape, or where fresh alluvium is deposited, or on steep slopes where erosion removes the weathered products and exposes fresh rock materials, the soils are good. It is a curious fact that the low, flat tropical areas need an accelerated erosion to strip off the 50 to 100 feet of weathered and impoverished materials that overlie the fresh rock that could supply the nutrients plants must have for adequate growth and production. Yet it is just the flat areas that are not subject to erosion.

There is a seeming contradiction in all of this. The tropical ran forest is the highest, densest, most varied forest on earth. If tropical soils are so poor, how can this be? No forest can grow luxuriantly on a soil that cannot provide nutrients. In areas of utterly impoverished soils in the tropics, despite optimum tem-

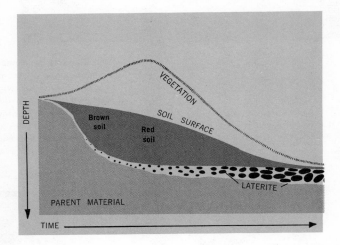

FIGURE 3-1   Soil Weathering and Time

Starting with unweathered material of any kind, the soil-weathering process can be illustrated as shown in the diagram, which reads from left to right. Any material subjected to surface weathering will at first improve in ability to support vegetation, and then gradually it will lose fertility as the weathering releases mineral nutrients that are removed from the soil by the rain water that percolates through the soil mass and drains away. Simultaneously, the surface of the land is being lowered by erosion, and redeposition of minerals is going on at depth. Iron and its related minerals, manganese and aluminum, accumulate at depth, and in time this accumulation may produce a heavy slag-like iron layer: laterite. When the land surface erodes down to this layer, an almost sterile land surface is the result. Characteristically, soils change color through time. Those starting from nonred materials are first gray, then brown, then red. Since heat and moisture speed weathering, tropical humid soils tend to age rapidly. Nevertheless, it normally takes one-half to one million years of stable weathering conditions to reach the stage of greatly impoverished soil with an immense laterite layer.

Mount Bromo, Java. *Active young volcanic materials keep the soils fertile, and even the streams coming down these volcanic slopes carry much nutrient material in solution. (Embassy of Indonesia)*

FIGURE 3-2   Cycling of Nutrients in a Tropical Rain Forest
The complex cycling of nutrients in a tropical rain forest is diagrammed in simplified form here. Except for the continuous action due to lack of seasons in the tropics, this cycle applies to vegetation in the seasonal climates too. The plant utilizes water and nutrients drawn from the soil and then utilizes solar energy to combine them to form wood, leaf, and fruit. Leaf, fruit, and branch falling to the ground decay, and the nutrient materials released, minus leakage to the ground water, are taken up by the tree roots and reused. Nutrients are also gained from the decay of rock material. As weathering progresses, the amount of weatherable rock minerals near the surface decreases, and the depth to relatively fresh rock increases. When the depth to fresh rock exceeds the depth of penetration of the roots, the system is precariously balanced, for the plants are running on the material in the cycle. Any large loss, as would occur if the forest were burned, might bankrupt the system.

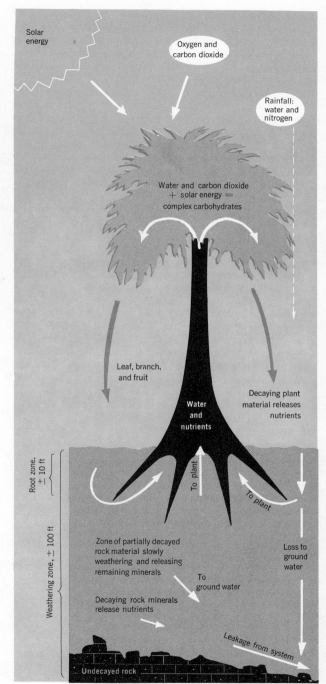

perature and rainfall, the vegetation is desertlike. There have been areas of luxuriant forest which, when the forest was cut down, were found unable to maintain any crops whatever. Furthermore, once the land was abandoned, it could not be brought to support a forest again.

This interesting situation arises because the tropical forests are cycling the nutrients that exist in their soil and plant system, and there is very little leakage out of the system. The minerals taken from the soil go to form branch, leaf, and fruit; these ultimately fall to the ground and there decay. As fast as their minerals are returned to the ground, the waiting network of roots takes them up and reuses them. Since the plants are using carbon dioxide taken from the air to make carbohydrates, and combining these carbohydrates with minerals, there is a slow enrichment going on that tends to balance the slow leakage of nutrients through the weathering process. Such a system can run for a long time, delicately balancing the losses and the gains but imperceptibly running down as the rocks are weathered to greater and greater depth and the surface mantle of weathered materials comes to contain less and less weatherable material that can release the minerals needed by the plant cover.

When man enters such a critically balanced system and cuts down the trees and burns the woody and vegetable material, he destroys the balance. The immense tonnage of woody organic compounds is converted to gases that go off as smoke and flame. The ashy residue is largely washed down through the soil mass and drained away in the ground water. Little is left except the insoluble end products of weathering, and little can be grown in this; the forest cannot reestablish itself.

This imbalance, of course, represents the extreme case. In most tropical rainy regions the forests tend to accumulate the organic and mineral materials and cycle them in the way described. Usually the soils have not reached extreme weathering. Consequently, when the forest is cut and burned, there is enough fertility to grow a crop or two and still enough fertility for the forest to regenerate and begin accumulating organic materials again.

Land Use

There are many ways of using the tropical forest lands. The simplest and undoubtedly the earliest was that of simple hunting and gathering, a way of life that can be practiced in every environment with the pos-

sible exception of the Arctic. There are a few people left in the world who still utilize the tropical lands in this way, for example, the Veddas of Sri Lanka, the Pygmies of the Ituri forest of Zaïre, and the people of the Yellow Leaves in northern Thailand. They wander through the forests, gathering fruits and honey, catching small animals, and getting some fish. But most of the tropical forest people of the world have long left this way of life and turned to farming. There are several kinds of forest agriculture, but they can be treated under two headings: shifting agriculture and permanent agriculture.

Primitive forest agriculture has always used the forest as a long-term rotation crop. The land is allowed to rest under forest cover for 5 to 20 years, during which time a large stock of material is accumulated in the plant cover on the land. Burning releases this material, and the ash provides a dressing of mineral fertilizer. The forest cover also shades out all the weeds. In a year or two, when the weak fertilizer effects have worn off, the land is again allowed to rest under its 5- to 20-year rotation into forest cover. This is a way of land use, of course, that dictates that there be 5 to 20 times as much land in the rotation reserve as is necessary for sustaining the agricultural needs of the inhabitants. Students of tropical agriculture generally agree that this type of land use is an efficient way for primitive agricultures to use a rain forest. While it seems wasteful of timber, this product itself has no market. The patchy forest clearings create little erosion risk and barely disturb the ecology of the land. But what is the determining force here: the physical environment or the cultural one? Could a tropical forest area be used in other ways?

There is a built-in potential crisis in the shifting agriculture pattern of land use. If the population grows, a point is reached where the land cannot be allowed enough time to recover between periods of use. Then the weeds get established, the soil becomes run down, and the situation deteriorates rapidly. At this point new land must be found, or the population must be reduced, or some new and more intensive way of using the land must be found. Few peoples have discovered new ways of land use, and the population density of most of the tropical forest lands of the world has therefore been fixed by the amount of forest available for rotation.

The wet-rice solution

A way of developing a denser and more permanent settlement on the wet-forest land was worked out in Southeast Asia. The rice plant was found to be capable of growing in standing water and of making a crop on soils so poor that they would not yield a return under any other circumstances. We are uncertain of the origin of rice. It does not belong in northern China or northwestern India and is surprisingly late to appear in the East Indian islands. Since its cultivation depends upon changing lengths of day—that is, seasons—rice cannot be an equatorial plant. Perhaps its homeland was southern China, in some area such as the Yangtze River Valley.

Rice is an amazingly versatile plant, for different kinds can grow under a great variety of conditions. One type can be planted in hillside forest clearings and grown without irrigation; another can be planted in shallow water, will elongate its stem as the floods raise the water level, and will develop a rice crop while floating like a rooted pond weed. This rice is harvested when the flood waters recede and the land is dry again. There is rice that will mature a crop in 120 days and rice that needs more than 360 days. In addition to the elongated rice grains familiar to us, there are both longer and more slender grain types and shorter and thicker types right down to nearly round grains. Such a variety of adaptation—long to short season, swamp to dry field—gives man great opportunity to select ways of handling his crop. Is this variability something man found in the plant and utilized, or did man take the plant and develop a wide range of uses? Probably it was a little of both.

Only Asia developed a grain plant to grow in a swamp! Although it is difficult to see just how this was done, it seems more likely that man started with a swamp plant than with an upland plant. It is conceivable that he then extended the swamp environment by simple dikes and ditches, eventually making whole flood plains and even mountainsides into artificial swamps for rice growing.

This development has made possible a dense permanent settlement on land that could not otherwise continuously yield a harvest, because the wet rice draws a good deal of its sustenance not from the soil but from the water. A rice paddy is a bit like a balanced aquarium, only more complex. In the rice field there are fish, frogs, insects, micro-organisms, and birds and animals to prey on them. In the warm shallow waters these life forms all live and die, and the rice plant profits thereby. It draws some sustenance from the often poor soil, some from the air, some from the water, and some from the by-products of the rich plant and animal community that shares its pond.

Sometimes the paddyfields are simply waffle-like, catching only the rain water that falls. Rain water is not pure water; it contains measurable amounts of impurities, especially the important plant nutrient nitrogen. Other fields are fed by streams, whose waters carry dissolved nutrients. Streams from young volcanic mountains, especially those from basic volcanic areas, may carry relatively large amounts of desirable nutrient materials. It therefore pays the farmer to irrigate his fields, even though there is an abundant rainfall to supply the water needed to keep the paddyfields flooded. In Southeast Asia, Java is a conspicuously densely settled area, and some, though by no means all, of this density is due to the fact that Java is largely composed of active, basic, volcanic mountains. Its streams are an extremely dilute but highly valuable liquid fertilizer because these young basic volcanic rocks yield measurable amounts of chemicals under active tropical humid weathering.

We are gradually becoming aware that Southeast Asia was an important cultural center that originated much that we think of as typically Chinese. As an example, note the list of contributions that linguistic studies show to have come from the south of China and Vietnam. It seems probable that rice was contributed by this area and that the whole complex of irrigation, terracing, transplanting, and harvesting probably originated there.

*Some of South China's Contributions
to Chinese Culture*[1]
Austro-Thai loan words in Chinese
higher numerals
chicken, egg
horse, saddle, riding
elephant, ivory
cattle, goat, sheep
bee, honey
garden, manure, plough
mortar, pestle
seed, sow, winnow
rice, sugar cane, banana
coconut, ginger, mustard
salt, steaming (rice)
metals (gold, copper, iron)
boats, rafts, oars
crossbow
fireplace
kiln, pottery
weaving, plaiting, basket

[1] Selected from Chester Gorman's "A Priori Models and Thai Prehistory," a paper given in 1974 at the International Congress of Anthropology and Ethnology in Chicago. Gorman drew upon P. K. Benedict, "Austro-Thai Studies," *Behavior Science Notes*, vol. 2, no. 4 (1967), pp. 275–336.

These words, but part of the whole list, suggest a flourishing advanced culture to the south of emergent north China from which wholesale borrowing went on. The situation is comparable to British borrowing nearly the whole Mediterranean culture complex.

Wherever and however discovered, the wet-rice method of using tropical lands is important, for it made possible permanent dense settlement on a scale not seen in the rain forest before this time. It is potentially useful in much of the world, but notice how little of tropical Africa or tropical America grows rice. Yet by utilizing Asiatic methods, these areas too could support similar agriculture. As always, each new idea potentially remakes the world, and as always, the potential is exceedingly slow in being developed.

Mechanization of rice production

It is often said that rice growing requires coolie labor, and that its spread is therefore not desirable. Rice-growing methods in the United States refute this. Rice growing began quite early in colonial America, where the low swampy land on the eastern coastal plain was well suited to this crop. The lands were diked and the crop produced with the use of much labor, as in the Orient. However, the scarcity of labor and limited European markets, as well as the higher return for other crops such as tobacco and cotton, led to the decline of rice production.

Note here the negligible role of the physical environment. For thousands of years the eastern United States coastal plain existed as it does now, but the Indians did not grow rice or any equivalent crop there. Nor can it be said that such a plant was not available in North America, for there is a plant here called wild rice. It is a grass of a different genus than rice, but its growth pattern is like the Asiatic wet rice. The American Indians harvested it, but never cultivated it; this is doubly interesting since they did cultivate corn. They had the idea of farming, but in the eastern United States, where they harvested both corn and wild rice, they lacked the idea of irrigation. Hence even in an ideal environment and with a ready-made plant whose use they fully understood, they failed to accept "the environmental invitation." There was no invitation, of course. The possibility lay mutely before them, and they saw it not.

The American colonists had contacts directly and indirectly with Asia, and from this source came the ideas. The environmental possibility for the foreign grain was seen. The ideas came in a package: Asiatic wet-rice cultivation with Asiatic rice. The colonists did not use the American plant, probably because the idea never occurred to them. They certainly would have had marketing difficulties if it had, for people are very resistant to new foods. Who would take up a strange new grain? Only recently has the American "wild rice" found a limited market as a luxury "specialty" food.

After rice growing on the eastern seaboard of the United States had declined into insignificance, an entirely new impetus was given to the rice industry by a quite accidental happening that serves to illustrate some of the complexities of the spread of ideas and the changing usages of landscapes. A Japanese who had come to America to study in a northern university traveled through Texas. He was struck with the extensive flat, hot, fertile, well-watered coastal plain. A thought occurred to him that was not likely to strike a cotton farmer entering Texas from the Deep South: this was ideal rice-growing country! He stayed there and wrote his family in Japan to sell everything, come to America, and bring some good strong field hands and some good seed. This was the beginning of the Texas rice industry that now occupies hundreds of square miles of the Texas coastal plain. Many more Japanese came, settling especially in the vicinity of Houston, where their names are to be seen on streets and houses and where their descendants today are more active in businesses and professions than they are in farming. Given the extensive contacts with Asia that we now know occurred at varying times over the past 4000 to 5000 years, one wonders why rice was not introduced earlier into other parts of America. It is a lot easier to ask the question than to provide the answer.

In America a transfer of methods occurred where the rice-growing area of the lower Mississippi met the wheat growers of the Middle West. Once the wheat farmers had mechanized their own production, they looked at the laborious hand methods of the rice farmers and saw a wheatlike grain, grassy straw, and the price of rice. If they could handle rice with their mechanized cultivators, seeders, harvesters, and threshers, they could produce rice cheaply. The wet soil was a problem. They proceeded to control this by cultivating the land when it was dry, drilling the seed in, then flooding the land by use of pumps. Harvesting was done by drying the land and utilizing slightly modified wheat combines and threshers. By these methods it was found that rice could be produced cheaply enough to compete with the "cheap labor" of the Asiatic area. Rice is now grown in the United States in a number of areas, including the desert oases of the Imperial Valley and the San Joaquin Valley of California.

An example of how these ideas can be applied elsewhere is found in northern Australia. This tropical savanna area has been a problem to the Australians. It

(Top) **Rice Threshing in Harvest Season, Japan.** *Advanced techniques, improved strains, and the heavy use of chemical fertilizers make Japanese farms among the most productive in the world. Japan's chief crop is rice, the country's staple food. The rice production per acre is the highest in the world. In the past decade Japan's harvests have been so consistently bountiful that a yield of around 12 million tons, once considered a more than bumper harvest, is now regarded as normal. Although Japanese farms are extremely small in size because of the mountainous topography, farming methods are continually being mechanized and modernized. Today Japanese farmers employ various modern farm machines designed to fit the needs and requirements of the small-scale farm. (Japan National Tourist Office)*

(Middle) **Spraying Insecticide over Rice Paddies, Japan.** *Helicopters have come to play a vital role in Japanese agriculture and are now widely used for spraying insecticides, weed killers, and fertilizers. The increased use of agricultural chemicals has contributed much to Japan's success in becoming self-sufficient in its rice production. (Japan National Tourist Office)*

(Bottom) **Wet-Rice Fields, Java.** *Notice the terracing of the slopes leading down to the streams. The clumps of trees, mostly around house and village sites, completely hidden by the trees, are all that remain of the tropical forest. (Embassy of Indonesia)*

is distant from their population centers, transportation has been meager, and the seasons are so extreme that during the dry season the streams cease to run and the grass withers away. It is poor land for cattle or sheep, or wheat or oats, but it is good rice land—flat, alternately flooded and dry. Immense tracts are available at very low cost to provide rice in demand in the world market.

Large-scale development of rice growing is now underway, and a 10 percent increase in world rice production is planned. Again one sees the importance of ideas. The Australian environment has changed little in the past 10,000 years. Irrigation rice farmers have lived in nearby Asia for perhaps 5000 years and on the adjacent Indonesian islands for at least 1000 years. Europeans with abundant knowledge have occupied Australia for 150 years, and they have looked for some way to use "the north country" for at least these past 50 years. Yet the final putting together of man and land and ideas comes only in the 1960s.

Should Asia now revamp the use of its rice lands and put tractors and harvesters to work instead of men? Perhaps, but surely not suddenly. The Asiatic rice paddy is a small field, adapted to a bullock and worked by hand labor. The fields are not laid out for tractors. The whole scale of the Asiatic operation is small, the labor input is great, and the yield per acre is large. The American system is the reverse. It could not be introduced into Asia without reshaping the land, reorienting the people, and almost totally changing the social and economic structure of the countries. Such changes can be made only gradually.

In Japan today the mechanization of small rice fields has begun with the introduction of heavy-duty rototillers. These are a one-man garden tool comparable to a power mower but designed to stir the soil of the flooded fields. They are relatively light and cheap and can be operated by one man on his traditionally small fields. Already rototillers have increased rice yields significantly. This innovation should release manpower from field labor, and this manpower can then be employed in Japan's growing industry while being fed from the greater amount of rice produced by fewer men on the land. Not every nation is ready for this step, for there must be growing industry to absorb displaced farm labor. Moreover, there must be a farm-labor force ready to accept, care for, and use wisely the new machinery, which is relatively expensive and complicated compared to either the old hand tools or the simple plows used with water buffalo.

We have been through a change of this kind in the United States. To use rough figures: 100 years ago this was a land of small farmers who produced a small surplus by using horses for power. There were 10 farmers for every city-manufacturing-commercial worker. Today this is a land of increasingly few, mechanized, large farms producing such high surpluses that our people are now about 95 percent urban workers and 5 percent rural. Asia could, and probably will, make a similar shift in the next 100 years.

Modernizing tropical agriculture: the Philippine experiment

The experimental work being carried on in the Philippines is so important for all of the tropical human lands that it deserves extensive review.

The potential for increasing food supplies in the tropics is very large, for the advances of modern scientific agriculture are only now beginning to touch these areas. Further, while the mid-latitudes are limited mostly to one crop per year, the tropics, due to the 12-month growing season, can grow anywhere from three to five crops. In order for the tropics to achieve this potential, however, more education, scientific research and its technological applications, and production of fertilizer must be introduced. In the underdeveloped countries of the tropics there is variable

opportunity for expanding the crop areas: huge in Brazil and generally more modest elsewhere. Most of the tropical people could easily increase the yields of single crops, and almost all of them could move to producing multiple crops on a given piece of land.

Rice is the basic crop of the Asiatic tropics, and it is said to feed more people than any other crop. Yet the yields are low, only about 1200 to 1500 kilograms per hectare in most countries. The production is very inefficient also. The average Japanese farmer spends about 1750 man-hours of labor in producing one hectare of rice, and this results in a low income even when the yield per hectare is high. In most of the tropics the rice farmer puts in only about half as much labor, but he gets only about one-third as much rice and ends up worse off than Japanese farmers.

The Ford and the Rockefeller Foundations studied this situation and collaborated in establishing an International Rice Research Institute (IRRI) in the Philippines. An international team of experts was assembled, put to work, and fully supported. Within six years they produced rice varieties and cultivating practices that gave yields ranging from 4 to 10 metric tons per hectare in regions where the average yields had been only 1 to 2 tons per hectare. Much of this success was due to the finding of a short and thick-strawed rice that did not respond to fertilizer by growing excessively tall and toppling over. One of the most interesting findings was that the farmers, as soon as they saw the productivity of the new rice, wanted it for their farms. There was no resistance to change. Resistance came in terms of taste. The Asiatic rice eaters are fussy, for they like certain kinds of rice and won't eat others. The rice breeders are ingenious enough to cater to these preferences, and out of the 10,000 varieties of rice available in the IRRI, they are manufacturing rice types to suit all tastes.

The next step has been to develop multiple-crop planting schedules. There is an enormous amount of solar energy available to work with. Los Baños in the Philippines has four times the cumulative degree days, the measure that the crop men use, than Ithaca, New York, has. This correlates very well with measures of the amount of dry matter produced in a tropical rain forest compared with that produced by a deciduous mid-latitude forest, for the differential is again four to one. To take full advantage of this huge solar opportunity, the crops must be kept moving and the land kept in use. These ends can be achieved by several means: throwing up the soil in ridges to speed drying, minimizing the volume of soil tilled and the number of tillage operations, and using fast-maturing crops. Crops that can be cut, but which will form sprouts (ratoons) again, can be used advantageously, for this eliminates one or two planting operations. Transplanting of slow-growing vegetables from starting beds to fields minimizes the tying up of the field. Some crops can be used in the immature stage—sweet corn for instance; intercropping whenever possible increases yields still further. An example of the kinds of crops that are being worked with is given in Table 3–1.

TABLE 3-1 High-Yielding Crops in the Philippines*

Crop	Days to Harvest	Range in Yields†
Rice	114–124	4– 6
Sweet potato	90–110	20–30
Soybeans	60– 90	2–3.6
Sweet corn	60– 70	40,000 ears
Sorghum 1	75– 85	5– 7
Sorghum 2	70– 80	6– 7
Sorghum 3	70– 80	5– 6

* Extracted from reports of the International Rice Research Institute in the Philippines.
† Yields in metric tons per hectare.

Perhaps the most startling finding of the IRRI was that 4 to 6 tons of rice per hectare could be produced by direct seeding on unpuddled soil. It has always been the practice for the rice farmer and his bullock to puddle the soil thoroughly before planting the rice. It is lengthy, exhausting and unnecessary work. Notice too that rice is not by any means the highest-yielding crop.

The sweet potato, from America, is one of the highest producers on a calories-per-acre basis. Many of the newer varieties are also very rich in β-carotene, a precursor of vitamin A, which is deficient in the diets of many people in the rice world. The tender tips of the vines are greens for humans, and the entire vine is food for livestock. The sweet potato is nutritionally a better food than the Irish potato. Why do we eat Irish potatoes, if we are fully rational beings? The answer is that we aren't; we are culture-bound beings.

The soybean is of great value since it is one of the richest sources of high-quality protein and the diets of much of the underdeveloped world is low in proteins. The soybean averages about 40 percent protein and 20 percent oil. In the Philippines it produces two to three times as much protein per acre per day as any other crop such as beans or peas.

Corn, again an American crop, can be grown every month of the year in the Philippines. It is eaten as we eat sweet corn, though the corn grown there is mostly what we call field corn. Sweet corns are incredibly more tender, juicier, and sweeter than field corn, and that the Philippines are just now getting to know the delights of sweet corn is one more example of the vagaries of diffusion. By some fluke they did not get sweet corn from Mexico and simply grew field corn—for 400 years!

Sorghum is now the third most important food crop in the world, and it is very flexible in its growth requirements, for it can stand either greater drought or greater moisture than corn. It has the great advantage in the tropics that it can be ratooned—cut and then let sprout up again.

By using these crops in sequences, several crops per year can be obtained from the same field. It is estimated that food production can be increased three-to fourfold if full advantage is taken of the long tropical growing season. If increases in crop yields similar to that in rice can be achieved, increases of from 4 to 16 times the present yields are possible on the better lands. All of this, of course, calls for water management to make possible continuous cropping. Data such as those given here do not lead to any great pessimism about the future of the tropics.

Cattle in the tropics: the King Ranch demonstration

The theme being developed is that the utility of land is determined by man, and man's determination is due in part to his technological status and in part to his perception of his opportunities. An excellent case of changing perceptions is found in the King Ranch expansion into the tropics around the world. The King Ranch is well known in America for its vast land holdings in Texas on which has been developed a superior breed of cattle that can withstand the heat and drought of south Texas. Oil and gas have been found on these extensive holdings, and the income thus derived has helped make possible the extension of this cattle kingdom on a worldwide scale. The result has been the utilization of tropical lands, generally thought to be of only marginal use, for the profitable raising of cattle.

The original King Ranch was 1,200,000 acres. One set of heirs took their share, but 860,000 acres remained, to which 110,000 acres have been added in recent years. In Australia the King Ranch holdings amount to 9,754,000 acres. They also have 7000 acres in Africa and 513,000 acres in South America. For this discussion, attention will be centered on the Australian holdings, but everything said about them is equally applicable to any of the tropical areas.

One of the keys to the King Ranch success has been the development of cattle that can withstand great heat, drought, and insect attack and yet can produce good beef. This type of animal was created by crossing the good British beef breeds—largely shorthorn —with the tropical heat- and disease-resistant Brahma cattle. The European breeds are not good for the tropics, and the Brahma is not very good for beef. The result of the cross was the now world-famous Santa Gertrudis breed, and it is thought to be able to withstand any tropical burden and still produce good beef at a high rate of gain per cow per year.

The breed was actually introduced in Australia, Cuba, and Brazil, but Castro aborted the Cuban experiment, and the Australian project has overshadowed all others including the South American holdings. Northern Australia is much like Texas. It is flat, hot, and subject to fitful rainy seasons that make a feast or famine situation for the cattlemen. The King Ranch took control of about 1000 square miles of land in the Barkly Downs area south of the Gulf of Carpentaria, on the dry edge of a tropical area of seasonally wet and dry climate. Two hundred Santa Gertrudis heifers and 73 young bulls were imported, and they were an almost instant success. Bulls sold at once for $2500 apiece. The cattle succeeded in the hot, dry north, withstanding heat, drought, and disease, and they produced steers that weighed 100 pounds more than competing breeds of the same age.

Over 4700 square miles of land near Brunette Downs were then acquired from people who were struggling against drought and heat with inadequate capital and ill-adapted cattle. Wells were drilled—120 of them at a cost of $12,000 each—and $2 million was spent on fencing. Today 50,000 head of cattle graze on the land. More and more properties were acquired by the King Ranch, and 14,000 square miles were brought under their control. Negotiations for an additional 1,300,000 acres in the Kimberlies were then undertaken. Few areas have been judged by geographers and others to present as formidable a challenge as this. There are thus tremendous implications here for the comparable areas of tropical wet and dry lands in Asia, Africa, and South America.

The expansion into the tropical rainy area is of equal importance and interest. The tropical rain forest of the Queensland peninsula had never been considered to be potential cattle country. The rainfall is 168 inches per year on the average, but a wet year may have over 300 inches. The land is covered with a huge forest and filled with vicious animals. The alligators and snakes attack, as do noxious insects; Australia is notable for its poisonous snakes, and their pugnacity adds to the peril. Clearing this land required the use of the largest tractors available in order to bulldoze the huge trees and prepare the ground. Costs were estimated at $30 per acre for clearing the land and setting up the fencing, roads, and housing. It was estimated that a 50,000-acre holding would be required to justify the scale of expense. This was arranged for through the Australian government, which has been highly cooperative throughout, for it was apparent that if the King Ranch projects succeeded, Australia had much to gain.

The mixture of legumes and grasses planted on these tropical wet lands were carefully selected from all over the world: Brazil, New Guinea, Africa. The grass grows 6 feet tall, and when burned over, it sends up new growth at the rate of 3 inches per day. No one knows what the true carrying capacity of such land is, but it is estimated that it will easily carry 30,000 head of cattle. The King Ranch project is demonstrating how the potential of huge areas of the earth—Brazil, as a classic example—can be made a reality.

The King Ranch has been discussed impersonally so far, but it is a highly personal matter. The key man has long been Robert Kleberg. By holding most of the descendants of the original owner together, he has been able to bring about remarkable developments around the world. Today oil wealth is frequently criticized. It is true that oil made possible much of what Kleberg accomplished, but lesser people took their oil depletion allowances and did nothing. The King-Kleberg combination used their wealth in such a way that both they and the world benefited.

The plantation system

Another way of using tropical lands is by means of the plantation system, the application of capital and skilled direction to the production in the tropics of a product destined to enter the world market. Examples of such products are sugar, bananas, coconuts, palm oil, and rice as produced in northern Australia. The origin of this system is often credited to the Portuguese, who produced sugar on the islands off West Africa by taking advantage of the tropical climate, using slave labor, adding their managerial and commercial skills, and profiting by the sale of the product in the European market. The Portuguese carried this system to Brazil, where they produced sugar under these same conditions. Other examples of the development of tropical crops under the plantation system have been the growing of bananas by the United Fruit Company in Central

America, the British and Dutch production of rubber in the East Indies, tea production in India, palm oil production in West Africa, and copra production in Polynesia. In each case the impetus (ideas, capital, and organization) came from outside and put the natural resources and native labor to use to produce a product for the world market.

The plantation system has acquired a bad name in this century because it has been linked with economic imperialism and identified with the use of slave labor. Like any human institution, of course, it can be and has been abused. The United Fruit Company in Central America is often cited as an example by those who dislike big business and plantations. Yet the company today pays high wages in the areas in which it works. It builds railroads and docks, brings about improved sanitation, and increases the national income of the nation within whose territory it operates. A company cannot succeed with sick and ignorant labor and cannot afford to have its skilled management die off. In its own interest it must sanitate its area and build up its labor force: tropical development results from such efforts. Productivity is necessary if tropical peoples are to have the adequate housing, sanitation, medication, and education that are the prerequisites of making these regions the potentially useful regions of the world that they could be.

Since the tropics are natural tree areas, the production of tree crops seems the natural best use of this resource. There are many such crops, and the plantations have seized upon some of them to supply industrial materials, food crops, and textiles. Palms supply many kinds of oils that are useful either as foods or as industrial materials, for example, as the basis for soaps. There are many tropical tree fruits, of which the banana, citrus, and avocado are but a tiny sampling. With today's easy and fast transportation, there is little reason why tropical fruits should not enter world trade. Tropical nuts are already popular, the cashew and Brazil nut among others. There are also many tropical vegetables that come directly from trees, such as leaves that are like spinach and flowers that are like cauliflower. Even textiles are available: not only is Manila hemp a tree crop (a banana), but cotton is naturally a perennial shrubby tree that we have converted into an annual. In the tropics it yields continuously for years.

Modern efforts to use the tropical forests have not been marked by much wisdom in the early stages. When tropical forest crops such as rubber were first grown in plantations in Indonesia, the land was cleared, the trees planted, and the ground kept free of weeds and cultivated just as it would be in a mid-latitude orchard. The result under tropical heat and high-intensity rainfalls was enormous erosion and great loss of fertilizer from leaching down through the soil. Belatedly, it was learned that the proper way to grow a tropical rain-forest tree was to grow it under tropical rain-forest conditions. Today the approach is never to break the forest cover.

A new plantation is set out by going into the forest and making a tiny clearing, perhaps felling one or two trees, and planting the sapling of the desired tree in the spot of light on the forest floor that was so created. With such clearings spaced 100 feet apart, and with gradual cutting back of the adjacent tree as the economically useful trees grow, the transition can be made from a forest of many varieties of trees of little economic value to a forest of trees of high economic value. The transition can be made without ever breaking the continuity of the forest, and the end product is a forest growing under seemingly natural conditions but differing from a natural forest in the great concentration of economically useful trees. It needs no cultivation; it has no erosion risk.

The Ford rubber experiment

The limitations of tropical landforms—and their characteristic soils, plants, and climate—can be illustrated by the experience of the Ford Motor Company in its attempts to produce rubber in the Amazon some years ago. The Ford Company was trying at that time to establish its own supply of rubber, hence to be free of the virtual monopoly in the East Indies under British and Dutch control. What was more natural that to turn to Brazil, the home of the rubber tree?

Brazil was glad to cooperate. The Amazon basin had become virtually nonproductive after the British and the Dutch rubber plantations in the Far East had driven Brazilian wild rubber out of the world market with their cheaper and higher quality production. If Ford wished to direct some of the millions of dollars in the rubber business into Brazil, its government could only be delighted.

Ford was therefore given almost carte blanche in choosing land in the Amazon basin. Its first mistake was to choose a tract of gently rolling, nearly flat upland back off the floodplain of the river. Under tropical weathering conditions, lands that are level and not subject either to erosion—which would expose fresh rock material—or to deposition by flood waters—which would add fresh alluvial material—will be steadily impoverished by the weathering process. These flat, forest-covered, uneroded uplands—free from river deposition—had been subject to the processes of tropical weathering for geological lengths of time. Such a site was guaranteed to produce poor soils.

The second mistake in the experiment was made when large areas of land were cleared, the wood burned, and the loosened and torn-up soil exposed to the tropical downpours and the strong tropical sun. The downpours started immense erosion; the hot tropical sun, beating on the soil that had for millennia been shaded by layer upon layer of forest trees, oxidized the remaining organic materials. The temperature of the soil was raised to heights so much greater than normal that the micro flora and fauna of the soil were virtually exterminated.

The third mistake was to plant in this desert—bare soil and blazing sun—seedling trees adapted to making their start under the shaded, moist, hothouse-like atmosphere of a rain forest. It was like planting a hothouse plant in the Sahara. The result was large losses of the transplanted trees.

The plantation struggled along with poor soil, erosion, soil impoverishment, and seedling losses. When the mistakes were finally realized, it was decided to start afresh. The Brazilian government remained cooperative. A new tract nearer the river mouth and on better soils was selected, and the trees were set out in forest clearings. But further difficulties developed. The trees that yielded such large amounts of rubber in the Far East, because they had been carefully selected for high yield by the Dutch and British, suffered tremendously from diseases when they were reintroduced into their homeland. Why should this have occurred?

The original trees shipped to the Far East had been seedlings, and the seed had been carried to Great Britain, where they were started in Kew Gardens. The diseases normal to the rubber tree in its homeland, and to which the trees had a high resistance, were all left behind when the trees were then sent on out to the Far East. The selection of trees over the succeeding years was for high yields, and since there was no disease load, there was no selection for the continuation of resistance to the plant diseases of Brazil. As a result, the trees were generally no longer immune to resistant. When then they were reintroduced into the environment where the diseases that had been associated with rubber trees for millions of years were very much a part of the scene, these nonresistant strains were rapidly struck down.

It would seem here that the physical environment was at least determining something. Man had to use his ingenuity now. The quality desired was the ability of the rubber tree to yield a large amount of latex from the trunk of the tree. The disease attack concentrated on the roots and the leaves. The solution, therefore, was to grow a native tree, graft onto the native root a high-yielding strain, and when this strain had produced a stem capable of carrying a graft, to graft onto it a top of one of the leaf-blight-resistant native strains. This tripartite plant then had a disease-resistant root, a highly productive trunk, and a disease-resistant top. Man had thus manufactured a plant to meet the disease problem in this particular environment. He could have done it by other means too, but selecting and breeding high-yielding strains of the naturally resistant native trees would have taken longer.

Yet, after overcoming all the problems of soil, disease, and climate, the Ford plantations still failed; the causes were social, economic, and technological. There is almost no labor supply in the Amazon: population density is extremely low, and the region has an evil reputation. Under boom conditions in the past, manpower had been dragooned into the area and cruelly exploited, and the loss of life under conditions of inadequate sanitation, housing, medical care, and income had been immense. Ford wanted to avoid such mistakes. Good housing, pure water, medical care, and adequate food at moderate cost, combined with wages well above national averages, were counted on to attract a labor force, but to no avail. Workers were reluctant to enter the Amazon, a death trap for labor forces of the past. Perhaps given a longer time, the successful operation of the Ford plantations would have convinced workers of the superior living conditions provided there, and there might have developed the type of stable labor force that a plantation requires. But a technological revolution intervened.

Synthetic rubber was brought into production at cost levels that forced natural rubber to face increasingly steep competition, with the probable eventual elimination of natural rubber as the major source of supply. At this point, the Ford Company cut its losses. After decades of effort during which they had learned a great deal and had invested many millions of dollars, Ford turned its plantations over to the Brazilian government and withdrew from its experiment in tropical agriculture. It would be very easy to use this as an example of the impossibility of accomplishing anything in the tropics; however, the errors made and the great costs and delays that resulted could have occurred in any other region. The technological change that led to the final decision to abandon the effort could equally have occurred in many another field. Indeed, another example is the former Dutch monopoly in quinine, produced from the cinchona tree (also a native of America taken to Southeast Asia and there improved), which was virtually destroyed by the production of a synthetic malaria drug that was as good as quinine in suppressing the symptoms of malaria and had fewer distressing side effects.

Change and displacement due to technological advance are not limited to impacts on tropical agriculture, though it may seem so because of the importance at present of plantation agriculture in the tropics. The steamboat replaced the sailboat; the auto, the horse and buggy. As these examples show, there are many ways to use a tropical rainy region. Most of them could have been used by any men anywhere if it is assumed that mankind is alike, all men are highly inventive, and all men maximize their use of their resources. These examples, especially those illustrating the high potential of tropical lands, should be kept in mind while reading of the various tropical lands and men's ways of life in them.

CULTURE AND CIVILIZATION IN THE TROPICS

Clearly, there is a considerable variety of wet to semi-wet tropical lands. The problem of man in the tropics has to be discussed with this variety in mind. If the wet tropics are defined as having at least a mean monthly temperature of 65° F. and sufficient rainfall for agriculture without irrigation, the hot wet climates have an area of 14,500,000 miles (see Table 3-2).

TABLE 3-2 The Extent of the Wet Tropics

Areas of Hot Wet Climate	Square Miles
Asia and the East Indies	3,000,000
Melanesia, Oceania, Australia	750,000
Africa	5,750,000
America	5,000,000

In the main there are thinly populated areas. In America, the average density of population is about 2 per square mile; in Africa, between 2 and 3; and in New Guinea, less than 1. Average figures can be deceptive, of course, for areas like Central America have spots of dense population in the midst of large empty regions. These are static figures, for the American tropics probably had much denser figures in the past. Yet it remains true that these hot wet regions are virtually empty in America and are only thinly populated in Africa. Outside Asia these regions have only 8 percent of the world's population on 12 percent of the earth's land surface, and generally speaking, the people are food gatherers or simple cultivators.

In the hot wet parts of Asia, on the other hand, one quarter of all mankind is crowded onto 8 percent of the land surface of the earth. Despite this crowding, these people are higher on the scale of civilization than those in the more thinly occupied parts of the tropics. At first glance, it would seem that the humid tropics need vigorous population growth. However, it is commonly held that the tropics are unhealthy—too unhealthy to support large populations.

Physiology of Man in the Tropics

One of the commonest beliefs is that white men cannot live productive lives in the tropics. This implies that there is a direct effect of the hot wet climate on man and that it affects the white race in particular. The usual implication is that both general health and mental activity are influenced. A brief review of man's physiology with special reference to heat seems the proper way to approach this problem, and it might be noted first of all that we have no data showing any significant differences between the various races in their reaction to heat and humidity.

Life processes are dependent on chemical changes; the human body burns food to produce energy. A fair analogy: the internal combustion engine in automobiles: the heat produced must be dissipated, or man rapidly runs into difficulties. The general term for this consumption of energy by the body is metabolism, and basal metabolism is that energy needed for bare existence—the working of the heart, lungs, and so forth—even when the body is at rest. Some energy is used even in digestion, but man uses the greatest amount when he works. Ninety percent of this energy appears as heat.

Man must normally maintain body temperature at about 98.6° F., for increasingly great variation from this is accompanied by discomfort, distress, and eventually death. Hence, when men work, they must lose large amounts of heat to the surroundings. If they are unable to do so, physiological failure will result just as surely as an automotive failure if the engine's cooling system fails. It is common experience that autos overheat more often when working hard, as on mountain grades or at high speeds in hot weather. We can expect the human body to react similarly, so we must first consider the human cooling system.

Heat loss can be looked at in terms of simple physics. Heat can be lost or gained by radiation. The warmth of a stove can be felt without touching it, for it is warmer than a human body and produces heat. If the stove was filled with ice, a hand could lose heat even without touching it—a process independent of the temperature of the air. Sitting in a room filled with warm air, but whose walls were cold, would be a very chilly experience because of the radiation heat loss to the walls, a not uncommon experience in winter in mid-latitude houses that are not sufficiently insulated.

Sunlight also acts like radiant heat. Its effect is quite complex because the different parts of the light (ultraviolet to infrared) have different qualities for penetrating the layer of the atmosphere or the skin, for reflection and absorption in relation to colors, and so forth. Solar radiation is heating, and in the tropics the nearly vertical sun makes its effects intense. Added to the already warm and humid air, solar radiation increases the heat being added to the body, which already has the problem of disposing of the heat that it is generating. Wide roof overhangs, large hats, and parasols are all sensible cultural reactions to this heat load. How many people native to the tropics had them? Relatively few.

Heat may also be lost by conduction and convection. Heat is carried by the blood from deep body sources to the skin; the warm skin can then lose heat by contact with the surrounding air, provided that the air temperature is below the skin temperature. Since skin temperature is lower than body temperature when the air temperature is about 95° F., this cooling process ceases, and at higher temperatures the body is heated rather than cooled. Conduction is not a very efficient cooling process. The heat transfer is from the skin to a thin layer of air in contact with the skin, which soon becomes skin temperature. Air movement facilitates the heat transfer by bringing new air in contact with the skin, accounting for the cooling effect of breezes and fans. If the air temperature was higher than the skin temperature, however, the effect of a breeze would be to increase the heat added to the body. But in most of the tropics the average monthly temperature is well below 90° F., and the highest temperatures ever recorded, at Panama and Singapore, are well below those annually experienced at such places as St. Louis, Chicago, and Baltimore. This is only part of the story, however.

Our most effective cooling system is based on evaporation, heat being dissipated by using it to evaporate water. In dry regions simple air conditioners work on this principle. Since bodies lose moisture through the skin and lungs, and the temperature of these areas is fairly constant, the rate at which they can lose moisture is most nearly measured by the absolute amount of moisture in the surrounding air. This process, like the conduction loss, goes from the skin surface to the layer of air in contact with the skin. Rapid replacement of this contact layer of air ensures a new absorbent layer to replace the saturated layer. Breeze, then, greatly increases the efficiency of this system, and the drier the air the more moisture it can absorb and the more efficiently the evaporative cooling system can operate.

The body is equipped to use these methods. Blood volume, the water content of the blood, and the amount of blood sent to the surface areas are all increased as the heat load on the body increases. This process brings more moisture to the skin surface, where moisture loss occurs in two ways. There is simple evaporation from the moist surface, or perspiration. In addition, man—though not all animals—has special glands that can greatly increase the amount of water brought to the surface of the skin to evaporate: the sweat glands. The lungs are also areas of water loss from the body, hence a cooling zone. Under warm conditions, people breathe a little more deeply, thus increasing the ventilation in their lungs. This also tends to increase the oxygen content of the blood, an important function, for the diversion of blood to the skin areas can decrease blood flow to such vital areas as the brain. Even a slight decrease in oxygen supply there will result in discomfort, lassitude, and headache and, if continued, could lead to greater distress.

Man reacts automatically to heat load. He does not have to think about his blood volume or perspiration, and even deeper breathing is automatic. Just as he curls up to keep warm when cold, he stretches to cool off: the principle involved is that of increasing or decreasing the area from which heat will be lost. Behavior changes too; people rush around in the cold and slow their movements in the heat. Can we say, then, that tropical people tend to move slowly and are less vigorous, or would it be more apt to say that they are simply making a normal efficiency adjustment? Observers report that tropical workers move more slowly but more efficiently. They get the same amount of work done with less energy output, and hence less heat is generated.

This raises the problem of how to evaluate the degree of strain imposed by heat and humidity. Stretching out, slowing down, breathing deeply, sweating, and eating less (man needs less food for energy for heating and gains a radiation advantage by being lean) are all perfectly natural adjustments. By sweating, the body can be kept cool for long periods even under the extreme conditions to be found in the wet tropics, just as long as the water intake is kept up—and water is available in the wet tropics. Only when these changes have failed to meet the heat load requirements can we say that a strain has been put upon man's physical makeup.

If the body cannot lose sufficient heat, a stress is set up. If too much dilation of the capillaries of the skin causes a lowered blood pressure, the circulation will be impaired, and a deficiency of oxygen supply can occur. The first area to feel this is the brain, where a slight change can produce a large effect. Judgment slips, lassitude occurs, then headache, then nausea, and finally unconsciousness—and, rarely, death. The milder forms of this—inertia, irritability, mental fatigue—are the things that tropical people experience. But mid-latitude summers are as hot and humid, or hotter and as humid, as weather in the tropics.

Physiological climatologists state that under conditions likely to be found in the shade in the tropics, the capacity of healthy acclimatized people is not likely to be strained. They report also that troops in the tropics with the heaviest and most constant work to do are the fittest and most contented so long as they feel that their work is of direct military importance, which brings us back again to the matter of attitudes discussed in Chapter 1. Experiments and experience have indicated that attitude often determines comfort or discomfort. Military units in tropical areas that showed "tropical deterioration" also showed signs of neurasthenia. The men complained of backaches, loss of weight, sleeplessness, and the like, yet physical examination showed nothing physiologically wrong. Doctors' offices in all climates are filled with such cases.

The wet tropics have their problems. They have many times the number of plant and animal forms that the mid-latitudes have. These include plants and animals that attack man and man's works. There are fungi, bacteria, parasitic worms, and termites in profusion. The constant sweating often causes prickly heat, which leads to scratching and possible infection.

However, the main idea developed here is that when we look at man's physiology in relation to heat, we find that man is well equipped to deal with the kind of heat load to be expected in the humid tropics. Since the tropics are the probable area of man's origin, this seems reasonable. We must look, then, at the cultural record with these facts in mind: man should be well adapted to heat and humidity.

White Settlement in the Tropics

The problem of the white man in the tropics may be studied by seeing what happened to some of his settlements there. What we learn is probably applicable to all of mankind, for we have no evidence of large differences between the major races in climatic responses. After 1500 Europeans spread over the world, but the way in which they settled differed from place to place. In Asia relatively few Europeans settled permanently. They came as traders, managers, and civil servants. The great masses of people already there supplied the labor force. In India in 1921 there were about 160,000 Europeans and over 300 million Indians. In the Netherlands East Indies there were less than 250,000 Europeans and over 60 million Indonesians.

This lack of large-scale, permanent settlement on the part of Europeans has often been explained as an effect of the tropical climate. It is true that a number of the Europeans who went to these areas began to drink to excess and to live with native women. Single men sent to outposts remote from their own society found solace in such means of "escape," no matter what the climate was. This phenomenon was more a matter of personal maladjustment, of course, than it was of climate, and to some extent resulted from a defeatist attitude about the tropics. The European whose life had been lived in cooler climates felt that the intense solar radiation might damage his central nervous system; he therefore never stirred without his pith helmet. Some even wore a strip of red flannel down their backs to protect the spinal column from excessive radiation. Everyone "knew" the climate was "bad." Everyone "had" to go home for six-month vacations to stave off "tropical deterioration."

Many missionary groups provided the contrast to this point of view. They lived in the hot wet tropics with their wives and children and took only infrequent, short home leaves, and they suffered no apparent tropical deterioration. Their motivation was sufficient to help them adapt to their new environment.

In the New World tropical settlement took a quite different course. For various reasons (new diseases, warfare, political and economic disruptions), the native populations were virtually swept away in many areas, many of which had been only moderately to thinly populated to begin with. At times Europeans settled on the land, but in part the needed labor supply was imported from Africa. As the New World recovered in strength, there sprang up a great mixture of races. Was this due to climate?

The West Indies as a Test Area in History

The West Indies are like a laboratory experiment, with varied races placed in isolated spots scattered over a tropical sea. Temperatures within this area are quite uniform. Note that most of the islands are close to the 75° isotherm for January. The whole Caribbean averages 80° in July. This uniformly warm humid climate with small annual temperature range is the typical tropical situation. The islands are also reasonably alike, being mostly volcanic peaks with limestone fringes derived from anciently uplifted coral reefs. Into this area the Europeans spread confusion and disease, and the native populations melted away. Then Europeans of various nationalities and Africans occupied the islands in various mixes, and they have remained there for 300 to 400 years. There has been ample time for tropical deterioration, if it acts with the speed usually attributed to it, or if it acts at all.

British colonial Barbados

By 1502 the Spanish had 12,000 Europeans on Haiti; later the British population on Barbados num-

bered 40,000, or 200 per square mile. Land was held in plots of 5 to 30 acres and worked by white men with white helpers. This was a family settlement of the land, and it flourished. After 1642 due to civil war in England —*not* a tropical climatic factor—immigration ceased. Thereafter the North American colonies were more attractive and an inflow of "undesirables" began in Barbados.

The new immigrants were white servants, indentured men, kidnapped men, and political prisoners from the Puritan Rebellion, and they were all treated worse than slaves. After all, an African slave cost his owner money and worked for him as long as he lived. An indentured man or political prisoner served out his time and had to be freed. His contract often called for giving him clothing and a set of tools at the end of his term of service. Financially, it was desirable to maintain the African slave, but to be less concerned with the fate of the European servant. In view of these conditions, the European working class left the islands, and more and more Africans were brought in. But what role did race and climate have to play in this development?

Actually the list of cultural forces at work is a long one. All of them worked against the small European landholder. High taxes hampered him, and wars ruined the economy of the whole islands, at the same

MAP 3-1   The Caribbean:
January Average Temperature
The Caribbean is shown with isotherms for the average temperature in January. All of the Caribbean area is warmer than the southernmost tip of Florida. Most of the Caribbean islands have January temperatures close to 75° F. This is the average summer temperature of the belt of states from Maryland to Missouri. Panama, with 80° F. temperature throughout the year, is comparable to summer in the Gulf area from Louisiana to Florida.

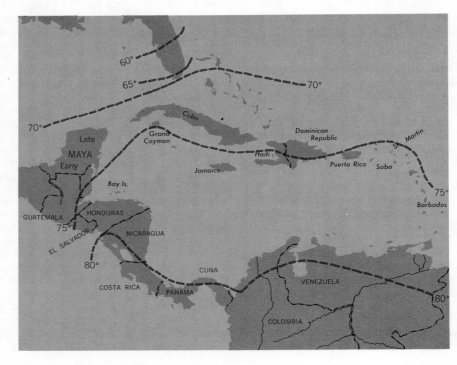

time creating such opportunities in buccaneering and pirating as to drain off much of the active manpower. Only large planters survived, and absentee ownership of large estates became common. Taxes, navigation acts, all patronage in the hands of the kings' ministers, troops quartered on the people—all of these things combined to drive the white settler off the land and out of the islands. None of these are physical-geographical forces or peculiar to the tropics.

Further, it cannot be said that the British settlers lived a wholesome life. Their houses were poor, damp, and infested with vermin. They wore heavy British clothing, well suited to the damp, cool British Isles but not at all suited to the humid tropics. They ate heavy food in large quantities, and since they did not need so large a food intake, they added fat, thus decreasing their ability to lose heat. They also drank heavily of the available liquor, which was rum, a by-product of the sugar cane production, the major industry of the island. This raw rum was near to poison. As a result, the death rate ran 10 men to every woman. There is no climatic-physiological basis for such a difference; the women simply drank very little, as was the accepted cultural pattern. A clear example of the forces at work is found in military statistics. In 1796 nearly 20,000 men were sent to the British West Indies; 17,000 died within five years! One could make a devastating attack on the tropics with such figures. However, look at the cultural setting. The men had been recruited from the poorly nourished slums of the British Isles and therefore represented poor physical stock. No attention was paid to hygiene and diet; they ate Old World rations—salt meat five times a week—and drank as much raw rum as they could lay hands on. The men lived in crowded quarters, ideal for spreading disease, and they wore the traditional heavy European uniform, bathing rarely. Would it have mattered much if they were in Florida, Washington, or Boston?

That none of this was inevitable is suggested by a description of the planter aristocracy in the British West Indies at the end of the eighteenth century. Significantly, the people were said to be taller and more slender than their British ancestors. The Jamaican historian who gave us this description knew no physiological climatology, yet he described a change toward the slender form of greater advantage for heat dissipation. This is reported in Australia in the twentieth century, and it is probably a nongenetic change. The body is fairly plastic and can respond to some degree to environmental forces, as we see in our society today where children are normally taller than their parents due to improved nutritional standards, especially in infant feeding.

These planters were also said to be graceful and agile men, less liable to local infectious diseases than newcomers. Their women lived calm, even lives, but as a result of little exercise they were also pale, spiritless, and languid. Their children were said to develop early, especially in mental ability, but to fail to keep pace later, a statement that turns out to be more myth than observation whenever closely studied. In fact, there is little in this contemporary account that indicates evidence of any climatic effect. Except for some desirable minor shifts in body form, the tropical environment had left these aristocratic planters untouched after two to three generations of living in the tropics.

Today Barbados has one of the greatest population densities of any place on earth. In 1971 there were 1548 people per square mile compared with a United States figure of about 50. The population growth rate is high even though the birth rate is not high for the region and thus indicates that health standards are high and death rates low. One of the surprising things about Barbados is that it was long a malaria-free island in a sea of infection. The physical environment provided optimum conditions for the establishment of malaria, for the winds, humidity, rainfall, and topography are all suitable. The western and southern part of the island have low-lying land with brackish water ideal for mosquito breeding, and there are other wet areas elsewhere on the island. After a period of much emigration by labor to work on the Panama Canal and the Cuban sugar plantations, a flood of workers returning home in the 1920s brought typhoid fever and malaria. The problem of the mosquito carrier, technically the vector, on Barbados is little understood. Probably no malaria carrier had been on the island previously because mosquitos cannot fly far, and the present vector, *Anopheles albimanus,* the principal malaria vector in the Caribbean, was introduced about this time by coastal schooners. The mosquito cannot fly the distance from adjacent islands and is unlikely to arrive in an open boat, but it can be carried in the hold of a closed boat. One mosquito will not do, of course. A considerable number must arrive at the same time and escape to an adjacent breeding area. All of this must occur relatively quickly —entering the dark hold of a ship, being transported to the island, and escaping to a breeding area—for the life of a mosquito is not long. These requirements seemingly account for the long delay in mosquitoes' reaching Barbados. Once in Barbados, the mosquitoes bred in the coastal marshes, flew up to one-half mile to settled areas, and spread malaria. By 1927 a virulent outbreak was under way. Once it was recognized, it was attacked head on with campaigns to control the breeding places, and by the end of the second dry season the malaria had been wiped out through control of the mosquito. The stage had been set for a long time, for there were coastal marshes, ponds, and other areas of stagnant water in and around fields, and for a period of years men had been returning from other areas loaded with malaria in the erythrocytes in their blood. All that was lacking was a mosquito population that would transmit the infection from one man to another. When *Anopheles albimanus* was introduced, probably by coastal schooners just before 1927, the circle was closed: environment, disease, vector, and man—and an epidemic resulted. Note, however, that with modern public health measuers in a well-organized community, the cycle was promptly broken and Barbados reverted to its malaria-free status. This provides us with an excellent example of the complexity of so simple a thing as a malaria-free condition being broken and reestablished, with the roles of culture in carrying and then controlling the disease.

Puerto Rico

Puerto Rico is a spectacular example of the inertness of the physical environment and the dominance of cultural factors at work on the landscape. Puerto Rico is a large island measuring roughly 100 by 35 miles, making it just a little smaller than Hawaii, the largest island of the Hawaiian group. There is a coastal plain on the north side of Puerto Rico, backed by limestone hills with solution valleys. Behind this lies the deeply dissected central plateau with elevations of 1500 to 2500 feet, and behind this the mountainous backbone of the island lies close to the south coast, where at times 3000-foot mountains are within 8 to 15 miles of the coast. Rainfall varies greatly, from less than 20 inches on the coast to over 120 inches on the mountain tops. Twenty inches of rainfall in the tropics is slight, for evaporation is high; the result is an arid landscape. Aridity is increased by the limestone, for water percolates down through this material and in time makes great solution holes and underground channels so that surface water tends to disappear underground. The temperatures are moderate in summer and winter: winter 75° F. and summer 80° F., with temperatures over 90° F. seldom recorded. The trade winds blow most of the year, and the steady breeze further diminishes the effect of the heat. Hurricanes coming principally in the fall are the most destructive natural force in the environment.

Into this idyllic setting the American Indians moved at some early time, and when Columbus arrived, the islands of the Caribbean were thickly settled by a farming and fishing people. Population estimates vary, but the figure of one million people living on the island of Hispaniola at the time of the Spanish discovery is supported by C. O. Sauer and gives an idea of the dense settlement in pre-Columbian times. The Indians cultivated primarily by heaping up the ground into mounds in which they planted a wide variety of crops: manioc, peanuts, maize, beans, squashes, sweet potatoes, and several others. It was reported that 20 men working 6 hours a day for one month could plant crops sufficient to provide bread for 300 persons for two years. Sauer has commented:

The people suffered no want. They took care of their plantings, were dexterous at fishing and bold canoeists and swimmers. They designed attractive houses and kept them clean. They found aesthetic expression in woodworking. They had leisure to enjoy diversion in ballgames, dances, and music. They lived in peace and amity.[2]

The Spaniards burst into this situation seeking gold. They found little in the placers (deposits of gold in the gravels of the streams) and soon exhausted the supply. Their impact on the natives was immensely destructive. They interrupted the native life, introduced new diseases, and soon the native population dwindled to some small fraction of the original number. The Spaniards destroyed the most important resource on the island, for without manpower there was very little that could be done with the land. They moved on to the mainland, where the conquest of the great empires in Mexico and Peru offered fortunes in gold and silver and lands still thickly populated with Indians. Only a thin sprinkling of Spaniards remained in the islands of the Caribbean; Negro slaves were soon brought into Puerto Rico as laborers, and the result was a Negro-Spanish-Indian mixture. Later Portuguese, English, and American settlers came, especially after 1815, when the island was opened to commerce with the ending of the Spanish trade restrictions. During the revolution for independence from Spain early in the nineteenth century, many wealthy Spanish families fled the newly created countries on the mainland to settle in Puerto Rico. They brought education, enterprise, and capital.

The development of the island was almost purely agricultural. In 1845 Negroes and whites were almost equal in numbers; since then, the number of Negroes fell steadily: 1802, 52 percent; 1860, 48 percent; 1920, 27 percent; and by 1971 Puerto Rico was grouped with the Latin American nations that are about three-fourths white. These figures probably reflect a shifting race classification, but they surely do not indicate any failure of the white race in the tropics.

[2] C.O. Sauer, *The Early Spanish Main* (Berkeley: University of California Press, 1966), p. 69.

(Top) **Oil Refineries in Puerto Rico.** *This is a classic example of industry versus environment. The waterfront is despoiled, beaches are ruined, and the air is polluted. Puerto Rico has found the refineries not to be big employers, and they are destructive of the amenities that attract tourists.* (Clarissa Kimber)

(Middle) **San Juan, Puerto Rico.** *This modern metropolis has an air of prosperity like that of Florida or California.* (Clarissa Kimber)

(Bottom) **Puerto Rican Landscape.** *This open-parklike landscape represents old fields and a moderately humid environment. The result is an open land with trees along the water courses. This is very much like the coast range country in central California.* (Clarissa Kimber)

The population increased 50 percent from 1900 reaching 1,500,000 in 1930. In 1960 it reached 2,350,000, and by 1973 it increased to 2,875,000. Birth and death rates are high, but both are falling now. Tuberculosis, malaria, and hookworm are prevalent, and schistomaisis is present in all the streams. The diet is poor in rural areas, primarily dried beans, corn, salt fish, and polished rice. Under United States influence the sugar economy was intensified to virtually a monoculture, which led to seasonal unemployment and undesirable corporate relationships with too much wire-pulling and interference with government. This rich tropical land had to import food.

In the first half of the twentieth century Puerto Rico was known as a "slum of the Caribbean." The explanation was simple: it had no resources. Its economy was primitive, and its population mostly illiterate and impoverished. Yet in the 1970s the island has a rising living standard. One-third of the entire population is going to school, the literacy rate and the average personal income are the second highest in Latin America, and a middle class is developing. The climate, the physical geography, and the geographic location both physically and in relation to the shifts of world power and development have not changed—the social outlook has.

Sugar cane, introduced in 1515, rapidly became the principal crop and has occupied the best of the land. Ownership was concentrated in large holdings, and the mass of the agricultural population was relegated to the poor and primitive uplands. The tendency recently has been for the acreage of sugar lands to decline and for the people to flow off the land into the cities. In viewing this scene in 1940, a soils man com-

mented that the land was seriously crowded, and with a population density of 500 per square mile, the population already exceeded any possible agricultural base. What, he asked, are we to do? Divide the sugar lands into small farms and turn to subsistence crops? He pointed out that sugar-cane land produced the maximum value per acre and that the wages earned by the laborers on the land exceeded the value of the food that could be grown on the land. He considered immigration to the United States, noting that it was already considerable, but still less than the population growth. Concerning industrial growth, he pointed out that the only resource was an abundance of manpower: no fuel, no raw materials.

Puerto Rico has an advantageous political status today. As a self-governing commonwealth associated with the United States, its people are American citizens, but do not have to pay United States income taxes. The island has embarked on a plan to attract private industry and capital, including a 10- to 17-year exemption from local taxes for desirable firms and the possibility of government financial aid. The tax impetus, combined with a stable government, a location within the United States political orbit, and a large supply of low-priced labor has drawn large amounts of capital to Puerto Rico.

In 1962, just as the industrial surge began to slow down, the petrochemical industry moved in. Hundreds of millions of dollars were invested in immense refineries able to handle crude petroleum and turn out basic materials for industries utilizing synthetic fibers, resins, plastics, insecticides, and surface coatings. Since that time refining has mushroomed. The number of jobs has been disappointing, however, for these installations are highly automated. Worse, they are sources of pollution of all kinds. They mar the landscape, their stench offends even insensitive noses, and the wastes discharged into the sea from the plants and from the oil tankers are fouling the beaches that are major attractions for the lucrative tourist business. Puerto Rico now faces difficult decisions. Shall it be a Belgium—a factory-dominated land—or shall it be a lovely tropical island with clear air and pure water? If Puerto Rico restricts the factories, where will it get its income? Can tourism alone support the people? Can tourism and industry coexist? Because industry came to the island for cheap labor and unequalled tax advantages, it would be easy to discourage business or at least certain kinds of business.

There is already a considerable ebb and flow of businesses. Some businesses take their tax-free years and then leave, but more stay. For example, in 1970 plant closings cost Puerto Rico 4600 jobs, but 43 new plants arrived, representing an investment of $300 million and an employment potential of 7000 new jobs.

The tourist business is now so important in the Caribbean—to Puerto Rico in particular—that it seems likely that the island governments will begin to limit industries that are environmentally destructive. Moreover, political unrest in the Caribbean has tended to focus on tourism in Puerto Rico. Air travel, modern hotels, proximity of a large reservoir of affluent people in the United States, combined with Puerto Rico's political association with the United States, has made tourist business a major source of income. Agriculture that was once the major source of income has been declining in relative importance. The island has shifted away from sugar, the crop that long dominated the economy. Some of the land that was taken out of sugar production is used for dairying. Industry obviously supports a great many more people per acre than sugar does, and it pays much higher wages, which allow a shift to higher-value foods such as dairy products. Dairying is a high-value-per-acre use and a high producer in tropical lands where the land is never out of production because of winter conditions. Coffee, which had been an inefficient producer, has been upgraded. Yields in Puerto Rico had been around 200 pounds per acre, as opposed to an expected 1000 pounds per acre from good trees on well-managed land, and a maximum potential is estimated to be 2000 pounds per acre. New crops are also being added. Pineapples, which seem indelibly associated with Hawaii in the American mind, are at home in the Caribbean and becoming an increasingly important commercial crop. Improving agriculture, burgeoning tourism, and explosive industrial growth make Puerto Rico the success story of Latin America.

The results are spectacular. San Juan is described as giving a feeling of affluence like Florida or California. Supermarkets, shopping centers, and housing developments for workers have been built. These are measures of the growth of masses of people who can afford to buy what we consider the ordinary necessities of life, but which for most of underdeveloped Latin America have been unattainable luxuries.

This has been a planned development. The kind of factory attracted and supported has been carefully screened to suit Puerto Rico's characteristics of few raw materials and a large unskilled labor force. The aim is to get away from monoculture: sugar, sugar, and more sugar, with a little tobacco and coffee. Clothing industries and electrical and electronic industries have been attracted, and the number of skilled workers has increased. Until heavy industry moved in, the Puerto Ricans were combining economic growth with a tropical paradise.

As income has risen, food consumption has gone up. Puerto Ricans now eat twice as much meat and dairy products as they did 25 years ago. The government is pushing farm production in an attempt to make the island much more productive of fish, grain, poultry, and meat. As the growth picks up headway, as education spreads out and incomes rise, there are increasing opportunities for economic enterprise.

All is not easy and rosy, of course. Change and growth and development are never accomplished without pain. Unemployment is still high; new factories have not yet soaked up the pool of labor. The factories tend to draw the villager into the big city. Here the often unskilled and illiterate poor form large slums. Some of the unemployed have flooded into such centers as New York City, where their low standards of literacy, hygiene, and differing customs have created problems. Nevertheless, the Puerto Ricans in New York are only repeating the experience of many another minority group that has come to our great eastern cities for economic opportunity and gained education and technical knowledge.

There has now begun a return flow of these people to Puerto Rico, bringing more new outlooks and abilities into their island. With the current program of schooling for everyone, a stable government aiming at attracting private capital, and governmental aid in the growth and development of health, education, and industry, Puerto Rico is experiencing a revolutionary development from the "slum of the Caribbean" to the "showcase of the Caribbean." The climate has changed, it is true, but it is the *social* climate, not the physical climate.

Puerto Rico's present economic status still leaves it poorer than any state in the Union, and the economic situation is not expected to overcome that level before 1990. Thus a revolutionary rate of change, starting from an advanced status (not a tribal situation) and aided by political and economic ties with the United States, is expected to take at least 50 years to achieve a bare United States minimum. This may be very close to the optimum rate of change achievable. It is significant that the changes here have been carried out under a program of governmental control with the encouragement of private capital and enterprise, and that under these conditions a good supply of private capital has come forward. For our inquiry into the future of the tropics and of the white man in the tropics, it is apparent that the races and race mixtures in Puerto Rico seem capable of awakening from centuries of "tropical lassitude" to "tropical energy" when given political and economic freedom and opportunity. The physical-geographical potential has always been there. It has been the economic, political, and historical forces that have varied, and the use of physical resources has changed accordingly.

Hawaii: a physical parallel to Puerto Rico

In 1940 when the United States assisted Puerto Rico with its soil survey, the Americans commented on the parallel to Hawaii: islands in the same latitude, in the Northern Hemisphere, similar climates, similar lack of basic fuels and minerals, and overspecialization on sugar. The Hawaiian Islands have a total area of 6451 square miles, which is one and one-half times the size of Puerto Rico. The islands are volcanic cones fringed with coral limestone, and the great elevation of some of the cones, up to 13,825 feet, assures high rainfalls and great variety of climate. Lee slopes are arid, windward slopes are wet, ranging from 20 to 600 inches of rain per year. This gives lush rain forest in contrast to arid steppe environments within a few miles, but this is true of Puerto Rico also.

When the Hawaiian Islands were discovered by Captain Cook in 1778, Europeans already had been living in the Caribbean for over two centuries. The islands were then densely populated by a people living much the same way of life as the native people of the Caribbean. They had a varied agriculture of root crops and tree crops and were adept at harvesting the seafoods of both the reefs and the deep sea. The impact of whalers, traders, and missionaries had the same sad result on Hawaii as the Spanish in the Caribbean. Disruption of the native way of life, combined with the introduction of new diseases, caused a melting away of the population. Measles, a mild childhood disease for us, was as deadly to them as smallpox is to us.

When development of the islands for sugar and pineapples created a need for a labor supply, one of the most varied assemblages of mankind ever seen in the world was created as people immigrated to Hawaii:

1876–1897	46,000 Chinese
1876–1913	17,500 Portuguese
1876–1900	180,000 Japanese
1901	6,000 Puerto Ricans
1904–1905	8,000 Koreans
1907–1913	8,000 Spaniards
1907–1950	130,000 Filipinos

The results are a vast conglomeration of racial blocks and racial mixtures, with the Japanese standing out for their resistance to intermarriage with other groups Nearly 50 percent of the population is of Oriental ancestry, and pure-blooded Polynesians now probably number less than 10,000. Hawaiians like to say that

Agriculture in Hawaii. *Here polyethylene mulch is being laid. Under the impervious sheet are soil fumigants and fertilizer. Weeds are suppressed, and moisture is retained. Pineapple slips are inserted at each white mark. In its large capital outlays, advanced technology, and large-scale operations, Hawaiian agriculture fits most definitions of plantation agriculture. The use of a mulch sheet in this relatively dry part of Hawaii is a reminder that not all of the tropics are naturally rain forests.* (Dole Photo)

beds that compose the mountains and is then tapped to supply the city. The outer islands (outside Oahu) are quite thinly settled.

The economy of the islands is dominated by tourism, construction, and the armed forces, followed by sugar and pineapples. Twenty-eight plantations controlling 225,000 acres produce the sugar. This industry was a major reason for importing foreign labor, and it still employs nearly 50,000 people despite mechanization of some of the work. The armed forces (military plus civilians) support 150,000.

Sugar has long been the largest crop in acreage and is fairly constant at 225,000 acres. Pineapples have been decreasing in acreage—now at 58,000 acres, down from 75,000 acres in 1960. Tropical fruits and flowers are grown, and some of the flowers are air freighted to the mainland. One million acres, three-fourths of the agricultural land, are devoted to raising beef. The big island of Hawaii is a vast cattle ranch; indeed it has one of the world's most extensive cattle ranches. Much of the landscape looks like Arizona or west Texas, for grass and cactus dominate the arid, stony land. Coffee and taro are declining, and rice is no longer grown at all.

Tourism has risen rapidly, just as in Puerto Rico: from 46,000 visitors in 1950 to 600,000 in 1965 to 1,600,000 in 1970 to 2,630,000 in 1973. It has now surpassed the defense industry as a major source of income and has fueled a building industry of immense importance. Recently a new tourist source was tapped: the Japanese. By 1970 more than 120,000 Japanese were coming to Hawaii. The Hawaiians, going all out to promote this business, now count the Japanese tourists as a major source of income.

PROFILE OF HAWAII

Area
Seven principal islands covering 6,450 square miles.

Population
831,000 resident population: 705,000 civilians; 56,000 armed forces; 70,000 military dependents. About 80 percent live in Honolulu on the island of Oahu.

Density of population
137 per square mile; 1200 per square mile on Oahu. Outer island population now exceeding Oahu's growth rate.

Tourism
2.63 million visitors in 1973; up from 32,000 in 1941 and 243,000 in 1959. Visitor expenditures total $890 million.

Construction
$696 million in 1972 and $904 million in 1973 with hotel construction predominating.

everyone in the state is a member of a minority group, for no single ethnic group is in the majority. They list their principal ethnic groups as: Caucasian, 38.8 percent; Japanese, 28.3 percent; Filipino, 12.2 percent; Hawaiian, 9.3 percent; Chinese, 6.8 percent. The present population is 832,000 people, with 80 percent living on Oahu, principally in Honolulu, where the density of population is 1200 per square mile. This is, of course, a very deceptive figure, for most of these people live in the Honolulu-Waikiki urban complex, and most of the island is thinly settled and occupied by great fields of pineapples and sugar cane. Large areas are dominated by two mountain ranges that are virtually unoccupied. These ranges perform a most useful function, for by intercepting the trade winds, they cause heavy rainfall, and this water becomes trapped in the volcanic

Military
Expenditures equalled $872 million.

Manufacturing
Value added by manufacturing rose from $58 million in 1939 to $453 million in 1971 (latest data available).

Racial Makeup
Caucasian, 38.8 percent; Japanese, 28.3 percent; Filipino, 12.2 percent; Hawaiian, 9.3 percent; Chinese, 6.8 percent; others (Korean, Negro, Samoan), 2.4 percent.

Land Use
Forest reserve, 1.2 million acres; grazing, 1.15 million acres; urban use, 156,000 acres. Unused open space: 50 percent of Oahu, 62.5 percent of the other islands.

Land Ownership
Federal government, 356,000 acres; state government, 1,585,-000; private owners with 1000 acres or more, 1,918,000; small owners, 270,219.

Sources
The Population of Hawaii 1973, Statistical Report 102; Land Use and Land Ownership Trends in Hawaii 1973, Statistical Report 98, both published by the Research and Economic Analysis Division of the Department of Planning and Economic Development of the State of Hawaii.

(Top) **Waikiki Beach, Island of Oahu, Hawaii.** *Waikiki's fame has nearly been its undoing, for now high-rise hotels and apartments crowd its beaches. Famed Diamond Head is in the background. The yacht harbor in the foreground contains somewhere between $3,000,000 and $5,000,000 worth of yachts, measuring rather neatly the amount of wealth in the islands.* (Hawaii Visitors Bureau Photo)

(Middle) **Cultural Crosscurrents.** *The Polynesians populated the Hawaiian Islands, but the impact of disease in the nineteenth century nearly destroyed the island population. Today here at Waikiki Beach the culture is American. And yet the old is still there in the twin-hulled boat, the catamaran, a type of boat developed by the Polynesians. This one is plastic hulled, dacron sailed, and has an aluminum mast, yet the basic idea is ancient Hawaiian.* (Hawaii Visitors Bureau Photo)

(Bottom) **Tropical Vegetation.** *Palms are almost the trademark of the tropics. These are coconut palms on a Hawaiian beach. Palms normally occupy sunny niches in the tropics: the beach, stream edge, or the forest edge. Many palms have valuable products such as palm oil and the coconut, but in Hawaii they are most important for lending the tropical atmosphere of this tourist paradise.* (Hawaii Visitors Bureau Photo)

The land ownership pattern in the islands exceeds in concentration anything known in Latin America. Eighty-five percent of the land is in the hands of the government, the big corporate landowners, the estates of the early settlers, and heirs to Hawaii's native royalty. Two estates own one-fourth of Oahu's 600 square miles of territory. There has long been agitation to break up these large holdings, but the situation created at Waikiki Beach—where wall-to-wall high-rise hotels block the view and preempt the beach—as a result of small holdings has changed the viewpoint. It is now felt that the large holdings are far preferable, since they make it easy to control development and to keep the uncrowded and unspoiled atmosphere that the islands recognize as one of their major tourist attractions.

The Caucasian population of the islands is a minority, and some fears were expressed when the islands became a state that the Orientals would take over the government. As it has turned out, the senators and representatives are all of Oriental origin, and many of the state and local officials are Caucasians. Further, given the tendency of Hawaiians to go to the mainland seeking employment and Causasians to flow to the islands, Caucasians may become a majority on the islands.

Hawaii was long touted as the model of race relationships, but it was not. The Japanese in particular stayed strictly by themselves and only married other Japanese. The Chinese were almost as closed a society. In 1915 only 11 percent of the marriages were interracial in this intensely multiracial society. By 1972 this figure had climbed to almost 39 percent, and interracial marriages were accelerating. The fact that Hawaii is now a racial melting pot makes it a bridge to the Far East, and many large corporations employ Hawaiians, since doors in Asia open to them when they would be closed to mainland Americans.

Parallels and contrasts with Puerto Rico are numerous. The native setting was similar and the catastrophic loss of life nearly identical, leading to massive imports of other races of men as a labor supply: Africans for the Caribbean, Asiatics for the Hawaiian Islands. Both areas developed sugar as a major crop, and both now have a rapidly expanding tourist business. Puerto Rico supplies a warm winter playground much closer to masses of people than Hawaii does, and its proximity to the mainland aids it in its industrialization. Even minor industries such as coffee supply studies in parallels and contrasts. In the hands of the Japanese on Hawaii it is a very high-yielding crop; in the hands of the Portuguese on Puerto Rico it is relatively poor in yield. Puerto Rico is densely populated; the Hawaiian Islands are thinly populated. Both of these lands are tied into the United States economy and have all the advantages of access to this superlative market for sales, investment capital, and so forth. Hawaii, with a longer association with the United States and its status as a state, is far ahead by most health, education, and welfare measures, but Puerto Rico after a belated start is improving fast. They form an interesting study of cultural differences and similarities developed on strikingly similar physical geographies.

In Hawaii the descendants of the first white families, many of them missionary families, obtained control of most of the land and still own it. Far from showing any evidence of "tropical deterioration," these families have proved to be energetic and canny businessmen capable of building great fortunes and holding them; no shirt sleeves to shirt sleeves in three generations, as the cycle of laborer to tycoon to laborer has sometimes been epitomized in "cyclonically stimulated" mid-latitude America. The contrast is unfair, of course, for some great families with long traditions of success are also found on the mainland. However, for those who expect tropical deterioration in white men in the tropics, the Hawaiian Islands are a disappointing case. No one is surprised, of course, that the Orientals flourish there, but that overlooks the fact that the Koreans are from a land no more tropical than our Middle Atlantic states and that the Portuguese, who also have flourished in the islands, are not a tropical people.

Cuba

Cuba currently presents the exact opposite of the Puerto Rican picture, even though its pre-Columbian and Spanish colonial history is much like that of the rest of the Caribbean. It is the largest of the Caribbean islands (10 times the size of Puerto Rico), has the largest forested area, and is rich in minerals (copper, nickel, manganese, chrome, and iron). Its population density is a modest 150 per square mile as compared to Puerto Rico's 650. If population is a resource, Cuba is richer, for it has 8,600,000 people, while its smaller neighbor has only 2,500,000 people. Cuba's population is 73 percent Caucasian, 14 percent mixed, and 13 percent Negro in origin. Cuba once had a flourishing winter tourist trade, which now goes to other Caribbean islands. As we have seen, this trade amounts to tens of millions of dollars annually.

Before 1958, when Fidel Castro took over the government, sugar occupied one-half the cultivated land and dominated the economy, accounting for 85 percent of the exports by value. So great a concentration of one crop influenced the way of life of the whole island. January to April was a time of furious activity for the sugar workers, but the rest of the year was a time of inactivity called *tiempo muerto*, dead time. Such imbalance in employment of the labor force is obviously wasteful in terms of productivity and destructive to the individuals involved.

Tobacco was the second crop. It occupied 3 percent of the land and supplied 6 percent of the export income. Havana was almost synonymous with fine cigars. In addition, there was a large vegetable business aimed at the United States market, and large cattle ranches, developed with American capital and skills, supplied a sizable export of beef. Cattle raising in the tropics is a growing business with a good future now that feeds, fertilizers, and appropriate stock are being found to fit the soils and climate; the pasture lands produce continuously, and there is no winter period with its costs for storage of feed and housing of animals. Cuba is well placed to capitalize on such an industry.

The Cuban economy is collapsing under Castro's regime. The role of technology in tropical plant-and-animal management is illustrated by Cuba's rice problem. Rice, a major food crop produced on the island, is subject to disease attack. The United States, prior to 1958, maintained an agricultural experiment station in Cuba and through application of modern knowledge kept the rice diseases under control. Since the closing of this station, the rice diseases have spread, and the crop is greatly reduced. The cattle ranches have been confiscated, and the stock largely eaten up. The preferred status for sugar sales in America is gone, and in the highly competitive world market the return on the decreased sugar production is slim. New sources of cigar tobacco have been found, and Cuba may never regain this business. Industry is closing down, and there is scarcity everywhere.

Rationing of food exists. Houses go unpainted, and there are intermittent shutdowns of water and electric services. Despite much propaganda about improvements in housing, there is a massive housing shortage. The sugar crop year after year is disappointing even though the whole economy is disjointed by the massive drafting of virtually all able-bodied workers to facilitate the harvesting. It is true that the gap between the rich and the poor has been narrowed, but this has been accomplished by impoverishing the wealthy without aiding the poor. On the plus side, there is more medical service than before, and more people are being educated though the quality of education has dropped. There is unemployment of a peculiar sort, for men simply choose not to work since their money buys so little. The state classes them as vagrants and attempts to round them up, but this is a clear admission of de-

clining morale. Some of this situation is due to Castro's ineptness, for he is said to implement even good ideas badly; for example, he herded thousands of students into schools before he had teachers. Moreover, Cuba since 1958 has suffered what is probably the greatest drain of all: flight of educated and productive people. The United States has gained thousands of doctors, engineers, educators, businessmen—the very people most needed for building a society. That much of what has happened in Cuba is typical of the socialistic planned society is suggested by similar happenings in eastern Europe, Chile, and elsewhere.

It is a common occurrence in history for a nation to decline: Egypt, Greece, Rome, Spain, to name a few, and the tendency often has been to invoke climatic change or some other physical-geographical cause. In the case of Cuba it is apparent that the causes are not physical-geographical but cultural. The results will change the physical landscape: shifts in land use, changes in investment allocation, and in this case deteriorating housing, declining yields of crops, lowering standards of living, and a brain drain that will force still further declines.

Panama

Everyone knows something about Panama and the building of the canal but usually only a little. Panama is a perfect example of a tropical rainy location at 9 degrees north latitude. The heat equator does not coincide with the true equator and passes through Panama, where the average monthly temperature stays at 80° F. through the year. The annual rainfall is 130 inches, but February and March are relatively dry. The area is low and swampy. (See Table 3-3.)

This area was once known as the "pest hole of the world." In 1903 the official description was "one of the hottest, wettest, and most feverish regions in existence. Intermittent and malignant fevers are prevalent, and there is an epidemic of yellow fever at all times. The death rate under normal conditions is large." In 1931 the description ran: "formerly a permanent regional focus [for yellow fever] because [it] existed at some place in it at all times; and from such infected places other places in the region were infected from time to time."

We know relatively little about the pre-European settlement of this region, but there are some indications that the Indians considered it a good land. Panama is characterized by a special style of art work in gold, while nearby people, the Cuna, had a form of writing. Agriculture, metallurgy, and writing all point to a developed society.

The Spanish record here, however, is one of terrible mortality. Colonies were attempted but failed, and many officials died in attempting the brief crossing of the isthmus to reach their posts in Peru. The California gold rush brought floods of people surging across the isthmus, and the loss of life was fantastic. A railroad was then planned, but the 1000 Negroes imported to build it were all dead in six months. Chinese were brought in, and they too died within six months. As though in a laboratory experiment, each race tried and failed, none proving to have any special racial resistance to tropical heat, humidity, and disease.

Then came the French to try building a canal, for they had succeeded at Suez. They lost 16,000 men from malaria, yellow fever, dysentery, and smallpox. Despite tragic heroism, the French failed. The "climate" could use up men faster than they could pour them in. But was it climate, or was it disease? Or do the two go together?

Then how did Indian civilization survive there? Interestingly enough, few of these fatal diseases were present before the Europeans brought them: both malaria and smallpox were imported into the New World. They are an excellent example of how man changes the environment, for without them the tropical lowlands of the New World would be much healthier. It is a mistake, moreover, to think of them as "tropical" diseases. Malaria once scourged the United States up into New England; its disappearance here is due mainly to better housing. Yellow fever and smallpox are also not limited to the tropics; they are held down in midlatitudes by public-health measures.

Finally to Panama came the American effort. The role of the mosquito in transmitting yellow fever and malaria had been isolated, and unlimited funds were made available. A zone 10 miles wide was drained, sprayed, and screened, and sanitary facilities and innoculations were instituted on a grand scale. The health level quickly approached continental United States levels: the "tropical" diseases were practically eliminated.

In the building of the canal the test of various ethnic and racial groups continued. Southern Europeans (Spaniards, Italians, and Greeks) were used in large numbers, and although they worked hard in any weather, they were difficult to discipline. Those who obeyed the sanitary rules remained healthy; those who did not became "victims of the tropics." Negro labor was also used in large amounts. Large numbers from the West Indies were brought in, but at first their efficiency was extremely low. As the effects of medical care and good food began to show, however, their work improved. The key to the situation again seems to have been neither racial nor climatic.

Three generations of Americans have lived in Panama, for there is a strong tendency for those who have worked there for years to stay there, and for those raised there to go to work there. Children thrive. They wear sun suits and go bareheaded, and the tropical sun somehow fails to "strike their nervous system." Most childhood illnesses prove to be mild, and some are lacking. As this shows, these children have not been victims of tropical deterioration.

The Canal Zone remains today a sanitary band across Panama, with reservoirs of disease in the native populations on either side. Health and vigor at Panama is related to knowledge and will power: the value of good hygiene must be known to be practiced, and knowledge of its value aids in the will power to keep up the necessary measures. The Canal Zone shows that the white man can succeed in the tropics, and it supports no racial selection for tropical life. It emphasizes again the ability of man to modify his environment. Europeans introduced diseases that made this the "pest

TABLE 3-3 Climate of Panama

Station	Latitude	Temperature (°F.) Mean	January	July	Relative Humidity* (percent)	Rain (inches)	Wind (miles per hour)
Balboa	9° N.	78.6	78.2	78.7	83	68	7.6
Colón	9° N.	79.8	79.8	79.9	83	128	10.5
Yuma	32° N.		54	91			
Miami	26° N.		66	81	78	60	8.9

* Relative humidity taken in early morning—lower during heat of day. Note that Miami in summer is hotter, almost as humid, and no breezier than Panama. New Orleans, from June through August, is hotter and almost as wet as Panama (temperatures of 80.6°, 82.4°, and 79.2° F. and rainfall about 6 inches per month). Balboa on the Pacific side is slightly cooler, drier, and less humid than Colón on the Caribbean.

hole of the world;" their descendants cleaned up a strip of it and showed that it could be a healthful, pleasant, productive part of the world where any race could flourish.

Civilization in the Tropics

We can turn now to the problem of civilization in the tropics. There have been and still are civilizations in the tropics, India being one of the oldest in the world. But Asia also had civilizations in Indonesia, Vietnam, and in the Americas: Mexico, Central America, and South America. About the time of Christ there was far more civilization within the tropics than outside them.

The Mayan civilization

In discussing tropical civilizations, the Americas are sometimes overlooked. In pre-Columbian times, *all* civilization was in the tropics: either in the arid tropics of coastal Peru or in the humid tropics (Mayan). The Old Empire of the Maya was in the tropical rainy area of Honduras, and the New Empire was erected in the similar climatic area of Guatemala and adjacent Mexico. Within this area there is actually considerable range of climate. Northern Yucatán is dry bush savanna, central Petén is a wet savanna, and the rest of the area is rain forest.

The Maya were preceded by the Olmec people. Our knowledge of these people is in its infancy, but so far it seems that they appear with the seeds of civilization about 1200 B.C. They had writing and calendars and began building ceremonial centers. Their portrayals of themselves show predominantly Negroid features, and one possibility is that these culture bearers were from Southeast Asia; according to the Chinese annals this area was at this time solidly Negroid, and increasingly Chinese scholars are recognizing the importance of south China as an early center of civilization.

Leaving aside the question of ultimate beginnings, by the time of Christ the Old Empire was flourishing in a rain forest. The next 600 years saw extensive cultural and territorial growth. Great civic centers were built; sculpture, painting, architecture, astronomy, and mathematics flourished. Thereafter came the Middle Empire, from A.D. 600 to A.D. 700, often described as a period of consolidation. It was followed by the great period from 700 to 900, when the arts and the empire flourished. Then there was the New Empire in the Yucatán savanna area, 900 to 1200, and the period of Mexican influence, 1200 to 1500. The Maya were still literate,

Mayan House on the Island of Cozumel, Mexico. *The thatched roof, the rounded ends, and the platform on which the house is built are all carry-overs from the ancient past. Over a thousand years ago the Maya built houses such as this. Today rural electrification is reaching into the back corners of Mexico. Change is apparent in house types, as can be seen in the background.* (Hilda Bijor for World Bank)

monumental city builders—perhaps best equated with Egypt in one of its great periods—when the Spanish arrived.

The Mayan empire crashed under the Spanish impact. Brutal conquest, destruction of written records, disruption of the priestly, learned, and ruling classes, and a melting away of the people from other causes brought an end to this civilization. It is easy to understand the disintegration of a culture whose traditions are smashed and records destroyed. Cultural declines have occurred in history for lesser reasons. The melting away of the people has received less attention, but we have already touched the probable reasons: plague, dysentery, malaria, yellow fever, smallpox, measles, and the dozens of other new diseases. What smallpox alone could do to a nonimmune population is illustrated by its effect on the Mandan Indians of North Dakota.

How remarkable were the Mayan achievements? They knew and used skillfully the corbelled arch, the only arch used by most of the world east of the Mediterranean at their time. They built vast pyramids on which they erected architecturally notable buildings, and their religious-civic centers rival in beauty and layout those of Egypt, Mesopotamia, Greece and Rome. Their pottery was tastefully decorated, and they did beautiful work inlaying and carving stone, bone, and wood.

The Mayan achievements in mathematics were equal to Egyptian efforts in Ptolemaic times. They had several calendars of very great accuracy. They knew the 365-day year, which they divided into 18 units of 20 days each plus a 5-day period. They also had a sacred year of 260 days, with 13 units of 20 days each. This sacred year they permutated with the natural year to create a 52-year cycle. If this general description of a very complex calendric system sounds intricate, it barely indicates how far advanced their astronomic and calendric knowledge was, for they also had moon calendars and Venus calendars, and they fully understood the interrelationships of all of them. Using the moon calendar, a date could be fixed in the "long count" (as opposed to short-term reckoning) that would not recur for 374,440 years.

They considered the length of the year to be 365.2420 days. It is actually 365.2422 days, though our own calendar is based on a 365.2425-day year. They had astronomical observatories, predicted eclipses and the heliacal rising and setting of Venus, knew the period of Venus' revolution, and were aware that their calendar had an error of 8/100 day every 584 days.

The Maya wrote in an ideographic form, which in degree of development resembles Chinese. They carved dates we can read and accounts we cannot read on their monumental buildings. They had books of paper, but few of these have survived, and we cannot read them.

What we have here, then, is a record of political, economic, and cultural achievement of the highest order maintained in a tropical rainy area far longer than many other cultures in the world survived.

The problem of Mayan agriculture and population

The Maya have long posed a number of problems. How could a simple, forest, shifting agriculture ever have supported such a civilization? How could such an agriculture be continuously maintained on the thin soils of a limestone plateau in a tropical rain forest? Was the great shift in location between the Old Empire and the New Empire—and the several spurts in development during the Maya's long occupation—due to changes in agricultural practices and the abandonment of areas that had exhausted their soil fertility or been overrun by weeds such as grasses?

Studies of shifting, forest-type agriculture suggest that it can support about 100 to 150 people per square mile, and perhaps this is enough to account for the Mayan civilization. However, there is some indication that the population may have been much larger than this. At Uaxactun, which was not a major city, a survey made through the surrounding jungle revealed so many house sites that if all were occupied at one time a population density of 1000 people per square mile would be indicated. Even if only one-fourth of them were simultaneously occupied, the density of population would have been well beyond the possibility of simple, forest, rotation agriculture.

The agriculture had to be efficient in order to support not only the people on the land but also the great numbers who produced the monumental buildings and works of art, as well as the scholarly and priestly classes who did government work. Further, the population figures given above describe a secondary civic center. The great cities of the Mayan area must have had larger populations and even denser settlement on the land. Therefore, there must have been a much more intensive land use in the Mayan area in the past. An Oriental pattern of husbanding is suggested—all waste materials being returned to the land, combined with all possible organic material from the adjacent lands, with the careful composting of this material so as to maintain the fertility of the fields. The question then arises: Once achieved, why would such an agricultural way of life be abandoned?

Intensity of land use is related to population density. Decreasing density of population leads to de-

creased intensity of land use. A classic example occurred in colonial North America. The colonists came
from areas that had developed advanced agricultural
techniques for utilizing the land intensively. This included crop rotation, the use of animal manures, the
use of vegetable matter for composting, and the beginning of the use of mineral fertilizers. In North America
when the colonial population found itself with a huge
amount of land and very little capital or labor, they
substituted extensive use of the land for the other two
factors. This led them to practice a shifting, forest-type
agriculture that their European cousins always described
in utterly scandalized tones because of its weedy, im-
permanent, primitive nature.

If, then, something had happened to the Maya
that suddenly shifted the relation of the population size
to the land available, it would be logical that they
would change their manner of land utilization. Quite
clearly a catastrophe struck the Maya in the appearance
of the Spanish and the European diseases. Between
wars and disease the Mayan population fell from a
minimum of 100 per square mile and from a possible
high of 1000 per square mile to about 1 per square mile,
its present status. It seems inevitable that under these
circumstances the agricultural methods would shift
radically toward a simple land-use system. The laborious maintenance of land in permanent production
would be abandoned for the much easier clearing of
a new patch of forest where natural processes had
restored the soil fertility during the period that the land
was allowed to rest and to grow up to a forest cover.
The more difficult thing to understand is the failure of
the Maya to rebuild a population density above the
present figure of 1 per square mile. Probably this is an
example of the dangers inherent in destroying a social
fabric. With the disappearances of learning, tradition,
and religion, accompanied by a tremendous loss of
life, a whole way of life disappeared, and a new way
of life with the drive to increase and be fruitful has
not yet arisen. Explanations that lean on soil exhaustion,
weed infestation, or climatic change seem unnecessary
to account for the record.

THE NEW WORLD TROPICS

A survey of the wet tropical regions of the New World
must keep in view what was, what is, and what could
be. Our view of the future is necessarily very limited.
We cannot see all the present possibilities; moreover,
we probably have difficulty seeing future possibilities
as well.

In the Americas the center of pre-Columbian
achievement lay in the wet tropical regions. Civilizations existed in Middle America, and village agricultural
settlement characterized most of the rest of tropical
America. We know little about the densities of population in these areas except that they must have been
high even by modern standards. Outside the civilized
city-state areas, population densities were lower, possibly somewhere between 10 and 20 persons per square
mile. Shifting agriculture can easily support even greater
populations. The disruption from the European conquest and the devastation of the new diseases so thoroughly removed the people from the land that we
probably underestimate how many people there once
were even in the areas of shifting agriculture.

The Old World parallel to the Mayan civilization is in India. Agricultural impulses arriving about
3000 B.C., from the Middle Eastern center bringing
wheat and from the Southeast Asian center bringing
rice, started India on the road to civilization. These
impulses started the great city-state civilizations in the
Indus Valley. Thereafter came the Aryan invasions, the
incoming of cattle-keeping Indo-Europeans about 1500
B.C. By 1000 B.C. great city-states comparable to the
Mayan city-states were flourishing in the Ganges Valley, and Indian civilization had its characteristics well
established. By this time there was a flourishing growth
of the arts, with especially notable accomplishments in
sculpture and in architecture. Indian music is a development quite outside the Western tradition and therefore difficult for Westerners to appreciate. It is nonetheless a fine art. India contributed our numbers, which
we call Arabic numbers because the Arabs were the
people who carried them to the Mediterranean world.
The value of this contribution can be most easily visualized if you attempt to multiply or divide any set of
numbers using the Roman numerals that were the common possession of the Western world until the Arabic
numbers arrived. The people of India are also credited
with the invention of negative numbers and the zero,
extremely difficult concepts that most of mankind failed
to achieve elsewhere. India is also noted for its philosophy and its being the birthplace of such major world
religions as Buddhism. The list of cultural contributions
could be lengthened, but the point should be clear by
now that under conditions of high heat and considerable humidity these people have maintained a civilization for more than three millennia. Of all the great
civilizations, only India and China have such a record.
It can hardly be used as evidence that tropical climates
are inimical to the growth and maintenance of civilization. In passing it should be noted that while India is
basically part of the ancient dark-skinned-race area of

the tropical world, there have been sizable contributions by north Europeans and Mongols. Still the major contribution must be credited here to a people variously described as originally Australoid and Negroid. India stands, then, as a refutation of the arguments of "no civilization in the tropics" and "dark-skinned peoples have never accomplished anything, and this is evidence that they never will." With that point established we may proceed to a survey of the tropical regions of the world.

Tropical North America

The wet tropics of southern Mexico and Central America are savanna and rain forest. Most of the people are subsistence farmers, clearing patches of forest and moving as the soil fertility declines. Corn, beans, and squash are the great staples, but other grain and root crops are grown. Cotton grows as a perennial. There is much fruit, and commercial production of cotton, bananas, coffee, and chocolate supply important exports. These are grown in plantationlike systems that vary from relatively small holdings in the El Salvador coffee area to the very large corporate developments of the United Fruit Company. In addition, Mexico has continued since colonial times to be an important producer of metals, especially silver, and more recently of petroleum.

Many of these lands are largely Indian. Mexico is about 15 percent white, 55 percent mixed, and 30 percent Indian. Since much of the classification is social rather than racial, it is probable that a considerably higher portion of the population is actually Indian. Guatemala, El Salvador, and Honduras are about 80 percent Indian and 20 percent mixed. Nicaragua and Panama are Indian, European, and Negro, in that order. Costa Rica is predominantly white in race. Over most of this area the men on the land are Indians living very much as they did at the time the Spanish arrived and often still speaking their own languages. They have been stripped of their native civilization, and the European civilization has not yet rooted and flourished. Although European domination has caused everyone who can to claim to be European, growing appreciation of the Indian accomplishment in art, architecture, agriculture, and government is beginning to remove these pressures.

The Caribbean area we have reviewed. Depending on the vagaries of colonial history, the major islands are largely white, as in Puerto Rico and Cuba, very much mixed as in Trinidad, or largely Negro, as in Jamaica or Haiti. Again, there is no climatic basis for such differences. In the Caribbean the pattern is largely subsistence farming for the mass of the population with plantation-type production of sugar, coffee, tobacco, cacao, and bananas supplying the export crops. Imports are typically fish, rice, flour, cloth, iron and steel, machinery, and vehicles.

While Puerto Rico demonstrates development via association with a great power, making the economic and political environment attractive to capital, and Cuba demonstrates the road to poverty by a different program, the Central American republics of Nicaragua, El Salvador, Costa Rica, Guatemala, and Honduras have formed a common market and are stimulating an economic revolution. Hundreds of factories are being built by local and foreign investors. While this has mostly affected the cities, it will inevitably have repercussions in the countryside.

The Central American Common Market began operations in 1963 and linked tiny populations such as Costa Rica's 1,900,000 with Guatemala's 5,800,000, El Salvador's 4,000,000, Nicaragua's 2,100,000, and Honduras' 2,765,000, to make a total market in the five countries of 15,565,000. The total area is only a little larger than California, and yet a unit of this size provides a market for factories of economical size, while tiny markets do not. The argument that the size of the country is a physical-geographical factor that determines the market size is not valid. The size of the country is man-made. It can be changed by combining countries or by retaining national boundaries but erasing economic boundaries.

The economy shifted rapidly from dominance by agricultural produce toward industry. As in Puerto Rico, much of this change has been fueled by foreign capital, and United States firms such as Alcoa, Standard Oil (New Jersey), Texaco, United Fruit, Westinghouse, IBM, and Sears, Roebuck show how broad the investment is. Perhaps of even greater importance is the change in attitude of the Central Americans. They too have begun to invest in local business, and young men educated in America now tend to return home to seek opportunity, instead of viewing home as too stagnant and too small for an ambitious man.

The experiment is young, and problems are inevitable. For example, there is not enough skilled labor, and too many landless peasants have rushed to the cities. So far the benefits of the boom have reached only a small percentage of the people. Only about 11 percent of the economically active population earns its living in industry. It is not that the millenium has arrived but that rapid progress is being made. Lands seemingly doomed by tropical heat and humidity have begun to move toward industrialization and diversification, rescued from dependence on one or two crops such as bananas or coffee. An increasingly stable economy aids political stability and attracts more investment capital. The growing per capita earning increases purchasing power, and this supports more industry. The growing middle class obtains more education and skills to build industry and buy industrial goods that encourage still more industry. Gainfully employed people are good customers and demand goods that require employing more people who in turn become customers. If the United States is a test, there is seemingly no end to this spiral, for we turn yesterday's luxuries into necessities and demand ever more production.

Tropical South America

Tropical South America presents a similar picture. Population density, on the average, is low. People are concentrated on the rim of the continent. Here there are areas of dense settlement and modern cities with populations over a million. The interior is thinly occupied by primitive Indian tribes and some mixed groups. The

people on the edges resemble those of the Central-American area in their racial makeup: Negro, Indian, mixed, and European. The largest Indian populations in South America are not in the wet lowlands but in the Andean highlands. Brazil is approximately 60 percent European, 15 percent Negro, 20 percent mixed, and the rest a tiny Indian population. Venezuela is predominantly mixed. The Guianas are mixed Indian and Negro. In all of Latin America, population declined under the impact of the Europeans, and a low point was reached around 1800. Only a fraction of the peak population that the land once supported—quite probably only half as many people, possibly only a tenth—remained. Since then the population has been growing rapidly. It doubled between 1900 and 1950 with the major growth appearing in the European and mixed groups. Today it amounts to about 300,000,000 and continues to grow at a rate of 2 to 3 percent each year. In recent decades the rate of growth has been 30 percent, or a doubling of population every 30 years.

The peripheral location of population in South America is striking. It is an excellent indication of the direction in which the colonial settlers looked. They reached the coast first. Since land transportation was nonexistent, their contacts, markets, and sources were

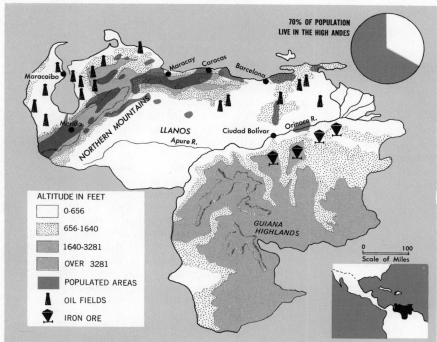

MAP 3-2   Venezuela
Venezuela has highlands in its north, and a disproportionate part of its population is concentrated in them. The vast Guiana interior is nearly empty. The great lowlying, well-watered interior, the Llanos, is thinly populated. Such population concentration in a few areas is typical of much of Latin America. The major resource now being exploited is petroleum, with iron reserves also becoming important recently. Such resources are bringing in a flood of money, some of which is going into the development of the country. The network of roads is unusually good for Latin America. The growth of the capital city of Caracas has been extremely rapid and has drawn people away from the land. Paradoxically, in this tropical land, well watered and rich in plains, not enough foodstuff is produced to feed the population. (Reprinted from Focus, American Geographical Society)

all in the Old World, and the sea was their highway. Quite naturally they stayed near the coast and looked overseas. This inheritance still marks the land and is only slowly changing.

Venezuela: oil and development

Venezuela is a typical Latin American country with a young, rapidly expanding population, highly concentrated in a few cities and coastal areas, while the vast potentially rich interior lies empty. It is also typical in its turbulent politics, history of dictatorships, and its racially mixed population. It is unusual in its immense income from oil and its recent increasingly rapid rate of development.

The climate, while tropical, varies from semi-arid in the northwest corner to strongly seasonal rainfall in the Llanos of the Orinoco to year-round rain in the Guayana area of the south. The Maracaibo basin is a lowland cut off from the rest of the country by a spur of the Andes that swings eastward to form a steep Atlantic face from Maracay to the delta of the Orinoco. The Orinoco drains a vast grassy inland area called the Llanos, and south of this are the Guiana Highlands, with one peak 11,000 feet high. There are vast potential resources: the well-watered plain, tremendous oil and iron deposits, and a large hydroelectric potential. The Orinoco is navigable by ocean-going ships up to Ciudad Bolívar, where the river is pinched between highlands, and thus produces a potential hydroelectric site.

The population pattern inherited from the Indian-Spanish period was concentrated in the Andes, but in the postcolonial period the country has slowly reoriented itself to face the Atlantic, with Maracay and Caracas growing in importance. After independence from Spain the country suffered from dictatorships, revolutions, lack of a basic educational system, and a stagnant agricultural economy. A major turning point came in 1914 with the discovery of oil in the Maracaibo basin. This provided the potential capital through taxation of the flood of oil for the development of the country. Quite understandably, there was considerable delay and distortion of the economy.

The economy until 1914 had been dominated by agriculture, with products such as coffee and cocoa providing a trickle of international exchange to pay for some imports. Within a few years petroleum supplied from 85 to 95 percent of Venezuela's exports, and an enormous inflow of capital became available. The struggle with unstable political conditions, primitive transportation, undeveloped school systems, and so forth seemingly was a long one, but in a longer view

of history it was actually short. In 1947 an iron and steel industry was begun, and the ambitious developments in the Guiana region indicate that heavy industry will continue to expand. Manufacturing in general is rising at a rapid rate, and it is expected that by 1980 the number of people engaged in manufacturing will equal those engaged in agriculture. In 1950 there was one person in manufacturing for every four in agriculture. The rate of change is extremely rapid, but the distance to go is great. In the United States, for example, 5 percent of the population is in agriculture. Nevertheless, Venezuela illustrates an almost optimum rate of change, made possible by the enormous income from petroleum. The fifty years required to get to a take-off point provide an astonishingly good record.

Population growth was also rapid: 2,400,000 in 1900, 5,000,000 in 1950, 11,500,000 in 1974, and increasing at a rate of 3 percent a year. Fiscal income in the same period went from $13,100,000 to $1,370,000,-000—a more than 100-fold increase. By 1972 Venezuela and Argentina had the lead in South America in gross national product per capita. Theoretically Venezuela should have been able to build a model state, but, in fact, dictatorships and inexperience have prevented a utopia. Roads, schools, and industries have come along with increasing speed. Education has been pushed especially since 1959, when half the nation was illiterate and 500,000 children had no schools. This problem has been attacked at all levels: building of primary schools by the thousands, expanding teachers' training to staff these schools, multiplying the high schools, and expanding university enrollment by nearly threefold. A major campaign of adult education aimed at teaching reading and writing also has been undertaken. To say that results will be apparent in a generation or two is not defeatist. The children in primary school today begin to influence the economy by their skills in 15 to 20 years, depending on the level of education they achieve, and the ones who take the longest training programs (engineers, doctors, lawyers, professors) will have the greatest influence.

Growth now is spreading into other regions and various economic endeavors. Rice, corn, sugar, and poultry production are being increased. The agricultural potential is enormous, and an ambitious growth rate for agricultural development of 8 percent per year is now aimed at. This is tied to distribution of land, rural electrification, extension programs, and other agricultural education programs. The timing is wrong. The world is moving toward large farms, scientifically managed and operated with large capital expenditure on heavy equipment and fertilizers. The time of the small family producer is gone; land distribution systems aimed at re-creating this type of production will not be efficient or provide good incomes, and the people will not stay on the land. The problem in Venezuela, however, is little different from that in the rest of the world. The tenant farmers have been paid incredibly low wages, and they now move to the cities as a means of escape. They are unskilled and often illiterate, and the industrial growth has not been rapid enough to absorb them. They move then from rural slum poverty to urban slum poverty.

Venezuela and the rest of Latin America, as well as most of the other developing nations, recognize this and are striving to industrialize. The problem is to industrialize fast enough to absorb the people off the land. As agriculture in Venezuela is modernized, a shift from 79 percent to 10 percent of the people on the land is anticipated. This requires far-reaching, simultaneous educational, agricultural, industrial, transport, and social changes, which demand immense capital investment. Venezuela is one of the more fortunate

Urbanization in Venzuela. *The barrio of San Jose is seen here in the foreground with the city of Caracas in the background. This type of contrast is typical of large cities in the developing countries around the world. Rural people seeking escape from the grinding poverty of a peasant life on the land flock to the cities. Unfortunately, most of them finally settle in slum areas on the outskirts of these growing metropolises. The contrasts between modern central cities and the surrounding slums is often extreme. (Ray Witlin, United Nations, for World Bank)*

countries because it has developmental capital, but nevertheless the struggle will be long and difficult.

The Guiana Highlands are an example of development of latent resources in Venezuela. The region has diamonds, gold, kaolin (clay for making porcelain), manganese, titanium, bauxite, and enormous iron ore reserves of high (58 percent) iron content. The hydroelectric potential of the area is enormous. One dam has an electric generating capacity of 1,750,000 kilowatts, which can be increased to 6 million kilowatts. Power generated at Niagara Falls is about one-sixth of this latter figure. This power is now being used for electric reduction of some of the iron ore.

A heavy industry to supply Venezuela is being built, and a city of 500,000 people is projected for a formerly empty area below Ciudad Bolívar. While the area is far from the major population center of the country, the Orinoco allows water transportation to the Atlantic and the population centers about Caracas.

Venezuela exemplifies much of the tropical South American problem, except that it has an enormous advantage in the capital income from oil and iron.

Brazil: potential world power

Brazil is larger than the United States excluding Alaska and contains half of the area and people of South America. However, 80 percent of its 100 million people live on or near the Atlantic Coast, and this leaves the vast interior thinly populated. Imagine the United States with 100 million people living along the East coast and only 20 million spread over the rest of the country, living on a primitive, subsistence, agricultural basis. Brazilian settlement has emphasized one particular vegetation type, the semideciduous forest of the tropical, rainy southeast coast. Now that type of land is almost gone, so they will have to use other types of land. This tendency to use one type of landscape to the near exclusion of all others is a common human approach to the land and will be seen again in the grassland discussion (see Chapter 7).

Brazil has the enormous, empty Amazon basin. As the map of landforms of Brazil shows, this is not as extensive a lowland as is often thought, for hilly uplands hem the river plain. The soils of the Amazon basin that are subject to overflow are moderately rich. The rest of the land is greatly leached and often very poor. While a wet-rice culture could support millions of people on a Chinese coolie level on the flood plain, this is not viewed as a desirable goal for Brazil. Extensive rice production on a California plan would be possible, but the Brazilians are not putting in the im-

mense capital this would require, despite the fact that the world demand for rice would assure a market. It is not a particularly hot land, and it does have rain well distributed throughout the year. It could produce rice, rubber, coconuts, palm oils, or many of the other tropical crops in demand today. That it does not is a quirk of Brazilian nature, not of Brazil's physical geography. The rivers coming through the hilly uplands to the Amazon present the same kind of hydroelectric potential that Venezuela has in the Orinoco. Yet the population map shows that the interior of Brazil is empty, and this is true whether it is flood plain, plateau, or hill land, whether there is hydropower potential, agricultural potential, or developable mineral wealth.

PROFILE OF BRAZIL

Population
103 million; about two-thirds Caucasian

Area
3.3 million square miles, about half of South America (The United States has 3.6 million square miles; Europe, 3.8 million square miles)

Gross National Product
$41 billion in 1972, up from $20 billion in 1962

Per Capita Income
$400

Education
50 percent illiterate

Participation
25 million of 100 million people in the money economy

Population Growth Rate
2.8 percent per year; expected population by A.D. 2000 to be well over 200 million

In northeast Brazil there is a well-populated area that has been a source of vast misery for its people. It is overcrowded in relation to its degree of development and is subject to periodic drought that sends swarms of people to seek a livelihood elsewhere. The landscape of this region often looks like a semidesert. Much of this is the result of man's abuse of the land—a familiar story. This was a wealthy sugar-producing area in the sixteenth century. Then the investment capital shifted elsewhere, the area became decadent, and the misuse of the land decreased the carrying capacity of the region while population increases aggravated the overgrazing and erosion. Only a quarter of the people can read and write. Today the Brazilians are applying all the latest techniques—contour plowing, insecticides, soil conditioners, and power machinery—in an attempt to revive the area, and they are revolutionizing the agriculture. If this effort succeeds, the area will have gone from prosperity to poverty and a damaged physical resource, back to prosperity again without any physical-environmental causation—indeed in spite of physical-environmental deterioration.

The Atlantic coast of Brazil from São Paulo north is tropical rain forest, and yet this is the locus of the major cities and the center of the population. The city of São Paulo has over 6,000,000 people, Rio de Janeiro has 4,250,000, and five of the other state capitals have over 500,000. Brazil thus has a high percentage of urban dwellers (55 percent) and a high concentration (57 percent) of population along its eastern seaboard, north of São Paulo and south of the Amazon's mouth. Thirty-five percent of the population is in the subtropical south. Very few, about 8 percent, are scattered over the vast interior. The Brazilian history of looking homeward across the Atlantic explains this far better than the physical geography.

Brazil is a land with vast resources of iron, oil, timber, manganese, industrial diamonds, and hydroelectric power. Its 100 million people occupy almost half of South America and produce coffee, sugar, rice, cocoa, citrus fruit, and cotton.

Brazil is alternately described as the land of the future and as a sea of despair—either statement could be true. The potential for becoming a major world power is all there, but the mismanagement at times makes it appear as if the potential will never be realized. Some of the historical and psychological implications of this will be dealt with later (see Chapter 5), and the present situation will be discussed here.

A pro-Communist regime mismanaged the government and worsened the inflation that has been virtually endemic in Brazil. A military coup overthrew this regime in 1964 and attempted to control the inflation and speed up the development of the country. Since World War II the United States has provided Brazil with more than $2 billion in aid, mostly in loans which must be repaid. Drastic declines in the prices of raw materials, the major part of Brazil's foreign exchange, at first made repayment of this debt difficult, but the growing scarcity of raw materials is now raising prices, and Brazil's position seems certain to improve.

Within the country there is an enormous disparity between the great masses of the poor and the few who are very rich. The Brazilians speak of the many people who are not in the money economy and the relatively few who are. This is not peculiar to Brazil but is a mark of the developing nations throughout the world—the sort of opening that the Communists exploit with maximum effectiveness. There are extreme

differences in the country as one might expect in a country as large as the United States. The overcrowded and abysmally poor northeast is the scene of maximum unrest. Rio de Janeiro is a center of growth at a leisurely pace, while São Paulo is a center of energetic bustle and growth, with the individual Paulista moving at a pace that astonishes and exhausts the more leisurely native of Rio.

São Paulo with 5 million people and a growth rate of 6 percent is the fastest-growing city in the world and the center of South America's biggest industrial complex. It is a city of skyscrapers, factories, and traffic jams, and has built a subway system and superhighway complex to alleviate some of its problems. It is an automotive center that ranks twelfth in the world and is exceeded in the Western Hemisphere only by the United States and Canada. There are varied industrial plants as well as huge investment by United States and European companies. It is estimated that about 75 percent of the people in the metropolitan area are in the money economy as opposed to about 25 percent in the country as a whole.

The descriptions of Paulistas stress their widespread ambition. Unskilled workers study to become skilled workers, bookkeepers study accounting, and accountants study business administration. One of Brazil's leading magazines wrote that other Brazilians cannot hide their dismay when they come to São Paulo. They inevitably ask, "Why do Paulistas run instead of walk?" The Cariocas, the light-hearted inhabitants of Rio, scoff, "All those people know how to do is work." Lest one leap to conclusions about environmental influences, it may be pointed out that the two cities are both in southern Brazil and in similar climates, with São Paulo having a slight climatic edge due to the city proper being 40 miles inland at an elevation of about 2000 feet. The difference in physical geography does not seem great enough to account for the psychological differences.

It is not just São Paulo that is moving ahead, for the federal government is pushing mightily to try to develop the long-ignored interior of the country. A new national capital has been built at Brasília in the remote interior. The cost has been in billions of dollars, and so far most of the governmental offices have placed only skeleton staffs there and continue to do business in the major cities. But construction is underway on more than 50 highways to interlink Brasília, the large coastal cities, the state capitals, and the frontier areas. The placing of a capital in a wilderness duplicates the placement of Washington, D.C. and Canberra in Australia, for both were deliberately placed in undeveloped areas.

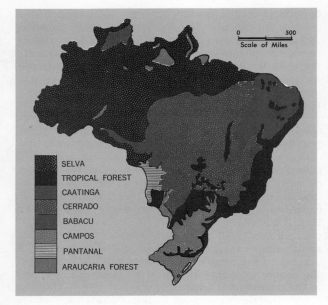

NATURAL VEGETATION

SELVA
TROPICAL FOREST
CAATINGA
CERRADO
BABACU
CAMPOS
PANTANAL
ARAUCARIA FOREST

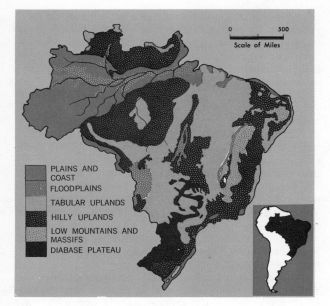

LANDFORM

PLAINS AND
COAST
FLOODPLAINS
TABULAR UPLANDS
HILLY UPLANDS
LOW MOUNTAINS AND
MASSIFS
DIABASE PLATEAU

Brazil has been increasing its agricultural production rapidly. This has been accomplished in large part by bringing new land into cultivation, largely in southern Brazil where soils can be farmed by using limited modern technology without short-term soil exhaustion. Beginning with slash-and-burn farming of partially cleared land and continuing with clearing until the land was ready for commercial cropping gave a 20-year yield expectancy before heavy fertilization had to be undertaken. But that kind of land is now almost gone, and further expansion must be into the areas of tropical soils where soil exhaustion must be expected in a few years.

Brazilian studies indicate that there are about 500 million acres of land with good potential for cultivated crops in the tropical area if an advanced level of technology is applied. They also report an additional 500 million acres of useful but less rich soils in the area. This can be placed in perspective by considering that Brazil is now cultivating 74 million acres and the United States is cultivating something less than 300 million acres. However, the use of the tropical lands of Brazil is qualified by the phrase "application of advanced technology." Only about 10 percent of Brazil's land is now being so farmed. Whole series of studies must be made to determine how to use fertilizers and crop rotations, how much machine tool work can be used, and what crops can be produced at what costs for what markets at home and abroad. The conclusion of the research agronomist working for the Agency for International Development was: ". . . Brazil possesses sufficient soil resources suitable to application of industrialized agricultural technology, that, if used according to their potential capacity, . . . [would enable it to] compete favorably with any nation in the world."[2]

The potential for tropical agriculture has long been discussed by experts. We have already seen examples—in the King Ranch's successful demonstration of cattle raising in tropical Australia and in the experimental work done by the International Rice Research Institute in Manila. Both sets of findings are applicable to all of the wet tropics of the world—in America, Africa, Asia, New Guinea—and should be borne in mind in examining the potential of the tropics as a whole.

Occasionally it is of value to give some thought to projects that can remake the earth. An American group has proposed such a project for the eastern portion of South America. The group points out that with a few low, earth-fill dams, the least costly type to construct, the entire waterway system of this part

[2] John C. McDonald, "Brazil's Untamed Cerrado: Potential Farms," *Foreign Agriculture*, vol. X (1972), p. 9.

MAP 3-3 Brazil: Vegetation, Landform, and Population

Some of the botanical wealth of Brazil is suggested by the number of vegetation types shown here. The selva and tropical forest are rain forests. The caatinga, cerrado, and babacu are scrub and thorn forests. The campos and pantanal are grasslands, and the araucaria forest is a mid-latitude forest.

The vegetational boundaries tend to cut across the landforms. The population is strongly centered in the uplands along the Atlantic coast. Note the absence of population in the interior and northern hilly uplands. Note especially the huge area of cerrado. This is the land that could readily be converted to cattle raising of the type illustrated by the King Ranch developments in Australia or to other agricultural uses.

(Reprinted from Focus, American Geographical Society)

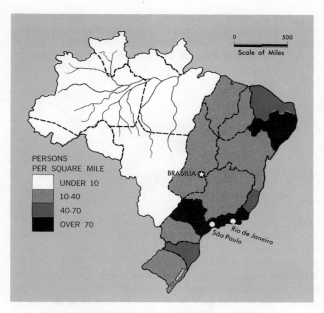

POPULATION DENSITY BY STATES

PERSONS PER SQUARE MILE

UNDER 10
10-40
40-70
OVER 70

of South America can be connected: from the Orinoco, into the Amazon, into the Madeira, thence into the Guapore, and from there into the Paraguay River and back to the Atlantic at Buenos Aires. This system would allow interior steamship service to the vast hinterland that is now so poorly served with transportation. The plan involves the construction of six dams and the creation of four lakes the size of Lake Ontario. In addition, a dam on the Amazon near Monte Alegre would create an inland sea the size of Texas. Since the Amazon alone has a flow equivalent to 14 Mississippi Rivers, the dam envisioned there would generate 28 percent as much power as is used in the entire United States at the minimal cost of 1 mill per kilowatt hour. The project is so vast, especially the plan for creation of an inland sea in the Amazon basin, that serious questions are raised about its climatic effect. One French engineer has calculated that the weight added there would slow the rotation of the earth by three seconds per year. The project is taken seriously, though the difficulties of negotiating the necessary international agreements guarantee that development will be delayed. The gains in transportation, access, power, fisheries, and other values have to be balanced off against great losses of land. For the moment, Brazil can easily afford the land loss and may decide that power supplies and access are worth it.

Brazilians are promoting an all-weather highway 4000 miles long to open up the Amazon Basin. They hope to settle land with people from the frequently drought-stricken northeast corner of Brazil and also to tap vast natural resources in the virtually uninhabited interior. The trans-Amazon highway stretches from Brazil's easternmost tip across the widest part of South America to the Peruvian border. The road runs parallel to but 200 miles south of the Amazon, and it crosses four rivers wider than the Mississippi. This is to be the principal road, but 9000 miles of connecting roads will form a network linking both north and south and the coast and the deep interior. The project has caught Brazil's imagination, and now tens of thousands of applicants are on file for land in the interior along the new roads. The government is granting long-term loans to build a house and pay for 250 acres, land clearing, and other costs, with preference going to landless farm workers. Towns are planned every 25 miles or so; each town consists of colonists in 50-family clusters. There are exciting prospects: production of cotton, cacao, coffee, rice, and cattle is being tested; Manaus, the glamour city of the rubber boom, is awakening; port traffic is heavy; hotels are full; and an airport has been built to handle super jets. Mining prospects are particularly promising. Some of the world's largest deposits of iron ore, bauxite, and tin have been found. Moreover, the discovery of gold, silver, platinum, manganese, diamonds, and rare metals is expected.

Amazonian Mineral Discoveries
Near Marabá: the world's largest deposit of iron ore
Trombetas River: an equally important bauxite (aluminum ore) discovery
Amapá territory: manganese
Pôrto Velho: tin deposits in alluvial materials
Petroleum: immense findings adjacent to the Brazilian borders with Colombia, Ecuador, and Peru, with probable extensions into Brazil

Interest and support for Brazilian projects come from all over the world. An American has achieved record rice yields; he is also doing well with oil palms and a fast-growing African tree that is useful in the wood pulp industry. Both the King Ranch and Swift-Armour have invested millions of dollars in cattle ranches in the interior. The Brazilians liken the whole movement to the homestead law in America that opened the West to a flood of settlers, filled the land, made it productive, and contributed so much to the making of America. Brazilians see themselves as a great power in the twenty-first century—just 25 years away.

The rest of humid tropical South America presents a similar picture. In Colombia 80 percent of the people live in the western third of the country in the mountains out of the tropics. The principal products are agricultural—coffee for export and corn for consumption at home. The principal imports are manufactured goods. Colombia, like Brazil, has an opportunity to make some drastic changes in the world by building a few dams. The divide between the Atlantic and the Pacific in the Choco Valley is so low that an interocean connection was discussed as early as the seventeenth century. Dams on both sides of the divide would justify their costs by the power gained alone. The addition of locks would make this a bypass of the Panama Canal. Since the cost of the dams would be borne by the hydroelectric power, the tariff on ships would be set very low: one suggestion was $.10 per ton as opposed to $100 charged at the Panama Canal. With the 400-inch rainfall of the Choco Valley, the hydroelectric potential would be enormous, and it would supply the three developing cities Medellín, Manizales, and Cali, which are just 50 miles to the west. The 1,600,000 inhabitants are doubling their consumption of electric energy every seven years, hence a market for the power is assured.

The South American countries within the wet tropics have great potentials for development, but handicapped by a colonial past, recent political instability, peculiar population distributions, and—except for overcrowded cities—too few people on the land, they have lagged in the underdeveloped nation status. There is considerable evidence now of growth, and if political stability can be maintained, such nations as Venezuela and Brazil could quite rapidly emerge as developed nations.

THE OLD WORLD TROPICS

Africa: The Problem of Cultural Lag

Africa south of the Sahara is, racially, Negro Africa, in the past thought of as a vast tropical forest inhabited by a culturally retarded people. The inferences drawn are usually that the tropics prevented development or that the race is inferior—or both. We must consider environment, race, and accomplishment.

The wet tropics of Africa are not as extensive as is usually thought, for North Africa is mostly desert and steppe, as is South-West Africa. Only about one-fourth of tropical Africa has dense forest, and large areas are covered with scattered scrubby trees set in brushy and grassy landscapes. This scrub area was partly caused by the action of man through his long-continued practice of slash-and-burn agriculture, but it also reflects the fact that much of Africa has only about 40 inches of rainfall—not a large amount for a tropical climate.

Much of tropical Africa is highland, because a short distance inland there is a rapid ascent to a plateaulike upland. This upland averages 1000 feet above sea level, even in the Congo basin, and in East Africa there are large areas of 5000-foot elevation or greater. Since the normal temperature decrease with elevation is 3° per 1000 feet, many equatorial African stations are 6° to 15° cooler than their latitude would suggest. Launde in the Cameroons has average monthly temperatures ranging between 70° and 74° F., and Entebbe, Uganda, right on the equator, has a range of 68° to 71° F. for its coolest and warmest months. These are April, May, and October temperatures at New Orleans, and they are much cooler than the five, hot summer months of that "subtropical" city.

Water power potential is very high because of the tropical rainfall on high mountains along the eastern side of the continent and because all the rivers have to reach the sea by descending from the plateaulike interior. This creates waterfalls that are among the world's largest potential sites for hydroelectric development. Resources of coal are small, and oil prospects

MAP 3-4   Africa

The location of the major peoples of Africa is shown here, together with arrows suggesting movements of people and ideas. The Bantu expanded southward relatively recently and had not yet reached South Africa when the Europeans landed at Cape Town. The extent of early sea trade with East Africa on the part of Egyptians, Indonesians, Chinese, and Arabs is just becoming recognized.

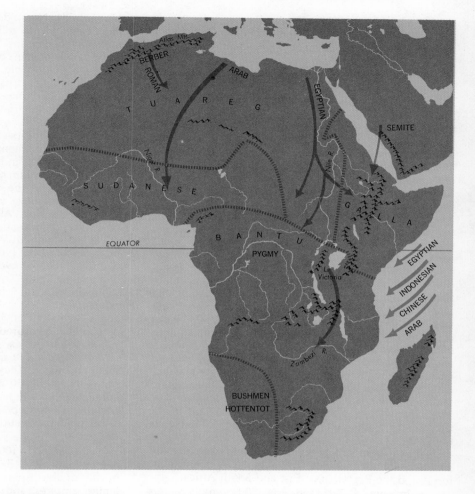

in tropical Africa are little known, but the central Sahara oil could, of course, be easily piped into tropical Africa. Mineral wealth is known to be great, for Africa is already a major producer of antimony, chromite, cobalt, copper, manganese, tin, industrial diamonds, and vanadium. In times past, tropical Africa has been a major producer of gold, but the major developments are now outside the tropical wet parts of Africa.

Tropical soils are often poor, and Africa does suffer from this. The relatively level land has allowed old soils to develop with the associated extreme weathering and leaching. Nevertheless, Africa has large areas of good soils, especially in the East African highlands, and coffee, sugar, palm oils, and other plantation crops have already been most successfully grown here.

From this description one cannot make much of a case for Africa as a geographically handicapped area. Even the tropical climate, if this were considered a handicap, is modified by elevation, and great civilizations have flourished in hotter, wetter, and in some ways poorer environments than this one. Further, in Negro Africa there are many kinds of environments: hot, wet lowlands, warm well-watered uplands, cool, moist highlands, grasslands, wooded and grassy savannas and rain forest, plains, plateaus, and mountains. It is a rich and varied environment well stocked with useful plants and animals. One may well ask why the Negroes have done so little with so good an environment, but first one must review what they have actually done.

All of Africa was held by people of Negro race until the end of the Paleolithic period, at which time Europeans appeared in northernmost Africa. Small, brown-skinned people of the Bushman type seemingly held much of Africa originally, but they have been progressively displaced by the expansion of the large, very dark-skinned Negroes, especially the Bantu, after A.D. 500. Migrants into the region were the Hamitic people who entered Egypt from Arabia and the Semitic- and Hamitic-speaking people of European race who entered Ethiopia. There was also a minor racial influence from Indonesia that introduced Asiatic race types to the east coast of Africa.

Cultural stimuli reached Negro Africa principally from the north and the northeast. As early as 1500 B.C. states modeled on the Egyptian–Middle Eastern pattern (priest-king, absolutist rule with courts, harems, nobility, and so forth) were developed by the Negroes in Nubia, and despotic states of typical African type derived from these models began to appear along the northern border of the western Sudan. By the eighth century B.C. the people of Meroe in the Sudan were smelting iron, surely due to ideas stemming from the Middle East and spreading into Africa via Egypt. After 1000 B.C. maritime trade across the northern Indian Ocean led to the introduction of Malaysian and Indian food plants into East Africa, where they were combined with the Middle Eastern and West African domestic plants to strengthen greatly the African agricultural base. People from Indonesia, principally Borneo, established settlements along the East African coast in order to engage in trade with Negro Africa. After A.D. 1 the empire of Ghana arose in the western Sudan, and the new food plants introduced from Malaysia allowed the growth of dense populations in the Guinea coast area, where the Bantu people had their homeland. It was the expansion of these people that thereafter progressively reduced the areas held by the smaller peoples of Africa: the Pygmies of the Congo forest and the Bushmen of East and South Africa.

Between A.D. 1 and A.D. 500 Semitic traders displaced the Indonesians in the East African trade, and the Indonesians moved to the island of Madagascar, establishing their speech there, where they remain even today a major part of the racial stock. After A.D. 500 milking belatedly began spreading into Africa from the north, although animals used only for meat and hides had been introduced much earlier. A parallel situation had occurred in China, where cattle keeping was introduced without milking.

The absence of milking in large parts of the cattle-keeping world has long been a puzzle. Recently it has been found that most of adult mankind cannot drink fresh milk. Only northwest Europeans can do this. All of mankind has enzymes in infancy that break down milk into digestible materials. Most of mankind loses the ability before adulthood. As a result most

Mongoloids, Negroids, and Amerinds cannot tolerate large doses of raw milk. Neither can 10 percent of the northwest Europeans. The resistance of the Chinese and many of the Africans to milk and milking then becomes more understandable.

It is not a total explanation, however, for milk that has been converted into butter, cheese, or fermented milk in any form is then readily digestible. It is just these forms of milk that are used in large quantities by the pastoral people. The African area is mixed in its treatment of this problem. Some Africans milk their cattle, and it has been found that there is a good relationship between milk tolerance and Caucasoid blood in the tribes of northeast Africa. Other African tribes have substituted the bleeding of their cattle for milking. Although the idea is abhorrent to us, it is a practical and useful way to obtain food from the animals without killing them. Blood is close to milk in its makeup, and bleeding the herd does not interfere with the breeding cycle of the females; moreover, it is possible to "milk" males and steers as well. It is so useful an idea that it makes us look a bit foolish for not adopting it.

The flow of ideas southward in Africa at this time is also measured by such stone buildings as Zimbabwe in East Africa. These stone buildings are the center of considerable controversy concerning who built them and when. They probably were built by Negroes between A.D. 1000 and 1400. There must have been contact with the Indian Ocean traders, for Chinese porcelain and Indian beads have been found in the ruins. Local gold production was the probable stimulus for the trade.

After A.D. 1000 the pastoralists who practiced dairying and the Bantu continued their expansions, mostly at the expense of the small Negroid peoples. States, empires, and kingdoms continued to rise and fall in West Africa, the Sudan, and East Africa, the areas most influenced by developments in areas bordering the Mediterranean and Indian oceans.

Progress in tropical Africa can then be summarized as follows: With the introduction of agriculture from the north between 4000 and 3000 B.C., a steady cultural growth began. Within 2000 years states and empires appeared on the northern margin, and from continuing stimulation through trade across the Indian Ocean an agricultural and population base was laid for the expansion of these culturally advantaged northern people at the expense of the nonagricultural, smaller Negro peoples. The picture resembles the spread of ideas among the American Indians: those having access to ideas reached the stage of empires and states and those too far from the areas of stimulus remained in the stage of hunters and gatherers. The hunters and gatherers near the centers of development who did not change were overwhelmed by the increasing numbers of the peoples who did change.

While we may ask why a people reaching the stage of empire between 2000 and 1000 B.C. did not go on to ever greater heights, we must recognize that the course of events in Africa is little different from that in most other areas, for no race has maintained an even course onward and upward.

The location of tropical Africa poses problems in terms of accomplishment. Is it close to or far from the important center of growth of ideas in the Middle East? It clearly is an area of cultural lag, but how much of this is due to distance, how much to physical-geographical factors such as climate and soils, and how much to

race? In crude distance central Zaïre, Great Britain, Inner Asia, and India are all equally distant from the Middle East. Great Britain and India could be reached by water and experienced early growth toward civilization. African contacts were limited to the margins of the continent, for no waterways penetrate it. Although the Sahara has never stopped contact entirely, it has greatly reduced the flow of men and ideas from the Mediterranean to Africa south of the desert. If cultural growth is in proportion to the length of time and the intensity of cultural stimulus, the Sahara must be seen as having greatly reduced the intensity of contact. Unusual amounts of time would be required to compensate for this. The vast desert west of the Nile River, with its long and difficult oasis routes, reduced the flow of men and ideas to a trickle. Even the Nile valley

Zimbabwe, Southern Rhodesia. *Few archeological sites have attracted as much attention as the series of sites in this area. They are built of dry laid stone without mortar. Of native construction, they represent some of the inflow of ideas into southern Africa in medieval and later times.* (Southern Rhodesian Department of Tourism)

seems to have been unsuccessful as a corridor for transporting ideas. In part this was because the Egyptian civilization spread very slowly up the Nile and cultural stimulus along so thin a thread was perhaps a pinprick instead of a shove. More time would be needed when contact is on a narrow front.

Africa's contacts have been punctuated by stormy interludes. If we start with the latest happening and work backward, we can see what the effect of changing contacts can be. When the Europeans entered the Indian Ocean just after A.D. 1500, they found a rich trade flourishing between East Africa and India. They disrupted it so thoroughly it was as if they had killed the goose that laid the golden egg. Production of gold and ivory and other products stopped, and not only did the trade die but contact with the people of interior Africa was cut off. Thereafter came a long period of confusion and isolation. Nearly three centuries later the Europeans finally penetrated the interior of Africa and began the full-scale development of roads, railroads, mining operations, and so forth. Were there similar setbacks whenever a new group dominated the trade with Africa? If so, Africa's slow progress despite the long-continued Indian Ocean trade becomes more understandable. The succession of Egyptians, Indonesians, Yemenite Arabs, Muslim Arabs, and finally Europeans suggests that roughly every 500 years a violent setback occurred.

To these disruptions of external sea-borne contacts we must add comparable disruptions in overland contacts and equally great cultural setbacks due to internal explosions. For example, the expansion of the Bantu from West Africa and the cattle-oriented culture of the mixed race groups from northeast Africa must have disrupted the culturally advanced people whose agriculture was marked by irrigation of terraced hillsides, reminiscent of the rice terraces of Southeast Asia. The course of cultural history is punctuated with these events. There is a parallel in the outpouring of the Germanic tribes and their overwhelming of France, Spain, and Italy, with a consequent cultural collapse so deep as to be called the Dark Ages. The course of events in Africa seems, then, to follow normal patterns, and the locale and race are only variations on a theme. The Africans were more isolated and lacked a Byzantium to keep the cultural flame bright while adjacent areas fell back into the darkness of barbarism.

It is common to use colonialism as a whipping boy for the troubles of underdeveloped areas. European colonial dominance of Africa had its seamy side, but it did bring development of resources, the beginning of a modern transportation system, and the preliminary stages of educational, governmental, and economic systems. The Africans were just beginning to enter this system as real, rather than passive, participants when the postwar burst of freedom struck. In many areas the colonial political units have splintered into tribal groupings, and often there has been savage intertribal warfare. In general, disruption of economic, political, educational, health and hygiene systems has been extreme, and Negro Africa, with few exceptions, can be said to have regressed in most measures of achievement. The causes seem perfectly clear. The people were ill prepared to take over their own affairs and too ready to bite the hand extended to aid them. If, in past times, similar collapses occurred, would not similar events have followed? Why then invoke race or climate as causes? We should be careful not to conclude that the difficulties at present prevent the possibility of political stability and a resumption of growth of development in the future.

Africa does have some handicaps, notably in tropical diseases. It is suspected that this may be the home of malaria, for the Negro race is notable for its specific resistance to this disease, which suggests long exposure and the development of a partial immunity through a blood abnormality called sickle-cell anemia. Africa is also the focal area of sleeping sickness, carried by the tsetse fly. This disease attacks man and his domestic animals. The wild animals are not killed by it but are infected with it and supply a reservoir of infection from which the disease is constantly reintroduced into human settlements. Most of the other tropical diseases are also present. Has the disease burden been unusually high, and has this been a contributing cause of Negro Africa's failure to advance?

Sleeping sickness is caused by a trypanosome that is carried by a fly. It sounds moderately simple until the details are examined. There are varieties of these flies, they live in different habitats, and they have different habits. Men may be more exposed to one type if life is lived in a forested area and to another type if the way of life is adapted to a savanna area. Changing the character of the vegetation, something that man does when he pastures cattle, practices slash-and-burn-clearing, or burns over territory, may limit or spread the fly. Brushy and woody landscapes tend to favor the fly, and many land uses promote this kind of growth. Men hunting in the savanna reduce the game that harbors the fly, and thus the chance of infection is reduced. In general the fly-borne disease is widespread in central Africa. It has been suggested that this disease may have been a major factor limiting the spread of civilizing influences into Africa and one of the contributing factors to the lesser role of Africa. With modern knowledge of the relationship between land use and fly infestation, it is possible to employ

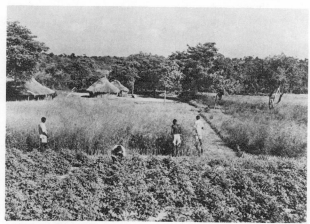

(Upper) **Food Distribution in Zaïre.** *Manioc, a New World crop, is being distributed by men in Western dress to women in native attire. Such mixtures of cultural origins and stages are commonplace in the world today. The vegetation is typical of the tropical forest.* (John Moss, Black Star, for World Bank)

(Lower left) **Well-Watered Area in Kenya.** *This area near Makueni in Kenya is being aided by loans for development from the World Bank. The round, thatched houses are typical of the area. The relatively heavy vegetation suggests a good water supply for the land.* (British Information Services for World Bank)

(Lower right) **Open Land in Kenya.** *Much of Africa is arid, and there is little rain forest. What remains is largely open land such as that shown here. This cleared land has been used agriculturally for millennia, and it owes its open nature more to the hand of man than to any natural cause. This is not arid land, though one has a feeling of a dry land because of the extensive clearing.* (Ivan Massar, Black Star, for World Bank)

some of the new techniques, such as releasing sterile male flies or perhaps the oncoming "third-generation insecticides" with their potential for limiting attack to single species of insect, to solve this disease problem that places so heavy a burden on tropical Africa.

The sleeping-sickness problem has been attacked with vigor by the colonial powers but with very little success. Great, organized slaughters of wild animals have been carried out in the hopes of reducing the host population. These have all failed, for the wild-animal populations rebuild, and the domestic animals continue to become infected.

Here we have an example of man plodding ahead in the face of vast difficulties instead of trying to make the means at hand accomplish the task. The idea of domestic animals has not been generalized. Domestic animals are cows, pigs, sheep, goats, ducks, and chickens. Giraffes, African elephants, zebras, gnus, antelope, and deer are not only not domestic but are generally said to be nondomesticable. If these animals could be domesticated, Africa would have sleeping-sickness-resistant "cattle." This idea seems not to have been seriously considered.

For part of Africa it has been suggested that it would be wise to use native animals as a protein resource. Students of the ecology of tropical Africa point out that many of the old tropical soils, when stripped of natural vegetation and farmed, either erode or harden into a rocklike iron crust, with the result that pastoralism often leads to erosion and declining yields. They suggest that the animals natural to the area are so suited to harvesting the native vegetation and resisting the local diseases that they are the logical means of production, and they call for the management of game animals as a crop. As they point out, there are 20 to 30 useful hoofed animals in Africa that are well adjusted to the environment. To disregard them and introduce cows, sheep, and goats is to both narrow the resource base and shift to poorly adapted animals.

The old tropical soils of Africa are poor in potassium, calcium, and nitrogen, but the vegetation and wildlife are adapted to these old soils. Deep-rooted leguminous trees that produce protein-rich beans are important feed sources, and giraffes can reach up and harvest them. The grassy areas are poor feed sources, for they grow on soils deficient in calcium, phosphorous, and nitrogen, and in the dry state these grasses are poor sources of protein. Since the grassy areas are in the zones of long dry season, they provide poor feed for six to ten months of the year. Only on the arid, less leached margins or on young volcanic soils is there grass that in the dry state is protein rich.

Each native animal occupies a special niche, harvesting his food his own way, and is physiologically adjusted to these mineral-poor soils. The giraffe has no competition in his treetop browsing, the hippopotamus rules the watery realm, the buffalo grazes the savanna, and the elephant forages everywhere except in the desert and the deep swamps. He harvests marshes, bamboo forests, dry bush, arid grassy savanna. He makes paths, digs water holes useful to other animals, plows the land by pushing over trees so that their roots bring up the subsoil. Elephants are easily harvested and highly edible. The potential beef animals can be listed in their equivalent in sheep: one giraffe equals 40 sheep; the elephant is the equivalent of 80 sheep; the hippopotamus equals 60 sheep; the wild cattle, called buffalo in Africa, are worth 15 sheep; and various antelope are rated from 3 to 12 sheep. It is a remarkable test of man's perception that he has not seen these possibilities that the physical environment flaunts before him. These animals need not be domesticated, sheltered, fed, or protected from endemic diseases to which they are immune but imported animals are not. They need only be managed and harvested. Harvesting wild animals is, of course, a very extensive use of the land and is only economic as long as land is cheap. On the other hand, domestication of some of these animals could lead to intensive land use. The real point is that there seems to have been no perception of the local animals as potential domestics until the mid-twentieth century.

This is a precise parallel in animal domestication to the situation described for wild rice among the American Indians. Cases such as these show how difficult it is to perceive the possibilities in the environment. What we do perceive is culturally conditioned, and our inability to see even obvious relationships is startling.

When the "nondomesticable" notion is examined, it is often found to be untrue. The African elephant is as domesticable as the Indian elephant, and the facts have been in the records for 2000 years; the Romans and the Carthaginians used domesticated African elephants. More recently, elephants in the tropical rain forest were tamed and trained as work beasts. Probably the goatlike antelopes and gazelles, the striped horses, and numbers of other African wild animals are potentially useful domestic animals that the environment displays before mankind, but man in his unimaginative way still imports animals that are nonresistant to disease.

The same story is true for the plant world. Africa is rich in plants but has supplied relatively few domestic plants to the world. The watermelon and the grain sorghums (milo maize, kaffir corn, and their relatives) are Africa's major contributions to the world's agriculture. Africa domesticates are sorghum, millet, cowpea, watermelon, gourd, okra, and yams; these have become widespread outside Africa. There are an equal number of plants that have only local importance. The problem of single or multiple origins of agriculture, and, if multiple, whether they are examples of stimulus diffusion (in this case the spread of the idea, but not the plant), is as yet an unresolved question.

Although Africa has made contributions to plant domestication, vastly more has been gained in recent times: maize and cocoa from America, bananas and rice from Asia. Two other African plant contributions, coffee and cotton, came not from wet tropical Africa but from highland Ethiopia. The record of Negro Africa looks poor, but the record for northwest Europe is no better, and for America north of Mexico it is much poorer. When it is considered that most of the peoples of the world contributed nothing to the agricultural heritage that is the real basis of all great cultural accomplishment in the past 10,000 years, the Negro contribution of some plants takes on added significance.

Negroes in America now have their destiny virtually in their own hands. What will they accomplish? What will this tell us about race and achievement? How can we use this knowledge to predict the probable sequence of events in other parts of the world? Note that in 100 years—from the time of the Civil War—American Negroes have slowly and then with increased opportunity accelerated their rise both economically and professionally. Negroes were said by some to be incapable of self-government, today they boast mayors of major cities and a Justice on the Supreme Court. In every area of learning growing numbers of Negroes are making important contributions. It is clear that opportunity rather than race has been the determining factor.

What kind of progress, then, are we to expect elsewhere, particularly in Africa? If we use America as our model, is it reasonable to expect a transition to take about 100 years? In Africa, with Negroes acting for Negroes, no segregation, and no carry-over from slavery, there may be more rapid progress. On the other hand, with a poor economic base and less political stability, there may be slower progress. It has been over 100 years since the mixed racial societies in South America achieved independence. At that time their economic and education base was then little better than that in Africa today. They splintered politically and only now do they seem to be on the verge of major economic and social success. Because the more successful countries have varied racial backgrounds, success cannot be attributed to race.

An assessment of black Africa's future may be to expect political turbulence similar to that in South America, with education and economies improving gradually at first and then accelerating, as was the case in America. All this change is economic, political, and historical in origin—not racial. Yet the retarded progress of Negroes in America and the people of Indian, Spanish, Portuguese and other varied descents in South America has been blamed on race. But what race? With Mexico and Brazil forging ahead, once more we see that race itself does not hinder or promote human achievement.

This review, then, should remove the picture of Africa as a vast tropical forest, occupied from time immemorial by a backward people. Instead, a picture emerges of considerable accomplishment by the Negro people, and this accomplishment was achieved in areas where proximity to and contact with the outside world would lead us to expect it.

Modern development

When modern development of tropical Africa began, its pattern was much like that of Brazil. Coastal areas were the focal points to which produce came to be shipped overseas. With this outward look, transportation systems developed as short-line routes feeding the nearest port; areas were not linked together, and there was no transportation network except for today's airlines. In part the flow of heavy goods such as the copper of Katanga has followed national lines rather than natural routes to the sea. In the present national fractionization of Africa, there is unlikely to be either the will or the immense capital necessary for the construction of a national rail network. This is not due to the physical geography, however, for the African terrain is not a difficult one for railroad construction.

Most of Africa remains today in the mixed economy of subsistence village dwellers who are just being drawn into the orbit of the world of commerce. Under colonial rule the extractive industries were started; mining began on a modern scale, with metals other than the precious ones sought. Agriculture was expanded into commercially valuable crops such as chocolate, palm oil, cotton, tea, coffee, and peanuts. The native populations typically had to be urged into production of crops of commercial value, for their outlook was comparable to that of the Yumas: no incentive existed to produce for the world market. However, this attitude is typical of all noncommercially oriented peoples, and only when desires for the amenities of the civilized world develop do they become willing to work for wages, to produce materials they do not themselves consume.

Tropical Africa has both advantages and disadvantages in the modern world. Agriculturally there is little that cannot be grown there, for the area ranges from coastal lowlands to highlands, from old acid soils to rich young volcanic soils, from high rainfalls to low ones, and from constant rains to seasonal showers. Perhaps the greatest disadvantages are the large extent of old soils, with the characteristic accumulation of iron and low amounts of soluble nutrients, and the broad belt where the tsetse fly makes cattle keeping impossible and human living risky. Yet agricultural chemistry can enrich soils, though at a cost, and it is not unthinkable that modern science can eliminate the tsetse fly. In the modern world of instantaneous communication, of spreading education, and of ever-growing commercial connections, Africa is unlikely to remain isolated, and despite predictable political travail, growth and development seem probable.

India and Southeast Asia: Land and People

Wet tropical Asia includes most of India, Southeast Asia, Indonesia, and southern China. This area holds representatives of all the major races of man. The people of northwest India are dark-skinned members of the Mediterranean race, while those to the southeast are Dravidians, a darker-skinned people often called Australoids. Probably all India was originally dark skinned and curly haired, and Caucasoid people entering the area from the northwest modified the original situation. The Burmese, Thais, Annamese, and Chinese are usually classified as Mongolian in race. Actually, the true Mongoloids are to be found in Inner Asia and northwest China. The original inhabitants of Southeast Asia were small black people. Relatively pure survivors of these people are found today in relatively remote areas such as Tasmania, the Andaman Islands, and the mountains of the Philippines. The people of Southeast Asia today are a racial mixture of Mongol and Asiatic Negritos. The people of the islands between the mainland and Melanesia are called Malays or Indonesians. They are an intermediate group, perhaps more closely related to the Mongoloid race than to any other. From New Guinea to the Fiji Islands the people are Negroid. Why there should be two centers of Negroid people in the world, one in the Pacific and one in Africa, is one

of the mysteries of the origin of the races of man. As suggested earlier, possibly a continuous distribution of dark-skinned people has been broken by relatively recent expansions of the Caucasoid and Mongoloid peoples. For comparative purposes it is useful to have in the tropical area of Southeast Asia and the adjacent island world nearly one-fourth of all mankind and representatives of all the races. These differences in way of life cannot be a result of the tropics, since the whole area is tropical and most of it has a monsoon-type rainfall. While one might expect racial uniformity in such a region, the earlier discussion of race origins has pointed out that relatively late movements primarily of European racial types into the northwest Indian area and Mongoloids thrusting into the southeastern area have modified the ancient race distributions. This gives an even better test of race accomplishment, for both the Europeans and Mongols are mid-latitude races. Moreover, there has been the persistent myth that the Europeans could not succeed in the tropics, yet in India the white race has played a major role in the northwest for at least 3000 years.

Monsoon India

Monsoon India is largely Hindu. The Muslims occupy mainly the drier northwest, although they have an outlying group in the Ganges delta, which became

MAP 3-5   India: Population, Landform, and Rainfall
This series of maps is focused on the population problem. The total population is large; it exceeds 1000 persons per square mile over considerable areas and is growing rapidly. The overall population density is 500 per square mile. This is 10 times that of the United States but only half that of Java or Holland. In the densely settled parts of India, the land is supporting two, three, and at times four persons per acre. Comparison of the population density map with the landform map shows that the high densities are on the margins of the Deccan Plateau. The rainfall graphs show the distribution of rainfall through the year. Notice that Hyderabad has no month with as much as 10 inches of rainfall and more than six months of very little rainfall, and that this is in one of the areas of low population density. That rainfall does not always explain population density is shown by the heavy population at Bombay and by the only moderately good fit of the population density map with the map of average annual precipitation. All the graphs show the strong seasonal concentration of rainfall that is characteristic of monsoon climates. (Reprinted from Focus, American Geographical Society)

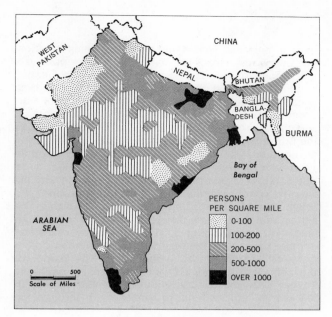

POPULATION DENSITY

PERSONS
PER SQUARE MILE
- 0-100
- 100-200
- 200-500
- 500-1000
- OVER 1000

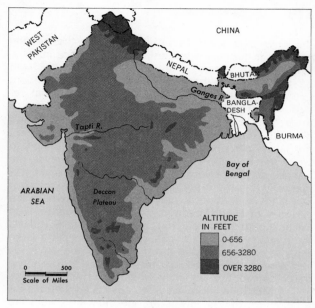

LANDFORM

ALTITUDE
IN FEET
- 0-656
- 656-3280
- OVER 3280

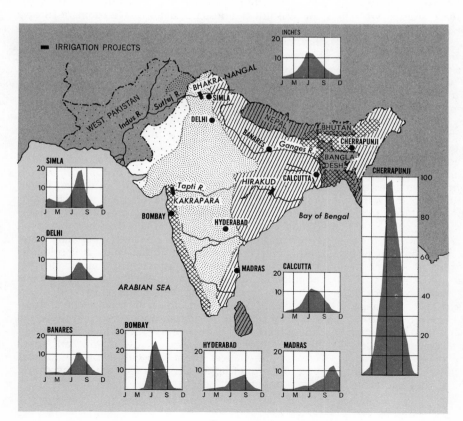

IRRIGATION PROJECTS

SIMLA

DELHI

BANARES

BOMBAY

HYDERABAD

MADRAS

CALCUTTA

CHERRAPUNJI

AVERAGE ANNUAL RAINFALL
BY MONTH

169

the basis for forming East Pakistan during the splitting of India at the time it gained its independence in 1947. With still further tragic bloodshed, East Pakistan split off from West Pakistan and became the independent country of Bangladesh in 1972. That the Muslim religion is strongly associated with dry lands is one of the accidents of history. Mohammad was an Arab, and his followers spread principally through the adjacent dry lands, perhaps because they were accustomed to and knew how to cope with such lands. However, the Ganges delta is a very wet land, and Muslims are numerically important not only there but also in such hot wet lands as Indonesia and the Philippines, where the currents of world trade carried their influence. The existing pattern of religion, race, and climate reflects the movements of trade and conquest through a large expanse of time. Muslim history began in the dry lands and spread widely through them, but as India shows, it also spread into wet areas when opportunity offered.

Monsoon India is sometimes called a subcontinent, although it is only about one-third the size of the United States. It is set off from the rest of the world by the Himalayas to the north, by deserts to the west, and by seas on its other sides. The Ganges-Brahmaputra plain in the northeast is a major area of fertile, well-watered land. Peninsular India is a triangular plateau with mountains on two sides. These bordering ranges intercept the seasonally reversed wind flows and are areas of high rainfall; the areas between them are surprisingly dry. The phenomenon of winds blowing in nearly opposite directions in two strongly marked seasons is called the monsoon. It is most strongly developed in Asia because this greatest land mass of the world extends here from low to high latitude, and on so great a land mass the seas can have little tempering effect. In the winter the interior becomes extremely cold, and the cold, dry air flows outward, bringing a dry season. In the summer the continent becomes extremely hot, and moist air masses from over the warm tropical seas flow in over the land, bringing heavy summer rains.

It is often said that India is a natural unit for a nation. A review of Indian history shows that this was not at all obvious in the past. The present degree of unity approximates the greatest that the peninsula has ever had, and it is far from perfect, as the existence of West Pakistan, Bangladesh, Kashmir, Nepal, Sri Lanka, and other independent states and countries shows. India as a political unity was a British notion and lasted only as long as they retained control. Religious beliefs proved far stronger than geography or economic advantage in determining the unity of this natural physical unit, once British control was removed. They had taken over an area that lacked any notion of being a nation, but this is not surprising. Nations also developed late in Europe, for it was Bismarck who put the German nation together in the nineteenth century, and even Great Britain is composed of Scots, Irish, Welsh, Cornishmen, and others, so regional loyalties still flare up and threaten to dissolve this political union. In India the British found a swarm of minor states with differing governments in frequent conflict with one another. Their lack of unity made them easy to conquer since one could be used against the other, the same way the Spanish conquered the Mexicans and Peruvians. The origin of civilization in India helps to explain this. The agricultural revolution first reached India through its northwest corner about 3000 B.C., and the next great impulse came from overseas to spark the growth of an early civilization in the Indus valley. The barrier of the Himalayas channeled land invasions through the Khyber Pass. The cultural impulses coming in through this corner took millenia to spread south through the peninsula.

Under Asoka in the middle of the third century B.C. much of the peninsula was united, but thereafter it was a shifting sequence of principalities and empires. This was not an unusual situation if we compare Europe under the Romans, the subsequent fragmentation of the "natural units," and the belated formation of the European nations in the form we now see them. Students of India state that until the Chinese invasion of India in the early 1960s there was still no sense of being "an Indian." Patriotic fervor has now created some feeling of nationalism among some of the people, although not among the remote villagers, the hill people, and the millions of tribal people in India.

The population of India is strongly concentrated in the Ganges River valley and on the well-watered sides of the Deccan Plateau. The dry interior of the plateau averages half the density of the wetter, lower, flatter parts of monsoon India. This is a geographically determined population location *if* one properly understands it. India's major population density is based on wet-rice agriculture, which demands abundant water and is most easily irrigated on flat lands. It is primarily this choice of a wet-rice agriculture that dictates the population concentration where the land and climate are most productive. If their major crop were sweet potatoes or some other crop with requirements different from rice, the maximum population density would undoubtedly be elsewhere. As it is, India produces six times as much rice as wheat. Jute in the Bangladesh area of the Ganges Delta and cotton on the Deccan and in the drier upper Ganges valley are the next most important crops.

Village India. *India is a myriad of little villages such as this. The man in this picture has a bicycle, a status symbol that he has acquired by working on a dam being built in the neighborhood.* (Jack Ling, UNICEF, for World Bank)

(Above) **The Taj Mahal, Agra.** *Reflecting the Muslim influence in India, the Taj Mahal is acknowledged as one of the architectural gems of the world.* (Embassy of India)

(Opposite, right) **Tiruvannaimalai Temple, Tamil Nadu.** *This temple is typical of the purer, earlier Hindu architecture. In its steepness, many levels, and great ornamentation it recalls some of the Mayan temples. The Chinese pagoda is thought to have evolved from structures such as this.* (Embassy of India)

(Opposite, far right) **A Contemporary Wood Carving in the Classical Style of India.** *Much of the ancient tradition of India is encompassed in this carving. The sacred cow is given a central place. The elephant-headed human is the god Ganessa, guardian of the village. Few arts have portrayed couples more sensuously than the classical art of India.* (Embassy of India)

India faces chronic malnutrition and recurrent famine. It is said that it is an unsuccessful famine that does not reduce the population sufficiently to alleviate the hunger of the survivors. With population increasing at the rate of 3 million people per year, the situation may seem hopeless, but it is not that simple.

First of all, India has twice as many cattle as the United States. They are used as draft animals that produce dung, which can be dried and used as fuel. Further, although devout Hindus cannot eat beef, they could eat the large amounts of feed that keep the cows alive. At best a cow is an inefficient converter of plant material into protein, but if the cow's protein is not used, the loss is total. In addition, the cattle dirty the city, injure people, and create a traffic hazard. Yet man here elects not to use the major potential of this animal but, on the contrary, to carry it as a heavy economic burden. The difficulty does not stop there, for the principle of *ahimsa*—life is sacred and man should kill nothing—discourages the killing of rats, monkeys, and other living things. Rats swarm through the villages, destroying food and materials and spreading disease, and the monkeys, considered gods, pilfer and destroy more food. It is difficult to assess the cost, but it is huge.

Secondly, it is difficult to say of any land that it is overpopulated; it is more realistic to view the people as underemployed. Indian peasant agriculture is inefficient: tools are light, scarce, and poor, and human energy levels are low. India's great political leader Gandhi is reported to have said that India's greatest handicap is the almost unconquerable inertia of her people. The will to accomplish something does not fill the ordinary Indian, and until it does, the land will remain overpopulated in the sense that the people will not make it productive. It is useful to compare densities of countries, for we often labor under quite erroneous notions. India's population density is 474 per square mile; the Netherlands, 900; England and Wales, 832; New Jersey, 939; Puerto Rico, 837; Taiwan, 1179. All of these lands with more than twice the density of population of India are more prosperous and some of

them are wealthy. It is not mere population density nor mere numbers that determines wealth or poverty; the manner in which that population is employed will be a greater determining factor.

Many of the reasons for India's nonproductiveness are attributed to its caste systems. While the caste system has been abolished legally, it continues to function. One cannot pass a law and thereby change attitudes ingrained into millions of people by millennia of usage. Nor does the law suddenly give education, motivation, or more than very limited opportunity. Those legally released are unequipped by tradition to take advantage of their new opportunities as such a change is resisted by the more privileged groups. The parallel in the United States is quite useful to consider. The Negroes were freed from slavery over 100 years ago, but accelerating change in their position only began about 40 years ago. Realistically, at least another 50 years will be required to approach equality of opportunity. India's caste problem is infinitely more complex and involves more people, who have at this time limited literacy and immense problems in political economy. To the individuals caught in the cultural web it must seem too much to bear, but the reality is that a century will bring improvement, and another century may perhaps see the problem reduced to a near vanishing point.

While this is unquestionably a major factor in its low productivity, India is in many ways a typical underdeveloped country. The majority of people are illiterate, subsistence farmers. While the rail transportation network is only fair compared to Europe or the United States, for it was designed by the British to serve external trade rather than internal communication, it is superior to transportation in most underdeveloped areas. (Compare Alaska, for instance.) To raise the Indians to a better standard of living, a whole series of changes must be made. The man on the land must improve his efficiency to the point where he can support several families in industry, commerce, and transportation. This more efficient farmer will need a larger farm. The people displaced from the land in this process must be absorbed by a growing industry. Industry demands more skills as it becomes increasingly mechanized, so the people must be educated to be useful in the city industries. Capital must be found to build industries; marketing channels and skills and transportation networks must be set up to distribute the goods and services created. In other words, generations of education, growth, and development are required. With a vast mass of people we cannot wave a wand and have simultaneously a revolution in attitudes, education, transportation, production, population distribu-

tion, and standards of living. If this can be said of relatively advanced India, how much more difficult is the African problem.

England went through extremely difficult times when it industrialized early in the nineteenth century. The United States today is struggling with the impact of automation in the factories and mechanization on the farm. Even with our enormously developed economy and society we are suffering from the resulting dislocation. Why, then, do we think that a few glowing phrases and a few billion dollars can accomplish what only time and desire can really bring into being?

When India gained independence, she made basic decisions concerning her development. Industry was to be planned by the government and controlled by the government, and new plants were to be largely owned by the government. The first needs were seen to be for heavy industry, and building a steel industry was given high priority. The rest of the world has invested an enormous amount to aid India. Despite such aid famine threatens annually. This problem arises from an explosive population growth: from 1892 to 1922 population increased by 12 million; from 1922 to 1952, by 113 million; in 1973 the population increased at the rate of over 1 million per month.

Few facts illustrate as well as the food situation in India the inability to predict over even short time spans the probable events in growth, development, and production. In 1965 the United States was sending one-third of its wheat production to India. Famine threatened; indeed it was freely predicted as an absolute certainty in view of the exploding population and former neglect of agriculture. Belatedly the Indians began to pay some attention to agriculture, to allocate some funds to fertilizer plants, and with the benefit of the green miracle—the use of wheat and rice plants with short, stiff stalks that convert more of the fertilizer input into grain instead of into stalks that get excessively tall and topple—the Indians had a surplus of grain in 1972. Surplus is a misleading term to use in a country where the bulk of the people live in unimaginable poverty, and the per capita food consumption is at least 25 percent below health requirements. Nevertheless, the 113-

India: A Land of Contradictions. *These three photographs show some of India's contrasts. (Opposite, top) Women carry baskets of earth on their heads to help build an immense dam. (Magnum for World Bank) (Opposite, lower left) A tractor pulls a sheep's-foot roller to compact the dirt carried. (Magnum for World Bank) (Opposite, lower right) A Sikh with tongs picks up a molten bar of iron at the Burnpur Works of the Indian Iron and Steel Company. (Magnum for World Bank)*

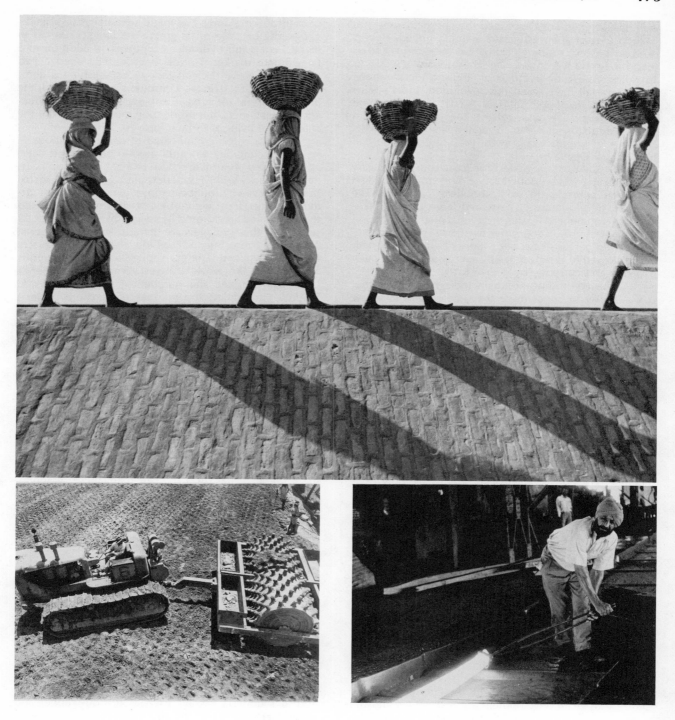

million-ton grain crop exceeded the approximately 100 million tons of grain estimated as India's need. Hybrid seed, irrigation, cropping throughout the year instead of just during the wet monsoon, security of crops due to irrigation instead of great risk of failure when the monsoons fail—all have aided this revolution. India's Minister of State for Food has said that at its current rate of increase India would outstrip the United States in grain production in seven to eight years. With its current rate of population growth, the only response can be: India will have to.

Despite pressing need for food and for industry that uses much semiskilled labor, an enormous effort has been poured into steel and heavy industry. The planned development has stressed large capital-intensive development in a land of great capital scarcity and vast labor supply. Labor-intensive small-scale production would perhaps have been more desirable; this path was followed in recent times by successful nations such as Japan and currently by Puerto Rico.

The controlling group in India has argued that it was better to concentrate their investment in a few areas where maximum return could be achieved rather than spread it widely through a vast agricultural sector. Heavy industry frees India from costly buying abroad and provides an industrial base suitable for military defense. However, the savings on foreign exchange are unreal if the cost of local steel is greater than the world market price. One review of the situation points out that whether rapid industrialization or agrarian expansion is more correct for India, the significant fact is that economic planning so far has brought neither prosperity nor efficient industrial production. Analysis of planning boards has shown that at the top levels where the decisions are made there are no technicians (engineers, scientists, and social scientists)—only politicians. Analysis of steel production shows that private industry such as India's Tata Works produces steel 15 percent cheaper than the government mills and wishes to finance its own expansion but is forbidden to do so by the government. The Soviet Union and Great Britain have each built steel mills for the Indian government. The unsophisticated Soviet mill has been described as "built by peasants for peasants," and the top management is Russian. It works. The British mill has sophisticated equipment and is operated by Indians. However, it is in difficulty. The British trained Indians in Great Britain for their jobs, but a high percentage were not put into the jobs for which they were trained, and Indian management demands that even minor decisions go through many people; therefore weeks or months of delay result. The contrasts with the Puerto Rican and Central American development are instructive, and the contrast with the successful development of Japan is es-

pecially striking. India has many advantages in physical geography because of its great size, large resources in metals, coal, and hydroelectric power, in addition to a large home market. Notice that Central America had to undertake the difficult task of establishing a common market to gain as consumers a fraction of the population that India has. People are potential producers and customers, but production must come first if the standard of living is to be improved.

India is too complicated to generalize about. It has the technology to build atomic bombs and actually exploded an atomic device in 1974, but most of its peasants farm with a wooden stick pulled by an ox. Rural India is 400 million people living in 500,000 far-flung villages, most without electricity and many linked to the outside world only by footpaths. Only about 15 percent of these villagers are literate. The gross national product is more than $37 billion, making the country tenth in the world. The per capita income, however, is $70, among the world's lowest. Since independence India has doubled the number of grade schools to 500,000 and tripled the number of teachers. Malaria has been cut from 85 million cases per year to 87,000. Despite all this, the considered judgment of those who have to deal with India is that it is bedeviled by bad management. The government is capricious, bureaucratic, highly socialistic, and suspicious of all business, especially foreign business. It is not an atmosphere such as in Puerto Rico which is likely to encourage foreign investment and the growth of industry that India so badly needs.

This is not to say that India cannot become again a great world power. In the remote past the Indians have been creative producers; by the severe test of high accomplishment, they are an able people. India is rich in resources: coal, iron, nickel, manganese, graphite, and other valuable industrial resources; and it is rich in people. If these were more gainfully employed, a great productive center would emerge. India's problems are not tropical, not physical-geographical, they are human-geographical: the outlook on man and land is a bit like that of the Yuma Indians.

The human factor in Southeast Asia

In Southeast Asia, India for the past 2000 years or more has exerted a great influence on Burma, Thailand, Cambodia, Vietnam, Malaysia, and Indonesia. The arts (architecture, dance, music, and the like) show this strongly. Southeast Asia is a land of forested mountains and seasonally flooded plains. Rice is the major crop, although in Malaya rubber equals it in tonnage and in Indonesia cassava exceeds it. Both rubber and cassava are American Indian crops.

It is sometimes said that in this region cultivation means constant struggle with the environment. Where is cultivation any different? It is also alleged that this area has the ideal climate for the rubber tree and thus became the world center of rubber production, yet the causes were quite otherwise. The rubber tree is a native of the American tropical rain forest: it would grow equally well in any similar climate. Africa, Melanesia, any of the American rain forest areas, and vastly more of Asia than actually grows rubber are equally well suited by physical geography. Indeed, if there were such a thing as a natural rubber-tree climate area, one would think that its American homeland would qualify.

Rubber is growing primarily in Southeast Asia, rather than in America, for political-economic-entrepreneurial reasons. The British and the Dutch, applying plantation methods combined with botanical knowledge and colonial control of areas well populated with sedentary, skilled agricultural people, took advantage of a rich market potential. The production of rubber from stands of wild trees in Brazil was inefficient, costly, and limited. Plantation agriculture was more efficient and less costly because of the closer spacing of trees, the better control of labor and production methods,

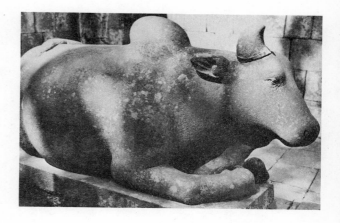

A Holy Bull in a Temple in Central Java (above) **and the Use of Indian Humped Cattle as Draft Animals in Central Java** (below). *These are examples of Hindu influence in Indonesia, which was especially strong between the eighth and twelfth centuries. Such Hindu influences are apparent in the art, architecture, religion, and various other aspects of the way of life in this area.* (Embassy of Indonesia)

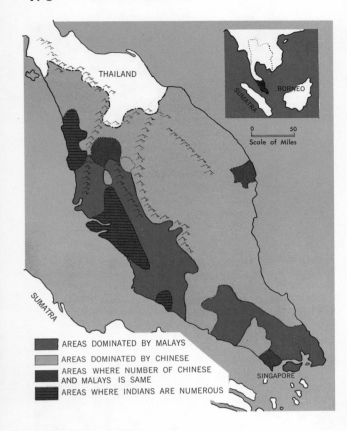

MAP 3-6   Major Ethno-linguistic Groups in Malaya

The Malays have been reduced to a minority position in parts of their country by the inpouring of Chinese, primarily, and Indians, secondarily. Indians are numerous in areas on the west coast; in these areas the Indian population and Chinese fraction are 40 to 60 percent of the total, often making the Malays there a minority group. The Indian- and Chinese-dominated areas coincide with the rubber- and tin-producing regions. Ethnic influx has changed Southeast Asia racially from Negrito to a racial mixture, usually called Malay or Indonesian.

AREAS DOMINATED BY MALAYS
AREAS DOMINATED BY CHINESE
AREAS WHERE NUMBER OF CHINESE AND MALAYS IS SAME
AREAS WHERE INDIANS ARE NUMEROUS

leading to a high quality of material, and the skilled plant selection of varieties yielding much more rubber than the natural trees. There were many physical environments where this could have been done; the human factors placed it in Southeast Asia.

In some areas the human factors have led to further complications. In the Malay Peninsula the tin mining and much of the plantation work have been done by Chinese. The native Malay did not choose to do this work; hence Chinese labor was imported. Many Chinese have stayed on, and much of the wealth of the area (factories, plantations, mines, banks, stores) is now in their hands. In Singapore they make up 80 percent of the population. There would seem to be no racial difference to account for this, for the Malays are a more tropical people than the Chinese. However, the Malays are less commercially motivated, and the hard-working, money-minded Chinese easily gained economic leadership. This situation is due neither to race nor to climate but to culture.

In the past there has been a great change in the racial composition of Southeast Asia, exemplified by a recent Chinese statement that south China was Negroid in race as late as the Chou dynasty (which ended 200 B.C.). The Chinese replacement of the native population in large parts of the Malay Peninsula shows how these changes occur. The Chinese intermarry with the local population to a considerable extent. They change the area culturally toward the Chinese pattern and shift the racial makeup toward the Mongoloid race. Something comparable to the present situation in the Malay Peninsula probably took place in south China after the time of Christ. In the past the Chinese spread to the Yangtze Valley, the Canton area, the Haiphong area, and now to the Singapore area.

In the period of colonial expansion after 1500, all of monsoon Asia went under the control of European powers except Thailand, which lost Cambodia to the French but managed to maintain its independence. Today all of these lands are independent, and their problem now is how to maintain their freedom. These countries were all more advanced than any of tropical Africa, yet literacy, a sense of political unity, and an understanding of the processes of government were not common in the masses of the people. At this crucial time in their national development, they are caught in the international struggle between the Communist world and the free world. They are in a precarious position, with only a fair chance to maintain law and order and to build up productivity and unity in their respective areas while remaining free from the Communist bloc, which desires to become the new colonial world power. Yet most of the former colonies of Southeast Asia, aware of their recent colonial past at the hands of the West, do not see clearly their present danger of a much more dangerous colonial future.

The delta lands of peninsular Asia

Burma, Thailand, and Cambodia are three delta regions of Southeast Asia. Burma is set off from India and Thailand by mountain ranges. The country has two

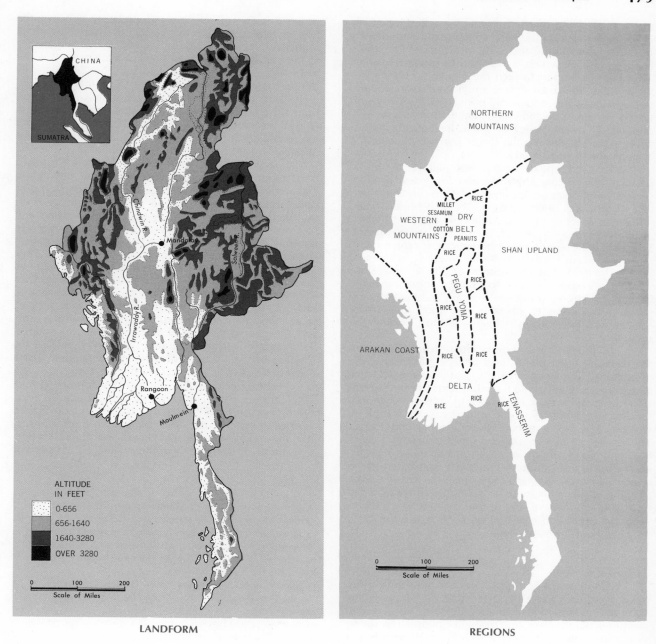

LANDFORM

ALTITUDE
IN FEET

0-656
656-1640
1640-3280
OVER 3280

0   100   200
Scale of Miles

CHINA

SUMATRA

Chindwin R.

Irrawaddy R.

Salween R.

Mandalay

Rangoon

Moulmein

REGIONS

NORTHERN
MOUNTAINS

WESTERN
MOUNTAINS

MILLET
SESAMUM
COTTON
PEANUTS

DRY
BELT

SHAN UPLAND

PEGU YOMA

RICE

RICE

RICE

RICE

RICE

RICE

ARAKAN COAST

RICE

DELTA

RICE

RICE

RICE

TENASSERIM

0   100   200
Scale of Miles

MAP 3-7   Burma: Landform and Regions
*Burma is a mountainous land. Most of the people and most of the agricultural production
are in the delta or in the dry belt, while the least are on the Arakan coast and in Tenasserim,
the rice-producing areas. (Reprinted from* Focus, *American Geographical Society)*

major valleys, and on these flood plains, and especially in the deltas, rice is grown. Since the population density of the country is only about 100 per square mile, there is a rice surplus, which is exported to India, Sri Lanka, and Malaysia, in part to feed the plantation workers of those areas. Southeast Asia is a region of great population contrasts, for moderately populated Burma is separated by only one mountain range from the densely populated Ganges Valley of India. In fact, the mountainous part of Burma, the larger part of the country, is quite thinly populated.

Thailand repeats the Burmese picture: the dense population is in the lowlands, and the major product is rice. Cambodia, on the lowland between Indochina and Burma, was once a part of Thailand under Indian influence. A series of petty kingdoms came and went; at times the area was united, but more often it was not. There are curious cultural combinations, for the religion is Buddhism, derived from India, where it has almost ceased to exist; yet the speech of the area is Sino-Tibetan, a language related to Chinese. India, however, has had the strongest influence on art, architecture, and custom. This complex situation is a good measure of the cultural crosscurrents that have long marked the region.

In Thailand as in Burma most of the population lives on the delta plains, and few people inhabit the mountains. The people of the plains use a wet-rice agriculture; the people in the mountains use forest rotation, the familiar cycle of clear-burn-cultivate-abandon. One of the products of the mountain area is opium, which can be grown in the remote clearings undetected. The market for this contraband product creates a value high enough in proportion to its bulk to make it profitable despite the transportation costs through roadless regions. In these back-country areas the people wear clothing of silk, a luxury to us but not to those who can raise, spin, and weave silk; to them, cotton cloth represents the expenditure of money, so they use the cloth they can themselves produce.

Indonesia

Contrasts in Southeast Asia continue into Indonesia. Here Java has a population density approaching 1000 per square mile. Sumatra, Celebes, Borneo, and the other islands have much smaller population densities, with large areas having only 5 to 10 people per square mile. Java, like Burma and Thailand, was strongly influenced by the Indian culture. Indian colonies that grew to kingdoms, complete with elaborate capitals, well-developed trade, and literate classes, marked this whole Southeast Asian area for more than 1000 years.

When it is considered that Indonesia lay between the two ancient culture centers of Asia and was powerfully influenced by both for more than 2000 and perhaps for as long as 4000 years, their level of achievement is not surprising. These influences had built up a moderately dense, agriculturally based population that produced high-value, low-bulk materials such as spices for world trade. The long, slow process of building a civilization is well exemplified by its modest success here over a long period of time, in an area well located to receive influences from more advanced neighbors.

Thus the Europeans who colonized Southeast Asia stepped into a "going concern." The population was settled on the land. They were accustomed to producing material for the world market. There were established governments, tax systems, capital cities, and so forth. The European powers simply took over the controls. These people had been in contact with traders from both East and West for thousands of years. They had already been exposed to most of the world's known diseases and did not melt away following contact with Europeans. There was no empty land for Europeans to move into, and there was no need to import black slave labor. As a result, Southeast Asia remains Asiatic, and the people are mostly descendants of the original inhabitants—with the exceptions we have noted of the Chinese in the Malay Peninsula and Singapore, where the influx radically changed the picture.

The effects of the colonial developments were very uneven. The Dutch on Java stopped the local wars, improved health and sanitation, and developed a series of export crops. They brought in American rubber and cinchona trees and developed highly productive strains. The efficiency of their management allowed

(Opposite, upper left) **Malay Family Scene.** *The Malays are a mixed race, as this family group shows. Notice that the house is built well off the ground. This is useful in tropical regions because it creates ventilation under the house. The practice is widespread in Southeast Asia but rare elsewhere in the tropics. (Mary M. Hill for World Bank)*

(Opposite, upper right) **Malay House Garden.** *Tropical gardens have a variety of food plants. The girl here is climbing to pick a papaya. (Mary M. Hill for World Bank)*

(Opposite, below) **Forest Clearing.** *This scene is commonplace in many of the tropical forested areas of the world. First the great tropical trees are felled, and then the area burned over. The logs that don't burn are left to rot. Planting is carried out between the stumps and the logs. The scene varies little except in scale when the clearing is done by men with stone axes or giant bulldozers. (United Nations for World Bank)*

AVERAGE ANNUAL RAINFALL

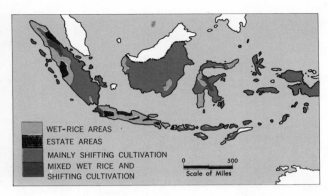

AGRICULTURE

MAP 3-8 Indonesia: Rainfall, Agriculture, and Population
*The relatively limited areas of wet-rice cultivation are notable.
That this situation is not determined by physical geography is
easily seen by comparing the agriculture map with the rainfall
map and considering that all the islands are mountainous and
many of them are volcanic in origin. It is a clear record of the
relatively late arrival of wet-rice culture. Population density
declines toward the marginal areas, as does general cultural
level. (Reprinted from Focus, American Geographical Society)*

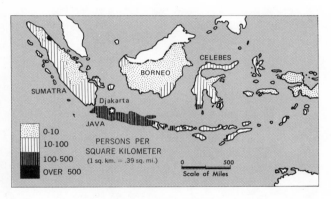

POPULATION DENSITY

them to capture a large part of the world market; in the
quinine market (derived from the cinchona bark) they
had a near monopoly. They also developed sugar, tobacco, pepper, and coffee. Programs of land use were
put into effect that left native populations undisturbed
in their land ownership while still drawing them into
the plantation production system as a labor supply.
Under Dutch management, at first cruelly exploitive
but later increasingly benevolent, the populations and
productivity of the Netherlands Indies expanded greatly.
The population of Java, for instance, went from 20 million in 1880 to 28 million in 1900 to nearly 50 million
in 1950 to 75 million in 1973.

Java's great density of population is in part related to its rich, volcanic soil. Much of Java is a string
of active, or recently active, basic volcanoes. The young
volcanic materials assure fertile soils. However, this is

not sufficient to explain the difference between densely
packed Java, moderately settled Sumatra, and nearly
empty Borneo. All of these areas experienced early
Hindu influences, all have good alluvial plains, and
most have some young volcanic soils. Distance is not
a negligible factor in Indonesia; however, Sumatra is
closer to India than is Java, so distance alone is not a
sufficient answer. The concentration on Java is probably the accidental result of a number of human decisions focusing attention here rather than elsewhere.

Indonesia is now an independent nation. In the
past the area achieved some unity under the Hindu
Majaphit princes and then again when conquered by
the Muslims. Most of the people there today are Muslim. However, 25 different languages are spoken, and
the culture level varies from civilization in parts of Java
to headhunters in unexplored interior areas of Borneo.

Illiteracy is high, technological knowledge slight, and capital extremely limited. Production is nearly all in primary materials such as quinine, pepper, kapok, rubber, copra, palm oil, tea, and tin. The costs of administrating a nation composed of islands strung across an eighth of the earth's equatorial circumference is enormous.

The uneven distribution of population is nothing new. Densely populated Java has not sent out floods of people to the adjacent thinly populated islands. Under Dutch management, although people were encouraged and aided to move to the less populous islands, the movement was small. The primary resistance was social. The Javanese villagers did not want to leave their kin and their settled existence in their beloved village to go and live among strangers in a strange land. The immense contrasts between Java with 1000 people per square mile and Borneo with about 10 were much more socially determined than physically determined.

At this time it is increasingly possible through air travel and radio communication to make a unit out of lands spread over a vast sea area such as Indonesia, but progress requires a technological development that Indonesia does not now have. To gain these skills an educational program is required that will be costly and time-consuming. The first generation of effort educates some and produces technicians and teachers to make possible a bigger effort during the next generation. The third generation can expect to begin to reap the harvest. All of this assumes governmental stability and steady economic growth.

Indonesia after World War II drove the Dutch out, and with them went a great store of managerial skill. The result was a decline in production. Then the Communists were welcomed, and this led to a brutal showdown when the Communists tried to take over the nation.

When the Communist coup in 1965 came within inches of taking over the country, it was headed off by the army, and General Suharto took the reins of government. The country was in a chaotic state and had huge foreign debts. With stability and a climate favorable to foreign investment established, capital began to rush into Indonesia, and if it continues, it may make Indonesia the industrial center of southeasternmost Asia. To the list of timber, rubber, and minerals there is now seemingly to be added oil, for the vast, shallow shelf on which the Indonesian islands stand offers rich prospects in this resource. Development is centered in the Djakarta area, a city of 5 million people. The country has a balanced budget, and export earnings of over $1

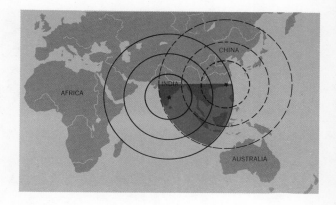

MAP 3-9   Indonesia in Relation to China and India
From centers on the coasts of China and India, circles 5°, 10°, and 15° of longitude in radius have been drawn. The area equidistant by sea from India and China is indicated by shading. This includes Burma, Thailand, Cambodia, Indochina, Indonesia, and the Philippines. These are the areas of culturally advanced people. Beyond this zone of double influence lie Australia and Melanesia. Melanesia, the closer of the two, received agriculture; Australia did not.

All simple correlations of this sort are, of course, suspect. Chinese trade goods reached East Africa in quantity, but no comparable trade developed with closer Australia. Indonesian people became culture bearers in their own right. They settled Madagascar across the Indian Ocean in one direction and greatly influenced developments in Polynesia in the other. Nevertheless, the role of location in relation to culture centers seems clear here. That the contacts in this zone were early is suggested by evidences of sea trade between the Persian Gulf and India about 2500 B.C. and between India and Ethiopia about 1500 B.C. Coastwise trading is probably considerably older than this suggests. The Indonesian cultural growth seems, then, to be the product of joint influences of India and China over a period of nearly three thousand years.

Even this concept may be greatly influenced by our limited knowledge of history. If Nam Viet (the area comprised of south China and Vietnam) was a vital center of cultural origins, then at some earlier time the flow of ideas was from this area to India and China and only reversed later in time. The parallel in the West would be Great Britain first influenced by elements from the eastern Mediterranean and later becoming a cultural dominant now contributing to the eastern Mediterranean.

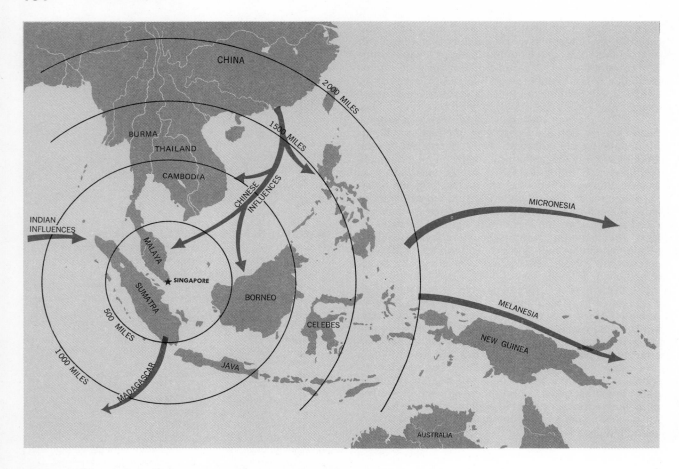

MAP 3-10   The Indonesian-Malay Area as a Center of Cultural Influence.
*Making Singapore central to this region, arcs mark distances at intervals of 500 miles. At 2000
miles from Singapore the transition is made from rice farmers with metals, developed trade,
and towns to nonmetal-using villagers in New Guinea and nonagricultural people in Australia.*

billion per year. It has severe unemployment problems associated with rapid population growth that adds about one and a half million people to the labor pool each year. Large numbers of these people flood into the Djakarta area seeking employment but are disappointed. The solution remains: migration to the outlying islands; industrialization, which requires time; increased agricultural productivity, more easily accomplished in the outer islands than in the overcrowded Java-Madura area; or population control, which requires education

and time. Despite the many problems, the area is a potentially rich one and has much space, many resources, and good organization working for it.

The Philippines

The Philippines provide an example of how difficult the process of change can be. The Philippines are another island chain scattered over a vast area of the Pacific. The total land area is 115,600 square miles and the population some 40 million, or a density of about 350 per square mile. Compare the United States, which has about 57 per square mile. Like Indonesia, these islands are mountainous, and their coastal plains are generally small. Chinese influence reached the islands early.

Numerous other people reached the Philippines and influenced their development, notably the Muslims, with the result that before A.D. 1500 the islands presented a great variety of peoples of differing cultural levels. In the mountains there were Negrito peoples living a hunting and gathering existence. In forested regions there were simple farmers practicing the ancient shifting agriculture so well adapted to thin settlements. There were also rice growers who had carried the idea of Asiatic terracing of mountainsides with them and put the land into permanent, densely settled, intensive agriculture.

When the Spanish reached the Orient, the Philippines became their base of operations. They sent out Mexican silver via the Manila galleon from Acapulco and obtained spices, porcelain, silks, and other Oriental luxury goods. Building a capital city at Manila, they controlled the islands from a few key points, developing the land relatively little. They did introduce the Spanish language, Christianity, and a pattern of large-scale landownership that has been called "the Iberian curse." Perhaps the most lasting Spanish achievement was in making this a Christian nation, for they managed to convert about 90 percent of the people.

When the United States took over these islands in 1898, the American educational system was instituted, and an attempt was made, in a limited way, to break up the large land holdings. The great mass of illiterate tribesmen in the interior, however, remained relatively untouched. The Americans did encourage the production of such crops as sugar, manila hemp, and copra, and began the exploration and development of mineral resources. The Philippines are a major producer of iron, chromite, and gold, even though the land is still poorly explored for mineral resources.

The importance of the time factor in cultural-change situations is well illustrated. The long nonintensive Spanish period of contact was followed by the more intensive contacts of the American period. Yet the American period lasted only from 1898 to 1946, less than 50 years, and World War II stopped much of the progress that had been instituted. This was too little time for extensive change even though the vast resources of the United States were deployed.

The Philippines, independent since 1946 and given preferential treatment in the United States market, are struggling to build a modern nation. Their problems are typical of underdeveloped lands. Sixty percent of the working population is directly engaged in agriculture, and more than 75 percent of the total foreign exchange earnings comes from the export of seven agricultural products. Analyses of land use show that little more than one-half acre of land per capita is used for producing food crops. This means that each square mile of land devoted to food crops is supporting about 1200 people. Manufacturing has been strongly encouraged since the time of independence, and today it accounts for 20 percent of the gross national product. Textiles and cement have been the most rapidly expanding industries.

In the earlier editions of this book it was pointed out that primitive agricultural practices produced yields of corn and rice in the Philippines that were fractions of what they could be, but it was easily within the capability of the country to increase the returns on its agriculture three- to fivefold. It was particularly stressed that the Philippine record of poor production was not due to poor soil or climatic limitations. Now, only five years later, the crop revolution that was conceived as a possibility has been realized.

Improved varieties of rice, and the so-called IR rice that was discussed previously, have been introduced together with the application of commercial fertilizers. The increase in yields has been so great that for the first time in modern history the Philippines have moved from being net importers of grains, especially rice, to net exporters. The changes have far-reaching effects. Expansion of the fertilizer industry has employed capital and labor, and it has required the building of roads and the development of a whole distribution network of dealers, distributors, salesmen, and so forth. The need for insecticides has also led to the development of chemical plants with similar requirements for distribution. Here, as in India, it has been found that the farmers did not evidence the resistance to change that had been anticipated. Quite the contrary, once the farmers saw the results of using the new seeds and applying fertilizers, they jumped at the chance to increase their yields. There have been difficulties, of course. Some farmers concluded that if a little fertilizer is a good thing, tons of it must be wonderful. In fact, too much fertilizer becomes a poison. Others have tried to shortcut the hybrid seed situation. As with hybrid corn, it is necessary to use the specially grown seed to obtain the best yields. It has been found in the Philippines that when the farmers saved grain from their harvest and used it as seed, they ran into rice disease problems that buying of new, specially grown

WATER

Tropical rice plant before panicles emerge.

Mature plant "lodged."

WATER

Modern semidwarf before panicles emerge.

Mature semidwarf.

(Above) **Syntha, a Traditional Rice Variety from Indonesia.** *Traditional rices in the tropics have long, weak stems and wide, droopy leaves. When fertilized, the plants produce heavy panicles, or grain heads. The stems cannot support the heavy panicles, so the plants "lodge" or fall over. The grain may dip into the paddy water or be eaten by rodents. The wide leaves at the top limit photosynthesis because they prevent sunshine from penetrating the leaf canopy. This type of plant produces few tillers. Yields are about 1 to 2 tons per hectare (1 hectare = 2.47 acres).* (International Rice Research Institute)

(Below) **Modern Semidwarf Rice Variety.** *The new semidwarf rices have strong, stiff stems. When fertilized, the panicles become heavier, but the strong stems hold the plants upright, not allowing them to fall over. Narrow, erect leaves permit more sunshine to penetrate the leaf canopy, increasing photosynthesis. The modern semidwarfs are high tillering. With proper management and inputs they yield about 5 tons of rice per hectare.* (International Rice Research Institute)

seed obviated. However, these are among the learning pains that accompany any new enterprise. What is astonishing is the doubling and tripling of yields with relatively minor changes in agronomic practices and the lightning speed of the spread of these new practices, with consequent huge changes in the food situation of whole nations.

The changes are far-reaching. It is already clear that the rice production pattern of the whole world will be affected. In recent years the United States has exported rice to 110 different countries, but many of these countries can now be expected to become either self-sufficient in rice or exporters of rice. Countries such as Thailand and Burma that have been heavy rice exporters are likely to feel a severe economic pinch. Countries such as Malaya and Sri Lanka that have emphasized export crops such as rubber and tea and have imported great quantities of rice are now aimed at self-sufficiency in rice. This will give them great gains in international exchange that can be used advantageously, but the countries from which they formerly bought rice will suffer considerable dislocation. The situation in the Philippines, then, is to be seen as only an example of the revolutionary changes that are possible in agriculture, not just tropical agriculture. Rapid changes in the world grain-marketing patterns are to be anticipated.

The Philippines are 60 percent forest covered with trees that are varied in type and valuable. The lowlands have extensive hardwoods: 2000 species, including much Philippine mahogany. In the mountains there are large, nearly pure stands of pine that produce valuable timber plus resin and turpentine. The lumber industry is growing both for local consumption and for export to Japan. There is also a wealth of minerals. The Philippines have been a major gold source and have large deposits of high quality iron ore that can be open-pit mined at locations suitable for shipment by sea. There are sizable copper deposits of relatively low grade, but gold, silver, and iron pyrites (useful for manufacturing sulfuric acid) add value to these deposits and make them commercially useful. The Philippines are also one of the world's leading producers of chromite. The greatest lack is their small amount of coal. Although oil seeps are known, there are no developed fields as yet. Manufacturing since 1950 has been developing rapidly, especially around Manila, concentrating on light industry producing soap, chemicals, rubber products, plastics, textiles, cement, wood, and paper. Hydroelectric power is being used on Mindanao to produce steel and manufacture fertilizer, and the undeveloped hydropower potential is large.

In a sense the Philippines have everything. Located, like Britain, on an island "protected from invasion," they are far richer in soil and minerals and not hampered by short growing seasons or drought. Moreover, they are surrounded by a sea rich in fish, which they still harvest to only a limited extent. Culturally enriched both by the Orient and Occident, they need only move ahead, but it will not be easy. Like so many other underdeveloped nations motivated by understandable pride, they may have cast off their tutors too soon. Education takes time; 40 years is not enough for the development of a broad educational base. If the Philippines do not advance as they might, beware the cry that the "tropical heat and humidity got them down."

The Negroids of the Pacific

Melanesia is the appropriate name for the group of islands stretching along the equator from New Guinea to Fiji. These lands are occupied by the Oceanic Negroes. In fact, there is a variety of Negroid people here: Negrito peoples, Papuans, and Melanesians. They represent three waves of mankind that moved into this area at various times in the remote past. The Asiatic-Indian-Malay occupation of Southeast Asia and Indonesia probably came after the Negro movement. The presence of remnants of a Negro-Negrito population in Southeast Asia—for example, on the Andaman Islands, in the mountains of Indochina, and in the mountains of the Philippines—supports this displacement theory, but it must all have occurred extremely long ago. This thesis is supported not only by the scattered occurrence of Negritos in relict position throughout Asia and by the Chinese statement that south China was Negroid up to the time of Christ, but also by the finding of a Negrito skull on Borneo dating back about 40,000 years.

Melanesia is culturally a low spot in the wet tropics. Nowhere did the people exceed an early Neolithic, agriculturally based way of life. Political organization remained on the village-tribal level; metallurgy was nonexistent, writing unknown, cities unheard of. Why? It cannot be the wet, tropical climate. Civilization flourished nearby in similar climates. It cannot be soil or landforms, for these were extremely varied. Indeed, the landforms include everything from temperate highlands to broad lowland flood plains, from volcanic high islands to low coral islands. Ideas rolled up to Melanesia and died away or surged past. Advanced political and religious ideas, elaborate trade networks, and skilled navigation techniques swept past Melanesia into the scattered islands of the vast Pacific.

The Old and the New in Rice Cultivation. *Although about two-thirds of the population is engaged in agriculture, the Philippines must import meat, dairy products, and cereals. The per capita income in the country is only a quarter of that in urban areas. Change is under way in cultivation, crop rotation, and plant improvement. The contrast in cultivation practices is shown here: carabao (above) versus rototiller (opposite page). (United Nations for World Bank)*

Even the inhabitants of Easter Island had a form of writing which they clearly owed to some cultural impulse coming ultimately from India.

It is impossible to answer the question of why Melanesia failed to progress. It was not the land or the climate that was at fault, for the land is rich and the climate wet tropical to cool highland. It would be easy to say that since ideas were available for a long period of time, the race was at fault. But how could one be sure of this? If the cultural outlook of the Melanesians was such that the cultural currents were repelled, the effect would be the same whether the inherent ability of the people was great or small. The flow of ideas that clearly penetrated the Pacific and reached America carrying the germ of civilization largely bypassed Melanesia. This is not an unusual happening. This is a chain reaction; thereafter, because they were cultural laggards, they would continue to be avoided. Note, for instance, that the Spanish, Portuguese, and Dutch avoided New Guinea but fought for possession of the Spice Islands (now the Moluccas in Indonesia). The choice in the past often was between developed economies and primitive economies. In a sense it still is; by far the greatest amount of trade and travel and con-

tact is between the developed countries of the world. Such routes often bypass undeveloped areas for centuries. At least we can be absolutely certain that the physical environment is the least likely reason for the lack of advance in New Guinea. The most likely reason is the pattern of high culture avoidance of primitive areas.

The potential for New Guinea can best be understood by reviewing the possibilities for plantation crops, expansion of food crops as in the Philippines, or of cattle raising as in Australia under King Ranch leadership. With the expanding world demand for food and other raw materials, areas such as New Guinea have an obvious opportunity. All that is needed is education, stable government, abundant capital, developed transportation networks, and the economic infrastructure of credit, wholesale and retail outlets, and so forth. Since none of these requisites are present, development will be slow. But none of these are physical-environmental. Neither are they racial. In the broadest usage of the word, they are cultural factors and their absence here is historically caused. As Puerto Rico shows, they can be overcome, but even under the best of conditions change will be slow.

Summary

The wet tropics show every stage of social development. There are simple hunters and gatherers, simple agriculturists, advanced agriculturists, and nations of advanced development. It cannot be said, then, that the physical environment has determined what happened. Why do we have the notion that the wet tropics depress human achievement?

Several unsubstantiated views led to this conclusion. First, it has been too long taken for granted that the heat and humidity act directly on man to decrease his activity and that the white race in particular is subject to this direct influence. But two sets of facts counter this notion. If the origin of man indicates his probable adaptation, he should be more at home in warm, humid climates than outside them. In addition, the hottest parts of the world are not the equatorial wet regions, but the low-latitude desert regions, and seasonally, the lower mid-latitude areas. Both of these hotter regions have been the scenes of notable human achievement. While the effects of extreme heat and humidity are undeniable, the record simply shows that they are not at their maximum in equatorial regions and have not deterred man even in the areas where these maximums are reached. As for the white race's ability to work successfully either physically or mentally in the tropics, the record clearly suggests that the white race can achieve in the tropics. Failures of white settlement are usually due to social, economic, and political factors. As for evidence of cultural lag by the Negroids, the facts fit the map of opportunity. Those in the mainstream of ideas became civilization bearers and cultural innovators and creators; those in isolated areas lagged. But this has been the case with all races and occurs irrespective of climate.

It is worth emphasizing that the leading nations of the world have at times been in tropical, humid regions. The center of pre-Columbian civilization was in the equatorial area and included lowlands from Yucatán through Panama. India has at times been a world center of civilization, and its colonies in Indonesia were advanced nations in their time. We see the world at a moment in history when the tropical regions are "underdeveloped." The conclusion we must reach from all we know is that this probably is a momentary situation and not one causally connected with the tropical climate.

The tropical problem is largely in our minds. We are a nontropical people; the tropics were and are strange lands to us. We have overplayed their difficulties and underestimated their advantages: the lack of seasons is *not* a disadvantage; it can be an advantage.

The tropics contain some of the largest areas of lightly populated parts of the world: tropical lowland South America, much of Central America and parts of Mexico, much of tropical Africa, most of Indonesia, all of New Guinea, and most of tropical Australia. This is a large part of the habitable world, for if the Arctic and Antarctic regions, the high mountain areas, and the as yet unirrigable deserts of the world are excluded, the areas just enumerated make up about half the potential for human occupation. The demonstration of the potential for such lands in multiplying the grain yields and producing high-quality protein food such as Santa Gertrudis beef is then of incalculable importance. Should even one area such as Brazil capitalize on this opportunity, the food situation of the world would be revolutionized. Even the modest gains from rice in the Philippines and India has demonstrated how far-reaching such changes can be.

The Wet Tropics. *Bananas almost epitomize the wet tropics. They grow only near sea level under conditions of high moisture. Here in the Guayas province of Ecuador the harvest has begun. Small plantations have sprung up all along the new roads built in this province. (Paul Sanche for World Bank).*

Review Questions

1. The United States, a land of high-cost labor, produces rice more cheaply than Asia, where labor cost is low. Explain why. What does this suggest for the future of rice production in Asia?
2. Discuss the lessons the Ford rubber-growing experiment affords.
3. What is the significance of the cultural level and population density correlation in the wet tropics?
4. Since Puerto Rico was able to accomplish what it did, why do the adjacent islands not do likewise? Is United States participation absolutely necessary for such development? Give reasons and examples. Consider also the Central American Common Market.
5. What does the Mayan civilization illustrate concerning tropical life?
6. To what extent can Brazil's present status be said to result from Brazil's personality? Can a parallel be seen in any other nation?
7. Discuss India's economic growth as compared with that of Puerto Rico. What are India's advantages and disadvantages? Which of these is physical geography? What is cultural (political, economic, psychological) in the actual situation?
8. Why are not the Philippines, rather than Japan, the Great Britain of the Pacific? Stimulating versus tropical climate perhaps? Presence or absence of basic resources?
9. Discuss Brazil as a potential world power.
10. What do IR rice and King Ranch cattle suggest about the future of the tropics?

Suggested Readings

Bartlett, H. H. "Fire, Primitive Agriculture and Grazing in the Tropics." In W. L. Thomas, Jr., ed., *Man's Role in Changing the Face of the Earth.* Chicago: University of Chicago Press, 1956. A broad-guage study with an extensive bibliography.

Blakemore, Harold, and Smith, Clifford T. *Latin America: Geographical Perspectives.* London: Methuen, 1971.

Gourou, Pierre. *The Tropical World.* London: Longmans, Green & Co., Inc., 1954.

Hirsch, J. "Comfort and Disease in Relation to Climate." In *Climate and Man: Yearbook of Agriculture, 1941.* Washington, D.C.: U.S. Department of Agriculture.

James, P. E. "Trends in Brazilian Agricultural Development." *The Geographical Review,* vol. 43 (1953), pp. 301–328.

Kimble, G. H. T. *Tropical Africa.* 2 vols. New York: The Twentieth Century Fund, Inc., 1960. Encyclopedic coverage.

Lambrecht, F. L. "Aspects of Evolution and Ecology of Tsetse Flies and Trypanosomiasis in Prehistoric African Environment." *Journal of African History,* vol. V (1964), pp. 1–24.

Lee, D. H. K. *Climate and Economic Development in the Tropics.* New York: Harper & Row Publishers, 1957.

———. *Physiological Objectives in Hot Weather Housing.* Washington, D.C.: Housing and Home Finance Agency, 1953.

Murdock, G. P. *Africa's People and Their Cultural History.* New York: McGraw-Hill Book Company, Inc., 1959.

Pollack, N. C. *Studies in Emerging Africa.* London: Butterworth's, 1971.

Price, A. C. *White Settlers in the Tropics.* Special Publications, no. 23, New York: American Geographical Society, 1939.

Robequaine, C. *Malaya, Indonesia, Borneo and the Philippines.* London: Longmans, Green & Co., Inc., 1958.

Sauer, C. O. *The Early Spanish Main.* Berkeley: University of California Press, 1966.

Simoons, Frederick J. "The Determinants of Dairying and Milk Use in the Old World: Ecological, Physiological, and Cultural." *Ecology of Food and Nation,* vol. 2 (1973), pp. 83–90.

Steel, Robert W., and Prothero, R. Mansell, eds. *Geographers and the Tropics.* Liverpool Essays. London: Longmans, Green & Co., Inc., 1964.

Stone, R. G. "Health in Tropical Climates." In *Climate and Man: Yearbook of Agriculture, 1941.* Washington, D.C.: U.S. Department of Agriculture.

Thorp, James. "Climate and Settlement in Puerto Rico and the Hawaiian Islands." In *Climate and Man: Yearbook of Agriculture, 1941.* Washington, D.C.: U.S. Department of Agriculture.

Wagley, Charles. *An Introduction to Brazil.* New York: Columbia University Press, 1963.

West, Robert C., and Augelli, John P. *Middle America, Its Land and Peoples.* Englewood Cliffs, N.J.: Prentice-Hall, 1966.

Problems of Humid Tropical Regions. Paris: UNESCO, 1958.

FOCUS

AFRICA: **Congo, Gabon, Ghana, Kenya, Madagascar, Mozambique, Nigeria, Resources of the Tropics, Rhodesia and Nyasaland, Rwanda and Burundi, Somalia, Tanganyika.**

ASIA: **Burma, Cambodia, Ceylon, India's Agriculture Problems, India's Agricultural Growth, India's Languages and Religions, India's Population Problems, Indonesia, Malaya, Resources of the Tropics III, Sabah, Thailand, Vietnam.**

CENTRAL AMERICA AND THE CARIBBEAN: **Guatemala, British Caribbean Confederation, Cuba, Dominican Republic, Mexico, Puerto Rico (1953 and 1963), Virgin Islands.**

SOUTH AMERICA: **Brazil, Guianas, Paraguay, Resources of the Tropics II, Venezuela.**

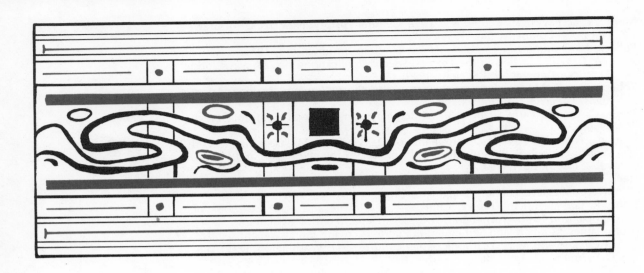

The Mediterranean Climate

CHAPTER FOUR

THERE are five small areas on the earth that have what is called the Mediterranean climate. These are around the Mediterranean Sea and parts of California, Chile, South Africa, and southern Australia. Most of the earth is either continuously wet or dry, or if it has a seasonal maximum of rainfall, this occurs in the summer. In the Mediterranean lands, however, summer is drier and winter is wetter. It is not only rainfall that sets these lands apart: their summers may vary from hot in the interior to mild on the coasts, but their winters are characteristically mild and sunny. They receive high percentages of the possible sunshine, summer and winter. For example, Los Angeles (presmog) received 90 percent of the possible sunshine in summer and 60 to 70 percent in winter. This combination of mild winters and much sun makes these Mediterranean climates winter resort areas today.

The Physical Setting

Typically these lands are located on the tropical margin of the mid-latitudes between latitudes 30 and 40 degrees and usually on the western sides of the continents. In these areas the seasonal shifts of the climatic belts of the earth place the land alternately under the desertlike conditions created by the subtropical high

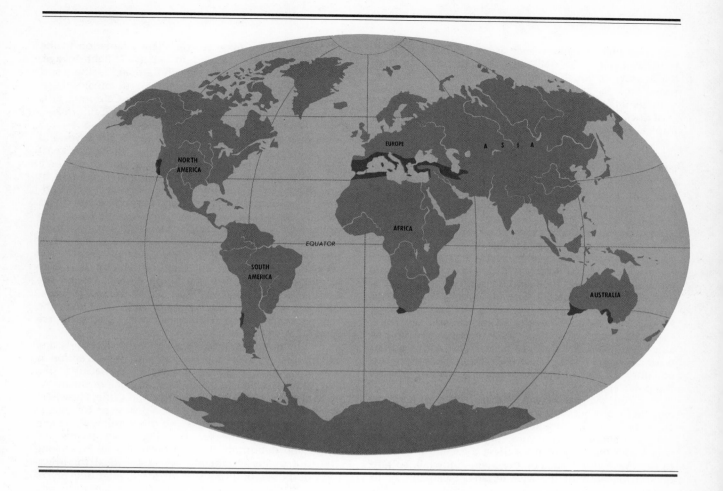

pressure systems in summer and the variable weather systems of the westerly wind belt in winter. Thus they have constantly sunny, hot, dry weather all summer long, with alternating sunny and rainy weather in the winter season. One can think of the Mediterranean Sea as having a Sahara climate in summer and a cool English or French summer for its winter climate.

The absence of cold weather is striking. Summers are warm, averaging from 70° to 80° F., but winters average only 40° to 50° F. The annual range of temperature is therefore small, and in some coastal areas where the sea helps to hold the temperature constant, the range may be only about 10 degrees. Inland, the temperature range may be higher—that is, much

hotter in the summer and somewhat colder in the winter. Classic examples are found in California where San Francisco normally has a cool and foggy summer with temperatures in the 70°s, while Sacramento, 90 miles away in the Central Valley, may experience temperatures above 110° F. in summer. In winter San Francisco has much bright, sunny, mild weather between spells of rain, while Sacramento may have spells of freezing weather. This variation is typical of the Mediterranean climates of the world. Summer temperatures on the coast are mild, inland hot. Winter temperatures on the coast are mild, inland cooler, but still warm for their latitude and with plenty of bright sunny days. The nearby oceans moderate the climate, and often high mountains such as the Alps or the Sierra Nevadas cut off the outbreaks of cold polar air that reach areas of much lower latitudes elsewhere. As a result, frost is rare in many Mediterranean areas. Its rarity tempts men into growing exotic plants, and when frost does come, it is often extremely damaging to such tropical crops as oranges and avocados.

Rainfall is generally less than moderate in Mediterranean lands, averages varying from 15 to 35 inches. Fifteen inches of rainfall is low for agriculture, except at great risk of drought. However, winter rainfall is efficient. If it came in the summer during high temperatures, the losses through evaporation would be so high that such rainfall would do little good. In the winter, on the contrary, evaporation is slow and a great deal sinks into the ground where it is useful to the plants. This winter concentration may be quite extreme; for example, at Los Angeles 78 percent of the annual precipitation comes between December and March. Rainfall increases steadily from the equatorial, dry edge toward the polar edge of these climates. San Diego has about 10 inches, Los Angeles 15 inches, and San Francisco 23 inches of rainfall. Such changes are typical along the coast of Palestine, from North Africa northward along the coast of Portugal, and in Chile from the equatorward end of the Mediterranean climate toward the polar end.

Most Mediterranean climates have mountains, and these profoundly modify the rainfall pattern. The air rising over the mountains is forced to drop much greater amounts of moisture; thus very high rainfalls occur, and the streams from these areas are important to the much drier lands below them. Snowfall, so rare on the lowlands of the Mediterranean climates as to be newsworthy when it occurs, is common in the mountains. A typical scene in many Mediterranean lands shows orange trees and palms on the coastal plain with snow-covered mountains in the background. These snow covers are of great value, not only for vacational skiing, but also for the period when the drought is greatest on the plains and the water most valuable to the crops.

Vegetation as well as climate is unusual. The plants in the drier areas are dwarfed, often thorny, and usually evergreen around the Mediterranean and in California. The evergreen oaks have stiff spiked leaves similar to the holly plant. California has a scrub oak that never makes a tree but only achieves a tall, bushy form. It forms a part of the brush-covered landscape typical of the drier part of the Mediterranean areas.

The seasonal rhythm of the Mediterranean plants is different. Growth starts with the winter rainy season, and flowering and fruiting take place in the early spring at the end of the rainy season. Plant adaptation to the summer drought is found in relatively light foliage, thick bark, and small leaves with hard and leathery surfaces. This type of shrubby growth dominates large areas around the Mediterranean and in California. It is called maqui in Europe and chaparral in California.

Poleward and on higher mountain slopes, the increasing rainfall leads to decreasing brush cover and increasing dominance of trees. Thus along the coast there is a progression from treeless San Diego to a region of scattered trees and grassland beginning at Santa Barbara and finally changing to the wet redwood-forest belt about San Francisco. Ascending the Sierra Nevada mountains east of Sacramento carries one from dry grassy valleys, into brush-covered (chaparral) foothills, on up into oak and pine mixed forest, and finally into heavily pine-forested, high mountain areas. Similar plant distributions are found all around the Mediterranean and in Chile and South Africa. Around the Mediterranean, however, man has so seriously modified the vegetation that the land looks drier than the rainfall would suggest: formerly tree-covered areas are often reduced to brush.

In these mountain-backed areas, lowlands and plains exist to only a limited extent. An exception is western Australia, where mountains are lacking. The largest areas of lowlands in these regions occur in the central valley regions of California and Chile. Elsewhere most of the land is sloping to steep or even to rugged. Rivers rise in the mountains and often follow short and steep courses to the sea, bringing down gravels and sands and depositing them in great alluvial fans along the foot of the mountains. These are ideal for irrigation purposes, both because of the natural delivery of water to the apex of the cone-shaped landform and because of the expanse of good alluvial soils that they present. There are problems, of course. Rivers in Mediterranean regions are subject to great fluctuations in volume between the desertlike summer, the mild, wet winters, and the spring periods of snow melt.

It is difficult to generalize about soils in Mediterranean climates. The rugged landforms can have little soil stability; hence they must have thin young soils on them. The small alluvial plains and valleys may have rich alluvial soils. In general the low rainfalls tend not to wash out the nutrients of the soils rapidly, but in Mediterranean lands there are many soils that are older than contemporary climates and have been strongly leached in past times of much wetter climates. Thus on the coastal plain about San Diego in southern California nearly 50 percent of the soils had their dominant characteristics determined prior to 10,000 years ago when the glacial climate brought much greater moisture to this area, now at the dry end of the Mediterranean climate. Such climatic change must have affected all of the Mediterranean climate areas, and many of them have old, acidic red soils that probably date to premodern climates.

One of these soils that is frequently mentioned is the terra rossa, or red soil. Strictly speaking, this is a type of soil developed on hard limestone. It is red, friable, permeable to water, and of modest fertility. The term is very loosely applied to reddish loess deposits in some areas, for example, in North Africa, for like many soil names, terra rossa has different meanings in different places. Most of the Mediterranean soils are brown soils with well-developed horizons and lime accumulating in them at modest depths, and they are moderately rich soils.

The five areas of Mediterranean climate are, then, much alike. Their similar climates have caused similar soils and vegetation to develop. While the details of their geology differ, all except Australia present a steep coast with a narrow coastal plain to the sea. For the student of cultural geography, they present a neat group of similar environments in which he can study the differing reactions of men. If the physical environment is the dominant cause of human behavior, there should be striking similarities in the ways of life in these regions. If not, then we must turn to other causative factors. In the investigation we should learn a great deal about how the various factors interact to create the differing ways of life that men lead in the various regions.

THE MEDITERRANEAN OF THE OLD WORLD

Some of man's earliest and greatest achievements occurred in the Mediterranean basin of the Old World. Here the Greeks wrote great literature and led the world in philosophy and arts. Here the Romans built a great empire. And here was the homeland of the Judaeo-Christian religion. Just how and why did this all happen? Was it because of the favorable climate that great men brought forth great thoughts? Was it because of the sheltering sea and the many islands that navigation and trade started so early and became so important? The answer to these questions in the past has often been yes, and the impression was that the favorable physical environment virtually created the cultural achievement. What is this environment like? What happened in it?

The Mediterranean Sea, literally the sea in the midst of the land, is divided into two basins. The western basin, lying between Italy and Spain, is the more open one; the eastern basin, lying between Sicily and ancient Palestine, has bordering seas to the north, the Adriatic, Ionian, and Aegean. The latter two seas, bordering Greece, are filled with islands. The Mediterranean Sea averages a depth of 5000 feet, but the depth at the Strait of Gibraltar is only 700 feet. Within this nearly closed basin, tides are almost absent. Winds, however, are often violent, especially in winter.

Around this basin the Mediterranean climate is found on the Palestinian coast, on the coast of Turkey, throughout Greece, extending up the Balkan coast of the Dalmation Sea, and covering all of Italy—except the Po plain to the north, which has too much summer rainfall to be classified as a Mediterranean land but which will be included here as a part of Italy, which is predominantly Mediterranean in climate. The climate extends along the south coast of France and follows the coasts of Spain and Portugal, but it does not include the interior of the Iberian Peninsula, which is too dry and in winter too cold, due to its northerly latitude and its high elevation, to be classed as a Mediterranean climate. Mountainous northwest Africa, coastal Morocco, Algeria, and Tunisia are also Mediterranean in climate. These last lands are the equivalent of southern California in climate, while northern Portugal is like San Francisco.

Cultural Achievements and the Spread of Ideas

Within these varied but similar areas bordering on one inland sea, what has happened in human history? This is a very ancient homeland for man. Stone tools belonging to some of the earliest of human cultures have been found in the desert areas of North Africa. Early types of fossil men have been found in Italy, Palestine, and Turkey. Some of the early great art of the Upper Paleolithic age has been found in caves in Spain. Most significantly, one of the possible homelands of agriculture lies at the eastern end of the Mediterranean, where the earliest known agriculture in the Old World is dated about 8000 B.C. This is a most interesting date, for it falls at the end of the glacial period and the beginning of the time when we are dealing with climates similar to the present climates in this area. This dating places us close to the recent-glacial boundary line, now precisely placed at 11,000 years ago, and it means that there must be a complete readjustment of all environmental thinking, for during the glacial period the climate of the Mediterranean basin and its neighboring regions was greatly changed. This glacial change has sometimes been credited with bringing the climate of France into North Africa. Much of what is now desert in North Africa, Israel (the Negev), and Arabia must have been at least well-watered enough to have been called Mediterranean in climate during the times of glacial accumulation in the north.

For our present purposes we should not try to start too far back. Our knowledge becomes very thin, and speculation plays too large a role. It is better to focus on the beginning of town dwelling and farming ways of life around 8000 B.C. in the Middle East. Out of this grew city-state civilizations and, relatively quickly, empires and dynasties. By 4500 B.C. Mesopotamia was well on the way to civilization; Egypt was clearly on its road to greatness by 4000 B.C. From these centers of early accomplishment, ideas radiated outward to adjacent lands.

The beginnings of civilization, then, were not in the Mediterranean climates, but in the adjacent hot and dry lands. Ideas, populations, and techniques spread from these lands into the marginal lands, and quite naturally to the edge of the Mediterranean Sea nearest the centers of the origin of ideas. Thus the Phoenician trading states became established on the coast of what was once called Palestine and is now Israel, Lebanon, and Jordan; for our purposes here, we shall refer to the area as Palestine. This does not make much of a case for the mild Mediterranean climate as the stimulator of ideas and accomplishment.

By 3000 B.C. there was a civilization developing on the island of Crete. The roots of this brilliant early center grew from the trade in grain, olive oil, and luxuries that flowed from the areas on the north side of the Mediterranean that produced raw materials to the centers of civilization to the south and east. The people on Crete were in a protected island location, in the midst of the stream of traffic, but their protected position lasted only until another group of people acquired an equal or better sea power. It was the stream of traffic that made the location of the island advantageous.

The spread of agriculture in the Mediterranean

The earliest expansion of agriculture seems to have been along the coast of Turkey and Palestine into Egypt. The fact that it spread to the adjacent islands of the eastern Mediterranean, as well as to the region of the Atlas Mountains in northwest Africa, clearly implies a well-developed maritime culture. Actually, we know that as early as 11,000 years ago the Greeks were importing obsidian from islands in the Mediterranean. Good boats capable of carrying sizable loads and overseas trade are implied. There must have been a relatively rapid flow of agricultural ideas down the length of the Mediterranean. This is an interesting example of the priority of sea routes over land routes for the long-distance spread of ideas. While every 20 miles on the land might place a traveler in jeopardy from a new group of people or a new tax collector, once clear of the land on the high seas he could go hundreds of miles with complete freedom.

The Cretans gained ideas from both directions and wealth from the trade. Their cities had indoor plumbing and many civilized features that later disappeared from the earth for thousands of years after the fall of these early civilizations. The Minoan civilization of Crete spread to the adjacent mainland of Greece, where it is called the Mycenaean. The inflow of ideas formed the basis for the city-states that grew up in the peninsulas and islands that make up the favored land of Greece.

Ideas from the Asiatic end of the Mediterranean Sea also formed the basis for the beginning of civilization in Italy. The Etruscans and the Romans used as building blocks for their civilizations such ideas as the alphabet, which shows clearly its origin to the east. They developed later than Greece, which was later in time than Crete. Beyond Italy lay Spain and Portugal and the North African Mediterranean lands, but they remained outlying colonial regions and suppliers of raw materials during this early period. The picture is almost deceptively simple: distance from the center of the origin of ideas, growth, and development in the Middle East was reflected by progressively later development to the west and by a dying out of the impulse as the distance exceeded certain limits. The map of cultural diminution, as distance from the focal center increases, is much like that for the American Southwest, where the Middle American area played the role of the center of growth.

The Greek Golden Age: Climate, Genius, or Culture?

The problem of accounting for the tremendous accomplishments in Greece between 500 B.C. and about the time of Christ has interested scholars for centuries. How much was due to physical geography? How much to race? How much to culture?

First, notice that race probably changed little from 1000 B.C. to the present, while culture rose to great heights and then receded. Race would therefore not seem to have been a critical factor. Second, it is clear also that climate changed little, if any, in this time period. Moreover, mountains, seas, rivers, and similar physical features all remained about as we see them today. It is hard to see these unchanging factors as causes of a truly immense cultural rise and fall. What did make Greece great? And what part, if any, did geography play in this?

For a time in history Greece was located very close to the centers of the greatest growth of knowledge then occurring on the earth. Ideas from these centers reached her shores by means of traders, colonists, and, quite probably, conquerors. Agriculture, at first simple but increasingly complex, brought a more assured food supply, and population grew. Men from the great cultural centers reached her shores by boat; shipping was thus introduced, and with it the idea of trade. Trade brought wealth and also an increased number of contacts with other peoples—hence more new ideas.

Sitting in a crossroads position, with new ideas flowing in from contacts in all directions and with wealth accumulating through trade, the Greeks were enabled to increase in numbers and wealth. Wealth makes possible some leisure. It is probably important that the Greeks were not organized into a great superstate and they had a class of free men with a degree of leisure, because they were supported in part by a slave class. Since the Neolithic revolution, with the great growth of population and the appearance of kingdoms and empires, such freedom for groups of men had been rare. In Egypt and Mesopotamia only a tiny fraction of the population had had leisure, freedom, and education. In Greece a significant percentage of the people had these gifts. How were these people to use their freedom?

Their interests were only partially deflected into intellectual activities. City fought city; government was often unstable, and good rulers were sometimes exiled out of jealousy. The Greeks were forced to defend themselves against incursions by the Persians, and they finally fell before the onslaught of the Romans. Through all of this, however, there ran a constant interest in ideas and a theme of individuality.

The Greeks did not put their efforts equally into all realms of knowledge. They were philosophers concerned with the nature of reality, but their religious thought was a confused, unsystematized collection of everything current. They were mathematicians skilled in certain kinds of geometry, but did little work in arithmetic and algebra. They were artists of extreme skill in certain kinds of architecture, but they hardly used the arch at all. The startling thing about Greece is the appearance of so many men of first rank in accomplishment in so small an area in so short a time in a few fields of thought.

Greek achievements in early classical times were in these fields:

Philosophy
Plato and Aristotle, 400–320 B.C.
Science
Euclid and Archimedes, 320–220 B.C.
Medicine
Hippocrates, 420 B.C.
History
Thucydides, 420 B.C.
Arts
Sophocles, Aristophanes, and Phidias (the Parthenon), all about 450 B.C.

Did the climate with its hot, dry summers lead to long, leisurely hours in the shade with discussion wandering through all the ranges of the human mind, with the consequent foundation of philosophy? Nothing like this happened in the other Mediterranean climates, which have equally hot summers. Did the strongly seasonal climate encourage men to work hard in one season and during the resting season use their leisure for mental activities? It did not happen elsewhere. Only in this one place at this one time did this particular cultural development occur. Why?

It does not seem too mysterious after all, for the contact with ideas was there. A particular social setting was there: leisure for an unusually large part of the population spread the opportunity to produce intellectually. But most important was the value placed on ideas in their culture. But not just all ideas, only certain kinds of ideas.

Could you predict the particular field of high accomplishment for the United States in the period from A.D. 1950 to 2000? I would try, with some confidence. Today in many a university the freshmen class is addressed by representatives of various departments. The physicists, mathematicians, and chemists can say to these students selected for mental ability: "If the brightest of you study with me and go on to take an advanced degree, the sky is the limit in your potential financial reward. You will be able to pick and choose between jobs in terms of where you prefer to live and the particular work that you wish to do. You can expect to start with a handsome salary." And although this is not said, it could be added that as an atomic physicist, or physical chemist, or mathematician, the populace will honor you. This is an important motivation because man's ego craves honor.

Greek Influence. *So overwhelming was Greek accomplishment in the arts that it conquered the country's conquerors. The Parthenon became the model for temples around the world. This Greek temple is located in Agrigento, Italy. The American equivalent is the Supreme Court Building in Washington, D.C.* (Italian Cultural Institute)

On the other hand, the representatives of such fields as philosophy, art, poetry, history (all fields in which the Greeks were well represented) can say to these same young men, "Come and study with me and take an advanced degree. We can probably get you a job, but you will have to go where the opening is. You can expect to start at a low wage and advance slowly. You will be viewed with a sort of tolerant and kindly respect for your work, but you can expect no adulation or admiring recognition. Further, no matter how good your work may be, most of you will have to settle for a lower monetary reward than men in the sciences."

Under these circumstances where do you think the best minds go? Most, though not all, of course, are deflected into the fields having not only high material premiums but also high public esteem as part of the reward. There is strong attraction in being in the leading field where there are the greatest discoveries that draw public attention. Just which area of learning is the most exciting in general esteem varies from generation to generation. The best minds that have the greatest natural gift for work in this particular field will then accomplish the maximum; these men we call geniuses.

To be recognized as a genius, it is necessary to be born with the right gifts in the right age in the right place. Had Socrates been born in Greece in 10,000 B.C., before the growth of ideas had made possible his philosophizing, he would have been lost to history. Were Socrates living in Greece today, would he be noticed? Probably not. The Greek location at that time in history, together with a particular bent of the Greek culture made possible the flowering of a particular set of minds. Had a man with a mind suitable for handling modern higher mathematics been born then, he might have perished unrecognized. The tools of his trade, the setting of ideas, the permissive tendency of the society were not ready for him.

So, for our society, it seems predictable that the twentieth century, especially the second half of it, will be marked by a flowering of genius in the physical sciences—and the climate and soil will have nothing to do with it. Our location in the world in place and time and the orientation of our culture set the pattern for this.

Location and Progress

It has often been noted that the center of civilization shifted northwestward through history: from the Middle East, to Crete, to Greece, to Rome, to France, to north-western Europe. This is, of course, a partial view. The spread of ideas from the Middle Eastern heartland was going in several directions: northeastward to China, eastward to India and Southeast Asia, and ultimately to America. From each new center there were secondary radiations. The course was radial and simultaneous in all directions. While ideas were spreading from Crete to Greece to Rome, they were also spreading from the Middle East to the Indus valley to India and beyond. The northward course of civilization is a one-sided view arising out of an overemphasis of the Western European on his own cultural history.

The last in time of the great Mediterranean powers were the Iberian states of Spain and Portugal. The maritime skills of the Mediterranean, the commercial tradition of the peoples of that sea, and the slowly increasing knowledge of other lands with exotic products of great value, all came together in these lands in the fifteenth century and set off one of the most explosive cultural changes of the world. But note the immense time lag: 2000 years later than Greece, 1000 years after Rome.

A great deal has been said about the importance of the location of Portugal and Spain on the Atlantic and the importance they gained when the Muslims cut the traditional routes to the Far East. There is some truth in this, but it must be seen in the right perspective. First, the position did not change, the historical situation did. The position remains pure physical geography; the situation is human geography. Neither is an absolute determinant, but it is clear that the situation is the ruling one.

For a time in history the Mediterranean Sea was the "lake" where Western civilization was growing at one end. Later this sea was the highway for the transportation of goods and ideas westward and northward. The exploration of the great oceans is often thought of as the turning point that left the Mediterranean powers in a position of great disadvantage. The focus was now on the Atlantic. All of this is true as far as it goes.

It is interesting that with the opening of the Suez Canal in 1869 the Mediterranean again became a crossroads of the world, but there was no resurgence of the Mediterranean powers. Italy did not seize upon her central position to control the Mediterranean Sea as of old, to become the dominant power of the world. Simple geographical position is not enough. Portugal, Spain, Italy, Greece, and Egypt, all great powers in the past, still occupy their ancient geographical position and again sit at the crossroads of the world. Control of the critical geographical points in the Mediterranean (Gibraltar, Suez, Cyprus, Sicily, and Malta) fell into the hands of the Atlantic powers, and the ancient Medi-

Use and Abuse of the Land in the Mediterranean. *The three pictures here show much of the story of the Mediterranean. (Upper left) Goats and the plow have accelerated erosion until there is often little but rock left. (United Nations for World Bank) (Lower left) A man is irrigating what seems to be just a rock pile. (United Nations for World Bank) (Right) Soil has been saved by piling up rock into walls to effectively stop the erosion. Such land abuse and just such immense investment in terracing and rock piling is found all around the Mediterranean, which was particularly vulnerable to erosion because of its mountainous nature and the prevalence of limestone with thin soil covers. These pictures are all from Lebanon, famous in the time before Christ for its magnificent cedar forests. Only tiny remnents now remain. (United Nations for World Bank)*

terranean powers lagged behind. Perhaps it is better to say that because they were lagging, the critical control points fell into the hands of others.

Today the growing importance of air travel continues to focus attention on this area, yet the Mediter-

ranean powers still lag. It is often argued that their geography is against them, that coal, iron, and oil are required for modern developments. But the economic development in countries like Belgium, where there is little coal and iron but extensive industry, disproves this theory. Japan after 1900 is an even better example of what a country can do in spite of poor environment if human energy is channeled along productive lines.

It is easy to overestimate the physical-geographical position of the Mediterranean. It would be of no value without people, or if the people were on the cultural level of the natives of Australia. Only when Portugal and Spain were occupied by an aggressive, commercial-minded, and adventurous people who had sufficient mastery of sailing and an idea to drive them to attempt great voyages did the raw fact of the physical location of the Iberian Peninsula come to have any great importance. Even then the importance of mere physical location seems almost accidental. As history shows, colonial empires around the world were also founded from other Atlantic states, such as France, Belgium, Holland, and Great Britain. All these people shared a cultural background that included shipping skills, commercial developments, and a taste for the luxuries of the Indies. These factors, more than geographical position, drove these peoples to their overseas exploration. This is not to say, of course, that the seaside position had nothing to do with the whole matter: the Swiss, Poles, and Austrians did not take to the sea to found vast colonial empires. However, seaside location only leads to colonial expansion overseas for those who have the idea of expansion in their cultural pattern. This idea was not important to the shore dwellers of California, Australia, or South Africa.

In the Mediterranean lands of the Old World it has been clearly demonstrated that the human material present is of the very highest caliber. The achievements of Greece, Rome, Spain, and Portugal, and of earlier peoples at the eastern end of the Mediterranean and at Carthage in North Africa show that gifted peoples could, when motivated, make these Old World Mediterranean lands great centers of creative thought, productivity, and world power. The same race is still there; the lands have changed but little. Perhaps the land is somewhat the worse for wear in places, but in other places endless labor has terraced and irrigated the land until it is better than ever. The travel patterns of the world have shifted back and forth: the means of transportation has changed from the sea to the railroad to the air, but the sea has always remained the basic bulk carrier of the world. In this area at the crossroads of the world, a gifted race has been resting on its oars. It is a clear example of a good environment with a tested race that is now relatively unproductive. Physically these are not the richest lands, but neither are they the poorest. Economically they are all now underdeveloped. They have not shared in the industrial revolution. The following review of the physical settings, current developments, and future prospects of these countries should be read with these principles in mind.

Palestine and the Israeli Experiment

Palestine is much like southern California. San Diego is the equivalent of Tel Aviv, and Los Angeles is located at Beirut's latitude. As in southern California, the narrow coastal plain is backed by a mountain range, beyond which lies a trough reaching below sea level—the Dead Sea, which is like the Salton Sea. Jerusalem has a location similar to the little town of Julian in southern California, perched on the mountains above the great, deep, arid trough to the east. The coast of Palestine is rainier than southern California, but the two areas have a steady increase in rainfall toward the north. Palestine's soils and forests have been used and abused by agriculturists and pastoralists for more than 5000 years; southern California's soils have endured less than 200 years of agricultural abuse. More arid California, therefore, has more vegetation and soil on its hills and more forest on its mountains than rainier Palestine.

Shifts of economic and political power at the end of the Middle Ages began to leave the Arab world aside. Soil erosion and deforestation impoverished the physical base for an agricultural-pastoral way of life, and the level of development sank very low.

The Jewish nation of Israel has now been established in this area. It is a relatively small expanse of land, about the size of Vermont, but much of it is desert—the Negev. Israel has been considered both as a Mediterranean land and earlier as a dry land. The old, worn, land, however, is being revived by the application of capital and skills from the Western world. Jewish refugees from Europe brought a flood of skilled workers and professional people. The Jewish community of the world has contributed capital. Hills are being reforested and plains irrigated, and even the Negev is being studied for resettlement. It is estimated that the area once held 5 million people. With modern technology it should readily support this many again; with industrial development it could support many more. The critical shortage is water, for both industry and population require large amounts, but desalinization may help solve this problem. The interesting point is that even a terribly abused land can be brought back into production; with will, skill, and capital any land

Reforestation in Israel. *The Jewish reclamation in Israel is a spectacular case of man changing the land. Too often over the past millennia man's use of the land has simply degraded it. In Israel the poorest land, the stony, eroded hillsides (upper left), has, have been reforested. Here we see both young forest growth (lower left) and tall trees (right), probably eucalyptus, representing one of the earlier plantings. Religious motivation and international financial support from the Jews of the world have made this possible. (Jewish National Fund)*

can industrialize, even if it lacks basic minerals and fuels.

Unfortunately the world is in turmoil, and the Arab world views the Jews as intruders. The future of Israel will depend not on physical geography but on political geography. What the Jews are demonstrating meanwhile is that the area here called Palestine could be another southern California.

Greece

Greece, the seat of remarkable mental achievements, is a mountainous set of peninsulas extending southward into the Mediterranean Sea. Greece is the size of New York plus Rhode Island, but the population of 8,800,000 is less than half that of these two states. The latitude is approximately that of San Francisco. The rainfall is somewhat greater than for central California, but the climate is much like California's. Temperatures in summer are hot, but the winters are mild with frost being infrequent on the plains, though snow may blanket the mountain tops. The land is rough, and there are few good alluvial plains, so only about 20 percent of it is arable. Another 20 percent is in forest, 20 percent is waste lands, and the remaining 40 percent is pasture-land, used by sheep and goats.

On this meager physical base one-half of the population make their living: a fair measure of the lack of industrialization. (Compare the United States figure of about 5 percent of the people in agriculture.) The industry is mostly small scale, dispersed among chemicals, rugs, textiles, leather, cigarettes. Wheat and flour, petroleum products, machinery, and most manufactured goods are imported. With half their population in agriculture, they still do not feed themselves, but they offset this by agricultural exports such as grapes, olives, and tobacco. These exports command sufficient price on the world market to make it profitable to sell their labor through these specialty crops and buy the wheat they need.

No nation in first rank today is primarily an agricultural producer. Greece's problem is how to increase industrialization, despite limited amounts of iron, silver, copper, nickel, lead, and zinc. There is no fuel of significance in the country, but petroleum is widely available today, and some of the major producing fields of the world lie adjacent to Greece. The greatest need is for education, organization, and capital. With these, there is no apparent limit to what Greece could produce in the future.

We might well ask why a nation that once led the world in so many artistic and intellectual fields, including such practical ones as mathematics, is now in a state of eclipse? The answer lies in history. While Rome reduced Greece to a vassal state, Greece continued as an important center of culture under Roman control. The Goths, however, devastated Greece twice; thereafter, with the fall of the Western Roman Empire, Greece was left in the outer periphery of Byzantine culture. Then came frequent conquest and changes of rule: Slavs, Crusaders, Byzantines, and then the Turks swept down on Greece. After centuries of incursion and the devastation that alien occupation and wars ensure, Greece finally emerged as a nation again in 1832. But it was a poverty-stricken, internally quarreling, foreign-dominated nation. Greece has been struggling ever since, with further setbacks in World Wars I and II. Perhaps we should ask how any people could have survived under such a load of misfortune, rather than asking what is wrong with Greece. It is a good land with a good people who have been burdened with 1500 years of invasion, occupation, and destruction. It is a small unit and would function better as part of an economic union.

If the developed nations are industrial, what would be the basis for Greek industrialization? Puerto Rico has demonstrated that a nation need not have local supplies of fuel and raw material for development in the present stage of worldwide transportation. Everything that the Greeks need can be found around the Mediterranean—from raw materials to markets—and everything is accessible by sea. Indeed, the geography of Greece, with its irregular coastline, facilitates shipment of raw materials to every major population center. The peninsular structure of Europe also provides easy access to the sea for all imports and exports. But the physical geography does not promise that maritime practices will develop; the geography alone cannot determine that Greece will become an industrial nation importing raw materials by sea and exporting manufactured goods by sea. That Europe juts into the sea is less important than the fact that it is thickly populated by one of the most productive people on the earth. It is the economically advanced nations that supply the greatest markets today. The Palmer Peninsula of Antarctica, the myriad islands of southern Chile, the Aleutian Islands are all accessible by sea, and the Aleutian Islands are in the same stormy latitude as Ireland and England. It is their lack of habitation, not their lack of approach by sea, that offers little possibility for either industrial development or markets. Greece, on the other hand, has people on its land, and, as with Puerto

Rico, these are the major resource. Puerto Rico and Greece both have mild winter climates with abundant sun and warm seas of translucent water, which offer an escape from the long, cold winters of Germany and Great Britain. In the United States many an industrialist has moved his business to southern California or Puerto Rico, drawn by similar possibilities of fun in the sun. At the moment, Greeks are supplying a labor force for northwest Europe. As in Puerto Rico, a return flow of people with a little capital and a lot of know-how may greatly influence Greek development. This is not necessarily good. Development too often means air pollution, crowding, traffic jams, and sundry ills of poorly planned growth. Already the industrialization of Athens has polluted the air with acids that are dissolving the marble of the Parthenon at an alarming rate.

Italy

The physical-economic base

Italy is a slim, mountainous peninsula, about 850 miles long and 150 miles wide. Plains make up about one-third of the land surface, and the valley of the Po (Lombardy), the only large plain, is about two-thirds the size of California's Central Valley. The mountains are relatively high: peaks range from 6000 to 9000 feet. Since they rise at rather short distances from the sea, they are marked by steepness and give Italy a rugged aspect. They also give the typical Mediterranean winter scene of snow-capped peaks rising within sight of the warm blue sea. Such snow- and rain-catching mountains are important in a land of summer drought. The mountains get increased precipitation and feed the streams that descend to the coastal plains; they are also areas of potential hydroelectric power developments.

A mountainous land more than 800 miles long must have a variety of climates. Southern Italy is warm in winter, while the Po plain is very cold. Summer temperatures vary not only with altitude, but with location on or away from the coasts and the ameliorating sea breezes. Rome is so hot in the summer that all who can, leave for the cooler hills. Only tourists visit Rome in the summer. The vegetation varies as the climate does. Southern Italy has the olives, palms, figs, and pomegranates that one would expect, while northern Italy is much too cold in winter for subtropical trees. The mountains are oak and chestnut forested at the lower levels and pine covered on their higher levels. Despite its mountainous land, Italy is poor in basic minerals and resources: oil, coal, and iron. Other min-

erals, except for mercury and sulfur, are not present in large amounts.

Italy can be compared in many ways with California. Italy is 116,224 square miles and California 158,693 square miles in area. They are almost exactly the same length, but California is a little wider in extent. Italy lies farther north than California: Sicily is at the latitude of San Francisco, while Rome is near the latitude of Eureka in northernmost California. This gives Italy much better rainfall than California but considerably colder winters also.

The Italians have done an astounding amount with this small and rugged land. They support nearly 54 million people on less land than California uses to support about 22 million. The Italians, with only one-third of their land in plains, have made about 55 percent of it arable by terracing the steep and sloping parts, or otherwise managing to bring it into agricultural use. California, with a greater extent of plains, has only 11 percent arable land. In Italy more than 20 percent of the population is engaged in agriculture, yet not enough grain is produced to feed the population. Wheat and corn are major crops, but wheat flour is still one of the necessary imports. Wine production is so high that Italy accounts for nearly 20 percent of the world's production. In olive-oil production Italy is second only to Spain, for vines and trees can be grown on steep slopes if hand labor is used. It is this kind of land use that raises the percentage of Italy's arable land.

Past greatness

In the past Italy has been the center of cultural and political achievements of first rank. It may well be argued that in the preindustrial past the geographic base did not need to be so large and the shortage of mineral resources was not as critical. While this is certainly true, it is less than a full explanation.

In part the success of nations has to be seen as a study in comparative advantages. At the time of Christ, Italy was a significantly large unit, with an agricultural base that supported a dense population for its day. Italy was then also one of the technologically advanced nations of the world. Great Britain, the center of the nineteenth-century equivalent of the Roman Empire, is approximately the same size as Italy with almost the same population. True, Great Britain has coal and iron, while Italy has little. But despite this, Great Britain has now lost its empire and pre-eminent position. All that would be needed to complete a test of the freedom from environmental control of these variations in accomplishment would be for Italy now to rise to an important position again.

The fall of the Roman Empire can hardly be blamed on Italian geography. The empire was built on population and organizational force based in Italy. By 265 B.C. the Romans had won control of all of Italy and went on from this base to build their immense empire. Thereafter came a long history of conquest abroad accompanied by shifts at home from a republic (265 to 33 B.C.) to dominance by emperors. By A.D. 476 a series of barbarian invasions had brought the empire to an end. The causes of the fall of Rome have been endlessly debated. Whatever they were, they were not due to climatic change, movements of the land, or even of soil erosion. The physical geography remained unchanged. Rome became corrupt internally, and there was a loss of the organizational driving force that had built the empire. The most conspicuous change was the shift from a republic, with elected heads of state limited to one term of office, to an empire ruled by virtual dictators.

The breakdown of the republican form of government was hastened by generals such as Sulla, who in 83 B.C. used his troops to defeat the popular party and seize his opponents' lands, which he then distributed to his veterans, who often neglected them. Thus productive lands went into lowered production or even went out of production. The tax situation in later Rome accomplished the same thing. Land was taxed, and if some land went out of production, the tax on the rest of the land was simply increased. This drove more land out of production. The material wealth of a nation is measured by its productivity; some of its strength is supplied by the number of its people, but its greatest source of power comes from the moral forces that check public and private immorality. Scholars agree that both public and private morality declined throughout the history of the empire. They describe also a weakening of the state as the republican vigor of the Romans declined. This process seems to have been related to the corruption that led to a lessening of production, which in turn led to increased taxation, which led to lessened production. But there is very little physical-geographical determination in this cycle of disintegration.

History in Italian Architecture. *Influences from the eastern end of the Mediterranean sparked the cultural growth in Italy. The Greek theater at Messina* (top) *on the island of Sicily is a measure of this influence. After the collapse of the Roman Empire and the Dark Ages that followed came the Renaissance, epitomized here by the Palazzo Vecchio* (bottom) *in Florence.* (Italian Cultural Institute)

The weakened empire finally collapsed and the Germanic tribes ravaged it from end to end. Production and population fell to still lower levels. Italy sank into a period of history where there was no unity: city-state fought city-state, neighboring countries contended over various parts of the peninsula. Spain, France, Austria, and other countries fought over and divided Italy again and again. During a momentary lull in the squabbling in the fifteenth century, the Italian state achieved greater prosperity and a measure of stability, a period that gave rise to the flowering of genius known as the Renaissance. At this time Italy was the foremost country of Europe in learning and the arts. Thereafter, the dreary recitations of wars, conquests, and the inheritance of different parts of the land by various dynastic houses of Europe resumed. In the early nineteenth century, following the American and French revolutions, a movement toward Italian national unity gained momentum. It was not until 1861 that the first Italian parliament was convened, and ten years passed before Italy was united.

Italy had not stood still throughout this period. The various city-states had been active centers of trade, wealth, and culture. Venice, for instance, was a city-state of immense importance in the Mediterranean, with sufficient power to have overseas possessions of its own. But if Italy is considered small for a modern state, a peninsula divided into city-states was even less adequate.

Unified Italy entered belatedly into the acquisition of a colonial empire, obtaining parts of northeast Africa late in the nineteenth century. After World War I Mussolini established his Fascist dictatorship, and the Fascists worked hard at the material buildup of Italy: reclamation of wasteland, modernization of agriculture, construction of highways, modernization of ports, and the encouragement of industry. Under extremely centralized, dictatorial control, the development was meager, only to be reduced further by depression beginning in 1929. To give his countrymen foreign exploits to take their minds off their misery, Mussolini led Italy into its conquest of Ethiopia, an activity that placed Italy more firmly on the side of Hitler's Germany and opposed to the French and British. This in turn led to Italy's disastrous role in the North African campaigns during World War II and the eventual destructive campaigns in Italy itself. The peace treaty at the end of the war stripped Italy of its African colonies and saddled it with severe reparations charges, for example, $100 million to the Soviet Union. The country thus emerged in 1945 wrecked by warfare, internally torn by strife between the Fascists and the Communists, and with little unity in the opposition.

Against such a background, the question might well be asked: What bearing has the geography had on the fact that since the end of the Roman Empire Italy has been one of the weaker nations of the world? Given a modicum of unity and stability can Italy accomplish anything in the modern world? Or is its land too small and poor to support its people? Must it continue endlessly to suffer from poverty while its people are exported as surplus?

Industrial renaissance

Since 1945 there has been rapid change, for the industrial revolution is finally reaching Italy. As a result, Italian engineering and business techniques are being exported all over the world. Italians have built pipelines from the Soviet Union to four satellite states in Eastern Europe, a power plant in Pakistan, roads, port facilities, and power plants in Colombia, Ecuador, Honduras, Mexico, Peru, and Uruguay, and are engaging in similar engineering enterprises around the world. This kind of activity brings in needed income for development at home. It is also a measure of a high level of human achievement.

Production in many lines of endeavor—from heavy industry to women's *haute couture*—has increased to the extent that the gross national product now exceeds $100 billion. Italy's merchant marine, practically destroyed during World War II, has meanwhile been rebuilt to fourth place in the world, and Genoa is now the Mediterranean's busiest port.

This kind of explosive growth has resulted in strong Italian gold and dollar reserves. American investment in Italy alone exceeds $1.5 billion. There are some 600 subsidiaries of major American firms, mostly concentrated in the industrial triangle: Milan, Turin, and Genoa. American methods are employed in a wide range of activities, including scientific chicken raising and ice cream factories.

The growth of industry has begun to change Italy. Italians eat twice as much meat as they previously did, and many now have cars. Traffic jams, noise, air pollution, and the need for many new costly highways accompany such growth. Italy's *auto-strade* equal the best roads in the United States, and are revolutionizing travel time through that long and mountainous country. More cars and more roads mean more work for people who can then buy more cars, which will need more roads. However, Italy's difficulties are not at an end. There are still at least 1,000,000 Italians unemployed, and the distribution of the industrial boom is not even. Northern Italy, long the industrial heart of the nation, is the most prosperous region; Sicily is the poorest. Italians still have a lower standard of living than French, German, Belgian, and Dutch workers. Italians are restless and complain about prices, but part of this is due to greater expectations. For the first time they can afford to buy a small car or choice cuts of beef. Now they are clamoring for better roads and lower prices.

The large labor pool in Italy is an advantage because other Western European countries are running into labor shortages. With the growth of the European Common Market, there will be increasing mobility of capital, and more industry can be expected to flow into Italy to take advantage of the pool of relatively low-cost labor there. This will inevitably lead to rising standards of living and increasing labor costs as the Italian worker pulls abreast of the standard of expectations in the rest of Western Europe.

Natural gas has been found in the Po Valley and is an important supplement to the hydroelectric power that has been able to draw water power from both the Alps to the north and Apennines to the south. Southern Italy has now also found natural gas, and combined with its labor force this should promote industrial development there.

Even Sicily, one of Italy's most depressed areas, has a combination of good land, rich fisheries, large sulfur deposits, and a reservoir of labor. Therefore it has a real potential for industrial development once the free flow of capital and goods within the Common Market reaches the country.

Italy has some curious advantages in this mid-twentieth-century growth. The lack of coal is of slight disadvantage, for petroleum is available nearby and easily transported by tanker. Coal is a depressed industry in other countries, where it can be obtained at less than cost because governments are subsidizing its production. Meanwhile technological developments in the mineral industries are making the use of coking coal less and less important. Even the absence of colonies has proved to be a blessing. The other Western European nations are losing their colonies, often along with the investments they have made in them or, as in the case of France, after having unsuccessfully poured out immense sums in the attempt to hold them. It seems reasonable to predict that in a unified Europe Italy may be one of the dominant states. The old geographical analysis that emphasized the lack of basic minerals, the limited amount of land, and the crowding population as the causes of Italy's impotence will once more have been confounded by the decisive force of human geography.

At present Italy is struggling to build industry and especially to reach the areas of greatest poverty:
the peasant hill farmers and the people of southern Italy. Industry is concentrated in northern Italy partly because of historical reasons; this is the area in immediate contact with industrialized Europe.

Southern Italy is a land of grinding poverty, of earth-floored rooms lacking running water or electricity, often one room for a family shared with their goat and chicken. It was not always so. When the eastern Mediterranean influences first spread to Italy, it was southern Italy and Sicily that benefited first and most. It is said that there are more Greek ruins there today than in Greece. Cities were built on the plains near the coast, and the grain, olives, and vines made the region desirable. In the time of terrible wars, after the fall of the Roman Empire, and during the struggles between Muslims and Christians, coastal towns were vulnerable to attack, and the people retreated to fortified hilltop positions. War, pillage, piracy, and the loss of people dragged off into slavery ruined the economy. When industrial growth began, it was in the north while the south stagnated. The government is now trying to rectify this misfortune by encouraging new industries to locate in the south. At Brindisi a $300 million petrochemical plant employs 4000 workers. At Taranto a state-owned steel plant has been built; at Bari, an electronics plant; a Gela, a petrochemical plant; at Syracuse, Italy's largest oil refinery. This illustrates the mobility of raw materials and the ability of tankers to move petroleum anywhere. Despite the building of new plants in the south, together with new roads, aqueducts to irrigate the land, and great increases in electric power production, southern Italians still flock to the northern industrial cities to seek employment. All too often, they join the ranks of the unemployed and gravitate into urban slums.

The attempt to end rural poverty in the toe of Italy in Calabria exemplifies the problems. It is an area subject to natural disasters such as earthquakes, landslides, and floods. It has 30 percent of its work force in farming. They have few tractors but many mules and goats, and they farm tiny terraced farm plots. Houses are small and dilapidated, without plumbing, and often with so little furniture that people sleep on the floor. It is almost as if one were stepping back into medieval times. Per capita income is half that of the rest of Italy, unemployment is about 10 percent, and about 70,000 people per year emigrate to seek a better life. Why, when Italy tries so hard, does this area continue to lag? The businessmen who have tried to run plants there say that is because the people are still peasants, not factory workers. They want to work when it suits them rather than on a regular basis. Further, militant unions have demanded pay equal to that of the more

Hill Towns of Italy. *Much of Italy is mountainous or at least hilly. In times past towns were built on the highest places for defense. Only belatedly are the Italians now moving out of the more inaccessible of these defense sites. The landscape is typically Italian, with its carefully tended vineyards and small fields, often on steep slopes.* (Italian Cultural Institute)

disciplined and skilled northern workers. But such demands simply remove the only advantage Calabria has: abundant low-cost labor. The result is continuing stagnation.

The industrial movement is much more successful than an attempted land development, where large blocks of land were divided and small farmers were aided with housing and equipment to become self-sufficient owners. This failed in large measure because the farm units were too small. The trend today is toward large units of land with heavy equipment and increasing efficiency of production.

A more successful if much more modest attempt at improving the lot of the small farmer has been undertaken as a public service by an oil company. They sent a young agronomy college professor into a group of hill villages in northern Italy. When he began asking questions about production of live trees, he was greeted with suspicion and hostility because the peasants feared that he was a tax collector. When he assured them that he was not, they told him what he needed to know. Olive trees clinging to odd bits of land produced enough fruit to provide a quart of oil every other year. He taught them to prune the trees so that their branches were spread out and light penetrated the leafy area. The trees then produced more each year. He convinced them by simple cost accounting that it was more expensive for them to grow wheat on their hill farms than it was to buy it from the lowlands, but first they needed a road to the village. He encouraged the villagers to get out with pick and shovel and build their own. The project had been discussed since the Middle Ages. But this act that had always been within the ability of the villagers was accomplished only under outside stimulus. Now chestnut wood and other products flow out, and insecticides, fertilizers, building supplies, and newspapers flow in.

This is a small undertaking. The area is described as 100 hillsides totaling 16,000 acres, inhabited by 8000 people. The cash income was $150 per capita. Two-thirds of the land is in forest, and one-fifth is unfit for cultivation even by Italian peasant standards where steep slopes are terraced and fruit trees such as olives are tucked into tiny bits of useful soil on craggy hillsides. On such land wheat is a poor crop and maize a better one. The professor persuaded 10 farmers to try hybrid corn at their own expense. The increase was so spectacular that the idea spread like wildfire. With olive trees as old as the tower of Pisa suddenly yielding two to four times more than before, with costly low-yielding wheat replaced by high-yielding hybrid maize, and with real connection to the outside world, production, ambition, and income rose.

Power in the form of chain saws and tractors increased twelvefold. Corn and meat production doubled. Marketable production rose four times as fast as for Italy as a whole. It has been so successful that the locality is now a demonstration area where courses are given on how to do likewise elsewhere. Thirty-nine communities in Sardinia are now being "advised," and similar units are at work in Portugal. In the United States this is what the county farm agent does, and we take him for granted.

This kind of development has great advantages for the underdeveloped nations of the world. It is not costly in terms of scarce capital. It moves people from a Neolithic peasant level of existence into the industrial ways of the twentieth century by introducing power tools, electricity, and modern machinery. It builds strength into the economy by expanding production and purchasing power among the masses of the people. As skills accumulate and labor markets expand, the people on the land already familiar with machine tools can supply a more skilled work force for the growing heavy industry. This program is in contrast to that followed in India, where the attempt has been made to jump directly into heavy industry with little thought about the source of skilled and semiskilled labor or of an adequate agricultural base to support the industrial sector of the population. The long-run future does not lie with the hill villages, or with the peasant farmers anywhere in the world, but a transition has to be made from their ancient way of life to that of the twentieth century. In Italy it is obvious that the steepest and poorest hill lands are being abandoned, along with the little fortified hill villages built to survive the days of political disorganization and banditry, and that just as elsewhere in the world, the people are moving off the land and into the industrial centers. Those who remain on the land are farming larger tracts of better land and producing far more than their ancestors did, and the surplus is helping to feed the people who are engaged in industry.

The Iberian Peninsula

The Iberian Peninsula is a varied land. The edges are Mediterranean in climate and the southern valleys are warm, but most of Spain is a plateau about 3000 feet in elevation with ranges of mountains towering above the plain. In latitude, altitude, and climate it is comparable to Turkey. In location, history, culture, and in part in climate, it is a Mediterranean land and will be discussed here. If Portugal is included, the land areas are comparable, but the population of the Iberian Peninsula is twice that of Turkey. Spain's plateau is semiarid and alternates between freezing cold in winter and blazing hot in summer. Madrid can have temperatures of 107° F. in July and −7° F. in January. This kind of range is like that of Reno or Salt Lake City, cities that share with Madrid interior location on plateaus with high mountains nearby. The mountains are forest covered, and in winter they are snow covered. Streams from these mountains carry large volumes of water in the time of snow melt in the spring, but they fall to low volumes in the summer period. Further, most of these rivers are deeply entrenched in the plateau so that their water is not easily distributed for irrigation. However, streams descending from well-watered mountains to semiarid plains are potential water power and irrigation resources, as the arid American West shows.

A land with its plateau averaging 3000 feet above sea level, liberally crossed by mountain ranges with peaks to heights of 11,000 feet, is mountainous highland country. Has this been a good environment or a poor one? Was it the highlands and the mountains that determined this? If not, what, then, were the causes?

Portugal shares the Iberian Peninsula with Spain, with the Atlantic-facing coast south of the northwest corner. In many ways it is like the California coast: few coastal plains, mountains near the coast, rainy to the north and increasingly drier to the south. But all of Portugal lies north of Monterey, California, and extends to 40 degrees, the northern boundary of California. In California this is the redwood belt: the coolest, wettest, least sunny part of the state but also the most drought-free part. Like Spain, Portugal has lagged behind Europe in the twentieth century. How can this be when in this same time the very similar land of California, peopled by similar Europeans, including many Portuguese, has risen so spectacularly? For convenience we will discuss Spain and Portugal separately.

Spain's historical background

Spain early showed signs of being inhabited by skilled people. In Upper Paleolithic times, 20,000 years ago, men were painting the walls of caves in Spain with the most advanced art that man had yet achieved. It compares favorably with some of the art produced today. Later, there is clearly African influence in the art of Spain. There is a constant repetition in Spanish history of her place as a bridge connecting Europe to Africa, for the slight break at Gibraltar has seldom stopped men from crossing in force in either direction.

In the sixth century B.C. Celtic peoples from France poured down through the Pyrenees and dominated much of Spain. About 100 B.C. the Iberians, racially Mediterraneans, entered Spain from the north of Africa and became a dominant people. Later, Germanic tribes poured in. Like the Strait of Gibraltar, the Pyrenees seem to have been slight bulwarks, for armies moved north or south through them, depending on the tides of cultural dominance. However, the Pyrenees are as good a physical barrier of their kind as one is likely to find. They present a high continuous wall extending practically from sea to sea. As with all such natural barriers, the most important factor is the differential in human forces on the two sides.

The Phoenicians traded with Spain, and Carthage founded trading cities there. Modern Cartagena is the descendant of New Carthage. The Romans could not allow Carthage to have the wealthy peninsula, and this was one of the major causes of the Punic Wars. The Romans eventually destroyed Carthage and developed Spain themselves. Under Rome the Iberian Peninsula was united, and Hispania became one of the most prosperous areas of Roman power. With the disintegration of Roman power upon the outbreak of the Germanic tribes in A.D. 409, Spanish unity was destroyed; it was not to be put together again for another thousand years.

Vandals and Goths made their way through Spain, and the Visigoths established a kingdom there for almost 200 years. Christianity came to Spain, but unity and prosperity remained mainly a memory. Then in 711 the Muslims arrived from North Africa and took all but the northwest corner of the peninsula; after an interval of mismanagement a strong ruler, Abd-er-Rahman, came to power in A.D. 756. His descendants established their capital at Córdoba and made this a city second only to Constantinople, the greatest city in Europe at that time. Muslim Spain at this time was a leader far in advance of most of northwestern Europe. There were more universities teaching more medicine and mathematics and stressing more literature and philosophy than were to be found anywhere else in Europe. When the Muslims fell to fighting among themselves, the Spanish Christian kings who had been steadily, though slowly, pressing the Muslims southward, began in 1031 a major drive to push them out of Spain. After 1212 the Muslims were limited to Granada in the south, and here they established one of the greatest and most splendid of all of the Muslim realms. This lasted until 1492 when, united by the marriage of Ferdinand and Isabella, the rulers of Aragon and Castille, the Spanish finally drove the Muslims out of the country.

Then came empire. Columbus tried to find the East Indies by sailing west but found the West "Indies" instead. The immense wealth of the Aztec and Inca empires, followed by the sustained production of the mines of the New World, then poured into Spain. The mines of Potosí alone produced at the rate of a billion dollars a century for two centuries. If Spain had been great in the past, this should have been the time of its greatest height.

A vast inflow of money may not be a blessing, for if it is not spent productively, it may simply create inflation. The sixteenth-century Spaniards had no mercantile tradition. Land was the basis of production, and its ownership was the mark of accomplishment. Too many of the returning conquistadors threw their money around in the manner of the newly rich and paid $20 for a new pair of boots when $2 was the going price. Money is easy to dispose of, as demonstrated today by some of the oil-rich kingdoms where hundreds of millions of dollars per year were squandered without anything to show for the expenditure. In Spain the influx of gold led to severe price rises and economic dislocation. There was little investment in a modern sense: new enterprises that would increase the national productivity and increase the capital. The money was largely expended on current living at an elevated level until the money was gone. Moreover, the chaos of international politics and kings prevailed. The Spanish monarchy became a branch of the Hapsburg dynasty and thus involved Spain in the Holy Roman Empire. Spain, thereafter, at times included parts of Italy and the Netherlands. Sometimes its kings spoke no Spanish, and at all times they were involved in wars with the French or the British or were squandering treasure trying to put down Protestantism in the Netherlands. The treasury was stripped to build an armada to attack England, and the immense fleet was lost to warfare and storm.

With the French Revolution not only was there further disastrous involvement in international affairs, but the germs of nineteenth-century liberalism, rationalism, atheism, anarchism, and socialism were introduced, and, given typical extremist development by various Spanish groups, these rival philosophies tore the precarious political structure apart. Weakened by internal dissension, Spain lost its colonies one by one between 1800 and 1830. This is not to be wondered at. Instead it seems remarkable that so small a country with its handful of people had had the energy and genius to hold so much of the earth so long.

Liberalism had fostered divisive forces, such as clericalism versus anticlericalism, monarchism versus republicanism, and separatism versus union. Spaniards, by their own account, are extremists, and they were wracked by such differences more than any other nation. For more than a hundred years, revolutions occurred at the rate of about one every three years. When civil war broke out in the 1930s, the Soviet Union backed the communist-socialist-anarchist group, and the Italian and German fascists backed the other. Spain became an international battlefield, and the situation proved to be the prelude to World War II. The country was torn apart, the people set against one another. The communist-backed group massacred priests, professors, landowners, and industrialists, and the little people on the land became cannon fodder for both sides.

General Francisco Franco's action defeated the communist-backed forces, but the country he saved was torn, impoverished, and bitterly divided. All the gold from the treasury had been seized by the Soviet Union, and thousands of Spanish children had been kidnapped and taken to Russia for indoctrination. Faced by the impoverished residue of a state that had long been wracked by internal strife, Franco established a dictatorship. Although bitterly criticized, his action served to give Spain 40 years of stability, the most it has had in more than a century. During this time universal education has been attempted, and for the first time in modern history the literacy rate in Spain is now approaching that of the other developed western European nations. Despite the fact that Franco saved Spain for the free world and by keeping the country out of World War II saved Gibraltar from falling into the hands of the fascists who had aided him in the civil war, he was condemned as a dictator. Spain was quarantined by part of the Western world, denied membership in the United Nations—though the Communist dictatorship was not denied membership—kept out of the European Common Market, and was for a time excluded from the aid that the United States has extended to the rest of the world so freely, even to the Communist nations.

Spain's physical geography and present status

This may seem like a long recital of political history. It has a purpose. What geography could overcome such fantastic mismanagement, such continuous war, and such destruction of men, facilities, and resources? If Spain, a mountainous highland country is today a poor land, are we to conclude that it is because it is a poor environment? Is it not more reasonable to recall that whenever it was given good rule, whether autocratic or not, whether Christian, Muslim, or pagan, it developed as a center of brilliant cities with great universities—a leader of the whole of Europe?

Further comparisons can make this even clearer. Spain has an area of nearly 200,000 square miles, and a population of about 35 million. (See Table 4-1.)

Spain is somewhat like California. Both have great variety of landforms and climate. Both have high year-round rainfalls in their northwestern corners, high rainfalls and winter snow covers on their mountains, a winter maximum of rainfall over most of their area, and large areas of rainfall of about 10 to 20 inches. By most measures Spain is richer: more arable land, less arid, and better mineral resources. California in 1500 lagged more than 10,000 years behind Europe, and Spain was one of Europe's leaders. California began its growth in 1850 and by 1950 was a leading state in the world's leading nation, and by the 1960s it was first in population in the United States. Spain, meanwhile, had stood still and hence declined in relative position; it is today about 50 years behind California.

Spain is today a land dominated by farming, and it is still a primitive agriculture. The principal crops are wheat and barley, and 75 percent of these crops are harvested and threshed by hand. The United States was on this level about 100 years ago. Plans call for mechanization of Spanish agriculture by 1980, and this is expected to reduce the percentage of the labor force in agriculture to 19 percent, compared to 5 percent of the American labor force.

There is a wide range of other crops: cork, olives, onions, grapes, and oranges are typical. Spain is one of the world's leading wine and cork producers. Livestock is important, with sheep being by far the most important. Mining is the other important source of wealth. Spain is highly mineralized, and despite the fact that mining has been carried on here for about 4000 years, it is still one of the major mineral-producing areas of the world. Coal, iron, copper, lead, and mercury lead a lengthy list of available minerals.

Spain's rich iron mines are major sources of supply for other European countries, which smelt the ore and manufacture steel products; Spain then reimports the finished products. Belatedly Spain has begun smelting some of its own ore and exporting pig iron and machine tools, thus employing its own people and extracting some of the potential profit from processing its own raw materials. Spain is not a have-not nation, yet it is today one of the world's poorer nations. How can this be? Is this the effect of the mountains and arid plateaus?

Spain has three times as high a percentage of arable land as Switzerland but only one-half the density per square mile or one-fifth the density per square mile of arable land. Switzerland's wealth is not all agricultural, of course. It is built on dairying, banking, and

(Upper) **Cave Painting in Altamira, Spain.** *This buffalo, painted nearly 20,000 years ago, has been preserved in a cave in northern Spain. It gives us a glimpse of one of the great periods of art dating from the Upper Paleolithic period.* (Spanish Ministry of Information and Tourism)

(Lower) **Roman Theater in Badajoz, Spain.** *The remains of an open theater on the outskirts of Badajoz in southwestern Spain is evidence of the Roman influence that has survived nearly two thousand years.* (Spanish Ministry of Information and Tourism)

(Opposite page) **Mosque in Córdoba, Spain.** *Moorish architecture in Spain is characterized by the elaborate pillars and arches shown here. Moorish influence is strongest in southern Spain, the home of the Alhambra.* (Spanish Ministry of Information and Tourism)

manufacturing (one-half the world's watch trade). As noted in this context before, banking requires a stable government and a reputation for integrity, and any nation could develop these qualities. Spain's agricultural base is both absolutely and relatively better than Switzerland's. Spain's potential for industrial development is much greater because of the presence of basic raw materials and a large working force with low wage demands, which could give Spain an important advantage during a developmental period. The main point here is that Spain, now rated as a poor nation and often described as such because of its physical geography, proves to be quite the opposite—it is a relatively well-endowed area in terms of physical geography.

Comparison with the United States again places Spain in an advantageous position in terms of percentage of arable land. Comparison with Texas, not a mountainous land, but more nearly the size of Spain, and an area with much arid land, again shows Spain to be well off in useful land. Further, in all of these comparisons, the number of people in Spain is relatively high compared with American densities. While there is great concern today about the growth of population, it must not be overlooked that people are the productive force

TABLE 4-1   Spain and Other Countries Compared

	Area (square miles)	Arable (percent)	Meadow and Pasture (percent)	Forest and Woodland (percent)	Other (percent)	Population Density		
						(millions)	(per square mile)	(per arable acre)
Spain	200,000	37	46	10	7	35	179	500
Switzerland	16,000	12	41	25	22	6.5	411	2500
United States	3,000,000	23	36	23	18	211	57	240
Texas	267,000	20	62	12	6	12	44	175
California	158,000	11	23	41	25	21	131	1240

of nations. In Europe today the countries with rapidly increasing production are importing manpower. Even Italy, the chronic source of overpopulation in Europe, is beginning to feel a labor shortage. Spain, then, is again in an advantageous position. It has its own manpower potential, but this still remains only a potential, because labor is increasingly useful only as it is educated, trained, and organized for production.

To a considerable extent Spain has made significant beginnings in these developments. Primary-school education in Spain is free and compulsory. There has been a period of freedom from wars, and there is now an improved international climate in regard to Spain. With greater access to international capital and markets, it is possible that Spain can, if its internal affairs are kept in order, rapidly rise again to a position of leadership. Spain has one of the world's greatest reservoirs at Salto de Aldeadávila. Certainly there is nothing in its physical geography that prevents development, and there is much of great advantage for such a development.

Spain has a long way to go for 30 percent of its population is engaged in agriculture largely on a plateau and mountain area that in rainfall, temperature, and elevation is somewhat like our central Great Plains, an area that we consider climatically risky for wheat farming. Although 30 percent of the population is so engaged, they produce only 20 percent of the national income. In recent years, especially since the United States extended military aid, Spain has been developing its industrial base. By 1968 it was eighth in the world in motor vehicle production, up from almost zero production 20 years ago. Cement, petroleum refining, shipbuilding, and steel are the principal heavy industries. Light industry is varied; Spain is sixth in the world in fishing. In four years (1968–1972) Spain went from twentieth to fourth in the world in shipbuilding.

In the decade 1960–1970 the gross national product jumped from $8.5 billion to $28 billion. The per capita income rose from $300 to $700, and illiter-

acy declined from 12 percent to 3 percent. In 1972 per capita income reached $1000 and is expected to reach $1500 in 1975. In 1970 1.5 million workers were employed abroad and were sending home $560 million per year as remittances. This figure is dwarfed by the flood of tourist dollars, now above $2 billion per year and expected to reach $3.7 billion by 1982. The experts shake their heads and wonder if the tourist business can continue to increase, but people find Spain beautiful, varied, historical, and relatively inexpensive, so they continue to flood in; as long as Europe's prosperity continues, there is seemingly no end to the supply. The tourist business is an obvious source of income, but tourists want more than scenery. They want certain amenities, they don't want to be cheated, and they want to be made welcome. To an astonishing degree much of Europe has looked on the tourist, especially the American tourist, as nothing more than a source of income. By law there is no tipping in Spain, a situation uncommon in the usual tourist center. The Spanish government has built and created inns over the country, often using ancient castles or noblemen's homes and often building at interesting or scenic spots. The rates are low and the service good—if you can learn to postpone your evening meal until 10 P.M. or later. The point is that the Spanish decided that the tourist dollar was important, and they set out to get it. Despite adverse publicity stemming from their civil war and its aftermath, they have diverted a treasure-laden flood into their poverty-ridden, underdeveloped land that had been relatively unknown to tourists.

Spain had more than $1 billion worth of foreign investment in the 1960–1970 decade. About $700 million of this amount came from the United States and was invested by over 90 separate American companies. The large supply of cheap and reliable labor and the huge internal market in which everyone wants a television set, a washing machine, and so forth—and for the first time it is becoming possible for people to buy

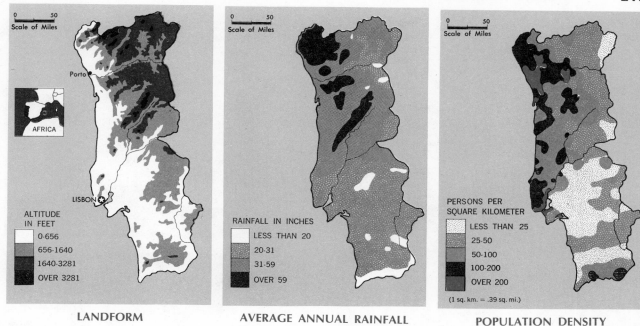

LANDFORM   AVERAGE ANNUAL RAINFALL   POPULATION DENSITY

MAP 4-1   Portugal: Landform, Rainfall, and Population
Portugal's highest land is to the northeast; its greatest rainfall is in the northwest. Its population is densest in the rainier areas. Rainfall in Portugal is high for a Mediterranean climate. Even the south, with 20 to 30 inches of rainfall, which comes in the cool season and hence is very efficient in wetting the ground, is well watered by California standards (San Diego, 10 inches; Los Angeles, 15; San Francisco, 20). The drier south of Portugal produces cork, wheat, and olive oil; the northwest produces wine and cattle. (Reprinted from Focus, *American Geographical Society)*

them—is attractive to business investment. While the economic boom has begun to slow a bit, there is little reason that it should not continue at a vigorous pace.

Spain has a foundation suitable for rapid building. Governmental structures and experienced administrators already are available. An ancient and respectable university system and a rapidly expanding school system are supplying educated and skilled people. Roads, cities, water supplies, ports, railroads—all are there even though in need of expansion. Spain's situation is not at all like the underdeveloped nations of sub-Saharan Africa but much more like Puerto Rico when that country began its spectacular growth. Spain has the advantage of even better socioeconomic foundations, a far larger and more varied and richer physical environment, and a larger internal market. In a sense Spain is belatedly shedding its heritage from the Roman

Empire and, leaping into the twentieth century, needs political stability for a few more years while the transition is completed.

Portugal

Portugal's history of development is much like Spain's. Mediterranean civilizing impulses reached it early, and trading cities were established on its southern coast at least as early as the first millennium B.C. There were probable settlements there in the second millennium B.C. Skipping a lengthy history that includes the development of a maritime fishery that probably had Portuguese fishing for cod off the Grand Banks of America before Columbus discovered America, we can pick up the Portuguese in the fifteenth century when they were the leading people in maritime exploration.

Portugal: Land of Contrasts. *Tiny Portugal has a long history of seafaring. Part of the sardine fleet on the south coast of Portugal* (left) *illustrates this long continued tradition. Portugal is also noted for its fine wines. Vineyards can be seen* (upper right) *with the vines pruned low, placed on high trellises, or even trained up on trees. On Portuguese highways one still sees ox teams pulling heavy solid wheeled carts* (lower right). *In the day of increasing use of diesel trucks this poses some traffic problems.* (Dan Stanislawski)

They rediscovered the Azores and the Canaries, and they pushed exploration down the coast of Africa until they finally rounded the southern tip and found the sea route to the Indies, the magical land of silk, spices, and gems. Then followed a rich period of Portuguese trading and colonizing. Portuguese trading enclaves in India (Goa) and China (Macao) gave access to rich markets. Colonies in Africa, Angola, Portuguese Guinea, and Mozambique have been bones of contention. The Portuguese maintain that they are integral parts of Portugal, these people are Portuguese citizens, and there is no racial or other discrimination. Their claims are gen-

erally valid. Neither in Africa nor Brazil nor at home do the Portuguese draw racial lines. Their legacy in Brazil is one of racial tolerance; in their African colonies voting rights are determined by literacy, not skin color. Whatever charges may be leveled at the Portuguese, racism is the weakest. As a tiny nation with economic problems at home, Portugal has been unable to proceed with the development of its African colonies at a rapid pace. The Africans, with much of the pressure coming from outside the Portuguese colonies, are determined that their countries shall be free. For some, the granting of independence is already in process.

Long racked by extremist groups and endless revolutions, the Portuguese called upon a college professor, Antonio Salazar, to manage them. He acted as a benevolent dictator until his death in 1970. He gave them what Franco has given Spain: a period of peace and stability in which to heal wounds and recover some national sanity. It was also a period of slow economic growth, perhaps inevitably. Measures of Portugal's lagging position could easily be seen 10 years ago in agriculture. Ox teams were commonplace, slowly pulling carts and wagons down paved roads along which modern diesel trucks thundered at high speed. People pastured goats on the roadside to garner each bit of grass. Men carried stable manure in baskets on their backs to tiny terraced fields on the mountainsides. The scene resembled something in China, with the traditional coolie labor utilizing every scrap in the environment to eke out a living. Men were conspicuously scarce in the landscape, for they had gone in great numbers to northern Europe to get wage labor.

Changes, epitomized by the diesels, were already under way. Industry was building slowly, but the tourist trade was bustling. As in Spain, the government was doing all it could to attract this golden horde. Excellent facilities at low prices in scenic and historic places were readily available, and courtesy overflowed. This development has continued, and the Algarve, Portugal's southern coast, long an isolated and impoverished area with little coastal fishing villages, has exploded into an instant Riviera. As a winter resort, the Algarve, 500 miles farther south, is the more attractive, and in summer it is at least as good. Currently luxury hotels are sprouting like mushrooms, and golf courses are spreading rapidly. Millions of dollars are flowing in as capital investment; hundreds of millions of tourist dollars will be spent there. Thousands of Portuguese will be employed, and the entire economy will get a boost. It comes at a fortunate time, for with the coup d'état in 1974, Portugal has already begun the move to relinquish control of its African colonies. This will relieve the country of what has been an increasingly costly burden. If Portugal then turns toward Europe and follows the Spanish model, we may see a renaissance in this Atlantic-facing land.

North Africa

Northwest Africa, the Atlas Mountains and the adjacent coastal strip, is also a land with a Mediterranean climate. The coastal part of Morocco, Algeria, and Tunisia is a piece of land 1200 miles long and about 150 miles wide, or twice as large as Mediterranean California. Its location compares with southern California, and the annual rainfall of 10 to 20 inches is also comparable. The area originally had much forest cover and undoubtedly was well mantled with good soils. The mountains are high enough to be good precipitation catchers and are often snow covered well into the summer. In an area of summer drought this is an enormous advantage.

Located at the extreme end of the Mediterranean, far from the ferment of ideas at the eastern, Mesopotamian end, only a few ideas—agriculture, cattle keeping, and irrigation—reached the area in pre-Roman times. A Neolithic way of life with agriculture, pottery, and domestic animals but a lack of metals was established. It has persisted with little change to the present. The Carthaginians based themselves on a few favorable spots on the seacoast and little utilized the inland region, so they had a minimal influence on the native peoples there, the Berbers. The Romans held a slightly deeper coastal area but again left the Berber tribes relatively untouched. Even the French and Spanish occupations of modern times have altered the old ways of life very little. The French impact of the nineteenth century was much greater because they moved thousands of colonists into the coastal strip of Algeria and substantially increased the productiveness of their coastal land.

The Spanish, with less to work with, barely touched the hinterland of Morocco. The land with its enormous potential is not empty; the population of Morocco, Algeria, and Tunisia is 33,500,000—nearly equal to that of the Iberian Peninsula. It is a rapidly increasing population, which formerly faced starvation in times of drought, and continues to exert a heavy pressure on the land as it increases.

The use of the land is largely agricultural. The most heavily settled lands are in the hill regions where irrigation and terracing form a way of intense land utilization. Domestic animals include cows, sheep, goats, and camels. The vegetation, formerly largely forest, has been immensely impoverished. Wood is cut for charcoal to supply household needs: every home has a charcoal fire for cooking and making tea, and it has been estimated that much more charcoal is used in tea making than in cooking. Wood is also used in house construction. It is destroyed by fires that are used in land clearing, which are sometimes deliberately set and sometimes "just happen." Finally the grazing animals, especially the goats, greatly deplete the remaining vegetation. Clues as to what the land might once have been are found in areas around shrines and holy places where the vegetation is usually carefully maintained, forming islands of beautiful forest in a sea of scrub. The arid and impoverished aspect of much of the land is an artifact: something made by man.

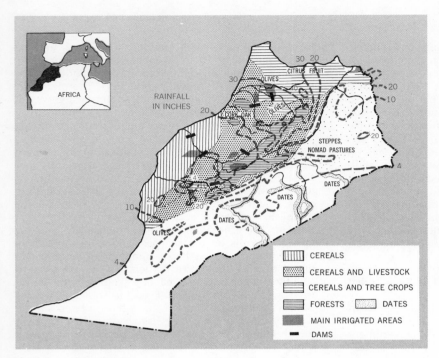

MAP 4-2 Morocco: Agriculture, Rainfall, and Irrigation

The monarchy of Morocco is larger than the state of California and has a population of 15,500,000 people. Its location is similar to that of southern California. The comparison extends to a steady increase of precipitation from south to north and a sharp increase toward the mountainous interior, followed by a sharp decrease into a desert interior. In California one-half the population is in the southern areas with 12 inches of rainfall or less. In Morocco most of the population is in the areas of 20 inches of rainfall or more. As in California, crops include cereals, citrus fruits, vines, olives, and dates. Morocco has the climatic advantages of southern California but lacks the politically and culturally determined locational opportunities of the New World area. (Reprinted from Focus, American Geographical Society)

Agriculture is carried on with primitive tools, and the work is almost entirely done by hand. The plows are little better than the light Roman plows of 2000 years ago, and the olive presses are like the ancient Roman presses. There is little mechanized agricultural machinery, except in a few recently developed European areas; grain crops are planted, harvested, and threshed by hand. There is little hydroelectric development, and a general state of economic undevelopment persists. This is a land that has stood still, or nearly still, for the past 4000 years. Cultural advance has been slight and has probably been more than offset by the deterioration of the land through deforestation and soil erosion through poor land management.

North Africa is not a poor land; it is well suited to grain, vine, olive, citrus, and fruit production. Nor is it mineral poor, for it is a major exporter of iron and phosphate. In sum it is not a land that must remain poor. But divided as it is, it is handicapped by internal dissension and lack of capital for development and unlikely to attract international capital while plagued with internal troubles. The immense pools of oils found in the Algerian Sahara and in adjacent Libya, combined with modern pipeline movement of oil, creates the possibility of industrial development. Abundant, potentially cheap mineral fuel offers a substitute for charcoal

in home cooking, and this in turn could relieve the heavy pressure on the forest resources, which in turn could help to reduce floods and soil erosion. The problems are not primarily in the physical environment; they rest in the millennia-long, fiercely independent, and unchanging outlook of the Berber and more recently the Arabic culture of the Muslims.

The location of North Africa gives it a great advantage in relation to Europe. It could play much the same role as California plays in the United States. Its subtropical location could be utilized to produce off-season fruits and vegetables that the highly productive industrial societies of northern Europe desire and could pay for. North Africa could be developed as a winter resort area for adjacent Europe. Physical geography demands, urges, offers these opportunities. Seemingly these possibilities will be very long in being realized. North Africa is Muslim in religion, and it is culturally and politically oriented toward the Arab world. It is fiercely anticolonial and nationalistic and therefore will probably not enter the European Common Market and share in the economic growth of Europe. In spite of its physical geography, it seems destined to remain in a preindustrial stage of development for some time.

Yet even here differences unrelated to the physical geography are apparent. Morocco is more open to

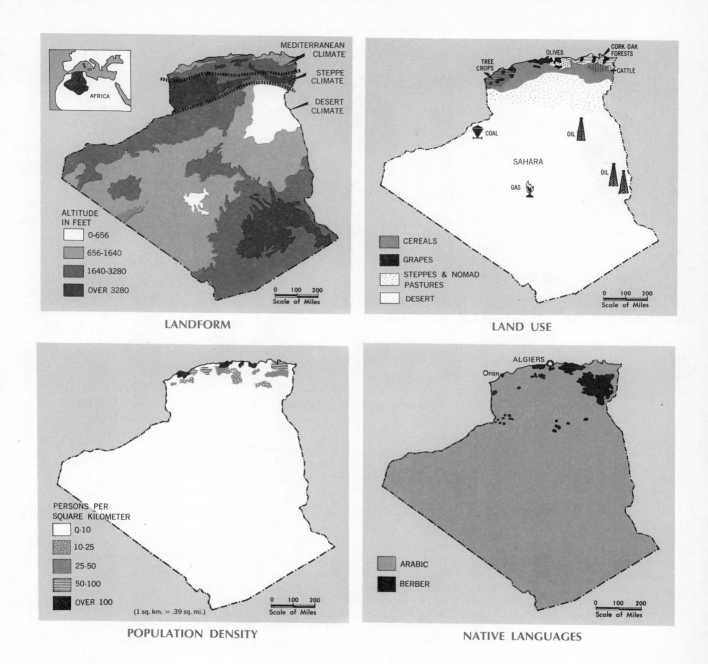

LANDFORM

MEDITERRANEAN CLIMATE
STEPPE CLIMATE
DESERT CLIMATE

AFRICA

ALTITUDE IN FEET
- 0-656
- 656-1640
- 1640-3280
- OVER 3280

0   100   200
Scale of Miles

LAND USE

TREE CROPS
OLIVES
CORK OAK FORESTS
CATTLE

COAL
OIL
SAHARA
OIL
OIL
GAS

CEREALS
GRAPES
STEPPES & NOMAD PASTURES
DESERT

0   100   200
Scale of Miles

POPULATION DENSITY

PERSONS PER SQUARE KILOMETER
- 0-10
- 10-25
- 25-50
- 50-100
- OVER 100

(1 sq. km. = .39 sq. mi.)

0   100   200
Scale of Miles

NATIVE LANGUAGES

ALGIERS
Oran

ARABIC
BERBER

0   100   200
Scale of Miles

MAP 4-3   Algeria: Landform, Land Use, Population, and Native Languages
*Algeria is two-thirds desert. The steppe and Mediterranean climate areas are larger than
Texas and much larger than all of California. Its 14 million inhabitants are mostly in the
better-watered areas. Significant recent developments include the discovery of vast oil
resources in the desert interior and independence from France, with the loss of about one
million Europeans. (Reprinted from Focus, American Geographical Society)*

221

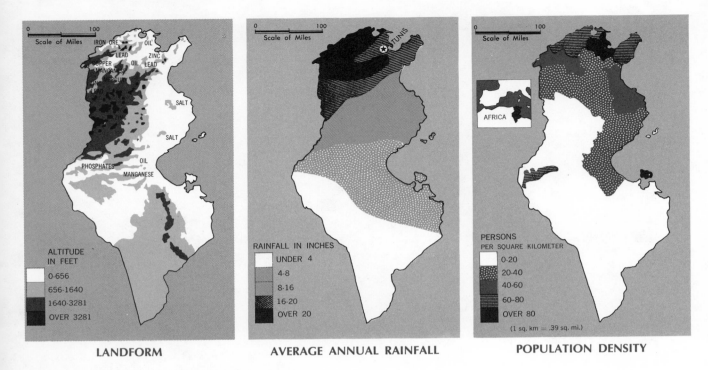

LANDFORM   AVERAGE ANNUAL RAINFALL   POPULATION DENSITY

MAP 4-4   Tunisia: Landform, Rainfall, and Population
These maps show the relationship of relief, rainfall, and population for this Mediterranean climate area in North Africa. Notice the abundant mineral resources, the very good rainfall, and the presence of mountains with their potential for water storage and hydroelectric development. (Reprinted from *Focus,* American Geographical Society)

European contacts than is Algeria, so it may advance more rapidly. Stimulus, growth, and development are dependent on contacts, and the unwillingness to enter into exchange situations is deadening. Japan closed its doors and stagnated. The Yumans rejected change and remained primitive. The most obvious thing for the North Africans would be a common market, which would combine the population and agricultural base of Morocco with the oil of Algeria and the minerals of Tunisia. The Muslims expelled the French, the Christians, and the Jews, thereby driving out the educated and enterprising minorities and alienating international capital. This is not a program that is likely to lead to development, even though this land is larger than and in many ways equal to two Californias plus Arizona, New Mexico, Nevada, and Texas; it has the climate of southern California, the oil of Texas, the arid waste and irrigable lands of Arizona, New Mexico, Nevada, and west Texas, plus mineral wealth. The possibilities are obvious.

THE NEW WORLD MEDITERRANEAN CLIMATES

Two of the five Mediterranean lands of the earth are very much alike in their physical structure. These are California and central Chile. The central parts of both areas have similar climates, similar temperatures through the year, and similar soil and vegetation types. In both lands there are coastal ranges that rise steeply from the sea leaving little coastal plain available. The mountains are not entered by the sea, so there are few harbors. Inside the chain of coastal mountains lies a great trough or central valley. Behind this trough rise high ranges of mountains: the Andes in Chile and the Sierra Nevadas in California. These mountains are continuous chains in both countries with relatively few passes through them. They are snow covered in winter and in spring the melting snow provides floods of water to the depression lying between the higher interior and the lower coastal mountain ranges. One of the few differences between the two countries is that the Central Valley of California is drained by two rivers that run from the ends toward the middle and join to reach the sea through the great system of bays at San Francisco, while in Chile there is no unified river system. Instead a series of rivers cross the central valley at right angles to its axis and reach the sea by separate mouths. This is a minor difference in comparison to the large number of similarities. Given this remarkably comparable pair of countries, how similar are their histories? And if physical environment is similar, what of race and cultural setting? These can be profitably studied in two periods: before 1500 and after.

California: The Indian Period

In California, which is now viewed as a golden land of vast opportunity and into which more millions seem to be destined to crowd each decade, what was the situation before 1500? Instead of a cultural center complete with great universities, immensely productive acres, and industry, there was one of the most culturally backward areas of the Americas—and indeed of the world. The Indians of Mediterranean California were called "Diggers" because they were gatherers of roots and seeds and hunted relatively little. There was plenty of game: deer were abundant, and buffalo, elk, antelope, and grizzly bears—as well as brown bears— were all present in moderate to large numbers, but this resource was relatively little used. Such small game as rabbits, quail, and squirrels were more important to most of the people.

There were many tribes in California, and it has been said that there were as many speech families in California as in all of Eurasia. A speech family is a language division for which there is no known relationship to any other language. The degree of difference can be grasped by recalling that languages as different as English, French, German, Latin, Greek, and Sanskrit are all closely related in the view of linguists. It is startling, then, that in California adjacent valleys had entirely separate languages. Was this the result of numerous separate peoples entering the area over a considerable length of time, each having an entirely separate language? Or was it the result of differentiation within the area over a period of time due to the isolation of the tribes in their separate valleys?

Since the similar situation did not create separate language families in Greece or North Africa, isolation in valleys cannot be the answer. If the number of speech families was the result of entry of different peoples, where are their language relatives? Only a few of the California tribes had any known language relations. This leaves the time factor. The prolonged debate about the length of time that man has lived in California now seems nearly settled. New dating methods place man in southern California as early as 48,000 years ago.[1] Between 70,000 and 30,000 years ago a continuous ice barrier stretched across Canada from the Atlantic to the Pacific and barred man's entry into the United States. If man was in southern California 48,000 years ago, he must have entered America before the ice barrier. Man was in California, therefore, for more than 70,000 years, and there was ample time for extreme language divergence. It is thought to take about 20,000 years for languages to diverge to the point that linguists are unable to distinguish any relationship. It is probable that the multiplicity of languages in California was the result of several of these factors. Different peoples, over a very long period of time, drifted into California and settled in limited areas where they remained with little movement, maintaining their own way of life unchanged.

Mediterranean California was an area of great cultural lag. There were only three spots of minor cultural advancement. Two of these are so clearly the result of stimulus from the outside that the very strong suspicion is raised that the third may be so also. The influences that we have traced from central Mexico reaching northward to the Pueblo country and the Yuma along the Colorado River barely reached into southern California. In San Diego county along the Mexican border the Indians spoke a Yuman Indian language and shared a few cultural traits that clearly point eastward across the desert to the Yuma and the Hohokam, and through them to Mexico. These include such things as pottery making and cremating the dead —putting the ashes in funerary urns. Agriculture failed to make the trip, although it could have if the idea of

[1] J. L. Bada, R. A. Schroeder, and G. F. Carter, "New Evidence for the Antiquity of Man in North America Deduced from Aspartic Acid Racemization," *Science*, vol. 84, no. 4138 (May 17, 1974), pp. 791–793.

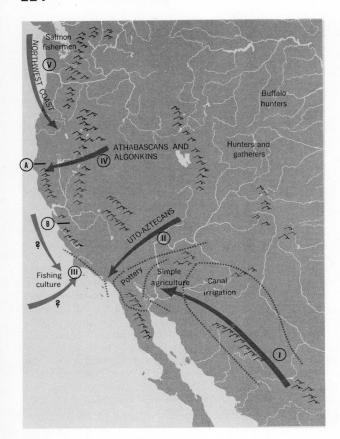

MAP 4-5   Cultural Influences in Pre-Columbian California

Five of many lines of influence are suggested here: (I) the flow of ideas coming up from Mexico, bringing irrigation to southern Arizona, floodwater farming to the lower Colorado, and pottery to southern California; (II) influences coming in from the Great Basin, indicated in southern California by the presence between San Diego and Los Angeles of a group of people speaking a Uto-Aztecan language; (III) the isolated occurrence of a fishing people equipped with hook and line, nets, spears, and plank boats in the Santa Barbara region. Many of their traits suggest relationships with either the Pacific Northwest or Polynesia. (IV) Small groups of Athabascans and Algonkin-speaking peoples in northern California show that the drift of peoples into California went on for a long time. (V) Influences from the advanced cultures on the northwest coast of America brought such items as the plank house and the dugout canoe to northwestern California. The southern limit of the dugout was about Cape Mendocino, A on the map. The finest possible boat-building tree, the redwood, continues to grow along the coast as far south as Monterey Bay, B.

irrigation had come with it. The gathering people, however, were not receptive to the idea of farming. Their whole way of life was geared to moving about the countryside, gathering the crops that came into ripeness with the differing seasons.

On the eastern side of San Diego county, through which area the idea of agriculture would have come, the people had a particularly rich and varied range of harvests. In the spring the yuccas and agaves were baked for 24 hours in an earth oven—a pit dug in the ground, filled with fire wood, and burned till very hot. When the fire died out in the pit, it was filled with yucca and covered to give slow baking. This produced a product that was very much like candied sweet potatoes. Later the mesquite bushes at the foot of the mountains yielded their immense harvest of beans. Still later, the piñon pines on the tops of the mountains produced their abundant harvest of oily, nutritious nuts. The coast ranges and valleys were sprinkled with

oaks that supplied a heavy harvest of nutritious acorns. On the coastal areas there were also wild grasses and edible roots, bulbs, and shoots. Why plant when the natural harvest was so varied and abundant? Why disturb the ancient family band organization that had made this a happy and easy way of life in a pleasant land for time beyond counting? Instead of asking why agriculture failed to spread here, we should ask, "Why did people ever take up agriculture?

Another area just beyond Mediterranean California that felt the wind of cultural advancement was northwestern California. Here there were good dugout canoes, not huge sea-going types, but good river boats, well shaped and useful. In northern California too there were some houses built of planks. These are traits highly developed to the north in coastal Oregon, Washington, and British Columbia. It is of utmost interest that these boats and houses disappeared in northern California long before arid, treeless southern California was reached.

Boats disappeared even before the great San Francisco Bay and the immense marshes at the juncture of the Sacramento River and the San Joaquin River were reached. In this vast delta, known as the Suisun Marshes, there was a whole world of reeds and channels filled with fish, water fowl, and with acres of starchy roots of the cattail ready for harvest. In this watery land a boat was needed, and a dugout would have been a most valuable possession. There was plenty of suitable timber in the adjacent hills and along the natural levees of the rivers running through the country. The ideas were available nearby in northwestern California where good dugouts were in use on the Klamath and the Eel and the Trinity rivers. But almost as if the Mediterranean climate was a deterrent to progress, these useful ideas failed to penetrate into the Sacramento–San Joaquin system. Instead the ancient primitive reed-bundle raft was used. It could, of course, be thought that a reed boat is the natural response of men living in a reedy swamp. A survey of the boats used in the world will quickly show that this is not so. There are many reedy areas where only wooden boats are used and some areas of scarcity of reed where only reed boats are used.

The third area of California that showed a spark of change from the simple hunting and gathering way of the adjacent coast a maritime culture that built good bara. In 1500 there was found on these islands and part of the adjacent coast a maritime culture that built good boats by sewing planks together, used nets and lines for deep-sea fishing, and had through these developments built up a sizable population. Although the arts of this culture were simple, they were definitely more advanced than those of the gathering folk of the mainland. The people carved artistic animal figures in the round and made tasteful inlays using shell on stone and bone.

Why was there this sudden appearance of a maritime skill in the California scene? Is it significant that it appears right where an island habitat made this desirable? Are we uncovering at last a natural response by man to an ideal habitat? These people had shell fishhooks, made in the same way that the shell fishhooks of the Polynesians are made, plus compound fishhooks, also much like those of the fishermen of the Pacific, and good nets. Boats, which we know they had in historic time, must have been present earlier to make some of this gear useful; carved stone replicas of boats have been found, indicating that boats were indeed being made. It is an open question whether this culture developed here in response to the special environment, or whether the plank boats, the shell and compound fishhooks, the peculiar pestles, and other Polynesianlike items indicate that a Polynesian or Asiatic people, such as the Japanese, or people from the northwest coast of America, reached these islands. If so, they came without bringing a whole society and persuaded a group of California Indians to initiate a maritime cul-

Redwoods, California. *The redwoods are old and huge, and this has led to the idea that they are slow growing. On the contrary, they grow very rapidly, and they are easily managed for timber growth. Similarly, although redwoods are now found only from the northern border of California to San Francisco, they could easily be grown in many other parts of the world. The long, straight, soft wood logs are ideal for dugout-canoe construction but were so used only at the northern end of their range.* (USDA Forest Service)

ture on this coast, where otherwise the sea had only been used for shoreline gathering. When we consider that for tens of thousands of years men had lived on the Mediterranean coast of California without making a move toward going to sea but did make such changes around the time of Christ when it is known that the Pacific was being peopled by sea-going fisherfolk, the possibility that influences from the Pacific may have reached California seems not too unlikely.

California before 1500 remained, however, a backward area. The blessings of having no seasons, mild winters and moderate summers, beautiful scenery, rich

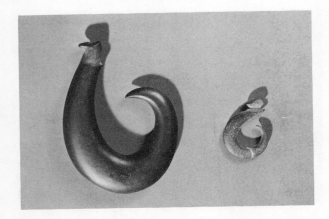

Fishhooks from Easter Island (left) **and California** (right). *The appearance of a specialized fishing culture on the islands off the coast of California poses interesting problems of cultural history. That this culture may not have been a simple response to the environment is suggested by the appearance of a number of specialized items duplicated by the highly skilled seafaring people of the Pacific. These fishhooks represent one such parallel.* (Courtesy of The American Museum of Natural History)

soils, water from the mountains available for irrigation, gold to be had for the taking, and oil to be had for the drilling, were wasted in the cultural world of the pre-Columbian Far West. The resources and opportunities were there, but the natives did not perceive them. Only at its southern end had the last gasp of Mexican influence stirred the cultural quiet. A similar faint ripple stirred northern California. And from a mysterious source the beginning of a martitime culture influenced the Santa Barbara region.

Spanish California

It is almost as striking a failure that the Spanish did so little with California. This area, after all, had a climate like the Mediterranean shores of Spain. The landscape of rich valleys with large evergreen oaks, mountains covered with brush in the foothills and pines on the highlands, and streams coming down to the warm valleys all should have looked attractive to the Spanish. And the Spanish did look at it. In 1540 their explorers entered the beautiful, finely sheltered bay of San Diego, one of the few natural harbors on the coast of Cali-

fornia. The Spanish explorers described the river running into it, the sheltering wooded headland, and the large expanse of sheltered water. They also noted the wild-seed-gathering Indians living there. They then explored up the coast, reporting on the islands off Santa Barbara and the fishing Indians there—people who had fine, light boats. The Spanish went on up the coast, missing San Francisco Bay but finding Monterey Bay, whose values they overestimated.

Yet California was not attractive to these Old World explorers. There were no civilized Indians among whom they could settle and whose productivity would support them. There was no gold to be looted from the natives—the coastal expeditions had seen no evidence of the wealth of gold that was later to be found in the interior of the state. There was evidence of petroleum as tar on the beaches, on the land in tar pits, and on the sea as great oil slicks. But the technology of Europe was still three centuries away from appreciation that this was a "resource." To the Spaniards California was a distant land with little to recommend it. Of course, they could grow wheat, raise cattle, and cultivate citrus and grapes there, but they had sufficient land for that at home or in closer places where production was already organized and where they did not have to transport such products half way round the earth. When there were empires in Mexico and Peru with gold to be had for the taking and more gold and silver in the mountains, why bother with a remote, undeveloped country? (This is the kind of world view that probably best explains cultural lag in places like Australia and New Guinea as well.)

The Spanish might well have left California untouched indefinitely had it not been for potential rivalry with Russian incursions into the area. For centuries Russian explorers had been reaching across the subarctic world, following the source of furs. They reached the Pacific and crossed over to Alaska in 1740. Their pursuit of the sea otter led them south into California and down to the islands off Santa Barbara. In 1812, to support this distant effort, they established a fort to the north of what is now San Francisco. Fort Ross on the Russian River marks this farthermost outpost of Russian expansion.

Spain's settlement of California, therefore, was a reaction to this appearance of the Russians in territory that Spain claimed, for only occupation could maintain Spain's claim. But two centuries had already elapsed before the threat of Russian occupation moved the Spanish to settle in California, and during that time the Indians had been left to follow their own simple way of life. In those 200 years the great Mexican and Peruvian empires had been conquered, looted, their rich mines exploited, and the pattern of Spanish colonial America had been blocked out. The developed and productive lands were all in the central part of the empire. The margins in both North and South America were the poor lands: New Mexico, Louisiana, Florida, Argentina, and Chile. Since there was little profit to be gained from these regions and from California, few chose to settle an undeveloped frontier where hardship and danger were the principal rewards.

But there was one group eager to go: the missionaries, who sought souls to save rather than wealth. In the true meaning of "all men are created equal" they considered the souls of the Indians as important as those of any other men. Therefore, while the motive of the crown was political, the motive of those who came to settle was spiritual.

The Spanish effort was well planned in light of the information on hand. Since it was known that California could be reached by land around the head of the Gulf of California, a land expedition with great herds of animals was sent by this route. It was also known that a good harbor was to be found at San Diego. According to the Spanish records of the exploration of 200 years earlier, there was another fine harbor at Monterey. Since the settlement was planned as a check to the Russians, it was ordered that the capital be placed at the northern frontier where the enemy could be more readily watched. From San Diego north, missions were to be placed about a day's journey apart. A ship's expedition was sent by way of Lower California, where supplies were picked up from the earlier missions settled there. It then went on to San Diego where the first new mission settlement was made in 1769. A *presidio* (garrisoned place) was built on the first height of land near water on the bay, but it was a poor site, too far from the harbor entrance to protect the harbor and too far down the river to assure abundant water for agriculture. The mission soon moved up the valley 10 miles to a location more accessible to an assured water supply. Although the presidio remained where it was first established, the harbor defenses were moved out near the entrance.

A string of missions was quickly established along the coast leading to Monterey, which was found then to be an open bay that afforded very little protection for shipping. The virtual inland sea of the San Francisco, San Pablo, and Suisun bays remained undiscovered until a party of hunters from the presidio at Monterey topped the hills to the east and looked down on this vast sheet of water. The role of environmental determinism and "cultural determinism" here had an interesting test. The immense sheltered system of bays with the Sacramento and San Joaquin rivers

San Diego Mission: The Bell Tower. *San Diego Mission, built in 1769, was the first of the missions in Alta, California. Its founding came almost 150 years after the conquest of Mexico and after repeated explorations had failed to interest the Spaniards in remote undeveloped California.* (San Diego Convention and Tourist Bureau)

entering into them gave perfect protection for shipping and excellent waterway access to desirable lands without the cost of building roads. Instead of a capital on an open cove on an isolated coast, the San Francisco Bay area offered a location for a capital at the focal point of a waterway system that could be the basis for development of the country. However, given the basis of a royal order as to the placement of the capital and the prior establishment of the capital at Monterey, the San Francisco Bay region was not then developed.

A mission was placed at the present site of San Francisco, but the capital, in spite of all its environmental disadvantages (location on an exposed coast in

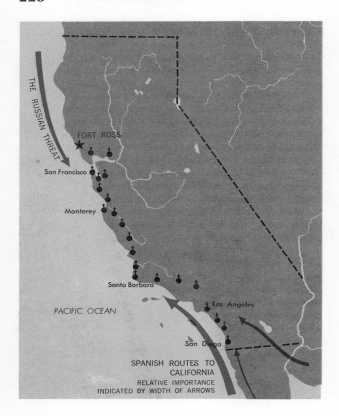

THE RUSSIAN THREAT

FORT ROSS

San Francisco

Monterey

Santa Barbara

PACIFIC OCEAN

Los Angeles

San Diego

SPANISH ROUTES TO
CALIFORNIA
RELATIVE IMPORTANCE
INDICATED BY WIDTH OF ARROWS

MAP 4-6   Spanish Missions in California, 1769–1823

The first missions in California were at San Diego, 1769, and Monterey (Carmel), 1770. Within 15 years there were nine missions. By 1823 there were 21. They were located along the coast, about a day's journey on horseback apart. The coastal location was selected because the principal contact with Mexico was by ship rather than overland. The distribution of the missions reflects the limited occupation and development of California achieved in the Spanish-Mexican period that ended in 1848. The breakdown of the colonial economy brought on by the confusion due to the Mexican revolution against Spain damaged the missions. Secularization in 1832 ended most of their work prematurely, for few of the Indians were yet sufficiently civilized to withstand first Mexican and then American pressures.

The location of the capital of the Spanish province of California at Monterey was influenced by erroneous reports of the original explorations, which described a fine harbor there near the area where the Russians, moving south from Alaska, were threatening Spain's territorial claims.

the fog belt, outside the main lines of movement in the region and lacking any easy communication with the interior) remained at Monterey. The glowing accounts of 1540 had described a fine harbor, and this had determined the location of a capital two centuries later, a location that was to remain fixed as long as the Spanish retained control of California.

There exists a romantic notion of Spanish California as a land of riches, but it was really a poor mission-frontier settlement. True, horses could be had for the asking, travelers (virtually none) could find open hospitality, and guests probably would not have touched a pot of gold left in the guest room, but what would gold have bought in such a region?

The missions themselves were productive centers, where fields were cleared, orchards and vineyards planted, and irrigation begun. The Spanish knew the Mediterranean climate, and they successfully introduced into this similar land wheat, barley, oats, olives, citrus, grapes, cattle, sheep, and horses. It was too dry for much farming at San Diego, but northward, as

rainfall increased, the crops were more productive, and the missions grew richer. Although grapes and citrus and other fruits did well and grain production was excellent, there was no market for their produce. Shipment to Spain was out of the question; citrus and other fruits would spoil, wine was bulky, and even grain was not profitable enough to send overland in a country that had no roads to ports where ships rarely called. And no Spanish colony, of course, in this age of mercantilism, could trade with any nation except Spain, any more than an English colony could trade with any country but England. The result was a meager handful of Spaniards in an undeveloped land, though the streams in the foothills of the Sierra Nevada were full of gold, and one of the world's greatest silver lodes lay just to the east at what was later called Silver City, Nevada. Spanish California was almost as poor, backward, and isolated as Indian California.

When Mexico won its independence from Spain in 1821, the missions were secularized and their vast holdings distributed to Spanish Mexicans. A trickle of trade opened up with the United States, whose vessels collected hides along the coast (an episode preserved for posterity in Richard Dana's lively book, *Two Years Before the Mast*). A few Americans looking for beaver wandered over the mountains and were promptly thrown in jail. A few settlers, such as Sutter, a Swiss,

Stockton Harbor. *Although the Spanish overlooked the natural waterways of the San Francisco bays and the advantages of the immensely rich delta region of the San Joaquin–Sacramento rivers, modern California has capitalized on them. Here the ships pass through America's Holland, the diked and drained peat lands of the delta that are said to contain 1000 miles of waterways. (Stockton Chamber of Commerce)*

managed to obtain land and settle down. But these few men were unable in the brief period between 1769, when the first mission was founded, and 1848, when the United States took over the territory, to do more than scratch the surface of this immense, rich land, inhabited by a gentle, simple, illiterate people.

American California

When California became United States territory in 1848, there had been little change. Climate, soil, vegetation, all were nearly as the Spanish had received them from the Indians. There were now horses and sheep on the land, but only in the southern and western part. The only significant change was in the makeup of the grasslands, where there had been a major shift from native grasses to introduced grasses that had come in as weeds in the Spanish packing materials. But this changed little; the grasslands had been there before. Only the kinds of grass had changed.

The United States acquisition of the lands of the Far West was not looked upon with favor by all. George Perkins Marsh, one of the greatest scholars of his time in America, who was then in Congress, viewed the action with alarm. Why take in those vast arid lands of the West? It was rumored that there was better land along the coast, but Marsh considered all of it to be useless to the United States. He said it was too far West, separated from the rest of the nation, and was virtually lost in the Orient. And he was not alone in his view. If it was of so little use to the Indians, the Russians, and the Spanish, how could it be useful to Americans?

Gold populates the state: from mining to farming

After the cession of this territory by Mexico in the Treaty of Guadalupe Hidalgo, the unexpected happened. Gold was accidentally discovered in the foothills of the Sierra just east of Sacramento. The excitement of the time is difficult to recapture. Thousands of men stampeded to California. Distance suddenly became a minor problem in proportion to the overwhelming drive for gold. Hordes set out by all possible means of travel to cross the continent. Indians, heat, drought, disease, cold, mountains, and rivers all took their toll. The Donner party made headlines by perishing in the deep snows of the Sierra Nevadas, and one group ran into serious difficulties trying to cut through the Death Valley country. Thousands died in more prosaic ways: by disease, accident, and drowning, to name just a few.

Another flood of mankind took to the seaways. Some went around Cape Horn. Counting the tacking of the ships, they probably traveled the equivalent of the circumference of the world at the equator. Another impatient group tried to shorten the trip by taking a ship to Panama, crossing the isthmus overland, and then taking a ship again on the Pacific side. Men dropped dead in 24 hours time from the pernicious fevers or were robbed and murdered in the crossing. If they reached the Pacific side, there usually were no ships. If distance, difficulties, and death could stop men, no one would have gone to California, yet thousands of men from all over the world poured in through the Golden Gate.

The gold fever was of almost unimaginable intensity. The ships dropped anchor in the harbor, and there most of them stayed for a year or so. The sailors jumped ship and took off for the gold fields. Soon the bay was a vast forest of masts rising above idle ships while the world cried for transportation to California.

There was gold, but not as described in the rumors going around the world. It could not be scooped up by the bucketful but had to be won by hard work. This often meant standing in icy water all day, digging down through sands and gravels to reach the bedrock where the gold would be found, if any. It was bone-chilling, back-breaking, wet, dirty work. At the end of the day there were no comfortable quarters, and food cost unbelievable prices. Gold dust and nuggets melted away as fast as the average man could dig them. A few lucky ones struck it rich, and some of them even got home with their gold. More lost it in gambling or were cold-bloodedly murdered for it.

Some men looked about them and decided that plowing, planting, pitching hay, feeding hogs, and all the other chores of farming looked less irksome than digging for gold. It was work that they knew, and with prices sky high, they knew there was more profit to be gained by cultivating the land than by digging in it. They might miss a bonanza, but they might never hit one anyway. Thus many men turned to the rich earth. The great Central Valley of California contains thousands of square miles of gently sloping alluvial lands. The Sacramento River Valley has good rainfalls; the winters are mild and, though the summers are blazing hot, there are many streams coming down from the mountains on either side. It was found to be an ideal land for barley and wheat. When the disillusioned sailors began returning to their ships, a whole fleet was available, looking for cargo to carry back East. Grain began to move, and the barley won prizes as the finest malting barley in the world. Grain was indeed profitable, even in this distant land. Some men became wealthy, and great farms began to grow. Near Sacramento one large-scale effort totaled 50,000 acres. Sheep became important in the 1870s and 1880s, and wool was exported.

Mining gradually became big business. The easily accessible placer gold soon was worked out, and the large gravel deposits were attacked by hydraulic operations. This consisted of using a super fire hose to wash down the gravel banks, sluicing the material through a flume with traps in the bottom to catch the heavy gold. This led to water management: the building of dams and flumes and the accumulation of a great deal of engineering know-how. The transfer of this kind of knowledge to the supplying of the water to the land was an easy step; all the required parts were there. The idea of agriculture, what plants were suited to the climate, even the idea of irrigation, were all present in the land. They already had been introduced, tried out on a small scale, and found to be successful by the Spanish. The Americans, with their characteristic vigor, made a large-scale production of it.

Orchards and vineyards were set out. The yield was huge, the quality fine—but where was the market? Fresh fruit could not last the trip via Cape Horn, and overland transportation in wagons was equally out of the question. They solved the problem by drying the fruit. In the hot, desertlike summers of this region, the fruit could be rapidly dried with practically no risk of rain spoilage. Dried peaches, apricots, pears, and prunes became specialty crops of different areas. At one time a large proportion of all the dried prunes of the world came from the Santa Clara Valley at the southern end of the San Francisco Bay. Dried fruit is no longer as important today because now swift means of transport and refrigeration make fresh fruit useful, but it was an important solution for its day.

This land that the Spanish had found so useless now grew rapidly in population, productivity, and wealth. San Francisco became a great city and Sacramento a large town while southern California lagged. For a time it retained the sleepy, sunny, relaxed heritage of Mexican California, and some wish that it had remained that way, for change is not always progress. Change, however, was in the wind. The nation was determined to bind the two coasts together. The railroad had been invented, and steam engines harnessed to the cars were revolutionizing transportation. The government sponsored military surveys to find routes to the West. Soon lines were reaching across the Great Plains, through the mountains, and down to the Pacific coast. The land so far to the West, lost in the Orient, was now linked to the rest of the nation, but the new railroads ran through largely empty land, and the coast that they reached was still thinly populated, only beginning to be developed.

Railroads, oranges, and cooperatives

The railroad entrepreneurs of the day plunged optimistically ahead, subsidized by government grants of enormous tracts of land along their right of way. The amount of land thus given to the railways has often been grossly exaggerated. Studies show that the railroads took only a fraction of the land they were entitled to. Further, they agreed to carry the armed forces and its supplies in perpetuity. As a result, the railroads did not earn as much as often believed—on the contrary, such an agreement cost them enormously in the long run. Great profits were made by speculators of the day, but the railroads were built, and they made possible a new era for California.

Now produce need not be dried: it could be shipped fresh. Now manufactured goods bought in California need not stand the extra cost of shipment more than half the way around the world. Now people could get to California swiftly, cheaply, and safely. The "geographic location" of California had suddenly changed. The railroads pressed for development, for they needed people at the far end of their line: they sold one-way tickets to California at unbelievably low prices.

A classic example of the economic development that followed the railroads was the citrus industry. Americans had not eaten many citrus fruits. Johns Hopkins' famed E. V. McCollum, however, discovered that one could eat all the calorie-producing food needed by the body to survive and still fall sick. Tiny amounts of certain other things were needed; these he named vitamins. Oranges were found to contain large amounts of vitamin C. The Californians had oranges and plenty of land to produce a great many more. More and more people could afford to buy oranges as the level of productivity, particularly in the Northeast, raised the average standard of living higher than the world had ever experienced before. Even though the East coast was at the opposite corner of this continent-sized nation, the railroads made possible the shipment of even perishable materials that far. The potential market was there, the means to reach it had been built, the product had been supplied by the Spanish, the motivation had been stimulated by medical research, and all that was necessary was the ambition to take advantage of these factors. But it is misleading to say "all," for we have already reviewed large areas of the world that have neglected innumerable opportunities. California did not, however.

Erosion in California. *These are gullies in denuded hill slopes in the foothills of the Sierra Nevadas. Some of the rounded, earth-covered, vegetated hills of California have started on the road to becoming bare, angular, rocky hills such as characterize many of the anciently cultivated, similar climatic regions in the Mediterranean.* (USDA Photo)

The shipping of oranges began on a small scale; it was profitable but filled with problems. It was still difficult to preserve the fruit for long periods of time, and the conditions in the distant market were hard to anticipate. Market fluctuations could do great damage to the individuals who had large investments in land and improvements with a relatively high overhead. Orchards had to be irrigated, sprayed, pruned, protected from frost, whether the price was up or down. Solution to these problems was found in cooperatives and individual inventiveness. Oranges were picked without tearing the skin at the stem end—that is, they were cut off the tree; they were washed and waxed to prevent decay; and they were packed in specially designed boxes. Heaters were invented to ward off the occasional cold waves that threatened the orchards, while specialists were called in to develop sprays, fertilizers, and irrigation techniques. What individuals could not do, large organizations could do.

Orange-growing cooperatives built large packing houses, bought their own box factories, and finally even bought the forests and saw mills to produce their own lumber for the boxes. They spent money on full-page advertisements in the leading home magazines to convince the whole country to buy California oranges, not only because they tasted and looked good, but because they were essential to good health. The orange industry of California became a spectacular case of organization and salesmanship that seized on an opportunity that had been latent in the physical environment and made a paying proposition out of it.

There are many sides to this, of course. The California Indians could have grown oranges, if they had known the fruit and how to irrigate the land, but their opportunities for marketing them were limited. The Spanish had the oranges and the irrigation, but their potential market was Mexico and the other Spanish colonies, all of which had their own oranges. It was a special circumstance that American California was a part of the United States and had the rapid transportation that made possible this development. In passing we should not forget that the orange is not native to the Mediterranean climate. It is not Spanish, for the Arabs brought it to Spain. It is not Arabian either; the Arabs had probably received this delicious fruit from their trade with India. The home of citrus fruits is in Southeast Asia. It is grown in Mediterranean lands more in spite of the climate than because of it. If there were not tariff and quarantine restrictions, our oranges could best be supplied by tropical areas such as the Caribbean or tropical Latin America, where there is no frost risk and irrigation cost. Orange growing is not a simple response to the subtropical climate of southern California. It is a very complex response in which the physical environment located within the political boundaries of a very productive nation, with a highly developed communications network, offered an opportunity that was seized and developed at a particular time in a particular way by a particular group of men.

Other agricultural opportunities of this kind followed much the same pattern. Off-season crops grown in California are sent to the winter-bound markets of the north and east. Lettuce, celery, carrots, cantaloupes, watermelons, and tomatoes are a few of the crops that move out of specialty areas in trainload lots. The specialty areas are of interest. When an area grows lettuce in large amounts, almost to the exclusion of other crops, it may seem reasonable to conclude that there must be something special in the physical environment to make this so. But this is seldom true. Most berry, lettuce, peach, or potato specialty areas have been developed because of the economies due to mass production, shipping, and marketing rather than because of special soil, water, and climate factors. A single lettuce grower can seldom assemble a carload lot of lettuce, but carloads move more cheaply.

Insecticides are expensive in small lots, but the local supply house can cut the price if it can buy by the carload. Most specialty produce centers could be in any one of dozens, if not hundreds, of spots. They are grouped for economic reasons.

Manpower and industry

California has now passed through its agricultural stage. The dominant role of agriculture ended with the vast influx of people that came in after World War II. There had been floods of immigrants before; in terms of original numbers, the days of the forty-niners brought the greatest number. The railroads brought a boom in the 1890s that was followed by a colossal "bust." World War II brought floods of soldiers who were either stationed in California or at least spent some time there on their way to the Pacific. It sometimes seems now as if they all came back and brought all of their relatives with them. A steady growth of population had long been underway, and it burgeoned when industries began to move west.

California is in many ways a have-not area. It has no coal, it has used up most of its oil and gas, and it has little iron ore. Light industry came first. The movie industry came because of the daylight and sunshine. The early movie capital was in the East, but film could be exposed only between 10 A.M. and 2 P.M. in good weather: California had more bright sun through more of the year and also had a greater variety of scenery in a small area. Does the change reflect physical environment? Well, yes, but only for a short time, because today film is not so limited, most movies are made on sets, and the film industry often uses other locations.

The airplane industry is a light industry that moved west, partly because of better weather for testing planes. In part too it was taking advantage of the growing pool of available manpower there. Originally a growing population attracted industry; now, growing industries are advertising for people—special people such as engineers, it is true, but people nonetheless who will also need doctors, tailors, plumbers, and electricians. Just when the population of California will level off is anyone's guess.

Problems in paradise

Problems have appeared in this paradise. At Los Angeles a particular feature of the physical environment has suddenly taken on a malignant aspect. The water off the coast of California is cold because the ocean current along this coast moves south along the continental border. The force due to the rotation of the earth drives the moving surface water to the right, away from the continent. To replace this water, deep ocean water moves toward the surface. This is cold water. The warm air masses coming toward the continent from off the Pacific must cross this cold belt and are cooled in their lower layers, creating an inversion: a very stable air situation, much fog, and a persistent, high-stratum cloud layer, without which southern coastal California would be blazing hot and very much more desert than it is.

This was all right until too many people with automobiles and too much industry moved into this situation. Then it was found that the inversion acted like a great lid over a limited amount of air. This air will support a certain amount of people and industry and will absorb only so much waste gas material. The Los Angeles area used up its quota of air 30 years ago. Now the citizens of that area are breathing second- and third-hand air, air that has been run through someone's automobile and through some factory furnace and through many other lungs. The air has become loaded with industrial waste products, especially with partially burned automobile fuels, which react with the oxygen in the air to form a variety of irritating compounds. The beautiful, sparkling clear air of the Los Angeles basin ceased to exist 30 years ago, and a yellowish-brown acrid substance that irritates the eyes, nose, and throat and damages the vegetation has taken its place. Once more man has changed the environment. Great effort and immense expense is now going into attempts to solve this problem. Industry has enormously reduced its pollution of the air, backyard trash burners have been eliminated, and now the automobile with its polluting exhaust is being controlled. Southern California, however, simply has a limit on its air supply, and no one has yet found a way to increase it.

Mediterranean California also has a limit on its water supply. Northern California has the water; southern California has developed most of its local supplies, reaching across the state to tap the Colorado River, and now up the state to take water from northern California. Arizona has fought California in the courts for the right to Colorado River water and won much that California has claimed for herself. Northern California is a bit cool to southern California's plan to tax the whole state to supply the billions of dollars needed to transfer water to the south. The vastness of the projected transfers of water is difficult to grasp. For Los Angeles to bring water from the Hoover Dam to its area is the equivalent of Baltimore getting its water

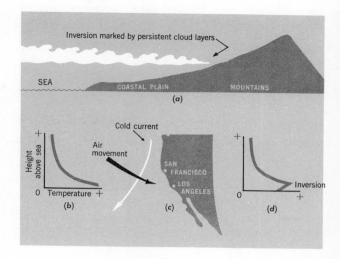

FIGURE 4-1 The Atmospheric Basis for the Smog Problem in Southern California

The general atmospheric situation at Los Angeles is shown here. Due to an inversion layer marked by a persistent cloud cover a few hundred feet above the ground, the mixing of large amounts of air is prevented (a). Only the air below the inversion layer is available for human and industrial use. The physical basis for this is as follows: The offshore current is cold due to upwelling of cold waters; the vast expanse of the Pacific to the west of California is warm. Air from over the warm Pacific has a normal decrease of temperature with increasing altitude, as shown at (b). Air masses are easily mixed in depth. However, when such an air mass moves over a cool zone, such as a cold ocean current (c), it is chilled in its lower layer and develops the temperature-elevation distribution shown at (d). Under these conditions the air becomes extremely stable, and the line of change of temperature increase to temperature decrease acts as a barrier to vertical air circulation. In effect this puts a lid over the Los Angeles plain.

from Lake Erie, or Boston and New York City getting water from Lake Ontario. In a sense it is a case of the physical environment offering a natural limitation to the number that can live in southern California, but southern Californians are willing to pay the cost to live there by extending those natural limitations.

The water problem is complex. The use of pumps on underground supplies has lowered the water level, and areas that had shallow wells now must lift water 100 feet or more while some areas have exhausted their underground supplies. Others near the

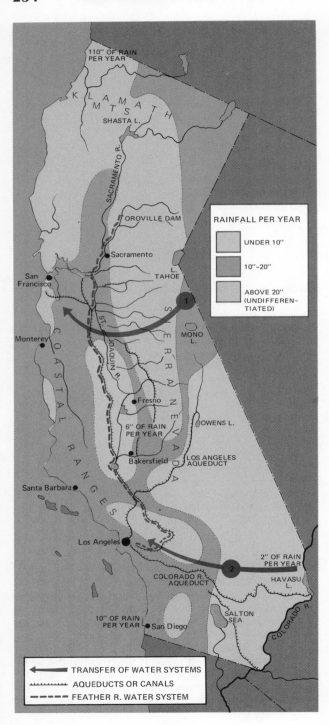

110" OF RAIN PER YEAR

K L A M A T H
M T S

SHASTA L.

SACRAMENTO R.

OROVILLE DAM

Sacramento

San Francisco

L. TAHOE

S   I   E   R   R   A

N   E   V   A   D   A

MONO L.

Monterey

C O A S T A L   R A N G E S

S A N   J O A Q U I N   R.

Fresno

OWENS L.

6" OF RAIN PER YEAR

Bakersfield

LOS ANGELES AQUEDUCT

Santa Barbara

Los Angeles

2" OF RAIN PER YEAR

COLORADO R. AQUEDUCT

HAVASU L.

10" OF RAIN PER YEAR

San Diego

SALTON SEA

COLORADO R.

RAINFALL PER YEAR

	UNDER 10"
	10"–20"
	ABOVE 20" (UNDIFFERENTIATED)

➤ TRANSFER OF WATER SYSTEMS
········· AQUEDUCTS OR CANALS
– – – – FEATHER R. WATER SYSTEM

MAP 4-7   California's Water Problem

Rainfall in California is highest in the north, but most of the people are concentrated in the drier south. The Californians transfer water all over the state: (1) from the Sierra Nevadas to San Francisco, (2) from the Colorado River to southern California. The most comprehensive water distribution program is the Feather River project and the associated aqueduct system. Water is carried from the world's largest earth dam at Oroville to the delta area where the Sacramento and San Joaquin rivers meet. From there water is distributed around the San Francisco Bay region. A 400-mile-long aqueduct carries water on down to southern California, crossing the Tehachapi Mountains at the southern end of the Central Valley. Six hundred miles of canals distribute the water to coastal towns, and several major reservoirs store water to control supplies.

sea have found that they were drawing salt water in through their underground wells and thus ruining their land by irrigating with sea water. The Colorado River water is loaded with dissolved salts and when used as irrigation water, evaporation leaves these salts to accumulate in the soil. Eventually this poisonous accumulation will reach deadly levels. In the midst of all this, Californians continue to dump millions of gallons of slightly used sewage water into the sea, and with it goes enormous loads of nutrient material that might better be returned to the land. Meanwhile millions of dollars are being spent in trying to find some economical way to desalt sea water. This now hovers on the brink of achievement. If accomplished, the human geography, not just of this region, but of most of the arid parts of the world, would be changed overnight.

The immense influx of population into California has centered on two areas: the Los Angeles basin and San Francisco Bay. Crowding has decreased the joys of living there. Smog burns the throat and eyes, and traffic clogs the freeways, while the freeways themselves absorb the land, encouraging the suburbanization of more distant rural areas. The beautiful orange groves that made the Los Angeles area a perfumed center of subtropical beauty with a backdrop of snow-covered mountains are largely gone, replaced by miles of developments, and the mountains can seldom be seen through smog. The smog also makes it impossible to grow many vegetable crops. In the San Francisco area the square miles of prune, peach, and pear orchards that were a delight during the spring blossoming have been replaced by square miles of subdivision. The huge influx of people is overtaxing the recreation areas: beaches are overcrowded, the Sierra Nevadas are overrun with pack trains, the desert is inundated each weekend by people with camping gear, land that was once unspoiled solitude now has a camping party in every canyon, while the landscape is littered with trash. Too many people can turn even a beautiful California into a mess.

Chile: California's Physical Twin

Chile is the physical twin to California, but not an identical twin. As we have already noted, the Mediterranean climatic sections in both countries fall in about the same latitudes, but Chile's zone is somewhat closer to the equator than California's. In both areas there is a coastal mountain range, then a central valley, and finally very high mountains. A minor difference is that the Andes are much higher than the Sierra Nevadas; the highest peaks in the Andes are over 20,000 feet, while the Sierra Nevadas reach only 14,400. In both cases the ranges are excellent precipitation catchers, but they are also formidable barriers to wheeled travel. As might be expected, the passes through the Andes are higher than those in the Sierra Nevadas: about 10,000 feet versus 7000 feet.

The coast ranges are about the same width and height in both cases: about 50 miles wide and 2000 to 7000 feet high. However, the California ranges are composed of series of steep ridges, while the Chilean coast range is a plateau. Both coasts are straight and often precipitous, with few good harbors. California with San Diego and San Francisco is far better off than Chile, whose best harbors are more like the open bay at Monterey. This is not an insurmountable handicap, for Los Angeles, lacking a natural harbor, built one. The coast ranges in California are broken at only one point—where the rivers from the interior valley pour out to the sea at San Francisco. In Chile, on the other hand, each major river crosses the coast ranges at right angles, reaching the sea through steep gorges, which are, of course, potential sites for reservoirs, water storage, and hydroelectric developments that California would be delighted to have.

The central valleys of the two regions are strikingly alike. They are both about 50 miles wide, and the Mediterranean climate sections of each extend for about 300 miles. In Chile the northern, dry end of the valley is cut off from the rest by a spur of mountains so that Valparaíso on the Aconcagua River is separated from the rest of the central valley, and a 2600-foot-high pass must be crossed in going from one section to the other. This is matched in California by

Tom Bradley, Mayor of Los Angeles. *Tom Bradley is the product of Calvert, Texas, a tiny town about 80 percent black. He was raised there during the time of segregated schools and before many of the disadvantages for the American blacks had been ameliorated. Ability and determination carried him to the top.*

the separation of the Los Angeles basin, the dry end of California's Mediterranean climate, from the Central Valley by the transverse ranges whose Tehachapi Pass reaches an elevation of 4000 feet. Perhaps elevation is the most significant difference between the two great valleys: Santiago in Chile's central valley has an elevation of 1700 feet above sea level, while representative cities in California's Central Valley have much lower elevations—Sacramento, 30 feet; Fresno and Red Bluff, 300 feet. Since temperature decreases about 3 degrees for each 1000-foot increase in elevation, this has important effects on the climatic comparisons.

The climate of these lands must be greatly varied because of coast, mountain, and interior valley conditions. In both countries the presence of an offshore current with a marked upwelling of cold water cools the coast and creates a strong inversion, with the potential smog problem that has already developed in California. The situation also creates the setting for an immensely rich offshore fishery, which is extensively used in California and until recently has been little used in Chile. Following Peru's lead, however, Chile is now beginning to harvest its rich fishery and claim control of the sea out to limits parallel to Peru's claims. Chile is now number two in fishing, second only to Peru, which leads the world in tonnage of fish caught. Sharp temperature change accompanies the change to the interior in both areas, but there are lower summer and winter temperatures in Chile's interior valley due to the increasing elevation. Chile, then, has moderate summer temperatures compared to the blazing heat of California's Central Valley. Santiago in Chile's interior valley has winter average temperatures about 46° F. and summer averages of about 69° F., which is closer to San Diego's equable winter 52° F. and summer 63° F. than it is to such interior California cities as Fresno. Finally, the north-south rainfall gradient in Chile is much steeper than in California. Coquimbo has 4.5 inches, Valparaíso 13 inches, and Concepción 30 inches; the comparable sequence in California would be San Diego 10 inches, Los Angeles 16 inches, and San Francisco 22 inches.

In both areas the great snow fields on the high mountains melt during the summer periods and supply vast amounts of runoff just when it is most needed in the hot interior lowlands. The same geomorphic processes in Chile and California have built alluvial fans sloping westward from the foot of the high ranges, with natural slopes ideal for distribution of this water by gravity flow over the great alluvial-filled valleys. In California, as we have seen, the Indian people did nothing with this agricultural invitation, and even the Spanish left the interior valley untouched. The Americans have introduced agricultural developments that make Central Valley one of the most productive areas in the United States. With the same physical-environmental opportunities, what has happened in Chile?

Despite the primitive conditions under which many of the Chilean people live, the population is growing: it doubled between 1885 and 1940. At present the population of the country is about 9,675,000, and 90 percent of these people are in the central valley—this is about the population of Los Angeles and San Diego counties. Two cities, Santiago and Valparaíso, have been growing and developing some manufacture, and population is moving to these centers. People are also emigrating from this relatively empty land to adjacent Argentina, suggesting that the number of people does not determine whether a land is overpopulated; the state of development determines it. Chile has the familiar problems of development. Increasing mechanization of its agriculture should free many hands for industrial and service work, and education and industrial development must parallel such changes in agriculture, or economic and social chaos will result. Already the flooding of illiterate landless poor into urban areas not equipped to employ, house, and educate them is creating severe social problems. But these are not Mediterranean climate problems, and they are only accidentally and not uniquely Chilean problems.

Why is this land that had a better start on the Indian level and an earlier start with Spanish development—and which physically is so much like California—still so poor? Why has it not, even now, matched California's achievements?

Cultural growth

Mediterranean Chile came within the area of influence of the Peruvian-Bolivian civilizations. The Inca empire even reached to the Aconcagua River valley in the northern end of the Mediterranean climate region, and the ideas of irrigation agriculture were introduced into the central valley of Chile, while simple agriculture of the forest clearing type extended to the island of Chiloé. This was a remarkable spreading outward. If the center of Andean civilization is considered to be between 10 and 15 degrees south of the equator, then irrigation agriculture was spread to 32 degrees south and simple agriculture was spread to 42 degrees south. The comparable spread in North America would have been for Mexican agriculture, with irrigation, to have reached into the Central Valley of California, and for simpler agriculture to have reached northern California. Had this been true, the history of post-Columbian America surely would have been greatly different.

In 1540 when the Spanish looked at California, they found it a beautiful land but one lacking in settled people. As we have seen, they then ignored it for over 200 years. A California with established village- or town-dwelling Indians irrigating their crops by use of the streams flowing into the Central Valley would have attracted Spanish settlement, as happened in Chile. This settlement surely would have led to discovery of California's gold and a period of intense activity, which would have made California an active part of the Spanish colonial empire instead of a remote territory vaguely known and claimed by assertion rather than occupation. Possibly the existence of a center of development to the north would have increased the flow of men and goods by land and sea, and thus the Arizona and New Mexico colony would have been strengthened and the United States might never have obtained the Far West, for the land-hungry Spaniards would have occupied it in force and valued it. In Chile they came looking for gold and empires; instead they found people settled on the land, irrigating the rich central valley. This moderately advanced and modestly dense settlement of people was attractive to some of the Spaniards. They took over this land, even though it was as distant from the center of their interests in South America as California was from their principal area of interest in North America. It is clear that the difference was developed land versus undeveloped land, and once again we see that it is men and ideas that make land valuable.

Santiago was established in 1541 and Concepción in 1550. Despite vigorous warfare on the part of the Araucanian Indians, the Spanish maintained their hold. Since the area lacked gold, the Spanish made the land the source of wealth and prestige, which in reality was only a continuation of an ancient Mediterranean pattern easily traceable back to Roman times. The land was soon divided among the few Spaniards to form great landed estates, some of which have been held intact to this day. Since there were few Spaniards, who, like most Mediterranean people were relatively free of race prejudice, they intermarried freely with the Indian population and created a mixed race. The Araucanians and the Spaniards were both proud, independent people, and this spirit has endured in the land.

The Spanish development centered in the Mediterranean climate area of the central valley. Here was a climate that they knew well, land that was easily and abundantly irrigated with a settled and agriculturally skilled labor force. To the north lay the great desert of Atacama, one of the most barren lands on earth. To the south lay forested lands held by warlike, less settled Araucanians. Major military force exerted against them at a later time found the task so difficult that a peace treaty rather than conquest was the outcome. On this early Spanish colonial frontier no such force was available, so the Spanish settled down to the de-velopment of a land that they understood. The result was such a concentration of development that 90 percent of the country's population is still found in this small area.

Of Chile's 9 million people, 30 percent are in the single province of Santiago, leaving half of the country uninhabited. California has a comparable situation in that one-third of the population is in the Los Angeles area. However, very little of California is uninhabited, and there is no agricultural population in the Sacramento-Stockton area in any way comparable to Chile's. The Los Angeles population concentration is industrial, and in all of California the agricultural areas are thinly, not densely, populated. In Chile the central valley has population densities of 125 per square mile, and near Santiago the rural population densities approach 500 per square mile. These extremely high figures reflect the ancient agricultural system that characterizes much of Chile—and most of the rest of the world where the scientific, chemical, mechanical, and industrial revolution has not yet penetrated.

Land ownership in Mediterranean Chile is also highly concentrated. Small farms (a 500-acre farm in Chile is considered small) are rare. In the central valley large properties, called *haciendas,* are only 7 percent of the total number of properties, but they occupy 90 percent of the land. In the Aconcagua Valley 98 percent of the land is owned by 3 percent of the property owners: 375 properties cover over 20 square miles in size, occupying 52 percent of the privately owned land in this region. The largest of these estates is about 600 square miles, an area 40 by 15 miles, roughly one-half the size of the state of Rhode Island. This is not quite the situation it seems, for these vast estates often have relatively small amounts of irrigable land and large amounts of dry and mountainous land. On the other hand, studies suggest that some of the large properties in Chile grossly underestimate their acreages to decrease tax payments. Air surveys have been carried out that are designed to check this. It is a notable step, however, for this could only be done if the large landholders themselves agreed to such remedial measures. During Salvador Allende's Marxist regime, land reform was one of the major goals. More than 1300 large farms were expropriated (many of them illegally by peasants who simply seized land). The land that the government took over was often converted into "agrarian reform centers," which in effect substituted one boss for a bigger one—the government. Disruptions led to a massive decline in farm productivity. The military regime now in control has returned illegally seized land and seems prepared to gradually break up the great estates.

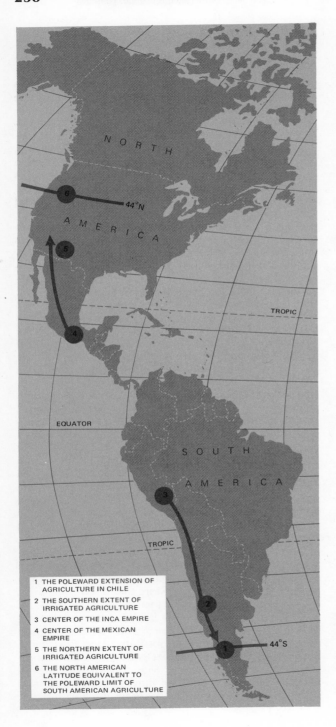

1 THE POLEWARD EXTENSION OF
  AGRICULTURE IN CHILE
2 THE SOUTHERN EXTENT OF
  IRRIGATED AGRICULTURE
3 CENTER OF THE INCA EMPIRE
4 CENTER OF THE MEXICAN
  EMPIRE
5 THE NORTHERN EXTENT OF
  IRRIGATED AGRICULTURE
6 THE NORTH AMERICAN
  LATITUDE EQUIVALENT TO
  THE POLEWARD LIMIT OF
  SOUTH AMERICAN AGRICULTURE

The great estates and the small farms are almost uniformly run on an agricultural system that is similar to the colonial one, with mechanization at a minimum and labor at a maximum. It is this prodigal use of manpower on the land that accounts for the dense rural populations. It is estimated that only one-quarter of the cultivated land is worked by mechanized units. The bulk of the farms rely on ox-drawn carts, and grain is hand sown, reaped with scythes, threshed by trampling horses, and winnowed by men pitching the straw into the air so that it will be blown away by the wind while the heavier grain falls to the threshing floor. Only one-quarter of the land, therefore, can be said to be in the twentieth century, with machinery replacing manpower on the land. This cannot be changed easily, for if men are displaced from the land, they must find employment somewhere.

The principal crops are cereals, with wheat predominant. This is the stage that California was in just after the gold rush a century ago. It is probably not far from the truth to say that Chile today is about what California would have been today had it remained subject to Latin American social, economic, and political traditions and economic and political ties. Although cereals predominate in Chile, only about one-tenth of the productive land is in grain in any one year; the rest is fallow or in pasture. This emphasis on livestock is again a Spanish heritage, and it affects the land use of Argentina and Uruguay in much the same way. In Chile animals are driven from winter lowland pastures to upland summer pastures, as in Spain and Switzerland.

The bulk of the people on the land are tenants on the haciendas, and their condition resembles that of serfs. As in medieval Europe, the men belonged to the estate and had both duties and rights. A meager living was man's lot, but it was an assured one. So long as a man stayed on the estate, he would never starve, and his sons could continue to work there after him. He might be legally free to move, but he was firmly bound by custom to the land. It is not easy to leave, for the tenant knows no other life, has no education that would equip him for doing a skilled work, and

MAP 4-8   Poleward Extension of Agriculture: Chile and California

A climatic cause has often been invoked to explain the absence of agriculture in California. The wealth of wild foods and the lack of intense and sustained contacts seem more likely to be the reasons. Political organization must also be reckoned with, for Mexico's influence died out rather quickly while Peru's influence was far-reaching.

usually is emotionally attached to the land where he and his forefathers have lived and served one great landowning family for 500 years. For cradle-to-grave security he gives up personal freedom and accepts poverty. Most of mankind since the spread of the Neolithic revolution has lived this way, and today much of the world is still in danger of bartering freedom for security under some form of paternalism.

Nevertheless, the Chilean agricultural system is productive. The proof is the number of men who live on the soil, very poorly by our standards, it is true, but relatively well by the world's standards. Each tenant has his own house, although it is unheated, with no running water, separate kitchen, or bath. Two acres or so about the house are his to raise domestic animals and to garden; in addition, he is assigned fields where he can grow his own grain. A Chinese peasant would think he had wealth indeed if this were his. On the other hand, a Californian might be taken to court if he put migrant laborers in dirt-floored, thatch-roofed, unlighted houses and told them to use the polluted irrigation ditch for their water supply.

All recent regimes in Chile have had difficulty in the area of agriculture. Eduardo Frei's Liberal government promised land reform. One farm economist reported that Frei's men spent 90 percent of their time on agrarian reform, trying to change the land ownership pattern. According to the economist, that is less than half the problem. Credit, technology, marketing, and irrigation are more important for they could have led to large agricultural increases. Land reform, however, is thought to be the magic key and sparks many a revolution. Allende's government tried to distribute land without improving the fundamentals of agriculture. The current military junta now faces the same problem.

This discussion has focused on the Mediterranean climatic area of north central Chile, neglecting the whole political geographic unit. Mediterranean Chile is one part of a country that has desert to the north and rainy west coast marine forested lands to the south. Development in these areas affects the area of winter rainfall. For instance nearly one-half of Chile's income derives from mining, and this is concentrated in the arid north. If the central part of the Great Valley is Chile's breadbasket, the north is one of its major cash accounts. Three great copper mines in the arid north supply about half of Chile's foreign exchange. One of these mines, Chuquicamata, developed by the Anaconda Company in the middle of a desert more barren than the Sahara, is the world's largest. It produces more than 300,000 tons of copper per year and has a 100-year supply in sight. Under United States ownership the mine paid three times the average daily wage outside the mines and gave the miners free housing, schooling, and medical care. Foreign ownership of copper, one of their major resources, has galled the Chileans. Under Eduardo Frei the government arranged to take over 51 percent interest in the major copper mines. Under Allende's regime the properties were seized without any recompense. Under the present military regime, payment for the properties has been negotiated, but Chile is retaining full ownership. With skilled management lost and production declining greatly, the problem now is to regain high-level production in this industry, which has accounted for three-fourths of the export income of the country in recent years.

There are also nitrate, iron, and sulfur mines. The sulfur is in the high Andes at elevations of 17,000 to 20,000 feet, an obvious hindrance to the development of enormous deposits. The nitrate is in the desert area, and the industry is now depressed because the electric fixation of nitrogen is competitive. Local use is low, however, despite the return that fertilization could supply. The mines also play a role in absorbing some of the country's labor supply, though mechanization limits this. In 1945 oil and gas were found in southernmost Chile, an inconvenient location in relation to the center of population. The oil is brought up by ocean-going tankers, and some of the gas is being liquified and shipped by tanker.

Chile has been pressing industrial development aimed at reducing the amount of money that must be spent abroad for manufactured goods and increasing its ability to employ its own people. This has been so successful that the number of people employed in manufacturing now begins to approach the number of people in agriculture who traditionally have earned minimal wages.

Chile's problems are not space or resources. There are vast empty areas in the central valley and even more in the forested south. The central valley alone is capable of feeding several times the 9 million people in the country, but under an extensive hacienda system it does not. Strenuous efforts are now underway to change this. School enrollment is up, schools are being built by the hundreds, housing is being built, and irrigated farm land is being greatly increased. Meanwhile the copper industry, whose taxes are a major support of the development program, is being redoubled. If Chile can maintain the pace, it can become the California of Latin America in more than physical geography.

SOUTH AFRICA
AND AUSTRALIA

South Africa's Region
of Mediterranean Climate

The part of Africa marked by concentrations of rainfall in winter and by summer drought is a triangular area on the southwest tip of the continent. The dimensions are roughly 200 by 200 by 300 miles, an area comparable to California south of San Francisco. Just as the Mediterranean climate of California is only a part of the state, in South Africa this summer dry climate is only a part of the entire country, which includes large arid and humid areas as well.

At the south tip of Africa there is a coastal plain backed by ranges of folded mountains, and behind them looms the high plateau edge of the continental mass of Africa. Aridity increases to the northwest, while rainfall increases to the northeast and soon becomes well distributed through the year. It is only the Cape region, then, that is Mediterranean in climate. Here the mountains parallel the coast, and their seaward flanks intercept moisture, but the interior valleys are drier; rainfall can be over 200 inches on the mountain tops, but under 5 inches in the valleys. As in all Mediterranean climates, the greatest problem is water, especially during the long, dry summer months, and the high mountain ranges with their abundant rainfalls are therefore invaluable. So far all is familiar. There is a desert "to discover," and a scenic, pleasant land to attract people: it could be California all over again. There are differences, of course.

The folded mountains create many valleys, but there is no great central valley as in Chile or California. The higher of these valleys are well watered, but the lower ones are dry. For example, east of Cape Town lies the 100-mile-long, 10- to 12-mile-wide Worcester-Robinson Valley where rainfall in the lower end is only 5 to 10 inches, on the flanks 20 to 30 inches, and at the head of the valley 30 to 40 inches. Permanent streams flow through many of these valleys, and the irrigation potential is large. There is a broad coastal plain, but here the rainfall is generally low: 10 to 20 inches—and the area is like the drier end of the Palestinian or Californian comparable climates. All in all, however, the region is the now familiar land of sunshine, mild winters, and opportunity for crops grown on the winter rains and for summer crops dependent on irrigation supplied by the rain-catching mountains.

As we have seen, these can be areas of great opportunity.

Man was as slow to see that opportunity in South Africa as he was in California. The earliest inhabitants of this part of Africa seem to have been the people known as the Bushmen—a small, yellow-skinned, slant-eyed, triangular-faced race of mankind. Quite clearly they once occupied much more of Africa and have only recently been driven from the well-watered parts into the arid environment that no one else wants, which has led to the curious and erroneous conclusion that some of their unusual bodily characteristics were developed for life in the desert. On the contrary, their history is that of long development in the high, cool, well-watered lands of East Africa and slow retreat before the onslaught of physically larger and culturally better equipped people, until today when they find themselves thrust into their ultimate refuge, the Kalahari steppe and the Namib Desert. As in other cases, men had the possibility of changing their ways to meet the thrust of new ideas or of clinging to their old ways. In an extraordinary number of cases, men of all races have chosen not to change. This is much like the story of the Indians in California, except that until the coming of the Europeans they were left undisturbed.

Early cultural lag

Like California, South Africa was far from the centers of growth and development of ideas and long remained in the hunting and gathering stage of development. However, ideas had reached the area before the Europeans arrived, and the ancient way of life of the seed gatherers and animal hunters was ending under the onslaught of another way of life.

The idea of cattle keeping, a form of pastoral nomadism complete with milking, had reached South Africa some time before the Europeans arrived. The Hottentot people, who had taken up this way of life, had moved into the Cape region and were competing for space with the Bushmen. Racially the Hottentot seem to be intermediate between the Negro people and the Bushmen and probably result from an incoming people mixing with the indigenous people. The use of a Mediterranean climate by simple hunters and gatherers was the original California land use, succeeded by ranching, a settled way of cattle raising. In South Africa a pastoral nomadism of the type usually associated with the steppe or desert areas developed, which suggests that pastoralism is less limited to the semiarid regions than relegated to them, as wheat growing is

(Upper) **Cape Town, South Africa.** *The landscape is typically Mediterranean: sea, narrow coastal plain, steep mountain slopes. The mountains behind the town are smooth sloped because of the intact vegetation and cover. Compare this photo with that of the Greek shoreline, where destruction of the vegetation has allowed loss of soil and reduced the land to bare rock.*

(Lower) **Landscape at Worcester, near Cape Town, South Africa.** *This location is in an interior valley, and the landscape would be completely interchangeable with locations in California or Chile. Here Frisian cattle are grazing on lucerne (clover).*

displaced from the more humid regions where it would actually do better.

European occupation

Into this scene of an already changing cultural landscape, two streams of people converged by the seventeenth century. The Portuguese early discovered the Cape region, but in 1510 a sizable group of them were slaughtered by the Hottentots, and the Portuguese thereafter centered their interest in the more settled and prosperous trading towns on the coast of East Africa. The lovely lands of the Cape region, with a climate like their homeland, were forsaken in favor of the steamy, more civilized areas offering greater and more immediate profits. Not until 140 years later, in

1652, did the Dutch decide that the Cape area made an excellent stopover point on the long voyage from the Netherlands to the East Indies, a conclusion precipitated by the wreck there of a merchant vessel in 1648. The amenities of a Mediterranean climate had almost nothing to do with the choice: the initial settlement was purely strategic, much like the motivation for the Spanish settlement of California. Thereafter the desirability of having provisions always available led to colonization, and by 1662 there were 60 free burghers (farmers, fishermen, and craftsmen) established there. Not only Dutch but German and other northern European people came to the colony. The scarcity of labor and the availability of land suggested extensive rather than intensive land use and led to a shift toward cattle culture, although wheat and grape growing early became important. Expansion into the interior was slow, for the population was small, and away from the coast the valleys became increasingly arid.

When the British took over the Cape region in 1795, they found a population of 16,000 Europeans (Dutch 50 percent, German 27 percent, French 17 percent, others 5.5 percent). In addition, there were 17,000 Malays, some Negroes, and no one knows how many Hottentots and Bushmen. An occupation of 225 years had not seen much European population growth, but it had seen the beginnings of the racial mixture that was to become the dominant characteristic in the life of the region. In 1806 the European population was 26,000, Hottentots 17,000, slaves 29,000; the Bantu were the enemy being held at bay along a borderland to the northeast. The frontiersmen were the Boers who lived as cattlemen moving over the land renting vast tracts at minimal rates. The British were anxious to establish a more permanent and productive settlement, so they brought in settlers: 5000 in 1820, 4300 in 1850, 12,000 in the 1860s. Yet after 350 years the number of people on the land was few, and the tempo and mode of life like that of Spanish California.

Agriculture and industry

South Africa might have lingered in a modestly developed agricultural stage had it not been for the discovery of great mineral wealth. The discovery of diamonds at Hopetown in 1867, and later at Kimberley, brought a flood of immigrants from all over the world. Then came the great gold discoveries on the Witwatersrand near Johannesburg in 1887, and again the immigrants numbered in the tens of thousands. What government planning and encouragement had not accomplished in centuries was quickly accomplished when

economic opportunity beckoned, and a large European population became established in the land. The mineral discoveries were not, however, in the Mediterranean climate region, and the great growth of population therefore centered in the northern part of the Republic of South Africa, of which the Cape area with its desirable Mediterranean climate was a less-valued appendage. The great modern city of Johannesburg marks the center of the mining community, while Cape Town marks the center of the development in the Mediterranean climate region.

In the Cape area, the real Mediterranean climate zone, the present development is strongly centered on agriculture. Wheat, long the dominant crop on the inland part of the coastal plain, is decreasing in relative importance. The commercial production of deciduous fruits, vegetables, grapes, and the export of wines is a very important part of the economy. The deciduous fruits require irrigation in the summer, but the mountains with their perennial streams flowing down the alluvial fans along the mountain flanks make irrigation feasible here just as in California and Chile. The area runs the curious risk of too warm a winter for the deciduous fruits. When there is not sufficient cold to make the trees shed their leaves and go into their normal dormant period, their seasonal rhythm is so disturbed that they may yield no crop the following years. Apples, pears, plums, peaches, and apricots are grown in large quantities for export. These are, of course, just the fruits with which California began her horticultural period. Vine cultivation is also important: grapes, raisins, and wines make a multimillion-dollar export industry. Productivity of the land expressed in yield of wine per acre in the Cape region shows 440 gallons per acre, as compared with Europe's average of 176 gallons per acre. Cape wines, grown under a stronger sun, have more sugar and less acid than European wines, but Europeans prefer their thinner and more sour wines. This is cultural determinism at work again: why should thin sour wines be preferred to full-bodied sweet wines?

Table grapes and deciduous fruits bring high prices and are shipped under refrigeration to European and United States markets. The Southern Hemisphere location of the Cape Town area puts its fruit on the market during the Northern Hemisphere winter. Some of its development was at first dependent on the empire-preference system that gave members of the British Commonwealth preferred status in the British market. Difficulties focused on the race problem caused South Africa to withdraw from the Commonwealth, but a market for its products still seems assured as the growing productivity of Europe is greatly increasing the number of people able to afford imported fruit and wine. Increasing prosperity in South Africa should eventually create local markets.

South Africa (the political subdivision, not just the Mediterranean area) has only 4 percent of the land and 6 percent of the people of Africa but much of the continent's developed industry. On a population base, the productivity of South Africa is nearly 10 times that of the rest of Africa. By 1964 South Africa's commanding lead in economic development could be stated in terms of the percentage of Africa's total productivity that was concentrated in South Africa: 60 percent of electric power, 50 percent of railroad stock, 48 percent of all autos, 54 percent of the farm tractors, 90 percent of the gold, 87 percent of the coal, and 75 percent of the steel. Ten years later the only major change is due to the discovery of oil in Libya, and the benefits of the capital inflow have yet to be applied to production, which is already far advanced in South Africa.

The race problem

Great stress has been placed by critics on the segregation program—apartheid—and too little stress on the difficulty of the problem and the strenuous effort that has been made to help the nonwhite populations. "Populations" is used in the plural for there is no unity among the nonwhites; even the Bantu are separated into tribal units with divergent languages and aspirations to separate national status. The difficulties are formidable. Nowhere in the world has a quick and easy way been found to bring people from the hunting-and-gathering stage into the twentieth century. The problem is only a little less formidable when one begins with pastoralists such as the Hottentot or tribal agriculturists such as the Bantu.

In each case there is illiteracy, complete ignorance of and consequent disregard of the simplest hygienic practices, and totally different cultural views on sex, business, and politics. We need not make value judgments concerning which system is best, although there is no reason that such judgments might not validly be made, to see that such discrepancies create enormous difficulties in a society. Those in favor of apartheid favor complete separation of the races, with each race having its own national-racial territory. Since the apartheid group is in the majority among the whites and the whites are in control, their program is being enacted.

The white sector of the population is not ignoring the needs of the other races, but is carrying on a massive educational, housing, and development program. This includes a housing program designed to produce tens of thousands of modern dwellings for the native people and the destruction of the slums sur-

Meadowlands, a Bantu Township South of Johannesburg. *These are some of the 130,000 family-size brick and concrete houses which are planned as a total project to be built to accommodate more than 700,000 Bantu. The total cost is expected to run $100,000,000. Rentals are to vary from the equivalent of $3 to $12 per month.*

rounding the major cities. For vast numbers of the natives, it is their first experience with running water, electric power, and such advanced devices as outhouses (such as characterized America in the nineteenth century, but persist as a sanitary arrangement only in our most remote rural areas today).

Fifty years ago the Bantu were an illiterate tribal people mostly living in mud huts and clothed in animal skins, but today many of them live in modern housing and wear European dress. They have their own newspapers, 80 percent of the Bantu children attend school, and there should be no illiteracy by the turn of the century. There are Bantu teachers colleges, universities, and medical schools so that the Bantu are becoming capable of educating their own. Most of the cost has been borne by the white population, but the product is a growing, educated Negro middle class with an increasing income and capability to pay for further development. The ultimate solution to the problem of race relations in the South African area remains to be worked out, and the fate of the region of Mediterranean climate about Cape Town will depend upon the outcome. The Cape area also is affected by the economic growth of South Africa and benefits from its connection with this powerhouse of industrial growth in much the same way that California has benefited from its association with the industrial East. If the race problem can be solved, the Cape area should have a bright future.

Change is certain to come to South Africa, and the first signs of it are already obvious. The Afrikanders, descendants of the Dutch farmers who settled the land three centuries ago, are uneasy about their position. They still want to preserve the identity of the white, the black and the coloreds, yet they still have a sense of missionary zeal and responsibility toward the other groups. They begin to have faint doubts about the wisdom and the morality of their course. The church leaders are increasingly critical of apartheid. Some of the businessmen see the restrictive employment practices as the cause of the critical shortage of skilled labor. American companies in South Africa are quietly beginning to pay the blacks and coloreds on the same basis as any other employee, and this puts pressure on all other industries to do likewise. The black African homelands such as Transkei are making progress. Throughout the republic 89 percent of the black children under 14 are attending schools. This is double the 1960 figure and more than three times the figure for the rest of Africa. It is quite predictable that as education increases, so will political awareness. The problem is how to make the change from the present system without blowing up the political economy that is responsible for the education and the rising standards of living. The danger is a Cuba instead of a Brazilian solution.

Australia's Mediterranean Lands

Australia has two areas of Mediterranean climate: one in the southwest corner around Perth and the other around the series of gulfs and peninsulas west of the southeastern corner of the continent, with Adelaide as the principal city. These regions differ from the other Mediterranean lands so far described because their mountains are less important. Western Australia has none, and the area around the gulfs has only the modest-sized Flinders Range, with general elevations of 2000 feet and with peaks reaching 4000 feet. Although the two areas have had similar development, they can best be discussed separately, except for the aboriginal situation.

All of Australia, whatever its climate, remained in the Stone Age until the British settlement. For people with a hunting-and-gathering economy, all environments are useful. Hence there is nothing significantly different about the way of life of the Australian aborigines whether one is discussing the tropical north, the seasonal south, or the arid interior. On so simple and unspecialized a level, there is something to gather and something to hunt wherever you are, and the fitting of economy to the land that marks the growth and development of economic life hardly appears at this level of human activity. This tends to undermine the common saying that man has increasingly won freedom from his environment, for in a very real sense the simpler people were less environmentally bound than we are.

The coming of the Europeans

Australia was first reached by Europeans crossing the Indian Ocean on a southerly route that brought them to the west coast of the continent. The discovery areas then focused on the southwest corner of the continent, an area much like California in climate, soil, and vegetation, and also in being occupied by a people of extremely simple culture. The reaction of these early seventeenth-century discoverers was much like that of the Spanish in California and the Portuguese in South Africa. They saw little value in the land and went on to the attractive hot and steamy tropics to the north of Australia where there were plenty of settled people who were producing spices of great value. Belatedly the British began the settlement of Australia, but from the east coast in 1787, beginning in climatic zones similar to the southeastern United States. Only later, almost reluctantly, did they settle in Australia's Mediterranean climate. For the Spanish, Portuguese, and English, a land without a settled and developed people was unattractive. Only a great change in economic potential such as the discovery of gold in California could lead to rapid settlement. Thereafter modern transportation linking California to the eastern United States assured continuous growth. A pleasant climate was not an important criterion then.

Western Australia

Western Australia as a political unit is a huge slice of the continent. Its southwest corner, known as Swanland, is characterized by winter rainfall that is an abundant 40 inches, decreasing inland until at the 10-inch line the boundary of the humid land is reached. The adequately watered, Mediterranean climate area amounts to 150,000 square miles. It is equal to California in area but immensely more useful for agriculture because it is flatter land. Most of Western Australia's 1 million people are in this humid corner. Compare Great Britain's 56 million people in an area of 94,500 square miles of less productive land for a measure of the emptiness of even the good parts of Australia. It is a low-lying, largely level land, for heights are only hills of some 1000 feet in elevation. Settlement began at Perth in 1828, and the first 10 years saw the arrival of only 2000 settlers. As in California and South Africa, wheat was the important crop, and in typical Australian fashion, sheep became the other important use of the land in the early period. Settlement has been aided by the attraction of gold, for in the 1880s gold was found at Kalgoorlie in the adjacent dry region. This brought the usual gold rush, and the flood of miners included men who then elected to settle on the land.

Development of the land has followed the rainfall belts. The rainier southwest has an apple-growing region, and along the coast of Perth there is a zone of orange growing and an area of concentration of vineyards. Inland, toward the drier regions, there is a wheat belt, stretching 450 miles north and south and extending 30 to 100 miles east and west, where farms run to 1000 acres. Rainfall here is from 10 to 20 inches per year. On the arid side of this belt the land is dry farmed; that is, part of the land is rested and allowed

to accumulate moisture for a year or two before a crop is planted again. Beyond that area, sheep are grazed out to areas having only 7 inches of rain per year, and further still lies the desert.

Iron ore has been discovered, and American capital has been pouring into this area of stable government and large opportunity. Japan's need for raw materials has proved to be a key to development of some Western Australian resources with American capital, a neat example of the interlocking movement of capital and resources and accelerated development. Western Australian iron mined by American companies goes to Japan, the third largest steel producer in the world. Japan is developing into a major customer for Australia's aluminum, salt, wheat, wool, and meat. As in California, the adjacent desert is being made productive in part by irrigation. Another large area of humid climate is being made productive at a cost of a penny per acre by applying trace elements to the soil. Millions of acres of this land, which formerly grew very little and sold for 55 cents per acre, now are productive. Such land will carry three sheep per acre in a rotation of clover, wheat, and sheep, and equivalent land in California would be worth several hundred dollars per acre.

Development has been slow up till now because the area has no great advantage in Australia. Australians have plenty of sunshine and warmth over most of their country all year round, and they can grow oranges, grapes, tropical fruits, or vegetables in the winter almost anywhere. There is, then, no tourist trade for Western Australia in winter, nor any possibility for the development of specialty crop areas to supply a winter-bound adjacent area. It is not surprising therefore that growth in this region has been modest; the area can be thought of as resembling California 50 years ago or South Africa today. This is not a backward area, for none of Australia is that. Perth is a modern city with some manufacturing, and its population of 703,000 equals more than half the total population of the state. The strong urban concentration of the population is a mark of efficient, modern agricultural production. Lightly populated Western Australia is simply a California attached to a tropical and subtropical country. This places it in an entirely different situation from a similar area attached to a mid-latitude country.

South Australia

The Mediterranean climate area centered on Adelaide is the smaller of Australia's two such climates. The 100,000 square miles of land with winter rain, year-round sun, and moderate temperatures would be a most welcome addition to most countries. In Australia it has attracted less than a million people, and more than half of them live in the principal city. Settlement here began with 500 settlers shortly after Western Australia: 1836 compared with 1828. With few men and immense areas of available land, the development was toward extensive use of the land, and the choice from the Australian viewpoint was wheat and sheep.

Australia's history is oddly intertwined with that of the United States. It was the breaking away of the American colonies that precipitated the settlement of Australia. The gold rush in California set off a comparable rush in Australia, for an Australian who had gone to California returned home still infected with gold fever and discovered gold there. By 1850, nearly 100 years after the beginning of settlement, Australia's total population was only 400,000 people, and development seemed to have slowed to a walk. But gold soon had it galloping, for in the four years after 1851 the population doubled and redoubled, and the land began to fill up. The biggest gold strikes were in Victoria, adjacent to South Australia, and the flow of settlers that followed the gold rush added to the men on the land around Adelaide.

It was a fine land that they settled. Peninsulas and gulfs provide pleasant land-and-sea contrasts, and the Flinders Range, even with its modest heights, adds to the scenery and catches much-needed moisture. Rainfall declines rapidly from an abundant 40 inches at the southern end of the Fleurieu Peninsula south of Adelaide going northward, to 30 inches, then 20 inches, and then more slowly to 10 inches; to the north thereafter lies the desert. The heaviest rainfall in the area, 47 inches, is on Mount Lofty behind Adelaide, but this is a modest amount compared to the immense rain and snow falls on the lofty ranges of California and Peru and the 200 inches on the ranges of South Africa.

As in Western Australia, the land use is adjusted to the rainfall belts. Apples are grown in the rainy area, oranges and vine crops are raised south of Adelaide in the 20- to 30-inch-rainfall zone, and wheat is grown on the dry side of the 20-inch isohyet. Beyond that lie the sparse grazing areas of the desert and semidesert. As with Western Australia, there is no great advantage in South Australia's favor. Oranges and grapes can be grown in much of Australia; deciduous fruits can be grown as well or better in adjacent areas. North Australia is warmer in winter than is South Australia, and when anyone wants sun, there is always the desert handily adjacent in all of Australia. Again, the Mediterranean climate when so situated has relatively little advantage, and when its summer drought is considered, it is actually at a disadvantage compared with the adjacent subtropical regions.

Summary

Australia's two areas of Mediterranean climate remain lands thinly occupied not for any fault in the land, but because in the present stage of Australia's development—with the whole land thinly settled—these sunny, mild-winter and warm-summer lands, with their high year round sunshine, must compete with adjacent lands that also have mild climates and high percentages of sunshine and are often better watered throughout the year. Nonetheless, the areas of Mediterranean climate are attractive and are bound to develop along with the rest of humid Australia, which continues to attract a strong flow of European immigration and to develop its industry. The future clearly seems to be one of steady growth on a high standard of living based on balanced industry and agriculture.

Summary

We have, then, five areas of land that share a common climate, sea-border positions, and, in four cases, rugged topography as well. Two of the areas, Chile and California, are alike to a quite striking degree. The question now arises: In these similar environments, just how much similarity of human activity has there been? Is there a Mediterranean way of life whose primary conditions have been determined by the similarities of these environments? The answer is no.

In the Mediterranean proper, cultural accomplishment was generally high, at times the highest that man has attained. The other two Old World Mediterranean climates were cultural low spots, and this is true also of the New World

Mediterranean climates. If the physical environment did not cause cultural uniformity or similarity, or stimulate much of anything in the way of human accomplishment in four out of five cases, we may well expect that crude physical environment (soils, climate, topography, position in relation to the sea) simply was not a major causative factor. What then?

We might well suspect race. Accomplishment in the European Mediterranean was primarily in the hands of the swarthy, graceful Mediterranean race with an assist from the closely related large, blondish neighbors from northern Europe, but accomplishment was followed by a long decline. It does not seem to matter what the race, for neither Australians nor Negroids (Bushmen first and Hottentots later) in South Africa nor the various types of men broadly classified as American Indians accomplished anything in this finest of all physical settings, if one judges by what the Greeks were able to accomplish. While it is true that the native Australians accomplished little and the primitive people of South Africa about as little, the American Indians in other settings proved themselves to be most able people. Race must, at least for now, be left in the limbo of unanswerable questions. We can do much more with the cultural factor.

It is clear from a survey of history that the initial cultural impetus in the European Mediterranean came not from the Mediterranean itself but from the adjacent Middle East. The spread of Neolithic and Bronze Age cultures around 4000 B.C., reaching out into the islands of the eastern Mediterranean, brought that land into the realm of high culture. The island of Crete was an early center of civilization; here the Minoan culture flourished about 2500 B.C. and came by 1700 B.C. to dominate the Aegean Sea. Minoan culture was clearly built in large part on trade between the Aegean and Egypt, and the main source of cultural stimulus lay in the much more ancient and advanced culture of Egypt. The major basis for advances in knowledge, then, lay outside the Mediterranean.

If we concentrate on Greece as the earliest seat of cultural growth in the Mediterranean (outside of Crete) and the center of some of the greatest human advances, we find that at all times Greece received ideas from adjacent areas. The beginnings of Greek cultural growth appear in such cities as Mycenae and Tiryns. Characteristically, growth was based in large part on local populations and imported ideas, apparently with a sufficient supply of immigrant skilled people to organize, teach, and eventually to get a local development going. It is difficult to see either a physical-environmental argument or a racial argument in this. It looks like a simple case of cultural diffusion. As we have seen, increasing distance from the center of influence in the Middle East is accompanied by a steady falling off of early great accomplishment.

The story changes little for the other lands with Mediterranean climate. Those near centers of ideas stood higher on the cultural ladder than those far from such centers. Since only one of the Mediterranean lands was really close to a great center of advancement, only one of them enters the list of early cultural innovators and centers of accomplishment. When the wheel of history turns, the positions of the countries shift. The Old World Mediterranean has been in an eclipse despite its proximity to the European mental powerhouse. That this may be only a temporary situation, if already centuries long, we have indicated by reviewing developments in Italy today. Spain and Greece have similar potentials. Italy, the more advanced in education and industry, should hold the lead. Spain, Portugal, and Greece could challenge Italy in a few decades if they continue their educational and industrial growth. North Africa seems to be second only to South Africa in the bleakness of its foreseeable future because of its sociopolitical situation. Expulsion of all Christian and Jewish minorities and withdrawal into the Arab world is not the kind of program that is likely to make northwesternmost Africa bloom. This potentially rich land, blessed with a start toward civilization as early as 4000 B.C., is seemingly doomed to stagnate still longer. Perhaps it is race? If so, the Caucasians are at fault. All of this, of course, is going on in the unchanging physical environment of the present world. The causative factors must then be nonphysical. The situation can be understood in terms of changing perceptions, with new ideas as the key to the changes.

In the New World the story is the same with the addition of the abrupt changes due to impact from Europe. California and Chile, as alike as two peas in a pod, have opposite histories. California starts low and after a slow start goes up the ladder of civilization. Chile starts well, but then goes nowhere. The changing world and the differing political associations and cultural heritages are clearly the causes of the differences. The future of each of these lands is predictable over a short time in terms of the people, their attitudes, and their situation in the world. California's future is unlimited because of its association with the other states, its large, highly educated population, and its access to the technology and capital sources of the world—combined with a passionate drive to be the best. It must still solve the problems of water supply, air pollution, and crowding. Chile's similar area has a small, poorly educated population, dependent on a small, underdeveloped nation with little access to technology and capital; more important, Chile seems to lack either the drive to achieve or the social mechanisms to accomplish its aims, so it may be destined to stagnate a while longer. In the Cape area of Africa men (the Bushmen) at first perceived that the environment was good for gathering the wild plants and catching the native animals. Shortly before the arrival of the Europeans, other men (the Hottentots) entered the land and had a different view of its possibilities, seeing it as excellent grazing country. Then came the Europeans, who saw the area as a strategic location on the route to India; once settled there, they saw other possibilities, for the rich soil and subtropical climate fitted their cultural patterns in allowing the growing of fruits and vegetables and the production of wines. Today, as part of the rapidly developing industrial unit in South Africa, the Cape area should have a bright future, but all of South Africa's future is clouded by racial problems of tremendous complexity. The future of Australia's Mediterranean lands is uncertain only because the part they will play in the total Australian development is uncertain. In general, prospects are good because of the educational level of the people and the degree of technological development.

We can state all this in terms of perceptions. At one time the men in the regions of Mediterranean climate lacked any idea of domestication of plants and animals, the utilization of metals, or the harnessing of water power, fossil-fuel power, and atomic power. Some of these areas were so located that new ideas reached them early, and they then developed their landscape with new vision. Of the five areas of Mediterranean climate in the world, the one first reached by such ideas was in the Old World, around the Mediterranean Sea. The four other lands of Mediterranean climate lagged behind. Only in Chile were there sufficient contacts to change the people's perception of the potential of their land; they shifted to agriculture, irrigation, and settled village life. In the other areas—California, South Africa, and Australia—the inhabitants continued to view the land as best suited to hunting and gathering and an unsettled way of life. When the Europeans spread over the world, their effect on lands of Mediterranean climate varied greatly, both in time and degree of change. They viewed Chile, a settled but non-gold-producing land, as a limited good. The first discoverers passed up the other three areas because they sought well-settled land that had gold, or spices, or some product of high value and low bulk that fitted their perception of an exploitable area. Only slowly did the Europeans, even those from similar climates, come to value these lands for their genial and productive climates. This changed outlook has only developed within the past century.

Review Questions

1. Would the explanation for the flowering of Greece about 500 B.C. have any validity in explaining the Italian Renaissance? Discuss.
2. Would the unification of Italy be a study relevant to the situation in Central America? What would be some useful comparisons?
3. Compare the North African area of Mediterranean climate with California. What part of California is it most like? Is it richer or poorer in basic resources? Will its development be like that of California?
4. Since California was physically much like their homeland, why were the Spanish not attracted to it at once?
5. Why are the superficial population distributions in Chile and California not really comparable?
6. In this period of population explosion everywhere and economic difficulty in England, why is there not the equivalent of a "gold rush" to the "golden land" of southwestern Australia?
7. Compare the cultural positions of the five principal areas of Mediterranean lands at the time of Christ, in A.D. 1500, in A.D. 1750, and A.D. 1950. Use chart or essay form. What is the significance of the similarities and differences?
8. Discuss Chile as an example of environment and man: pre-Columbian, post-Columbian, and modern.
9. Discuss South Africa's resource potential and its racial tension. What outcome do you foresee?

Suggested Readings

Allbaugh, L. B. *Crete: A Case Study of an Underdeveloped Country.* Princeton, N. J.: Princeton University Press, 1953.

Barbero, G. *Land Reform in Italy; Achievements and Perspectives.* F.A.O. Agricultural Studies, no. 53. Rome, 1961.

Caughey, J. W. *California.* Englewood Cliffs, N. J.: Prentice-Hall, 1953.

Hanson, E. P. *Chile: Land of Progress.* New York: Reynal & Hitchcock, Inc., 1941.

Karmon, Yehuda. *Israel: A Regional Geography.* New York: Wiley-Interscience, 1971.

Mikesell, Marvin W. *Northern Morocco: A Cultural Geography.* University of California Publications in Geography, vol. 14. Berkeley: University of California Press, 1961.

Myers, J. L. *Mediterranean Culture.* London: Cambridge University Press, 1944.

Semple, Ellen Churchill. *The Geography of the Mediterranean: Its Relation to Ancient History.* New York: Holt, 1931.

Stanislawski, Dan. *The Individuality of Portugal.* Austin: University of Texas Press, 1959.

Suber-Caseaux, Benjamin. *Chile: A Geographic Extravaganza.* Translated by Angel Flores. New York: Hafner Publishing Co., 1971.

Walker, Donald S. *A Geography of Italy.* 2nd rev. ed. Advanced Geography Series. London: Methuen, 1967.

Way, Ruth. *A Geography of Spain and Portugal.* Advanced Geography Series. London: Methuen, 1967.

East Coast Mid-Latitude Forest Lands

CHAPTER FIVE

THE middle latitudes, roughly from 30 to 60 degrees away from the equator, are moderately to extremely moist lands. On the west coasts of continents in these latitudes the westerly wind belt with its cyclonic storms is present through the year. The flow of the wind and the structure of the storms bring precipitation. On the east side of continents in these latitudes the cyclonic storms bring rains, especially during the winter season, and are reinforced during the summer by monsoon tendencies, the inflow of air onto the continents from over the warm tropical seas. The result is precipitation well distributed throughout the year, generally ranging between 20 and 60 inches annually, with 40 inches the usual amount. In the cooler parts of these lands, even 20 inches is enough for forest growth; therefore wherever there is sufficient rainfall,

and as long as at least one month of the year is above 50° F., a forest cover is found over these mid-latitude lands—with rare exceptions.

Under cool, moist, forest-covered conditions, weathering proceeds at varying rates. In the cooler north, where there are long cold winters, the rate is slower than it is in the warmer, short-wintered southern parts. The tendency is reinforced in the north, where glaciers once rested on the land and left either bare rock or fresh debris. Since the great retreat of the glaciers happened about 10,000 years ago, that is the maximum time that weathering can have occurred in formerly glaciated regions. The difference is easily seen in the eastern United States. Weathering penetrates tens of feet deep into the solid rocks in the South, and the rich coloring of the soil materials, often

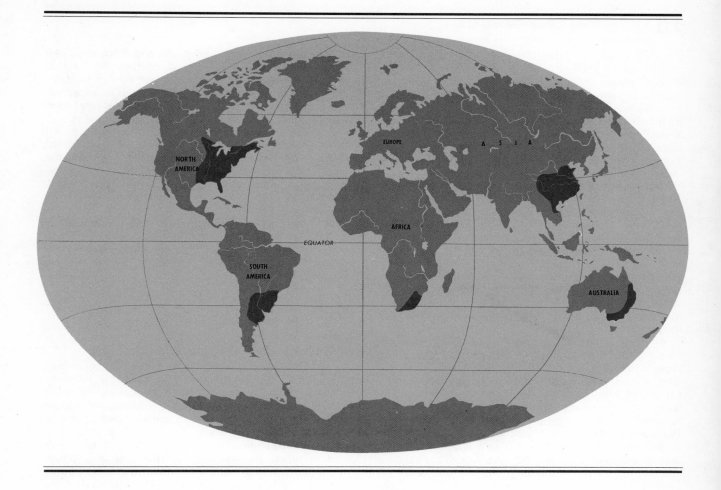

a bright red, expresses the degree of chemical action. In the North the soils are not deeply weathered, the colors are never bright red, and the rocks are only weathered to slight depths. The rounded forms that the glaciers gave to the bedrock masses that they over-rode can be seen in the rocks in the Hudson River Palisades, and the smooth polish that the glaciers gave to the granite rocks can still be seen and felt in many places in New England.

All other things being equal, younger soils are richer in humid climates because of the tendency of the weathering process to remove the soluble nutri-ents. The longer this weathering process goes on, the poorer the remaining mass of material becomes in the chemicals that plants need for growth—a phenom-enon discussed in the section on the soils of the tropical

rain forest (see Chapter 3). It applies to *all* forest cli-mates. Within the mid-latitude forest climates, then, the older soils will be poorer than the younger ones. Those weathered under the wetter and warmer cli-mates will have more of their soluble nutrients leached out than those weathered for less time in the cool, formerly glaciated regions.

There are two quite different mid-latitude forest climates on the two sides of the continents. Those on the eastern sides of continents have continental-type climates marked by hot summers and cold winters. Those on the western sides of continents have mild, oceanically modified winters and summers. The physi-cal similarities between the areas of mid-latitude forest climate will be seen much more clearly if we discuss these two kinds of mid-latitude lands separately.

The eastern United States is an excellent example of an east coast mid-latitude forested country. Rainfall ranges from a 60- to 80-inch high along the Gulf coast, through a 40- to 60-inch belt that covers most of the South and East, declining to a 20- to 40-inch belt that extends west to the 100th meridian and takes in the Great Lakes region. Most of this area was once forest covered. Except for the hill country of the Appalachians, it is a low-lying area dominated by great riverine plains.

The region has considerable variation in weather and climate. The South has long, hot, humid summers: New Orleans, St. Louis, Charleston, Baltimore, and even New York have spells of hotter humid weather than any equatorial location can expect. The winters are of varying intensity too. In the South, winters are mild and snow is rare—it seldom stays on the ground longer than a day or so. In the southern Great Lakes region the winters are subarctic, daylight becomes short, temperatures fall below freezing and stay there for considerable periods, and snow may cover the land for periods of weeks.

This description of weather and climate applies to a number of other areas of the world as well. South America has a block of land that is very much like our East, both in physiography and in climate. From southern Brazil through Uruguay, including southern Paraguay, and extending south to nearly 40 degrees in Argentina, there is a great Atlantic-facing lowland that has much the same type of rainfall and seasonal change as we have just described. China has another major area of mid-latitude forest climate. Again, the rainfalls are greatest in the south near the sea and drop off toward the north and the interior. The south of China is like our South; the northwest and adjacent Manchuria (now part of the People's Republic of China) are like the Dakotas. Climatically, then, China is closely comparable to our East. This surprises some people, for China seems to have acquired the label "tropical" in many peoples' minds. China differs from the United States and South American counterparts in its physiography: the south of China is hilly and mountainous; the plains are in the northeast.

Australia's east coast is similar. Its climatic range is much like that of the east coast of the United States, but the extent of this type of climate is very limited. It forms a strip along the coast and ends to the west of the mountains that parallel the coast. There is also a small region of this type of climate on the southeast side of Africa.

All of these areas share similar physical settings to some degree, and some of them are comparable to a high degree. Rainfall, seasonal regimes, and variation in climates both north and south and from the coast inland are similar. With rare exceptions the land is, or was, covered with forest. Under similar rainfalls, climates, and vegetations, similar soils can be expected. Again, we have a laboratory for examining what man has done and is likely to do in these similar areas.

The population map of the world shows that these lands differ greatly in density of population: China is densely populated, the United States and the South American area have moderate population densities, and the Australian environment is thinly populated. If we looked at the population map for the year 1500, the Chinese density would still stand out, while the United States and South American densities would fall to low levels, and the Australian density would fall still lower. Another 500 years is likely to see still further changes in these patterns of density, both in relative positions and in absolute numbers.

The Eastern United States

The Indian scene: population and culture

The eastern United States in the pre-Columbian period had a modest Indian culture varying in intensity, tending to decline from a cultural focus in the south toward a less developed north. Corn, beans, and squash were the staple crops, raised on lands cleared of forests. As elsewhere in the world, the forest was used as a long-term rotation crop to clear the land of weeds and to renew the soil fertility. The level of development had reached the stage of town dwelling, pottery making, and broad field cultivation that approached the early stages of civilization. Little was lacking except political organization into larger, more stable units. Towns of considerable size existed, populations were dense enough and had enough organized productivity to free much labor for monumental works, and men with sufficient political and architectural skills had appeared to direct this labor force in large-scale works. Some of the earth mounds built by the southern Indians rivaled the pyramids of Egypt in their bulk. The great mound at Cahokia, Illinois, was shaped like a truncated pyramid 1000 feet on its base and 100 feet high. The level of development was comparable to predynastic Egypt. We can only speculate on the progress that culture would have made if it had been undisturbed by European contact.

The degree of development and number of people in the eastern United States in pre-Columbian time have been greatly underestimated. The usual claim is that 350,000 Indians cultivated less than 1 percent

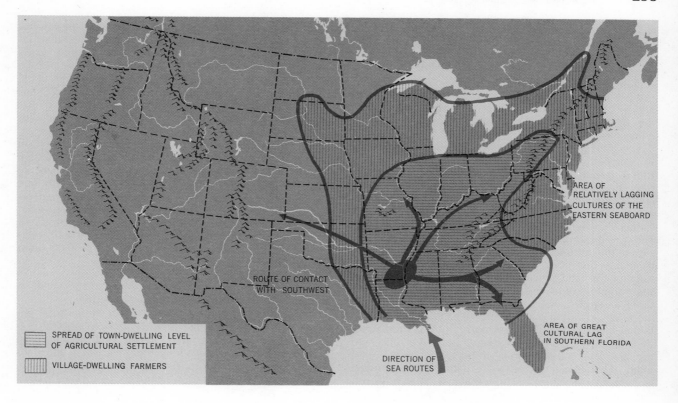

MAP 5-1 Cultural Level in Pre-Columbian United States

Within the central area outlined agriculture was the pre-Columbian way of life, and villages grew to the size of small towns or larger. The peak of cultural achievement lay in the lower Mississippi valley. Art, pottery styles, truncated pyramid mounds, and agricultural minutiae all indicate that the lower Mississippi was the focal area into which ideas were pouring. Continuing work has failed to find a route for this inpouring through the west Texas area. It seems increasingly probable, therefore, that the major contacts were by sea. That the routes were not primarily from the Caribbean via Cuba, however, is suggested by the low level of culture in pre-Columbian southern Florida, as well as by the focusing of cultural change patterns on the lower Mississippi.

of the land available for maize growing. The simple gatherers of California are credited with twice the population density of the farmers of the eastern United States. Both in terms of ecology and of culture this is contrary to what might be expected. The anthropologists have tried to explain this seemingly low population in the East as the result of warfare. While there undoubtedly was warfare, it did not hold down the populations in great areas elsewhere in the world, so why should it have done so here? Instead the populations of the East were probably much greater than they

have been formerly estimated. There is some support for this view in a 1568 traveler's account of a trip from Mexico to upper New England. Towns en route of up to one and one-half miles across were noted; such urban centers would make vast pyramids like Cahokia more understandable than a picture of a relatively empty land.

Analysis of the distribution of buffalo in the eastern United States reveals a similar story. Prior to the time of European contact, buffalo were seemingly absent east of the Mississippi River. Shortly thereafter they flooded into the wooded East, and the colonial landscape was dotted with areas called Buffalo Spring, Buffalo Falls, and Buffalo Creek. Buffalo are grazing animals and require grassy pastures; their presence therefore indicates extensive pasture. Why were they not present before the Europeans came? Why did they appear in the interlude between the disappearance of the Indians and the taking up of the land by the white men?

The answers may be something like this. The Indians had cleared large areas of the land. When the Indian population was greatly reduced by Old World disease, vast areas of cleared field were abandoned and gradually returned to grass and eventually to woods. This man-caused grassy environment suited the buffalo, so they moved into the virtually empty land. The pre-existing agricultural fields would also have suited the buffalo, which found maize and beans to its liking. But since a relatively dense Indian population on the land considered buffalo a succulent dish, the buffalo could not invade the humid East. The buffalo's opportunity appeared when man inadvertently prepared a suitable environment and removed the Indian threat. The buffalo's opportunity was short lived. In a state of nature the forest would have reclaimed the grassland in due time, or the Indian population would have reestablished itself with people immune to European diseases. In this particular case, however, the culturally advantaged, disease-immunized European population moved in to take permanent occupancy. They removed the rapidly expanding forest and slaughtered the bands of buffalo. For a time the land reverted to the agricultural state it had experienced at the hands of the Indians, but when the industrial revolution came to the United States, the growth of population and the development of the land were both intensified.

It may be that contacts with traders and fishermen, long before the colonists, brought a deadly flood of disease to the Indians of the East. Perhaps the populations seen by the first colonists were only the few who survived these epidemics. At Plymouth the Indians claimed that there had formerly been many more people, but that a great epidemic had nearly emptied the land a few years before. There is corroborating evidence too: the colonists wrote that they cultivated the already cleared, unoccupied land of old Indian fields. This would bear out the buffalo diagnosis.

The impact of man on the environment should be expanded to include the impact of new diseases. The effect of the lack of immunity to such a disease as smallpox has been documented. The reduction of the Mandan from 2000 to 200 in one winter is one example. The reduction of the Easter Island population from around 3000 to 111 is another. Before the days of vaccination, individuals who had had the local diseases and survived were then immune. Only children or the rare adult who had not yet been exposed came down in the recurrent epidemics. But since the children made up only perhaps 10 percent of the population, this cannot explain the large reduction of population that the Mandan or the Easter Islanders experienced. For even if 90 percent of the children died in an epidemic, only a 9 percent reduction of the total population would occur. Smallpox introduced into a population never before exposed might well kill 90 percent of its total population. Smallpox, of course, was only one of a great many diseases that the Europeans inadvertently imported, clearing the land of its native population in America and elsewhere. The serious effects of these diseases have not been recognized, and our judgment of the occupancy of the land before 1500 often has been erroneous.

Agricultural evidence of cultural origins

It is easy to trace the origin of the cultural impulses that created the semicivilization of the pre-Columbian eastern United States. Man had lived in the Americas for a very long time. He was well established as a skilled hunter of horse, camel, ground sloth, mammoth, and mastodon long before the last glacial retreat. About 10,000 years ago, his way of life began to change: the big game animals were becoming scarce or were extinct. Thereafter the beginnings of agriculture appear. Some local plants were taken in hand; the sunflower was developed for its seeds and tuberous roots. Soon the Middle American plants appeared, first corn and squash, later beans. Immediately one knows something of the direction from which ideas were reaching the eastern United States. These tropical plants could only have been introduced from the south. They probably came from Mexico, though the exact route has never been demonstrated. This leads to the suspicion that it may well have been by water—where no evidence would survive.

Too few questions have been asked about this strange agricultural beginning in the southeastern United States. Was this an entirely independent beginning? Or was it the result of stimulus diffusion? If so, from where? From Mexico? Then why were corn, beans, and squash not introduced directly? From Europe? An unthinkable thought up to now. However, Roman contacts with America and other earlier Mediterranean contacts seemingly should alert us to some such possibility. Of special interest is the revival of interest in the finds of European artifacts in Tennessee and Kentucky.

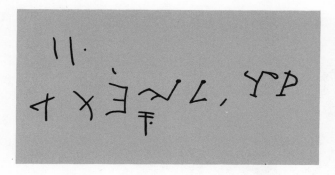

FIGURE 5-1   Bat Creek Inscription
The Bat Creek inscription records the presence of Jews in the southeastern United States in the early centuries A.D. The inscription is one of several records that we have. Coins dating from the same period have also been found. Moreover, a festival celebrated by the Yuchi Indians in that area resembles the Old Testament festival of the booths. The evidence stands as an example of a failure of diffusion. Metallurgy, wheels, and writing, all well known to the Jews, failed to spread to the Indian people who absorbed the Jewish settlers. (From Cyrus H. Gordon, The Book of the Descendants of Dr. Benjamin Lee and Dorothy Gordon, *Ventnor Publishers, 1972)*

Coins and an inscription all dating to the first and second century A.D., and clearly of Jewish origin, have been found over a period of 150 years—and blandly ignored. They clearly signal Mediterranean influences in America.

There are several striking things about the renewed interest in these finds. They do not stand alone, for a Roman pottery head found in Mexico dates to this same period. What became of these people? What effect, if any, did they have on America? Where is the European record? Did they too bring diseases? Is there a break in American cultural history about this time—a tremendous interruption suggesting a loss of life such as happened in Europe with the Black Death or in America after 1500 with the introduction of smallpox? Or a sudden surge in knowledge and skills such as contacts with the advanced civilization of the Mediterranean might have brought? If we can't answer these questions, how can we know what the roles of physical geography, physical man, and cultural man are? After a century of silence, the questions are being asked, but we haven't the answers yet.

By using botanical data we can make a tentative map of the time, place, and sequence of influences from the south. In passing it is clear that other influ-

ences were coming in from the north, but they were relatively minor in comparison with the current of ideas that flowed in from the south bringing temple mounds, religious cults, pottery forms, and agricultural plants. The most revealing data are found in the corn plant. We know that maize accidentally crossed with a wild-grass relative in Mexico, providing great variability, as the result of which the Indian was able to bring about the many varieties of corn that we now have. We know that in this process the tiny original corn ear was gradually increased in size and in the number of rows of kernels. We know that the original corn had a hard flinty kernel and that later other types of corn with soft kernels and mixed hard and soft starch in their kernels came into being. When we plot the distribution of these traits on a map of the eastern United States, we find that the early types of corn are on the periphery of the area, and the closer we come to the Lower Mississippi area the more evidence there is of the later types of corn.

This kind of evidence can be expanded. When the original primitive corn crossed with the wild grass, certain genetic markers were transferred. In the nucleus of each living cell are chromosomes, threadlike bodies along which the genes that determine the heredity of the plant are arranged. Each plant has its own characteristic number of these chromosomes: those of the original corn were plain; those of the wild grass with which the corn was crossed had knobby lumps on the chromosomes. In the plant crossing, some of this knobbiness was transferred to the maize plant, and as time went on, more and more of this material was transferred to the maize plant. Knobbiness, then, is to some extent a measure of the degree of mixture and of time.

This knobbiness pattern works out interestingly on a map. The original corn began spreading northward before there was much, if any, of this mixture. Later, forms with a little mixture started north, and still later, increasingly mixed forms started the trip. Those that started earlier had more time to be adapted to increasingly short seasons and cold springs. If the frequency of knobs on the chromosomes of maize in the eastern United States is plotted on a map, distribution suggests that the influences from the south were entering somewhere in the Lower Mississippi area, spreading up the Mississippi valley, across the coastal plain, and moving northward. This pattern is very much like that made by the distribution of some of the archeological remains such as the great mounds and therefore is a further illustration of the importance of location in relation to the major center of cultural growth and development. For the pre-Columbian eastern United States, it was the Gulf region that was nearest the center of

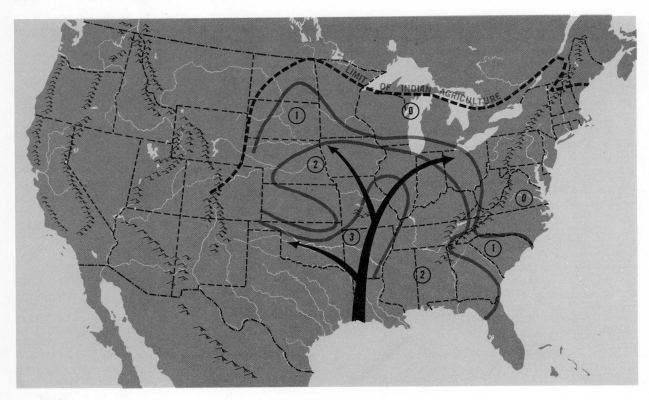

MAP 5-2   Knobbiness of Maize Chromosomes as Evidence of Cultural Movements
*The numbers refer to knobs on the sixth chromosome of maize, which can have either
no knobs or knobs in one, two, or three positions. The earliest condition is presumed to be
that of no knobs, with increasing knobbiness measuring additions of new genetic material
later in time.*

*The data, then, suggest that the first maize to enter the eastern United States was
knobless. Further, the direction of introduction is from the south, centering on the lower
Mississippi valley. This is also the location of the peak of native culture in the eastern United
States. The building of great mounds and many other evidences of town-dwelling ways of life
in pre-Columbian eastern United States fit this pattern.*

*The routes of movement of culture that the chromosomes of maize suggest are natural
ones. The flow seems to have been up the Mississippi River to the Ohio and the Missouri.
A further flow seems to have crossed the Great Plains via the Red River and via the Arkansas-
Canadian river system to the Southwest. It is notable that the later types of corn failed to
reach New England, the Middle Atlantic states, or southern Florida. While at first glance this
seems to be a simple case of agriculture following the path determined by the physical
geography, note the failure of this type of agriculture to spread northward through the piedmont
and coastal-plain provinces of the Middle Atlantic states, an environment richly suited to this
type of agriculture.*

growth and development, and the degree of influence from this center diminished to the northeast, leaving New England a relatively backward area.

The northeastern part of the United States was in one of the focal areas of pre-Columbian contacts with America; its history proves that contacts do not automatically stimulate cultural growth. There are Roman accounts of the North Atlantic, which they called the Cronian Sea, and of a series of large islands, one of which, Cronusland, seems to be Greenland. A cache of Roman coins found on Iceland bore dates around the second century A.D. Pythias sailing from Marseille had described the frozen northern sea and the phenomenon of midnight sun even earlier, about the end of the third century B.C. Since America can be seen from Greenland, and Greenland from Iceland, and since the early navigators were often storm driven far beyond their intended land falls, it seems most probable that America was touched by Mediterraneans at this time.

Although the record is exceedingly dim, the point is that the contact was too slight to be effective: no American Indian plants were brought home, and no traceable American ideas went back to Europe. There are suggestions of Mediterranean contacts with the Caribbean about this time, with ideas seemingly carried to America and at least some return flow, for an American plant, a pineapple, is portrayed at Pompeii, and a Roman style pottery head has been found in Mexican archeology. Since slight contact in the North Atlantic produced no such evidence of exchange, one would judge that a mere extensive contact must have occurred in the tropical Atlantic than in the North Atlantic. Proximity and currents of wind and water would lead one to expect this. Anyone sailing out through the Strait of Gibraltar enters the Atlantic in the zone of northeasterly trades and the North Atlantic drift, and if disabled or blown away, is delivered directly to America. In the North Atlantic the ocean currents and normal wind currents are from America toward Europe. However, there are infrequent periods of easterly gales. The Norse record these as taking them to America even against their will. But these are rare; the trade winds are regular. One should lead to more frequent contact with America than the other.

The next contact we know about is with the Norse, whose sagas tell of the accidental, storm-driven discovery of America. They give useful sailing directions for reaching America and tell of repeated voyages to the mainland, including deliberate sailing to and re-

FIGURE 5-2 A Discovery That Failed
The eight runes shown here are from Heavener, Oklahoma. The figures and the letters are the keys to an elaborate cryptogram that conceals the date A.D. 11 November 1012. Gloria Farley has long championed the authenticity of the Heavener inscriptions, and Mongé and Landsverk have proved their antiquity. The significance of this inscription is that the Norse reached far west, and other inscriptions show that the Norse were still in the United States late in the fourteenth century. Despite this lengthy contact, the wheel, domestic animals, and metallurgy were not learned by the Indians. (Courtesy of Gloria Farley)

occupation of houses built on earlier trips. The exact location of some of these sites may never be known, but evidence of Norse occupation now being found suggests that settlement was more extensive than the scanty saga accounts had led us to believe. Runic inscriptions are found all over the northeastern United States and as far west as Oklahoma. From the cryptographic work of Alf Mongé it has been found that many of them contain hidden names and dates. Many of them have been decoded, and they reveal that parties of Norse with priests traveled widely in America and stayed in the vicinity of such unexpected places as Heavener, Oklahoma, for up to 10 years. The great lack is further evidence in America and Europe.

The Norse settlements on Greenland flourished from A.D. 1000 to nearly 1500, and it is as likely that they were destroyed by European pirates as by climatic change, though the theme of climatic change has been favored. Further, the Norse casually mention that the Irish had settlements on the American coast and that one of their men who disappeared became a leader of one of the Irish settlements and was baptized. The record, then, is one of dimly seen contacts with northeast America beginning well before the time of Christ. A quickening of these contacts is documented about A.D. 1000, and this contact was never broken, for the Norse settlements continued until the time of Columbus. American Indians, especially Eskimos, arrived intermittently on the coast of Europe. The arrival of one birchbark canoe with living occupants is recorded with the comment that this was the type of boat used on the coast of *bacalaos,* the codfish coast, meaning the northeast coast of America. Europeans must have been familiar with the northeast coast of America to recognize a birchbark canoe.

There are several important points: the inability of the north Europeans to "discover" America even though they had known of this land for millennia, but most especially the failure of proximity and touch-and-go contact to transmit knowledge and stimulate cultural growth. Sparse contact of the type hinted at in the Roman and Greek accounts seemingly had no results. The Norse contacts had startlingly little. No European plants were carried to America and, even more notable, the idea of domestic animals did not spread. This is especially interesting since we know that the Norse brought their cattle with them; therefore the Indians saw domestic animals in use. Moreover, the Norse did not take back maize, beans, squash, pumpkins, or even tobacco. One of the few accounts of learning describes a captive European teaching the Indians to make a fish net. Almost the only known biological transfer was the European sea snail (*Littorina littorea*), which apparently hitchhiked to America on a Norse ship and became established on the American shore around A.D. 1000.

This was completely reversed in the period of European settlement. The area of greatest proximity to the center of cultural and economic growth and development was along the northeastern seaboard. American cultural growth and development began here with the first colonies and spread almost concentrically for the next 300 years. Irregularities in this early period reflect a mixture of physical and cultural forces. Colonization followed lowland water routes for the first 250 years, but the Spanish to the south, the French to the north, and the Iroquois Confederation in New York were even more influential forces than landforms.

The general conclusion is that the importance of the physical environment was very modest. New England's dominance in the realm of American thought and culture has been greatly overstated. The Middle colonies and the South made greater contributions than it has been fashionable to acknowledge. For instance, Maryland's contribution to the concepts of religious tolerance and freedom probably exceeds any similar contribution by any other colony. The South still leads in the number of presidents it has supplied, despite its smaller population and the handicap due to the Civil War and its aftermath.

The dependence of colonial America on Europe eventually came to an end. Settlements expanded across the continent, the center of population shifting steadily southwestward, until it is now beyond Chicago. Centers of industry, education, and political power have

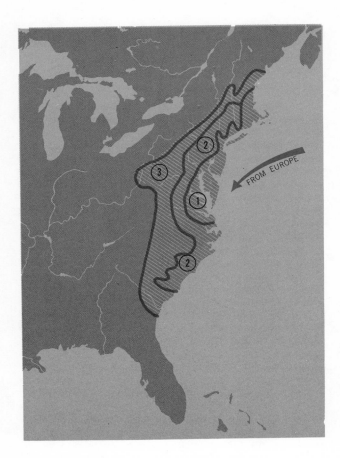

MAP 5-3   Colonial Settlement in North America
In the early decades European settlement focused on the north and central Atlantic coast of the United States. Area (1) was being settled by 1700. Area (2) represents the reach of settlement by 1750. Area (3) marks the expansion to 1775. Proximity clearly played a large role. It was not the only factor, however. The presence of Spain in Florida and France in Canada also played a part, as did physical geography. Lord Baltimore's colonial effort in Newfoundland was shifted to Maryland because the Newfoundland climate was found to be too severe for English tastes.

shifted with it, though with a lag determined by such cultural factors as resistance to change and the cost of change. Moving or abandoning a factory is a costly business; moving a university is also, but the biggest resistance is probably cultural. Harvard is still acknowledged as one of our greatest universities. It is in New

England, the area formerly dominant in America that is now losing its relative importance. But who would think of moving Harvard from its "yard"?

The momentum of the early start of the Northeast has kept it going for a full century beyond its time of vitality. While much of northernmost New England has now reverted to wilderness, southern New England is still maintaining momentum from its past. In this progression from a backward area in pre-Columbian times to an area of great growth and development during the eighteenth and nineteenth centuries to a modern period of decline, the physical environment of New England has remained constant. It is the same land, with the same climate, occupying the same place in the physical world. But the cultural world has pivoted around and around, and this change has been decisive. New England developed *in spite of* its physical environment; it continues *in spite of* both detrimental locational and physical factors.

American greatness: environment or culture?

This development and decline of New England leads us to the next environmental consideration: How much did the physical environment of the United States have to do with the greatness achieved? It is commonly held that the United States became great because of its extremely fortunate physical geography. Since the West only belatedly became a part of the United States and is only now being fully developed, that greatness must have been founded in the East.

The United States started with vast forests, enormous tracts of good land, the world's greatest coal resources, excellent fisheries, abundant iron ore, and immense petroleum reserves. Notice however that these resources come to have value at different times: fish and land in early times, timber later, coal still later, and petroleum very late indeed in the history of the nation. Even the land that the infant nation first developed was poor, for the rocky soils of New England and the sandy soils of the Atlantic coastal plain are not America's best. It has also been alleged that New England has one of the most stimulating climates in the world. Statements reciting these advantages often suggest that any country with the same resources as the United States could have achieved similar success. The land would have produced a great nation, runs the argument, under anyone's management. Therefore let us review what did happen.

Settlement began at various points along the East coast of America early in the seventeenth century. This is an oddity. America had been "discovered" in 1492. Why did the British wait a century before beginning settlement? We do know that they did not particularly want North America. In Hakluyt's *Voyages*, published in the sixteenth century, there is a discussion of where colonies should be sought. Not in latitudes similar to Great Britain's, ran the argument. Instead, tropical colonies that would produce what England could not and that England could supply with what they could not well produce were desired. What use is it to have colonies that produce wool and wheat? England produces those items. How can one build a trade that way? However, the tropical areas were largely in the hands of such powerful nations as Spain, Portugal, and the Netherlands. The British had to take what they could get, and there was a section of the North American coast, handily adjacent to Britain, that was open to settlement.

Settlement took several directions. The Virginia colony was one of gentlemen adventurers. They might have been splendidly successful if there had been an empire to conquer; instead they found themselves in a fever-ridden swamp among village-dwelling Indians. There was only a moderate opportunity for trade, and no treasure of gold to be had for the looting as in Mexico, nor even any rich mines to be discovered as in California. The colony faced hardship, starvation, dreadful loss of life from disease, and Indian attacks. It was a most difficult situation. Reading the accounts, one wonders why the colonists and their British backers persevered. New England was no better. There may have been less fever there, but the winters were much harder.

In both colonies the wheat did poorly, for it was not adapted to such a climate. After all, the climate of the eastern United States is not at all like that of Great Britain, as the colonists always remarked. They spoke of the violence of the seasons. Winter was hard and in New England, very long. Summer was hot and humid and marked by violent thunder storms. Plants used to the gentle rainy summers of Great Britain were far from at home in this land of sudden, violent weather changes, with a supertropical summer to the south and a subarctic winter to the north.

An example of the geographic ignorance of the times is illustrated by Lord Baltimore's attempts to found a colony. He first sought and obtained a grant for a colony in Newfoundland. Since this land lay in the same latitude as England, he expected it to have the climate of England. A very well-planned party was sent out in the early summer to have time to erect buildings before winter. By winter they were snugly housed, but that season proved so long and so severe that the colony was abandoned. Thereafter a petition was filed, and finally granted, for a colony near the Virginia colony. This new settlement was the beginning of the area that was to become the state of Maryland.

The New England colonists worked hard at making a living out of their land. Fisheries helped the food supply, but the bulk of the people worked the land, clearing it of woods and rocks. A trip through New England today still reveals the fruit of these labors, the stone walls that were the result of laboriously picking up the stones in the fields, hauling them to the field borders, and stacking them. It was stupendously hard work. The land was far from "free." It had to be won from the forest, defended from the Indians, and freed from the burden of rock that the glaciers had left. Even then the growing season was short, and the risks from early and late frosts were very high. Yet New England was at one time largely cleared and farmed. Today it has largely gone back to forest. Even after the immense effort at improving the land, it was simply marginal land climatically, and soon left behind by the developing centers of population in America. The greatness of New England was *in spite of*, rather than *because of*, the physical environment. Perhaps it was the challenge of a difficult environment that shaped the New England character, but more likely their stern Puritan consciences had even more to do with it.

The other colonies had little better beginnings. In the Middle colonies and down the coastal plain the early settlements were on the sandy coastal soils. We think so little of many of these poor soils today that much of this coastal plain is abandoned and has reverted to timber. Sands are poor for agriculture. They had the advantage for the early colonists that they could be cultivated with light and simple tools, but labor shortage and abundant land led the Middle colonies into bad agricultural habits. The poorness of the soil and the early dependence on tobacco made a system of forest slash-and-burn, plus the early abandonment of the land, seem like an economic way of land use. The land was never cleared, stumps were left standing, and cultivation was by hoe. This labor was accomplished at first by indentured servants, and soon thereafter by slaves.

Visitors from Europe were shocked at the miserable, weedy state of the agriculture. People moved so frequently that it hardly paid to build a good house. In one of the Middle colonies a law had to be passed against burning down houses just to get the nails out of them to use at the next plantation. Tobacco did not prove extremely profitable. Although some people made money on it in the early days, as more land was put into production, the price dropped. After a relatively brief period of good returns, therefore, a long period of very small profits followed. Meanwhile, the poor soils were further impoverished by the tobacco and the accompanying poor agricultural practices. Nor

in all of this were the forests of any great value. Trees were weeds; they had to be felled and burned to clear the land. Timber was not profitable enough to ship to so distant a market as Europe, except for specialized uses like timbers for masts, but there were not even many of these—the millennia-long Indian custom of annually burning the forest had left few tall, straight trees. Mast trees were found largely in swampy, fire-protected areas. Man, again, had modified his environment.

New England found a way out of some of this difficulty when the industrial revolution began. Close contacts with Great Britain paid off when the know-how of making spinning machinery was brought to New England. Here there were a number of favorable factors. The natural increase of the population had created a labor force, and the land was so poor that it was easy to get some of the people to come to work in the factories. The small streams coming down out of the glaciated uplands often went over falls and rapids and could readily be harnessed as water power to support mills. Mills at this stage were small, and streams now not considered worth harnessing to light a house ran a mill. The South was producing cotton, and New England spun and wove it. The young nation had little industry. New England plunged into supplying the need.

As always, the physical environment was playing its role as the backdrop for the culturally determined action. Even as stated here, the physical environment's role is overplayed. The initial utilization of machinery in New England was a historical accident. It could as well, or better, have begun in the Middle Atlantic or southern states that had as good or better water power opportunities and better raw material supplies.

(Opposite, top) **New England Woodlands.** *Today large parts of New England have reverted to forest, and much of the remaining open land is used for pasture. This is a land of thin and rocky soils and short summers. Although land of this kind is cultivated in other areas of the country, much of New England remains in its natural state.* (Maine Department of Commerce and Industry)
(Opposite, lower left) **A New England Church.** *The simple architecture, the tall spire, and white painted wood suggest the Pilgrim heritage of this area.* (Maine Department of Economic Development)
(Opposite, lower right) **New England Winters.** *Long and snowy winters in such areas as Maine shown here were a time of staying indoors in the past. Today paved roads and snow plows give freedom of movement. Sports such as skiing and snowmobiling have changed the traditional image of the stillness of winter.* (Maine Department of Commerce and Industry)

It is one of the curiosities of history that Samuel Slater who brought from England the plans for mechanizing the production of cloth considered the cotton-producing South the logical place to introduce the machines. However, the southerners were doing very well and had no interest in the spinning machines. They were planters, not manufacturers. Middle Atlantic colonies also rejected the newfangled ideas, and as a last resource the ideas were sold in New England. While the manufacture remained there, the tendency was to explain it as a natural thing: New England had the water power, the stimulating climate, and so forth. Today the cotton manufacturing business has largely moved to the South, without the benefit of any changes in climate or in the distribution of the water resources of the nation.

After the Revolutionary War the population expanded over the Appalachians, posing transportation problems. The interior rivers flowed to the Mississippi and so led to the Gulf, but Americans did not control this vast inland area. The Ohio River ran to the Mississippi, and the Mississippi was alternately in French or Spanish possession. The British and French held some of the adjacent land, and the Spanish held the rest. There were no roads, and in the days of wagon transportation, only materials of immense value in proportion to their bulk could stand the cost of overland transportation. This situation was the basis of the Whiskey Rebellion in the 1790s. The only way people of the interior of the country could get their bulky, low-value corn to a market was to concentrate it to high-value whiskey. When this was taxed, they rebelled.

Strenuous efforts were made to overcome the problem. Canals were dug, or at least projected. Attempts were made to put canals across the Appalachians, and a canal linking the Chesapeake and Delaware bays was discussed in George Washington's day. The first railroad built in America was laid across this narrow neck. But all such efforts are costly. A better endowed land would have supplied the necessary waterways at much less cost than the Americans had to bear.

The early colonists had no major iron ore resources. The bog irons of Maryland so important in colonial times are so poor that no one touches them today. Coal was no resource until later, and oil was not even dreamed of. Yet the bumptious colonists freed themselves at the end of the eighteenth century and 150 years later were the world's great power. They had, of course, fought Great Britain in a second war, fought Mexico and annexed the West, bought the whole Mississippi Valley from France and Alaska from Russia. But these are evidences of the strength that had already been engendered in the colonial East, which was no great natural gift.

The land that they eventually came to control, extending from sea to sea, was none too good. A full third of it is dry land. Only 40 percent of it is considered agriculturally useful land. As we saw in the section on the Mediterranean lands, it was the driving of the railroads west and the vigorous capitalizing on the advantages of being linked to the East that made California what it is, instead of another Chile. America *is* what Americans have *made* it.

The role of freedom

The Europeans early in colonial times remarked on the air of freedom that permeated America. Fewer men seemed to accept the role of someone else's inferior: everyone believed he could become whatever he wanted. As one European traveler commented with astonishment, anybody seemed to feel free to have a coat of arms painted on his carriage.

The drive was to own land and produce. Land ownership for various reasons never went into the hands of a few. In New England the settlement was from the beginning one of small holdings. In the other colonies this was true, though to a lesser extent. In Maryland large grants were made. However, large estates often were accumulated from the buying and selling of various properties, and at the end of each generation inheritance tended to break the estate up into a number of smaller estates. After 1862 the distribution of the western land by the government in 160-acre lots assured a very broad ownership of the land. It created the possibility that anyone with the strength and the will to go out and make a start in life as an independent, self-employed worker could do so. This set free what is probably an unprecedented amount of human energy. It also tended to distribute the population over the land, for the free land was always on the frontier.

In most of the world for most of civilized time, conditions have been the opposite. The structure of human society after the Neolithic revolution has been characterized by a tiny percentage of the people in a privileged position, with the rest of the people in a menial position. One of the miracles of America is that we broke away from this pattern. In part, the great disparity between nobles and commoners in the past was a matter of productivity. It took many men producing only slightly more than they needed for bare existence to support one man in a way of life that did not produce food, clothing, or tools. It is difficult to be a learned man if one's time is taken up with growing potatoes and hunting game for the table. Yet until the productivity of the working man went up, a hundred men had to toil so that one man could have leisure.

But the American accomplishment was based even more on the idea of individual freedom. Other nations by this time, in fact for a long time earlier, had had the opportunity to strike out on such a new line. They were sufficiently productive. However, the inertia of a going social system inherited from the past prevented change. The abrupt transplantation of groups of Englishmen into America placed them for a time in considerable isolation from each other. The various colonies worked out somewhat different ways of government, but each based their ideas on the liberties guaranteed by the Magna Carta and such ideas as popular representation and self-determination. The businesslike New Englanders drew up a contract, the Mayflower Compact, explicitly stating the purposes of their group and the rights and duties of each member. This was America's first written constitution.

The Marylanders, escaping from religious persecution, instituted religious freedom. This has been curiously misrepresented in the American history texts of the past. The religious dissenters who fled to New England fiercely demanded that in their colony everyone believe as they did, yet New England is often spoken of as if religious toleration began there. Among the early colonies religious toleration was in fact the innovation of the Marylanders. But the point here is that each colony was experimenting in some isolation with a set of ideas that they had brought from Europe. Among the most important of these ideas were the rights of individuals, the right to have a say in government, the right to own land or to go into business, and the right to limit what the government could do.

In most of history, a man could be hard working and intelligent as a genius, but if born a share tenant or a serf or a peasant, his chances were slim of ever having a chance to be more than an ordinary farmer. Education was for the very few at the very top, and position, honor, and opportunity were similarly limited. If a society elects to allow only the ability that appears in 10 percent of its population to develop, it will get a much smaller return than a society that attempts to give all its people a chance. It is impossible, of course, to give everyone exactly equal opportunities, but one of the things that has made America great has been the idea, firmly fixed very early, that the way was open to all—and to a remarkable degree it was.

Much of this has also been said many times. We therefore have two opposing themes: one, that the physical geography of America made it great; the other, that it was a set of ideas that made America great. Some people have put them together: a set of favorable ideas placed in the most favorable imaginable environment made America great. However, the physical environment does not deserve even this much credit. It seems probable that just such a development as America has had could have occurred in any one of a number of environments. Conversely, a settlement of this country by some other mixture of people would surely have led to a different kind of development of the country, and most probably would not have led to the enormous productivity that underlies our high standard of living, vast educational system, and relatively leisurely life.

We can explore this idea, much as we did the idea of the growth of culturally different "worlds," beginning with a uniform world. In this case we will imagine that the British settled in Brazil instead of in the United States. To complete the story, we will also imagine that the Portuguese settled in the United States instead of in Brazil.

What Happened in Brazil

The Portuguese discovered and exploited Brazil. Unlike the Spanish, they did not find great native empires laden with gold and silver. They found, instead, a tropical country occupied by native people with a simple way of life. They found an immense river up which they could sail three-quarters of the way across the continent. Along the Atlantic coast low ranges faced the sea, with only a narrow coastal plain, and many small streams descended these hills to the sea. To the south the climate was temperate, not unlike the American South. Along the western highlands, the climate was temperate because of altitude. The vast interior was principally plain and gently rolling land. Almost the whole of this land was well watered, but like the United States, it was an undeveloped resource to which the energetic Portuguese had to apply themselves to make a productive land.

The Portuguese had long been producing sugar on tropical islands under large-scale production methods using Negro slave labor. They used what we now call the plantation system. A tropical setting such as Brazil gave them a chance to expand this type of industry, and they transferred it to America. The Portuguese were the managers and the armed force. At first they used Indian labor, but as the local supply diminished, they sent armies deeper and deeper into the interior for more. When rounding up Indians for a labor force became more expensive than importing Africans, they did the latter.

Sugar was a booming crop, commanding a staggering world price. Its value at this time can be measured in part by remembering that when settling one of her squabbles with England, France elected to give up all of Canada rather than surrender one tiny sugar-producing island in the Caribbean.

Exploration inland soon uncovered gold, and during the eighteenth century Brazil produced 44 percent of all the gold in the world. Thereafter came the rubber boom, then coffee. Early Brazilian history has aptly been described as one bonanza after another.

Why has Brazil failed to develop, fill its land with people, produce sufficiently to give its people a high standard of living? It had, and has, everything. It is a larger territorial unit than the United States, without Alaska, and it is one of the best watered lands of the world.

Arable land and production in Brazil and in the United States are as follows:

Brazil
80 percent is potentially productive.
2 percent is used, therefore food must be imported.
90 percent of the inhabitants live on or near the Atlantic coast.
United States
60 percent is waste land.
40 percent is potentially productive—we feed ourselves and
    export food.
The population is well distributed.
World
25 percent is arable (average).

Brazil uses only 2 percent of its land and does not even feed itself, yet 80 percent of its land is potentially productive as compared with 40 percent of the United States and a world average of 25 percent. Brazil has abundant iron ore, a near monopoly on industrial diamonds, known, but little exploited ferroalloy minerals, some coal, and immense hydroelectric potential. Although high-grade petroleum was found in the upper Amazon years ago (and the Andean piedmont is still a rich prospect), Brazil imports petroleum products at great cost.

Brazil by all measures is potentially one of the richest nations on earth. But under a boom-and-bust, plantation-slave agriculture, with broad ownership of the land concentrated in the hands of a few, it has accomplished the minimum with these goods. If the United States might be said to have succeeded in spite of its physical resources, Brazil might be said to have failed in spite of its physical riches. Perhaps it would be fairer to say that too much easy money has delayed the development of Brazil.

These statements so often provoke outright disbelief that it may be well to quote as an authority one of the American geographers who has made a lifetime study of Brazil, and who, in turn, quotes the Brazilians themselves. The Portuguese

. . . were attracted less by the prospects of earning a living by persistent toil than by the opportunities for speculative profit. As one Brazilian writer puts it, the ideal was "to collect the fruit without planting the tree." Whereas some of the peoples of America have been led by force of circumstances to be content with less spectacular returns from more intensive forms of economy, the Brazilians, with their huge land area, and their small numbers, are still seeking new ways for the speculative exploitation of the treasures stored up in nature.[1]

Could it have been otherwise? It will be remembered from the earlier discussion of Brazil that changes are now under way.

British Brazil

Suppose the British had colonized Brazil. Not having a tradition of sugar plantations, let us suppose they began a settlement of the land just as they did in North America. From study of the British colonial efforts in the Caribbean at this time, we can conclude that the climate would not have deterred them. Given the politico-economic climate of the American colonies and an equally good human material, they should have fared well. They could have produced sugar on small farms under a freeman's husbandry, as indeed they did successfully in the Caribbean whenever they had the opportunity. Tobacco, rice, indigo, cotton, and maize are all at home in Brazil. There seems to be no reason that a healthy growing British population should not have developed along the Atlantic seaboard of Brazil at least as easily and probably more prosperously than it did in America.

A British assessment of Brazil would probably have emphasized the cooler south. As the population expanded, it could have spread into the vast, rich interior. Given the rapid send-off from sugar, capital would have been available to develop the machine textile industry which their home contacts would have brought to them. The small streams coming down the flanks of the coastal mountains would be far better than the streams in New England. There would be no winter freeze, and the rainfall would be large in amount and well distributed through the year. Cotton growing could

[1] From *Latin America* by Preston E. James, copyright © 1942 by Western Publishing Company, Inc., reprinted by permission of The Bobbs-Merrill Company, Inc.

well have been linked with spinning and weaving, just as in the United States. This could have been the foundation for a broad industrial and commercial development.

As the population spread inland, gold, diamonds, iron, coal, and oil would all have been found. There were no deserts and mountains for the railroads to cross, and the river system would be more extensive and useful than that in the United States. There were temperate highlands, a seasonal south, a wet tropical north, and no dry and nearly useless desert dividing one-third of the land. There would be no reason why, under an American constitution and with the growth of American free enterprise, population should not have spread evenly over the land, and agriculture, industry, and commerce should not have flourished. There is nothing that has been accomplished in the United States that the same people could not have accomplished more easily in Brazil, and there are numerous opportunities in tropical agriculture in Brazil that do not exist in the United States. Indeed, a sobering thought is this: Life might have become too easy, wealth too easily accumulated by a few, and a pattern established resembling what actually did develop in Brazil. The theme would not be changed greatly. Rich resources do not necessarily make great nations, but ideas may.

A Portuguese United States

What might the Portuguese have done with the United States? The Portuguese assessment of this country would probably start with coastal exploratory trips. They would note the near-tropical South and Gulf coast. To tropical plantation experts these low wet coasts would look attractive. To the north they would note that the low coastal plain narrowed, winters increased in intensity, and north of Long Island, soils were thin, sandy, and increasingly stony.

In the Gulf coast they would discover the mouth of a great river. Occupation of its mouth would give control of the interior. Similarly, occupation of Florida's east coast would provide control of the sailing route to the homeland. Given their background of tropical agriculture, they could be expected to place their major settlements from Georgia, around Florida, into the Gulf coastal plain. They would establish plantation agriculture, rapidly use up the Indian labor supply, and begin importing African slaves.

Their Indian hunting would take them into the interior. Up the Mississippi they would find rich plains. To the north would be cold lands of little attraction. New England would have been left as a fur-producing

wilderness. To the west the increasing aridity and the vast mountains would make that area seem useless. In all probability they would, like the Spanish, consider the Pacific coast of the United States too far and too poor even to occupy, except possibly as a mission frontier. Hence, like the Spanish, they would probably find no gold. Nor would they have any world monopoly in anything like rubber or coffee. Their plantations would lack even so rich a product as sugar, for the United States is distinctly marginal for sugar cane. They probably would have specialized in suitable crops such as indigo. Nor would the Portuguese United States have access to the new ideas of the industrial revolution. These were not developing in the Iberian world.

So the Portuguese United States would logically be an agricultural nation, perhaps far poorer than modern Brazil. Twentieth-century viewers might hypothetically assess the situation as follows:

What could you expect? The land is one-third arid. There *may* be some resources there. We hear of possible iron deposits up on their northern border. But of course that is too far away from the center of things to be useful, and besides it is impossible to work in that area during half the year because of the intense cold. Geologists tell us that there should be oil, and one test well indicates that there are really high-grade resources there. But, unfortunately, they have never been developed. There is some coal. But what can you do with a country that has a partly frozen north, whose good land in the West is separated by mountains and deserts from the good land in the East? They only have one good navigable river and it is full of shoals. It is just a mediocre lot of land, poorly arranged, and *that* is why the Portuguese United States is a second-rate land!

And what might the explanation of British Brazil be?

Why, how could such a land be *other* than populous and productive? Look at the magnificent, gently rolling terrain! Consider the well-distributed rainfall. No other large land even approaches it in potential arable land. Then consider the great inpourings of capital from sugar, gold, rubber, coffee, and industry. Of course, they have been ingenious in making their meager coal match their rich iron. But with a near monopoly on industrial diamonds, with quantities of the necessary ferroalloy minerals, with vast forest, immense hydroelectric power, and the world's greatest, most navigable river, how could they fail to be a great nation? What's that? Something about freedom, rights of individuals, capitalistic free enterprise? Don't be ridiculous! Those British Brazilians simply fell into a vast natural resource. Any system would have succeeded!

There is a danger that this presentation be twisted into an attack on the Portuguese. However, the Portuguese have a relatively admirable record of peace and stability and a great record of exploration and colonization, indicating organizing and creative ability of high order extending far back in time. Consider that their homeland is only about the size of Tennessee. They have proved themselves able people, but by nineteenth-century American standards they have been handicapped by a very restrictive sociogovernmental system. This system is not strictly Portuguese; it is quite simply pre-industrial European. It has survived in Portugal, Spain, Italy, and eastern Europe. The slow tide of change is just reaching some of these lands. Note, for example, the recent developments in Italy discussed earlier (Chapter 4). Recall also that Brazil is currently forging ahead and its Portuguese inheritance of complete racial tolerance has kept it free from the terrible race problems of South Africa and difficult ones of the United States.

Comparative forecasts

Today Brazil is in a ferment that includes railroad building, hydroelectric development, industrial development, expansion into the interior, and a general dynamic drive that reminds one of the United States in the nineteenth century. Should Brazil finally turn its attention to "planting the trees" and releasing human energies on a broad front, rather than drawing primarily from a few families, in the twenty-first century it could be the home of a numerous, prosperous, educated people with a standard of living at least equal to that of the United States. Brazil, like much of Latin America, is plagued by inflation and unstable politics; this discourages foreign investment and hampers growth and development. The future depends on the solution of these problems, not on limitations in the physical environment. Brazil can be populous or relatively empty, rich or poor, with education and opportunity widely dispersed or restricted to a few, a first-rank or a third-class world power, all depending on the social solutions that are adopted. The geographical potential is large, as it has always been.

The forecast for the United States is only slightly different. Since our physical base is no better than Brazil's—it may actually be argued that it is poorer—the enormous discrepancy between what we have done in our physical environment, compared with what the Brazilians have done, must be due to some cause other than physical environment. In the United States it clearly was the release of human energy from a very wide slice of the total population which, coupled with widespread settlement of the land, a broadly based educational system, and a rapid population growth, led to a steady increase in productivity in both material and immaterial things and thereby to the world's highest standard of living.

The gist of this is that the United States is not the only place where people could have achieved great things and that we should take care not to change the fundamentals, for if this imaginative comparison is correct, the American physical environment will support us in the style to which we have become accustomed only as long as we maintain the cultural outlook on man and land that made the American development possible. Should we adopt some set of institutions that would suppress the burst of individualistic achievement that has marked America, it is conceivable that we could lose our position of world leadership in productivity in both material and cultural fields. A shift in cultural values such that Brazil would come to surpass the United States in standards of living and educational and cultural achievements is not unimaginable, and all of this could occur without any change in physical environment.

Discussions of this sort are most difficult to hedge about with all possible disclaimers. It is not contended that everyone in the world should copy the people of the United States, nor that being a great world power is the most desirable aim in the world, nor even that the materially rich life of the United States is the ultimate goal in life. Nor is any slight on the Portuguese or the Brazilians or the Latin Americans in general intended. Insistence on a proper distinction between what is physically-environmentally determined and what is socially determined is the aim of these comparisons.

On the other hand, there is no intention to underplay the American achievement or to portray the United States as a barren haven of material goals. While it is true that our productivity has made possible better houses with more indoor plumbing and television sets than in any other country, the same productivity is the basis for more doctors and public health measures, which give our people about double a primitive man's span of life—a period of years that can be spent in obtaining and profiting from a lengthy education made possible by a productive society, whose educational products in turn make the society still more productive. Further, the American society is rich in artists, musicians, writers, and spiritual leaders, and these also are the result of a productive society that both supports and produces such people.

The productivity that has made all this possible is not due to our land's being the world's best but to our cultural heritage, which led to the world's broadest-based release of human energies. The product of this release of energy has pervaded all fields of human activity—spiritual, cultural, and economic. The opportunities for others to do likewise is virtually unlimited, especially as technology advances. This is the point of the Brazilian-American contrast. A corollary is that America, should it change its system, could become another example of a nation that rose to greatness and then declined.

The Eastern United States Today: North versus South

If now we return to the eastern United States to look at it in more detail, we find considerable difference between the North and the South. The eastern, humid, originally forested United States is not physically uniform. The soils to the north are young and relatively rich. In some areas they are richer than others because the glaciers brought to them ground-up limestone rock. In Iowa, it is possible to plot the limits of particular glaciations by the fertility of the soils and by the need for drainage. The greater fertility is found in the younger glacial soils, and the greatest drainage is needed on the same areas because of the irregularity of the land that is a heritage from these same glaciers. In other areas, as in New England, the irregular topography is not accompanied by so large a blessing of lime-rich soil, and the product is poorer. These northern areas are also short seasoned. This is a handicap for crops because of the risk of early and late frosts. It is a handicap for forestry, for the amount of tree growth is related in part to the length of the growing season. It takes nearly a century to produce a saw log, one capable of yielding boards, in our Northeast.

The South has no broad area of recently renewed soil fertility; many of its soils are old and leached. But the growing season is long and the winters mild: crops that cannot be grown in New England flourish in the South. Trees that need a century to reach saw-log size in New England need only 50 years in the South. Cattle that have to be carefully housed all winter in New England, with feed expensively harvested and stored, can live outdoors all year round in the South; and in the deep South they can graze on green pasture throughout the year. Carefully managed pasture in the South can carry one cow per acre per year. New England probably needs four times this amount, and in the arid West it may take 40 times as much land to carry one cow for a year. The physical environment is really there, and given a particular choice of way of life, it has its real influences.

Pasture management in our South is a particularly good example of what man can do. In 1860 in beef production the South averaged about 6 pounds of live weight gain per acre per year. This was on the poor native grasses with the poor cattle of the day. In 1900 carpet grass was introduced and productivity increased fivefold. In 1960 lespedeza and common Bermuda grass were introduced and yields doubled again. Since then, Bermuda grass has been improved and fertilizer used more efficiently, and there is talk now of obtaining an unbelievable 1000 pounds of live weight gain per acre.

That the role of the physical environment can change with changing cultural outlooks, however, is well illustrated by the use made of the hill lands in the humid eastern United States in the past and in the present. In the pioneer stage the hilly lands of the Appalachians were viewed as good land. In New England the hills were cleared of rock, and small homesteads covered the land. When men were aiming at a subsistence from the land and little more, when the land was worked largely by hand and even plows were little used, these steep lands were adequate. Small pieces of good land sufficed for crop lands, the surrounding rough lands supplied wood for fuel, timber, and were a source of game.

As farming shifted from subsistence to commercial and as mechanization crept into agricultural production, this situation changed completely. Now land is worked in large plots, and the large machinery requires large, flat expanses. The high yields and low costs achieved through heavy capital input for fertilizers and machinery sent the small hand-producer right out of business. We can produce rice by machine cultivation more cheaply than the Oriental labor can produce it. We can produce chickens and eggs under mass production methods at less cost and of higher quality than can the farm wife who counts her labor at zero. At the same time, the wants of the hill farmer have increased; he is no longer satisfied with mere subsistence. He wants a radio, television, and running water in the house, store-bought shoes and clothes, and a car or truck and the requisite gasoline. The little patch of land in the hills cannot supply these things. Hence, throughout the Appalachians and the Ozarks, the land that once supplied a subsistence living is being abandoned. Yet it is the way of life, not the land, that has changed.

The textile industry has moved out of New England to the South. Was it for reasons of climate? In the South, factories could be built more cheaply, and they needed less heating in the winter. On the other hand, they need air conditioning in the torrid southern summers—or at least many people are coming to think so. Closer examination of the forces leading to movement

MAP 5-4   Soils of the Eastern United States

These soils fall into four age groups depending upon their relationship to the glaciers and the glacially fluctuating sea levels. (1) The weak podzols are very young, for they have developed in the areas freed from an ice cover within the past 10,000 years. There has been little time for weathering processes to impoverish them. (2) The gray-brown podzolic soils are somewhat older but not yet greatly weathered. (3) The red and yellow soils are in the nonglaciated regions and are quite old, strongly weathered, and often poor. (4) Along the coast are other soils usually developed in relatively poor, sandy materials. Their age and degree of weathering are related to sea levels of past times. In general the higher soils are older and poorer. On the western margin the approach to the arid lands is marked by the appearance of the grassland soils: prairie and chernozem types. The red prairie soils are the older and more weathered of the grassland soils. The planosols are moderately old soils with strongly developed subsoils.

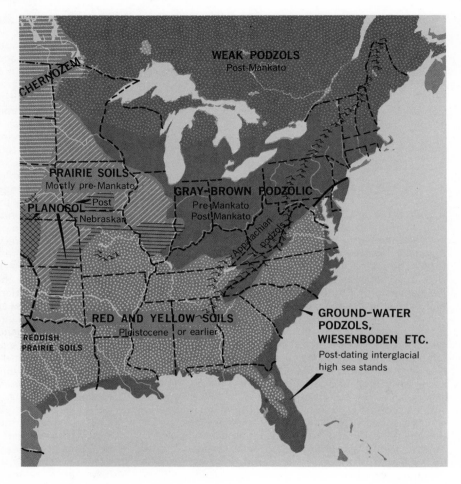

CHERNOZEM

WEAK PODZOLS
Post-Mankato

PRAIRIE SOILS
Mostly pre-Mankato

GRAY-BROWN PODZOLIC
Pre-Mankato
Post-Mankato

PLANOSOL — Post
Nebraskan

Appalachian podzols

RED AND YELLOW SOILS
Pleistocene or earlier

REDDISH PRAIRIE SOILS

GROUND-WATER PODZOLS, WIESENBODEN ETC.
Post-dating interglacial high sea stands

shows that much of it was due to the attraction of a large, stable, relatively low-cost labor force in the South as opposed to costly labor in New England; also tax incentives were offered businesses in the South while there were high taxes in the Northeast. The labor force in the South existed because of the shifts in agriculture that were being brought about. As the southern hill farmers are drawn into the commercial life of our day, they move down out of the hills into the factory towns. This ready-made labor supply is an enormous asset. It is the same kind of asset that built the New England textile towns 100 years before. The abandonment of the New England hill farms started earlier. Today it has progressed so far that many areas of New England no longer have a large pool of labor. With industry moving out, unemployment is created. Some of the labor

also moves. The remaining labor pool is small, and often older, for it is the young people who move. The result is an unattractive labor situation. New England, described 50 years ago as the epitome of all that was good in America, is distinctly losing its place. But it will not sink to the level that it might have remained in if the accidents of discovery and settlement had not made it an important area in early America. There is too much invested there now. Boston is not likely to wither away. Yet, had America been settled from the West or from the mouth of the Mississippi, New England would probably never have emerged from its forests.

This is not to say that New England is now a land without a future. Its harsh winters are less of a handicap than they once were. Central heating, paved

(Top) **Forestry in the South.** *Vast areas of the southeastern United States have been turned into forests, displacing the subsistence farmer and the sharecropper from the land. This has helped to stop erosion, regulate stream flow, and end a poverty level way of life, but it has also depopulated large areas.* (Georgia Pacific Corporation)

(Lower left) **Timber Production in the South.** *Note how much can be accomplished simply by good forest management. The trees are of the same age, but the one on the right has been grown with man's help and the other under haphazard "natural" conditions.* (Georgia Pacific Corporation)

(Lower right) **Improving Trees through Selection.** *This amounts to domestication, or at least the first step in that direction. The seedling at the left is better than average, but the one on the right is obviously growing four times as fast. Since domestication of plants goes back 10,000 years, why have we only now begun to domesticate forest trees?* (Georgia Pacific Corporation)

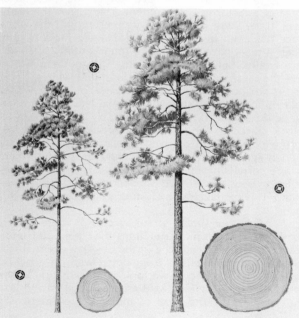

roads, and the automobile have changed our reaction to the winter very greatly. Indeed, with the growth of interest in winter sports, the area has gained in attraction. Even its loss of population has increased its value as a vacation area. For those who long to escape the social pressures and confusions, and the crowded and fume-laden population centers that grow bigger and more artificial year by year, areas such as the cool green-in-summer, snow-mantled-in-winter hills of New England are a place of blessed relief. But again, the values of the natural landscape are being assessed not only in terms of themselves, but in terms of the man-made setting to which they find themselves now peripherally attached, instead of centrally located as they were in the past.

Meanwhile the South has belatedly begun to forge ahead. One of the best things that has happened to the South has been the boll weevil. This highly destructive pest entered the South · from Mexico in 1888. By 1920 it had spread throughout the cotton-growing area of the South, and the cotton industry began moving out of the South to the arid West. This was the straw that broke the back of the cotton monocrop system in the South. The South was forced to do what agriculturists had long been urging: to diversify, to grow crops less destructive to the soil, less corroding to the social structure, and more productive economically. This change, combined with the growing industrialization of the South, has started a revolution. In a very real sense, the South in the mid-twentieth century has finally entered the industrial revolution.

It is a far-reaching revolution. Agriculturally, the land is highly productive because of its long growing season and its extensive areas of plains suited to mechanized agriculture. It is now one of the most important cattle regions, for the reasons mentioned earlier. Its forestry can be put on a sustained yield basis, for the rate of growth is high enough to allow harvesting before so great an elapse of time that any reasonable counting of interest charges on investment has eaten up all the possible gains. The labor supply, freed by the mechanization of agriculture and the collapse of the subsistence farming way of life, has become an asset. It is too often unskilled and uneducated labor, but it has some value in industries not requiring technical knowledge. Even here the South is being aided by the mass migration of blacks to the big cities of the North and West. This transfers the enormous burden of a mass of uneducated and unskilled people over more of the country. The South is in the midst of social and economic revolution, and by many measures of economic growth it is one of the most rapidly devel-

oping parts of the nation. Why was it ever otherwise in the past?

Causes of southern lag

Explanations for the backwardness of the South have often stressed the physical geography. Common misstatements are: the area is better suited to agriculture than to industry; or, the depressing effect of heat and humidity and the absence of the stimulus of seasonal change doom the area to low mental effort. It is certainly true that the South has enormous advantages in agriculture over the North's short season and over the West's aridity. It is strange to argue that any area possessed of the abundance of hydroelectric power found in the South, and possessing some of the major minerals such as sulfur and petroleum, is ill-equipped to do well in industry. Especially is it hard to dismiss such a situation as Birmingham, Alabama, where cooking coal, iron ore, and limestone are found almost side by side. This gives Birmingham the lowest assembly cost of raw materials for iron- and steel-making of any industrial site in America. Nevertheless, the South has lagged. Perhaps there is something to this climatic argument after all. Or perhaps there are other factors at work, and this is just another of the numerous examples of a potentially good region standing relatively still. We certainly have seen enough examples of this.

In some areas we have studied the question of race as a factor in development. That is not the problem in the South. Two races are present, but the white race has been in charge. If there was a racial failure, then it must have been due to the inferiority of southern whites. But this is ridiculous. The southern whites are the same racial stock as the northern whites. There is an obvious correlation. Northwest European whites placed in the cold, strongly seasonal North were energetic developers of trade and industry. The same racial stock placed in the warm, weakly seasonal South became planters and decayed into nonintellectually inclined agrarians. This is one of the classic cases where a fine correlation may indicate virtually no causal correlation between the two phenomena so closely correlated. It is particularly interesting to twist and turn this a bit.

First, our survey of the success of white settlers in much more tropical situations than the South indicated that there is nothing in the physical environment that stops white settlers from doing well for a period of centuries. Second, if we look at the situation in the pre-Columbian period, we find a reversal of the cultural argument. The peak of Indian achievement was in the South, and culture declined toward the north with New England being toward the lower end of the scale of human achievement among the eastern Indians. There is no biological basis for an argument that the differences between the American Indians and the Europeans are such that the climate that stimulates one is deadening to the other. A much simpler explanation is found if we look at proximity to ideas. The stimulus of Indian America was coming primarily from the South. For colonial America, the stimulus was coming from the Northeast. In each case the area closer to the center of ideas led in cultural accomplishment.

This explanation can be overdone, however. In colonial America the South was a progressive region that produced more than its share of national leaders and cultural advances. The retardation of the South follows the Civil War. This is worth considering, for again and again the situation is found where a country or area was far in the lead of the rest of the world, and thereafter fell far behind. Often the period of loss of leadership followed devastation and war. The story of the South illustrates how such an event can affect a region and how long the effects may last, and how our judgments about an area are warped if we look at the momentary situation without considering the events that have conditioned the present use of the land. For the South, one could well substitute Italy, Spain, Greece, India, or any one of dozens of other cases.

It is now more than 100 years since the Civil War. The South is still a distinctive part of our nation, until recently marked by greater emphasis on agriculture than most of the country, by a single-party system, and by entrenched segregation. The first breaks in this structure have appeared, however. The South had emphasized agriculture before the Civil War, and it was the industrial might of the North that was decisive in the war. The South was a devastated and defeated nation in 1865. Major cities were destroyed, and manpower was at a low ebb; the mass of illiterate slaves were freed into a devastated, impoverished, and economically hobbled region. If nothing worse had happened, the South would still have had a difficult period of rebuilding. The worst was yet to come, however.

The North occupied various southern states for three to 10 years after the fighting stopped. In contrast to our treatment of defeated European nations in later wars, we did virtually nothing to help the starving South. In addition to a loss of $4 billion through the freeing of the slaves, the South had $2.5 billion in worthless bonds and currency, and on top of this, United States Treasury agents seized millions of dollars worth of cotton, land, and other property that they alleged had belonged to the Confederacy. The blacks were organized by the northern carpetbaggers and made to vote the straight Republican ticket. Under the leadership of the worst elements from the North, the South was robbed and beaten and left in debt. A debt of more than $100 million was accumulated by the southern states between 1866 and 1874. In one southern state, the carpetbaggers organized the blacks and elected a northerner for governor. On an $8000-a-year salary this governor accumulated $1 million in four years. The southerners were not in charge, and there was almost nothing to show for the squandering of this vast sum of money.

The problem of black voting in the South must be seen against such a background. They did vote for 25 years after the Civil War. Their votes were straight Republican while the northern armies were there. Thereafter they voted straight Democrat, with the particular candidate who could buy the most getting the most votes. The problem was that of an illiterate mass that could not be expected to take up the heavy responsibility of wise action in so difficult a field as government.

The obvious answer to this problem was to educate the blacks. The Freedman's Bureau created 4000 schools for blacks in the South. This Bureau was supported by taxes on cotton. Out of each $4 that it collected, it spent $1 on such projects as relief and schools. Incidentally, no aid was given to the South for building schools for the whites. Our public-school system is a relatively recent growth, and during the time that it was expanding most rapidly in the North, the South was struggling merely to live. At that the South did what it could to create a dual school system and to give some education to both races. The low status of southern education even today is a measure of the problem that they faced. As late as 1936, the South, with one-sixth of the nation's income, had one-third of the nation's school children in a dual school system. The result was that a white teacher received only one-third the pay she might expect in a state like California. Primary and secondary education were limited and poor, and graduate schools and medical and dental schools almost nonexistent. Schools are expensive, and advanced schooling is immensely expensive, but the South was our poorest area.

The South had its economic base virtually destroyed by the war and the freeing of the slaves without compensation. After the war, cotton was heavily taxed and given no protection in the world market. Northern manufacturing enterprises, meanwhile, were heavily protected. The resources of the South were bought up at distress prices by northerners. Pine land, more than 5 million acres of it, sold for as little as 25 cents per acre. The southern railroads were bought up and freight rates were so fixed that southern industry could not compete with the North. So vicious was the rate setting on freight charges and steel prices that it was cheaper for a businessman in New Orleans to buy his steel from Pittsburgh than from nearby Birmingham. These discriminatory rates were ended in 1945.

Seventy-five years after the Civil War, more than half of the South's farm families were tenant farmers with an average yearly income of less than $75 per person. Only 3 percent of the farms had indoor plumbing. More than 2 million families had malaria, and child labor was common. Wages in industry were about two-thirds those outside the area. Cotton had impoverished the people and led to the greatest soil erosion problem in the country. The South was a rural slum.

Revolutionary changes in the South came with World War II. The textile industry had been shifting toward the South to take advantage of the large pool of labor there. Abundant timber, hydroelectric power, oil and gas, sulfur, and water made the area attractive to light industries such as the synthetic textiles, plastics, and chemicals. Meanwhile, farming was changing. Cotton was being displaced by diversified agriculture, and mechanization of agriculture was coming in. This was displacing the tenant farmer, creating a continuing flow of manpower off the land and into the towns available for industry; this was the same situation that had existed in New England 100 years earlier and had supplied one of the impulses for the industrialization of New England.

By the 1950s the South was booming. Factory employment was growing four times as fast as in other states. Increase in number of cars and trucks was 50 percent higher than in the rest of the United States. Between 1950 and 1960, personal income rose from about $1000 per person to $1700 per person. Between 1955 and 1965, the South gained 850,000 factory jobs, while the Northeast lost 328,000. In the South during this period, there was a decline in the number of factory jobs in only one state (West Virginia); in the North, in two-thirds of the states; in the rest of the United States, in only one state (Kansas). This growth continued into the mid-1970s.

Between 1950 and 1960 white population grew one-third faster in the South than in the rest of the states, but the black population grew at only one-fifth the rate of increase in the rest of the states. This figure represents the migration of blacks from the South into the big northern and western cities where they found better opportunity. Between 1950

TABLE 5-1   Migration of Blacks in the United States: 1955–1970

	From South	To South
1955–60	402,000	123,000
1965–70	378,000	162,000

Source: U. S. News and World Report (Feb. 26, 1973).

and 1965 2 million blacks left the South. The responsibility of educating and employing them has in considerable measure been transferred to the other parts of the nation. This has come at the very time when the growth of income in the South has made possible the building of a good school system, and when the industrialization of the South has offered an opportunity for these people to be employed productively. The blacks who have remained in the South are not all impoverished laborers; today there is a growing middle class of professional businessmen and even a few millionaires. Conditions in the South are so changed that now blacks are moving back. Disillusioned with the North, and especially with the insecurity of the big northern cities, many young college-educated blacks are returning to the South where they are amazed to find that many people don't lock their doors at night. The desegregation of the 1960s combined with economic growth has made the South an attractive place for many blacks to live.

The opportunity for the industrial growth, the diversification of agriculture, and the accompanying rise in health, welfare, and education were inherent in this geographical region at all times. Notice how the race-and-climate argument has changed. The same white race, in large part the same families, is still dominant in the South. The monotonous heat has had even longer to work on them. Now, however, they are vigorous entrepreneurs, busily developing their area at a much more rapid rate than in the rest of the country. At this very time New England, the land of stimulating climate, has gone into a period of very slow growth, and in upper New England, of absolute loss of population and loss of productivity. The physical geography and especially the climate have changed little. Within the past 50 years the temperature of mid-latitudes has risen somewhat. This means that the South is somewhat hotter in summer and New England a little less frigid in winter, and still New England declines in relation to the rising South.

The ills of the South cannot all be attributed to the Civil War and the unjust treatment the region received thereafter. The southerners were plantation minded and far too interested in cotton for their own good. Like most men through most of time, they tended to keep alive the system that they knew. They faced an enormous task of rehabilitating their land, readjusting their society from slave to free, and working out a new economic and social order. It would have been difficult enough under the best conditions, but the war and the attendant crises made the situation worse.

In a wider view of the history of land use, consider now the effect of the Civil War. The South is a part of the United States, and the Confederates were our own people. Despite this, we treated that area in such a way as to retard its development for 75 years. Only after more than a full century do we see it making a recovery that promises to make it a leading part of the nation within 25 more years. Consider what the situation might have been had the area not been a part of our nation, with people speaking our own language and having our own religions and sharing our outlook on life. The North might well have oppressed them more heavily and longer. If this had been the case would we be explaining the backwardness of the South in terms of physical geography? Or would it be in terms of race? Or perhaps religion? If the South were not a part of the United States, would it be as successful as it now is? Its sources of capital, its markets, and many of the key people in its recent spurt of growth have been drawn from the West and the North. They could operate freely within the boundaries of the United States. Had the South been successful in the Civil War it might have made a more rapid initial recovery, but its enormous potential as a part of a continent-sized republic would have been lost.

Thus within the mid-latitude forest environment of our East, a series of dramatic cultural growth and land-use changes has occurred in a relatively brief time. Indian America had its greatest cultural centers along the Mississippi River and in the South. Population density and cultural achievement declined steadily northward and eastward reflecting the importance of Middle America as a source of stimulation.

The colonial period saw the reversal of this: the eastern seaboard became dominant. Although differing economic and social systems marked the various colonies, both North and South prospered and produced notable men in all fields of endeavor. Then came the Civil War, and the South was reduced to poverty. The geographical explanation offered in the early twentieth century was that the climate caused this. At mid-century, the South, without benefit of climatic change, sprang into a race for leadership in the nation. If the pattern of the eastern United States of the future looks more like that of Indian America, will we now conclude that the South is an environment that is superior? A review of what man has done in similar climates elsewhere in the world may help answer that question.

South American Mid-Latitude Forest

The South American area with a climate like the eastern United States centers on Uruguay and includes adjacent parts of Brazil, Paraguay, and Argentina. Equivalent latitudes are Saint Augustine, Florida, and Porto Alegre, Brazil; Buenos Aires and Cape Hatteras or Memphis; Bahía Blanca, Argentina, and Philadelphia. The Brazilian highlands decline into plains in southern Brazil; the rest of this climatic zone in South America is characterized by extensive plains. Most of the area is drained by the La Plata–Paraná river system, which can be thought of as somewhat like the Mississippi in its location and potential function. As in the United States, aridity increases to the west and seasonality increases poleward. Within one climatic area, largely on one landform type, under the domination of Mediterranean peoples, remarkably unlike nations developed. A brief review of southernmost Brazil and adjacent Uruguay, Paraguay, and Argentina will bring out these differences.

Southern Brazil, Uruguay, and Paraguay

The south of Brazil has varied areas of cultural settlement. In order to settle the extreme south of Brazil, especially the state of Rio Grande do Sul, German immigration was encouraged in the nineteenth century. The settlement was made in an otherwise sparsely settled country, and the settlers brought their northern European patterns of crops, fields, houses, and animals. Consequently this area looks different from the rest of Brazil. Note the sequence of causes and results. The area was a thinly held frontier to which the Brazilians wished to establish their claim. A convenient source of desirable settlers was found in Germany. Southern Brazil is the least tropical part of Brazil. Its development has been in a mid-latitude pattern. It would be extremely easy to turn this into an environmental-deterministic argument: "The mid-latitude of Rio Grande do Sul was the only area suited to northern European people. Here it was that the German settlers found a stimulating seasonal climate. Naturally then, it was here that they settled. And quite naturally this has become one of the dynamic centers of Brazilian development." However, the causes for settlement here were political rather than climatic.

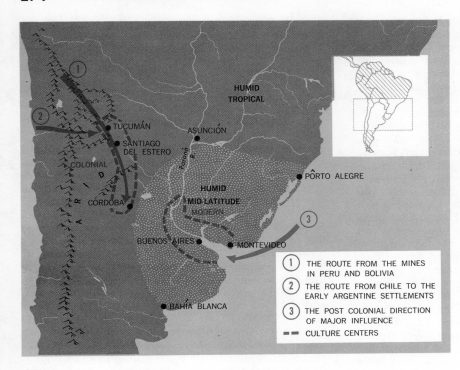

MAP 5-5   South American Mid-Latitude
Forest Lands

The South American mid-latitude forest lands center on the La Plata estuary and the Paraná River. They include most of Paraguay, the southernmost part of Brazil, Uruguay, and part of Argentina. In Argentina much of the land was in grassland at the time of discovery, and it is thought that much of this humid grassland was the result of fire driving the trees out of their natural habitat.

The colonial approach to the area was from the west via Chile and from the highland areas of Bolivia. The early centers of growth and development were consequently on the dry margin of the area. With the breakdown of Spanish controls, the focus of the area came to be the La Plata estuary. The two great cities of the region, Buenos Aires and Montevideo, developed there.

Later other Europeans emigrated into this area. For instance, there is an Italian area. Here the houses and crops and field patterns are more like Italy. Where the Germans raise hogs, the Italians are more noted for their vineyards. Note again the importance of cultural tradition. These settlements are set down in the midst of Portuguese settlements. Each has its own distinctive type. The environment has not demanded that there be one use of the land. Southeastern Brazil, even though it is climatically like our South, has not developed like our South, nor even like adjacent Uruguay.

Uruguay is approximately half the size of California. It has a population of about 3 million, nearly one-sixth that of California. Since our country is more urbanized than is Uruguay, the density of rural population is to be thought of as much like that found in rural areas in the eastern United States. The settlement of this country was by Spanish people in the colonial period, and the population today is nearly 90 percent European stock. Interest was in livestock; thus grain, cotton, fruit, and many other potential crops are relatively little developed. Only about 7 percent of the land is cultivated, nearly 60 percent being in pasture. As

with most of the Latin American countries, a high percentage of the population is in one large city. Montevideo, with 1,300,000 people, has nearly one-half of the population of the nation.

Uruguay offers free education to its citizens and has a high literacy rate (65 percent). Further, there are pension plans, child care, free medical care, and other social programs. As elsewhere, the costs and abuses of these systems have proved prohibitive, and the country is experiencing financial difficulty. As a geographic unit, Uruguay is small but extremely well endowed in topography and climate. Mineral resources are known but little explored or developed. Uruguay is in a stage of economic development similar to our South in the period of cotton dominance; only in Uruguay, cattle ranching is king.

Paraguay is a land of a mixed Indian and European race, with more than 90 percent of the population either Indian or part Indian. It has a tropical to subtropical, humid climate, similar to lower Florida. Much of the land is flat, and the amount of potentially useful land is high. Citrus grows virtually as a wild plant. Yet the land is little developed, and population density is only about 17 per square mile. Paraguay, the

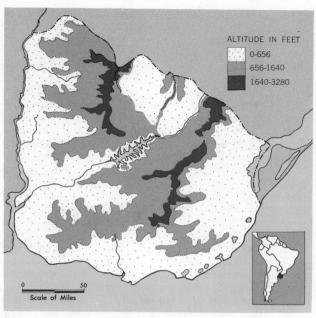

ALTITUDE IN FEET
- 0-656
- 656-1640
- 1640-3280

0   50
Scale of Miles

LANDFORM

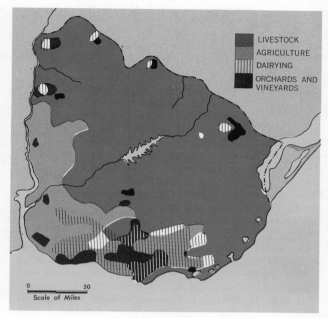

- LIVESTOCK
- AGRICULTURE
- DAIRYING
- ORCHARDS AND VINEYARDS

0   50
Scale of Miles

LAND USE

MAP 5-6 Uruguay: Landform, Land Use, and Population

Uruguay is a land of little relief, as shown on the landform map. It has abundant rainfall well distributed through the year. With such a physical base the whole country could be developed for intensive agriculture. Instead cattle raising dominates the area, and a modest population is fairly evenly spread over the country. (Reprinted from Focus, *American Geographical Society)*

same size as California, has only 2,750,000 people. Interior location, poor transportation, and absence of coal, iron, and any developed oil resource (although oil is thought to be present) are the usual reasons advanced for the lack of development of such a nation. Comparison with Switzerland or Belgium shows that these are not sufficient causes. The Paraná River is navigable and leads to the Atlantic. Undeveloped transportation and idle resources, illiterate people, and restrictive government are not parts of the physical environment. A nation need not have coal, iron, or oil to be a prosperous and productive nation. Paraguay is a

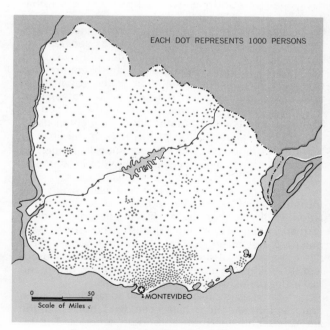

EACH DOT REPRESENTS 1000 PERSONS

0   50
Scale of Miles

MONTEVIDEO

POPULATION DENSITY

classic case of low-level use of a potentially rich piece of geography.

Argentina

Argentina has long been the most advanced nation in this area. The Spanish development of this land, so like the eastern United States in its physical setting, was to reach down into the Argentine grassland and develop it as a reservoir for the beef and mules needed in the mines in Bolivia and Peru. Argentina was approached from Spain via the Caribbean, through Peru, and across the Andes. Entry to the grasslands was from Tucumán and Córdoba. Use of the sea lane that led so easily from Spain down the Atlantic to the La Plata estuary was prohibited until the time of the American Revolution. Here again men worked in defiance of what the natural setting demanded.

In 1776 the restrictions against direct sea trade were removed. At that time a viceroyalty including what is now Bolivia, Paraguay, Uruguay, and Argentina was created. If this political unit had survived to become a state at the time of the breakup of the Spanish empire in America, it would have been a nation of the size of the United States east of the Mississippi, with a comparable climate and with comparable amounts of navigable rivers and well-watered plains. The possibilities that have gone unfulfilled are forever influencing the development of the present. Too often the results of such historical accidents are then read as geographical determinations. It was the human decisions made at the time of the collapse of the Spanish empire that isolated Paraguay. The orientation of the colonial period made Uruguay a tiny ranching nation rather than a farming nation. Argentina was similarly molded by Spanish interests and has only slowly reoriented itself. The political fragmentation of this part of the Spanish empire has hindered, and will continue to hinder, its economic growth and development. Note how important are the cultural versus the physical causes.

In Argentina, as on the American plains, the introduction of cattle soon led to a large wild population in the grasslands. As in America, the Indians learned riding from the Spanish and soon were mounted horsemen. As in the United States, these fierce horsemen of the plains became a military problem. It took major armed force to put them down and make the plains useful to the Europeans.

As an aside, it is the role of ideas that is again important here. There had been wild horses in the grasslands both in North and South America. The Indians had eaten them all up. There were wild camels present, and the idea of using them as domestic animals was present in adjacent Bolivia. But they were not so used in the plains of Argentina. When the North American Indians took up the trait of hunting from horseback, they used the bow and arrow for killing their prey and developed very short, powerful bows for use on horseback. Not so in South America. There the Indians used the bola, a set of stone weights attached to the end of strings to be whirled around the head and thrown to entangle the prey. We know the bola to be a very ancient implement, and we know the bow and arrow to be relatively recent in its appearance in the Americas. It seems probable that the bola was still the major implement, not yet fully replaced by the bow when the Spanish arrived. In Argentina, custom, not need or efficiency, decreed that the bola rather than the bow should be used on horseback when pursuing game, while in the United States the bow was used.

After 1776 Buenos Aires, which had languished as long as trade came largely from the Pacific, became the active center of the country. The development of modern Argentina has centered on this city that has grown up in the La Plata estuary with access to this great inland waterway, on the sea with access thereby to the lanes of the entire world, and on the edge of the vast rich grasslands of Argentina, the pampas. In 1776 its population was only 20,000. By 1800 it had a population of 40,000 people. Today metropolitan Buenos Aires is a city of 9 million.

The settlements of Argentina, then, were as if the settlement of the United States had started on the West coast, centered on a mining economy—a thing that could easily have happened had history been slightly different. The East then would have been developed from some points such as Sante Fe and Denver. The first use of the East would have been to supply livestock raised on the Great Plains. Later, the port of New Orleans would have gained importance. With this sequence of events, New England probably never would have been developed to important levels. The development of the United States would center rather on New Orleans. This is roughly comparable to what did happen in Argentina. The early cities were at the foot of the mountains in the dry interior at oasis locations: Tucumán, Córdoba, Santiago. Only after the Spanish grip began to slip and the colonies began to go their own ways did Argentina reorient itself to the Atlantic. The major city has become Buenos Aires, at the mouth of the Rio de la Plata, the great bay at the mouth of the Paraná River. The La Plata is as long as the Mississippi River, has nearly four times as great an amount of water, and is potentially an enormously useful artery for transporting freight.

LANDFORM

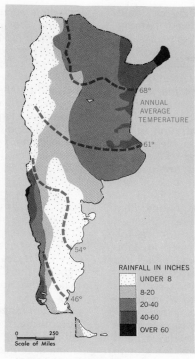

AVERAGE ANNUAL RAINFALL
AND TEMPERATURES

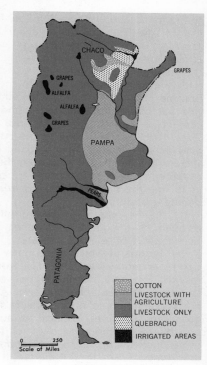

LAND USE

MAP 5-7   Argentina: Landform, Rainfall, Temperature, and Land Use
Argentina has a plain facing the Atlantic, and the spine of the Andes forms its western boundary. Rainfall is abundant in the northeast and declines toward the south and west; the southern Andes get heavy rainfall because they are in the westerly wind belt: The dry area to the east of the Andes is due to their intercepting the rain-bearing winds. The steady decline of temperature toward the south parallels the type of temperature decline northward along the East coast of the United States. Land use reflects physical geography, history, and economic development. The irrigated oases in the northwest of Argentina owe much of their development to colonial times. The great emphasis on cattle also reflects the colonial and Spanish background. The general zoning of land use within this cultural-economic orientation is fitted to the rainfall pattern. (Reprinted from Focus, *American Geographical Society)*

Argentina remained a nearly self-sufficient cattle-raising country until the mid-nineteenth century. By then, the growth of industry in Europe had created a market for foodstuffs to feed the population now drawn into manufacturing and gaining a higher standard of living, and thus able to import foodstuffs including such costly materials as beef and mutton. Under European stimulus, and largely with European financing, railroads were built to bring the produce of the interior to the ports, and great slaughtering and refrigeration plants were built to process the cattle for export overseas in refrigerator ships. Grain production became important, and corn, wheat, wool, and meat were the major exports.

The Latin tradition was for ranching, and a few families held large blocks of land. Immigrants supplied the farming labor—a steady inflow from Europe, especially Italy. The country evolved a mixed farming-ranching system geared to the world markets. The strong European basis of the nation was reinforced by the steady inflow of immigrants. As the wealth of the nation grew, industry began to grow. Urban growth became phenomenal. Nearly one-third of the population has come to be located in the four major urban centers of Buenos Aires, Rosario, Córdoba, and La Plata. Partially this is due to the concentration of the land in large estates, and partially it is a measure of the worldwide shift of man from the land to the cities as the efficiency of the world's agriculture improves so as to free the mass of mankind from the task of producing foodstuffs.

In the middle of this century, Argentina went through a politico-economic upheaval under a dictator, Juan Perón, that set it back severely. This period was followed by one in which the country struggled to stabilize its economy, reestablish its worldwide trade, and resume its program of internal development of manufacturing. Momentarily, however, Argentina stands as a classical example of the fragile nature of politico-economic structures. Political turmoil has made it impossible to get the economy back to its former levels of productivity. Strikes, assassinations, slowdowns, and government crises are not the fuel that fires the growth of gross national product. In 1972 Perón returned from exile in the hope of unifying the country once again. His death in 1974 leaves Argentina's future uncertain. The physical potential is good; the population has been drawn from nations of great accomplishments. Argentina's future is as great as its people have the will to make it.

The South American equivalent of our eastern United States, southern Brazil, Paraguay, Uruguay, and northern Argentina, is physically well endowed. It is politically divided, and the degree and kind of development is clearly related to the political present and the historic past. The differing cultural landscapes developed in similar climatic and physiographic settings again express the dominance of the role of culture.

China

China, firmly fixed in most minds as a tropical country, is actually very much like the United States. It has a warm subtropical south, a strongly seasonal north, and an arid west. Its slightly more southerly latitude is compensated for by its edge-of-a-great-continent position. The cold air of Inner Asia brings very cold weather to north China when it pours out in the winter. This area,

including such outer territories as Tibet, Mongolia, Sinkiang Province, and Manchuria, is a vast country nearly one and one-half times the size of the United States. However, just what the political fate of these outer territories is to be remains in question. The heart of China is east of the 100th meridian and south of the 40th parallel. This area is about half the size of the United States, and climatically it is much like the humid forested eastern United States, although the shape of the land is different. The plains are to the north, in the latitude of northern Florida to Maine. To the south lies a rugged hill land with much of the land between 2000 and 5000 feet in elevation.

Racial and cultural origins

There are two Chinas, north and south, which differ in race, language, culture, climate, and land form. They are nonetheless bound together in a national history and share a unique writing system. The east-west trend of the Chin Ling Shan Range divides China and makes the north-south contrasts more marked than they are in the United States. The Chin Ling Shan mountain range is long and high enough to deflect the outpouring of cold air from Inner Asia from a southerly course, giving southern China mild winters. Northern China is subject to invasion by masses of cold, dry air. Thus, in winter the north becomes cold and dry, and the south remains mild and rainy. In summer both are hot and wet. Some of the differences between north and south China are summmarized in Table 5-2.

TABLE 5-2　North and South China Compared

	North China	South China
Major rivers	Hwang Ho	Yangtze
Land	Great delta plains	Few small plains
Rainfall	Under 40 inches	Over 40 inches
Crops	Wheat, millet, soybean	Rice, tea, mulberry, bamboo, tung tree
Animals	Ox, camel	Water buffalo
Racial features	Taller, lighter color	Smaller, darker
Cultural characteristics	Slower in mental action, quicker to take physical action	Quick in mental reaction, subtle in business

What is the reason for these differences? The predominance of rice in the warm, humid south and of wheat in the drier, colder north seems a reasonable physical-environmental response. But does the warm,

MAP 5-8   Cultural Contacts with China and the Location of Cities

Two major routes of contact with China are indicated. The path shown by the arrows is based upon Heine-Geldern's map of the spread of ideas to China about 800 B.C. and eventually to a major capital city at Dongson on the Tonkin delta. This ancient contact route through the continental interior served not only for the flow of the cultural ideas that stimulated the growth of China but also later as an important trade route. The inland location of cities such as Peking, Chungking, and Nanking illustrates the dominance of this continental orientation. The vast growth of sea trade after A.D. 1600 shifted the orientation of China toward the sea. Tientsin, Shanghai, and Hong Kong illustrate this outward-looking interest. Currently Chinese orientation is again toward the continental interior. It is probable that such shifts of emphasis were known to ancient China too, and that there was orientation toward the sea-trade lanes in early times.

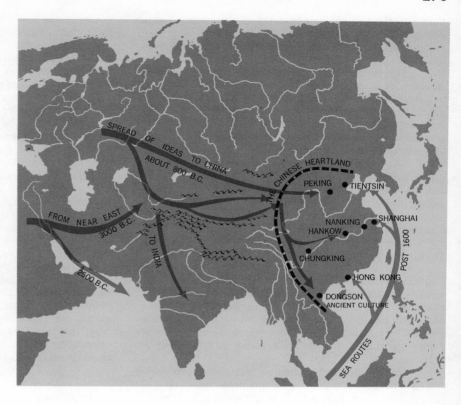

humid south versus the cold, dry north also account for the difference in physical makeup of the people?

Further, most descriptions of the Chinese distinguish between the northerners and southerners in mental characteristics. We have seen that there are reasons for some bodily changes from hot to cold climates. We are familiar in our own country with the clichés about the slow, easygoing southerner and the quick, driving northerner. However, in China it is the southerner who is the quick-witted, shrewd businessman. Can it be that climate has opposite effects on different races? Or is this a case of culture history rather than geography being the cause? Or, is this just one of these clichés such as "Englishmen have no sense of humor"?

Just as there are two Chinas in terms of physical geography, there are also two Chinas in terms of cultural history. It is the joint effects of the two sets of facts that have made two Chinas. The beginnings of this division go back almost to the beginning of the human occupation of Asia, for as earlier discussion

brought out, man reached China very early by a southern coastal route and was living near Peking about 400,000 years ago.

Other people spread from the Middle East after the third glacial period. They followed a route through West Turkestan that led via the Selenga River toward Lake Baikal. During the last part of the glacial period, Europeanlike people—with most of the cultural equipment well known to us from cave excavations in France—arrived in force in the area from Lake Baikal to the northwest of China; this was about 20,000 to 30,000 years ago. We see, then, an early tendency for two lines of expansion toward China. People from Southeast Asia moved up from the south, swinging eastward around the Tibetan highland. People from the European side of the Eurasian continent persistently flowed eastward along a route that led from the Caspian area toward the Baikal area. It is quite possible that the physical differences in the Chinese—tall, lighter-skinned northerners versus the smaller, darker southerners—stem in part from these ancient differences, a conclusion reinforced by the recent Chinese statement that south China was occupied by Negroes, persumably Negritos, as late as the Chou Dynasty, which ended just before the time of Christ.

This mixture of northern and southern influences is complex and becomes increasingly so as we learn more about Chinese origins. Scholars now speak of proto-Tungusic (northern), proto-Turkic (northwestern), and proto-Tibetan (western) influences as well as three comparable influences from the south, which are lumped under the general heading of proto-Thai or Yüeh. The Yüeh were maritime people with Indonesian connections, and they occupied southeast China around 480 to 221 B.C. They were wet-rice agriculturists; the distinction between the rice eaters of the south and the wheat eaters of the north persists today. Chinese culture owes to these rice eaters the dragon myths, ancestor worship, and a great deal more that we think of as typically Chinese (see Chapter 3). On the other hand, influences from at least as far west as the Caucasus Mountains are evident in Chinese bronze art as early as 1000 B.C. This was only part of the inflow of ideas from the west through Inner Asia. The Chinese borrowed ideas right and left (or, more literally, from west and south) and wove them into their own unique blend, the culture that we know today as Chinese civilization.

The differentiated lines of influence have continued. North China begins in the great bend of the Hwang Ho about 2000 B.C. This is a cold, dry, interior location near Mongolia. One could well expect Middle Eastern rather than Southeast Asian cultural influences here. The basic grain is wheat, the domestic animal is the ox; these are Middle Eastern, not Southeast Asian traits. Metallurgical arts and design elements, among other things, suggest that during the late Bronze Age, cultural influences from the Caspian–Black Sea area were extremely important in China. The existence of an empire, called Dongson, that spanned the zone from the Black Sea to the coast of south China has been reconstructed from archeological remains. The Mongol invasions of Genghis Khan are only a late example of such movements from west to east. The Great Wall of China is the physical proof of the Chinese attempt to block off repeated invasions.

From the northwestern corner of China came political and population expansion down the Hwang Ho onto the immense delta plain lying north and south of the Shantung peninsula. Soon, these men spread over the Chin Ling Shan mountains into the Szechwan basin. This sheltered, warm interior, moderately high basin became a major population center. Chinese political domination spread, and with it went the development of a distinctive Chinese culture, marked by a

sense of the importance of the family, the strong development of ethical systems, and a respect for learning —and all of this based on an intensive agricultural system that made possible one of the densest populations the world had yet seen.

The original people of south China were seemingly very different. On Taiwan the aborigines are small, dark people; if intermixed with north Chinese, the result would be much like the present south Chinese. In the mountainous parts of southwest China there remain today 20 million tribal people sometimes described as primitive Caucasoids, of which the Lolos and the Miaos are examples. Quite possibly they are remnants of some ancient racial variant of man that has been there for tens of thousands of years. They probably intermixed with the expanding northern people to enter into the formation of the typical southern Chinese. The use of rice and water buffalo is characteristic of these people, just as the use of wheat and oxen characterize the northerner.

These hill people were not all primitives; some of them used pictographic writing and had domestic elephants, a distinctly Indo-Burman idea. These elements are evidences of the stream of ideas coming from the south into China. It seems possible that the orientation toward business in south China may be related to such influences.

Population and land use

China's population is a double problem. No one really knows how many Chinese there are, for no modern census has been taken. The last census, which was taken in 1953, reported 583 million people. The Communist Party announced in 1966 that the population was 700 million. The United Nations estimated the population at 787 million in 1973 and growing at a rate of 13 million per year. A West German expert just back from the People's Republic of China—as it is now officially called—told of stringent birth control activities, and predicted that the next Chinese census would show a surprisingly small total population. Abortions are not stigmatized; they are performed at government clinics and are common. Marriage before age 27 or 28 for males and 25 or 26 for females is discouraged. Couples are sent to work in differing parts of the country and often see one another only every few months or perhaps only once a year.

With the birth rate slowed down and the grain production up, the race between people and food is at least slowed. It is generally accepted that China has increased its production of food grains by 28 million tons to 240 million tons in five years. This is somewhat more than China's average wheat import during the last decade: 22 to 24 million tons a year. This increase is largely due to increase in fertilizer use: from 7 million tons in 1965 to 13 million tons in 1970. The Chinese produce about half their fertilizer needs. The U.S. Department of Agriculture estimates that the best that China can do over a 50-year period is to multiply its food output two and one-half times, but the probable population will be three times what it is today.

Nevertheless, parts of China are still only thinly to moderately populated. Therefore, the heavily occupied parts have very high population densities. Taking the area of China to be 3,800,000 square miles, an estimate that some 260 million acres are under cultivation indicates that only slightly more than 10 percent of the land is cultivated, or less than one-half acre of cultivated land per person—one of the lowest figures in the world. Part of this is due to the land and part is due to the Chinese pattern of land use. The Chinese have developed an intensive agriculture. They can well be said to be gardeners rather than farmers, for the land they farm is cultivated with hand tools. The labor input is high: it takes 26 man-days of labor to produce an acre of wheat, Chinese fashion—in the mechanized agriculture of the United States it only takes one. Only a small piece of land can be cultivated by one family in the Chinese way. Therefore the yield must be high if the family is to be supported by the land, and only the best land can do this. Hence, the type of land used in our extensive, power-rich, agricultural system would be useless to the Chinese—useless, that is, unless he changes his way of farming. Again, it is the Chinese methods, not the Chinese land, that dictates this land-use pattern.

Thus China is a land climatically much like the United States or southeast South America; physically it is more rugged, and its densest population is very spottily distributed in the country because of the requirements of its unmechanized, hand-labor, intensive, self-sufficient agricultural base. There have been two great lines of racial and cultural influence that have persisted since the beginning of human history, and it is generally thought that it was the northern stream of ideas that initiated the population and political development that gave rise to the dominant pattern of Chinese culture.

Modern China

China in the nineteenth century was reduced to a near-colonial status. Until now it has participated in the industrial revolution only in a very limited way, and its accomplishments in recent centuries have been minor. Is this due to the land, the people, or to some other cause?

It surely is not due to any racial failure. The Chinese have a most distinguished list of inventions to their credit despite a relatively late start. After all, dynastic China only began about 2000 B.C. while the Middle East was well underway by 4000 B.C. However, gunpowder, the compass, paper, porcelain, printing, paper money, and a well-developed civil service are only a few of the important achievements that flourished in China before they were developed elsewhere in the world. Their art is acknowledged to be extremely fine, and their literature is equally developed. To what do they owe their present relative backwardness in comparison with the West in the nineteenth and most of the twentieth century?

It is not the "land." China is rich in basic resources: it has the world's largest reserves of tungsten, antimony, and tin; and its coal, iron, copper, lead, and aluminum reserves are among the world's largest. In this sense China is not a "have-not" nation. China is one of the many nations in the world that were not prepared to share early in the industrial revolution. The expansion of Western Europe came during a period when the Chinese were stagnating in a withdrawn state, sure of their superiority and contemptuous of those they considered relatively barbaric white devils. If we consider that Marco Polo was called a liar for his reports to Europeans on the unbelievably advanced state of China in his day, we can see why the Chinese also refused to believe that the Western people were civilized.

However, the Western Hemisphere was then entering the ferment of the industrial revolution, and the nations in a position to share the current ideas and disposed to use them were about to forge a commanding lead in world affairs. The humid, mid-latitude United States at this time had a fluid society harboring few fixed, traditional barriers to new developments. The mental climate in humid, mid-latitude China was just the opposite: it was at a peak of development of a socially fixed structure, and just past a peak of cultural-political achievement. Change was resisted at the very moment that it could have been advantageously embraced. China had abundant cheap labor and water power. These resources could have been combined with the ancient Chinese silk industry and the new spinning and weaving machinery to establish the basis of an industrial-commercial society. Everything was available in the physical environment.

As need for more power arose, and coal became important, they would have had ample supplies: iron for heavy industry is abundant. When hydroelectric power became significant, they had available such immense resources as the Yangtze River gorges. In the great gorge of that river, the Yangtze drops 300 feet

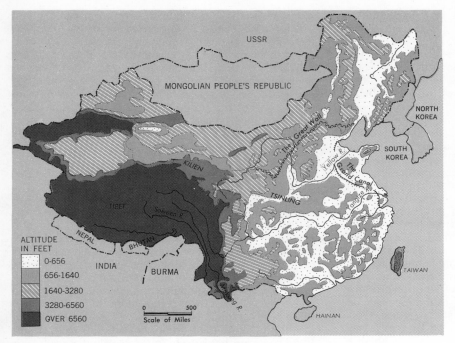

LANDFORM

MAP 5-9   China: Landform, Railroads, Water Power, and Minerals
The very weak rail network shown here is a measure of China's slight industrial development. Its actual situation is in sharp contrast to its potential, as expressed on this map by the widespread major coal deposits. Note also the wide distribution of known major deposits of other minerals important to industrial growth and development. The strong rail and industrial installations of Manchuria are due to Japanese influences prior to World War II. (Reprinted from Focus, American Geographical Society)

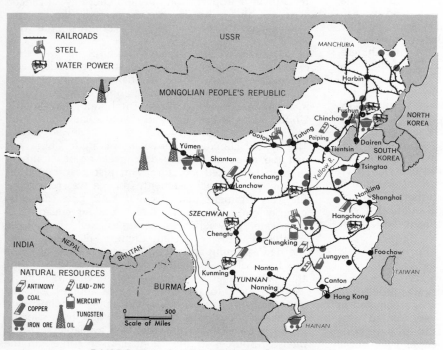

RAILROADS, WATER POWER, AND MINERALS

in 20 miles. At one point the gorge is only 600 feet wide, and the variation between high and lower water is 200 feet. A miniature Boulder Dam would make this one of the greatest power sites in the world and would level out the fluctuations between high and low stages, thus changing a dangerous and costly river passage into a placid lake ideal for water transportation.

If industry had developed, manpower would have been attracted into the cities, and machinery would have replaced manpower on the land. Extensive land use would have made more land useful. The need to feed the growing cities would have required development of a vast transportation system. Had the start been made, nothing in the physical environment would have seriously impeded development, and many items favorable to that development would have been found. China could be a land of great productivity with a high standard of living.

However, the course of events led another way. When its problem was to maintain the fertility of the land at a level that would support a sizable population with a surplus to support some effort beyond mere subsistence, China made a great advance. A pattern of life that allowed very dense, continuous settlement on the very best lands was worked out. A surplus of food was produced sufficient to support a tiny aristocracy of government, learned people, and a small merchant class. China then became set in this pattern. Most countries and most people throughout history have become "set in their ways." The progressive nations of today are in danger of becoming the has-beens of tomorrow. China's solution of her land-use problem was a brilliant one for its day, and it gave China a leading position in the world for millennia. Then Europe entered its period of exploration, mercantile expansion, colonialism, and industrialization, and China became a semicolony to Europe.

The colonial period in China created a situation somewhat like that of Brazil, where contact with the European powers by sea forced a strong development of the port cities. This is opposite to the trend of the older order in China. Peiping, Nanking, Hankow (now part of Wuhan), and Pahsien (formerly Chungking) are all interior cities; they represent the former internal focus of the country and the importance of contacts to the west via the old caravan routes across Asia. With the shift to dominance of sea-borne trade, especially in the nineteenth century, seaports replaced the interior cities in importance. Shanghai, with 10,800,000 people in the city proper, is the world's most populous city. Tientsin is twelfth, with 4,300,000 people.

Industry had barely started in pre-Communist China. The country's principal exports before the Communist takeover were cotton textiles, tungsten, and pig

bristles. China's coal is in the interior; its great metallurgical wealth is in the mountainous south. Mountains and abundant rainfall give that area an enormous hydroelectric potential. If these factors have any influence, and if China is to develop industrially, it is to be expected that the south may be the land with the greatest future. However, as we have seen, political and cultural orientation persistently outweighs mere physical geography. China's present turmoil only partially conceals the fact that its interests are in the Pacific basin, and in the long run the coastal cities seem likely to dominate.

Communist China

The People's Republic of China, the country's official name, is burdened with about 800 million people, most of whom live on a primitive subsistence-type economy. The present weakly developed industrial base is saddled with an armament program and an atomic development program that saps strength from the more basic needs. The Chinese steel industry turns out 20 million tons of steel for its 800 million people, while the United States turns out 120 million tons for its 200 million people. Power use is low; the railroad network is sparse; the road mileage is estimated at 350,000 of which only 1000 miles are paved. There are only 300,000 cars, trucks, and buses to service the nation, and most of them are reserved for the army; transportation is still dominantly in the wheelbarrow and rickshaw stage. Girls pull plows and women pull carts, for not only is the number of tractors inadequate (1 for each 2500 acres) but even domestic animals useful for traction are scarce. Accomplishments result from rigid controls over the masses which allow reducing their share of national income and the concentration of the minuscule surplus into a few areas such as armaments. It is a pattern that was widely practiced in antiquity to support an elite while building pyramids in Egypt, ziggurats in Mesopotamia, or terraces in Peru. Ultimately China's strength rests in her people. Most of them are farmers who are being hindered rather than helped in their productivity by vast collectivization plans. The results are epitomized as follows: each Chinese farmer feeds himself and two other persons; each Russian farmer feeds himself and six other persons; each American farmer feeds himself and 29 other persons.

In 1962 a Chinese writing from Shanghai described the Chinese Communist rulers as the famine makers, for although floods and droughts had been common in Chinese history, never before were there continuous expansions of disaster areas: 1950, 13 million acres; 1954, 29 million acres; 1958, 78 million acres; 1959, 107 million acres; 1960, 148 million acres. This is a record suggesting an economy in increasing disorder, and relatively easy to understand as descriptive of a land like China where a very tightly knit system of land use and an immense population allow few mistakes. Methods applicable elsewhere were tried: the Russian collective farm system was tried and it failed; then plows were tried but found inapplicable, and after costly crop losses the peasants reverted to hand tools. The government ordered close sowing (more seed per acre) for half the crop land, but did not increase the fertilizer application, and the result was a net loss in production.

The government then had hundreds of thousands of acres of cotton, hemp, tea, mulberry, peaches, oranges, litchi nuts, and bamboo razed and put into wet rice, wheat, and potatoes, but production was slight for the soils were usually not suitable for these crops. Over the millennia the peasant had found the crop suited to each patch of soil—wet rice and litchi nut trees have very different requirements. Further, if the mulberries are torn out, the village silk industry is ruined, and without bamboo hundreds of useful items cannot be made. A peasant people on the land represents a close adaptation of man, crops, and seasonal activity. Changes must be made with understanding of details, not by sweeping edicts. However, the famine makers worked in other ways. In 1959 the peasants in Honan Province reported locusts in dangerous numbers, but the authorities ordered them to concentrate on weeding. Two months later the crops in two counties were eaten up overnight, and in a panic 1,300,000 peasants were hurled into combat with the locusts, but it was too late. The losses included 1 million acres stripped in Honan, damage to 5 million acres in adjacent states, and 100,000 farm animals killed by airplane-spread insecticides.

There have been great campaigns concentrating on first one thing and then another. In the 1950s canals, dams, and dikes were important: 40 billion man-days were invested. By the end of the decade it was admitted that the big dams were most useful for impressing visitors, but many of the canals had been dug too deep and so drained the land and created drought; others had flooded land and salinized it; and the excessive canal building had led to enormous water loss by evaporation and seepage. Water control is a delicate business requiring knowledge of rainfall distribution, stream behavior, soil characteristics, and local conditions. This kind of coordinated knowledge was not applied. Next came the dream of Great Green Walls, a 1000-mile windbreak against the desert winds extending from China to Korea. This is very like the American

plan in the 1930s to plant a "shelter belt" from Canada to the Gulf of Mexico along the arid margin of the plains, a notion quite out of step with the realities of botany and climate. In China the "Green-up China in 12 years" program involved human seas of people swarming over mountain ranges, planting trees, but moving on to leave 90 percent of the infant trees to perish. Meanwhile millions of man-hours of precious labor withdrawn from other endeavors, including education, were lost.

PROFILE OF THE PEOPLE'S REPUBLIC OF CHINA

Population
760 to 800 million, 80 percent rural.

Area
3,700,000 square miles, slightly larger than the U.S.

Gross National Product
Between $80 and $120 billion (U.S. $974 billion, Japan $245 billion); 30 percent goes to capital goods and the military.

Per Capita Income
China has 3 percent of U.S., 6 percent of Japan.

Living Standards
Low but rising. Average wages in industry $270 per year.

Resources
Soils are poor. Coal, iron ore, oil, tin, tungsten in ample amounts.

Industry
Slight but expanding in steel, chemicals, cement. Military production gets top priority, but consumer goods are being produced. Bicycles are produced in large numbers. Steel output up from 5 million tons in 1957 to 20 million tons in 1971.

Outlook
Doctrinaire Maoist Communist. This may well bring on severe economic dislocations comparable to those of the years of the Cultural Revolution 1967–1968 or the Great Leap Forward in 1958.

It is exceedingly difficult to evaluate China's present accomplishments, for they are clouded in contradictory reporting. In the profile given here the salient points would seem to be that while the living standard is low, it is rising, and that while industry is slight, it too is growing. The growth of steel output is an interesting double measure. It is a significant increase, but it is far behind the goals set a decade ago. Japanese reporting on Chinese metallurgy, however, credits them with significant advances in technology, and with the governmental emphasis on steel production further advances are expectable. While consumer goods such as watches, radios, and bicycles are beginning to be available to the people, food is still rationed, cloth difficult to obtain, and housing of such scarcity that the usual arrangement is for one family to have one room and a dozen families to share one kitchen.

Further insight into the status of development in China is given by Table 5-3, comparing the economy of China with the United States. The tenfold disparity in gross national product is the best overall measure; the other data give the specifics. China produces one-sixth as much steel, 4 percent as much petroleum, and so on down the column. Even in manpower the figures are deceptive for their armies lack the support of sophisticated weaponry and transportation that our forces have. The low level of activity at Chinese airports almost looks like a blessing to anyone who has been "stacked" for up to an hour over some large American city while his plane waited for a chance to land.

TABLE 5-3   **Economies of China and United States Compared**

Economy	China		U.S.
Gross national product	100	$ millions	1000
Steel production	20	millions of tons	120
Petroleum	140	millions of barrels	350
Electric generating capacity	16	millions of kilowatts	350
Cement	13	millions of tons	65
Airlines*	400	transport aircraft	2700
Military			
Men†	2.6	million	1.9
Air force	180,000	men	720,000
Nuclear weapons	20	missiles	1,710
	120	bombs	50,000

* Peiping has 70 domestic and 18 international flights per week.

† Not equipped to United States standards.
Source: U. S. News and World Report (Feb. 28, 1972).

Thus after discounting the nuclear weapons developments and the relatively modern air force, one is dealing with a nation that is closer to the eighteenth century than it is to the twentieth. As in India, this is still a land of thousands of little mud-walled and thatch-roofed villages with dirt streets that are dusty in summer and muddy in winter. The farmer still lives as his ancestors did. Breakfast is a bowl of millet or rice porridge. Lunch is a ball of cold rice. Dinner is rice, or wheat or millet and some vegetable and a soup made

of roots or grasses. Fish or a snared bird is an occasional addition, and pork is a rare delicacy.

In the cities rents are low and food is cheap but housing is scarce. Such necessities as soap, matches, and medicine are always in short supply. Everyone dresses alike, men and women. As one Italian described it, China today impresses one as a single immense poor country whose poverty is respectable and proud but implacable. The interruption of education by the Red Guard activities and the Great Cultural Revolution may have set education back considerably. In a country that is short of engineers, scientists, technicians, and managers, this is critically important. This is the reason for the comment in China's profile on the effect of doctrinaire Maoist Communist outlook. With such doctrinaire interruptions the upward movement of the economy can be set back at any moment. An American with long experience in China commented that while China is now making progress, the present state of affairs has not lasted long enough nor consolidated itself sufficiently to warrant the conclusion that it will be permanent. Both Chairman Mao and premier Chou En Lai are aged and ill, and a transfer of power may lead to renewed turbulence.

Interruptions of life are the order of the day. Millions of families are broken up on orders from the government, and husbands and wives are often separated. In 1969 possibly 25 million people were ordered to leave the cities and go to work in the rural villages. This included all kinds of people, including hundreds of thousands of men who were being trained for factory work but were surplus workers when the factory program was cut back. Dumping millions of people on the rural villages placed a great strain on their food resources, and the people sent to the villages were often not fitted for agricultural work.

The people of China, wearied with wars and destruction, welcomed the Communists. But they have found that "the incorruptible men of the party" are greedy, ambitious, power-hungry individuals just like all the warlords. The Chinese peasant is busily seeking freedom to run his own farm, raise his own family just as he always has. It makes one wonder about the dictum that history never repeats itself.

The other China

The same people in a similar climate in the same time period have utterly different accomplishments. The Chinese Nationalists on Taiwan have had American aid, it is true, but mainland China could have had financial support from international sources and originally did have Russian technical and financial aid, an assist that was thrown away for doctrinaire political reasons. The story of Taiwan, on the other hand, is one of an open door to ideas and investments, and success 15 years after most observers were predicting failure—either absorption by the People's Republic or a permanent costly drain on the United States. Instead, it is now independent of United States economic aid, though still dependent for defense against invasion.

Taiwan is a rugged island of 13,808 square miles, four times as large as Puerto Rico or less than one-fourth the size of Florida. It is densely populated: compare its 1179 people per square mile with Japan's 758, China's 222, and the United States' 57. Two-thirds of the island is rugged hill and mountain land with peaks up to 12,000 feet. A plain on the west supports 80 percent of the population. The climate is subtropical, somewhat comparable to Florida, with mild winters and warm but sea-tempered summers. In Taiwan the peaks of high mountains are sometimes snow covered, but in the frost-free lowlands citrus fruit, sugar cane, bananas, and pineapples can be grown. The rainfall is heavy and well distributed through the year, and this plus high year-round temperatures have resulted in leached and impoverished soils. Nevertheless, the development of the island was agricultural, and before World War II Taiwan was the third largest producer of pineapples, most of which went to the Japanese market. After 1937, when development of the large hydroelectric potential was begun by the Japanese, the power was used primarily for refining aluminum for the Japanese market. This strong tie to the colonial power was typical of Taiwan's development.

In 1955, although only one-fourth of the total land area was cultivated, three-fourths of the gainfully employed were in agriculture, and farm produce amounted to 80 percent by value of the total production of the island. Rice, sugar cane, and sweet potatoes dominated, with 85 percent of the rice locally consumed as opposed to only 50 percent locally consumed during the Japanese-controlled period. In the past years Taiwan supplied Japan's needs as follows: 90 percent of the sugar, 95 percent of the tea, most of the pineapples, a large part of the rice, besides a wide range of items such as bananas, tobacco, and hemp. This is an excellent example of the economic distortions frequently associated with colonial status.

At the end of World War II there were immense disruptions and a wholesale reorientation of the economy. Nearly 500,000 Japanese were repatriated, and about 1,500,000 Chinese refugees came over from the mainland. The population tripled between 1905 and 1950 under Japanese development, and the population explosion has continued to the present. The problem in 1955 was: given this density of population, limited agricultural land, and rapid population growth, what can be done? Except in scale, their problem was similar to that of mainland China. The mainland had far more natural resources and developable land, but the problems with space and population were greater.

Today, Taiwan is booming, a showcase for other agriculturally based underdeveloped countries. Industrial production, which had been growing at a rate of 10 percent a year, hit 25 percent in 1972. Agricultural output had doubled in ten years, and real income had also doubled. Two revolutions are underway: industrial and agricultural. Agricultural land rents were reduced and farmers were aided in 1958 with low-interest payments to buy their farms over a 10-year period. More than 300,000 farm families became landowners, and after 10 years had passed they began banking a phenomenal 30 percent of their cash income and still investing heavily in farm buildings, fertilizers, and machinery. Interplanting, rotation, and diversification often yield three crops per year.

The industrial revolution is even more important. The major assets are abundant cheap labor, attractive tax rates, government policies favorable to foreign investment, and a connection with the United States. All of this is parallel to the case of Puerto Rico. The United States has primed the economic pump at a rate of $90 to $100 million a year, but this has now ended.

The Chinese immigrants from the mainland brought business experience. Using the $1.4 billion of United States economic aid, $400 million of which went into such basic investment as power plants, communications, and transport, a consumer-goods and light-equipment-oriented industry has been built. In 1972 foreign trade alone reached the $5.7 billion mark, and a goal of $10 billion was set for 1976. Per capita income, which was $25 in 1945, reached $375 in 1972, and the gross national product climbed to $6.5 billion. Such development would have been impossible without aid from the United States, but the real points are that these developments can occur, that they have been accomplished here, that they have not been achieved in other areas such as the Chinese mainland with the same race, better geography, equal time, total freedom from "colonial domination" but with a closed-door policy.

The Taiwan development will now get a severe testing, for the United Nations has given China's seat to the mainland regime, and the United States has stated that the settlement of the Taiwan-mainland China problem is one for the Chinese to settle, presumably peacefully. The Taiwanese have taken these changes philosophically. They are now playing down the talk of returning to the mainland, and are concentrating on the development of Taiwan. They may at some time and under some mutually satisfactory arrangement join the mainland government, or they may declare themselves an independent nation, as so many groups have done, especially in Africa, in the past few years.

They remain an inspiring success story. They have mastered what is called the first-stage industries (textiles, light manufacturing with heavy emphasis on relatively unskilled labor) and are now moving into plastics and more sophisticated light industries and planning the move into heavy industry such as iron and steel. The land reform program is a model for other developing nations. A major difficulty on the horizon is labor shortage. In this densely occupied land this situation exemplifies again that overpopulation is less a matter of numbers of people than of degree and kind of economic development.

PROFILE OF TAIWAN

Population
15 million; 85 percent Taiwanese.

Area
13,886 square miles; one-quarter the size of Florida.

Gross National Product
$6.5 billion; 18-fold increase since 1952.

Per Capita Income
$375; doubled in 10 years; four times that of the People's Republic. $1000 expected by 1980.

Urbanization
60 percent in cities.

Industry
Growing explosively and now moving from light industry to heavy industry; 78 percent of its exports are industrial goods.

Labor
Shortages appearing for industrial growth.

Agriculture
Successful land reform program carried out with farmers now sharing in prosperity.

One can say that the differences between the two Chinas are not race, climate, or physical geography. It is not even Chinese culture (strong family, hard work, thrift, premium on learning, and so forth) for this heritage has been the mainspring on Taiwan just as it could have been on the mainland. The dominant forces here have been cultural and political ideologies. Fundamentally, the contrast has been between the socialist-communist goal of maximizing state ownership and central planning versus limited central planning. Taiwan has followed a Puerto Rican style of minimal central planning combined with a maximum effort at creating a climate favorable to attracting international capital. Capital is essential for development and must come either from internal savings or from external investment sources, or, preferably both. Such countries as the People's Republic of China and the Soviet Union have undertaken to raise their development capital from within by rigid controls in order to channel the limited surpluses into desired areas of development. India has followed a similar course to a lesser degree. Puerto Rico and Taiwan have planned their development also, but with lesser centralization and control and with great emphasis on attracting international capital. While these successful examples are small units, and it might be argued that large units require more planning, it has also been pointed out that the larger the unit the more difficult and inefficient central planning becomes.

The contrasts in growth and development go far beyond mainland China and Taiwan. Japan, to be discussed later at more length, has come from near total destruction of its industry to a position as the number three industrial power in the world. American aid and protection, which has removed the burden of heavy military expenditures from their economy, plus statesmanship and immense hard work by the Japanese have made this nation a world power. South Korea is described as Asia's Cinderella. It inspires other Asians even more than Taiwan or Japan for they feel that if Korea could make a comeback, any country can. Fifteen years ago Korea was prostrate, virtually destroyed by the war. Today Seoul is a modern city of 5,900,000 people with factories spread for miles in every direction. The people are well dressed, American investment after years of hesitation is flooding in, land ownership —formerly divided into innumerable tiny plots—is being consolidated, and pumps are being used to raise water for efficient irrigation. The Philippines, on the other hand, are not nearly so bright a spot. Political unrest continues, the economic growth rate lags, and the birth rate soars. But, as noted, the miracle rice has not only averted a food crisis but has moved the Philippines into self-sufficiency. President Ferdinand Marcos is pressing the development of rice, roads, and schools, and has turned the military into "nation builders." They are out building highways and irrigation systems, developing ports and forest lands.

The gist of this surge in Asia is that the American postwar program has paid off. The Communists have been contained in Korea and Vietnam, have lost out in

Indonesia and Malaysia, and economic and political power is growing among the free Asians vastly more rapidly than it is among the Communist Asians. Despite the heartbreak of Vietnam, the development within South Vietnam even during the war has been prodigious: roads, bridges, improved agriculture, the beginning of industrial growth, and all the rest of the infrastructure necessary for development. Meanwhile a thorough-going land reform has been carried out in South Vietnam. If real peace is achieved, South Vietnam will be more prepared to follow the Korean or Taiwan course than either of those countries were at the beginning of their period of modern development.

Australian Mid-Latitude Forest

Another land with an area of east-coast mid-latitude forest is Australia. Australia is the size of the United States without Alaska. However, the population is only 13,335,000. Forty percent of the continent is desert and another 40 percent is semiarid, but aridity alone is not enough to account for only one-sixteenth as many people as in the United States. Time is perhaps more of a factor.

Australian settlement by Europeans was long delayed. The Portuguese and Dutch discovered the continent and explored its margins. They found the land to be occupied by a primitive people who had curly hair, dark skins, flat noses, beetle brows, and beards. They were not Negroids, for Negroids are not bearded. They were not white in color, nor were they Mongoloid. They are therefore usually placed in a separate category termed "Australoids."

For the development of Australia their racial classification was unimportant; it was their cultural status that was decisive. They were simple hunters and gatherers. They had no fixed residence and virtually no houses. They produced nothing that the Europeans wanted. To the north lay the Spice Islands (now the Moluccas) where goods of low bulk and high value could be obtained in trade; or, if settlement were made, where settled agricultural people supplied the labor for a profitable colony. Native Australia offered no inducement, and so European settlement came even later than it did in the United States—where, as we have seen, there was a 100-year delay after the discovery of the New World. Australia saw a nearly 300-year delay.

Captain Cook surveyed the coast in 1788 for the British. The well-watered coastal areas were charted, and colonization began shortly thereafter. The Amer-

ican colonies were 150 years old and had won their independence before even a penal colony was established in Australia. Australia was to get a poor start as a penal colony, and not until the discovery of gold in 1851 brought a flood of people did the country begin to develop. Australia was settled as follows:

1788	Penal Colony, Sydney
1824	Brisbane
1828	Perth
1835	Melbourne
1836	Adelaide
1851	Gold rush to Victoria

Australia's development was similar to that of California: no early advantage despite much good land; a gold rush bringing population in the mid-nineteenth century; than rapid growth and development. (See Table 5-4.)

TABLE 5-4   Population of Australia, 1788–1966

1788	1,500
1800	5,000
1820	33,000
1840	200,000
1850	400,000
1860	1,100,000
1890	3,000,000
1920	5,000,000
1950	8,500,000
1960	10,400,000
1966	11,500,000
1974	13,335,000*

*Three million immigrants, half of them British, have entered Australia since World War II.

A country that developed in the second half of the nineteenth century started in the railroad era: Australia began with the building of railroads. However, Australia's development was from several centers, each of which was independent of the others, much as the North American colonies were originally independent of each other. The rail lines were not integrated and were of differing gauges, hence rolling stock could not go from one line to the other. All goods had to be reloaded at transfer points, and this put a heavy cost on transportation within the country. Movement of goods, focused on the major ports, emphasized the dominance of urban centers, a phenomenon generally of the nineteenth and twentieth centuries. (In Australia, there are only the nineteenth and twentieth centuries.) Urbanization so characterizes this land that half of the total population lives in four cities: Sydney, Melbourne, Adelaide, and Perth. One-fourth of the total labor force is in factory employment. Most of Australia, including even the humid lands, is virtually empty.

The British settled the humid east coast. They avoided the tropical north, found the interior to be forbiddingly dry and the southwest to be good, but far from the rest of the developing part of the continent. The east coast is to be thought of as like the eastern seaboard of the United States, but lacking our southern coastal plain. One end of the coast is as tropical as the Caribbean, the other end is mid-latitude forest land comparable to Virginia. It was the seasonal, mid-latitude areas that the British settled first, and they have developed the most since. One could easily argue that the physical environment determined this; however, note that a tropical people might well have selected the north coast, a Mediterranean people the south coast, and an oasis agrarian people the Central Basin.

From Brisbane, Sydney, and Melbourne, settlement pushed across the dividing range. This situation is the equivalent of the American expansion across the Appalachians, but instead of a broad area of well-watered plains, the Australians found a narrow belt of lightly forested land, then park landscape (trees scattered through grasslands), and then the brushy-grassy lands of the desert. Exploration soon showed that the heart of the continent was desert.

The eastern side of the interior is called the Central Basin. Here there is an artesian water supply that makes the area more useful than the 10- to 20-inch rainfall in a subtropical position would lead one to expect. The southern part of this area has a river system capable of extensive development for irrigation. The Murray River rises from the Snowy Mountains in the southeast corner of Australia where high annual precipitation, heavy snow falls, and summer melt-off supply a steady runoff. The Darling River is the northern part of this system and also carries a useful amount of water. The drainage toward Lake Eyre is too erratic to be useful: Lake Eyre is normally dry, though in exceptional years it becomes an inland sea. The mid-latitude forest-land environment, then, is from Brisbane south to Adelaide, and reaches only a short distance over the dividing range.

Modern Australia

Sheep, cattle, and wheat have been Australia's principal exports. Cattle production is concentrated along the humid coast, while sheep and wheat are centered on the edge of the arid land and reach into the desert—until it takes 15 acres of land to support one sheep.

Two world wars have had much the same effect on Australia as they had on the United States. In 1918 the United States was still dependent on European machinery, dyes, chemicals, and similar goods. The war cut off the German sources and greatly reduced the amount that could be obtained from Great Britain and France. This forced local development; Australia's experience in wartime has been similar, and it is following the same course, again with a time lag. Iron and steel, automobile, and other heavy industries are now developing. In some ways, Australia today is much like the United States in the early 1900s, but with the super-cities of the mid-twentieth century.

Australia faces serious problems. Wool is in a position of declining importance; competition from synthetic fibers is cutting steadily into the market formerly dominated by wool. Sheep are losing their advantage as a dual-purpose animal. Their fate illustrates man's use of his environment. Goats were once the preferred food animal. When woolly sheep came into existence, the advantage shifted to this dual-purpose animal, and mutton came to be preferred to goat meat. If wool ceases to be a valued product, we might shift back to goats as a major meat source, and we could then use larger areas of dry land to raise goats than we ever could for sheep. We may also try to breed a woolless meat-sheep. Thus we see that custom, not the animal best suited to the environmental requirements, is likely to decide the issue.

Technological changes, such as the displacing of wool by synthetics, are common in our day. Rubber and quinine are also examples of how technological changes affect the production pattern of the world. Australia is fortunate that this change in textiles is coming at the very time when it is changing to a more industrialized society in which there will be less need for an export crop to supply the foreign exchange to pay for refrigerators, stoves, and automobiles.

Australia's east coast, with a humid and forested climate, was settled by a people of the same race and with the same cultural background as those who moved into the similar climatic zone in the United States. There are the same huge modern industrial cities, supported by a thinly occupied countryside and linked together by rail, highway, and air. The developments show variation in detail. Australia's rural areas are more thinly settled than those of the United States, though we are rapidly approaching the Australian condition. Simultaneously, Australia is industrializing and urbanizing in a way parallel to the United States pattern.

(Opposite, above) **Snowy Mountains Hydroelectric Power and Irrigation Scheme.** *Dams store water for irrigation in summer and for hydroelectric power. Some of the water is diverted to the arid interior.* (World Bank)

(Opposite, below) **Tree Removal.** *The immense forest growth, which characterizes northern Australia, is suggested by the log being dragged out of the forest by the tractor.* (World Bank)

(Above) **Clearing the Land.** *The clearing of forest land for cattle grazing and other uses is accomplished by tractors with huge chains for felling giant trees.* (World Bank)

Before the conclusion that race and environment must naturally create such a landscape, however, we must review other factors and other lands, for these peoples share a British cultural heritage as well as similar climates.

In the comparable climates elsewhere in the world, cultural developments were more or less like the American and Australian developments depending on the extent to which the countries were more or less influenced by Europe and the industrial revolution. Thus Argentina, strongly influenced by British enterprise, is Australian in its well-developed agricultural base, to which there is now being added a growing industrial base. Adjacent Uruguay has hardly begun its industrialization, and Paraguay has not done much even with its agricultural potential; the differences become even more marked when China is considered. While most of the areas of the world with this type of climate are currently moving ahead, the Chinese mainland is progressing slowly. Thus, within this one type of climate in the past 500 years, China has fallen from a position of world leadership into a state of relative decline; Australia has emerged from a stage of life that disappeared from most of the world 50,000 years ago and become a modern, increasingly industrialized nation, while the United States and South America have undergone large-scale changes between these two extremes.

Summary

Racially these four parts of the world with similar climates, soil, and vegetation were occupied before 1500 by American Indians, Australoids, and Mongoloids. Their cultural status varied. They range from extremely primitive aborigines to town- and village-dwelling agriculturists to sophisticated civilized urbanites. After 1500 race changed almost completely in the United States and in Australia, changed variably in South America, and hardly at all in China. In the United States, cultural change went from Neolithic town dwellers to advanced civilization. In South America change followed political rather than natural boundaries: from little or no advance in Paraguay to great advance in Argentina. China has stood still while the rest of the world has advanced, but now, as the People's Republic, is making progress. Australia is off to a late but strong start. Such enormous variability in similar physical environments underlines the relative unimportance of simple environmental factors such as climate, vegetation, and soil, and the overwhelming importance of the cultural-historical factors. Race seems to supply no basis for causal explanations. This continuing failure to find a racial answer suggests that race *is* no answer. Different men at different times brought differing views of how to use these similar landscapes, and the results varied according to their perception of how the land should be developed.

Review Questions

1. Discuss the probable population situation in the eastern United States before 1600. Why would it be surprising if at that time the population there was less than in California? What is the probable cause of the very low estimate of the population of the eastern United States?

Main in the Fall. *East coast mid-latitude lands are natural forest lands. Hills with extensive forests such as these are found in most of the lands of this category.* (Maine Department of Commerce and Industry)

2. To what extent is it true that the people of the United States made a good thing out of a poor physical environment rather than the country became great because of the wealth of the land? Whichever is true, why are there two contradictory ideas?

3. If one says that Brazil and the United States are what they are because of the personalities of their people, what philosophy of geography is being expressed? Could there be a geography based on the personality of nations? How would it compare with physical-environmentalist geography? Possibilism? If British institutions (law, parliaments, and so forth) were important in the United States, could there be an "institutional" analysis of geography? What would make such studies geography rather than political science or history or psychology?

4. Why did New England lose the textile industry? Is this the only industry that moved out of the area? Of the causes underlying the changes, how much is physical geography?

5. Why are the physically similar regions of southern Brazil, Uruguay, Paraguay, and northern Argentina so different in other respects?

6. Why are there "two" Chinas? Assess the role of race, physical environment, and culture.

7. The text stresses China's difficulties during the period from 1950 to 1967. What is the situation today?

8. Which is the more remarkable: that Australia was colonized so late or that Europeans bothered with it at all? Why had Asiatics not colonized it?

9. Why is the question of European contacts with eastern North America before 1500 A.D. of significance to geographers?

10. How many parts of the world of the size of the United States could you isolate and say, "Physically there is another potential United States." Need these all be climatically similar? What limits development in these areas?

Suggested Readings

Chang, Kwang-Chih. *The Archeology of Ancient China.* New Haven: Yale University Press, 1966.

Gordon, Cyrus. *Before Columbus.* New York: Crown Publishers, 1971.

Harris, Chauncy D. "Agricultural Production in the U.S.: The Past Fifty Years and the Next." *The Geographical Review,* vol. 47 (1957), pp. 173–193.

Higbee, E. C. *The American Oasis: The Land and Its Uses.* New York: Alfred A. Knopf, 1957.

Jefferson, Mark S. W. *Peopling the Argentine Pampa.* American Geographical Society Research Series, no. 16. New York: American Geographical Society, 1926.

Lillard, Richard G. *The Great Forest.* New York: Alfred A. Knopf, 1947.

McKinney, John, and Bourque, Linda. "The Changing South: National Incorporation of a Region." *American Sociological Review,* vol. 36, no. 3 (1971), pp. 339–412.

Mongé, Alf, and Landsverk, O. G. *Norse Medieval Cryptography in Runic Carvings.* Glendale, Calif.: Norseman Press, 1967.

Pendle, G. *Argentina.* London and New York: Oxford University Press, 1961.

———. *Uruguay.* London and New York: Oxford University Press, 1963.

Prunty, Merle, Jr. "Land Occupance in the Southeast: Landmarks and Forecasts." *The Geographical Review,* vol. 42 (1952), pp. 429–461.

Raine, P. *Paraguay*. New York: Scarecrow Press, Inc., 1956.

Sauer, C. O. "The Settlement of the Humid East." In *Climate and Man: Yearbook of Agriculture, 1941*. Washington, D.C.: U.S. Department of Agriculture.

Scobie, James R. *Argentina, A City and a Nation*. 2d ed. New York: Oxford University Press, 1971.

Taylor, G. *Australia*. Skokie, Ill.: Rand McNally & Co., 1931. Includes chapters on New Zealand and neighboring islands.

Wu, Yuan Li. *The Economy of Communist China; An Introduction*. Praeger Publications in Russian History and World Communism, no. 166. New York: Praeger, 1965.

FOCUS

China's Agricultural Resources, China's Resources for Heavy Industry, Formosa, Argentina, Uruguay, Changing South (U.S.), Korea.

West Coast Mid-Latitude Forest Lands

CHAPTER SIX

THE west sides of continents between 40 and 60 degrees of latitude are in the westerly wind belt throughout the year. Winds blow from the ocean onto the land, bringing with them oceanic, tempered weather conditions. Since the oceans are relatively cool in summer and warm in winter, the lands so influenced have winters mild for their latitude and cool summers. A measure of the oceanic effect on climate can be seen by comparing the mild climate of Great Britain with frigid, barren Labrador. The January temperatures are, respectively, 40° and 0° F. Although both countries are in the same latitude, winds from the oceans flow over Great Britain while winds from the continent flow over Labrador; hence the latter is much colder.

The westerlies are characterized by cyclonic storms. The winds are fresh from the oceans, the great reservoirs from which the air gains water to moisten the land. These western coasts are therefore very well watered: precipitation is 40 to 60 inches or more. Where highlands near these coasts force the moisture-laden air masses to rise over them, the precipitation may go up to 200 inches per year. The continuous moisture supply and the long growing season allow for immense forest growth. There are rain forests in these latitudes that rival the great tropical rain forests.

Very high rainfall and no dry season lead to active leaching of the soil. The great forests supply masses of dead vegetation, and the low temperatures

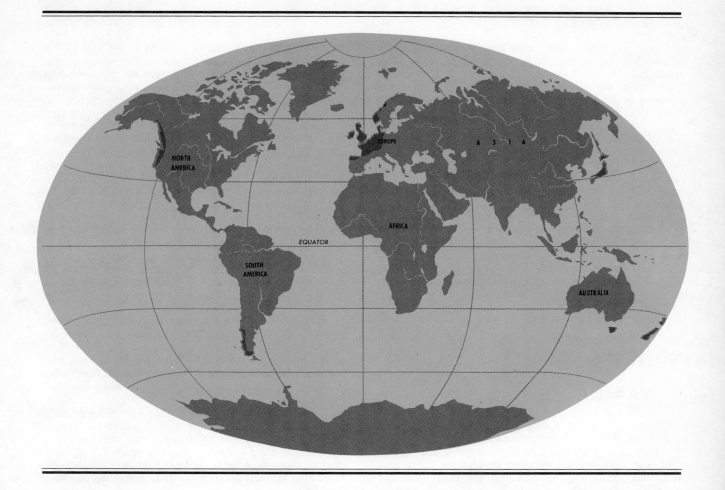

through the year prevent rapid decay. A thick mass of acid, partially decayed vegetable material then comes to cover the area. In boggy places partially decayed plant material accumulates in thick masses and forms peat. It is notable that slightly decayed plant material does not accumulate under conditions of warmth and moisture nearly so much as it does under these cooler, moist conditions. This suggests that our coal deposits accumulated under cool wet conditions.

The four major areas of this climate in the world are northwest Europe, the northwest coast of North America, the southwest coast of South America, and Tasmania and New Zealand in the South Pacific. All these lands have mountains on their coasts, and only

northwest Europe has extensive areas of plain. For convenience Japan is also included here. Japan is an island world off an east coast. Air masses reaching these islands come from the west across the sea. Their temperatures are modified so that they arrive warmed in winter and cooled in summer. Japan is also mountainous, as are the rest of these maritime-dominated areas. Therefore, despite the colder winters of Hokkaido and the warmer summers of Japan's southern islands, Japan will be included with the mid-latitude areas dominated by oceanic air masses.

Mountains in latitudes of 40 to 60 degrees, a position to get much precipitation, are bound to have a great deal of snow in winter. All mountains in these latitudes get some snow, west coastal ones get much snow, and higher mountains in these latitudes have perpetual snows. There are permanent ice caps; glaciers reach the sea in southern Alaska, and there are glaciers in the mountains of southern Chile, southern New Zealand, Tasmania, and Norway.

In the past all of these lands were heavily glaciated. The ice scoured the soil off the uplands, carved deep gorges whose bottoms extend below present sea level, and let the sea enter these mountainous coastlands for long distances. Such glaciated, mountainous, coastal lands characterize Norway, Scotland, the Panhandle of Alaska, British Columbia, southern Chile, Tasmania, and the west coast of the South Island of New Zealand. Once again, then, we have a set of lands widely distributed over the world with many physical-environmental features in common. What were they like and what are they now like in human occupation? How much of their similarity or disparity is due to physical environment, and how much to the play of cultural forces?

NORTHWEST EUROPE

Prehistory

Northwest Europe has a long, well-known history. When the ice last lay over Scandinavia and extended south of the Baltic Sea, man had already learned to use fire, he had dug snug houses into the ground for protection from the cold, and he was unquestionably wearing clothing made from the skins of the animals he was hunting. The Neanderthal men lived in western Europe about 60,000 years ago at the height of the last ice age. Then the ancestors of some of the modern Europeans appeared on the scene. The Cro-Magnon, Brünn, and Boreby people, still represented in modern populations of northwest Europe, appeared about 20,000 to 30,000 years ago. This was the time of the wonderful cave paintings and the engravings on bone of spirited scenes of their days. We know from such evidence that these people were skilled hunters of mammoth, reindeer, horse, deer, and bison.

As the ice retreated, the reindeer hunters had to follow the reindeer northward out of France into Denmark and Scandinavia if they were to continue to be reindeer specialists. Some of these men elected to stay at home. They shifted to eating new animals appearing with the increasing forest cover of the land: horse, bison, and deer.

Europe was a changing world at the end of the ice age. The melting ice fed the seas, the seas rose, the melting ice took an immense weight off the land, and the land rose. Depending on the rate of such movements, land was rising above or being overwhelmed by shallow seas. At the height of the ice age there was no English Channel, and much of the floor of the North Sea was land. The Rhine, Seine, and Thames rivers were the headwater streams of a large river flowing northward to the sea. Modern Europe took on its present shape when the Baltic ceased to be a fresh-water lake and became an arm of the sea, the North Sea ceased to be land, and the English Channel came into existence to make England an island. All of this happened within the last 10,000 years, and much of it within the last 5000 years.

The modern European climate was established 10,000 years ago and has fluctuated since only to a minor degree. By then the influences that were to change Europe from the realm of hunters and fishermen had made a start on their way to this distant land—distant, that is, from the center of growing civilization in the Middle East, and hence a retarded, backward area. Agriculture, the basis of the Neolithic revolution, was spreading along the axis of the Mediterranean, and from various points there it began to spread northward into northern Europe. By 2000 B.C. it reached the English Channel.

When agriculture spread into the cold wet lands to the north and west, it met serious problems. Wheat and barley belong in the warm lands and need hot weather for good growth and dry weather for harvest. Marine west-coast areas are wet and cool, and there is no dry season. Even today it is only southeastern England that is wheat country. Ireland, Scotland, northern Germany, and Scandinavia are lands of oats and rye.

Imaginative reconstructions of how oats and rye came to be important domestic plants have portrayed them as weeds that in time came to supplant the crop. Oats flourished where the climate was cool and very moist, hence in Scotland and Ireland. Rye dominated where summers were short and winters long and severe, hence in Scandinavia and northern Germany.

When man introduced agriculture in northern Europe, he sometimes changed the environment drastically. At other times technological changes even more drastically changed his assessment of the environment.

In England, for example, the early farmers stayed on the thin light soils of the uplands. Their simple plows could manage them, but they could not work the heavier clay-rich soils of the lowlands. The early agricultural settlement pattern of England, then, was cul-

Jostedal Glacier, Norway.
This glacier descends from a highland icecap into a valley that was filled by the glacier at the peak of the glacial period. Except for the cattle, this scene could be in Alaska, southern Chile, or southwestern New Zealand. (Norwegian Information Service)

turally determined. The concentration on one kind of landform and one soil type was determined primarily by the tool, not by the physical environment. Later, when the modern type of plow was introduced, the richer soils of the lowlands could be used. The forests that had been a haven for wolves and outlaws were cleared, and population centers grew up in their stead that dwarfed the old upland centers. Today the ancient roads and towns on the uplands have declined to such an extent that many of them have to be reconstructed from archeological studies. Aerial photographs sometimes reveal the old field patterns, town plans, and roadways on the uplands. Many of these date to Neolithic and Roman times and were abandoned after the Germanic invasions brought in the heavy, earth-turning plow that made the heavy, richer clay soils useful.

Another agricultural revolution in Europe occurred when a new plant was introduced from America in the seventeenth century. The cold, wet ground and cool, short seasons of northwestern Europe are not the best grain lands. When the so-called Irish potato was introduced from South America, land of marginal utility suddenly became productive. The potato not only grows in cool, moist, short-season conditions, but it gives a heavy yield per acre. This revolutionized the agriculture of the north German plain and was so important in Ireland that many people came to think of the Irish potato as truly Irish. The potato that made modern Ireland possible is an artifact, a domestic plant, that changed the potential of the landscape as much as the heavy Germanic moldboard plow did.

Northwest Europe early was influenced by ideas coming from distant centers of cultural growth. There was placer gold in Ireland and tin in Cornwall, and there is a half-hidden history of extensive exploration and prospecting of these regions by early Bronze Age people of the eastern Mediterranean who sought copper, tin, and gold. Somehow they found their way to the British sources by 2500 B.C. and developed them. This would seem less remarkable if we knew all the steps. Trade is ancient. In Neolithic times Baltic amber was traded to the Mediterranean, and the rich metal deposits of Spain were discovered early. Fishing probably has supported the people of Europe for more thousands of years than is commonly thought, and fishermen move about to a considerable degree. Quite

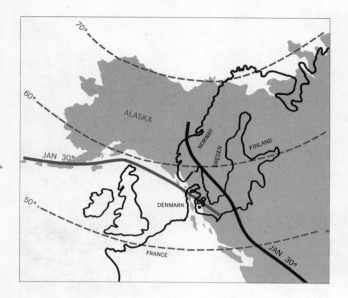

MAP 6-1   The West Coasts of North America and Europe
The west coast mid-latitude forest lands of North America and Europe are shown here at similar latitudes to stress the high-latitude position of this type of climate. The isotherm for 30°F. for January is added to show the average winter temperature of the two areas.

probably the metal-seeking people of the Mediterranean simply followed well-established land and sea routes to the North Sea region where they found the minerals that they valued. The Irish gold and Cornish tin led to long-sustained influences.

Ireland and adjacent southern England became a cultural center in the north. Stonehenge and similar monuments are some of the tangible evidences of these early contacts. Much of the doubts as to the age of the monuments has been dispelled by carbon-14 dating, which places them about 2000 B.C. Most recently outlines of bronze implements typical of the eastern Mediterranean Bronze Age have been found carved into the great stones.

This is an example of how we get to know things, for the paths to knowledge are often indirect. Dating by radioactive carbon, for example, depended on developments in atomic physics. Knowledge is often belatedly gained from overlooked obvious sources. The carvings of Bronze Age implements persisted for nearly four millennia but were not noted as such until the mid-twentieth century. From such dating we learn of the extremely long-continued influence from distant centers in the eastern Mediterranean, and we learn that distance dilutes effect. Thus, despite millennia-long influences that gave this northwestern corner of Europe a distinct opportunity, relatively little happened.

When the Romans reached the English Channel, they were reaching better-known territory than had been the case in France. The ports of southern Britain and adjacent Ireland, they said, were well known to them because of the ancient and extensive commerce carried on from those places. Northwesternmost Europe, then, has had a very long maritime history and through that activity has maintained contact with the center of civilizational ferment at the east end of the Mediterranean Sea for at least 4000 years. In that case, why was western Europe so long delayed in its ascendency? One must remember, of course, that the collapse of the Roman Empire and the subsequent pillaging by the Angles, Saxons, and the Norse set the entire region back enormously.

Cultural Lag in Northwest Europe

It has been argued that the climate of northwest Europe is the optimum one for human mental efficiency, but it has been so for approximately 10,000 years. Moreover, for the past 4000 years people with advanced ideas and skills have been reaching northwest Europe in important numbers, enough to stimulate the building of Stonehenge about 2000 B.C., and enough Roman influence in the 400 years between A.D. 43 and 443 to build cities, roads, and baths for Roman England and walls to contain the unconquerable Scotch and Welsh. Where was the climatic stimulus when for 400 years the Romans made potentially available the advanced knowledge of the Mediterranean?

The northwest European states are often described as having become maritime powers because of their advantageous position on the Atlantic at the time that world commerce broke out of the Mediterranean Sea. It is a fact that 90 percent of the earth's land surface is in the half of the earth centered on France. This land includes 94 percent of the world's population and 98 percent of the world's industry. It is said that for these reasons no other area of the world has so advantageous a position as Europe, but how much of this is determined by physical geography and how much by culture? The land-sea distribution has not changed greatly in the past million years, or even slightly in the

Stonehenge, England. *Stonehenge reveals the effects of Mediterranean people on England in the second millennium B.C. Modern mathematical studies show that it was a center for calendric studies of solstices and eclipses. It is a monument to the flow of ideas, engineering skill, and astronomic and mathematical knowledge to this land far from the center of the growth and development of ideas in those distant times.* (Courtesy of The American Museum of Natural History)

last 5000 years. The industrial concentration centering on Europe has come about in the last 200 years, and its dominant position, quite possibly, will have ceased to exist in another 200 years.

The notion that world commerce burst out of the Mediterranean about the time the northwest European powers began their rise is untrue. World commerce never had been contained within, or restricted to, the Mediterranean. World commerce plied the Indian, Pacific, and Atlantic oceans long before A.D. 1500. As we have seen, even the Mediterranean people were moving outside the Pillars of Hercules as early as 2000 B.C. to reach England. How much farther they may have reached we have yet to learn, but it is probable that they reached America. A Roman artifact in America, the appearance of American domestic plants in the Mediterranean area in pre-Columbian time, and a Semitic inscription and coins in Tennessee are but samples of the evidence of European contact with America.

Further, the northern Europeans were in contact with adjacent America from some time before A.D. 1000 and seemingly were fishing off the Grand Banks nearly continuously thereafter. In every way the climatic stimulus and Atlantic Ocean-edge explanation for the rise of northwest Europe fails to prove its case.

In a way the people of northwest Europe can be placed in parallel with the long-retarded Africans or with the Yuman Indians. They were not oriented to civilization; they were too far away for massive impacts

of civilized knowledge to reach them and force a rapid change. Change therefore came slowly and was subject to costly setbacks such as the Germanic outpourings in the post-Roman period.

The parallels with the American Southwest are interesting. Heavy civilizational influence reached southern Arizona. Canal irrigation began, and monumental structures were built. The Casa Grande near Phoenix is in a sense a Stonehenge, for not only is it a multi-storied building but slits in its walls oriented for celestial observations suggest cultural influences from distant centers of civilization. However, the ideas did not spread far and fast. The Yumans were, like the Scotch and Welsh, adjacent but resistant to change. Finally the civilizational impulse was greatly set back by the incursion of a raiding people: Germanic tribes in Europe, Athabascan tribes in America. The parallel can be carried even further. The Spanish influence on the Southwest was like the Roman influence on Great Britain; London was a distant colonial outpost comparable to Santa Fe. The Roman influence lasted for 400 years, the Spanish for 300. The effects were astonishingly small in both cases. It does not seem to have mattered whether the climate was arid or humid, warm or cold. Distance to an outpost among barbaric peoples seems to have been the controlling force. Later, new conditions led to rapid cultural advance. In both Europe and North America the physical environment set the stage for culturally determined events.

In Europe the Germanic outbreaks and the collapse of the Roman Empire initiated a cultural setback equivalent to the collapse of the Hohokam culture and the retreat of the Pueblo culture in the Southwest (see Chapter 2). This initiated the European Dark Ages, usually portrayed as far blacker than they were. The interesting socioeconomic system—feudalism—came to dominate the scene. Crop yields were low, but they probably always had been. Even under the three-field system in which some of the land rested and the cattle pasturing tended to aid the renewal of fertility, the return was low. Wheat gave only 6 to 10 bushels an acre, as compared with modern yields of 40 to 50. Population had to be in proportion to the ability of the land to produce. Further, this was a time of many small, warring political divisions, and much good land was kept out of production for political reasons.

However, trade never stopped. Old knowledge was kept alive in the monasteries, and new knowledge was brought by the traders. A great burst of new knowledge accompanied the increased contacts outside Europe that came with the Crusades. In the seventeenth century the middle class was rising with the growth of commerce, and the nation-state was beginning to develop in Europe. The little feudal units were forged into large units until even Germany, the last of the major units, was hammered together. Around 1760 the industrial revolution was underway, and one of the crucial periods of human geography was initiated. Notice that all of these changes—from pre-Roman barbarism through Roman colonialism into the political fragmentation of the Dark Ages, into the nation-state of the eighteenth century, and into the industrial revolution that has shaped the present—took place on a single geographic scene. The physical geography did not change one meaningful bit.

Antecedents of the Industrial Revolution

The industrial revolution is usually placed as beginning in England about 1760. It is marked by a shift from cottage to factory work and is clearly related to shifts in the use of power and machinery, for prior to this time the use of power was very limited. For nearly 2000 years men had very slowly been moving toward the application of nonhuman energy to the heavy daily work that mankind always had to do. The moving of loads across the land was eased by a machine composed of wheels and a frame—a wagon. But this re-

quired costly roads, and for a very long time no one thought of any way to apply anything except animal power to this device. Movement of goods over the water was more efficient, for not only was there an excellent load-carrying machine in the ship but man rather early learned to harness wind power by means of the quite clumsy square sail. He then was astonishingly slow to learn the use of fore and aft sail that allows a ship to tack into the wind.

The greatest power-need of early man was to grind his daily bread. By solving this problem, man made major advances that finally led to the broad application of power to all kinds of industrial needs. Grain can either be pounded or ground into flour, and men very early divided in their methods. Neither motion is easily converted into the circular motion characteristic of wheel-based machinery, but the grinding motion finally was converted into a circular grinding, long done by hand. This change from grinding with a back-and-forth motion came in Greco-Roman times. Thereafter there was a very long and little-known history of solving the problems of applying power to wheels to grind grain.

FIGURE 6-1   The Water Wheel
The water wheel was one of the earliest inventions that led to the industrial revolution. A stamp mill for crushing ore is shown here. Water is brought in a flume (1) to the water wheel (2), whose drive shaft (3) lifts the heavy timbers with hammers on their ends (4) and drops them on the ore (5). Such machines were used by medieval miners and persisted in the American West into the twentieth century. (Adapted from Charles Singer et al, eds., A History of Technology, vol. II, Oxford University Press, 1956)

The Nora Water-Lifting Device. *The history of the mill is long and complex. From the rotary grain mills there developed donkey-powered water mills. Here a Nora, introduced to Portugal by the Arabs, is shown. It is somewhat modernized in the gearing, but the principle is the same. It was also in the Middle East that the windmill was first developed. It was the experimentation with wheels, gears, and varied forms of power that set the scene for the industrial revolution.* (Dan Stanislawski)

These long, slow, halting periods when problems are gradually solved are the critical periods of history, but all too often they are hidden in the historical addiction of telling about kings and empires. Then, when the fruit of these hidden eras of discovery bursts forth in something like the industrial revolution, we are mystified at their "suddenness." Beginning with Roman times, water was harnessed by wheels, often to grind grain. There were innumerable problems to be solved such as controlling the water. The wheels were at first directly driven by the current; later the wheels were set on edge and at first undershot and then overshot. This required gears to change the plane of rotation from the vertical water wheel to the horizontal milling stone. Gears could not be very satisfactory until metal was commonly in use. Steady, though microscopically slow, progress was made along these lines for centuries. Among other inventions that came along was the windmill, which was first made in Persia in the eighth century A.D. The windmill also required gears. In summary, these inventions and adaptations, being made over a wide area in Europe and the Middle East, laid the base for the industrial revolution that was to take place centuries later.

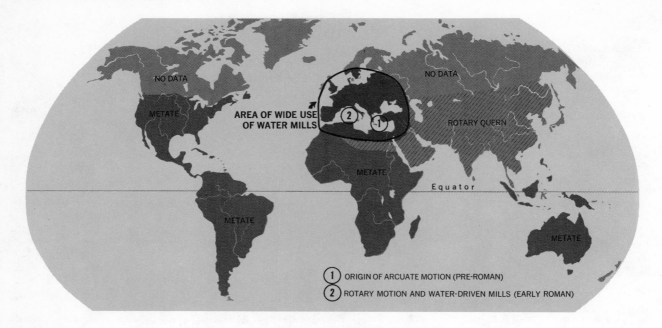

MAP 6-2 **The Growth of Skill in Milling in Relation to the Industrial Revolution: The Role of Antecedents**

The ancient problem of reducing hard foods to palatable form was first solved by the use of the metate, or saddle quern. This simple device consists of a stone slab with a smaller stone rubbed back and forth on it by hand to grind grain. The motion is approximately that of scrubbing on an old-fashioned washboard. In place of the back-and-forth motion, the Greeks introduced an arcuate one by adding a handle and a pivot to the upper stone. The Romans completed the development by making the motion fully rotary, thus introducing the principle of the wheel and making the process more easily harnessed to power. Soon thereafter donkeys were used to turn larger grain mills than men could turn, and before long the Romans had harnessed water power by means of water wheels to grind grain. Western Europeans applied these sources of inanimate power to many tasks such as sawing wood, cutting marble, and so forth. It was out of this millennia-long experimentation, the development with mills and wheels, and the application of machinery to the daily chores of mankind that the industrial revolution arose. Only in Europe and the Middle East were the antecedents present, and thus in no other part of the world was this event as likely to have occurred. Notice on the map that we lack data on use of the metate in the Arctic areas. These lands were occupied by hunters rather than seed gatherers; the metate may really have been absent, or we may simply lack evidence of it. For simplicity, use of the mortar and pestle, which in many parts of the world replaced the metate in ancient times, is not shown. It played no part in the development of the mechanized mill and can be viewed rather as a self-limiting advance, for the pounding motion was more difficult than the back-and-forth motion to change to rotary motion and hence to wheels and mills and machines.

Notice that by the application of the principle of antecedents we can eliminate large parts of the world as possible centers of origin for the industrial revolution. All of South America, North America, tropical Africa, Australia, New Zealand, Polynesia, and Siberia are out of consideration because these areas lacked the necessary ingredients. Some areas such as the Americas had agriculture and consequently the necessary economic base. In parts of Indian America the complex loom—with all the parts found in the European looms—was present; however, the development of the idea of the wheel was lacking. There were no potters' wheels, rotary grindstones, or even wheeled vehicles, though it must be noted that some of the Indians most certainly had the principle of the wheel, for wheeled toys have been found in their graves.

Equally interesting is the appearance in Peru and southern Mexico of the mano roller. The typical mano is flat sided. It must be moved back and forth, so a good deal of energy is expended in creating friction. A roller, which is a form of the wheel, uses rotary motion. In its use on a grinding slab the roller is moved with a to-and-fro motion somewhat like that of the mano, but friction is reduced, and the crushing action of the heavy stone roller "grinds" the grain. While this looks like a significant advance, it is again a dead end, for the to-and-fro motion is not readily harnessed by rotary machinery. In passing, notice how slow we have been in applying the principle of the wheelroller. Only in the mid-twentieth century have we shifted from back-and-forth motion in painting with a brush to the application of the principle of the wheel in the form of a roller.

In America the principle of the wheel was used in toys and in grain grinding; hence the necessary basic ideas were present that could have led to an industrial revolution. Nonetheless, the rotary principle was not elaborated, and without it the harnessing of power and its applications to do the work that man had always done by hand were hindered. If progress was blocked for agricultural, civilized, mechanically skilled Indian America, how much less chance there was for the other areas listed here! It is worth stating these distributions in reverse: in the Old World only in the belt from Europe to China and India were the necessary ingredients for the industrial revolution present before A.D. 1500.

Theoretically the shift to a much broader application of machinery to production could have come at any point in this vast area. In fact, it appeared first in Great Britain, giving that country such a head start that 200 years later it is still one of the leading industrial nations of the world, while large areas of the globe have yet to begin to enter the industrial era. The cultural lag remains even in these times of enormously speeded-up communication. But why Great Britain? Was it a stimulating climate and a channel-protected position, or was it a combination of poor climate and soil with proximity to the sea that led to the taking up of trade and the consequent contact with ideas? None of these factors has changed significantly in the past 4000 years. Yet after 1760 a change began in Great Britain that was to affect the entire world. This change was not due to climate, because there were several such climates. Neither was it caused solely by antecedents, for these were far more widespread than the beginnings of the industrial revolution. Nor was it due to race, for many Europeans had the same opportunity, but only the British took advantage of it. In a broad sense, the change was not even cultural since all the northwest Europeans had the same culture. What, then, was the cause for the change?

The scene was set, the requisite preliminary steps had been made, all the parts were at hand. It was only a matter of time until the parts were put together and the first spinning machines were made and harnessed by means of the already existing methods. As the repeated cases of simultaneous inventions of such things as the telephone, calculus, and anesthesia show, when a certain state is reached, the invention follows. But simultaneous inventions are always made within one cultural-geographical area and spring out of a similar background of ideas. The idea of natural selection, for example, as the basis of the changing of life forms occurred to two Englishmen, Charles Darwin and A. R. Wallace, who were well acquainted with the line of thought in biology that had been developing among the northwest European scientists for a hundred years. It was one of the accidents of history that two Englishmen rather than some continental scientists are credited with the idea. This in no way detracts from the role of the individual, for Darwin and Wallace were notable men, but the nature of their accomplishment was culturally shaped. This is the way in which the broad-scale application of power to industry must be viewed.

The beginning of the industrial revolution came in the spinning and weaving industry. In the consumption of manpower the preparing of cloth came second only to the never-ending task of grinding the grain. This was a most tedious and time-consuming task. Few today realize what was involved. Wool had to be clipped from the sheep, washed to remove the oily lanolin, carded to get all of the fibers lying parallel, then spun into thread; after these steps the threads had to be woven into cloth. All of these were originally hand processes, which gradually have been mechanized. Spinning wheels were an early, moderately complex machine. They pointed the way toward the possibility of mechanization, and the rotary spinning motion was a relatively simple one to harness. Looms had long been in existence and had been growing more and more complex for at least 2000 years. From simple hanging warps and all finger work man had progressed to warps fixed in a frame with shuttles to carry the thread and systems of attached strings that alternately raised one set of warps and then another. These were difficult motions to harness.

British Leadership

After 1760 men in Great Britain began to focus their attention on these problems. A whole series of solutions was reached. It then became possible to apply power to machines and to increase the rate of production of cloth many times over. This could not be done in each cottage, for power was at first water power and later steam power, and these installations were too cumbersome and too costly to make at every worker's home. The workers therefore had to come together at specific points instead of working at home. This was the beginning of the factory system, which led to the growth of the factory town and eventually to the large-scale movement of people off of the land into the towns. Since the increase in productivity of these mills allowed the British to undersell the world in cloth, their trade and commerce prospered, and their industrial cities grew. Great Britain's lead in manufacturing became so great that the country used its manpower in manufacturing and commerce and imported its food. Grain importation became important for Great Britain as early as 1760 and increased thereafter.

This is doubly interesting, for the means to make large-scale grain movements had marked Roman and even earlier civilizations. In Great Britain about 1760, however, the need to import grain was probably less than it had been. New ideas about crop rotation and manuring had come into being, and ideas about chemical fertilization were being promoted. The ability of the British to feed themselves was increasing to levels higher than ever before. However, the choice was to use the land to produce wool and to put the labor force to work in factories. The comparative advantage of machine-woven goods was so great that the British could make more by exporting cloth and importing grain.

It is a common thing that once an advantageous start has been made in a new line of endeavor, momentum quickly builds up. More and more effort and brains are poured into the new type of work, and further innovations are made. The present burst of discovery and invention in nuclear energy and rocketry and associated fields is an excellent example. In Great Britain the success of the application of power to the first simple spinning machines led to greater and greater inputs of human inventiveness. These led to further successes, which led to still further efforts. Yet these inventions were limited to one area—the applications of power and machinery to familiar tasks—and limitations on human inventiveness are apparent even then.

The spinning jenny dates to 1764 and the power loom to 1785. Water power was proving inadequate to supply the necessary power. The steam engine was invented in 1769 and was in factory use by 1785. At first these engines were fired by wood, but between the use of charcoal in smelting and the demands of the steam engines, Great Britain was rapidly being deforested. Coal had been used in pre-Roman days in Great Britain, but for a long interval it was used very slightly. About 1650 it began to be used as home fuel. Its use for smelting was long hampered because unprocessed coal is not ideal for use in smelters, and it was not until 1859 that coal was roasted to produce coke. This seems like an inordinate delay, especially when it is remembered that wood was customarily heated to produce charcoal for furnace use. The transfer of this idea to treating coal in similar fashion to improve its qualities in the same way for the same industry illustrates the difficulty with which men produce ideas, and this is more notable since it occurred at a time when men were invention minded.

Each of these ideas changed ways of living and thus changed the cultural landscape of Great Britain, and each of these changes gave that country an added advantage over the remainder of the world. This advantage enabled it to pour still more time, money, and manpower into further improvements of industry. The cloth industry required power, and this led to machines; machines required metal, and this speeded up mining. Machinery was applied to mining both of iron and of coal, and this led to still further demands for more kinds of machines and more metal to make machines and for machinists both to make and to run the machines. Meanwhile, shipping had to be expanded to bring in grain and raw materials and take out the manufactured products. The process, once started, expanded like a chain reaction that reached far and wide through the whole way of life of the people. Once Great Britain had the start, all of the cultural forces thrust it forward. The factors of distance, cultural resistance, and, to some extent, deliberate efforts on the part of the British, kept other nations from getting an equally early start.

It is clear that the beginning of the industrial revolution could have been in any number of places in the European world. Within that area it seems to have been in the nature of a historical accident that the beginning was in Great Britain. Once the process was started, the momentum of the movement toward industrial development swept it into an increasing lead, while the forces of distance and cultural disconnection prevented other nations from making a similar start. That the forces of cultural relationship could be greater than mere distance is well illustrated in this case by the fact that it was not the most adjacent countries that most quickly became industrialized. It was the culturally close, but geographically distant, country of the United States that led off, with adjacent Germany a close second.

Great Britain, by virtue of its start in industrialization, rapidly became *the* leading nation of the world, a role that it was to hold until the middle of the twentieth century, when, with the loss of its colonial empire, it receded to the place of *one* of the second-class world powers. This is an amazing accomplishment when we consider that the combined areas of England, Ireland, Scotland, and Wales only equal in size the state of New Mexico.

Great Britain is now an urbanized, manufacturing nation that produces only a fraction of the food necessary to feed itself, even though annual food production is greater today than it was a hundred years ago. On their little bit of land the British sustain a population of 56 million people. This is a population density of 598 per square mile. The ownership of agricultural land is in the hands of relatively few people. In Scotland 600 people own 80 percent of the land. In England and Wales about 2500 people own one-half of the cultivated land. In Latin America such ownership of land is called *latifundia*. It is severely attacked by social critics because it is considered one of the causes of Latin American lack of progress. It would appear that such land ownership has in some way failed to hold Great Britain back from progress. The lesson seems clear. It is not the land ownership pattern but the economic pattern that determines whether such land ownership is good or bad. In Great Britain the major employment of the people is not on the land but in the factories, while in Latin America there are not yet enough factories to employ the people. Latin America may find that it is easier to add factories than to dis-

tribute land. Further, the movement in the whole industrialized world is away from the small landholding that has characterized man's use of the land for thousands of years to large mechanized farms of very high productivity per man on the land. This is distinctly the American pattern and to some extent the British pattern.

The British development can be seen, then, as being little related to the "stimulating climate," its position near the center of the land hemisphere of the world or at the center of the manufacturing part of the world. The climatic stimulation had a belated effect on men in Great Britain and seemingly has lost some of its effectiveness in this century. Moreover, the position near the center of the land hemisphere did not have any effect until the arrival of the industrial revolution. The industrial revolution was a British creation, and it put Great Britain at the center of the industrial world. The urbanized, capitalized, industrialized Great Britain of today is what the British have made of their land. Its rainy, overcast, foggy climate with its long, dark winters may be stimulating to some, but it would be very depressing to anyone used to the sun. The factors that made Great Britain great were the human forces released within that country by the developments in the fields of government, economics, and science. British greatness is little dependent on the land; it is enormously dependent on human factors.

Since Great Britain is not growing today at the same rate as much of the rest of the world, it must be concluded that a change has occurred, not in the physical geography but in the cultural geography. Several things have been taking place simultaneously and are often mentioned by those who try to explain the decline. Great Britain has been through two great wars, lost her colonial empire, and shifted her governmental structure from a predominantly capitalistic free enterprise system toward an increasingly socialistic system within which the tax burden is so distributed as to virtually destroy extreme private wealth.

Something in Great Britain has changed, and it is not the climate or the country's position at the center of the land hemisphere. Wars can hardly be the answer, for Great Britain has survived numerous wars as severe as and of greater duration than those of this century, and countries more devastated by the wars of this century, such as Germany, Belgium, and Italy, have all recovered, while it has not. Although the loss of its colonial empire was a blow, this was probably not enough to account for the matter. Colonies can be costly, and Great Britain has borne large costs both directly and indirectly. Giving dominions such as New Zealand preferential treatment amounts to paying more

MAP 6-3 Ireland: Landform, Population, and Farm Size
Ireland is a land of low relief. Notice how little land is over 1500 feet elevation. The rainfall is high and the winter temperature moderate for the latitude because of its oceanic position. It is a small land, about the size of West Virginia. With 100 people per square mile, it is a relatively empty land compared with adjacent Great Britain, which has 600. The population is concentrated on moderate-sized farms to the southeast; the northwest is in relatively large holdings with few people. (Reprinted from *Focus*, American Geographical Society)

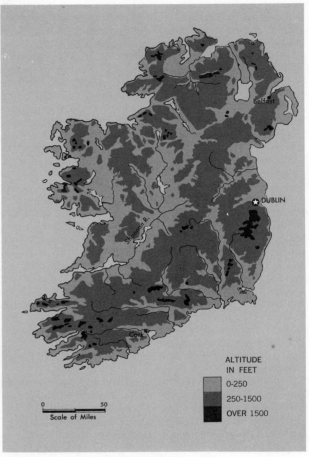

ALTITUDE
IN FEET

0-250
250-1500
OVER 1500

Scale of Miles 0   50

LANDFORM

for dairy products than they cost in the open market. Devotees of the capitalistic free enterprise system will, of course, be impressed by the parallel between the growing socialization of Great Britain and the destructive taxation levied on the capitalist class, with the attendant decline in Great Britain's fortunes.

Other forces have also been at work. The empire was wracked by growing nationalism, and the British solution, a commonwealth of nations, did not work. The British had stayed out of the original Common Market in the belief that they could make the Commonwealth a viable scheme. By the time they discovered it was not workable, De Gaulle was barring them from the Common Market, although they have since joined. Great Britain, the very nation that led in the industrial revolution with the inventive new machines and new ways of organizing labor, fell behind in just these areas in the postwar period. It ceased to invest heavily in new machinery, and union leaders insisted on wasteful practices that led to an estimate of British industry being 10 to 15 percent overmanned. This situation is compounded by the British class system, which tends to limit opportunity—educational and otherwise—for those not born into the right social strata. The system succeeded in producing the leaders to run an empire, but it did not produce the managers and engineers needed in modern industry. Indeed, in Great Britain there has been some disdain for "the trades." When all of these factors are added on to the drift into state socialism, the change from a first-class power to one of uncertain rank becomes understandable. To complete the study, compare Japan, a nation that has moved in the opposite direction.

Space limits the treatment here of northwest Europe. To discuss all countries would reduce each one to a few paragraphs. We have treated Great Britain at some length because of its leading role in the industrial revolution. Ireland will be discussed as an example of population explosion, disaster, and stagnation. Germany exemplifies what can happen to a nation split in two and managed by competing systems. The Netherlands are the classic case of man against the sea, and Norway will be briefly reviewed as a test of the common notion that the land determined that men should take to the sea.

Ireland

Ireland's lack of development, despite its being a part of Great Britain for so long, deserves notice. Its climate is equally stimulating, if cool, cyclonic, storm-dominated climates are stimulating. Why did the industrial

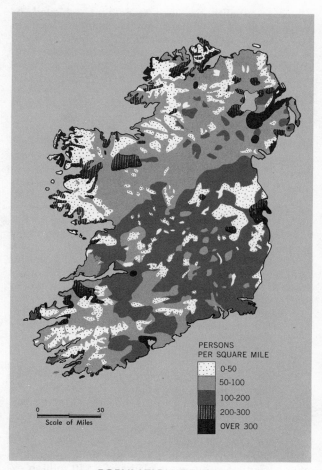

POPULATION DENSITY

PERSONS PER SQUARE MILE
- 0-50
- 50-100
- 100-200
- 200-300
- OVER 300

0   50
Scale of Miles

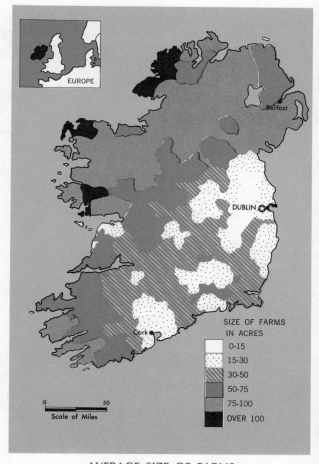

AVERAGE SIZE OF FARMS

EUROPE

Belfast

DUBLIN

Cork

SIZE OF FARMS IN ACRES
- 0-15
- 15-30
- 30-50
- 50-75
- 75-100
- OVER 100

0   50
Scale of Miles

revolution leap to the United States but not to Ireland? Surely it is an oddity that the Irish have had to emigrate to find industrial employment.

The Irish beginnings as civilized people go far back. They were one of the centers of the Mediterranean influences that reached northwest Europe as early as 2000 B.C. In Julius Caesar's time the Irish were still an important, seafaring, advanced people. The Irish setback came during the Norse outbreaks, for Ireland's location in the sea left it subject to pillage by rampaging Norsemen. The great monasteries with their ancient libraries—early centers of learning—were sacked and burned. Thereafter came the period of English conquest and colonialism. Only someone who has read of the 1000 years (roughly A.D. 900–1900) of English conquest and rule of Ireland with its endless revolutions and

reprisals can understand the almost pathological hatred that the Irish have for the English. Ireland was not industrialized because it was a rebellious colony that was held by force. So great was the problem that the English at times tried what today would be genocide. It was a bitter, black history of fighting that went on intermittently for seemingly endless time. Learning, science, and industry do not thrive on this sort of fare.

Toward the end of the seventeenth century there were about a million people in Ireland. Population thereafter grew rapidly. There were a number of causes, but one of the important ones was the arrival of the Peruvian potato. This plant does well in cool climate and short seasons, and it produces an enormous yield of carbohydrate food per acre. When the Irish were being harried by the English, one of their most vulnerable spots was their grain fields. Since these grains yield lightly, the acreages had to be large. The potato, on the other hand, yields hugely, and small fields could easily be hidden. Wheat yields about 30 bushels per acre while potatoes yield about 350 bushels per acre. The balance of forces are difficult to weigh. The introduction of the potato was a cultural adoption, and the climate was suitable. The result of the cultural and physical factors set up the explosive spread of the potato in Ireland. Lest it be thought that adopting a useful plant is obvious and natural, it may be useful to mention that in France, Germany, and Russia the potato was introduced only with great difficulty. In France royalty had to hold potato banquets—from potato soup to potato desserts—and wear a potato flower as a boutonniere to aid in popularizing this new plant. In Germany, the plant was viewed with suspicion, and the first farmers to grow it did not know what part was edible, when to harvest the plant, or how to prepare these newfangled roots for food. In Russia, the notion got abroad that the potato would sterilize those who ate it, and the troops had to be called out to put down the rioting farmers. Men do not easily change their basic food habits and quickly take up new plants. The Irish experience was an unusual one.

Once potato growing gained acceptance in Ireland, it became a national mania. Vast areas were planted with nothing else. Such a situation creates a paradise for any form of life that can feed on potatoes, and in such an environment the pathogen can grow explosively. The large population that had come into being completed the scene for a disaster. When disease hit the potatoes, a horrible famine occurred. The population of Ireland in 1845 is thought to have been 8,500,000. In 1851 it was 6,550,000. Within a period of six years two million people had died or emigrated. (Ireland's population is placed today at just under three million, with approximately one and a half million people in Northern Ireland.) Emigration was particularly heavy to America, which was industrializing and in need of labor, so the Irish flooded into northeastern United States. For the following hundred years in Ireland a pattern of emigration and late marriages has led to a lowered population and to a very poor economy.

Since 1922 southern Ireland, five-sixths of the land, has been allowed to move away from union with Great Britain to an independent status as the Republic of Ireland. In this time industrial growth has been marked, and it is clear that Ireland is in process of becoming an industrialized nation. Ireland, then, becomes a clear example of the failure of mere proximity. The complex interplay of nationalism, religious wars, and the disaster of the potato famine combined to prevent either industrialization, such as Belgium achieved, or commercialized agriculture on the Danish pattern. In a day when large markets greatly aid economic growth, as exemplified by common markets in Europe and Central America, the partition of Ireland and southern Ireland's withdrawal from the United Kingdom seems contrary to their own best interests. It provides one more example of the human will rejecting a possibility that exists in the physical environment.

Ireland's admission to the European Common Market joins the country to a huge market and gives it access to large amounts of capital. Ireland could now play a role something like Puerto Rico in relation to the United States. If Ireland were to open its doors to the inflow of capital and industrial know-how, and if European industry were to decide that the supply of Irish labor was an asset, rapid development could follow. In this day when raw materials are moved freely around the world and a rapid shift to nuclear power is expected, there is no obvious reason why Ireland should not follow other nations lacking basic resources other than manpower and the will to use it. It would not take long for industry to absorb the surplus labor in Ireland, for if there were one potential worker in each family of three, Ireland would have a labor force of about one million, most of whom are already gainfully employed. Should Ireland then increase its population?

Current developments tend not to stress the most labor-intensive development. For instance, the United States is investing in Ireland's developing pharmaceutical industry. In the decade 1963–1973 Irish exports of chemicals rose from less than $2.5 million to $85 million. Medicine exports rose from about $1.25 million to nearly $30 million. One major American pharmaceutical company is planning a $10 million expansion near Tipperary and the employment of 400 workers. Now that the Republic of Ireland is independent of Great Britain and a member of the European Common Market, it should be able to develop exports to enhance its economy.

Ireland today is torn by partition and terrible civil war. It has been the curse of the Irish that they have usually been divided, fighting each other as often as the invader. The present conflict stems from British rule with settlement of Protestants in Catholic Ireland in such numbers that they became a majority in the north. When the Republic of Ireland won its freedom from Great Britain, the British insisted that the predominantly Protestant north be partitioned. The Irish insisted that Ireland is Ireland, and because it is predominantly Catholic, it should not be partitioned. The notion that a country be partitioned on a religious basis seems strange in the twentieth century until one recalls the situation in India. But Ireland is not India, and the south of Ireland actually has had a good record of religious toleration. Not so Northern Ireland. The Catholic Irish there were treated approximately as the Negroes were in the American South: virtually disenfranchised, last hired and first fired—a pattern exactly like that in Boston where 20 years ago "help wanted" ads frequently carried the phrase "Irish Catholics need not apply."

The situation in Ireland is now acutely inflamed. As was frequently the case in Irish history, there is great power meddling. It appears that the Soviet bloc is currently supplying arms to the Irish revolutionaries, just as in times past Germany and even France did. In many ways Ireland is an oddity. It is indeed a natural unit, but so is the Scandinavian peninsula, the Iberian peninsula, and India. These people all partitioned themselves, not seeing geographic unity as being important in their political-economic-cultural world. The Irish, on the contrary, believe fiercely in the unity of Ireland, and seemingly there will be no peace until there is unity.

Cultural consciousness and national pride are interesting qualities and cannot be ignored. Of all the Europeans, perhaps the French and the Irish are those who have these qualities most. Note how strongly the French in Canada cling to their identity. Even more striking are the Cajuns, French Canadians forcibly moved to Louisiana by the British in a brutal attempt to solve the French Canadian problem. Despite everything they have maintained their distinctive culture for centuries. A cultural consciousness, a national pride, a feeling that one piece of earth is uniquely, irrevocably, perpetually and indivisibly Ireland is a force to be reckoned with. Mankind is dominated by ideas: their presence, their absence, their strength. Ireland is a classic case of an extraordinarily powerful set of ideas that centuries of adversity have not stamped out but rather seem to have inflamed. Perhaps the most powerful force in Ireland is the idea that there is an Ireland.

Germany: A Nation Divided

Germany, for centuries France's principal enemy on the continent, is a classic example of a late starter and a strong finisher. The Germans were the last of the north European peoples to be organized into a national unit. German expansion had long been eastward into Slavic territory, and this had added the low, wet plains east of the Oder River to Germany. Germany then came to have two distinct parts: northern Germany, a land of plains draining to the North and Baltic seas, and southern Germany, a hill and mountain country.

South Germany was the closer to the centers of ideas in the Mediterranean. Trade routes from the north Italian area crossed through the Alpine passes and kept the south German area in contact with the advances in the Mediterranean world, while northern Germany, through its merchant and fishing contacts, maintained wide contacts around the Baltic and North seas. The industrial revolution began in England in 1760, but it was not until about 1850 that industrial growth began in Germany. Significantly it began in northwest Germany. The delay is of great interest: the distance was small, and the contacts between the peoples were direct and frequent.

It seems difficult to account for this technological delay in any way except in terms of the lack of interest of the Germans at this time in a factory-type system of industry. They had many highly skilled enterprises. For example, the art of German map making is centuries old and has long been one of the world's finest, but this is typical small-scale industry, using artisans to produce a low-bulk, high-value product. Further, it was centered in southern Germany, the area in close contact with the north Italian industrial centers with similar outlooks. As with everyone else, the Germans had to grind their grains and spin and weave their fibers. Once England had shown the way, they could have made the same start and followed much the same line of development. Nevertheless, there was a lag of about one century.

German industrial development gained a great impulse through the unification of Germany and the successful war against France in 1870, which gave Germany control of the iron of the Alsace-Lorraine— a district that could be combined with the great coal resource of the Ruhr valley. This was also the time when the coking of coal was invented to improve its value for smelting. Yet all of these things can be overstated. The political boundary drawn to give Germany Alsace and Lorraine was drawn with strategic defense in mind, not to gain the iron. Further, the iron ore was of little use because of its phosphorus content, and the solution of the problem of making it useful lay ahead in time.

The good coal had been there for a long time; its usefulness had been potential for thousands of years. Coal for household heating had become important in England by 1650 and was in use in France by 1715. The Ruhr valley, with its very high-grade coal and iron ores nearby, located in a country with contacts reaching southward into the Mediterranean and north-

westward around all of northwest Europe and in the world's most stimulating climate (according to the environmental determinist thesis, see Chapter 1), should have been the center of the industrial revolution, instead of lagging behind for a century. The story is again that resources and other physical factors are secondary in importance to cultural factors.

Once Germany began its industrialization, its development went at a phenomenal pace. The cultural elements on which to build were there. The educational base was broad, there was a widespread, diversified, small industry, and trade and commerce were ancient callings. Well-developed routes up the Rhine, over the Alpine passes, and down the Danube had long been in existence. When national unity, with purposeful direction of the nation's growth into industrial lines and away from the dominance of agriculture, began in the 1860s, Germany sprang from a minor position to a first-rank industrial nation within 50 years. The change was as dramatic as the emergence of Japan from its long isolation into the modern world.

Germany had an advantage in its late start because the latest improvements could be incorporated in its new industry. England in contrast had an investment in going plants that were on their way to obsolescence but which it was reluctant to scrap, and this reluctance to make necessary improvements, along with the punitive tax structure of the post-1900 era, were factors that cost that country its world position. After each of the wars in this century, Germany has made a phenomenal comeback. In each case energetic application of human skills and rebuilding with the most advanced and newest equipment have allowed the Germans to jump back into an advantageous competitive position.

(Upper) **Coal Mining in Germany.** *Germany was late in unifying and in fully entering the industrial revolution. However, it has developed rapidly, building upon its assets—a broad base of skilled workers, manufacturing knowledge that had long existed there, and its many natural resources such as coal—and has become a major industrial nation. Changing technology is now making Germany's coal less important, but the skills and social energy of the German people keep them in the forefront among productive nations.* (German Information Center)

(Lower) **Ceramic Works, Westerwald, Germany.** *The handwork on steins shown here is an example of the continuation of a traditional craft of earlier times. Drawing from such traditional manufacturing, Germany was able to develop skilled workers and experienced industrialists.* (German Information Center)

It is too often said that Germany's advantages are enormous. It has the greatest coal reserves in Europe, vast areas of lignite, and the valuable Stassfurt salt deposits. This omits the fact that Germany is short of oil, iron, copper, and a long list of essential materials. The scale and intensity of Europe is well exemplified by the Ruhr area. This industrial zone is 45 miles long and 15 miles wide and has 5 million people in its nearly continuous urban development. Several individual United States centers exceed this in size, equal it in population, and approach it in production. German soil is poor in the north and limited by the rough nature of the land to the south. Less than one-fourth of the country is suitable for agriculture. The pre-World War II area was only equal to that of the state of Texas. Crippled by partition between the Western powers and the Soviets, with its capital city divided between these powers and isolated in the Communist zone, with the eastern provinces given to Poland including the industrialized Silesia areas, West Germany today has nevertheless come back as an industrial power. It is a classic case of man over environment.

The role of economic and social systems has seldom been more clearly demonstrated than in the different courses of East Germany and West Germany. The division of the country, intended originally only as a matter of convenience while a peace treaty was worked out, has turned into a protracted division of the country, due to the Soviet seizure of every pretext to make of East Germany a colonial area subject to its exploitation. East Germany was the more agricultural part of the country, yet it contained important industrialized areas in Saxony and Silesia. Under Soviet domination there has been a continuous flight of people from East Germany, which has drained off particularly the educated and skilled part of the population. Recovery has been slow and production low. Although the Soviet need of German productivity has created an endless market, the fixed prices and controlled production in a socialist planned society have stifled productivity instead of stimulating it.

West Germany adopted an aggressively free enterprise system after the war. Despite the loss of nearly half of its territory, occupation by the Western Allies, immense destruction during the war, and the loss of many of its leading scientists to both the East and West, cities have been rebuilt and industries reestablished, and now the general situation can only be described as prosperous.

Among other problems West Germany had to absorb over 10 million refugees. Although this was an enormous immediate problem, industry, in its explosive growth, has absorbed this nearly one-quarter increase in population and even run into labor shortages at times of peak employment. West Germany in 1970 had more than two million foreign workers: Yugoslavs, Italians, Greeks, Spaniards, Turks, and Portuguese.

As the data in Table 6-1 show, the population in West Germany is growing while that in East Germany remains static. The gross national product in West Germany has grown by $53 billion while that in East Germany has grown by $12 billion. Electric power in West Germany has increased by 113 kwh while that in East Germany has increased by 26 kwh. Refrigerator production, a measure of standard of living, has gone from one to two million in the West, but the East has yet to reach even one-half million units. This disparity is much greater than it seems, for a mini-refrigerator costs only $80 in the West while it costs over $320 in the East. The West Germans have a huge lead in every aspect of industrial output. The farm picture is similar. Per capita farm production rose by 40 percent in the West between 1965 and 1970 while it rose only 16 percent in the East. The East German farms have been collectivized; the West German farms are privately owned.

These disparities extend into other areas of the economy. West Germany has traffic jams and struggles with the problem of parking spaces. East Germany is still in the motorcycle and motor scooter era. West Germany had 1300 miles of super highways 20 years ago, has doubled that mileage now, and plans for 5000

TABLE 6-1   West Germany versus East Germany

	West Germany		East Germany	
	1960	1970	1960	1970
Area (sq. mi.)	96,000		42,000	
Population	55,000,000	60,848,000	17,241,000	17,000,000
Gross national product (billions of dollars)	$103	$156.5	$25.4	$37.7
Electric power produced (billion kwh)	118	231	40	66
Refrigerators produced	1,000,000	2,300,000	139,000	366,000

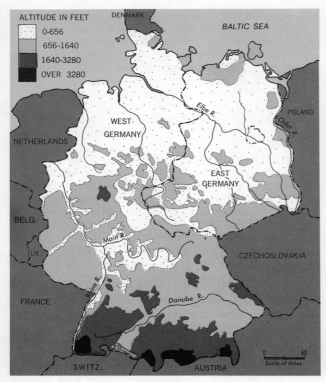

LANDFORM

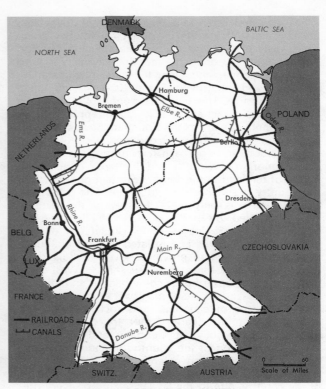

TRANSPORTATION SYSTEM

more miles of autobahn. East Germany had 900 miles of autobahn that it inherited and has built no more. Housing in East Germany is old, dilapidated, and scarce but cheap. Food is costly there; meat is 20 percent higher than in the West, and eggs are twice as costly. Only in education does East Germany come out ahead: the children of the poor have a better chance to get a higher education, though it may not be as good as in the West. Germany is an excellent test case, for at the starting point it was one race, one language, one culture, one nation. It was arbitrarily split in two, with two economic-political systems applied. As statistics show, one has flourished considerably more than the other.

The Netherlands: Man against the Sea

The Netherlands is the classic story of men against the sea. It is also, however, a success story illustrating that nations need not be huge, that great

population density need not spell poverty, and that an industrial nation need not have great resources of basic materials such as iron, coal, and oil in order to succeed. This tiny country of 14,140 square miles—about the size of Massachusetts, Connecticut, and Rhode Island combined—has a population of 13.5 million people, a density of 900 people per square mile. The Netherlands is about 10 degrees farther north in latitude than New England, comparable to the latitude of northernmost Newfoundland, but because it is on the east side of the Atlantic where the westerly winds bring ocean conditions inland, it is no colder than Rhode Island in winter and much cooler than Massachusetts in summer.

The outstanding accomplishment of the Netherlands has been its control of water and reclamation of land from the sea. Without the dunes and dikes the largest and most densely populated part of the country would be flooded. The area of the Netherlands took shape during the glacial period when the sea level was far below the present. Alluvium from the continent was swept far out into what is now the North Sea, and great dune belts were built up on this alluvial sheet.

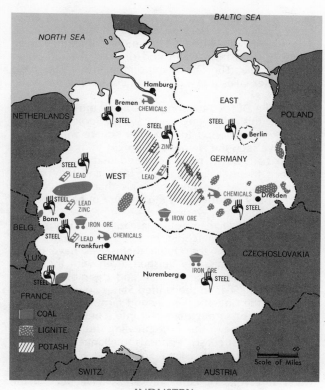

INDUSTRY

MAP 6-4 Germany: Landform, Transportation, and Industry
Germany is a land of low elevation with a great coastal plain to the north and modest highlands to the south. The relatively flat land has allowed Germany great freedom in laying out its railroad network, and the fact that Germany developed industrially in the railroad era has led to a dense network. The major rivers of Germany are navigable and carry a great deal of freight. In addition, the glacial edge ran east and west, creating east-west lowlands, which the Germans have utilized as connecting links between such rivers as the Ems, the Elbe, and the Oder.

In the postwar period under an enterprise system West Germany has rebuilt its industries and become one of the world's most prosperous areas. East Germany, despite its good resource base, has stagnated under the restrictive planned programming of the Communist regime. Note also on the map the location of major industries at ports such as Bremen and Hamburg where there are no notable natural resources. (Reprinted from Focus, *American Geographical Society)*

When the ice melted and the sea level rose, the sea flooded in over the lowlands, and the great dune belts were left standing as long series of islands. The Rhine then began filling sand and clay behind the dunes, and peat deposits formed in the shallow-water bodies. After 700 B.C. the encroaching sea began to sweep away some of the peat. Man increased this effect by digging peat and enlarging the areas of water. As early as 1200 the Netherlanders began building dikes to trap sand and silt and build up the land; this land reclamation has continued over the centuries.

Reclamation of land from the sea amounts to nearly 3000 square miles and is about 20 percent of the national territory. The really large-scale reclamation came after 1900 with the increasing availability of power machinery. The biggest undertaking was the turning of the Zuider Zee into a series of polders (reclaimed areas) with the remaining water area being a fresh-water lake. An even more ambitious plan would be to reclaim the Wadden Zee, but conservationists are opposed to this because the area is an important wild life sanctuary. Many polders are below sea level, and the excess water must be pumped out. The lowest

polder is 21.7 feet below sea level. Windmills—so firmly fixed in our minds as Dutch but actually invented in the eastern Mediterranean and then used widely as a grain-grinding device (recall Don Quixote tilting against the Spanish windmills)—were employed in the Netherlands to lift water from the polders; today this work is done by diesel or other power-driven pumps.

The population of the Netherlands is concentrated in the cities: 65 percent live in towns with more than 10,000 inhabitants, and 47 percent live in the Amsterdam-Rotterdam area. This is the result of the concentration of industry in the Rhine mouth area where the overseas shipping is concentrated. The rural population is declining as agriculture is mechanized, and it is this population that is feeding the growth of the cities. The so-called West Holland conurbation is today virtually a single city about 45 miles long and 40 miles wide. There is no single center, for this area includes Amsterdam, Haarlem, The Hague, Rotterdam, Utrecht, and Hilversum. The growth of this center has occurred in a hundred years. In that time the population of the Netherlands has gone from three million to 13.5 million, while population in the area of the conurbation has gone from one-quarter million to four million. The great impetus to the growth of the conurbation has been the ports of Amsterdam and Rotterdam that form major gateways into Europe.

Rotterdam is itself a virtual megalopolis. Factories are crowding into the area to the extent that the Dutch are concerned over the maintenance of tree belts and the amenities of the countryside. The growth extends 60 miles away to the Antwerp region in Belgium which had $1 billion of industrial development between 1960 and 1970. Rotterdam has been receiving industrial investment at the rate of $100 million per year. One hundred sixty million people live within 300 miles of this region, while only 50 million live within a comparable radius of New York. Enormous investments are being made in harbor improvements, including deepening, so that they will be able to handle containerized cargoes and receive the supertankers that are already carrying some of the world's cargoes. The immense dredging and filling operations also make more land available for expansion of industrial plants.

It is startling to find in a land with so great a density of population that so much of the area is still farmland. Land in general farming makes up 26 percent of the land; specialized horticulture takes up 5 percent; pastures occupy 46 percent; woods, heaths, and dunes account for 13 percent; and only 10 percent of the land is in buildings and roads. The principal crops are potatoes, sugar beets, and wheat, though most of the wheat for bread is imported. The cattle industry is very important, as the figure for pastureland indicates. The very labor-intensive horticulture features the production of vegetables, fruits, and flowers; Dutch tulip bulbs are shipped to the United States by the ton. Perhaps the outstanding point in the figures for land use is the relatively tiny amount of land used for housing, employing, and conveying (roads, railroads) the enormous number of people in this crowded land.

Although the idea of closing in the Wadden Zee is currently discouraged by ecologists, the Dutch have gone ahead with the so-called Delta Project. This was precipitated by the great destruction caused by a huge storm tide in 1953 that flooded 400,000 acres of land and drowned 1800 people. Too many miles of dikes were necessary to follow every inlet of the complex delta, and the Dutch boldly set out to close off the mouths of the estuaries. This amounts to shortening the coastline by 450 miles. Salt water is prevented from infiltrating the country, and isolated parts of the country are now connected with the rest. The losses are shrimp, oyster, and mussel fisheries of the estuaries, but the gains far outweigh these losses. One of the great gains lies in the control over the fresh- and salt-water situation. All industrial nations today are plagued by enormous demands for fresh water. Coastal areas find when they heavily pump the deep aquifers that they soon pump out the fresh water and draw in the salt water from the sea. The Dutch impoundments create huge fresh-water lakes that aid them in controlling the water problem.

The Dutch situation poses many intriguing questions. Is this what the world would look like if the population exploded to 1000 per square mile? Those who have seen the Netherlands might think it wouldn't be too bad. But it must be noted that the Dutch do not feed themselves from their land, and they import a great deal of the raw materials that they work up in their busy factories. Still the amount of open space—farm, pasture and, waste land—is impressive. Another question is how many other areas of the world offer possibilities for snatching hundreds of square miles from the sea? As the Dutch themselves have noted, they are not necessarily finished, even with their own territory. Would it be feasible and desirable to close off the Baltic Sea? In the United States could or should we close the Chesapeake Bay and farm most of it, and meanwhile, by turning part of it into a huge fresh-water lake, solve some of the more pressing fresh-water problems of the East coast? Proposals to change parts of San Francisco Bay, Long Island Sound, the Hudson River below Albany, and other such areas either into fresh-water reservoirs or cultivable areas have been made from time to time. How many more opportunities of this sort are there in the world and why, with the Dutch example before us, do not more people do likewise? All of the old fundamental propositions come to bear here: need, necessity, perception of need or opportunity, invention, and diffusion. It is a situation that merits much consideration, for there clearly are more potential Netherlands in a world beginning to worry about food and space than we realize.

Norway

Norway is almost the classic case of a far northern marine west-coast land. Norway lies in the latitude of mainland Alaska; the parallel of 60 degrees latitude passes through southernmost Norway, near Oslo, and through Alaska at the base of the Aleutian Island chain and the Kenai Peninsula. There is no land in South America so far from the equator. Tasmania and the South Island of New Zealand are far nearer the equator than these lands. The human assessment elsewhere in the world is that lands so far poleward are virtually

Rune Stone, Rök, Östergötland Province, Sweden. *Runes are the characters of the alphabet used in early Scandinavia. Their history is obscure, but they are related to the early alphabet systems of the eastern Mediterranean. They are evidence of the important flow of cultural stimuli from that region to northwestern Europe which continued for millennia. Stonehenge and Hadrian's Wall are other monuments to this cultural path.* (Swedish National Travel Office)

potatoes, oats, and cabbage, grow well. The Norse have turned to the sea for fish and to shipping as a means of obtaining foreign exchange. The forests supply them with wood and pulp for export either in pulp form or paper. In an environment that mankind has in recent time largely avoided, the Norse have built a good life and maintained a national existence. It could well be said that they have done so in spite of the environment. It is about equally true to say that they owe their advance to their proximity to the advanced developments in northwest Europe.

Consider, for example, developments in the similar areas elsewhere. The people of the northwest coast of America did indeed take to the sea and come to depend heavily on the rich fisheries there. At the same time, however, they turned their backs on the land, so agriculture was not developed. Until the present neither the Canadians nor the Americans have done much with these lands. Southern Chile languished in pre-Spanish time and remains underdeveloped today despite the presence of timber, water power potential, rich fisheries, and most recently oil at Tierra del Fuego. The southwest side of the South Island of New Zealand, again very much like Norway, was in the past and continues today to be an almost totally empty land. The Norwegian development, then, becomes a special case. Only there did men utilize this mountainous, forested, recently glaciated, fishery-rich type of coast in this way. The longer history of strong contacts with adjacent developed lands provides the obvious reason for the Norwegian development, and the comparative notes on the nearly identical lands confirm the judgment.

The Minerals of the North Sea

During World War II the Dutch found some very heavy petroleum on their sea margin. Immediately after the war it was found that there were huge natural gas deposits under the North Sea adjacent to the Netherlands, and the proved reserves there today are so great as to place the Netherlands in fourth position in natural gas deposits. These discoveries set off a flurry of exploration of the sea bottom by all the adjacent nations. But before exploration could more than begin, the problem of the ownership of the adjacent sea bottom had to be decided. There is virtually no law of the sea—an oddity in a world where over two-thirds of the surface of the globe is water. Beginnings have been made, however, and on this basis the nations around the North Sea divided up the sea floor. Little went to Germany, and Norway would also have received very little

worthless. Enormous Alaska has less than 337,000 people, and most of them live in the southward-extending coastal section, the Panhandle, that lies in the latitude of Denmark. Norway, on the other hand, has a population of over four million. While this population is mostly in the south, there are settlements all along the coast to beyond the Arctic circle.

The Norse have a poor physical environment in many ways. Minerals are scarce, coal and oil have been virtually nonexistent, and areas of good soils are best described as rare. The uplands are granite from which most of the soil was stripped by the glaciers, and the lowlands are few and limited in size. The growing season is short, and the summers are wet; because of the persistent cloud cover, relatively little of the potential sun can get through to the land. Few crops, principally

Henningsvar, Norway. *This photograph shows the sparseness of land in Norway and the country's great dependence on the sea. The steep snow-covered mountains are obviously of little use agriculturally. The drying racks with their burden of fish indicate the importance of the product. Would this scene be duplicated in Tierra del Fuego, a similar land?* (Norwegian Information Office)

if a strict interpretation of the nascent law of the sea had been adhered to; a long, deep arm of the sea cuts Norway off from the shallow North Sea area that it would otherwise have been able to claim. Norway insisted that its claim to that area be recognized, and the other nations conceded the point. The undersea drilling is now proceeding feverishly, and oil and gas are being found in significant quantities. Belatedly Norway has found important oil supplies on its share of the North Sea bottom. These quite unexpected finds alter the energy requirement picture of the North Sea countries, but not nearly as much as one might expect. The growth of energy demand is so great that even these new supplies have no great life expectancy. They should, however, tide these countries over until nuclear power comes of age.

Europe: Reorganization of Space

The European Common Market

A gigantic change in the economic and political geography of the earth is underway in Europe through the formation and expansion of the European Common Market. The original six nations have been joined by three more, creating a unit that will be a world power of startling proportions. The mere rearrangement of economic ties promises to remake part of the world's trade pattern and increase vastly the standard of living of most of western Europe.

The United States has been the single largest trading country with exports of $43 billion in 1970. The six-nation Common Market (West Germany, France, Italy, Belgium, Luxembourg, and the Netherlands) had exports valued at $89 billion, 32 percent of total world trade. To some extent this immense productivity in Europe is the result of American aid during the rebuilding period after World War II and of American investment and input of industrial know-how in more recent years. The European recovery has been so vigorous that in some respects the United States finds itself at a disadvantage in competing with this enormous new power that owes so much of its strength to a simple reorganization of space. With the addition of Great Britain, Denmark, and Ireland, the productivity of this unit grows to truly impressive heights. An industrial giant that will be second only to America has come into being.

The United States has a gross national product approaching $100 billion. The Soviet Union has just over half of this productivity, and Japan, the third-ranking world power in the present alignment, has not

(Left) **The Oseberg Longship.** *With these long, low, swift ships the Vikings ventured not only to northwest Europe but to the Mediterranean as well. For trading they used deeper, slower, more capacious ships and undoubtedly reached the United States many times. (Norwegian Information Service)*

(Upper right) **Harvesting the Sea.** *The Norwegians, along with the people of other Atlantic-facing nations, have long gone to sea for much of their protein. The boats they've used have become larger and more mechanized to the extent that today the capacity of the many fishing fleets to extract fish from the sea has exceeded the supply of the sea. (Norwegian Information Service)*

(Lower right) **The Fjords of Norway.** *For some nations the fish from the sea are just a supplement to the diet, but to nations like Norway, much of whose land is mountainous, rocky, and snow covered, the sea is of vital importance. The fjords are important as shelter for the fishing fleets and for attracting tourist traffic. Landscapes such as this can also be seen in Alaska, Chile, and the southwest corner of the South Island of New Zealand. (Norwegian Information Service)*

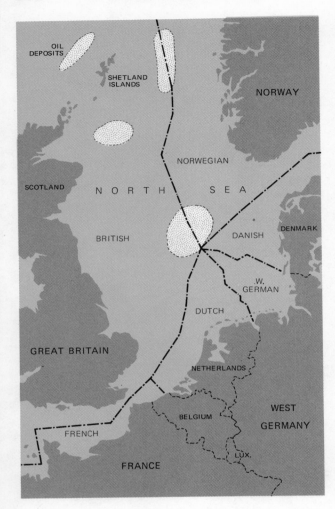

MAP 6-5   Oil Discoveries in the North Sea
The countries bordering the North Sea divided the ownership of the sea floor as shown here. Exploitation of the sea floor led to the development of a new area of international law. Here international boundaries were projected seaward to the midline between nations, with claims not to go beyond the continental shelf. Norway claimed an exception because it is cut off by a deep oceanic canyon. Negotiations resulted in the present divisions. Notice how critical a minor displacement of some of these boundaries would be. Fortunately, the boundaries were set before drilling began.

Exploration for oil and gas led to discoveries in the areas circled. The cost of exploration has been high: 10 times that of drilling on land. Drilling platforms had to be built to withstand waves 90 feet high and winds blowing up to 100 miles per hour. Billions of dollars have been invested by more than 200 companies, and huge finds have been made. Because the oil is adjacent to the consuming centers, finds have been highly profitable despite the high costs of exploration and development. Great Britain expects to meet half its needs as a result of its wells and also save $1.5 billion per year in foreign exchange. The natural gas available in the area can be used for central heating in British homes. Scotland is becoming the Texas of Great Britain, with Aberdeen the center of the booming offshore oil industry. So great are the modern demands for energy that these fields are expected to last only until the end of the century. By that time atomic energy or other energy sources should be developed to supply our energy needs.

quite $20 billion. The other European states, notably West Germany and France, then follow. With the nine western European powers combining, they emerge as the second power in the world, outranking the Soviet Union. Should Spain and Portugal and perhaps Switzerland be included—and it is difficult to see the geographic or economic logic of excluding them—this power block would become even more significant.

The effect of the Common Market on the welfare of the units is especially notable. Since 1958, when the Common Market was organized with the original six nations, trade among them has increased sixfold. This is twice the rate of growth of world trade. With figures such as these one begins to feel the enormous extent of cultural control over human productivity, well-being, and advancement.

The United States has worried considerably about the impact of the Common Market on its trade with the European area. The situation is extremely complicated, for there are numerous trade agreements, tariff laws, and other controls over trade. The Common Market tends to favor trade with one another. This is particularly true in the agricultural field where subsidies keep prices very high but agreements keep American products out. This is partly due to the member nations' desire to keep their farmers happy and partly due to the effects of the Green Revolution, bringing the enormous increases in agricultural productivity that are now so strongly influencing world food output.

United States exports to the Common Market are declining in the area of raw products and "nontechnological-intensive" products such as textiles, shoes, and manual typewriters. On the other hand, our exports are growing in the "technology-intensive" areas: computers, nuclear reactors, and scientific instruments. The developed nations, then, have their greatest trade with each other. It is not possible to sell many computers in tropical Africa, whereas the European market for them is huge. We can expect that increasing Euro-

MAP 6-6   The European Economic Community

The European Economic Community, or Common Market, is currently comprised of nine nations: Great Britain, Ireland, Belgium, Luxembourg, the Netherlands, France, Italy, West Germany, and Denmark. The early years of the EEC led to extraordinary growth in trade among the member nations, but in recent years economic and political differences have multiplied. The admission of Great Britain in particular has been accompanied by extensive economic problems. As we have seen, the difficulties of uniting nations, even those closely related in language, culture, and economy, are truly formidable.

pean productivity will mean increasing, not decreasing, trade with the United States. At this point we should take a look at the pattern of world sea trade. The greatest flow is between the developed nations. The second greatest flow is that of raw materials from the underdeveloped nations moving to the developed nations. This movement of goods is now so huge that should it stop, the whole economy of the world would be thrown out of gear. A measure of the importance of world shipping is the fact that the freight charges are more than twice the value of the yield of fish from the ocean. The United States, Japan, and the Common Market countries are so dependent on this flow of materials that none of them could tolerate even a brief interruption.

PROFILE OF THE EUROPEAN COMMON MARKET

Membership
France, West Germany, Italy, Belgium, Luxembourg, the Netherlands, Great Britain, Ireland, Denmark

Population
255 million

Area
715,000 sq. mi.; one-fifth the size of the U.S.

Gross National Product
$615 billion; 60 percent of U.S.

Trade with United States
27 percent of U.S. sales abroad: $11.3 billion
23 percent of U.S. purchases abroad: $9.4 billion

The Common Market is a loose federation of states with its headquarters in Brussels. All decisions must be unanimous, and this has made negotiations and planning very cumbersome. The movement toward political unification has been full of fits and starts. West Germany, once strongly for it, is now lukewarm, while Italy remains supranational. France is suspicious of West Germany, and its switch from opposing to sup-

porting the admission of Great Britain was made to counterbalance West Germany. This all suggests considerable difficulty in gaining political unity. However, the United States went through exactly this kind of struggle at its beginning, for the colonies were jealous and suspicious of one another, and a federal government was achieved with difficulty, despite the far greater linguistic, cultural, and social uniformity in America. In 1972 the unanimous decision requirement had caused a logjam of important decisions, some postponed from as far back as 1966. With more nations the situation can only become worse. Expectably a European parliament will be formed with limited powers, probably purely economic at first. When one sees the great difficulty of uniting one group of nations, all sharing related cultures, one realizes how enormous the problems of world organization actually are and how complicated it becomes to find solutions to these problems.

Changes are rapidly occurring in a number of fields. Agriculture is being revolutionized by investment in farm machinery, and the productivity of farm labor has greatly increased as a result. Just as in the United States, there are fewer men on the land who are producing in greater quantities. The Common Market is self-sufficient in many agricultural products, and it has a surplus in a few. However, it still imports large amounts of food and feed from the United States. Restructuring of land holdings from tiny fragmented plots to larger, more compact economic units has increased both efficiency and output. Food production is increasing more rapidly than the population, a notable accomplishment in this far northern land (France is in the latitude of New England) and quite a contrast to tropical lands in sub-Saharan Africa, India, and mid-latitude China where food production lags.

The Common Market and its industrial growth have led to enormous population movements within Europe, which may be the greatest force in ultimately leading to a unified western Europe. English is becoming the common second language, styles of dress are growing similar over large areas, business is interlocking and expanding across international boundaries. Frenchmen are drinking Scotch and Coca Cola; Italian refrigerators sell well in France. West Germany and France have huge numbers of foreign workers.

Remittances to homelands are important in such places as southern Italy, and the migrant workers are learning skills and accumulating capital. The immigrant workers get the same pay as the nationals, and they often qualify for family allowances, unemployment, and other benefits.

In the lands donating the labor the general reaction is one of enthusiasm. The industrialist feels that he stands to gain a trained and educated labor force at no cost, trained faster and better than would be possible in the infant industries of his own country. A Portuguese farmer is quoted as saying that he had one son in West Germany, one in Switzerland, one in Lisbon, and one at home in a cork-baling factory but soon leaving. He was glad that they had the opportunity to get away, though he hoped that one or two might return. Italy, the great source of labor, has sent 2,500,000 people "north": one million to France, 400,000 to West Germany, 200,000 to Belgium, 20,000 to Luxembourg, 15,000 to the Netherlands. In parts of Portugal and Spain more women than men now labor in the fields, and in Italy where the process of draining the manpower away from the fields has been under way longer, a great deal of land that was formerly farmed now lies abandoned. Nevertheless, the remittances to Italy amount to a few hundred million dollars per year and are a major source of foreign exchange.

There are problems and strains, of course. When Italian refrigerators sell well in France, French manufacturers cry for protection; when French autos sell well in Italy, Italian manufacturers ask for protection. West Germany and France tussle over the price of agricultural products for which they have differential costs of production. Nevertheless, together the Common Market countries, with their highly educated and productive people spread over a wide range of terrain and climate, have a broad agricultural base and many of the minerals needed for industry. With access by sea to the entire world, this group is a potential world power and seems closer to uniting economically than anyone had dreamed. If they do unite, the now scattered power of Europe will become a focus of energy that will make this western peninsula of Asia one of the most powerful pieces of geography in the world—because of the rearranging of the human geography. All the possibilities in the physical and human geography have always been there. Their potential has been recognized for decades, but the realization of the potential is only now slowly and painfully moving toward realization.

Summary

We may summarize the developments in the marine west-coast climate of Europe as follows. Unquestionably the development would have been utterly different had the course of human history been other than what it was; that is, there is nothing in the climate or the location that made this inevitable. The beginning of civilization lay in the Middle East. The Mediterranean provided a seaway leading west. This opened up the southern side of the peninsula of Europe to the penetration of ideas, setting the stage in northwestern Europe for the beginning of the industrial revolution. This technological change could just as well have come about almost anywhere in the whole European area, the Mediterranean, the Middle East, India, or China. That it began in Great Britain can be compared to a home run being hit by one man rather than another in a world series game. The stage was set, many men were prepared, they were at the plate, ready and waiting, and someone was due to hit sooner or later. Just who made the hit is not entirely chance; some individuals are more able than others. But without the proper setting they would never have a chance to display their ability—not even Babe Ruth or Socrates, in his own field of accomplishment.

Once the start was made, many series of events were set in motion. Some of these had momentum; many had feedback. Machines required metal; this promoted metallurgy. This in turn increased the efficiency of machines that required more metals and so on. The long sequence of events from early beginnings in the Middle East, with radiation outward of these ideas, set the stage for a unique growth in human history. Its almost accidental correlation with one climate has led to the grossest hypothesizing about stimulating climates and the role of favorable natural resources. Comparisons with what occurred in other marine west-coast lands can serve to give a better perspective.

THE AMERICAS

The Americas have two areas of marine west-coast climate. The northwest coast of North America, from 60 degrees north to northern California at 40 degrees north, is cool in summer and wet throughout the year. It is a narrow strip because the mountains parallel the coast and are close to the sea. Beyond the mountains the rainfall drops rapidly due to the rain shadow effect, and the seasonal temperatures increase due to the blocking of the oceanic influences. The identical situation prevails in southern Chile from the tip of South America at 55 degrees south to just north of 40 degrees south. This climatic physiographic resemblance extends into the common past glacial history of these two lands. There are sizable glacial remnants in both: ice-scoured valleys now drowned by the sea (fjords) and small amounts of flat areas with good soils, plus a dense cover of rain forest. In both areas the flatter land with better soils is at the equatorward end: Puget Sound and Willamette Valley in North America and the southern end of the central valley of Chile in South America. Poleward of these flatter, soil-mantled regions in both areas the mountains rise out of the sea. This gives an island-filled sea with innumerable channels, inlets, passages, and potential harbors. This fjorded, island-guarded coastal effect is much like the Scottish and Norwegian coasts.

The Northwest Coast of North America

The pre-Columbian period

In northwest America in pre-Columbian time the coastal lands from 60 to 40 degrees north were used by a sea-faring people, and the center of their development was approximately the Puget Sound area. Their culture area was bordered by the Eskimos to the north

and trailed off in development toward the simple tribes of California to the south. Tribes with the most advanced form of this culture were the Tlinkit, Kwakiutl, and Haida.

These tribes were skilled, woodworking people. The vast forests supplied them with immense straight logs of soft woods, such as spruce and cedar. From these they made large sea-going canoes. They also split planks off the standing trees for making houses. House frames of huge posts and girders and ridge poles were often maintained at several sites, and when the people moved seasonally, they took the planks for roofs and walls with them. These are the people who made the totem poles, so well known to everyone. It is less well known that the totem poles as we know them were a post-European development. Steel axes and adzes made it possible to produce in gigantic size a type of carving that until then had been made in small scale. Ceremonial staffs grew to super telephone-pole size. Their woodworking also extended into house furniture. Spoons, boxes, and combs were made of wood, beautifully carved according to a rich art tradition. Did the abundance of a suitable wood determine all of this?

The sea supplied a rich fishery, an area of immense salmon runs, rich halibut fisheries, great shellfish resources, and seasonal abundance of other fish such as the sardine-sized oulachon. With a seashore deeply indented by the sea and with innumerable islands forming endless sheltered passages, inlets, and harbors, what could be more natural than taking to the sea? Under these circumstances the northwest Europeans also went to sea. We have traced much of what the Europeans accomplished back to ideas coming from the Mediterranean and often ultimately from the Middle East. Where did the marine west-coast people of the American Northwest get their ideas? Or did they develop their own in response to the environment in which they lived? Here we can make a comparison with New England, especially Maine. This area has a rocky coast with innumerable islands, bays and inlets, a great forest cover, tremendous salmon runs (into the nineteenth century), and huge fisheries along the shore and on the banks. The physical environments of New England and the northwest coast region of the Pacific are remarkably alike, the wealth of fisheries perhaps somewhat greater in New England. Yet the New England Indians did not build strong canoes, develop a great fishery, build snug plank houses, or create great art in woodwork comparable to that developed by the Indians of the Pacific.

It was noticed long ago that many of the things that the people of the Puget Sound area did were repeated in New Zealand. The fish hooks of the Ameri-

cans, for example, resembled those of the Maori of New Zealand. Both groups built great sea-going dugout canoes, both built great community houses with post-girder-and-ridge-pole frames with plank covering, both groups were great wood carvers, and both carved the posts of their houses in similar grotesque art forms. The environments were both marine coastal climates, with all the similarities of forest cover and availability of large logs, and both people were oriented to the sea. But why should their compound fish hooks be identical, and oddest of all, why should their art forms be so alike? There are further similarities: for instance, both people built similar fortified villages and indulged in great give-away contests to create personal prestige.

The list of similarities becomes so great that we must ask whether we are looking at environmentally conditioned results that have led to such similarities. Or are we looking at transmission of ideas? Geographers have paid relatively little attention to these two peoples, for most geographers concentrate on the present-day scene. Although the anthropologists have been interested in the cultural similarities, they have used a geographic reason to deny any probability of exchange of ideas between these two people. They point out that the two areas are almost at the opposite ends of the earth, a distance much too vast for men equipped with mere dugouts to travel.

This is a curious bit of reasoning. First, these two peoples are on the same ocean, they are seafarers, and the seas are better highways for long-distance travel than the land. Second, their "mere dugouts" were often 50 feet long, and some of the New Zealand canoes approached 100 feet in length. These are not mere dugouts, they are ships. The *Niña*, of Columbus's fleet, is

(Upper) **Totem Pole from British Columbia.** *The totem pole is a series of animal and human forms, stacked one above the other, and used as heraldric devices. The New Zealand form is exactly the same thing and includes details of artistic style that are identical to those on the Northwest coast of America.* (Canadian Information Office)

(Lower) **Maori House-Post Carvings, New Zealand.** *These are strikingly like the Northwest coast Indian totem poles, which were also originally house-post carvings. The likenesses are obscured by such stylistic details as the elaborate use of decorative scrolls on the Maori carvings as compared with the simpler forms used by the Northwest coast Indians. Certain details suggest the close relationship between these art forms. The protruding tongue, shown on the second post from the left, is continued today in the Maori haka dance. Protruding tongues are also a regular feature of the totem-pole art.* (Courtesy of The American Museum of Natural History)

thought to have been only about 40-ton capacity and not to have been entirely decked over. Further, once men take to the sea, the inevitable storms are bound to create castaways, and the ability of man to survive at sea is so great that he will reach virtually all the adjacent shores of the seas. As an extreme case, the Eskimos have repeatedly reached the British Isles from some point in the American Arctic, probably Greenland, in their tiny kayaks. Japanese junks, some of them bearing living crewmen, reached the northwest coast of America repeatedly in the nineteenth century. From inscriptions on South Pacific islands we know that mariners from Java sometimes reached them or were drifted far out into the Pacific. Beyond that, Thor Heyerdahl and others have demonstrated that purposeful long-distance voyages were easily within the means of the people of the Pacific.

Perhaps the resemblances between the New Zealand and American northwest-coast way of life, art, architecture, and tools are due not to either group having voyaged the length of the Pacific but to their both having common origins in Asia. Studies of New Zealand art indicate that it is relatable to Asiatic art, specifically to early Chinese art forms. Studies of the American northwest-coast culture show that it is filled with Asiatic traits, and their art has the same foundation. Among the Asiatic traits are specific types of raincoats, special kinds of armor, and mythological accounts of the coming to them of people from the west.

Perhaps the clearest case is the parallel between the whaling techniques of the American northwest-coast people and those of the Asiatic coast. In both areas whales are hunted with poisoned harpoons, and the poison used is identical. During both the preparation for and the hunt itself, identical taboos are practiced by both peoples. While it may or may not be thought natural to hunt whales, it is very unusual to kill them with poison. That the same poison should be used is suspicious, but when the same nonfunctional magical practices are used by both people, a clear case of spread of ideas is before us.

Once the door is opened to the spread of ideas, the American northwest-coast people, with their highly developed sea culture, their fine art, and peculiar customs, turn out to be a transfer to this part of America of a way of life that was highly developed along the coast of Asia from Japan to the Kamchatka Peninsula in ancient times. The transfer was from one climate and one physical environment to another. The common factors were the sea and the rich fishery. A similar area lay in New Zealand. People coming from the Asiatic coast eventually reached this area too. Since both groups started with a similar stock of ideas, including how to achieve their art work, the results turned out much alike.

There are differences, however, for the Maori of New Zealand were farmers and the people of the Pacific Northwest coast were not. Did they both start

Old Indian Settlement at Tanu, Queen Charlotte Islands, British Columbia.
Notice the totem poles, the use of planks, the huge logs used as frames, the beach location, and the heavily timbered landscape. (Courtesy of The American Museum of Natural History)

as farmers, and one eventually lost the skill? Or were they both nonfarmers, one gradually gaining the skill? Did physical geography play any role in this? For example, did the people of the Pacific Northwest coast lose their agriculture because their plants were not suited to the cool northern coastal climate of the Panhandle of Alaska? If their homeland was tropical, this could well have been true, for the cool, cloudy, marine coastal climate of the Northwest is ill suited to tropical plants. On the other hand, tropical plants could be spread relatively easily across the tropical Pacific, but problems would be met when they finally reached subtropical New Zealand where the Maori were able to remain agriculturists primarily because they obtained the American sweet potato.

If the Haida had had access to the Irish (Peruvian) potato, would they too have been farmers? The Irish potato does well in the marine west-coast climate and could have been the basic starch crop for a civilization, as it has nearly become in northwest Europe. In passing, it is worth commenting that the wild relatives of the Irish potato extend up through the mountains of western America into the Pueblo region and were gathered there and eaten but never cultivated. This is another example of the failure to transfer the idea of domestication to the most obvious plants, and as we have seen, it is equally applicable to animals. Yet had the idea of the utilization of this prolific tuber spread into Mexico and on north to reach the Pacific Northwest, the whole basis of the Indian culture there might have been different. This root crop failed to spread through the seed crop area of Mexico, and here one sees a tendency for men to cling to patterns of farming to the extent that seed growers and root growers tend to be separate. The people beyond Mexico, once the Mexicans rejected root crops, were then blocked from the utilization of this idea even though the raw material (the potato) was present and the need for a plant suited to high altitudes (in the Rockies) and high latitudes on the northwest coast was also present.

Adding or losing skills upon movement into a new physical environment is not uncommon. However, the problem of how much of the change is to be charged to the new environment and how much to man's perception of this environment requires careful assessment, for as the examples above show, access to ideas can be extremely important. Many of the classic buffalo-hunting Indians had been farmers before they got the horse and traded hoeing corn for harvesting buffalo. The Athabascan-speaking Navaho, once hunters of the northern forests, became farmers when they entered the American Southwest. Men lived in the Southwest for millennia without agriculture, then they

adopted corn, beans, and squash. Today we utilize the land with these crops, plus small grains, plus cattle. Clearly men can change, but their changes are first dependent on whether they want to change, and secondly on what choices are actually and perceptually present. If they have no background in pastoralism, they will not perceive a herding possibility, even if suitable animals are present, and they usually are. In passing, the Maori called the flightless birds of New Zealand moa. This is the Polynesian word for chicken, and this suggests that the Maori knew of chickens but failed to think of moas as potential domestics and so missed the opportunity to domesticate a bird larger than an ostrich that grazed like a goose. Actually there were numerous intermediate-sized moas with other feeding habits. The Maori called them all chickens but treated them as game. This is not strange but commonplace, as we have seen in Africa with domestic animals and with the wild rice in northeastern America. It is comparable to our calling the American wild cow a buffalo (a domestic animal in much of the Old World) but dealing with it only as a wild animal. There is probably no major land area outside the Antarctic that lacks both plants and animals suitable for domestication, but only in rare instances has man perceived the opportunity. This is the kind of evidence that suggests that domestication is not a "natural" idea, one that man would be expected to evolve into frequently; rather, it is so difficult an idea that it is not to be expected as a frequent independent invention.

To return to the agricultural Maori and the nonagricultural people of the northwest coast, since they shared so much cultural background, it is probable that they also shared a tropical agricultural background of vaguely Polynesian type. The northwest coast was not at all suitable to taro, breadfruit, and coconuts, and the distance from the nearest agricultural base was so great that the agricultural tradition was lost. The Maori situation was different; with the cultural accident of the addition of a useful plant, the agricultural tradition was kept alive. They saw their land in a different light. True, they approached their land from the subtropical end, and the northwest coast of America seems to have been approached from the polar end.

If an Asiatic group equipped with taro had landed on the coast of Oregon, would the case have been different? None of the Asiatic oceanic crops would have flourished there either, and yet for other crops Oregon would have been a splendid opportunity. To a man equipped with boats and fishing skills and somewhere in his background knowledge of subtropical agriculture, the northwest coast of America was a good fishing area, but useless for agriculture, that is, the only kind of agricultural opportunity that he perceived in an environment filled with tuberous roots, succulent shoots, nutritious seeds, and delicious berries, many of which he gathered, but none of which he domesticated. Once again the use of the land was a culturally determined choice of the possibilities seen.

This cultural contact or spread of ideas also answers the question of why there appears in the marine west-coast area a sudden heightening of cultural level when there is a general decline in cultural level moving away from Middle America. This area is much farther from Middle America than California or the Great Basin. Could it be argued that it is due to the stimulating climate of the marine west coast? The evidence seems to indicate that it is due to increasing proximity to the Asiatic centers of cultural growth. Perhaps the first contacts were made by storm-driven fishermen, some of whom returned to their homeland to report a good area for settlement. Probably they found this rugged, wet, coastal area very lightly held by relatively simple, shellfish-gathering people who were easily displaced. Thereafter they developed their own culture, probably with minor additions from time to time from renewed contacts with Asia.

The modern age

This rich area, this strip of coast, is the climatic analogue to Norway, Denmark, Great Britain, and most of coastal France. It has an immensely rich fishery, great timber resources, known mineral resources, stupendous hydroelectric potential, sheltered coastal waters, and vast natural harbors. According to one theory, it has the most stimulating climate that man can live in, especially modern Western man with his knowledge of good housing and central heating. Yet it is one of the empty parts of North America.

The coastal strip of Washington and Oregon today is thinly populated; the population of these states centers in the interior valley paralleling the coast. British Columbia's population centers on the south end of Puget Sound at Victoria and drops off very rapidly to extremely low levels both along the coast and inland. The American settlement along the coast of Alaska amounts to little more than a few villages located at fishing and mining centers, yet this Alaskan area has coal, copper, gold, water power, timber, and fisheries that vastly exceed the Norwegian resources. Today Norway is a thriving nation with four million people, while Alaska is an underdeveloped state with 337,000 people. The causes are not difficult to find. Until it achieved statehood in 1959, Alaska in American hands was treated, to use the Alaskans' words, as a colony.

However, until now there has been so much greater opportunity in the other states that there has been little incentive for Americans to develop Alaska or for large numbers of people to go there. Indeed, the course of our development has featured progressive abandonment of good land in equally good climates and much closer to consuming centers. For instance, in recent decades a tier of states from North Dakota down through the center of the Middle West has either stood still in population or, in extreme cases, even lost population. Nor is this by any means the whole picture. Within states such as Maine large areas have never been effectively settled, and large areas of New England that were once settled and cleared at great cost have been abandoned. If good land close to the industrial heart of America is being abandoned, why should Alaska develop at all?

A major impulse toward development of Alaska today stems from Japanese needs. Japan is one of the world's most advanced industrial nations and must import nearly all raw materials. Alaska has oil, coal, copper, iron, timber, fish—and Japanese capital is pouring millions of dollars into the development of these resources in order to supply its economy. Indeed, Japanese demands are leading to the development of an economic empire in the Pacific Northwest from Washington through British Columbia to Alaska. Vancouver is building a seaport on a mudbank, Roberts Bank. It is planned to have 50 berths with a depth of 65 feet, enough to take very large sea-going vessels, and multitrack rail lines are being installed to service this new port. The spur to the immediate development is the Japanese market for coal. Contracts have been made for delivery of 51 million tons of coal to Japanese steel firms over a 15-year period. With this much business in sight the Canadian Pacific Railway is building special coal trains to deliver the coal to the bulk-loading facilities at the docks. The bulk loaders will pour 6000 tons of coal per hour into big cargo ships. These developments are symptomatic of Japan's boom and are a measure of the resources available in the Pacific Northwest and the scale of the movement of goods by sea today.

New pulp mills have been built at Prince George in British Columbia, where nearly $240 million in capital has been invested. Railroads are being pushed into formerly untapped areas causing booms in mining, logging, and pulping. Hydroelectric power is being generated at huge sites such as Bennett Dam in British Columbia and exported as far away as Los Angeles. Food-processing plants find that the Japanese are good customers. As one Pacific Northwest businessman correctly observed, about half the world's population lives around the Pacific Basin, and they all have to be fed,

(Upper) **A Farm in Mantanuska Valley, Southern Alaska.** *The soil is very fertile loess recently deposited by winds from the outwash of the glacier visible in the right-hand corner. The conditions that exist here—exceptionally fertile well-drained soils in a sheltered valley near a sea that is warm for its latitude—are found also in southern Sweden and Norway.* (Division of Economic and Tourist Development, Alaska)

(Lower) **Cattle Ranch on Kodiak Island, the Innermost Island of the Aleutian Chain.** *Nearly treeless and lying in the latitude of Denmark, this large island has only recently attracted settlers. The ranch shown here has built up a herd of 200 Angus cattle in a few years. Angus cattle are a Scottish breed widely used for beef. A winter scene such as this could be found also in Montana or Wyoming, but those continental interior states have greater temperature extremes.* (Division of Economic and Tourist Development, Alaska)

this area became one of the cultural high points in America north of Mexico. With the shift to European development of the continent, beginning in the East and belatedly reaching the West, the West has seen a delayed development. The American Northwest (Oregon and Washington) is now beginning to grow at an accelerating rate. Alaska now can take advantage of the sizable population centers that are developing on the Pacific coast and transportation services that are ever increasing, particularly with the impetus of the petroleum developments. But notice again in all of this that the physical environment has remained passive. The wealth of the area has always been there. Man's assessment and decision about how and when to use it have been the variables.

Chile

At the opposite end of the Americas is southern Chile, a land much like the northwest coast of America. Here there is extensive good land from 40 degrees south, including the island of Chiloé. These latitudes are the equivalent of Oregon and northern California. As in Oregon and Washington, there is an interior valley parallel to the coast with much good flatland. South of the island of Chiloé the land is mountainous with islands rising out of the sea, creating a landscape like that of the coast of British Columbia and the Panhandle of Alaska. The southernmost extent of this area reaches to the tip of South America, about 55 degrees south. This is closer to the equator than Sitka, Alaska, but the climates are much the same, and Sitka is considered to have a mild climate. Fifty-five degrees is close to the latitude of Glasgow in Scotland, Copen-

clothed, and supplied. Oil from Prudhoe Bay in northernmost Alaska is being brought down to the areas where it is needed. This will cause major refineries to be built, and at some time in the future one can expect local industrialization to begin. Everything except a major labor supply is there.

The story is the familiar one of differing assessments of the best use of the land. Given the cultural setting of the Americans of the pre-Columbian level,

hagen in Denmark, Belfast in Ireland, Newcastle in England, and Hamburg in Germany. This area is not a land of impossible climate. Judged by what the Europeans have been able to do with it, the climate is a most useful one—or as the physical-environmental determinists would have it, the most stimulating on earth. What, then, we may ask, has man done with this promising land of Chile?

Pre-Columbian Chile

In pre-Columbian time this was an undeveloped land. It had received some influence from the Inca empire, and there was a beginning of agricultural village settlement at the extreme northern end of this woodland area.

South of the island of Chiloé the land was occupied by groups of the simplest peoples in the world. The way of life was dependent on the sea. Shellfish were a major source of food. They were gathered from the intertidal zones and fished up from deeper waters by means of a split stick; sometimes the women dove for them. Sea birds and their eggs were used; sea lions were clubbed if caught on the shore. Inland hunting and gathering of plant foods was minimal. For travel on the water a bark canoe was used, not like the birch-bark canoe of the North American Indians but one made of a single sheet of bark, bound together at the ends, and propped open with sticks. It was so fragile that it could not be beached. The occupants had to get out in shallow water, and the canoe was tied offshore so that it would not be injured by contact with the land. Yet whole families traveled in them. Along this coast, composed of mountains rising out of the sea, the sheltered passageways let so weak a vessel function and made other types of travel difficult. The same environment contained trees whose bark could be removed in huge sheets and thus could have been used to make far larger and stronger bark canoes. Such large trees could have supplied logs for dugouts similar to those that the people on the northwest coast of America used or finely made plank boats such as the Norsemen used. When it is said that this is an area favoring water travel, it still leaves open what kind of water travel man may invent.

The climate of this coast is a cold and wet one. As we have noted, the climatic analogue of Tierra del Fuego is Sitka, Alaska, and the northwest-coast Indians with their good clothing and snug houses were the occupants of the Sitka area. The southern Scandinavians built warm houses and dressed to withstand the cold of this type of climate. Yet the people of Tierra del Fuego had virtually no clothing and next to no housing in contrast.

Darwin was astounded in his trip through Tierra del Fuego to see a boat come out to his ship carrying a family. The boat was leaky, the day was stormy, and sleet was falling. Yet the occupants of the boat did not have shirts or trousers or shoes or robes: they were practically naked. A woman was along, and she had a tiny naked infant in her arms; ice water dripped on it from the sleet that caught in the mother's hair and melted. By our standards the whole boatload should have died of exposure, especially the infant. If the environment could urge, compel, encourage, or suggest anything, it should have shouted at these people to put some clothes on. Somehow they seem to have been deaf to the obvious.

Is it possible that there was no material for clothing in their environment? This cannot be true. Bird skins can make a robe; feathers engaged in the strands of string can be made into puffy ropes, and these can be woven to make a soft, warm blanket, as was done in the Great Basin area by a people on almost as simple a level of living as those of Tierra del Fuego. Further, these people killed seals and thus had hides that could have been used to make capes, pants, moccasins, and leggings. The limitation was not in the physical environment, therefore, but in the cultural environment. Despite what is to us a seemingly imperative need for clothing for mere survival and an absolute need for clothing for any comfort, these people simply did without anything but an occasional skin worn loosely as a cloak. Here again we have necessity failing to mother inventions.

Their ability to withstand cold was phenomenal. They were often wet, for rain was frequent, and if it was not raining, the grass and brush were wet, and there were streams to cross. In these circumstances clothing loses some of its advantage. Wet clothing is long in drying while skin dries quickly. Wet clothing is cold, and if body heat is used to dry it, there is some question as to whether it is worth the drain on body temperature. One explanation for the Scots' addiction to kilts is that since the heather is almost always wet, anyone walking through it is going to be wet to midthigh. Trousers would then be soaking wet and would be wet for hours after. Kilts would be above the wet heather, and within minutes after entering a house legs would be warm and dry. Nevertheless, the Scottish herdsmen who live in this Tierra del Fuego climate felt a need for warm wool clothing, long cloaks, and snug houses.

The Tierra del Fuegians did not bother with much of a house. They built shelters of pole frames, thatched with grass, over which they might also throw a seal skin or two. This slowed the breeze down to a draft and delayed most of the rain. A smoldering fire

Natives of Tierra del Fuego. *These people, who live in a climate similar to that of southern Alaska and southern Scandinavia, have lightly built houses and do not wear heavy clothing. Their ability to withstand cold, including swimming in sea water at near-freezing temperatures, is explained by their high basal metabolism, which is about 150 percent that of Europeans.* (Courtesy of The American Museum of Natural History)

in the middle of the hut supplied some heat and a lot of smoke. It was noticed that the large fire we find comfortable caused the natives to retreat a long way off. They could not stand so much heat.

This has led to the suspicion that they may have been somewhat different physiologically from the rest of mankind. Belatedly, for these people are on the verge of extinction, an expedition was sent to study them and found them to have some adaptation to cold. Their rate of heat production when at rest is half again as high as ours; that is, their basal metabolism is 150 percent that of Europeans. This allows them to withstand exposure that would be fatal to us. Again note that as with the natives of Australia, this phenomenon of physiologic change to meet environmental stress is found among the least culturally equipped people. Culturally equipped mankind interposes such artifacts as clothing, housing, and heating between himself and direct environmental impacts.

Among the Tierra del Fueguians the women could withstand the most cold. In obtaining shellfish in winter, it was the women who dove for them. At Tierra del Fuego in winter the sea water is just above freezing. Because of its salt content, sea water freezes at 28° F. instead of 32° F. At times it is so cold that rain water (which is fresh water) resting on the broad leaves of the seaweeds freezes, forming a glaze, as we sometimes see in higher mid-latitudes in winter. The women nonetheless swim through the kelp and dive for mussels. They can stand only about 20 minutes of

this, and then they must be taken ashore and warmed at a fire. That the men cannot do this at all is of considerable interest. In our own society it is noticeable that women dress less warmly than men. Women are thicker bodied than men, and their fat accumulation tends to be distributed in a layer under the skin. Men tend to accumulate their fat in the visceral area. The thicker, fat-insulated body of the woman is much better suited to withstand cold than is the larger surface area, uninsulated body of the man. It is a good measure of how close to the limits of human tolerance the Tierra del Fuego climate is that this relatively small difference plays so decisive a role.

The introduction of better boats, clothing, improved housing, and better weapons for hunting and fishing did not cause a great increase in population among these people. Instead a wave of disease that accompanied these things reduced these people to the vanishing point. From a few thousand people scattered along a thousand miles of the coast of Chile, they are now reduced to a few hundred people, and most of them are now of mixed parentage. This is another of the grim stories of the impact of the whole gamut of mankind's diseases on a people who have lived for vast periods of time almost completely out of touch with the rest of mankind, hence without any immunity to even the commonest mild diseases.

Just how long these people have lived in this area is not known. Excavations in caves in Tierra del Fuego show a long period of occupation, with carbon-

14 dates going back to 6000 B.C. It seems probable that men were living in this area for longer than that, for man was surely in southern South America for far longer than 10,000 years. He may have been living in the marine west-coast type of climate of that area even during the last glacial period. This would explain the simplicity of his culture and the physiological adaptation that he has undergone.

Modern Chile

As we have seen, when the Spanish conquered the Inca empire, they quickly explored the adjacent regions also. The beautiful, rich central valley of Chile with its favorable water supply and equable, homelike (Mediterranean) climate attracted only a thin settlement. What is mere land when gold and silver and great cities and masses of people can be seized and put to work and taxed? Besides, Spain is a small nation: there was not enough manpower to settle the new lands. The Spanish penetration southward down the coast of South America was rapidly played out. The rich, well-watered, and forested country of the southern end of the central valley of Chile was left to the Araucanian Indians until the end of the nineteenth century. The southern end of the central valley was finally pacified when armies were systematically used. Thereafter German settlers were brought into this wet, forested land. They have introduced northwestern European ways of life including important dairying development. In this rainy area Valdivia is a modern city, flourishing despite an annual rainfall of 100 inches per year.

The densely forested, very wet, mountainous seashore of southernmost Chile was left untouched. Even in this century the Chileans do almost nothing with this land, which now has fewer men than it had before 1500, for the native people have now virtually disappeared. This is a land much like our northwest coast, where we have done almost as little. Both these lands are richer in minerals, forests, and water-power potential than their European counterparts. The disparities are the result of different histories, different present positions in relation to the great productive and consuming centers of the world, and differing viewpoints of peoples with separate economic traditions.

Only at the southern tip of this land, at Tierra del Fuego, has there been any development. Formerly there was some sheep raising here, and more recently gas and oil have been found. Chile needs such fuel, and it ships the oil by tanker to its population center in its region of Mediterranean climate. The gas could be compressed and shipped, but the costs of the installations are too high, and the Chileans have declined

international capital. They could also send gas to southern Argentina by pipeline and import gas from northern Argentine fields across the Andes, but they find this politically unpalatable and simply burn off the gas. With the highly mineralized Andes, great hydroelectric potential, petroleum and natural gas, and enormous forest resources suitable for timber and pulp, this region has all the ingredients for major industrial development, except people. Tierra del Fuego is in the same latitude and the same wind belt on the same side of an ocean as Great Britain, and it has comparable winter but much cooler summer temperatures. It is in a zone of great storminess. If cool, stormy lands are the optimum for mental work, this should be one of the most productive spots on earth. What reasons might explain why this is not the case?

Today, when the principal movement of man is out of marginal lands and into urban agglomerations, the probable course of development for southern Chile is one of slow and very slight development of the potentially rich resources. Overpopulation is a meaningless phrase in southern Chile. A nearby empty land awaits the coming of modern man, but man's view of his overpopulated earth is not one easily to tempt him to move into these empty lands.

LANDS IN THE SOUTHWEST PACIFIC

Tasmania

Among the other areas of marine west-coast climate is the island of Tasmania. It lies in the latitude of northern California and northern Portugal. It is a mountainous land, covered with forest, with rainfall well distributed throughout the year—not as excessive in amount as that found along the marine west coasts of America but more like that of northwestern Europe. From this description one would expect it to be a favorable land, but its geographical position has been against its development. From Tasmania it is about 7000 miles to South America, 8000 miles to San Francisco, 5000 miles to Tokyo, 4000 miles to Singapore, and 6000 miles to Bombay.

Yet man reached Tasmania fairly early. The original inhabitants were a small, dark-skinned people, probably of Negrito origin. They had a level of cultural development somewhat like that of the Tierra del Fuegians. As with the Tierra del Fuegians, contact with the European diseases proved fatal. Further, when

(Upper) **Tasmania, Derwent River at New Norfolk.** *Beautiful Tasmania has a climate like that of Ireland, southern England, and northern France. Water power potential, timber, minerals, and good land are still not enough to attract people and industry and lead to development, even in this time of world population growth and depressed conditions in Great Britain, the preferred source of immigrants. Tasmania, although the size of Ireland, has only one-eighth the population of that empty land.* (Australian News and Information Bureau)

(Lower) **Cradle Mountain and Crater Lake, North Central Tasmania.** *This area, a national park, is noted for its rugged beauty and numerous lakes.* (Australian News and Information Bureau)

populated. There are two towns, Hobart with 100,000 people, and Launceston with 50,000. The bulk of the island remains in forest, and lumbering is one of the principal industries. There is some stock raising, mining, and farming. Copper, tin, silver, lead, gold, and zinc are mined. Wheat, oats, peas, hay, potatoes, and fruit are the chief crops. This is a land as large as Ireland, blessed with dependable, well-distributed rainfall, mountains creating hydroelectric potential, immense forests, and a whole list of valuable minerals including coal. Therefore the lack of development and population is surprising. Without population there will be no development in this remote land. Distance seems to have outweighed all other factors, but distance means less and less in this world of rapid transportation. Adjacent New Zealand has not been deterred by mere distance.

However, Tasmania is not developing parallel to transportation growth. Instead the dispersed population of a century ago is ending, and people are moving down out of the interior hilly country. This is what we have seen in most of the industrializing world. In Tasmania the exodus is removing the few people who once lived in the interior; it is selective, and those remaining show up on intelligence tests as distinctly below average. As always it is difficult to separate cultural, physical, and biological factors. Since we are dealing with one race in one limited physical environment, those factors can be considered constant. This leaves cultural and biological variation within one uniform group. Are the lower intelligence test results due to biological selection? Or are they due to the increasing isolation of the hill people, with decreasing educational and cultural advantages? It is a typical problem. If we can assume that the intelligent are the more enterprising, it is to be expected that they would move out as conditions worsened. However, decreasing population

the British introduced sheep, the Tasmanians saw them as an added food source. The British thereupon exiled the last Tasmanians to put a stop to the destruction of the new economy, and the last native was soon gone. The British were then free to expand their sheep industry.

Tasmania has an area of 26,304 square miles, the size of Ireland. The population is under 400,000, one-eighth of the population of Ireland, which is also thinly

below certain limits creates a differing social order. Value systems tend to shift from bookish skills to frontier skills, and mental tests measure bookish skills.

Tasmania is one more example of the unevenness of the human occupation of the world. Its extreme isolation in the past allowed the survival of a people who preserved a Lower Paleolithic way of life, a way of life that disappeared in most of the world 50,000–100,000 years ago. British occupation of the island came at the same time as their occupation of Australia. The course of development has paralleled that of Australia, but Tasmania has been peripheral to Australia in the same way that the Alaskan Panhandle has been peripheral to America. Modern land-use tendencies are leading to concentration of its population in a few urban centers and the extensive rather than intensive use of the land. Tasmania's future is tied to that of Australia. As population and industry in Australia grow, Tasmania should also develop. However, despite its having the "stimulating" climate of northwestern Europe, it seems much more likely that the Mediterranean and east coast mid-latitude forest climates are likely to remain the dominant areas in the Australian-Tasmanian world. Tasmania must be viewed as a potentially rich land, undeveloped in spite of, rather than because of, its environment.

New Zealand

New Zealand is more isolated than Tasmania, yet it is prosperous and growing and has one of the world's highest standards of living. New Zealand has a marine west-coast climate. The North Island, lying between 35 and 40 degrees, is to be compared with the coast of northern California. The South Island, between 40 and 47 degrees, is to be compared with the coastal strip from northern California to British Columbia. The European counterparts are from northern coastal Portugal and Spain to Brittany in France. The total area of the two islands is slightly smaller than Great Britain plus Ireland, or about equal to the state of Colorado.

The Maori

The case of the Maori in New Zealand is an interesting story of the impact of man on the environment. When the Polynesian people reached New Zealand perhaps one thousand years ago, they found a land untouched by man. It was well covered with vegetation and populated almost exclusively by birds. This was the result of the great isolation in the sea, for only forms that could fly could reach this land. With

The Maori in Transition. *Two men wear traditional costume, but with wrist watch and war medals, for a ceremonial dance, the haka.* (Embassy of New Zealand)

no ground-dwelling predators, the birds evolved to fill the ecological niches open. Some became small, flightless feeders on worms—the kiwi. Others—the giant moa—became immense grazers of the grasslands, the largest ground-dwelling bird that existed, larger than the ostrich.

The Maori entered this land from the warm central Pacific with equipment suited to life on tropical islands. They were entering a subtropical land mass of far larger size than they had been accustomed to. Many of their familiar plants would not grow there. The coconut, breadfruit, and mulberry, the source of their tapa cloth for clothing, could not stand the seasonal climate. The sweet potato succeeded only if handled with care. For a time this made relatively little difference, for the supply of flightless birds, especially the huge moas, supplied an abundance of meat. The Maori became strongly specialized in hunting, and their settlements were often inland and well distributed over both the

North and the South Island. However, the eruption of a fierce predator such as man into an animal population that had a reproduction rate unadjusted to heavy predation had a familiar result. The moa became extinct. Man had again decisively altered his environment. This case may be closely parallel to what happened in America about 10,000 years ago when men with great hunting skills entered America and an accelerated extinction of large game animals occurred.

The Maori then shifted back toward the coasts and again became fishermen and farmers with an agriculture based on the sweet potato. Since the sweet potato is a tropical plant, it did best on the subtropical North Island, and the Maori center of population came to be located there.

The decision was only partially determined by the physical geography. Primarily it was based on choice of retention of an agricultural way of life, and the lack of plants suited to the strongly seasonal climate of the South Island. Had their basic crop been wheat, their primary area of settlement would probably have been on the drier, grassy plains of the east side of the South Island.

These are the people who are strikingly similar in many cultural details to the Indians of the northwest coast of America. Their art, house types, canoes, and many other items suggest some common origin. The two fundamentally similar peoples ended in different corners of the world, each of which offered varied physical environments. In North America the people settled on the wet, west-facing coast, ignored the drier interior valleys, and did not spread equatorward to occupy the subtropical coast. In New Zealand they occupied the subtropical north, used the drier plains to the south only slightly, and ignored the wet, west-coast region almost entirely. The contrast is heightened here by the common race and culture in similar environmental settings.

To the Maori, New Zealand was first a vast hunting ground, then a poor farming area, then good fishing grounds; to the Europeans, the assessment of useful land is utterly different.

New Zealand today

How can it be that this bit of land, nearly in the center of the water hemisphere of the world, is one of the most prosperous spots in the world? As we have seen, climate alone cannot account for this. Location, which seems to have worked so great a hardship on our own northwest coast, the Chilean south coast, and Tasmania, seems somehow not to have damaged New Zealand—despite its even greater isolation. Is the land

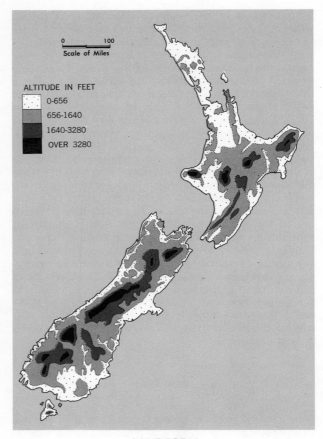

ALTITUDE IN FEET
0-656
656-1640
1640-3280
OVER 3280

LANDFORM

that much better? Even the very practical consideration of situation within the British Empire is common to Tasmania and New Zealand, but Tasmania languishes while New Zealand flourishes.

New Zealand is mountainous. Plains are limited in extent and, except for the northern part of the North Island, are strictly marginal to the mountainous interior. The population is strongly concentrated on these plains. However, away from the four major cities of Auckland, Wellington, Christchurch, and Dunedin, the density of population, even on the best lands, is only about 50 per square mile. The mountainous interiors have less than five people per square mile. On the South Island there is little population on the very wet, western side and a concentration of population on the drier, sunnier plains of the east side of the island. Rainfall on the islands is heavy; large areas in the North Island and virtually the whole west side of the South

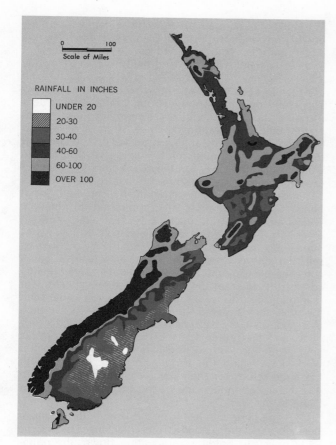

AVERAGE ANNUAL RAINFALL

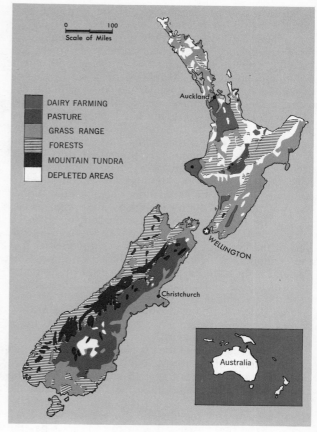

LAND USE

MAP 6-7 New Zealand: Landform, Rainfall, and Land Use
The very large extent of New Zealand's dairy pasture and grassland is shown on the land-use map, and the landform map shows that much of this land is in steep hill country. The immensely rainy southern alps stand out in the rainfall map; there is an absence of railroads in this region, indicating the absence of development there. (Reprinted from Focus, *American Geographical Society)*

Island receive 80 inches of rain per year. Such heavy rainfall on mountains produces an enormous hydroelectric power potential.

New Zealand is rated as having 2 percent arable, 47 percent meadow and pasture, 27 percent forest and woodland, and 24 percent waste land, mostly high mountain land. Such figures are deceptive because they contain the human decision as to how the land is to be used. The nearly one-quarter of the country that is useless because it is rugged mountain land is about the

only purely physical measure. What is meadow and what is forest depends in large part on which category man chooses to emphasize. In New Zealand the emphasis is on livestock. Much land that elsewhere would be left in forest has been converted into pasture, and much land that would be used by other people for field crops has been used for very high-yielding pasture lands. If the western half of the states of Washington and Oregon were used in comparable fashion, it is probable that more than half of their land would be in meadow and pasture, instead of the very high percentage that is left in forest.

New Zealand has only three million people. Of these the Maori, the original inhabitants, make up 160,000, a figure that might be compared with Great Britain's population of 50 million on a set of islands of comparable size in a similar climate—with the climatic advantage definitely being with the warmer land of New Zealand. New Zealand has managed her native population well, and the 160,000 Maori are rapidly being assimilated into the political and economic life of the country. The immigrant population was largely British, and they brought with them high levels of education and technology. Immigration is controlled to continue this type of population, and the educational policy within the country maintains these high levels.

The economy is based on modern mechanized farming: 90 percent of the country's exports are products of the livestock industry. This has gone so far that the economy has been said to be based on grass. New Zealand is the world's largest producer of mutton, and most of it goes to London. Wool is a natural associate product, and this largely goes to Great Britain. There is a large dairy industry, and butter and cheese are exported in such large amounts that milk products make up one-third of the exports.

The expansion of the grasslands has been pushed at the expense of the forest. (New Zealand could have been discussed in the grassland section as well. The choice was made on the basis that this is a naturally forested land that in most respects is more like Chile, Great Britain, and the rest of the marine west-coast lands, hence best discussed with them.) Grass has the great advantage that rolling land can be put into pasture, and if properly managed, there is very little erosion. However, the New Zealanders have at times pushed their clearing of forest onto such steep slopes that they have set off accelerated erosion. In the main, however, they have made a vast, rich grassland out of much of their islands. They have imported grasses chosen from the world's available plants for their nutritive value and suitability to the soils and climates of New Zealand. These grasslands are carefully managed and heavily fertilized. In areas where cattle did not do well for a long time, they have now discovered that the difficulty was caused by the absence of a minute amount of a necessary nutrient: cobalt. Additions of this mineral to the soil have changed a large area in the volcanic plateau of the North Island from a virtually sterile area into a highly productive cattle-producing region. This is typical of what an educated and vigorous people can do with a land.

New Zealanders have a welfare state with a curious background. The original plan, as set forth by one of the early governors, was to create in New Zealand another England, with the land held in large estates by a landed gentry. The price of land was artificially high in order to keep the land in the hands of the wealthy. This could only be done by strong central government planning and control, which among other things deliberately held immigration down. A struggle developed between those aiming at an aristocracy on the land and the people who wanted land of their own. The people won and got the land, but they continued the planned economy and strong governmental controls, which then became a system of governmentally supplied social benefits and governmentally controlled production. Today agricultural production is controlled through marketing boards. Most produce is sold to a governmental board at a fixed price, and the board in turn sells, largely to Great Britain, also at a fixed price. A considerable amount of New Zealand's governmental costs are paid from the profits that these government marketing boards realize from the difference between the price they pay and the price they receive. Through these boards the government controls the land, its produce, and the return that the individual gets from his land. With the assured British markets the system has worked well, and the people are quite content with the large return for their labor—a return sufficient to give New Zealanders one of the highest standards of living in the world.

The largest industry in the land is the fertilizer business, and the second is meat packing. Due to the early policy of holding immigration down, the population is now too small, and labor shortages appear on all fronts. In much of the world the mechanization of agriculture is releasing labor, and countries are struggling to absorb this manpower in industry. In New Zealand the already highly mechanized agriculture is itself labor short, and there is no surplus of labor to feed potential industrial growth. Within the country the struggle is between the urban population and the rural population, each wanting the economy controlled in its favor.

Two-thirds of the New Zealand population now lives in cities, a situation typical of the advanced nations of today, and it would appear that in time more industry will develop. Industrial development is handicapped, however, by the shortage of labor, the small internal market, and high production costs, in part related to the heavy burden of social benefits for workers. The largest city is Auckland with 690,000 people. This city is unusual in that it does not tower over the other cities of the country in the way that Paris dominates France, London dominates England, and New York dominates the United States. New Zealand has only moderate power resources. The hydroelectric potential is large, but the building of dams and transmission lines is expensive and characteristically will increase in cost since the best sites are always developed first. Volcanic steam in North Island has been tapped, and in 1961 a natural gas field was found on the east side of the island. New Zealand's immediate future does not seem to be industrial, however.

Currently there is some worry over the entry of Great Britain into the European Common Market and the loss of New Zealand's preferential marketing there. In view of the increasing standards of living in much of the world and the concentration of much of the economic growth in such highly urbanized areas as northwest Europe, there should be a permanent and growing market for New Zealand's meat, butter, and eggs. Moreover, New Zealand's location at the center of the water hemisphere—the exact opposite of Great Britain, which occupies the center of the land hemisphere—should theoretically be a severe handicap, but this has not been the case. Produce is shipped around the world to Great Britain. This is less startling than it seems, for sea transportation is the cheapest there is. Even with the advantage of a preferred position in the British market, New Zealand's success is a triumph of man over distance. If New Zealand were to lose its British connection, this might place a severe strain on its economy, but the changing economic world may compensate for this. Japan's industrial population is unable to feed itself yet is increasingly prosperous. New Zealand is a logical source of supply, and Japanese manufactured goods could replace British goods in New Zealand's international exchange.

Japan

Climatic comparisons can never be absolute. Japan is included here with the west-coast marine climates because its total setting is more like the lands that we have been discussing than it is like any of the others. Like Great Britain, Japan lies just off the coast of a continent from which the impulses of civilization came. Like the island dominion of New Zealand, most of Japan lies between latitudes of 32 and 44 degrees. As in New Zealand, this creates sharp climatic differences from north to south. Hokkaido, the large northern island, has a climate with summer-winter contrasts

(Upper) **Sheep, South Island, New Zealand.** *The rugged mountainous spine of South Island is seen in the background, with the relatively dry land in the rain shadow of the mountains (the east side) also clearly shown. Water from these mountains is used to irrigate the lowlands on the east side of the island.* (Embassy of New Zealand)

(Lower) **Queenstown, New Zealand.** *This winter scene illustrates New Zealand's mid-latitude location. Queenstown, on the South Island, at 45 degrees south can be compared with Portland, Oregon.* (Embassy of New Zealand)

about like Newfoundland and the coast of Maine. The southern islands have summer-winter contrasts like those of our deep South. Climatically, then, these lands are more like our east coast, but the island setting, with its limited size, mountainous structure, and limited amounts of plain, makes this land much like New Zealand and Great Britain. In size Japan is about equal to California, about one-third larger than Great Britain or New Zealand.

Prehistory

Japan was settled early by people with a Southeast Asian Lower Paleolithic culture. Just as a long sequence of invasions from the continent brought ideas to Great Britain, later people brought the cultural advances that were developing on the Asiatic mainland to Japan. These included Upper Paleolithic types of stone flaking, followed by pottery making and finally agriculture. The Japanese took from the Koreans the Chinese style of writing as well as their governmental forms—a continuation of a millennia-old sequence of movements of ideas. Among these ideas was Oriental-intensive, garden-farming techniques that make possible the development of dense, agriculturally based populations. This was apparently introduced by the people that we see dominating Japan today. They were preceded in the islands by another race, the Ainu. These interesting people are swarthy, bearded Caucasoids, whose origin and history remain obscure. They have been gradually displaced by the modern Japanese over the past 2000 years, and today the Ainu hold about the same status in Japan as the American Indians do in the United States. They are now limited to the north, principally the island of Hokkaido. If the racist argument is followed, should we conclude that since the white race never did anything here, it never will here or elsewhere?

The Japanese population growth was limited by its agricultural outlook to the relatively flat and fertile plains where rice could be advantageously grown. Since Japan is extremely mountainous, this means that relatively little of the total land surface could be used. It is estimated that only 20 percent of Japan is potentially useful agricultural land and that only 16 percent is actually so used. More than half of the cultivated land is in rice. While the overall population density of the country is high, the density of people per arable acre of land is enormous: 3000 people per square mile. Population densities in much of Japan are 500 people per square mile, and even in Hokkaido there are areas of this density, though most of the populated part of that cold northern island is about 250 to 500 people

per square mile. Japan's population is now 110 million, five times that of California and nearly 40 times that of New Zealand.

The present population cannot be fed from the land despite an intensive agriculture with immense input of labor and fertilizer, resulting in the highest yields per acre in the world. It remains an inefficient agriculture, however, for the amount of labor per acre applied to the tiny fields is costly by our Western standards. Even with extreme application, Japan can raise only about 80 percent of its agricultural needs.

The emergence of modern Japan

When Japan decided 120 years ago to emerge from its self-enforced withdrawal from the world, there already was a moderately dense agricultural population on the land. Japan possessed a developed fishery, an educated upper class, and a system of writing, and it had considerable tradition of trade and commerce. The highly structured society was adapted to modern production; the agricultural society remained little changed. The upper social level added the industrial and commercial development of the Western world that suited the land and society. Labor was abundant and cheap, and water power was available due to the high precipitation on the mountainous lands. One of the important early industries put these resources to work on the traditional silk industry of the Orient. Western mechanization made Japanese silk fabrics available in quantity and at attractive prices. This is like the British and American beginnings. As in these countries, the growth of industry and commerce stimulated a whole series of industries.

Japan has some coal, copper, and sulfur, though the coal supply has proven insufficient and there is little iron ore. Nevertheless, Japan has developed such heavy industries as iron and steel production, shipbuilding, the manufacture of machinery, and an automotive industry. They are the world's most efficient ship builders and are now building the largest ships ever seen. Their automobile industry has grown, and their cars find a market in the United States, which prides itself as being the world's leader in this field. Their raw material sources are overseas: iron ore from Alaska, South Africa, and west Australia, and other raw materials from equally widespread areas. As with Puerto Rico, their greatest resource is their industrious population. Anywhere in the world, raw materials are easy to get; skilled and productive manpower is scarce.

One of Japan's greatest assets is the combined governmental, industrial, and labor attitude. Big business of the type that would be called monopolistic in

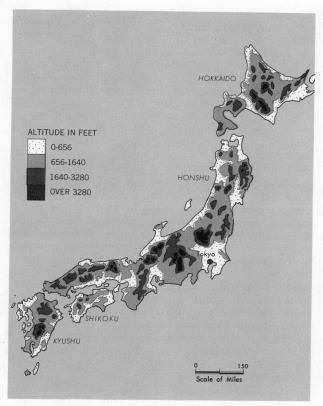

LANDFORM

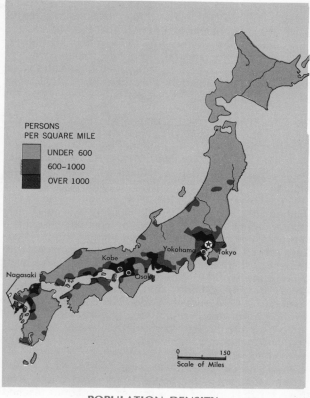

POPULATION DENSITY

MAP 6-8   Japan: Landform and Population
*The concentration of population on the small coastal plains of the south is clearly
shown here. The mountainous interior and the colder north are much less densely occupied.
Hokkaido, the northern island, is climatically like northern Vermont. The climate of Tokyo
resembles that of North Carolina and the climate of Kyushu, the southernmost island, that of
Alabama. With the population concentrated in the lowlands, more than half the land is left in
forest. (Reprinted from* Focus, *American Geographical Society)*

America is not frowned on in Japan if what it is doing
is considered to be in the best interest of the nation.
Combinations are encouraged to enter special fields; far
from hindering, the government then aids the development. The labor situation would shock an American
union boss. Workmen join a company and expect to
remain with that company for life; at the same time the
company accepts the lifetime responsibility for its employees. Workers reporting to work in the morning may
gather to sing the company song and wave company
flags in unison, after which they all plunge in and work
for the good of the company as if they were on a

college football team. We have nothing comparable to
it and find it extremely hard to compete with such
gung-ho productivity. Too often, not only in America
but in other countries, the workers and the employer
are arrayed against one another, and the results are
hardly to be described as conducive to productivity.
Japan's workers are highly productive, and they make
Japanese goods highly competitive.

The growth from a deliberately isolated and retarded nation to a major world power in little over
a hundred years by copying what others had done has
often been treated as a joke, as if it indicated some

inability to do more than imitate. It indicates, on the contrary, an extraordinary ability to assimilate useful ideas. Japan is the classic example that the human attitude determines whether ideas will spread to a given area. Japan deliberately chose not to change; when it did decide to change, it accomplished in decades what often has taken millennia for others to do. In Japan's case the highly structured society with effective control in the hands of a small group helped to make this possible. It was the fact that Japan was already civilized, literate, and politically organized that was decisive. In contrast, the Yuma were illiterate, economically undeveloped, and had virtually no political organization.

Japan's population has grown from about 50 million in 1910 to 65 million in 1930, 85 million in 1950, 93.6 million in 1960, 98.2 million in 1965, 100 million in 1970, and 110 million in 1974. The land is now producing at its maximum. The Japanese have always fished to supplement their diet, and in this century power applied to fishing boats has extended the range of fishing over all of the Pacific. They have been among the world's leaders in developing refrigerator ships, floating cannery ships, and deep-sea fishing methods. At the end of World War II they lost some of their northwest Pacific fishing grounds to the Soviet Union, largely through the Soviet maneuver of extending any temporary agreement in their favor into a permanent arrangement. After a decade of rapidly increasing fishery harvests, the world yields have turned downward. Japan, the Soviet Union, and others who have sought proteins from the sea have about reached the limit that the sea can provide.

Japan is now supporting a large part of its population through industry. It seemingly has little opportunity to expand its agricultural base and is doing an immense job of harvesting the sea. Where, then, is Japan going to turn for the resources to support its growing population? An attempt at building a colonial empire was repulsed. Emigration is never a solution for expanding populations; hence Japan must take care of its population growth primarily at home. In the main, support must come through further industrialization and further expansion of world trade.

It is worth noting, however, that there are probably some untouched agricultural resources at home. While it is true that Japan cannot be developed in the New Zealand pattern, would it not be possible to use some of the 65 percent of Japan that is in forest to develop intensive pasturage of the New Zealand type? While the native grasses may be poor for cattle feed, grasses can be introduced and rich pastures created. Or what would the Swiss assessment of the high mountain slopes of Japan be? Is it not likely that they might think of them as little utilized? The combination of population pressure and increasing knowledge is already moving the Japanese mountain people toward stronger development of forestry, and a major dairying industry is now developing on Hokkaido.

However, Japan's population has already outrun any likely land development program for a self-sufficient food supply. What then of the future with its predicted immense population? Quite possibly it will never come to be. In the rapid rebuilding of Japan since World War II, industrial and commercial power has spread more broadly through the nation than it ever did before, and there is an increasing middle class. The probable course of events is going to be the gradual decrease of the peasants on the land, the increasing growth of the urban industrial population, and with this a declining rate of population growth. In an expanding world economy with more and more people demanding greater numbers of goods and better able to pay for them, there seems to be ample opportunity for an energetic people to utilize almost any base to build an industrial and commercial economy that can supply these goods in return for the industrial raw materials and foodstuff that are needed. As we have seen, there are today potential butter-and-egg areas lying almost totally unused in various parts of the world. Already developed areas such as New Zealand may find Japan a major customer, for industrialized Japan needs foodstuffs, and unindustrialized New Zealand needs manufactured goods.

Japan is another Great Britain with a population dependent on skills exported in the world market. Growth is now well beyond any possibility of making the land yield the foodstuff and raw materials that could supply the people or the industries. This situation is by no means unique. Japan's uniqueness lies more in this development's occurring in Asia, an area that has not otherwise entered into the industrial revolution to any appreciable extent. Japan is an example of the potential free movement of ideas and the ability of men with ideas to make much from little.

(Opposite, above) **The Rush Hour in Tokyo.** *This scene could well take place in New York. The industrial success of the Japanese has led to both intense crowding and extensive pollution—problems typical of technologically developing countries. (Japanese National Railways for World Bank)*
(Opposite, below) **Japanese Family.** *The traditional Japanese family is still the mainstay of the country. The family shown here lives on a farm on the northern island of Hokkaido. Notice the absence of chairs and shoes and the generally sparse furnishings. This is part of a planned development, with 540 square feet allotted for living space. (United Nations for World Bank)*

Japan: A Land of Change. *Japan's extensive modernization in recent years is symbolized by this high-speed train streaking past sacred Mount Fuji in the background.* (United Nations for World Bank)

Summary

A survey of these lands of similar environments and physical opportunities reveals the usual disparities between what could be and what is. The northwest European peoples, with opportunity beating on their Mediterranean and Atlantic doors for something between 5000 and 6000 years, developed industrially in the middle of the nineteenth century and for a time seemed on their way to world domination. Hidden behind the apparent suddenness of this eruption lay the millennia-long seepage to the north and west of the ideas of agriculture, writing, and machinery. Through all of this time northwest Europe occupied the land center of the world, England held her channel-protected island position, and the climate changed in no meaningful amount. Europe, after leading the world for centuries, has recently lagged. A new alignment of economic and political ties gives promise, however, of setting off another burst of energetic development that could lift it back to world leadership.

In the Americas, proximity to Asia made it possible for fishing people to reach the northwest coast of America and start up a minor culture center there on the pre-Columbian level, but when the Europeans took over, this was an offside location. Our exploitation of the region has set back the aboriginal development, and we have only belatedly begun to develop the potential resources of the area. In South America the comparable area was so far from the centers of ideas and especially out of the effective reach of any tradition of use of this type of sea and land that virtually nothing happened there. Neither was this sort of land of any interest to the Spanish; hence the Chileans have no tradition for the use of such land. The usual explanation given for its lack of development is that its position is against it. That this is inadequate is shown by comparison with the development in New Zealand, the most out-of-the-way land in the world. Similarly, Tasmania's slow development cannot be attributed to poor location, nor does its record enhance the argument that these cool, marine, cyclonic climates are enormously stimulating for mental work. And, finally, the Japanese accomplishment, in a short time period on a relatively poor island in the mid-latitudes, is to be seen as similar to that of Great Britain. Back of the spectacular burst of Japanese accomplishment since the middle of the nineteenth century lie centuries of growth of traditions, learning, and political structures that formed the real basis for the events that so impress us. In this Japan, is also like Great Britain.

In these mountainous, coastal, rainy lands, representatives of all the races of mankind have lived for millennia. Despite the myth that these are the world's best climates for mental effort, in no case were they centers of early cultural growth. Every case of cultural growth can be shown to have been due to the arrival of cultural stimuli, relatively late in time. Those areas that received such influences flourished; the others did not. Western Europe, starting late, gained world dominance. Japan, starting later, has jumped to Asiatic preeminence and has provided us the interesting case of an early white race failing to progress in this "stimulating" climate. The explanation, of course, is to be found in the socio-political climate. This should counteract any tendency to judge the Negrito failure to progress in the Tasmanian climate where the European is lagging a bit, even in the mid-twentieth century. Location, opportunity, and receptivity—not raw physical environment or race—emerge as answers.

West Coast Mid-Latitude Forest Land. *At the poleward ends of all these lands mountains and sea meet. Snow is always close by—the glacial past still evident in the landscape. Skiing, a winter sport native to the Scandinavian lands, has become a year-round pastime for those who can travel to the Southern Hemisphere during the Northern summer. Outside Europe these scenic lands are little used at their poleward ends. (ENDESA for World Bank)*

Review Questions

1. Discuss Great Britain as an area of cultural lag. How does this affect the "stimulating climate" argument?
2. Discuss land use, population density, and the good life in the world in terms of Dutch development.
3. With growing demand for raw products around the Pacific, what do you predict for Alaska and British Columbia?
4. What are the prospects for a unified Europe? What would be some consequences?
5. Since the physical environment and ways of life, including even art forms, on the northwest coast of North America and in parts of New Zealand are alike, why should we not conclude that similar environments evoked similar responses?
6. Why did Irish potato growing not extend north? What might have been some consequences if it had?
7. If the world is overcrowded and we are about to run out of space, how do you account for thinly populated Tasmania, Chile, northwest-coast America, and the west coast of South Island, New Zealand?
8. Japan is poor in industrial resources and already unable to feed itself from its own land. What is the solution? Are there parallels in other lands?
9. If land for food production is so needed, why don't the Dutch enclose the Wadden Zee and the people in the Americas their coastal bays?

Suggested Readings

Butland, G. J. *The Human Geography of Southern Chile.* Institute of British Geographers Publication no. 24. London: G. Phillip, 1957.

Dempster, Prue. *Japan Advances: A Geographical Study.* New York: Barnes and Noble, 1967.

Evans, E. E. *France: A Geographical Introduction.* London: Christophers, Ltd., 1937.

Fleure, Herbert J. *A Natural History of Man in Britain: Conceived As a Study of Changing Relations Between Men and Environments.* The New Naturalist Series, no. 18. London: Collins, 1951.

Freeman, Thomas W. *Ireland: a General and Regional Geography.* 3d ed. New York: Barnes and Noble, 1965.

Hance, W. A. "Crofting Settlements and Housing in the Outer Hebrides." *Annals of the Association of American Geographers,* vol. 41, no. 1 (1941), pp. 75–87. A description of the survival of ancient ways of life in these communities so close to the center of growth and development in northwest Europe.

Landheer, B., ed. *The Netherlands.* San Francisco: United Nations, 1943.

Mayne, Richard, ed. *Europe Tomorrow: 16 Europeans Look Ahead.* London: Chatham House, 1972.

Stamp, L. D. *The Land of Britain, Its Use and Misuse.* London: Longmans, Green & Co., Ltd., 1948.

Taylor, G. *Australia.* Skokie, Ill.: Rand McNally & Co., 1931. Includes chapters on New Zealand and neighboring islands.

Van Veen, J. *Dredge, Drain, Reclaim: The Art of a Nation.* The Hague: N. V. Nijhoff, 1955.

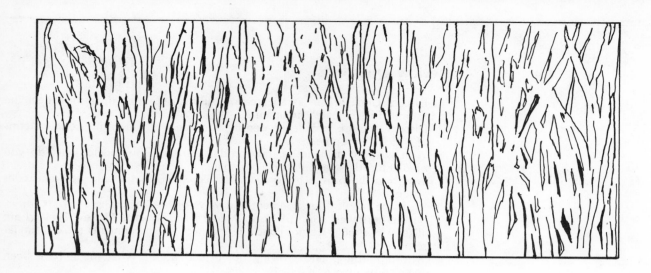

The Grasslands

CHAPTER SEVEN

THE rich, flat lands of Iowa, Kansas, and Nebraska were originally grass covered. Today we use them for grain and cattle production and proudly count them among our richer agricultural areas. How could it be otherwise? Grasslands are endowed with rich soil. Wheat and corn are grasses. Where could grasses grow better than in a grassland? Cattle are certainly dependent on grass. So again, what could be more natural than for man to seize the grasslands as a superior resource, ideally suited to grazing animals and growing grain? However, such a judgment is based on our present view and our Western culture. A survey of the use of grasslands through time and space will show that man has found other uses and had differing views of the grasslands. Our own outlook is perhaps the most interesting because of the changes that it has undergone, for we once called the great grassy center of the United States the American Desert.

This chapter departs slightly from the plan of discussing the world by climatic types, for there is no single grassland climate. Instead there are tropic and mid-latitude grasslands. They all share location between the humid forested lands and the arid lands, great climatic variability as they experience series of humid years followed by series of desert years, a predominance of grass cover, and usually a flat to rolling terrain. They are, then, a type of environment with much in common, even if some grasslands have equable tropical

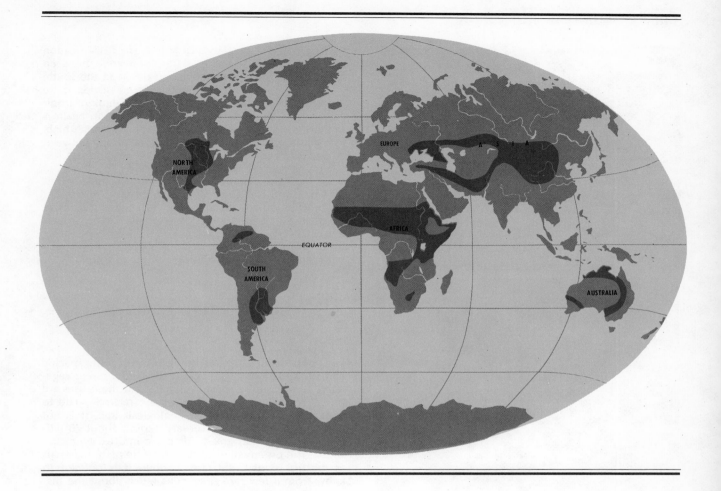

temperatures and seasons and others are characterized by the seasonal temperature extremes that are typical of mid-latitude location.

Location

The grasslands of the world characteristically are found in the transition areas between the more arid climates and the fully humid climates. Thus the North American grasslands are in the center of the continent extending north and south. This is a zone into which the moisture-bearing winds from the Atlantic and the Pacific pene-

trate with difficulty. The winds from the west are blocked by the mountains, and the winds from the east and south are deflected by the westerly drift of the air in these latitudes and by the distance that they have to go to penetrate the continental interior. Primarily it is the rain-shadow effect of the mountains that creates the strong north-south orientation of the grasslands.

The pampas of Argentina are similar mid-latitude grasslands. They lie in the rain shadow of the Andes and between the more tropical and rainier areas to the north and the more arid land to the south. The other major area of grass in mid-latitudes is in Eurasia. These grasslands start in Hungary and extend through south Russia, through Siberia south of Lake Baikal, and

terminate in the grasslands of Manchuria. This location along the 50th parallel is a transition zone between forest to the north and steppe and desert to the south.

The tropics also have extensive grasslands, quite contrary to the expectation of those who think of tropical lands as tree covered. The typical location again is in the transition between the zones of equatorial, high, year-round rainfall and the belt of great aridity between 20 and 30 degrees of latitude. The New World has less tropical grassland than the Old World. The llanos in Venezuela are a minor area. The southeastern quarter of Brazil and the Gran Chaco of Paraguay are other major areas of tropical grassland. Africa has truly huge areas of grasslands forming a great horseshoe around the relatively small equatorial rain forest. The other major grasslands of the world are found in Australia between the humid margins and the arid interior.

Origins: Botany, Zoology, or Man?

The close association of the grasslands with specific climatic situations suggests that they are a direct vegetative response to a specific climate, but the causes are probably more complex. Grasses are relatively late to appear in the long span of geological time. It is not until the opening of the Tertiary period, about 70 million years ago, that grasslands came into existence.

The preservation of such delicate plant materials as grasses is relatively rare. Further, it is difficult to know from a few rare specimens much about the dis-

(Upper) **Farm Land, Saskatchewan, Canada.** *The vast extent of flat land suitable for mechanized grain harvesting is evident in this aerial view.* (Information, Canada)

(Middle) **Indian Head, Saskatchewan, Canada.** *Wheat fields and grain elevators present a standard sight in the North American wheat belt.* (Information, Canada)

(Lower) **Fencing.** *Controlling the movement of stock is an ancient problem. In New England stone fences were used, in northwestern Europe, hedges, in the eastern United States in colonial times, split rail fences. The plains had neither stones nor timber and little that resembled the traditional hedging materials, but technology provided wire, and with Yankee ingenuity barbed wire was invented. It is now found around the world in all kinds of environments and used for many purposes other than controlling the movement of stock.* (U.S. Forest Service)

tribution of a grass or to learn such things as whether it was a scarce plant, growing along the edge of streams and in forest openings, or a plant found dominating vast areas of the earth's surface. It is from fossil animals that we learn when the grasses have come to be dominant plants over large areas. When animals appear with their teeth specialized for the grinding of grasses and their legs specialized for rapid fleeing over open territory, we know that the grasslands are present.

It has frequently been said that man has had much to do with making the grasslands. Properly stated and properly understood, this is certainly true. The grasslands only partially fit the climatic boundaries. In the United States grasslands extended east of the 100th meridian, which approximately marks the border between arid and humid in our continent. Along the 40th parallel the grassland reached east of the 90th meridian, as far east as Chicago. This is land with adequate rainfall to support a dense forest. The Argentine grasslands lie in a climate that is like our southeastern United States climate where a forest cover is possible. The adjacent Uruguayan and Paraguayan grasslands, even though they contain a sprinkling of trees, are climatically such that a complete forest cover is to be expected. In Africa the great bulk of the tropical grassland is in the savanna climate. In India, Southeast Asia, and Central America such climates have forest covers. In Africa most of this climate has only scattered trees and much grass. Such observations suggest that something besides climate is at work in creating grasslands. There are several possibilities: plants, animals, and man.

If the plants available are very different, there may be considerable difference in response to the similar climates. In one part of the world there may be grasses better suited to taking over wetter areas. A wet grassland, a prairie, may then be found and would therefore be botanically caused. This aspect of the grassland problem has been relatively little explored. The possible effect of the animals of the different parts of the world in creating or expanding grasslands has also had little discussion. Perhaps it is the least important cause, even though animals do have effects on vegetation. However, most of the data that we have of great vegetative changes caused by animals are based on observations of the effect of domestic animals. Much of mountainous Northwest Africa would be forested if it were not for the activities of man herding goats and camels. In the discussion of the American Southwest the effect of the Navaho sheepherders was described as turning grassland into desert. Quite probably the effect of wild grazing and browsing animals is similar though less drastic. Man, by herding and holding his animals in restricted areas, increases the effect beyond

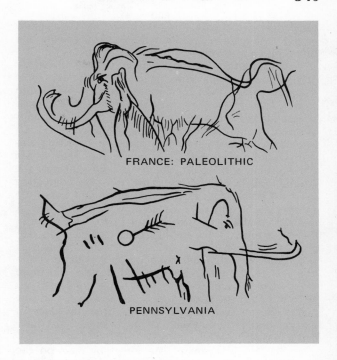

FRANCE: PALEOLITHIC

PENNSYLVANIA

FIGURE 7-1   Mastodon, Mammoth, and Man
That man and mammoths lived in America at the same time was long disputed. Now we know that man hunted these great animals in America just as he did in the Old World. Shown here are two drawings of mammoths. The one on the top is from the Upper Paleolithic period of Europe. The drawing on the bottom was carved on stone by Indians in Bucks County, Pennsylvania. A similar carving on shell has been found in New Jersey. Paintings and carvings of mammoths have also been discovered in the West. [From Anthropological Journal of Canada, vol. 4, no. 3 (1966).]

that of wild animals—who are freer to move when the grass or the browse is depleted and would usually move on before the damage became severe.

Is it significant that Africa, where the greatest number, both in kind and quantity, of grazing and browsing animals have survived into the present time, is the continent with the greatest amount of tropical grassland? An equally great number of varieties and perhaps equally great numbers of similar animals were once found in the Americas. American camels, horses, elephants, many forms of antelope, and a large number of other grazing animals became extinct within relatively recent times. We suspect that man has had a great deal to do with this extermination. What effect would this relatively sudden removal of grazing and browsing animals have on the vegetation?

It must have had some effect. Grass that was not grazed down would grow higher. Fires that started in the grass would then burn more fiercely; this in turn would tend to destroy woody perennials that were adapted to withstanding the lesser fires of shorter stands of grass. The grassland would then tend to spread. Or trees that had been severely pruned by browsing animals would be released from this burden. They would grow bushier and would have more limbs left near the ground and could better shade out the competing grasses. This suggests that in some areas removal of the animals might shift the balance in favor of the trees. A dramatic example of such a shift in England came in the 1960s when a disease killed most of the wild rabbits. Areas that had been grassland throughout English history became unpenetrable scrubby brushland within five years. Other shifts such as less disease and more rabbits would favor the grasses. We do not know just what did happen. We can be sure, however, that any major change such as the removal of many kinds of grazing and browsing animals in America must have had an effect.

What about man? The effect of man on the grasslands has been argued vehemently. Unquestionably man changes the vegetation of the earth. He may clear the forests and plant grasses such as wheat and corn; he may remove the native animals and introduce his own. These may be fenced and kept on limited pastures, or they may be herded widely over the landscape. We have abundant evidence that all of these activities change the land that is so used. We have been altering significant amounts of our humid East to make of this naturally forest area a grassland for the grazing of cattle. This is a radical change indeed. Earlier it was noted that in the East Indies some cattle-raising tribes, by the frequent setting of fires, have converted considerable areas of tropical rain forest into tropical grassland. This is certainly an extreme example of man's ability to change the landscape and to create grasslands. However, from the field of zoology we can demonstrate that extensive grasslands existed long before man. Clearly man did not create the grasslands; at most he may have expanded them, and the extent to which he actually did so is the real question. The answer in areas where we see grasses reaching far into lands where humid forests are climatically possible seems to be that man played a significant role in altering the nature of the environment in these areas and his major tool was probably fire.

Grassland Soils

Whatever their origins, grasslands have soils with certain characteristics. They are deeper and more humus charged through a greater depth than are forest soils. This is due to the different growth characteristics of trees and grasses. Most trees are relatively shallow rooted. A tree overthrown by the winds usually shows a broad mat of roots that penetrated the ground a relatively short distance. The reasons for this are that the nutrients are often concentrated near the ground surface by the accumulation there of the decayed organic material, and the moisture during the summer growing season is also concentrated there. In the forest lands of the eastern United States it is strikingly clear that after spring the ground is rapidly dried out and the trees are largely dependent on the shallow wetting of the ground by summer rains. The trees drop a load of leaves annually, and whether they are evergreen or deciduous, this load is placed on the surface. Only at long intervals, on the death of a tree, does the rotting of roots add organic material to the deeper layers of the soil. The movement of the surface organic material downward is slight. Plowing of the ground by the overturning of trees is a centuries-long process, and the activity of burrowing animals is probably much more significant. The total effect, however, is to leave most of the organic material concentrated in a shallow zone, as can be seen by inspecting almost any forest soil.

On a grassland almost all of these characteristics are reversed. At the beginning of the growing season the soil is wet to some depth and then dries out as summer comes on. The grasses are deep rooted for their size and have an immense net of hairlike roots, which seek moisture at considerable depth. Tall grasses may send their roots down to a depth of 6 feet or more, which is deeper than most tree roots. The annual grasses die down at the end of each rainy season. This adds a plant material to the surface of the soil to decay and become humus. Simultaneously the root system that permeated the soil to a depth of 6 feet dies, adding a load of organic material throughout the soil mass to this depth. Perennial grasses replace their roots at relatively high frequencies, and the old roots decay and add humus at depth in the soil.

In addition, most forest soils are found in wetter lands rather than the grasslands. There is therefore more leaching of the soluble nutrients than in the grasslands. The capacity of a soil to hold up the soluble nutrients that the plants need is in part determined by the presence of the organic compounds that can absorb and hold these materials. These organic materials are much greater in quantity and better distributed in the grassland soils. Hence, even in the grasslands growing in humid areas where leaching tends to remove the nutrients, there is greater fertility than in the woodlands.

There are, of course, differences between the soils of mid-latitude grasslands and tropical grasslands and between grasslands on the edge of the arid lands and those that invade the humid lands. Tropical grasses tend to be perennial, so they add less humus to the deep layers of the soil. Under the higher temperatures of the tropics the decay of the surface organic materials goes on much more rapidly, making it more readily leached out of the soil. The tropical grasslands tend, then, to be low in organic material.

The contrast between the soils of the arid and humid parts of the grasslands is well illustrated in the United States. The soils in the humid eastern extension of the grasslands are subject to the leaching action of 30 to 40 inches of rainfall every year. This is sufficient to maintain a permanent water table and a movement of water down through the soil into this water table to drain away and reappear as springs feeding streams. The soluble nutrients tend to go with this water. There is, then, a steady impoverishment of the soils, with the most soluble nutrients moving out first. This includes the calcium, nitrogen, phosphorus, and potash that are the nutrients needed in greatest quantity, and these humid grasslands then often show deficiencies in these materials. This is readily measured by their lack of basic materials, particularly calcium, whose absence leaves these soils with an acid reaction. Only the abundance of humus saves these soils from rapid depletion and reduction to the fertility of the forest soils.

Westward with declining precipitation, the degree of leaching drops. Near the 100th meridian the boundary is met where the amount of precipitation is no longer able to move through the ground to maintain a permanent water table, which means that the ground is annually wet to some depth and thereafter dries out. The moisture moving down through the soils during the wet season will carry soluble chemicals down with it. When the water evaporates from the soil, the minerals are left behind. Water lifted back up by the plants to be transpired from their leaves brings up only a small part of the minerals that have been carried down. Consequently soluble chemicals tend to accumulate toward the base of the zone of average wetting. This is easily seen in soils of a few thousand years age by the accumulation there of lime in visible amounts. With the passage of great amounts of **lime** the accumulation increases. Very old soils may have immense amounts of lime, and at depth they may be cemented to a rocklike hardness.

The amount of humus accumulation must also vary with these climatic conditions but in a different way. Under the 30 to 40 inches of rainfall in the wet grassland, the prairie, the grasses grew quite tall. In Iowa the first men who saw the tall-grass prairie described it as being so high that a man on horseback seemed to be moving along with no visible means of support. Westward the grasses steadily shortened. It is customary to divide our grasslands into the tall-grass, short-grass, and bunch-grass regions. The short grass still forms a continuous sod, as in a lawn. The bunch grass no longer forms a continuous cover, but as the name implies, it grows in clumps with some space between the clumps. These shifts are related to the amount of moisture available. With abundant moisture grasses of greater vegetative growth dominate and send their roots deep to tap the stored moisture. As the moisture decreases and there is less moisture stored in the soil, grasses of shorter growth, both in length of their growing season and in the height of their stems, come to dominate the scene. As the moisture falls still lower, the grasses can no longer grow close together in sharp competition with each other but must spread out to have enough moist ground available for their moisture needs.

The climate, the vegetation, and the soil in the American grasslands came to make up the grassland environment through a complex of adjustments. The relationships are even more involved than have been described here. For instance, the geological materials in various parts of the American grasslands are different in composition and in age. To the north the land was glaciated; this brought fresh materials into the area, and these contained much more of the soluble nutrients. The time since the end of the glacial period has been too little to leach out much of this mineral wealth. The southern grassland soils under similar climates are less rich because their material has not been renewed so recently. In Nebraska there is an extensive area of blown sand, called the Sand Hills. Sand is generally poor in nutrients, water-holding capacity, and resistance to wind erosion. Then too the rate of accumulation of humus varies with temperature. Much more organic material remains in the soils to the north than in the warmer soils of the south.

The high plains along the foothills of the Rockies are formed by the alluvium washed out from the mountains. Most of this vast apron of alluvium is now cut through by rivers that are flowing below the level of this plain; hence new material is not being added. The soils are therefore aging; in Texas, on the Llano Estacado, the land has been in such a condition for so long that the accumulation of lime in the subsoil has reached geological proportions. The soil is underlain at shallow depth by an immense cemented layer that now resists erosion and tends to preserve the smooth, flat, semiarid grassland landscape. The other grasslands of the world will be discussed in less detail than our consideration of the grasslands of the United States. If they were described in the same detail, similar differences would appear in all of them.

The grasslands are a distinctive environment. Shade is scarce; the sun shines down from a sky that is relatively cloud free. Grassland is close to the desert in the total amount of sunshine received. As in the desert, the wind velocities are relatively high and constant, for there is less vegetation to break up the flow of wind. The air is clear, and since there are few obstructions to the line of sight, it is possible to see great distances. The grass provides an immense food supply to grazing animals, and the grazing animals support the carnivores that feed on them. The grasslands are therefore a relatively rich environment in plant and animal resources, built on the rich soils.

What have these environments done to man? Or perhaps we should ask what has man done with these environments.

MAN IN THE GRASSLANDS OF THE UNITED STATES

The grassland of the United States is one of the most extensive mid-latitude grasslands of the world, and it has played a large part in American thought. It has been the home of the buffalo, the buffalo-nickel type Indian, the cowboy, and other heroic American stereotypes, such as the western marshal and the cattle drive, that have provided the content for song, story, motion picture, and television drama. The impact of the grassland has been described and debated, and at times it has been presented as almost a personified force that molded men and cultures to meet its conditions.

Hunters, Fire, and Change

Man has lived in the American grasslands for a very long time. We know his history there with some certainty from a time beginning toward the end of the last ice age. At that time the grasslands were considerably more moist than they are now. The tall grasses that are now found to the east of the 100th meridian must have extended far to the west, perhaps right to the foot of the Rocky Mountains. This was the time when the Great Basin had large lakes in areas that are now desert, and the southern deserts had a grassland appearance that may have approximated that seen today in the short-grass area of the Great Plains. The animals in the grasslands then included the American elephant, giant buffalo, and horse. The men of that day were skilled hunters and seem almost to have specialized in the killing of this large game. This is, of course, logical, for a single kill of a large animal supplies a huge amount of meat. We know too little in detail of this time to do more than mention that the animals indicate that the area was grassy, although it may, with the increased rains of that time, have had more trees, especially along the water courses, than we now see there. We know that the men hunted the biggest of the game animals and that presently these large game animals disappeared. We therefore suspect that man played a large part in the extinction. Man may also have been modifying the flora as well as the fauna, for the driving of big game animals in grasslands often is done by setting fire to the grass to stampede the animals over a cliff or into a ravine. In this way some are killed and some are injured and then easily finished by the hunters. Such kill sites are known in the Great Plains.

Fire is the enemy of trees and the friend of the grass. Grass germinates in the spring, utilizing the moisture stored in the soil during the winter as well as the rains of spring. Maturation is rapid, and seed is dropped by the onset of the summer drought. If fire then burns up the grass, the seed is safely lodged in the ground. Trees need a period of years to mature and produce seed. Too frequent fires, fed by sufficient grass to make a very hot fire, can kill the trees. The greater the frequency of fires the less chance for the trees. Lightning-set fires would occur at a "natural" rate. They would start most easily, run the farthest, and burn any given area most frequently in the drier areas and would have diminishing opportunity to start and run as the humid areas were approached. A balance would finally be struck toward the humid side of any arid zone, where the frequency of fire would be too low to kill out the trees; the shading of the trees would be dense enough to shade out the grasses so that they could not provide sufficient growth to kill the mature trees. All of this would happen without man.

Man could increase these effects. If the frequency of fire is important, man setting fires would add to the natural frequency and the fire effect would be increased. We have a number of studies of the effect of man and his working with fire. In the south of Texas great, grassy prairies were found by the first settlers. This is a land of 30 to 40 inches of rainfall, sufficient to support a dense forest. When the Texans put cattle on this land, they multiplied to an enormous extent. It is estimated that Texas had 100,000 head of cattle in 1830, 330,000 head in 1850, and 3,500,000 in 1860. A very large proportion of these cattle were in southern Texas. The crowding of so many cattle on the grasslands led to the appearance of woody and brushy forms of vegetation, and the prairie began to turn into a brushland. Under the protection of the clumps of brush, trees began to grow. In the thorny brush there was less food for the cattle, and they were very much harder to round up; hence the cattlemen found the land decreasing in value to them. They found ready buyers for the land in the oncoming wave of farmers, who, with an entirely different view of the land, cleared the brush and planted crops.

Analysis of these shifts suggests some causes. When the cattle population became heavy, the tall grasses were grazed down. Therefore when fires got started—whether natural or man-made—they drew on a limited amount of fuel. They could not build up the raging inferno of heat that the lightly grazed, 4 to 6 feet high, natural tall grasses could create. With the lighter intensity of fire, the heavy brushy vegetation such as the mesquite could make a start. In the past the fierce fires fed by tall grasses would have burnt it out. Thus through man's introduction of cattle the balance was shifted through overgrazing toward the lessening of the intensity of fire, and this was to the advantage of the woody vegetation. The question naturally arises: How did this humid land, naturally suited for tree growth, ever become a grassland? The answer must be that fire was the agent. Whether natural fire or fire increased through the activity of man is not readily answered. Surely man increased the frequency of fire. This must have been effective and further enlarged whatever grassland existed.

In the Ozark region a similar story can be found. Between St. Louis and Springfield there is a low dome of limestone, deeply eroded by stream action. This hilly upland is now largely forested. This was not its condition when American pioneers first saw it. The first accounts of the land describe much of it as parklike. Trees were scattered through a grassy cover. Trees, of course, grew thickly along the water courses and on rocky knobs, but so much of the country was open grassland with only scattered trees in it that it was possible to take a wagon and drive across country with some ease. This would not be possible today because the trees now form a closed forest over most of the area. What has happened?

We are sure of one thing: we have removed the Indians. We know further that it was the Indian custom to burn the land annually, for they had the idea that this increased the amount of game on the land. We are, or were, largely woodsmen: we have the fixed idea that fire is evil; we have everywhere taken drastic action to stop fires. In much of the West this fire prevention effort has brought on virtual states of war between the fire wardens and the range men. The range men angrily maintain that fire drives out the woody vegetation and increases the grass that their animals need. It is a fact that with the disappearance of the Indians and the forceful prevention of fire instead of its propagation, the aspect of the Ozark country has changed from an open, grassy, parklike landscape to one of forest-covered land. For the cattleman this is a change for the worse; for the forester it is for the better.

Significant shifts in the animal life have accompanied the vegetation change. The Ozarks were described in the early accounts as a sportsman's paradise. The land was thickly populated with deer, bear, turkey, quail, rabbit, and a wide range of other game animals. Today turkeys are scarce, bear and deer only a fraction of their former numbers, and the same holds true for the quail and rabbit. The reasons are closely tied to the shift in vegetation. A closed forest is poor in food for most animals except squirrels. The other animals need the browse that is found on the edge of woods, the berries and seeds that are found in the grassy meadow, and the nuts that come in greatest quantities not from trees standing in closed forest but in open stands where the competition for light and moisture and nutrients is not so severe.

This process has been most strikingly documented for some oak trees that were grown for genetic studies at the Missouri Botanical Gardens. These oak trees, planted on the parklike grounds of Shaw's Gardens, the intown part of the institution, matured rapidly and produced huge crops of acorns at least every two years, with minor crops in the off years. The trees that grew under closed forest conditions in the woodland at Gray's Summit, where the experimental forest is, produced so few acorns that they were useless for genetic studies. Since acorns are a highly nutritious food that supports pigeons, bears, deer, and turkeys, it seems that primitive men were wise to burn the forest sufficiently to keep it open, parklike, and rich in game.

Our needs change. When we wanted timber, we suppressed fire, and the Ozarks began to revert to a closed forest. More recently we have decided that we want more beef and that the Ozarks could be good grazing land. Currently men are using airplanes to kill the trees with toxic sprays, burning the deadened areas, sowing grasses from planes, and thus converting forest into pasture. With these methods square miles of land can be changed rapidly to suit our needs and at a reasonable cost. Because the environmentalist movement is opposed to chemical poisons, destruction of trees, and burning the land (all good management procedures when properly applied), it is usual to leave belts of trees along roads to preserve the wooded aspect of the landscape. The point is that by managing the land we can have whatever we want. The Indians wanted game and used fire to create a grassier landscape. We long considered fire a destructive thing, overlooking the fact that it can be an ecological tool, and now we find that defoliants plus fire plus aircraft allow constructive landscape changes.

Indians in the Grasslands

The grassy plains had only limited attraction for the Indians. They were too dry and too susceptible to being swept by raging grass fires to be useful for agriculture, except along the major river valleys. Probably the Indians would have had great difficulty cultivating the surface of most of this land anyway. The grass formed an unbroken tough sod. This is not easily turned into farmland without such tools as the plow and such power as horses, oxen, or tractors. In the prewhite period, then, the Great Plains had limited value for the Indians. The tribes around the margins made seasonal trips into the area to conduct buffalo drives. Either by fire or by skilled driving ending in a stampede, they drove the buffalo into ravines, over a cliff, or into a marsh. The Indians could not easily go out on the plain and approach a herd or even a single buffalo. They are large beasts, quick as a gigantic cat and dangerous to a man on foot. No one approaches range cattle on foot either, and there are many not very funny stories of hunters forced by "domesticated" bulls to seek safety in a tree.

The Indians got horses from the Spanish, together with the idea that they could ride these beasts, and the whole outlook of the Indians toward the plains changed. Instead of being an area of marginal utility, occupied by relatively few people, they became an area of great attraction. Riding a horse, an Indian could approach a buffalo with considerable safety and kill with some ease. Moreover, the problem of finding one of the herds was greatly simplified, for the horse gave man increased mobility. Suddenly the great herds drifting slowly about over the vast sea of grass in the center of America became an endless supply of meat, hides,

and sinew. The land, with no change in the physical environment but only with the addition of the art of horseback riding, changed from one of meager opportunity to one of great opportunity.

All the other ideas needed for the utilization of the horse were at hand. The northern Plains Indians already had the tepee, the familiar pointed, conical tent—a marvelous house that was portable, easily taken down or set up, cool in summer, and relatively free of smoke in winter. The Indian women set the flaps at the top according to the wind to make the fire draw well. The shape of the tent was chimneylike: in winter this made the smoke rise and go out the top; in summer, with the sides raised slightly, air flowed in around the bottom, and hot air rose and went out the top. When there were only women and dogs to haul the tent, it could not be very large. Men, of course, carried little more than their weapons when camp moved, for they had to be ready to kill game or repel the enemy. When the horse was added, much longer poles could be moved from site to site; further, the horse made it possible to kill more buffalo for more skins for larger tepees.

It is worth noting that not all the Plains Indians had the tepee. Some of the southern Plains people built grass-thatched houses. The Mandan, and in the past other Plains tribes, built large houses with log frames and covered them thickly with earth. This land of scanty timber did not dictate the house type: grass was available, skins were available, and sufficient timber was available along streamways for houses for the small population. Houses could also have been built of adobe bricks, stone, or sod. It is too easy to look at what is or was and conclude that it had to be or was "natural."

It is worth considering also how useful the tepee could have been in other areas. It would have been equally good as a home in any part of America where people had a supply of animal skins, which means most of America. Hence it becomes clear that the tepee was not made because this was the only way to meet the cultural need of a movable home in an environment that offered little except skins for wall construction. It was a clever invention made by people who elected to carry their home with them instead of making a new one of grass or sod each time they moved. They had still another choice, of course: to move less and to have one or several permanent houses. The choice was a cultural one. It was not, as it has too often been portrayed, a physical, environmentally determined one. Furthermore, the distribution of this type of tent in Siberia suggests that it may be an ancient invention made in some environment far from and very different from the American grasslands.

(Left) **Tepee of the Plains Cree, Saskatchewan, Canada.** *The tepee is the classic Indian house. This type was found in the northern Great Plains and in part of the adjacent Arctic forest area. Similar dwellings existed in adjacent northern Asia. In other adjacent areas in America, however, even in such similar environments as the southern Great Plains, entirely different types of housing were used. Compare the grass lodge of the Wichita Indians.* (Museum of the American Indian, Heye Foundation)

(Right) **Wichita Indian Grass Lodge under Construction.** *This type of dwelling could have been built in many areas of America, and almost any other type of housing used elsewhere in America could have been built here.* (Smithsonian Institution, Bureau of American Ethnology)

The Indians had the bow, the lance, and the shield, although they probably had had the bow and arrow for little more than a thousand years and had hunted with the throwing spear before that time. Originally their bows may have been long, but for use on horseback they used a short bow, just as our cavalrymen developed short-barreled rifles to replace the long rifle. The Indians also evolved a way of killing the buffalo from horseback. They trained special horses to run alongside the buffalo; the hunter then leaned over and fired at point-blank range. The Indians and their mounts both knew their task. Once the herd was sighted, the Indian did not have to guide his horse. The hunter, therefore, had both hands free for his bow. Once alongside the buffalo, the Indian timed his shot for the instant when the buffalo was stretched out in a gallop: this spread the ribs and gave the hunter a chance to drive his arrow through the lung cavity without touching a bone. Occasionally an Indian drove an arrow clear through one buffalo and dropped another running alongside the first.

This was a thrilling business. It had all the excitement of a football game, with the difference that all the men played and there were seldom any spectators. Great care had to be taken in the approach to the buffalo, and this was one of the few times that the extremely independent Plains Indians submitted to some discipline. Certain Indians served as police to see that no one broke ranks so as to scare the buffalo prematurely or to get a special advantage. Once the "run" was on, it went for miles, and a good hunter on a fine horse might kill several buffalo. The yield was tons of meat, dozens of hides, and almost more horn and bone and sinew than could be used. The Indians made good use of the material. The hides were stripped off, and·the carcass was cut up at the kill site. In the camp the meat would be stripped from the bones, cut into strips, and dried. The dried meat might later be pounded up and have fat poured over it: this is the famous pemmican of the Indians. The bones were then often boiled in order to get the marrow out of them.

When the horse was taken up, this way of life was so attractive that many of the tribes in the areas bordering on the edge of the plains gave up farming and moved out to become hunters. This is one of the few cases in history of farming village dwellers giving up their settled way of life for the wild nomadic life of the hunter. The whole period was a brief one, however. The Indians of the southern plains began to ride at some time toward the end of the seventeenth century, and in less than a hundred years all of the Plains Indians had horses. In another hundred years this way of life had come to an end because Americans had killed off the buffalo. In between these dates lies the history of the flourishing of the great buffalo-hunting Indian culture. The Spanish almost certainly were the source of the horses and the habit of riding them, and it needs to be emphasized that it was the trick of riding the horse that was important. The Indians had had horses before but lacked the idea that they could be ridden; they ate the horses instead. In time, of course, horses became valuable as a means of transportation.

European Views of the American Grasslands

Since the Spanish were horsemen and mid-latitude men and had wheat growing and cattle handling as a major part of their background, what was their reaction to this rich grassland?

The Spanish explored the grassland very early. Several long expeditions from the Gulf area and the Southwest reached out into the plains. De Soto entered the plains from the southeastern direction, and Coronado penetrated the plains from our Southwest. Both expeditions probably reached the vicinity of Kansas. They reported on the immense extent of level, grass-covered land and the presence of herds of buffalo with Indian people who made their living hunting them. However, they made no move to settle these lands.

This is not surprising. As we have seen, there were very few Spaniards available, and they were two continents to seize and to hold. There were also settled Indians possessed of great wealth of gold and silver and developed lands in the Middle America. In the North American grasslands there were only wild Indians and undeveloped lands. Not surprisingly, the Spaniards elected to pass up this area, just as they did California. The few settlements made were along the southern border of the grasslands and at the mouth of the Mississippi, more to hold the area politically than to try to settle and develop it. It was probably from these settlements that horses and riding spread to the southern Plains Indians in the seventeenth century.

Once the Indians had the horse, the Spanish spread into this area was made infinitely more difficult. Instead of a thin population of people of limited mobility and considerable poverty, the area came to have more people, and these were a mounted, mobile, aggressive, well-fed people who took to warfare as a special form of sport. For such tribes as the Apache, the Comanche, and the Kiowa, the settlements in Texas and Mexico became the source of readily available horses, guns, and captives. The Spanish—and later the Mexican—frontier was actually driven back by the Plains Indians. The Spanish can be said to have made no use of the vast grassy heartland of America, an area that was for centuries nominally, but never physically, in their possession.

Other Europeans also looked at the grasslands. The French had spread across the North, following the Great Lakes waterways that led them westward. Their explorations then took them down the Mississippi, and they followed the major rivers out toward the Great Plains, but their interests were concentrated in furs. These they obtained primarily from the forest Indians; hence they were deflected by their interests away from the grassy plains.

In the early nineteenth century there was a sizable settlement of French-Indian people along what is now the boundary area of Canada and the United States, just west of Lake Superior. From this point spring and fall hunting parties were organized to go into the plains for a supply of buffalo meat. They hunted in the rich agricultural area that is now Minnesota. But the agricultural traditions of the French were a thing of their past; in this new land they approached the plains as the Indians did, as a source of buffalo to be harvested seasonally, and they made no attempt to farm these fertile grasslands.

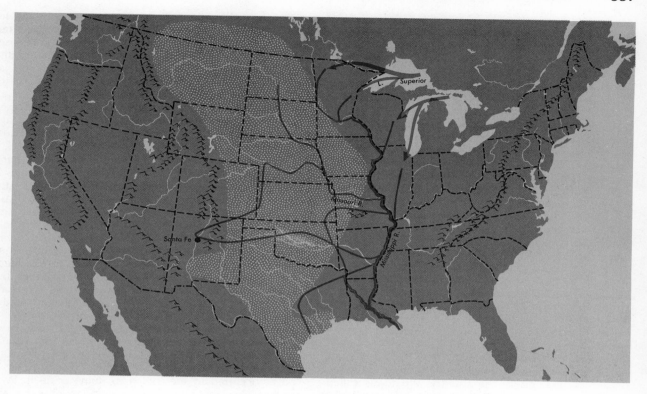

MAP 7-1   French Explorations in the Grasslands, 1673–1750
For the French the Great Lakes led naturally into the grasslands and into the Mississippi. Although they traveled frequently on the Mississippi and developed a major settlement at St. Louis, they explored the grasslands relatively lightly. Their major penetrations were in the north, just to the west of Lake Superior, where they made considerable use of the area as a major source of buffalo meat. In the main they stayed in the woodlands.

The American Approach

The American approach to this land really has to be described as plural. The first arrivals were mainly fur trappers, and they sought beaver. The plains had few, so these men drove on west to the mountain areas where beaver lived. The first settlers viewed the grassy plains as a poor land to be crossed as rapidly as possible. Their path through the plains became "the Oregon trail," traveling over the fertile prairies, our present corn belt, and the level, short grassland, our present wheat belt. They followed the Platte River to South Pass through the Rockies to reach the headwaters of the Snake River and went all the way to the Pacific Coast to find another forested area that was like the eastern region of the United States.

These early Americans had behind them a long tradition of forest dwelling, farming, and cattle raising, for this had been their way of life in Europe. It was continued in the forested eastern United States, where their whole economy was geared to a great abundance of wood. Trees told you whether the land was rich or poor. The trees that you felled supplied the timber for your house and barn, the rails for your fences, and the wood for your fires. Without wood, how could you keep stock out of your fields or keep the fire going in the fireplace for warmth in winter and for cooking the year round? A land without timber must be a poor land and, to one accustomed to unlimited amounts of wood, an impossible land. Better to go 3000 miles to Oregon than to try to do anything with this impossible grassland where timber was lacking, water was scarce, the grass fires raged across the land faster than a man could run, and the buffalo supported a series of Indian tribes whose avocations were horse stealing and warfare.

Eventually changes were to come that would alter this outlook. But the good, rich soils and the immense wealth of some of our richest agricultural regions were to those early Americans, even to the traditional farming people who first saw them, no resource at all. The "good" of the environment lay fallow awaiting another generation of men with new ideas and new techniques. In large part the techniques already existed, but the applications had yet to be made. The physical environment was to continue to lie fallow until a people came who looked on it not as a repulsive desert but as an attractive possibility.

The West was added to the nation in a series of purchases and settlements and wars: the Louisiana Purchase made in 1803, the Oregon country obtained by settlement with the British in 1846, Texas brought in by annexation in 1845, and the Southwest won by the cession from Mexico in 1848. With the Oregon settlement and the subsequent gold rush, America suddenly found itself with a large population separated from the rest of the country. What was called the Great American Desert (the grassy plains, the Rocky Mountains, the Great Basin, and the Sierra Nevadas) separated this Pacific population from the rest of the nation. In addition, there were the mobile horsemen of the plains making settlement or occupation of that area difficult, if indeed it was even desirable.

Opinions varied on what to do about this situation. Some thought the distance made regular contact too difficult. Others advocated bringing in camels to supply transportation across the deserts and to give to the Indians so that they would become herding people instead of hunters. It was reasoned that beneficial relations could then be established with them, such as the peoples of the Old World had worked out with the nomadic peoples of Inner Asia. They could supply camel hides, wool, and cheese in return for manufactured articles. The camels were actually introduced, but they were used mostly by the army for transportation in the southern desert route to the West through Texas and Arizona. Although they were successful in their way, they were rapidly displaced by the railroads. They had been introduced a few centuries too late.

The camels flourished in the arid West. Eventually stockmen shot them to keep them from multiplying and eating up the food on the range that the cattlemen desired for their stock. Were this not so, there would be wild camels in our West today—and this, like the horse, would have been only a reversal of history, for there had been plenty of camels in the past. The Indians had had a hand in their extermination, then the white man reintroduced them, but with new ideas: domestication and transportation. Before the Indian could take up the idea of using them as pastoral beasts, however, the quickening pace of change was to sweep away the old way of life and almost sweep away the Indian with it.

It would not have been a quick and easy thing to convert the Indian to a pastoralist. The change is not a simple one, for it involves an entirely new attitude toward the land. The keeping of a domestic beast, the milking, the gathering of the wool, spinning, and weaving are all difficult new ideas that have not only to be learned but be fitted into a way of life. Given time, the Indians could readily have done this. They had the horses for herding, the grasslands for the feeding, and the custom of moving about. Other tribes, the Navaho, for instance, show that such people can readily take up herding, shearing, and weaving; for the Navaho, the domestic animal was to be the sheep.

Curiously, little thought seems to have been given to the buffalo as a domestic beast. This animal is admirably suited to the plains environment: it supplies delicious meat, fine hides, and some wool. Yet it never occurred to the Indian to domesticate it. Why bother when there were millions available for the taking? The white men all stressed the ferocity of the beast and its utter untamability. Those who have seen these animals in zoos, parks, or in Yellowstone are likely to have been more impressed with their cattlelike appearance and action. No large animal is safe; even large breeds of dogs turn on their masters at rare intervals and kill them. For quite a time the greatest risk on our farms were the bulls and Jersey bulls had a particular reputation for meanness. The extent of injury and loss of life due to domestic bulls has only recently given way to injury and loss of life due to tractors.

We came to the Great Plains and found the buffalo a wild animal. Like the Indians before us, we continued to think of the buffalo in this way. When a thoughtful move was made to try to use these grasslands for pastoral nomadism, two patterns of thought emerged: first, that these were desert lands; second, that for desert pastoralism the camel was the proper animal. This illustrates the dominance of men's ideas over the realities of mere physical geography. The grasslands were not desert. There were already highly suitable animals in the grassland for a pastoral way of life, but it was easier to think of an already domesticated animal and invent reasons for not using the native

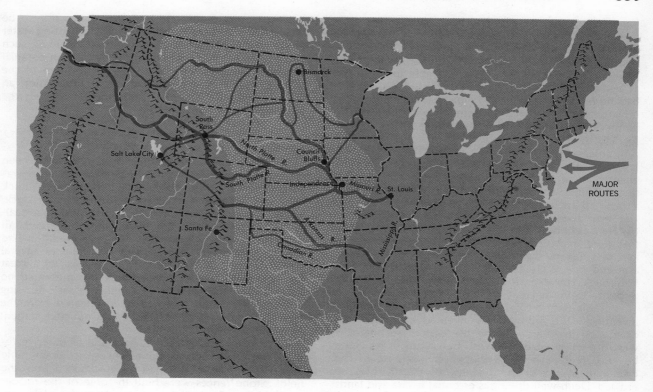

MAP 7-2   American Explorations in the Grasslands, 1800–1850
The Americans came to the grasslands later and approached them on a broad front along the Mississippi. The routes of exploration and eventual travel were the rivers, with the Missouri playing the major role. The Oregon Trail started up the Missouri, then followed the Platte River to South Pass, continuing along the Snake River to the coast. The great migration to Oregon went right across the grasslands, and the settlers saw nothing there worth stopping for. Early American exploration and settlement largely avoided the Spanish-Mexican territory until that land was taken into the union.

animal than to see the grassland and its native animal for what they were. In passing, it should be noted that the buffalo is a Eurasian animal that entered America relatively late in time, so much like the European cattle that its bones are barely distinguishable even by an expert zoologist. It is readily tamed and does well behind barbed wire on managed pastures.

Developing the grasslands

A series of inventions was introduced just after the Civil War that transformed the grassland and doomed the Indian buffalo-hunting way of life: the six-shooter put the Indian at a military disadvantage, and the railroad provided transportation to markets. The deep-well drilling technique made water available where it could not be reached before, and the windmill lifted the water. Barbed wire supplied the fencing for fields. Transportation doomed the buffalo and moved the crops to market. The self-polishing steel plow made it possible to break the tough sod of the prairie. Since these were all needed on the plains, were they invented there in response to the environment? At best only one of them can qualify.

The six-shooter was invented by a New Englander named Samuel Colt. As is often the case, he had the greatest difficulty getting acceptance for his invention. Eventually a number of them got into the hands of the Texans, who had been having their difficulties with the Indians. With a single-shot rifle, and at most a single-shot pistol, the Americans were at a decided disadvantage against the Indians, who might easily carry 20 or 30 arrows and be able to fire them at a rate of 10 or more a minute. The Indians, in fact, had perfected the art of charging the settler, drawing his fire, and then charging on in to drive their arrows home at point-blank range while the settler was reloading. And it did no good to jump on a horse and run, unless it was the fastest horse on the prairie. The Indian would chase anyone fleeing and easily pick his enemy out of the saddle with his 14-foot lance. The first encounter of men equipped with six-shooters with a band of Indians proved disastrous for the Indians. The men fired their rifles and then charged. To the Indians' astonishment, they whipped out short guns that fired again and again. The Indians were routed and lost half of their men. The pistol was a needed arm and one well suited to close-quarter fighting on horseback. However, it was still not invented in direct response to the need.

This is true of all but one of the rest of these inventions. The railroad was invented in Europe and developed in America along the eastern seaboard. It was to revolutionize the West, but it can in no sense be said to have been developed in response to the challenge of the great open spaces that had to be crossed. Deep-well drilling is essential in the arid lands where the water supply may be 50 or 100 or more feet below the surface of the ground, but it far precedes the American occupation of the West and in fact was a European invention. The drilled well is a 6-inch or wider hole bored by either a rotary or plunging type of bit and cleaned out by a long slender bucket with a valve at the bottom. These devices were taken up in the American plains, and they were used very extensively to get water in regions that were otherwise arid, but they were not invented there.

Water is readily lifted by a bucket on a rope over a pulley for distances up to 40 or 50 feet, if the water needed is only to be used about the house. If water is needed in larger quantities, for watering stock and irrigating a small garden, lifting the water by hand hauling on a rope is a very tedious business. For lifting the quantities of water needed, the Americans adapted the windmill. This is particularly useful in such treeless lands as the plains. There the absence of surface roughness does not cut down greatly on the movement of the air, and the wind velocities near the ground are, like those at sea, about 10 percent higher than they are over forested land. But where did the windmill come from? The usual reply is Holland, but the earliest recorded ones are in Persia. They came to be widely used all over Europe, not only for lifting water in Holland but much more commonly for doing such work as grinding grain. It was widely used in colonial America. The idea was taken up by the men of the grassland, and after considerable experimentation an efficient, metal, fan-bladed form was developed: the familiar windmill of our West.

Fencing

The one invention that seems to have been made in the grassland in response to a real need was barbed wire. The problem was a pressing one. Stock had to be kept out of the crops. To have good stock, it was necessary to keep the animals of desirable qualities separated so that they would not interbreed. All of this required fencing. The American tradition had become one of rail fences. This could be of several types, but they all had in common the use of a great deal of wood. In the forested East this presented no problem; on the contrary, it supplied a way of using some of the wood that had to be cleared from the land. In the plains there simply was not that amount of wood available.

Fences can, of course, be made of other materials. The New England stone fences are an example of a use of material that had to be removed from the fields. Stone fences were built in some of the rockier parts of the plains; however, much of the plains are not stony. Soil, which is everywhere, can be piled up either between board forms or as sun-dried bricks to make an earth fence. This was also tried, and it gave rise to the Americanism "as ugly as a mud fence." Earth walls were expensive in labor costs and melted down under the washing of the rains. Another means of fencing is by hedges; this is a highly developed art in England and in parts of France. A wide variety of plants can be used, and when properly planted, reinforced with stakes and branches in weak spots, and carefully tended, they make excellent fences—but they have some disadvantages. They use water, and they occupy space. In the plains environment, however, hedges also cut down on the destructive action of the wind and shelter the homestead; they catch drifting snow and thus accumulate moisture.

Considerable effort was expended on hedging, and attention turned to the Osage orange. This hardy native tree is thorny and hence useful in holding cattle, but there were numerous other possible hedge-making plants. To the south and toward the arid side of the plains the mesquite and the cactus could have been used. Hedging could quite possibly have solved the problem of fencing the grassland if the invention of barbed wire had not come along just at the time that the hedge work was in the experimental stage in the United States. There were then a number of ways in which the fencing problem could have been met. The actual solution was different.

The industrial revolution had made metal more readily available. Wire-making machinery had been invented, and wire was relatively cheap. Fencing with wire had been tried, but the animals soon learned that they could simply push their way through the wire. A number of men began experimenting with putting prickly materials on the smooth wire to stop the animals from escaping. At De Kalb, Illinois, a solution to the problem was found, and our modern barbed wire evolved. This came in the 1870s, and barbed wire rapidly became the dominant fencing material in America in all climates. It is unesthetic, unfriendly, uninhabitable by bird and beast (unlike hedges), and so damaging to stock that it is never used on good horse farms. It is not demanded by the physical environment, even in the grassland. It is the dominant fencing material of our day for economic reasons only.

Settlement

The grasslands of America were the last of our good lands to be settled. After the Civil War the railroads were thrust across the Great American Desert and on across the Great Basin to link the Pacific Coast with the rest of the nation. The railroads spelled doom for the buffalo and the Indian and salvation for the Texans. The buffalo herds were slaughtered by the hide hunters, for the buffalo was so docile an animal that a skilled hunter could locate a herd, hide in the grass at a considerable distance, and using a heavy rifle, shoot down 20 or 30 before the animals finally ran. Thus one rifleman could keep several skinners busy. The hides were moved by wagon to the railroad and thence went east to the market. Buffalo robes and buffalo coats became common items of wear in the East. But the slaughter was at a far faster rate than the reproduction rate, and within a few years the buffalo were on the verge of extinction. Only energetic action by a few individuals saved a small number to form the present small herds.

With the buffalo gone the Plains Indians had their means of livelihood cut out from under them. Harassed by the army and weakened by unaccustomed diseases, white man's liquor, and hunger, they were soon rounded up and put on reservations. The great grassy plains were emptied of both their dominant animal and the men who had based themselves on the harvesting of this animal. This void did not last.

The Texans had returned defeated from the Civil War and depleted in manpower to a land that had stood still or declined in productivity during the hostilities. Only one thing had increased: cattle. The semiwild, long-horned, lean and tough cattle had multiplied into millions, especially in the southeastern grasslands of Texas. There was no market in Texas, but as the railroads reached west, the Texans hit upon the idea of driving the cattle north to reach the rail lines. This started the saga of the cattle drive, so thoroughly exploited in literature and by television. It brought money to Texas, and it brought cattle to the rest of the plains, so recently vacated by the buffalo and the Indians. When the price was not right, the Texans moved their cattle out into the grassland adjacent to the rail head and held them there for a better market. The range was then open; no one claimed its ownership. Men simply put their cattle on it and held it. For a period of a few decades this system persisted. Then ownership had to be legalized by taking up homesteads and establishing a right to the land.

The agricultural settlers moving into the grasslands were often far ahead of the railroad, and they faced many difficulties. The settler, arriving from some point of departure in the wooded East with a wife, children, a team of horses, and a wagon loaded with equipment, looked about him at this land that was his "for free" under the Homestead Act of 1862. Except for trees along the water courses, it was treeless, and the water courses had water only in the winter and spring. The land was level and the soil was rich, but it had a continuous cover of grass. Along the humid edge of the grassland the grass might stand 4 to 5 feet high. While this was fine for the stock, it posed a terrific fire hazard. Once on fire, whether by act of nature or man, it was a threat to man and beast, to field and house. The first thing to do was to plow a fire break around the spot on which the homestead was to be placed, and this was not easy.

The sod was so thick and tough that a team of horses could not readily pull a plow through it. The old-fashioned plow with a steel tip but a wooden moldboard—the wide blade of the plow—did not slide through the sod but dragged heavily. It usually took teams of oxen to pull the plow through such heavy going. An invention that soon greatly aided the farmer was the all-steel plow. The steel mold-board polished as it went through the ground, so the friction was cut down, and the plowing went more easily.

When the plowing was done, the sod was still there, turned up in ridges with its roots to the sky. But this was sod, not clods. Clods will break down under rain and frost or under harrowing, but sod will not.

The fields plowed in the first year could not be planted in wheat. Farmers often walked along the furrows, chopping a hole in the sod and planting some crops such as potatoes. It was not until the second or third year that the land could be harrowed smooth and a grain crop put in.

Meanwhile, what was to be done about a house? There were no supplies of logs for cabins. There was no mill for sawing planks. But the whole tradition of these settlers was to build a wooden house. True, brick was also known, but brick making demanded a large supply of wood for firing the brick. Further, some sort of house had to be created rapidly, for winter in the plains required some shelter. The shelter could have been an Indian house—tepee, earth lodge, or grass-thatched lodge—but that was Indian housing and not for white men! The interesting solution was to take the sods, treat them like bricks, and lay them up in walls. The roof was then formed of timbers, covered either simply with canvas off the wagon, or planked with boards from the wagon and then covered with more sods. The result was a thick-walled, weather-tight but often damp house that was serviceable for a few years.

The psychological impact of this new flat, tree-less environment on the settlers is most interesting. They had come from the forested lands and usually from hilly lands. This flat, treeless, windy land was strange to them, and they longed for the hills of home. The letters and diaries of the time are filled with this longing. The women felt it most, for they were confined to the sod house and its immediate neighborhood. They had no one to talk to other than the children and their husbands. The nearest farm was likely to be a few miles away, two or three hours travel by foot or by wagon. The wind blew eternally, the heat in summer was stifling, and the winter was not only cold but a time of violent change—of cold waves and of blizzards. The women sometimes lost their minds in the isolation, hard work, and strain.

All of that ended with the first generation, for the children raised there never knew the hills and woods. The plains with their brassy skies and their wild winters and violent summers were home, and they loved them. To them, now plainsmen, the woods would be stifling because they hemmed one in and cut off the view. A plainsman is accustomed to seeing for miles in any direction, and it is disturbing to him not to be able to see. This is a theme that runs through all of man's reaction to his environment. Home is where you live, and home is right. All the rest of the world is different, and not quite as good. The desert man loves the desert, the forest man the forest, the hill man the hills, and the plainsman the plains.

The settlers took up the grasslands. The humid section became part of our present-day corn belt, the land out to the 100th meridian became the wheat belt, and to the west of the 100th meridian the land was dominantly cattle land. The railroads increased in number, and the growing population in the East took the increasing flow of grain and meat produced in the area. Thus the land increased in value. The 160 acres that the government gave to each settler amounted to values of hundreds of dollars per acre. This has sometimes been called an unearned increment. It is too bad that those who toss this phrase around could not have been on hand to endure the grass fires, the Indian raids, the plagues of locusts, the loneliness, and the absence of all the things that most people cherish—neighbors, a doctor within call, schools for children, and the like. True, the great rise in value benefited the second and third generation more than the first. But is it sensible to deny to the children of these pioneers the right to the product of their parents' sacrifice and work?

Climatic Risk

The grasslands are lands of natural climatic risk. This risk increases as one goes farther west toward the arid lands. Climate is an average of conditions met with over a long period of years. The humid lands are those that in almost all years have sufficient rainfall to keep the ground water table charged, the streams running, and the trees growing. The arid lands are those where water rarely runs, the ground water table is nonexistent, and trees grow only in exceptional spots because most years are marked by extremely meager rainfall. The grasslands of America lie in between these conditions. Sometimes they have a year that is truly arid, even desert, in its low amount of rainfall. At other times they may have a year that is humid, and if that were continued for many years, the grassland would turn into a forest. Wet and dry years in such lands come in spells. A series of wet years makes possible wheat growing far out into the land that is normally too dry for such crops. If the price of wheat is high, the land will be plowed. If there are a number of such years, much land may be plowed up in the hope that the streak of wet years is going to last just a little longer. Eventually the balancing sequence of very dry years comes. The wheat then fails, and the land dries out. Since the winds continue to blow, the land, lacking any protective cover of grass, begins to blow away. The Great Plains have seen at least two major periods of blowing in the brief period of American occupation.

The first period came at the end of the nineteenth century. The settlers had moved in during a wet spell, and they plowed, planted, and harvested. Then came the dry years, and the land blew out from under them. The settlers abandoned the land and either went on west or returned to the humid, more dependable East—and the land went back to cattle. World War I brought sky-high prices for wheat, together with a number of wet years in the plains. Again the land was plowed up. The inevitable drought came in the 1930s, and the land blew away. Dust clouds darkened the skies at midday in St. Louis and Chicago, and the dust could be seen clear to the eastern seaboard. People poured out of the area like refugees from a natural disaster. It was, yet it was not. They had created the difficulty themselves; the disaster was only natural in the sense that in the environment of the plains, winds are persistent, droughts are inevitable, and land left bare will blow away in the drought years.

The environment had been there unchanging for 10,000 years. In the realm of the buffalo and the Indian such disasters were infrequent. It is a chilling thought, however, to learn that even without plowing, the plains can blow out. The sandhills of Nebraska are the remnants of a period of immense blowing at some time in the past. Archeological excavations have revealed that there have been times in the not too distant past when the Great Plains, though unplowed, became so devegetated that there was immense wind erosion. Seemingly this could only have been caused by droughts of greater intensity and much greater duration than any that we have seen in the hundred years for which we have good records of the area. We can expect at some time to see such a drought in the plains again.

Modern Commercial Farming

On the eastern, humid side of the American grassland, the land has been made immensely productive. Since agriculture is still the basis of our economy and America is helping to feed the world, the working of a productive unit should be examined. This becomes more important since less than 5 percent of our people are farmers today; most of the remaining 95 percent have little understanding of the work of the vital 5 percent that feeds them.

The movement in America is to larger farm units; to the extent that agrarian reform in other lands tends to fragment ancient large-scale holdings into tiny units, it is moving against the economic tide. In Iowa today farms average 200 acres, and many successful family farms are 600 acres. The traditional American farm was the 160-acre homestead—land free for the taking and developing. This was in the past larger than needed for the support of a family in the humid East and too small in the arid West. Today 160 acres is seldom an economically sized farm, even in the humid regions. What is an economic size changes with technology, politics, and economy, and these all change through relatively short spaces of time. Thus in an unchanging physical environment man's appraisal would change.

In the 1920s a 160-acre farm in Iowa valued at $12,000 ($75 per acre) was adequate. In the 1940s such land cost $100 per acre; in 1950, $200 per acre; in 1960, $250 per acre; and in 1970, $400 per acre. This represents a large rise in costs which came at a time of need to increase farm size for economic operation. Many men sold small farms and left for urban employment, and the men remaining enlarged their operation. The investment in land, buildings, and machinery for a farm today amounts to a figure between $250,000 and $500,000. The risks are huge, for not only is the farmer exposed to all the market variations normal in any business, he is also at the mercy of the weather and susceptible to epidemics of plant and animal diseases. Farm income is low by any measure. Comparable capital and effort would give greater returns in almost any other type of enterprise. The farmer continually increases his output, but he seldom has a proportional increase in income.

Production and management have changed greatly and continue to change. Crop rotation to maintain fertility is gone, for fertilizers make it possible to use the same land for the same crop year after year if that is desirable. The disappearance of the horse ended the hayfield, and cheap, effective fertilizers have eliminated the grass, legume, grain rotation. Now it can be corn, corn, corn—indefinitely. Yields have risen from around 30 bushels per acre in 1900 to 100 per acre in 1970. Test plantings yielding over 300 bushels per acre indicate that the ultimate potential of breeding, fertilizing, and pest control for increasing yield has not yet been reached. The costs of such production are high. Fertilizer for one year may cost $10,000, and one tractor may cost a like amount (a 600-acre farm needs two tractors). Other $10,000 items are a self-propelled combine for grain and a picker-sheller for corn. A $25,000 investment in machinery is modest for such a farm, and this gives a good measure of the difficulties facing underdeveloped nations in modernizing their agriculture. This kind of operation requires capital and management skills utterly different from those found among the simple peasant farmers of the world.

On the other hand, the Iowa-type family farm has great flexibility, which corporate and government-controlled farms do not have. When plowing or harvesting demands it, the tractors can be kept going 15 or 16 hours per day with wife or children spelling the farmer, without problems with time clocks, overtime, or union regulations. Similarly, if pigs are born at inconvenient hours such as 2 A.M., the farmer serves as the midwife. They are his pigs and money in his pocket. They are just "dollars for the boss" to the hired hand or "pork for Moscow" on the collective farm. Production results not just from capitalization but also from motivation.

The increasing farm size combined with mechanization is displacing men from the land, and this is expressed in America by the loss of population in one-half of all the counties in America in the census decade of 1950–1960 and again in 1960–1970. Over a longer term we could say that once we were 90 percent rural and 10 percent urban, but now the figures are more than reversed. Moreover, the end is not in sight, for estimates of a further decline of 25 to 50 percent of the farm population have been made. We now have more food being produced by fewer men who operate large blocks of land with increasing capital investment and managerial skill.

The eastern side of the grassland is known as the corn belt. To the west of that lies the wheat belt: spring wheat to the north, where it is too cold for the wheat to survive the winter; winter wheat to the south, where wheat can be planted in the fall to grow slowly through the winter and then mature quickly in the summer. West of that area, cattle is king, but the cattle business has also changed. The vast holdings have become scarce, even though the average holding is large. The cattleman may move about over his domain with his pickup truck, pulling a light trailer carrying his saddled horse and perhaps his cattle dogs. He needn't stay out in a line shack overnight, for he can drive out 20 or 30 miles to a distant pasture, check his cows, and return home comfortably. Even the windmill is disappearing. It required a drilled well, was subject to storm damage, and had to be constantly maintained. Today a bulldozer will scoop out a hollow in a drainage way and build a wide overflow channel to take care of storm waters. Once built, the pond so created, known in the Southwest as a tank from the Spanish *tanque*, may require no maintenance for 20 years. There may be a million of them in the grasslands now, and they are beginning to affect the ecology of the plains by supplying ponds for ducks, water for wildlife, and water for cattle.

New plants have been introduced, notably the African grain sorghums. These are very drought resistant and high yielding, and enormous tonnages of grain go to the feed lots—a new blotch on the landscape. Cattle are penned in enormous numbers, fed heavily to get the maximum rate of gain together with the desired "condition" (fat percentage), and then slaughtered. The feed lots are highly efficient and operate on high-volume–low-cost principles that make them unbeatable by the small-scale cattleman. The man who raises calves sells them at the local auction house to the buyers, and the cattle go by truckloads to the feed lots. Tens of thousands of cattle may be concentrated in one feed lot to be fed by machine. The production of urine and manure is enormous, and the stench in wet weather is almost insupportable. The by-products filter into the ground water and may pollute a whole area. Regulations are now being created to control much of the offensive by-product of this highly efficient cattle management. As usual, we stumble into advantageous ways of doing things, find that they have drawbacks, then slowly make the necessary corrections.

Meanwhile, the cattle business is also shifting. The longhorn of the West was mostly skin, bone, and horn. The cow that displaced it was the Hereford, a British breed of vastly greater efficiency with far better taste. Then the Angus cattle were brought in, and recently a whole wave of other European breeds has been introduced: Charolais, Anjous, Brown Swiss, as well as the heat-resistant Brahma. But the most important development has been in the scientific breeding of cows for good milking, fast growing, and concentration of growth in the edible areas. We are actively redesigning beef cows for our needs instead of simply taking what nature gave us. The new cow will probably be about twice as efficient in producing beef as the longhorn and about twice as good to eat.

Summary

The American grassland changed at the time of the retreat of the glaciers from a well-watered land inhabited by elephants to something like the grassland that our frontiersmen saw. For the Indians it was a good hunting ground, but it was not a good farming land. This meant that the density of its population was limited. The forest-hunting Chippewa had a population density of one person every 2 square miles. Quite probably the density of the hunters of the plains was somewhat less. The density of farming people, such as the Huron, was about 10 per square mile. The Plains Indian hunters, then, were numerically, culturally, and militarily insignificant in the prehorse days. With the introduction of the horse, the man-land situation changed drastically. Now man was able to harvest the buffalo with vastly increased efficiency, and the Great Plains became an area of attraction, growing population density, interesting cultural growth, and military power. This development was cut short by the expansion of the Americans. With the removal of the buffalo the Indian was reduced to a reservation existence: a miserable, settled existence for men who had been as free and fierce in their occupation of the plains as the hawks in the sky.

The American assessment is notable for its early rejection of the plains. The idea of what the land was like was sufficient to prevent men from "seeing" it, even when they stood right on the rich lands of the corn and wheat belts. Then changing ideas, more men, and the use of a series of inventions, most of which were neither new nor invented in the grasslands, made possible the profitable occupation and development of these lands. In all of this the physical environment remained little changed, while man changed and changed again in his way of using this land.

THE ARGENTINE AND AUSTRALIAN GRASSLANDS

The Argentine Pampas

In South America there is a great grassland around the La Plata estuary. Climatically it is like the United States prairie, that is, it is humid enough to support tree growth. And like the North American grasslands, aridity increases westward toward the mountains. Another similarity is that it was the home of hunting Indians who, in times past, had known the horse but elected to eat this animal rather than domesticate it. The idea of domestication simply did not take root among them, although they had a greater opportunity to get this idea in pre-Columbian time than the Indians of the North American grasslands did. The people of the Andes had domesticated the llama, a little American camel, and they used it as a beast of burden. The idea that animals could be domesticated and controlled, and that thereby a surer food supply could be won, never took root among the Indians in the grasslands of South America.

When the Spanish explored southward over the Andes, they discovered this great grassland. As in North America they found little use for it in terms of the few Spaniards available and the vast numbers of men and acres there to be conquered and controlled. Once the mining period in colonial New Spain was underway, however, there arose a continuous need for mules to work in the mines and beef to feed the miners. The

grasslands were then utilized as an area to produce the vast amounts of animals needed. As in North America the contact between the Indians and the Spaniards then became close enough so that the Indians learned about riding horses and obtained horses to ride. As in North America there rapidly arose a way of life characterized by men virtually living on horseback, extremely mobile, and making an easy living from the animals of the great grasslands. As with the Indians of the northern continent these people became fierce fighting people, proud, able, and formidable.

The parallels are striking, and it is easy to read into them the "natural responses to the environment." Given the similar cultural additions in the similar environments, certainly there were similar responses, but there are some interesting differences also. The South American Indians lacked the buffalo. They had smaller game: the ostrichlike rhea and the camellike guanaco. Although they hunted these animals on horseback, they did not develop the short bow for this; instead they used the bola, an arrangement of a number of stone balls attached to the end of leather thongs. When these are thrown, they entangle the legs of the prey and throw it to the ground. It is difficult to know why this difference in weapons used on horseback came about. Perhaps it was because the bola was still in use in South America at the time of the change to horseback, and the habitual weapon simply became the one that was used. These differences can be multiplied. For ease of carrying burdens the Plains Indians never thought of the wheel; instead they used two poles dragged behind a dog, and later they used the horse. Their South American counterparts never developed this type of conveyance, nor did they build housing in the style of the tepee. The similarities, then, do not go very far in detail.

Eventually the Indians of the South American plains were reduced to reservations much as the Indians of the United States were, but only after a long and costly series of campaigns. The cattle owners then occupied the land unmolested, but the pattern of land ownership was entirely different from the United States pattern. Land was granted in huge estates, in the Spanish manner, which a historic-minded Spaniard might describe as something derived from the Romans. The result was the huge landed estates of Argentina that are still the features of the occupation of the Argentine pampas. The land is now utilized for cattle and grain, principally corn, wheat, and flax. Development has depended on the coming of railroads, which can move the produce to markets, a development that came as a result of British capital and energy, just as British capital provided early investment in the United States railroad system. It was British drive in South America

Gaucho. *The gaucho is to Argentina what the cowboy is to America. In this scene the middle man has a knife in his belt, essential equipment for a gaucho. He is pouring hot water into a gourd containing maté, the Argentinian equivalent of hot tea, to be drunk through the tube from the top. Wrapped around the waist of the man standing to the left is a bola, a rope which has two balls for weights at the ends. The baggy pants, tall boots, and two-gallon hats (as contrasted with the Texan 10-gallon hat) complete the gaucho's working equipment. (Panagra)*

that created the modern slaughter houses, built the refrigerator ships, and led to the upbreeding of cattle so that the kinds of beef produced would be salable in the European market. The development of the Argentine pampas was dependent on ideas and capital from outside Argentina. The land, however, remained in very large blocks in the hands of the original great families. Thus, though the environment was also similar, the details of this development show differences, despite the fact that the settlers of both areas were

Europeans who faced similar problems, who came to use many of the same devices to overcome these problems, and who came also to produce many of the same products to put into the same world markets. The resemblances that do exist are hardly surprising in view of the many cultural causes that are common to the two areas and the fact that modern developments occurred at about the same time.

The Australian Grasslands

Other grasslands show other patterns of use and change. In Australia the grasslands that surround the desert, forming the transition to the humid zones, were occupied by simple hunters and gatherers. European contact largely removed these people, with no period of horseback hunters intervening. Why the difference? The land is well suited to mounted hunters, and the large kangaroos can be hunted much as the American buffalo were. The Australian natives lacked the bow and arrow and used the spear thrower instead. However, this weapon could have been used from horseback fully as easily as the bola could, and as we have seen, the Indians of the pampas managed this miracle of juggling. The Australians also used the boomerang, the curved throwing stick; this would have been a powerful weapon if used at close range from horseback. The environment was suitable for horsemen, there were animals to be hunted, and the men had weapons that would have been useful on horseback. It looks almost like a case of human failure.

It is true that the Australian natives were among the most retarded in cultural development of all the people of the world, but as we have seen repeatedly, there are numerous possible reasons for this state. Here isolation and initial lag had combined to leave them so far behind that later people simply swept past them. When the British finally settled in Australia, they occupied the land relatively swiftly. There was no period of a century or two of contact along a grassland frontier as there was in both the United States and in Argentina. There simply was not enough time for the Australian natives to learn the mysteries of horse domestication and riding and adapt these new ideas to their nomadic hunting way of life. Hence in this similar physical-environmental situation the "human response" was utterly different. Even given the presence of similar people (Europeans) and introducing similar (though not identical) cultures into similar environments already occupied by nonagricultural people, the results were greatly different in the Australian case from those in

the two cases in America. Time, as has been suggested for several other situations where people did not do what seems to us to have been the obvious thing, is one of the necessary ingredients for learning—and for social change.

The modern Australian development of their grasslands is very much like the western United States. There is the same transition from predominance of grain growing on the humid margin to predominance of sheep and cattle raising in the more arid lands. Similarly, as the aridity increases, the ranches become larger and the ranch houses more separated, ultimately by miles and miles of empty land. The barbed wire, the windmill, and, in times past, the six-shooter also played their roles. Today the railroad, the truck, and the radio play similar roles in both areas. Why should the result be different? If similar people having similar cultural backgrounds are placed in similar environments, is not a similar landscape to be expected? That such a climate need not produce such a response is best illustrated by taking lands completely outside the European sphere of influence and looking at what happened there.

PASTORAL NOMADISM: EURASIA AND AFRICA

The Idea of Domestication

The use of lands that are neither quite desert nor quite humid but are often covered with grass for the herding of domestic animals is an invention of the people of the Old World. We know relatively little about just when and where the idea was first put into use. As with so many ideas, it seems to have begun somewhere near the Middle Eastern center of growth of ideas, population, and civilization. We know that agricultural beginnings there date back about 10,000 years. We know too that domestic animals' bones appear relatively soon after men started to settle down in agricultural villages.

The Eurasian domestic animals include the horse, cow, buffalo, pig, duck, goose, chicken, donkey, camel, goat, and sheep. Not all of these come from one area. The chicken is from Southeast Asia, the donkey may be North African, the horse and the cow probably came from southwestern Asia, and the home of the domestic camel may be Arabia. We do not know whether some of these domestications are entirely separate or if the stimulation of domestication in one area led to domestication of a similar animal in another. Could this be the case in the relationship of the donkey to the horse and the cow to the water buffalo? Did the desert people try raising horses and, finding that it was a difficult and unsatisfactory business, turn to the donkey instead? Or did donkey-using people turn to the horse when they found that it was swifter? Did the people of the hot wet tropics try the common cow, find that it did not do well under conditions of high heat and humidity, and therefore substitute the water buffalo?

For these things to happen there must first be the idea that animals can be domesticated. Also, as we have seen, there must be a great deal of time for the idea to take root and spread; still more time is required for alternatives, additions, and substitutions to be made. In America the available animal, the buffalo, was not adopted as a domestic animal in the two centuries that the Indians had the idea of the domestic horse. The turkey, which was a domestic animal in Mexico, was not domesticated in the United States by the Indians; yet these Indians had considerable contact with Mexico for at least 2000 years before the discovery of America and took up art forms, ceremonies and domestic plants from that area.

In the Old World things took a different course. The idea that wild animals can be taken into captivity and managed—the idea of domestication—came into being very early. It had ample time to spread far and wide and be tried out on a wide range of animals—even the elephant. It is worth repeating that all of the animals domesticated in the Old World were or had been available in the New World. The Indians had known, and in some cases eaten, the American camels, elephants, horses, turkey, wild pigs, sheep, and goats. Except for Middle America, where we suspect that Old World influences had been strong for more than 2000 years, none of these animals were ever domesticated. And time was apparently too brief to allow the spread of the idea of domestication from the Old World to the grasslands of America, where they could have changed the whole cultural landscape with incalculable results on the course of history.

In the Old World these ideas did take root in the grasslands that surround the arid belt that runs from North Africa through the Middle East and across the inner part of Eurasia. It is here, and particularly in Inner Asia, that the use of the grassland landscape through the domestication of animals reached its highest development. Details of this distribution suggest something of the possible areas of origin of this trait of domestication. The idea never reached southernmost Africa, it was relatively late in reaching the East African area, and some highly adapted animals such as the

camel were introduced very late to North Africa. The camel seems to have been brought into that area by the Romans. Further, with the possible exception of the donkey and the guinea hen, there seem to be no native African animals that were domesticated.

The Eurasian Pattern

The truly vast grassland of Eurasia stretches from the vicinity of the Black Sea eastward to Mongolia. This is the home of the highest development of pastoral nomadism. The people in this area had the greatest variety of domestic animals, controlled the greatest area, and dominated it and its bordering regions more frequently than did man in any other grassland area. They figure as the epitome of the picture of man moving over the grasslands with his herds. They had the horse, cow, sheep, goat, and camel as their principal animals.

These Eurasian nomads practiced milking more than any of the other herdsmen. It is only because of our European traditions that we view milking as a natural way to utilize domestic animals. Before the European expansion, milking was unknown in the Americas and much of Africa and not practiced in China or Australia. It is not unlikely that milking, prior to its spread into India in 1500 B.C., was restricted to the Eurasian zone north of the Mediterranean and the Himalayan barrier and west of the Gobi. It is obviously an invention that was made once, somewhere in this area, and spread to those people who had domestic animals and were receptive to the idea (recall our picture of invention and change in Chapter 1). The Chinese had domestic animals but did not take up the idea of milking. This is usually thought to indicate that milking developed considerably later than the spread of domestic animals. It is a further case of the slowness of the spread of ideas, even when the necessary antecedent idea is present.

Within Inner Asia the nomadic herders worked out a way of life in which the herds were moved about over the grasslands to harvest the natural vegetation. Man in turn harvested the animals. The harvest took several forms: milking to extract a part of the animal's production without destroying the animal; shearing sheep, another such device; and slaughtering animals, a drastic means of harvesting that is a bit like killing the goose that lays the golden egg. Nevertheless, the carrying capacity of any range is limited, and usually the bottleneck is the winter pasturage—a greater limit in the Inner Asian area than in many of the other grasslands because of the severity of the winters. The Asian

MAP 7-3 Oasis Routes of Inner Asia
Routes linking the humid sides of the Eurasian continent have persistently followed the flanks of the mountains where the water supply was assured by oases fed by runoff from the highlands. Man's earliest expansions into the area followed these routes, and overland trade developed very early. Chinese dynastic beginnings in the northwest corner of China reflect the arrival of important cultural impulses from the Middle Eastern hearth along these routes, and in late dynastic times, 3500 years later, the Chinese emperors were importing elephants from India via these routes rather than by sea from India. We see these routes today at a time when sea and air travel have greatly reduced their importance. In the past they were among the world's most important thoroughfares for transfers of knowledge.

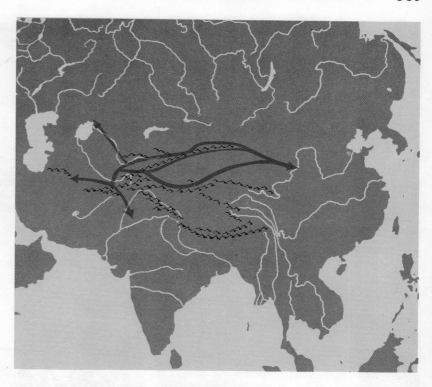

grasslands are distributed along the latitude of 50 degrees. This is more northerly than the Canadian border, and it is at this latitude in America that the maximum cold is encountered. In Asia, a much greater continental mass, still greater winter cold is met, and getting herds through the long, cold winters poses difficult problems. The Asiatic herders in many cases had oases as operating bases; these were occupied in the winter, and the herds were pastured on areas that had been left free of grazing during the summer period. Normally some agriculture was also carried on at the oasis, and sometimes fodder was stored up to help carry the herds through the winter. However, there usually had to be some reduction of the herd because the normal increase tended to produce more animals than could be safely carried over the long winter period of reduced forage.

During the summer various animals were milked. While most of us accept the milking of cows and perhaps, if prodded, of goats and even sheep, it still seems strange to us to milk camels and horses. Actually once the idea of milking has been tried out and found to be an excellent one, it is difficult to find a logical stopping place. Why not milk elephants? Or horses?

Horse milk is nutritious, and horses are effective animals in the grasslands. Milk, unfortunately, does not keep well, but this disadvantage was overcome by the making of butter and cheese. Since butter does not keep well without refrigeration, at least in its solid form, the solution was to melt butter and strain off the solids, retaining the oily material. This clarified butter, or ghee, does keep well. The pastoral nomads had worked out a whole series of such solutions to the problems of harvesting the grasslands with animals and then using the products of the animals for their needs. It will be recalled that most of adult mankind cannot drink raw milk. These soured, fermented, and otherwise processed milk products get around this difficulty and point to the absence of milk in China and parts of Africa as a cultural decision, not a biological one.

The horse provided transportation for speed and for war, while the camel was useful for slower transportation and carrying heavy burdens. The horse provided milk, and the camel provided milk and even wool. In the extreme cold of Inner Asia the camel is quite at home with a heavy coat of wool grown in the fall and shed in the spring. This is a very good quality wool, and since it is shed, the animal does not have to

Ulan Bator. *This is the capital of Mongolia where most Mongolian families live in "gers." These are sturdy, roomy tents with felt covers which are movable and warm even in the intense cold of Inner Asia. In the background are the roofs of Buddhist temples with the characteristic upswept eaves. Here there is a meeting of the nomadic life style of Inner Asia and the settled way of life of the Chinese. (Nolton for UNICEF)*

be sheared. Goats are milk and meat animals; sheep are wool and meat animals, and the Asian pastoralists even have special varieties of sheep that store fat in their tails. This is biologically comparable to the storage of fat in the camel's hump. In extreme cases the tails become so fat that the sheep have to be supplied with little wheeled carts so that their tails will not drag on the ground. It is this type of variety and specialization that indicates the degree of concentration on this pastoral way of life and hints at its considerable antiquity.

There are many other specialties. It was in this area that felt was probably invented. This is an ingenious way to make cloth, the principle being to beat the fibers until they are interlocked to form a strong sheet of material of uniform thickness. This process bypasses the laborious steps of carding, spinning, and weaving. A simple method practiced by some of the nomadic people was to take a large sheet of felt, spread a layer of wool on it, roll the sheet up around a pole, and then drag the cylinder around behind a horse so that the continuous rolling and bumping pounded and kneaded the material until the fibers were interlocked. This clever invention seems to have been made but once. It supplied materials suitable for shoes, clothing, tents, and saddle blankets, and we still use it for slippers and hats. It is a measure of our reluctance to try new things that we did not make vastly wider use of this valuable technique in the days before mechanized spinning and weaving.

Environmental Determinism

The nomads of Inner Asia have left a mark both on history and on thought. Repeatedly they have thrust out into the lands adjacent to them. The Mongol invasions of Europe reached well into central Europe. One group settled in the plains of Hungary, implanted its Asiatic language, and left just enough trace of Mongoloid ancestry for Hungarian babies to have a significant frequency of a bluish spot at the base of the spine at birth. This is the so-called Mongoloid spot. The nomads also swept into China at various times and even established dynasties there. Probably this type of movement has gone on throughout the time since the addition of the horse to the pastoralists' culture shifted the advantage to them as decisively as the introduction of the camel into the western Sahara shifted it to the pastoral people there.

These movements have given rise to one of the most extreme environmental-determinist schools of geography, and this has in turn influenced much thinking in history, economics, and related social fields. As a young man, Ellsworth Huntington, whom we mentioned in this context earlier, accompanied an expedition into Inner Asia. As they traveled through the country, they repeatedly found the remains of great cities lying in ruins in arid country. Huntington's conclusion was that the climate had changed and that the cities had fallen because of these climatic changes. He came to think of the repeated sweeps of the Mongols out of Inner Asia as the result of alternate buildup of population in the years of good rainfall followed by hardship and migration in the years of drought that inevitably come in these grasslands. He therefore came to speak of this pulsating movement as the "beating heart of Asia." It is a poetic and compelling thought: the Asian heartland filling with people in an expansive period of abundant moisture with its accompanying rich grass resources and then contracting during a dry period, pumping out streams of fierce nomadic mounted horsemen.

The attraction of such an explanation is that it is so simple and direct. Huntington's theory so gripped him that he vigorously assembled data to bolster his idea, instead of inspecting the data to see if they supported or denied his idea. Huntington's mass of evidence has been investigated by many scholars, and in almost all cases the physical evidence of droughts as the causative factors of the great known Mongol outpourings has been found to be erroneous. The Russian studies of the evidence from lakes and streams and other climatic records simply do not support Huntington's conclusions. A cultural change such as the introduction of the horse seems to be a more likely cause for the shift from oasis-based city dominance to pastoral nomad dominance. It is not improbable. Both the camel in the Sahara and the horse in the American plains created comparably drastic changes.

Huntington went far afield at times to bolster his ideas. For instance, he went to California and measured and counted the rings on the giant redwoods. Assuming that wide rings meant wet years and that thin rings meant dry years, he constructed climatic curves for a 2000-year period. He then compared the sequence of good years and bad years in California with the recorded history of Eurasia and concluded that there was a good correlation between climatic swings and historical happenings. This is an astounding conclusion for anyone to reach, for we know that the tree rings of California show almost no correlation with those in adjacent Arizona. It has been noted that in the United States, when one area is having unusual wet or cold weather, another part of the country is getting unusually dry or warm weather. If differences of these magnitudes are true within parts of one continent, how would it be possible to correlate such changes from continent to continent? This is not to deny that there are long-term changes of climate such as the glacial to interglacial shifts that have affected the world.

It is even more important to keep in mind that the human changes we can see in the full light of history are seldom necessitated by environmental changes. The Navaho and Apache invasions of the American Southwest were not linked with climatic changes. The movement of the farming folk out of the woodlands into the grasslands to take up hunting when the horse was introduced was culturally caused, not physically-environmentally caused. Expansions of peoples and the establishment of kingdoms and empires have a very large individual factor. The role of a Genghis Khan is probably more important than a swing in climate, and he is more likely to be able to muster men, equipment, and supplies for a vast military undertaking during times of good climatic conditions than in a time of poor climatic conditions. It is not necessary, however, to reverse environmental determinism here; it is simply worth noting that climatic shifts in the grasslands, lying as they do between the arid and the humid lands, are to be expected on a decade to decade basis. Great droughts of unusual extension, alternating with unusually good and long periods of moist conditions, are to be expected within some relatively long span, perhaps every century. If the outbursts of the pastoral nomads were primarily caused by physical-environmental conditions, they should have occurred at least once a century, but in fact they happened much less often than that. Other factors that may be involved would include the contemporaneous occurrence of weak government in the bordering humid lands and a period of strong leadership and unity in the grasslands; these might or might not correlate with climatic periods of good or poor conditions. The Germanic eruptions into the Roman Empire are an example of what may occur in any area at any time, with culture the cause rather than the physical environment.

In the other areas, where grasslands bordered on humid lands, there was no comparable movement of people from the grasslands into the settled lands. Prior to the introduction of the horse the American grasslands seem to have generated no force that struck at the settled peoples of the eastern woodlands. Neither is there any such history in the South American grasslands or in the Australian or African grasslands. Yet in each of these areas there were grassland-woodland contrasts, with the inevitable wet and dry sequences. What gave the people of the Inner Asian grasslands their peculiar ability was their possession of a way of life based on domesticated animals, particularly the mobility of the horse. Nor can the physical-environmental determinist escape by saying that once they had the horse, the environment then operated on the given culture to produce the observed sequence of invasions. As we have seen above, the evidence when critically examined simply does not support such a conclusion.

Ghost towns can come into being for a number of reasons other than climatic change. Mining towns are classic examples: when the ore is exhausted, there is often no reason for people to remain in the area, so the city is abandoned. Another common cause is the shift of a major route. Trade routes may shift because of political changes or technological changes: camels to railroads, sail to steam, a route around Africa to a route through the Suez. Such shifts are commonplace in history and are rarely due to climatic change. The

Egyptian closing of the Suez Canal to Israeli shipping caused the establishment of a port at the head of the Gulf of Aqaba and the revival of town sites along the route leading to the Mediterranean Sea. This route was important in antiquity, and cities then stood in the desert to serve it. They were long abandoned, but they are now being rebuilt. North Africa under Roman rule was a granary with rich cities. After Roman times it declined disastrously due to the collapse of authority and the incursion of the camel-using nomads who had not previously occupied the area. With stability imposed by the Italians and French after the mid-nineteenth century, these North African areas were rebuilt, and cities again prospered. Under modern political tensions the cities may be less prosperous, and their future hangs in the precarious balance of conflicting political forces—not on the variations of the precarious climate of this borderland between the humid and the arid.

The probable cause of most of the city ruins found in Inner Asia can be thought of in this way. The ruins are along the ancient caravan routes, situated where there are rivers coming down from the mountain passes. Under stable political conditions, with a flourishing trade and the construction and maintenance of canals and aqueducts, they could be flourishing today. There are exceptions, however. In some areas rivers have shifted widely across deltalike regions, leaving the former oases miles from the present area of river discharge, but this is not climatic change. Cities in arid regions run another risk: irrigated lands in arid and semiarid regions tend to accumulate soluble salts that are brought to the land by the irrigation water and left there when the water evaporates. The land slowly accumulates these materials until the soils are poisoned with them. When all the land that can be irrigated at a given site has been poisoned, man has no recourse but to move. While this is, in a sense, a physical-environmental cause, it is primarily a cultural cause, for if the land had not been so used, it would not have lost its fertility. If properly treated—an excess of water applied and drained away to some sump—the land can be used indefinitely, even under irrigation. It is bad land management that destroys the soil, and management is culturally, not physically, determined.

These ideas of environmental determinism are dealt with at this length because few ideas have had so great an impact on so many fields of thought as the notion that the violent climatic changes of Inner Asia were the driving force behind the great Mongol invasions that have been so important in Eurasian history.

Much of the grassland of Inner Asia is now under Soviet control, and patterns of land use are changing in keeping with Soviet aims. There are many parallels to the American experience. The trans-Siberian railroad is comparable to the transcontinental railroads of America, except that the distances are vastly greater. Resources such as iron, coal, and oil have been found in or on the borders of the grasslands, and cities and factories have been built there. Nomadic peoples have been encouraged, a bit forcefully, to settle down and enter the labor force. A vast effort is underway to bring huge areas of semiarid lands, the so-called virgin lands, into cultivation. As in America this has led to vast wind erosion and crop failures followed by more limited effort, more drought-resistant crops, and better tillage practices—with some success.

One finds in Inner Asia, then, the greatest development of pastoral nomadism with a whole series of special cultural traits. The culmination of this way of life was in the great empires of the khans, empires that controlled for a time most of the immense area from eastern Europe to the Pacific. It was an empire of great wealth and learning, with its capital city a sea of tents (*yurts*) that was moved about the central Asian grassland as the needs for pasture dictated. From this mobile capital the khans ruled a major part of the world. All grasslands had the same physical geographic potential: grass, open space, and potential domestic animals. Most grasslands had civilizations adjacent to them: Mexico for the United States, Bolivia for Argentina, Carthage and Rome in North Africa. There could have been several Inner Asian developments, but the role of cultural history restricted the possibilities to one.

The African Pattern

The grasslands of Africa, like all large regions, are complex. They grade off to the north into the Sahara, in the center into rain forest, in the east to the forests of the highlands, and in the south either into the Mediterranean scrub of the desert or the Kalahari region. They mainly occupy the margin between the arid and the humid, as do the grasslands of the rest of the world. One can thus say that the primary determinant of the location is climatic. There are numerous modifiers in this distribution. In many areas where tongues of trees extend far into the grasslands and around villages where there are extensive stands of trees, it seems that trees could grow all over the area if some cause did not intervene. Often the cause is fire, which is decreased near villages by the heavier grazing of the land. Fire is also the probable cause of some of the very sharp boundaries between the grasslands and the forest —where the fierce fires that sweep the seasonally dry grasslands meet the year-round greenery of the trees. Elsewhere soil differences enter the balance struck between trees and grasses, for the trees do better on the well-drained lands, and the grasses thrive on the seasonally waterlogged lands. The final balance struck between trees and grass involves climate and soils, fire and man, and grazing animals—all factors in the complex forces that shape the actual landscape.

In Africa it is the grasslands that are the principal home of the pastoralists. Cattle, sheep, and goats are the common animals, with some camels and donkeys in the north. Horses are rare throughout. The origins of these animals show complex sequences. Sheep in Africa are goatlike in that they are lanky and hairy. The goats are often very small, wiry animals able to browse and survive on very poor land. The cattle depicted in much of early African art are long-horned types, but later short-horned breeds were introduced, probably from the northeast of Africa. Later much zebu (humped-backed) cattle of Indian origin were also introduced. The zebu strains are excellent for meat, disease resistance, and their ability to thrive on the harsh grasses of the tropics. Throughout the cattle-keeping parts of Africa the mixing of these strains has given rise to new kinds of cattle that are adapted to the local conditions of soil, climate, and vegetation. The tendency of western European or American cattlemen is to try to introduce improved strains of western cattle. The results are usually disastrous, for the pampered stock from other areas do not thrive on the siliceous, low-nutrient grasses of tropical African grasslands and are not resistant to the local diseases.

African cattle keeping in the grasslands is usually done on foot since the concept of the mounted cowboy is largely foreign. While many of the African people, in the grasslands and out of them, milk their cattle, sheep, and goats, some of the African herdsmen bleed their animals. The latter custom seems strange, if not repulsive, to people of other traditions. Actually it is a quite reasonable way to draw off some of the animals' renewable resource for human use. Milking does this by utilizing the lactation period of the animals and has the disadvantage of being tied to the animals reproductive cycle. Bleeding an animal is little different from milking it, and it is wholly independent of the breeding cycle. In practice no animal is bled too frequently, and the drain is no more exhausting than our making a donation to the blood bank. The fault is in our perception of the problem (to obtain food from the animal without killing it), and we tend to forget that our custom of milking is as repugnant to the Chinese as the African custom of bleeding is to us. Bleeding may be the superior system since it does not involve the problem of most of adult mankind's inability to

Ethiopian Boy Herding Cattle. *Cattle keeping spread south through Africa in part following the East African highlands from Ethiopia. The landscape is typical of much of East Africa. The cattle here are both humped and straight backed.* (Kay Muldoon for World Bank)

metabolize milk. If bleeding cattle is so advantageous, why don't more people do it?

The greatest difficulty with cattle keeping in Africa is that cattle are considered not a source of income but a source of wealth. Thus a man tries to accumulate cattle, just as in Western society a man tries to accumulate money. A man purchases a wife with such tender or meets the emergency expenses that arise in any family by drawing on his stock of sheep and goats. This leads to emphasis on numbers rather than quality, and it puts a heavy burden on the land. A man may have many more goats or sheep than he can possibly use, and they may graze the land so heavily that they virtually destroy the vegetation and set off excessive erosion. The economics also present difficulties.

While the aim in Western society is to raise a young animal to maturity as rapidly as possible and market it as soon as it has finished its period of rapid growth, the Africans hold their animals until they are about to die of old age and reluctantly slaughter them only when they are on the verge of a natural death. This is uneconomic in our view of things and very hard on the land in anyone's view. The Navaho in the United States have exactly the same view of their sheep, goats, and horses, and the impact on the land of their swollen herds is equally destructive. In both areas the problem has been attacked in several ways, usually unsuccessfully, for the non-Western peoples simply do not accept our value system. It is a clear case of differing perceptions, for to them, many animals on the land are a good thing, just as we feel that many dollars in the bank are a good thing. All of this is in process of change as new generations come along with changing outlooks. In Africa the growth of markets for meat products, the rising demand for protein by people with the means to pay for it, and the tendency for tribal ways of life to break down are gradually changing the old perceptions. As this happens, we can expect to see a reduction in the numbers of livestock, increase in the quality of the remainder, more rapid turnover of the stock, and an improvement of the grassland. The future is forecast in part by the tendency even now to import grasses from around the world for adaptation to African grassland needs. As the Africans' perceptions of values change, their societies shift from tribalism to nationalism and from primitive barter economies to modern money economies. As standards of living change, it is logical that the "natural" grasslands of Africa will also change, and man's perceptions will be reflected in the physical landscape fully as much, or perhaps more, than the physical landscape will be discernible in man's outlook.

Summary

The grasslands show the usual wide variety of man on the land. They have contained a wide range of cultures, from the primitive hunters and gatherers of the Paleolithic level through pastoral nomads, simple hunters, specialized hunters, farmers, and today great urban centers. The greatest contrast in the use of the grasslands was caused in the early periods by the presence of domestic animals in the Old World and their virtual absence in the New World. The dog was the only domestic animal in the New World and the Australian grasslands in pre-Columbian times. The introduction of domestic animals into the New World gave rise in both North and South America to a period of nomadic hunting cultures based on the horse. In both cases they were shortly displaced by the expansion of the European cultures. Interestingly the grasslands we value so highly as agricultural lands were among the last lands that we have taken into cultivation. Yet these grasslands were in part, even though a minor part, caused by man's activities in the use of fire over a very long period of time. The areas least changed so far are the Old World grasslands where pastoral nomadism dominates. The African grasslands remain in the hands of the Negro cattle-and-garden people, and are as yet little changed. Inner Asia is changing as the Soviet Union attempts to settle the pastoralists and make factory workers and farmers of them. This represents a return to building cities at the oases, a completion of the cycle from cities to nomads to cities without physical-environmental causation.

Review Questions

1. If horses were common on the American grasslands, and if the Indians hunted buffalo for at least 10,000 years before the Spanish arrived, why did the Indians not learn to ride horses for hunting buffalo? What general principles are illustrated by their failure over this long time span, as compared with the rapid spread of horseback riding in the eighteenth century?
2. The American grasslands are greatly prized today for raising grain and cattle. Why did the early settlers value them so little?

Moving the Flock. *The pastoral peoples of the Old World traveled over the land following the rains and the change in seasons. Theirs was a particular way of life dependent on the domestication of animals. It developed nowhere else despite the fact that there were suitable lands and animals and men with similar needs. (Pierre Streit, Black Star, for World Bank)*

3. To what extent could the Iowa type of farm management be exported to the rest of the world?
4. If American-style farming spreads, what will be the effect on rural-urban population distribution? Will this be good or bad? Why?
5. Compare the development of the grasslands in Argentina with their development in the United States.
6. Why was development of the grasslands in Australia so different from their development in the United States?
7. What are some of the possible explanations for the fact that Inner Asia has been alternately a center of political power and an area of weakness?
8. Why is the natural fertility of grassland soils so high?
9. Why did pastoral nomadism become so important in Eurasia and not appear before 1500 in Australia and North and South America?
10. Discuss the role of antecedents in the American settlement of the grasslands.

Suggested Readings

Bartlett, H. H. "Fire, Primitive Agriculture, and Grazing in the Tropics." In W. L. Thomas, Jr., ed., *Man's Role in Changing the Face of the Earth.* Chicago: University of Chicago Press, 1956.

Bogue, Allen G. *From Prairie to Corn Belt; Farming on the Illinois Prairies in the Nineteenth Century.* Chicago: University of Chicago Press, 1963.

Budowski, Gerardo. "Tropical Savannas. A Sequence of Forest Felling and Repeated Burnings." In C. L. Salter, *The Cultural Landscape.* Belmont, Mass.: Duxbury Press, 1971.

Gourou, Pierre. *The Tropical World.* London: Longmans, Green & Co., Ltd., 1954. Contains discussion of tropical grasslands.

Jefferson, Mark. *Peopling the Argentine Pampas.* New York: American Geographical Society, 1930.

Malin, J. C. "Man, the State of Nature and Climax: As Illustrated by Some Problems of the North American Grasslands." *Scientific Monthly,* vol. 74 (1952), pp. 1–8.

————, *The Grasslands of North America; Prolegomana to its History.* Magnolia, Mass.: Peter Smith, 1967.

Shantz, H. L., and Marbut, C. F. *The Vegetation and Soils of Africa.* Research Series, no. 13. New York: American Geographical Society, 1923.

Stapdon, R. G. *A Tour in Australia and New Zealand: Grasslands and Other Studies.* London: Oxford University Press, 1928.

Thornthwaite, C. W. "Climate and Settlement in the Great Plains." In *Climate and Man: Yearbook of Agriculture, 1941.* Washington, D.C.: U.S. Department of Agriculture.

Webb, Walter Prescott. "Geographical-Historical Concepts in American History." *Annals of the Association of American Geographers,* vol. 50 (1960), pp. 85–97.

————, *The Great Plains.* Boston: Ginn & Company, 1931.

Wedel, W. R. "Environment and Native Subsistence Economies in the Central Great Plains." *Smithsonian Miscellaneous Collection,* vol. 101 (1951), pp. 1–29.

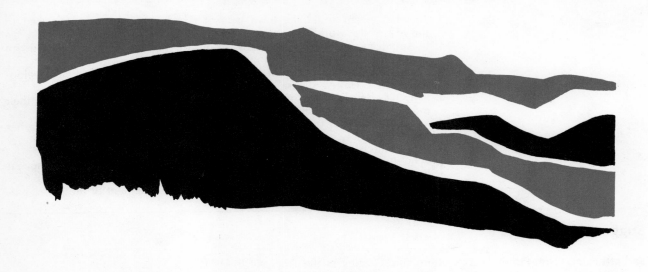

Mountain Lands

CHAPTER EIGHT

EW parts of the world have as much real effect on man as the mountains, and yet even their effect can be overestimated. It is often said that mountains are barriers to travel and mountaineers are backward because of their isolation, as well as fierce and warlike. Yet mountainous areas may have no population or be highly populated, and they may have high civilization or very low levels of cultural development. Again it seems likely that the causes of this variety are rooted largely in the different attitudes people brought to the mountains. The problem is to assess the effects of the physical mountains and the physical-cultural man.

Mountains are a landscape form; they are not a climatic type. There is a wide divergence in what people call mountains: all but the southern end of the

Appalachian Mountains are more hill-like than mountainous, while the Black Hills of South Dakota are really mountains. It is difficult to draw a line between what is a mountain and what is a hill, but one method is to call the landscape mountainous if the differences in elevation are great enough to cause vertical zoning of vegetation. This works well in the tropics and subtropics where a few thousand feet are required to accomplish this zoning. But in the Arctic and subarctic a few hundred feet are sufficient to cause sharp differences in the distribution of the vegetation. For most people local differences of more than 2000 feet elevation are impressive enough to warrant calling the landforms mountains. In such areas as the American West the local difference may be much greater. Denver, for instance, has an elevation of 5000 feet, and the moun-

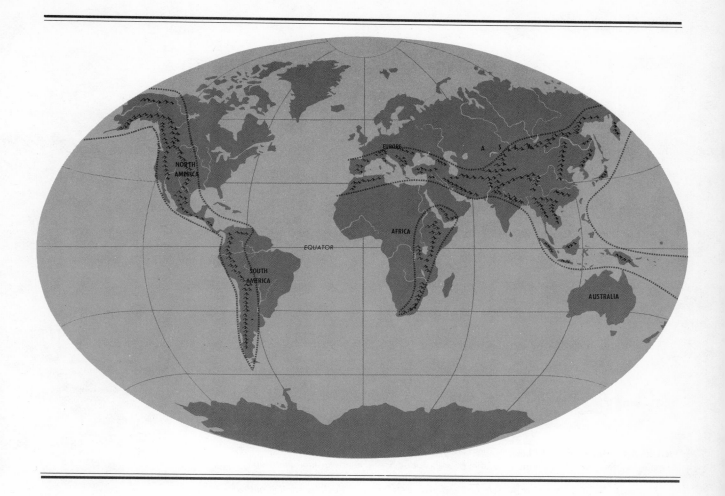

tains on the outskirts of the city rise abruptly to 10,000 feet. But less majestic masses than our Rockies can still be called mountains. Most of the areas to be discussed here will have at least 2000 feet of elevation change in short distances and will also exceed 5000 feet for their peaks. Such areas are truly mountainous, as contrasted with most of the Appalachians or the English Pennines, which, by the definition being employed here, are only hill country.

Mountains have strong temperature, pressure, and moisture relationships. The average decrease in temperature as the mountains rise through the atmosphere is 3.5° F. per thousand feet. Hence at 10,000 feet on the equator the average annual temperature is no longer an equable 80° F. but a chill 45° F. Air is compressible, and the air at sea level is relatively dense.

One-half of the atmosphere is below 17,500-feet elevation. In high altitudes, then, it is necessary to breathe much more air to obtain the necessary oxygen, as would certainly be true on the Tibetan plateau, which averages close to 1700 feet elevation.

The effect of mountains on the distribution of precipitation is also very strong. Mountains force moisture-bearing winds to rise in order to cross them. Rising air expands in the higher altitudes. Expanding air cools adiabatically at a rate of 5° F. per thousand feet. Since the capacity of the air to hold moisture is related to its temperature in such a way that cool air cannot hold much moisture, the rising air mass must drop moisture, and the side of the mountain mass facing the wind will therefore have relatively high rainfall. In tropical rainy zones this rainfall can be very great; in

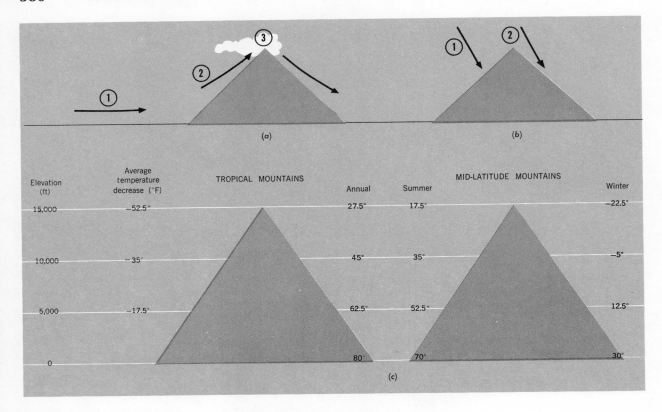

Elevation (ft)	Average temperature decrease (°F)	TROPICAL MOUNTAINS		MID-LATITUDE MOUNTAINS	
			Annual	Summer	Winter
15,000	−52.5°		27.5°	17.5°	−22.5°
10,000	−35°		45°	35°	−5°
5,000	−17.5°		62.5°	52.5°	12.5°
0			80°	70°	30°

(c)

FIGURE 8-1   Mountains and Climate

Climates in mountains are extremely variable over short distances due to elevation and slope exposure. At (a) the effect of wind direction is diagramed. With the wind flowing toward the mountain mass as at (1), a forced rise will be caused as at (2). The result will be cooling of the air mass due to its rising and expanding. This process will normally lead to the formation of clouds and precipitation, concentrated on the windward slope. Over the crest the wind will descend, be heated by compression, and will not only cease to drop moisture but will begin to pick up moisture. The two sides of the mountain will thus have strikingly different moisture conditions. This is strongly developed in the western mountains of the United States.

These conditions can be either increased or decreased as a result of the exposure of the slopes of the mountains, or of a given valley, to the sun's rays. At (b) the rays at (1) are vertical to the valley slope. This is the condition of maximum intensity of insolation. At (2) the rays are striking the opposite slope at a very low angle, and the insolation is dispersed over a greater area; hence there is less energy per unit area. Slope (1) would be warmer and slope (2) cooler. Whether this would be desirable or undesirable would depend on whether the slope were in mid-latitudes or tropical latitudes or on the rainy side or the dry side of the divide.

Temperature distributions to be expected for tropical and mid-latitude mountains are shown at (c). Notice that only one set of figures is given for the seasonless tropical situation but the summer and winter temperatures are given for the mid-latitude situation. Average annual temperatures are meaningful in the tropics. In the mid-latitudes, however, it is not the average temperature but the extremes of temperature that tell the story for crops, animals, and man.

arid zones it may be relatively small, but its effect in terms of vegetation, plant life, soils, and animals may be very large.

When winds descend the opposite slope, they reverse their effect. They are now drying winds, for in descending they are compressed, and as a result their temperature rises. They can now hold more moisture and will pick up available moisture from the land surface rather than moisten it. Since the effect of a mountain range in intercepting rain on one side and cutting off or decreasing the moisture on the other side is very much like the effect of the sun flooding one side of a mountain range with light while the opposite side is in the shade, the area deprived of rainfall by the effect of a mountain range is said to be in the rain shadow of the mountain. Where parallel ranges are arranged at right angles to the flow of rain-bearing winds, alternate wet and dry strips of land are found. In the Andes it is possible to go from desert coast up into forested heights, down into an arid interior valley, and back up onto a forested height again. Somewhat the same thing happens in the American West. The western coastal mountains have high rainfalls on their seaward-facing slope and dry areas to the east of their crests, and rainy conditions are met again on the west-facing slope of the next range, with another dry area to the east of that range.

The mountainous zones of the world are fairly simply arranged. The Pacific Ocean is rimmed with mountains. Since there are many volcanoes in this rim, it is sometimes called "the fiery rim of the Pacific." The other great mountain axis of the world runs from northern Spain across Eurasia to meet the Pacific mountain rim in East Asia. Where these two mountain systems meet, the greatest mountain mass of the world is formed: the Tibetan plateau with its associated mountain ranges. The only sizable mountainous areas outside these two zones are the mountains of East Africa, which include the Ethiopian plateau and the East African highlands. Mountains, then, extend from the Arctic to the Antarctic and are found in all the climatic zones that we have so far discussed.

Wandering Continents and Growing Mountains

Decades ago Alfred Wegener, a biologist, noted that the plants and animals on the two sides of the Atlantic and Pacific tended to match up. He postulated that the continents were all joined together until late in geo-

logic time, after which they had split up and drifted apart. His idea was rejected scornfully, for no one could explain how the light crustal pieces (the continents) could drift.

The obvious plant and animal pairs were accounted for by invoking climatic change that made polar regions tropical and so allowed alligators and pit vipers to migrate between Asia and America by way of chilly Siberia and Alaska. Those who advanced these theories had never been exposed to elementary physical geography, for it is known that as long as the earth sits in space with its axis inclined 23.5 degrees toward the plane of its ecliptic while revolving around the sun every 365.75 days, the earth must have seasons; poleward of 66.5 degrees there must be periods with no sunshine and intense cold from lack of solar warming. Recent work in geophysics shows that the earth's crust can be thought of as being composed of a series of platelike pieces. These pieces move due to slow convectional currents within the earth, which can be thought of as boiling very slowly. Where the convection currents well up, the earth's surface cracks, and the surface plates tilt and slide away. Where these titanic forces thrust plate against plate, or continent against ocean floor, great buckling and folding occur, and this gives rise to mountain belts. The mountains thus measure the movement of the earth's crust. They are young and growing features of the earth and zones of earthquakes and volcanic activity. Once the immense forces within the earth thrust them up, all the degradational forces begin to pull them down. Ice, wind, water, and especially gravity, work to reduce them. The mountains that so impress us are, then, relatively young features on the face of the earth. The old, enduring features are the great low plains of the earth in such regions as the Great Lakes–Hudson Bay area where ancient crystalline rocks of enormous age are found. The crust of the earth is unstable and in motion, and the crumpled edges of its parts are the mountainous evidence of the instability in geologic time of this earth that seems, and generally is, so immutable in human time.

Variability of Physical Elements

On world maps of geology, soils, vegetation, or climate, mountains are normally put into a special category. This is because it is virtually impossible to make a generalized map for these features in mountains. Mountains vary in their geology: they may be largely granitic, as are the Sierra Nevadas, or very complex in

(Upper) **Rogers Pass, British Columbia.** *This example of rugged higher mid-latitude mountain land is in about the same latitude as the mountains of southern Germany.* (Information, Canada)
(Lower) **Copper River Railroad in Alaska.** *Trees appear here only at the lowest elevations. This multimillion-dollar rail line was built to tap rich copper ores in the area, and it was abandoned once the ore was exhausted. Compare this with the situation at Potosí in the Andes.* (Alaska Travel Division Photo)

their makeup, as are the Rockies. All have much rock exposed, making them easy to prospect; the concentration of mines in mountainous areas is as closely related to this rock exposure as it is to any special localization of minerals in them. Soils vary so greatly from slope to slope that they cannot be mapped on a large scale. The unifying aspect is that all mountains are relatively young features on the earth. Since there cannot be much stability of weathered material on their steep slopes, old soils are found only where there are flat areas of considerable age.

Vegetation in mountains is also variable over short distances. Not only is it zoned vertically because of the changes in temperature that come with changing altitude, but it varies greatly depending on slope. In the northern hemisphere in mid- and high latitudes there is a large difference in the expected temperatures on the north and south sides of any slope. The south side is exposed to the sun's rays at a much steeper angle. The most efficient angle of exposure for the receipt of solar energy is 90 degrees. A mountain slope in mid-latitudes may create a 90-degree exposure in an area where the latitude would never lead one to expect such a concentration of energy. The opposite condition (the northern slope in the northern hemisphere), however, is that much less exposed and is therefore that much colder.

This has variable effects. If low temperature is critical for plant growth, the south-facing slopes will be the more heavily vegetated. This can have striking effects on the soils. More vegetation may mean that the products of weathering of the rocks may be held more securely on the slopes. A deep soil mantle and a smoother landform may result. In a semiarid situation the hot southern-exposed slopes may be too dry for optimum plant growth. This is a common feature of the arid American West. The greater plant growth, soil moisture, soil depth, and smoother landforms may then be on the north-facing slope. At times the absence of heavy plant growth on south-facing slopes may lead to greater erosion on such slopes. Then an area may come to have steep north-facing slopes with heavy vegetation on them, while the south-facing slopes will be marked by gentle inclinations and light vegetation. Since soil and soil moisture and plant type and abundance will all vary with these changes, it is easily seen that in mountainous regions there will be vastly more complicated patterns of plant, soil, animal, and climatic variation within short distances than is to be expected in flatter terrain.

In the tropics the high mountains have important areas of vertical zonation. Near the equator there are mountains tall enough that their bases have tropical rainy climates with the characteristic soils, vegetation, rainfalls, and animal life. However, with a decrease of 3.5° F. in temperature with each thousand feet of increase of elevation—the normal amount of decrease around the world—this tropical vegetation must soon give way, for it is notably temperature sensitive. A climb of 3000 feet in a 20-mile trip in a mountain range leads to a temperature change of 10° F. This is the summertime difference between Maine and the Carolinas or the wintertime difference between northern Florida and Washington, D.C. Temperature changes within 20 or 30 miles that would require hundreds of miles of travel on the flatter parts of the earth are the striking characteristics of mountain areas.

Tropical mountains therefore have very strong, vertical plant zonation in addition to the complex pattern created by varying exposure to sun and to rainbearing winds. Typical tropical rain forest seldom extends upward more than 2000 feet. The vegetation changes steadily upward, and at about 8000 feet coniferous trees often appear; at 10,000 feet the familiar broad-leaved trees are disappearing from the landscape. Really high tropical mountains often have shrub cover such as laurel below the snow line. The snow line on the equator is found at about 15,000 feet. In a vertical distance of 3 miles, then, it is possible to find virtually an equatorial-to-polar range of temperature.

The word climate has been carefully avoided in the last sentence. A tropical mountain zone with a mid-latitude type of average temperature would not be like a mid-latitude area in climate, for there would be little or no seasonal change on the tropical mountain. At Quito, Ecuador, for instance, the temperature stays very nearly—delightfully—springlike the year round.

Mountains in mid-latitudes lack the tropical and subtropical types of climate. It is as if the lower part of the tropical mountains were cut away so that one starts in the cooler climates of mid-height on the tropical mountains and proceeds quickly into the coniferous forests, then into the meadows just below the tree line, and thence into the permanent snow fields. High-latitude mountains may have their tree line at sea level and their snow line only a thousand feet above. Both of these mountain types have large seasonal changes, unlike the tropical mountains.

Mountains, then, offer the most varied kinds of environment for man. They vary with altitude and latitude with some degree of uniformity. High-latitude mountains, except for ruggedness of terrain, are not much like tropical mountains, unless the tropical mountains reach very high altitudes. But all mountains share the characteristics of great variation not only in vertical arrangement but in detailed differences from slope to slope depending on exposure to sun and to rainbearing winds. In no other part of the world was man challenged by a greater variety of conditions—of plants, animals, soils, and exposed mineral wealth—in so small an area. Unlike the plainsman who might never see anything but a vast flat grassland with a limited number of kinds of plants and animals and one type of climate, the mountaineer's environment assured him a wide range of experiences.

MOUNTAIN LANDS IN THE TROPICS: AFRICA AND ASIA

There are vast mountains in the tropics. In North America there are high mountains in the south of Mexico and in adjacent Guatemala and Nicaragua. In South America the immensely high Andes, extending from 10 degrees north of the equator, must be considered tropical mountains at least down to the tropic of Capricorn at 23.5 degrees south. In Africa there are very high mountain lands in Ethiopia and in the East African lakes region. In Asia the only high tropical mountains are in the Malayan area and especially in the great islands such as Sumatra, Celebes, New Guinea, and the Philippines. Note that by starting with tropical mountains we have narrowed the amount of variation in the physical environment.

Southwest Arabia

If we start in the Old World and recall that the center of origin of civilization was in the Middle East, applying the principle that ideas tend to spread first to the nearest areas, it becomes apparent that there are no tropical mountain areas at all near. The closest are those in southern Arabia, where the highest point approaches 10,000 feet in elevation. The highlands of Ethiopia are across the Red Sea in northeastern Africa. They are not only at a considerable distance from the Middle Eastern center, they are separated by the formidable Arabian Desert on the Arabian side and the Nubian Desert on the African side. Both of these highlands are in the dry belt. Their height provides for moisture catching, but only their upper slopes, and the better exposures of these slopes, have adequate moisture for agriculture.

The high southwestern edge of Arabia has at times been a center of some cultural growth. It holds an important position near the mouth of the Red Sea,

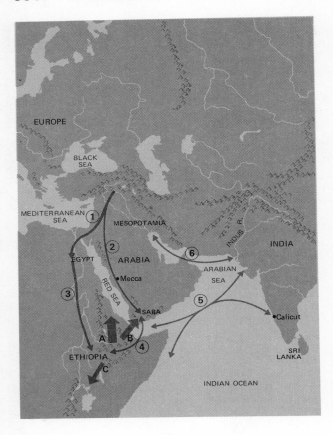

MAP 8–1
Cultural Influences and Reaction in the Tropical Mountain Lands of East Africa and Adjacent Arabia

These mountain lands are the only tropical highlands near the early cultural developments of the Middle East. Mesopotamian influences reached Egypt (1), Arabia (2), and eventually Ethiopia (3). Note the cultural reaction between Saba and Ethiopia (4). Sea trade routes in the area of the Indian Ocean influenced East Africa (5, 6). The wide arrows indicate Ethiopian reactions: strongest toward Egypt (A) and Saba (B), where there were centers of culture, and much weaker toward East Africa (C), which was less developed.

for sea trade from the Indian Ocean area coming toward the Mediterranean passes through the narrow strait there. The flow of ideas carried by traders was sufficient to lead to the development of agriculture, cities, and at times even states of some minor importance. Its history can be divided roughly into kingdoms of unknown extent and importance before 1000 B.C.,

the kingdoms of Saba and Ma'in from 1000 B.C. to the time of Mohammed, and Muslim kingdoms to the present. Mohammed is used here as a time marker since his life so profoundly changed this area of the world.

It is noteworthy that the kingdoms of Saba and Ma'in were centered on the highlands of the southwest where the increased rainfall gave some opportunity for agriculture and better grazing and hence denser populations; this area is present-day Yemen and South Yemen. The Queen of Sheba (Saba), as recorded in the Bible, came to Jerusalem to visit Solomon about 950 B.C. Mohammed also was a native of the northern fringe of this region. Because of his origin there, southwest Arabia for a time became the center of the Mohammedan power. However, Mecca was too far offside to persist as the capital for a movement that was sweeping west across Africa to Spain, northward into the Balkans, and eastward toward India. The capital was first moved to Kufa in Mesopotamia and later to Damascus in Syria. Location has worked against southwest Arabia, and despite some favorable physical geographic factors and a moderately good location beside a major sea lane, it has lagged behind the rest of the world.

It is worthwhile to examine this judgment. In the time of Solomon, location did not work against this area; in the time after Mohammed it did. Yet Arabia neither moved nor had a change of climate during this time. There were vast economic and political changes in this period, however, and these changes controlled the situation. When the East African trade was of major interest, adjacent Arabia was favorably located. When the shift of interest turned elsewhere, it was unfavorably located. After the opening of the Suez Canal in 1869, an important part of the world's shipping focused on the Red Sea, and the location became favorable again. Great Britain seized Aden (now part of South Yemen) in the south and held it to forestall others from gaining control of the canal. Meanwhile, the cultural level and dynamic drive that had once characterized the region had receded until little but ancient history and crumbling ruins of great dams remained to remind us that the potential of the land and the position were still there. Perhaps technology played a major role, for in the days of tiny rowing ships with crude sails Aden was an important stop.

Until very recently the Suez Canal was closed, and the sea lanes of the world were reorganized to meet this emergency. Moreover, today's huge ships and supertankers will be unable to pass through the canal. It appears now as if its closing accelerated developments in shipping that will prevent the canal from ever again regaining its old importance. Yemen, held but not

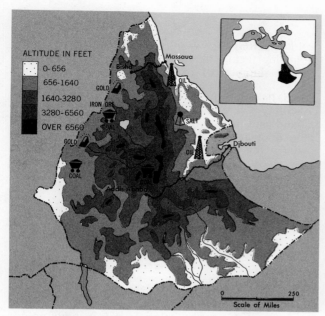

LANDFORM, MINERALS, AND RAILROADS

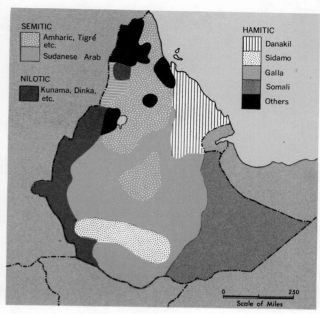

LANGUAGES

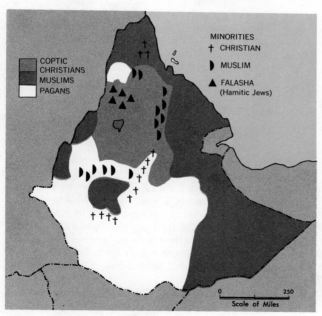

RELIGIONS

Map 8–2 Ethiopia: Landform, Languages, and Religions
The maps present the complex physical and cultural geography of Ethiopia. The well-watered high mountains and the arid plains of the north and east and the humid lowlands of the south and west are clearly shown. The low state of economic development is measured by the scarcity of railroads and highways, in sharp contrast to the presence of major mineral resources.

The language and religion maps show the imprint of history. The Nilotic languages are associated with Negroid people. The Semitic languages reflect later influences from Arabia and Egypt, while the Hamitic languages are related to ancient Egyptian influences. Religions parallel the linguistic pattern. The Hamitic and Semitic people tend to be Muslims. The ancient Coptic Church is strongest among the Amharic and Tigré Semitic speakers. All of this linguistic, religious, and associated racial diversity is concentrated in an area the size of Texas. (Reprinted from Focus, American Geographical Society)

developed by the British, now seems to have lost its crossroads position due to the quarreling between the Israelis and the Egyptians. With intertribal quarreling and the serious political problems of the area, this land seems destined to drift yet awhile.

Ethiopia

Across the Red Sea from ancient Saba lies Ethiopia. This highland country is slightly larger than Texas and Oklahoma combined. Mountains rise to 15,000 feet, with

385

many peaks in the 5000- to 13,000-foot range. Therefore the land has moderate temperatures despite its tropical location. The lowlands are, of course, steamy tropical or hot desert depending on their location. Most of the population of Ethiopia is in the highlands. Because of its location and the range of its elevations, Ethiopia has an enormous range of vegetation and animal types; most of the African flora and fauna are represented somewhere within its varied geography.

This relatively inaccessible land has a more complex history than one might guess. Its original language was Hamitic, originally the dominant speech group of North Africa. Today the dominant language is Amharic, a Semitic language, which suggests Arabian influences in Ethiopia, for Arabia is the home of the Semitic language family. The earliest civilizational influencies reaching Ethiopia came apparently from adjacent Egypt. In the eleventh century B.C. the Ethiopians broke away from Egypt, but now with considerable additions of knowledge. By the middle of the eighth century they conquered Egypt and dominated it for the next two centuries. This, of course, meant that even greater masses of advanced knowledge were transferred to Ethiopia. Ethiopians entered into trade with the Arabians. According to tradition their greatest king, Menelik, was the son of Solomon and Sheba.

Christianity was introduced in the fourth century A.D., and in the sixth century persecution of the Christians led to Byzantine intervention and the Ethiopian conquest of Yemen, the land of the ancient kingdom of Saba. This initiated a period of extensive trade and contact with such distant lands as Byzantium in one direction and Ceylon (now Sri Lanka) and India in the other. Then came the rise of Mohammed, the loss of Yemen, and the isolation of the Ethiopians by the Mohammedan expansion over North Africa and all of the Middle East, an isolation that has now lasted for at least a thousand years.

There are many interesting points. Although Ethiopia seems geographically isolated, it has repeatedly been drawn into the realm of advanced civilization. Each time it has profited greatly by the inflow of ideas and has reacted vigorously. In all cases the inflow of ideas has come from the north where civilization was growing; the reaction, similarly, was primarily oriented toward these centers of growth, development, and productivity. Little effort was exerted in a southerly direction toward the undeveloped lands and peoples of tropical Africa. This is parallel to the flow of trade in the world today. The major traffic is not necessarily to adjacent lands but is greatest between the most highly developed lands. It is one more example of the fact that it is not mere location, distance, or even difficulties such as deserts and mountains that are the determining factors in the cultural-geographical growth and development of a land or a people.

Ethiopia has again been touched by the outside world. Italy seized and held the kingdom briefly before World War II. New ideas have again flowed in, and air and land transportation have increased efficiency so that Ethiopia is potentially less isolated than it ever had been in the past. Are we to expect that history may now repeat itself? Could Ethiopia again rise to be an important force in the world?

Ethiopia has nearly twice the land area of France. Although its present commerce features hides, skins, grain, coffee, spices, and gold, its known resources suggest other possibilities. High rainfalls on high mountains indicate great potential water power. Considerable deposits of gold, coal, rock salt, sulfur, phosphate rock, copper, lead, iron, mercury, platinum, tin, petroleum, mica, potash, and tungsten are known, although only a few of them are mined commercially. Such an array of mineral resources is possessed by very few countries of this earth. Ethiopia's lack is therefore not of the physical goods of this world. Her present status is the result of a thousand years of isolation, of virtual absence of education associated with nearly complete lack of productivity. The potential for a repetition of its past history is certainly there.

The East African Highlands: Distance Becomes Important

South of Ethiopia are the East African highlands, which include Uganda, Kenya, Tanzania, and part of Northern Rhodesia, now called Zambia. In addition to this great highland area centering about the rift valleys of East Africa, there are outlying tropical lands in Angola and South-West Africa. Civilization slowly spread toward these lands. Yet the climate is no barrier. These are tropical highlands with the moderate temperatures and the varied vegetation and animal life that are found in all such areas.

Some ideas did work south. Agriculture, domestication of animals, iron working, pottery making, politically structured societies with kings, nobles, and commoners—to name a few important ones—all spread south into this area and beyond. These formed the basis of a Neolithic way of life: agricultural village settlements with occasional development of full-fledged kingdoms complete with capital cities. On this basis a moderate population density was established by the

MAP 8-3   Cultural Currents in Africa

The striking features of this map are the dominance of the east-west cultural flow in Africa north of the equatorial rain forest and the focusing of influences from west, north, and east on Ethiopia. The relatively weak flow to the south followed the East African highland. Most of the movements that we can detect depended on the acquisition of new cultural means to better utilize the land. The West African crops were dry country plants. The forest regions of Africa were not opened up for agricultural dominance until the arrival of Malaysian and American crops carried across the Indian Ocean by traders. The first of these crops arrived before the time of Christ. It was utilization of these crops that triggered the Bantu expansion in Africa.

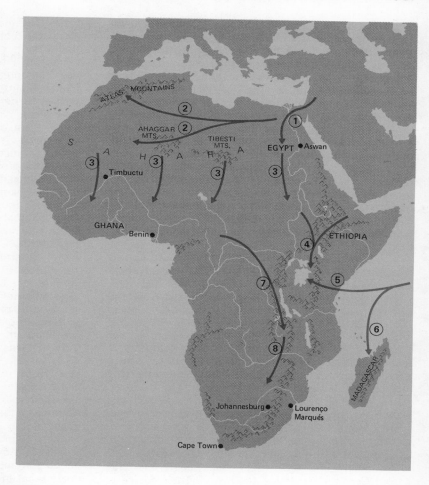

5000–4000 B.C.

1. Plant and animal domestication spreads to Egypt.

4000–3000 B.C.

2. Domestication spreads through the Sahara.

3000–2000 B.C.

3. The idea of plant domestication spreads southward; local plants, especially sorghums, are domesticated. Animal domestics from the Near East are retained.

2000–1000 B.C.

4. Ethiopian expansion takes place.

1000 B.C.–A.D. 1

5. Semites invade East Africa, and possibly east Asian plants are introduced via the Indian Ocean sea lanes.

6. Outrigger canoes and Polynesian speech probably reach Madagascar about this time.

A.D. 1–1000

7. Bantu tribes expand into East Africa.

A.D. 1000–1500

8. Continuing Bantu expansion leads to conflict with Europeans entering South Africa.

beginning of the European colonial era. Under benevolent European control, with tribal warfare and cannibalism diminished and the diseases of both men and animals increasingly brought under control, population has increased. In some areas the population can best be described as dense; around Lake Victoria in the East African highland, for example, sizable areas have populations of more than 500 people per square mile.

If the highlands of Ethiopia could repeatedly give rise to kingdoms, why did the equally good, or even better, highlands of the East African lake region so rarely give rise to anything even remotely resembling civilization? The physical environment is excellent. The proximity to Ethiopia provided potential access to ideas. Again we face the question of the success of the Negro as a possible carrier of civilization. It would seem easy to choose the racial answer, but this is in fact too simple.

It should be noticed that all of black Africa was bypassed for a long time in the period of western European expansion. However, we have seen that such choices are often made neither on the basis of the physical geography nor the race of mankind in question. Often the decision is made in terms of location and, equally important, the stage of development of the people in the area. This is like Spanish California: it was too distant in proportion to what it had to offer—good land, good climate, but undeveloped people. This may be said to have described Africa.

The flow of ideas in Africa was considerable. By 5000 B.C. agriculture with cattle keeping had spread from Mesopotamia to the Nile and by 4000 B.C. had spread throughout the Sahara. Some of these cattle and farming people also entered the Ethiopian highland.

Before 1000 B.C. an Ethiopian people, the Caucasoid Cushites, expanded southward. Their choice is of particular interest. Since they were a highland agricultural people, their expansion was through the East African highlands. After A.D. 500 milking spread into Africa, and the Cushites were one of the peoples who received this idea and passed it on to the south. The Cushites were then largely displaced from the highlands by the expanding Bantu-speaking tribes about A.D. 1000. The area where the Cushite influence was strongest and where racially mixed remnants of these people still exist is in northern Tanzania, where the whole landscape is different from the rest of black Africa. The land is in permanent agricultural production and is fully occupied; there is no brushland lying fallow, which accompanies a shifting type of agriculture and is common throughout most of tropical Africa. Instead there is a developed, cultivated landscape.

The mountainous northeast corner of Africa as a cultural center becomes understandable, then, when the currents of cultural connection are considered in relation to the peoples, crops, and trade routes of the day. The spread of ideas down the mountainous side of Africa illustrates that highlands can be highways as well as barriers, depending on the nature of men and their ideas.

Most of Africa, highland and lowland, is now in turmoil. Tribalism versus nationalism, dictatorship versus democracy, capitalism versus socialism, and the enormous problems of illiteracy, lack of capital funds, and inadequate transportation are some of the handicaps that hamper most of the area. For part of Africa the tsetse fly or leached tropical soils can be named as causes that place difficulties in the path of development, but this is certainly not true of the East African highlands.

In terms of our present assessment of the world, the African highlands are unusually well off. They are in moderate climates relatively disease free due to the high altitudes in an equatorial region, and they have good soils, partly the result of volcanic activity, considerable potential water power, and known mineral wealth. Such a combination should make the region advantageous for economic growth and development. Even sizable populations to supply the manpower needed for development are present. However, physical geography is passive. Action, such as development or nondevelopment, is a sociopolitical problem. The turmoil of the present suggests that a rapid development at the rate experienced by Japan is not to be expected, for while the physical base of the African highlands is better than that of Japan, the political fragmentation and social turmoil create an offsetting "climate" that may be the controlling factor for the next few decades.

In all climatic areas Africa is tending to fragment along tribal lines. In the process old scores are being settled by massacring some minorities and expelling others. The situation is somewhat like that of India at the time of partition, except that the civil service built by Great Britain has tended to carry India through her difficult transition. Africa may be better compared to Latin America, which also fragmented politically when it threw off the colonial yoke and then too often fell into chaotic dictatorships. After a hundred years Latin America still struggles for political stability and economic growth. Africa was less prepared for independence than was Latin America, and no one should be surprised if a prolonged difficult period of development lies ahead for Africa.

Ethiopia, a Racial Rainbow. *Ethiopia is a linguistic, religious, and racial mixture of peoples. The means and the extremes are shown here. The man (upper left)* is a Europoid type who looks as if he could also be Turkish. *(Kay Muldoon for World Bank) The girl in native dress (upper right)* is from near Lake Awasa south of Addis Abbaba. She is a classic Negro type. *(Terence Spencer, Black Star, for World Bank) The young women in a telecommunications class (below)* are an intermediate racial type. *(Johannes Haile, National Photo Studies, for World Bank)*

Tropical Mountains of Southeast Asia: Cultural Highways

The tropical mountain lands of Southeast Asia present a somewhat similar picture. They too are distant from the early centers of civilization. As civilization spread eastward, it reached India and then China. As these great centers of culture developed, their influences came to overlap in the Southeast Asian area. Influences from India reached across the Bay of Bengal not only to adjacent Burma, but to Thailand, Malaya, Sumatra, Java, Borneo, and up to the Philippines. Beginning about the time of Christ, this civilizing force established kingdoms of large size in these areas. Such great ruins as Angkor Wat testify to this influence from India, as do also the dances of Bali and much else in this whole area. Chinese influences reached down the coast and to the offshore islands. Chinese traders were voyaging to India before the time of Christ and probably for a very long time before Christ.

In this crossroads of cultural contacts there could have arisen a mighty center of civilization. If mountains in the tropics, because of their beneficial cool climates and great variety of soils, plants, animals, and minerals, are advantageous, we might expect to find a highland civilization here, but this did not develop. The reasons are not hard to find. The agricultural outlook of the civilized people was strongly oriented toward lowland gardening with the rice plant as the most important of their food plants and with tropical lowland root, fruit, and spice plants also important. This focused attention on the plains, not the highlands. As a result the Southeast Asian highlands are relatively empty.

In the mountains, shifting agriculture is the principal way of life, This is the familiar forest-clearing, garden-type cultivation with abandonment of a piece of land after a year or so for a new forest clearing with its renewed fertility. In Southeast Asia a large variety of plants are available, including both root and grain crops and many tree crops. The pig is the important domestic animal. On this basis a village-dwelling way of life that did not advance either to towns or cities has long been maintained in the mountains, while the flow of civilization moved around the mountains, following the lowlands of the valleys and the coasts.

This is not to say that highlands are always barriers. It has been demonstrated in the past 2000 years that highlands may be highways. The peninsula of Southeast Asia, including Burma, Thailand, Laos, Cambodia, and Vietnam, was originally Mon-Khmer in speech and strongly Indian in culture. Speech in this area is now mostly Sino-Tibetan, though the people are racially and culturally little changed. A review of the history of this region shows that through the past 2000 years relatively simple mountain-forest farmer-folk have worked their way southward along the mountain chains and outflanked the lowlanders. Repeatedly they have been able to take over government controls and have imposed their language.

Perhaps the mechanics of this aggression resembled the Aztec takeover in Mexico. The mountaineers were probably used as mercenary troops and by their strategic position were able to stage a military coup. Our main interest here, however, is the fact that the mountains served as their highway, not as barriers to men's movements. Even this deserves to be stated more fully and carefully. The mountains that for centuries had been barriers to the lowland-oriented Mon-Khmer people, who absorbed culture from the lowland-oriented people of India, were corridors for the movement of the highland-oriented Sino-Tibetan people. It was the cultural orientation, not the physical structure, that was decisive.

Eastward into Oceania levels of civilization decline. The center of India's influence was on the nearest islands, Java and Sumatra. The farther islands, Borneo, Celebes, Timor, and others, were less developed. This probably repeats an ancient pattern of the spreading of ideas eastward from the continent, with successive waves in part limited by man's ability to cope with the physical geography and by the degree of the penetration of the preceding ideas necessary for the establishment of the next set. Here, as in Africa, a point is reached where the necessary ideas did not penetrate or did not have time to engender developments sufficient for the next steps to be taken. We meet this line in New Guinea, where the Oceanic Negro people held the vast island (10 times the size of Puerto Rico) and a series of adjacent islands. This part of Oceania is known as Melanesia.

New Guinea: Distance and Lag

The highlands of New Guinea are extensive. They are estimated to cover 12 percent of the land area and contain 40 percent of the people. The people live in the valleys of the highlands, often at elevations above 5000 feet, which means that they live in a mesothermal climate with high daily ranges of temperature due to their high altitude and with the possibility of frost at the upper end of their habitat. Rainfall is high, averaging from 80 to 100 inches per year and more. Much of the mountain land below 8000 feet is grassland. This is thought to be the result of clearing and burning of the landscape by the natives, for detailed reporting of the life of the natives makes clear that they destroy large amounts of forest and eventually convert large areas into grassland. This is also widely reported from the interior highlands of Borneo and elsewhere in Southeast Asia. There are very aggressive grasses in this area, which seemingly accounts for the great speed at which this type of landscape is created. Frequent man-set fires in these grasslands perpetuate them. As in all mountainous regions the soils vary greatly from slope to slope and from one kind of rock formation to another.

Although the population of New Guinea is often called Negroid, it seems more closely related to the Australoid group than the Negroids of Africa. Man has been present in adjacent Indonesia from the opening of the Middle Pleistocene, perhaps 600,000 or so years ago. There is a long evolutionary series on Java, but just when man managed to pass the deep-water straits that divide the New Guinea–Australia area from the Indonesian area we do not know. Considering the smallness of the water barriers and man's ability to cross small water gaps, an early crossing is probable.

Cultural contacts are apparent from the plants used and the arts practiced by the New Guinea highland people. They have not only Asiatic plants and animals such as the pig, but most surprisingly some of the highland people have as their principal crop the American sweet potato. They also have and use American tobacco. We have no real knowledge of when these American plants reached here, but they show that even the remote highlands of New Guinea felt the flow of ideas from both Asia and America. On the basis of these ideas and materials, the people of remote, interior, highland New Guinea advanced to a primitive, Neolithic, village-dwelling way of life.

These people lived in small triblets, usually continuously at war with their near neighbors. It is of interest that a shifting agriculture supported populations in local concentrations of 100 to 300 per square mile, but with large empty areas in between these pockets. In these areas of concentration sufficient food was produced from a relatively small amount of land to leave a great deal of leisure. Much of this was spent in head hunting or simple warfare. The warfare was an activity in which a thousand warriors on a side might prance and feint all day long with only a few casualties on each side, although men occasionally were killed outright in battle, and more died of infected wounds. It was a satisfyingly dangerous sport.

Fighting, weapon making, and such skills as making women's skirts and house building were in the hands of the men. They also aided the women in the heavy work of preparing fields. Once the fields were cleared and thoroughly prepared, they were turned over to the women, whose duty it was to plant and cultivate and harvest. Families lived in small compounds, with the one or more wives of a man each having her own sleeping hut. Cooking was done in a large communal building, but even there each woman had her own fireplace. The men lived and slept in the men's house.

Most of the preliminary steps toward building the basis of a civilization had been taken. Agriculture had been introduced, population density had built up, leisure existed because there was an adequate food supply—but the surplus energy was not spent productively. Social structure was very simple: might made right. If a man could not protect his women, pigs, and fields, he deserved to be robbed and have his women assaulted. Status was based largely on prowess in warfare against outsiders and on brute force applied inside the tribe. Wants were few, and they were supplied largely from the local scene.

To change this culture, social advances were needed. The energy dispersed in warfare needed to be channeled into useful production, and the simplest way to accomplish this would have been an increased trade, which would have required the production of a surplus of some material for trade. Trade also requires stability and some degree of security for travel. However, the highland interior of New Guinea was so remote that only since the 1950s have we discovered valleys with thousands of settled farming people, who in the mid-twentieth century had never even heard of a white man. This was not due entirely to mountains. The people of the coast were little more advanced when first seen by Europeans.

It is not difficult to think of uses for these highlands. Their agricultural possibilities are considerable; they would be natural places to produce coffee, tea, and similar commercial crops, and they could be utilized for cattle production. Rainfall of 100 inches and more in the highlands, well distributed throughout the year, assures the presence of large amounts of potential hydroelectric power. The area has been little explored for minerals, though important gold fields have been found. Its potential must be said to be largely unknown, but it is likely that it is quite great.

Highland New Guinea has a better physical-geographical base than the Andes, which nurtured a civilization at greater elevation and under conditions of considerable aridity, or the Tibetan highland, which also reached the level of civilization at very high elevations.

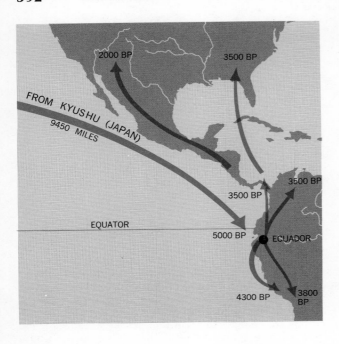

MAP 8-4   **The Introduction and Spread of Pottery Making in America**
The dates given (BP = before present) are based on carbon-14 and other absolute dating methods. The route across the northern Pacific is hypothetical and simply follows the winds and currents. The distance is given in land miles along the great circle (shortest route). Once pottery making began in America, seemingly on the coast of Ecuador, it spread somewhat radically. Of particular interest is its early extension into the Andean highlands, suggesting the prior presence there of a more settled and culturally advanced people. Notice also the relatively early arrival of pottery in the southeastern United States. A route across the Caribbean is suggested, and this is not improbable since the original bearers of the pottery-making idea were boatmen. All such constructions are subject to change as new data are accumulated, but if pottery was introduced at some point such as the coast of Ecuador, this is the pattern of spread that might be expected. Further, it is most unlikely that only pottery making was introduced and spread in this fashion. (Based on data from Betty J. Meggers, Clifford Evans, and Emilie Estrada, Early Formative Period of Coastal Ecuador, Smithsonian Press, 1965)

New Guinea's failure both in its highlands and its lowlands is due to its geographical location. Distance from the centers of ideas worked against an early start toward civilization, and its cultural lag further delayed growth. Nevertheless, the area was well into the agricultural village-dwelling stage of development and ripe for the town-dwelling stage when the Europeans erupted over the world. The specialized needs of an expanding Europe—high-value materials, such as spice in Indonesia, where proximity to idea centers had led to an earlier start, or open land where a more isolated people melted quickly from the area, as in Australia— were not present in New Guinea. Whether in its moderate-temperatured uplands or its steaming lowlands, New Guinea has lingered on in its Stone Age.

Summary

The tropical mountain environments of the Old World may be summarized as follows. The only mountainous areas that had growths to the level of civilization were Ethiopia and Indonesia, and in Indonesia civilization centered in the lowlands rather than the highlands. In both cases these lands progressed because they were reached by impulses from the great center of civilization in the Middle East. The other tropical mountain areas were beyond the reach of these influences, so they became cultural laggards. Thereafter

civilization passed them by in preference for other lands, even if they were much more distant lands.

NEW WORLD TROPICAL MOUNTAIN LANDS

In America the story is quite different from the Old World, for the tropical mountains lie in the center of the civilized area. The Toltec-Aztec civilization developed on the plateau of Mexico, and the great ancient series of civilizations of Peru and Bolivia were centered on the high plain, the Altiplano, of the Andes. Between these two peaks of American Indian civilization there was a conspicuous tendency for high cultures to be strung out along the mountain chains that linked the two centers. It must not be overlooked, of course, that this was not the only location for civilization in pre-Columbian America. The Maya lived in a tropical lowland, and part of the Peruvian civilization lay in the desert oases along the coast. Nevertheless, there is a conspicuous correlation between high culture and the tropical mountains.

What is cause and what is effect here? Did the mountains provide a better opportunity? Or did culture simply begin in such a position that the tropical mountains were in the favored location? We have seen that in the Old World the answer was largely one of access to the ideas streaming out from the center of civilization in the Middle East.

Was there something particularly different in the New World? For instance, were there diseases in the lowlands of the New World that tended to place a greater advantage on the highlands? This cannot be true, for many if not most of the scourges of the tropical lowlands of America seem to have been introduced in the post-Columbian period. This is certainly true of malaria, yellow fever, and the bubonic plague, to name only three. The tropical lowlands of America actually seem to have had greater freedom from adverse disease conditions than did those of the Old World. If this is a cause that favors the development of the tropical highlands, there was less reason for man to concentrate in the highlands of America than in the highlands of the Old World. Perhaps the location of the beginning of civilization in America was the key factor, and we may well consider one of the areas for which we are beginning to have some data as to the origins.

About 5000 years ago the people on the coast of Ecuador were primarily shellfish gatherers and fishermen. This focus on sea foods tended to establish relatively fixed settlement areas at good fishing and collecting spots where water was available and sheltered waters facilitated the launching of boats. Just such an area marked the southern coast of Ecuador. Here the people could gather the fruits of both the sea and the land by simple means, and the variety of plants and land and sea animals was such that they did not have to move frequently. They were sedentary fisher-folk even though they were following a hunting and gathering way of life, which elsewhere required considerable seasonal movement in pursuit of harvests of seeds, roots, shoots, and game.

On the coast of the south islands of Japan a very similar way of life existed, except that pottery had been in use for a few thousand years. Shortly after 5000 years ago the people on the coast of Ecuador suddenly began to make pottery. Was this due to invention or to diffusion? Comparison of the pottery made in Ecuador with that made by the sedentary fisher-folk in Japan at this same time shows that the techniques of manufacture and the minute details of decoration are extremely similar. There are similar designs made by similar techniques—rocker stamping, finger grooving, zoned punctating, and cord impressing. The relationships are so extensive that one people doubtless learned from the other. Since pottery making is thousands of years older in Japan that in America, it seems clear that it was the American Indians who were learning. The establishment of pottery on the coast of Ecuador has been explained

by the combination of a chance arrival of drifting fisher-folk among people leading a similar sedentary way of life. There are difficulties in supporting this hypothesis. For example, there were similar people leading similar lives all along the Pacific coast of both North and South America, and the chance that drifting fishermen from Asia would land on the American coast must have been a thousand times greater in the north than in the south. Still we can take this event, even though not fully explained, and use the results.

From this center pottery making spread both north and south, perhaps partly along the coasts, and very early it spread inland and up to the Andes highlands. Why would pottery making spread into such an area as the high plateaus of Peru even before it spread into the irrigable valleys of the coastal desert, an area later to become the seat of civilization? If the sedentary way of life helps explain the acceptance of pottery on the coast, is this also the reason for acceptance in the highlands?

A clue comes from the study of agricultural plants. The coast of Peru built its civilization on borrowed plants. The highlands had their own. This suggests that the mountain people of Peru had started on the road to civilization earlier through plant domestication, and they were probably already sedentary village farmers and thus receptive to the inflow of new ideas of which pottery is the tangible measure. Unique plants clearly domesticated in the highlands and indicating an early agricultural interest include ulluco, oca, topinambou, cubios, achira, yautia, and nasturtium. Only one, the nasturtium, is widely known outside the high mountain areas of the Andes. For us it is a flower, while for people of that area it is the tuberous root that is eaten. Data such as these, plus botanical data, prove that these plants were domesticated locally. Later manioc, sweet potatoes, Irish potatoes, peanuts, maize, and beans were added to the agriculture of the area, and as none of these are native to the Andean plateau, they testify to the inflow of ideas. While the pinpointing of a local center of plant domestication may help to explain the early flow of pottery making to the high plateaus of the Andes, it leaves unexplained the origin of this agriculture. It is the only complex agriculture of high-mountain origin in the world, and we have as yet no explanation for its origin. For the present, then, we must take this agriculture as a given circumstance that aids us in understanding the early acceptance of a flow of ideas from Asia arriving by sea in Ecuador and epitomized by pottery making. A similar case occurs later when the Spanish went where there were settled people and avoided such areas as California, where there were only primitive people. At this time our knowledge of the actual happenings is still so incomplete that the explanation offered here is only a hypothesis. While it gives a plausible answer, this does not necessarily make it the correct answer.

With the appearance of pottery making by 3000 B.C. apparently marking the beginning of the arrival of ideas from overseas, the stage was set for increasingly rapid cultural growth. Ideas spread not only down the coast into the irrigable valleys in the desert but also quite rapidly up to the highlands. The simple farming communities grew into theocratic states by 1000 B.C. These were characterized by sizable populations organized around religious centers of such magnitude that both large populations and organized political structures were needed to plan and carry out the constructions. One of the earliest and most brilliant in art and architecture was at Chavin in the northern highlands of Peru. Influences from this highland center can be traced widely, especially down to the coast to the growing centers in the desert coastal valleys. Temples, the use of gold and silver and their working by welding, soldering, annealing but perhaps not yet smelting, and the appearance of the true loom are advances that mark this period. It is also the time when maize appears in these South American cultures, marking widening contacts and the gaining of ever more useful ideas from Mexico. In the long period from 1000 B.C. to 200 B.C. the early states continued developing. In the various regions, coastal and highland, and from one valley to the next, differences arose as each developed its own personality in the arts. New features were added such as blow guns and pan pipes, of possible Asiatic origin, suggesting a continuing inflow of ideas. Other features were more probably developed locally. This was a time of population growth and developing political structures, leading to the organization of larger states.

Shortly after the time of Christ an empire arose on the plateau to the south of Lake Titicaca in Bolivia. The purest expression of the art and architecture of this group is found at Tiahuanaco, now a vast ruin consisting of monolithic gateways, enormous courtyards and temples, and platforms often constructed of huge stone blocks cut to shape at a quarry 3 miles away and transported to the city site. Some of these blocks weighed 100 tons, indicating that the engineering ability of the people must have been of a very high order for their day. Such huge stones were locked together by bronze clamps for which there are no antecedent forms in America. They duplicate Greek forms for which there is a long developmental series and are but one of a long series of such Mediterranean traits in the Andean region about this time. It seems quite possible that there was a new inflow of stimulating ideas and that a possible source of such ideas was the Mediterranean world. Then came a long period of breakdown of empires, fragmented states, and the building of new empires. The last of these was the Inca empire with its center at Cuzco, Peru. It was put together by a series of gifted rulers in the period between A.D. 1000 and 1532 and destroyed by the Spanish on their arrival.

This sequence is in many ways like that of the other areas of the world, especially the Middle East: the long period of hunting and gathering extending back into the misty beginnings, the little understood origin of the domestication of plants, the consequent growth of villages, towns, states, and finally empires, with the ebb and flow of conquest and collapse. The American example is unusual in the dominance of the highlands in the process, and the truly great height at which some of the sites are located: 12,000 feet for Tiahuanaco. It is as if the major centers of growth in eastern Asia had been on the Tibetan plateau rather than in lowland China and India. The persistent dominance of the cultures on the enormously high Andean plateaus makes the argument that human achievement is hindered in the lesser heights of the other mountain areas of the world seem inadequate. The Inca empire is sufficiently well known from historical sources that it can be described in some detail as an example of a highland society.

The Inca Empire

When the Spanish arrived in South America, they found the empire of the Inca comparable in some ways to the Roman Empire. It was momentarily in the throes of internal difficulties resulting from a struggle over the inheritance of the kingship. The Inca were an obscure people who began building an empire centered on the highland near Cuzco, Peru, about A.D. 1000. The Inca empire was but the latest of a series of such empires in this region. The Inca had a succession of able rulers, and in 500 years they united the highland and the coastal lowland from southern Ecuador to northern Chile. This domain includes most of what is now Peru, Bolivia, and Chile, excluding the interior hot, wet lowlands and the cold, wet southern parts of Chile. The north-south extent of this empire was about a thousand miles. It was connected by two systems of roads, one along the coast and the other lengthwise of the highlands. Messages could be sent rapidly great distances at night by a system of signal fires relayed from one mountain top to another. At regular intervals along the roads there were stations housing runners.

The Inca empire was an example of benevolent socialism—a planned society. In the Inca empire every person had cradle-to-grave security. The Inca had an ever-normal granary storage system that saved supplies in good years to tide the people over the inevitable lean years. They also had highly developed conservation practices that included periodic roundups of the little wild camels—llamas and alpacas—with selective slaughter to maintain the herds at levels desirable for the vegetation of the region. That a given range can carry only so much wildlife and that it is wasteful not to harvest the excess often is thought to be a twentieth-century wildlife management idea; however, in the planned society of the Inca, it was well established long ago.

When a young couple married, they were allotted land at the annual division of land; when they had a child, their landholding was increased. The people of a given settlement worked communally on the state lands. After those fields were tilled and planted, they then individually worked their own plots of ground. Harvest was organized in the same way. The government share was partly stored locally and partly transported to other areas where there might be need. It sounds like an ideal system, and in a way it was. However, there is a price for everything; in the planned society the price is freedom. In the Inca system the individual had an absolute minimum of freedom. He could not freely choose his profession, where he would live, or whether he would be single or married. The state decreed that all should marry and saw that they did. A man was a farmer or soldier or roadworker as the state decreed.

This tightly organized state with absolute authority over life and death and all intermediate activities centered in one person, the Inca ruler. Under the ruler's leadership the people achieved development of both the arid lowland and semiarid highland. Through the use of labor that could be put to work without any thought of cost, vast temples, palaces, canals, and land-terrace systems were constructed. Cities that extended over several square miles were built, canals many miles in length were dug, whole mountainsides were terraced, and the population grew. Arts were encouraged, and some of the finest work in textiles and pottery that the world has ever seen was produced.

The Spanish in the Highlands of the New World

The Spanish burst into this highland during the frenzied search for instant wealth that motivated their discovery and conquest of the New World. Between 1500 and 1520 the Spanish carved out a foothold in the Carib-

Andes Highlands, State of Mérida, Venezuela. (Upper) *At elevations above 8000 to 10,000 feet the forest is replaced by grass and brush. In pre-Columbian South America only the potato was grown at such heights, but after 1500 wheat was successfully introduced. (Lower) Cultivation with primitive, ox-drawn plows was originally introduced by the Spanish and represented an enormous technological advance in fifteenth-century America. Retention of this method in the twentieth century shows the cultural lag in such areas. Similar plows are still used also in lowlands. (Embassy of Venezuela)*

bean and probed the edge of the continent. In 1520 they began their penetration of the continents of North and South America. In the next 20 years they blocked out the areas that they wanted to develop. They sent ships on expeditions into the Pacific and scouted the coasts north and south. Then they sent land expeditions through eastern United States to some point in the Great Plains, such as western Kansas, and they sent another up the west coast of Mexico, through the Southwest, and out into the plains. Southward they crossed the Andes, found the Amazon, and sent a ship down its length, out to sea, and northward to the Caribbean shore of South America. They found that the two great centers of population and wealth were in Mexico and Peru. Between them lay lands with moderate cultural development, wealth, and population. North and south of this central area they found less population, less cultural development, and no gold or silver.

There were few Spaniards, so they could not physically take and hold all of two continents with the handful of men available. They chose to seize those areas where there were mines that could be made to produce more of the wealth they found there. Thus the Spaniards also went to the mountains—perhaps somewhat in parallel to the Asiatics who had preceded them.

In the mountains of Peru the Spaniards for the first time met the effects of truly great altitudes. It must be remembered that in this time there were no means for measuring the heights of mountains. Unquestionably the snow-covered Andes were high, but the Spaniards had seen snow-covered mountains in Spain. However, they found that they had strange problems to face in these South American mountains that no one seems to have experienced in the European mountains. The first accounts were greeted as unlikely stories. The Spaniards said that as they climbed these mountains, they found areas where the air made them seasick. Their choice of a parallel sickness was not bad, for the illness that they described was characterized by nausea and vomiting, just as in seasickness. However, it was well known that the sea's movement was in some way related to seasickness, and not even the Spanish storytellers claimed that the mountains moved. As a matter of fact, they gave a very good account of the phenomena for their time. In the words of one, Father Acosta, in abridged form:

I have wanted to say all this in order to describe a strange effect which the air or the wind which blows has in certain lands of the Indies, namely that men get seasick, yea, even more than at sea. There is in Peru a very high mountain range. I had heard of the effect of the air of this mountain and I started out prepared as best I could according to the information which is given out there by those they call old hands or guides. As I went up "the stairs" as they call the highest part of that range of mountains, I was suddenly seized with such a mortal anguish that I was of a mind to throw myself off the horse onto the ground. I was seized with such retchings and vomitings that I thought I should give up my soul, for after the food and phlegm came bile and more bile, this one yellow and the other green, until I came to spit up blood. If it had gone on, I believe I would most assuredly have died, but it only lasted a matter of three or four hours until we got down a good long way and came to more healthy air-temper.

It is not only that passage of the Pariacaca Range which has this effect but the whole great mountain which runs for more than five-hundred leagues in length. Wherever one goes through it one feels this strange distress, though more in some places than in others. It is felt more by those who go up from the seacoast than those who ascend from the heights of the plains. Animals feel this also.

I hold the op. ion that this region is one of the highest places on earth. In my view the snowy passes of Spain and the Pyrenees and the Alps of Italy are as common houses compared to tall towers. I am persuaded that the element of air there is so thin and delicate that it does not provide for human respiration, which needs it to be thicker and more tempered. I also believe this to be the reason for the stomach's being so greatly upset and the whole man put out of balance.[1]

This is both a good description of the effects of mountain sickness and a penetrating analysis of the causes of this malady by a man who had none of the advantages of modern chemistry, geography, or physiology at his disposal. Nevertheless, he recognized that it was something in the nature of the atmosphere at great heights that was the cause of this distress. What we would say in modern terms is little different. With the decreasing pressure that accompanies increasing altitude the density of the surrounding air decreases. At 17,500 feet, one-half of the atmosphere is below you, and one must breathe much more air to obtain an adequate amount of oxygen. There may, in addition, be complex physiological reactions to the decreased pressure. These effects are noticed by individuals at different elevations. Many people experience nosebleeds, headaches, and nausea as low as 10,000 feet. Thereafter the percentage of people affected rises sharply. Very few people born and raised at low elevations can go as high as 13,000 feet without experiencing mountain sickness. Those raised there, however, are not troubled by the elevation. In Peru men live permanently at elevations as high as 17,000 feet, but this is probably about the limit for any man.

This raises numerous questions. How did the first men penetrate to such great heights? Why did they do so? How long does it take for lowland men to adapt to these elevations?

Some of these questions are not answerable with any great certainty. All life forms tend to spread wherever they can. Baboons in Africa live from sea level to 10,000 feet in elevation. Men everywhere have used adjacent mountains as part of their hunting and gathering range because the diversity of plant and animal products makes them attractive areas. Further, in times of drought they have unfailing water supplies. Some of the plainsmen would in time remain in the foothills. Soon some of the foothill dwellers would be permanent residents in the mid-heights. By this process men would work their way up to the greatest heights that offered any possibility of food. Thus several generations would pass, and the men involved would gradually get used to the conditions of higher elevations. Man can become fitted to very high elevations; in the Andes men live up to heights of 17,000 feet, but this is exceptional. The bulk of Peru's population lives on the 12,000-foot plateau, and most of the people live nearer 10,000 feet above sea level.

Physiological problems at high altitudes

These are not altitudes that men enter easily. The Spanish experience again supplies most interesting data. It should be remembered that at least some of the Spanish came from the Spanish plateau, which averages 3000 feet in elevation. Nevertheless, when the Spanish went up into high elevations, they met great physiological problems. Reproduction virtually ceased, and this was not limited to men: bulls, horses, dogs, and particularly cats were affected. In one case of 12 sheep taken to 12,000 feet, only 6 reproduced. Some of the infertile ones were sent down to 7500 feet, and some of them then bore young. It was only after some time that children were born and much later that children began to survive. Then, as the Spanish priests noted, it was usually those who **were part Indian**. As Carlos Monge—who has studied the effects of altitude in those high regions—says, there is today no purely white race in these altitudes. The people present are 60 percent Indian, 30 percent mixed, and 10 percent European, with the faculty of resistance to great altitude carried over from the Indian parentage.

This does not mean that none of the Spanish were able to reproduce in these high elevations. The effect was variable. The effect of the anoxia (lack of oxygen) was in some cases complete and lasting sterility. In other cases, however, the effect was a temporary one, and after some time the individual's physiology was able to adjust to the condition, and fertility returned. Selection was abrupt for some, for those who could not reproduce saw their genetic line extinguished.

Monge presents interesting data on the fertility rate among the present Peruvian population. The population is now in two areas: about 66 percent is in the highlands above 8000 feet and 28 percent is in the lowlands below 800 feet. This is partially due to the physical geography: the lowlands can be irrigated; the foothills are extremely arid, not readily irrigated, and hence have relatively few people—only 6 percent of the population; and the highlands have more moisture and on the Altiplano have areas that can be irrigated. For the coastal lowlands the number of children per thousand women between 15 and 45 years is 144; for the foothills, 174; and for the highlands, 164. At first glance these figures suggest a large effect by elevation with optimum conditions in moderate altitudes: out of the heat of the lowlands but below the cold of the highlands.

There could be numerous reasons for this pattern. It could be related to the relatively small population density. Most infectious diseases spread in proportion to the number of contacts between infected people. The less dense the population, the less the frequency of contacts, the slower the spread, and the lower the disease rate. It is especially the infectious dysenteries and such things as measles and smallpox that kill the children in primitive societies. Conditions for their spread are at an optimum in crowded cities and towns. These are most common on the coast and in the highlands of Peru.

In the highlands the relationship between women of child-bearing age and infants under one year suggests very good reproduction rates between 7000 and 12,000 feet, but there is a considerable drop off above 12,000 feet. The high rate of reproduction of the highland people is just what the Spanish reported. But why then should there be a decrease above 12,000 feet? Is this an elevation effect? Those living at such very high altitudes may be the poorest of the Indian people, and the result may be nutritional. Of course, it may also be strictly physical environment: at some point the increasingly lower oxygen content of the atmosphere

High-Altitude Man. *This man is sitting at 15,000 feet in the Andes west of Lima, Peru. Few people can exist at such altitudes, but the Peruvian highland Indians can work at this elevation and at times have built shrines at even higher elevations. At those high, dry altitudes the bodies of those sacrificed simply mummified.* (Peruvian Transport Corp. Ltd. for World Bank)

must go beyond the possibility of man's adjustment. In the transition zone there would be physiological disturbances, and these would appear in such physiologically critical areas as reproduction.

Recent physiological studies tell us something of the mechanism of this sterility. At very high elevations women are unable to maintain sufficiently great oxygen concentrations in their blood to supply fetal growth. Women with certain heart defects have the same difficulty at low altitudes. Races adapted to high elevations maintain higher oxygen levels through increases in red blood cells, lung capacity, and circulation.

When the rewards were high enough, the Spanish lived in the heights. At Potosí, Bolivia, one of the greatest silver mines of the world was found: in the first two centuries of Spanish operation there $2 billion worth of silver was produced. This vast wealth attracted 20,000 Spanish and a working force of 100,000 Indians. The elevation was 13,600 feet, too high for plainsmen. The Spanish paid the price; for 53 years no Spanish child born there lived. For a long time the Spanish sent their wives to lower altitudes to bear their children, but even then the children often died when brought up to the high altitudes.

The on-the-spot reporting of the Spanish priests is again worth quoting. Here is Father Cobo's sixteenth-century account, shortened, but in his own words:

Those Indians who are born in high and extremely cold regions may be raised and survive better than those born in temperate and hot regions. The Indians are healthiest and multiply the most prolifically in the cold air-tempers, which is quite the reverse of what happens to children of the Spaniards, most of whom when born in such regions, do not survive. It is most noticeable that those who have half, a quarter, or any admixture of Indian blood are all raised with the same loving care as the pure Spanish children and yet, the more Indian blood they have, the better they survive and grow.[2]

These difficulties were so great both for the Spanish and for their domestic animals that they soon moved their capital down from the highlands, where the bulk of the people and the mines were, to Lima, where they were physiologically more at ease. Interestingly enough a hundred years later the Spanish commentators found this a bit difficult to understand. By that time domestic animals had undergone selection and adjustments to such high altitudes, and the valley of Jauja where the first capital had been placed was declared a fertile, productive, and healthful area for man and beast whether of European or American origin. Nevertheless, the cultural factor of resistance to change has maintained the capital at Lima, despite the fact that the coast is the minor population center of the country. This phenomenon, of course, is not unknown in other places: Annapolis, for instance, is the capital of Maryland by a similar historical accident, and the capital is maintained there despite considerable inconvenience. One question remains: if the Spanish had so much difficulty, how had the Indians survived so well?

Highland empire of the Incas

The Incas, beginning about A.D. 1000, developed a great empire based on the Altiplano. From these great heights they reached down into the arid coast lands and subjugated the lowland riverine oases toward the Pacific, but they left alone the steaming lowland jungle folk to the east. Two problems arise here. How did the highlanders manage to conquer the lowlands? Why did they avoid the humid lowlands and seize the arid lowlands? As usual, the question that we are asking here is: to what extent was this choice physically determined and to what extent was it culturally determined?

[2] C. Monge, *Acclimatization in the Andes*, pp. 31–32.

Megalithic Stone Work, Ruins of an Ancient Inca Fortress, above Cuzco, Peru. *The carving of huge stones so that they fit tightly together is relatively rare in the world. It is found in the Mediterranean, in Peru, and on Easter Island. Does this mean that ideas spread or that men reinvented this building technique several times? (Exxon Corporation)*

The Inca conquests eventually took in all of the irrigation-based city-state oases of the coast. The Incas never penetrated to a like degree into the hot wet lowlands along the foothills of the Andes. Since these were two utterly different climates, was the reason chiefly climatic? If the people of the Andes are physiologically adjusted to cold, they would have particular difficulty in tropical heat. This difficulty would be considerably greater in the hot wet climates than it would be in the hot dry climates. This was discussed in the opening sections of the book. It will be recalled that in hot wet climates the cooling by evaporation of perspiration is greatly slowed down.

It is quite probable that the people of the Altiplano are adjusted to cold, and quite possibly they have a slightly different metabolic rate than most of mankind. It has recently been found that the Tierra del Fueguians and the Australian aborigines have such physiological adjustments, and much the same thing is reported for these people. One of the Spanish priests noted that the people always felt warm—their hands were warm even in very cold weather. He commented that they could lie down on the ground and sleep comfortably even in a snow storm that covered them with as much as 6 inches of snow—as incredible as this may seem.

However, there are alternative explanations that suggest that a single-factor explanation is, as usual, probably wrong. In the desert areas the population was concentrated in the riverine oases, so it was vulnerable to attack. These people were also highly organized and productive, hence susceptible to conquest and worth taking over. The forest dwellers of the Amazon, on the other hand, were scattered, mobile, shifting agriculturists with little political structure. They were hard to hit, hard to hold, and had little that the Incas wanted. The combination of difficult physiological-climatological problems and sociopolitical problems seems to be the answer for the Incas expanding in one area much more than in another.

The Incas knew all about the effects of changing altitude; they knew that highlanders sent to the lowlands were prone to sickness. We know now that such illnesses resulted from exposure to new diseases, the lack of acclimatization to high heat, and perhaps pressure and oxygen changes. Combined, they made the highlanders highly susceptible to pneumonia, chronic bronchitis, lung abscesses, and tuberculosis of the lungs.

The Incas guarded against the effects of changes from one climatic zone to another with great care. When they attacked the lowlands, they prepared two armies. One was held in reserve so that if the campaign either in the hot wet lowlands or the hot dry lowlands exceeded two months, the first army could be replaced and returned to the highlands before physical deterioration set in. When people were conquered, the Incas often dispersed the conquered people and replaced them with loyal subjects. Care was taken to send highland people to highland locations, and lowland people to lowland locations. To the Incas the people were a major resource, and pains were taken to preserve them. They were perfectly aware of the fact that if they transplanted highland people to the hot lowlands or lowland people to the highlands, they would sicken and die. Indeed this became a form of death sentence. A man could be condemned to exile in a climate so different from his homeland that he could be expected to die. And many men did die under the less humane Spanish rule.

Spanish mismanagement of highlands and lowlands

The Spanish king issued decrees that the Inca policy be followed; unfortunately, he was far away, and the Spanish on the spot were greedy for gold. In spite of the king's orders, therefore, lowland Indians were taken to the highest areas for labor in the mines, and highland Indians who had to pay their taxes in gold or silver had to go to the lowlands to work for wages. In 1556 the viceroy reported that of 300 men who went down to the coast for work, not 20 would ever reappear in his homeland. He issued orders that this was to be stopped. Negroes were requested for labor in the hot lowland mines; it was observed that Negroes sent to the highland mines simply died. In the midst of all this turmoil the Spanish told the viceroy that they were certainly not going to work the mines themselves, and if he wanted production to continue, the Indians would have to be used. They recommended that the Indians so used be kept at elevations to which they were accustomed: lowland Indians for lowland mines and highland Indians for highland mines. There was lengthy legislation that all tended in the same direction: the Indians must not be moved radically from one climate to another. Even the tax laws were changed so that the Indians who lived in areas where gold and silver were not produced were allowed to pay in wool or cotton.

Despite royal decrees, the bitter protests of the Spanish priests, and the efforts of some of the officials, however, the Indians were moved about radically and abused terribly. The result was a catastrophic loss of life. This can be seen in the comparatively empty desert oases that contain the ruins of great cities and in the equally impressive ruins in the highlands. The Spanish had here entered an area of civilization, great mining wealth, and immense contrasts in climate. The contrasts were so great and concentrated in so small a lateral space that men adapted to one climate could be moved into another so drastically different that the results would be physiologically devastating. This is one of the most extreme of the environmental contrasts and limitations found among the varied environments of the world. Even then cultural wisdom could have prevented such a waste, for the Inca civilization had dealt with the same problem and handled it with little difficulty.

In these lands today the same problems exist. Mining is carried on at immense altitudes. For instance, the copper mines at Cerro de Pasco are even higher than the silver mines of Potosí. Labor still moves up and down the mountains seeking employment. Monge, the leading student of the effects of high altitude on man in the Andes, complained bitterly as late as 1948 that little attention was even then paid to the great cost in human life in moving men from the highlands to the lowlands. Of all the areas of the world, these very high altitudes provide the clearest cases of human physiological adjustment to extreme physical conditions.

Bolivia Today: The Potential

Although Bolivia and Peru began with great empire, great wealth in precious metals, and with a relatively dense population of politically stable people organized

Andean Lands. *The Andes are the home of great cities, primitive peoples, and all the stages in between. Cali, Ecuador, (below) is shown here ablaze with lights, like any modern city representative of the Western world.* (CHIORAL for World Bank) *The women in the marketplace (upper right) in Otavalo, Ecuador, represent the persistence of an ancient American custom: the market. From central Mexico through Bolivia the market is not only an economic affair but an important social event.* (United Nations for World Bank) *In Bolivia, on the waters of Lake Titicaca (lower right), 12,500 feet high and the world's highest navigable lake, an Indian boy is sailing one of the earliest forms of watercraft known. The boat is made of bundles of reeds and has a worldwide distribution, often surviving among primitive peoples.* (United Nations)

into a model socialist state, they have gone down the scale of cultural and political status. Under the mismanagement of the Spanish profiteers, the population was reduced to a fraction of its original strength. Estimates of Bolivia's pre-Spanish population go as high as 10 million. Probably an 80 percent reduction in population occurred under Spanish rule. The present population of five million is estimated to be 50 percent Indian, 35 percent mixed, and 15 percent white. After the Spanish empire collapsed, the old Inca empire was broken up: parts of it went to Peru, Bolivia, and Chile. From a rich, stable empire this part of the world declined into a poor, unstable society wracked by revolutions, limited education, and low productivity. In cases such as Bolivia, the situation seemingly is drifting from bad to worse.

Bolivia today is one of the most backward nations in America: only Haiti is poorer. Two-thirds of its people cling to the 12,000-foot Altiplano, a landscape so barren that it has been likened to a lunar landscape. Yet this was the base for an early civilization, that of Tiahuanaco whose ruins are one of the major archeologic wonders of America. Today 70 percent of the people are illiterate, life expectancy is 32 years, and for most of the people the future seems hopeless. More than 70 percent of the people are farmers, but they cultivate only 2 percent of the land and produce too little to feed the country. The mines have accounted for 90 percent of the foreign exchange earnings, but the tin ores are about exhausted. Oil has been found, gold mining is increasing, some population is moving to the more productive tropical lowlands, and education is slowly reaching more people. As a consequence the measures of economic growth are beginning to move upward after the centuries of stagnation that followed the catastrophic declines of the sixteenth century.

Bolivia's problem is to find outlets for its people's potential productivity. This requires energetic development of education in order to train individuals who have the genius to develop products, organize production, and capture markets—but this amounts to changing the whole Latin American culture pattern. Free-enterprise entrepreneurship is totally foreign to their whole way of life. It is the way of life, not the highland interior position of the country, that is hindering development.

Bolivia may well be compared with Switzerland:

	Area (square miles)	Population	Density per square mile
Bolivia	412,000	5,385,000	13
Switzerland	16,000	6,550,000	411

Could Bolivia do what Switzerland has done? Switzerland is noted for its political stability, which aids in making it an international banking center. Surely this lies within the power of the Bolivians to accomplish. If Bolivia had a stable government and banking houses of proved repute, then international capital, and especially Latin American capital, could well flow to it. Physical geography does not prevent this.

Switzerland's major agricultural effort is in dairying. Bolivia has a vastly greater land area, and much of it is potential cattle-producing land. If New Zealand from its position in the midst of the world's oceans can enter the international market with dairy products, wool, and fabrics to compete with Switzerland, surely Bolivia could do the same. We think almost automatically of the Swiss success in the making of watches. Small nations often specialize in some highly skilled product. The Dutch, for instance, are the most important diamond cutters of the world, and there are probably countless opportunities of this sort in the world, many of which could, through the application of human energy, be developed in the physical geography of Bolivia. One of Switzerland's important industries is tourism. Switzerland is scenic and hygienic, and Swiss service is among the world's finest. Bolivia is more scenic, for its mountains are higher, its snow is equally good for skiing, and it has, in addition, tropical lowlands, exotic primitive people, incredibly magnificent archeological ruins, picturesque lakes such as Titicaca and Poopó, and immense valleys that plunge down the eastern slope of the Andes becoming more tropical by the mile. Now that skiing grows in importance every year, the South American snowfields are attracting attention, for their Southern Hemisphere location means that there is skiing in the northern summer. A measure of the growth already underway is that an international ski meet was held in the adjacent Andes of Chile in 1966. But Bolivia is mostly unhygienic, and the services that tourists cherish are in the most rudimentary stage of development. These are human failures in the face of potential resources that another people might well make a source of wealth. The block is not in the physical geography or in the physical men—it is, as usual, in the cultural setting.

The Antioqueños: A Highland Success Story

In Colombia there is an interesting development that illustrates that a rebirth of the highlands as a center of energy and cultural growth is possible. Here in the

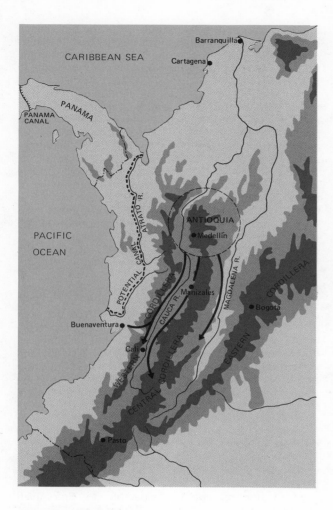

MAP 8-5 Antioqueño Settlement in the Andes
The original Antioquia was the mountain territory on both sides of the Cauca River, centering on the city of Medellín. After 1790 settlement expanded down the western flank of the Central Cordillera to Manizales, which was reached by 1850. Vigorous expansion after 1850 followed the high mountain flanks down the Western Cordillera and both flanks of the Central Cordillera but not into the Cauca and Magdalena river valleys. The linear expansion along the mountain flanks and the avoidance of the hot low valleys show the tendency of man to stay in the kind of environment to which his culture is adapted. A wet-rice cultural expansion might have been just as linear but restricted to the valley bottom. Changes are probable if the Atrato valley is developed. Low dams on the two rivers, one flowing into the Atlantic and the other into the Pacific, would generate enormous hydroelectric power and simultaneously allow a canal development competitive with the Panama Canal.

highland valleys of the northern part of Colombia, in the province of Antioquia on the slopes of the mountains flanking the Cauca River, is a group of people of Spanish descent who have been growing in number and expanding throughout the highlands. They are noted for their energy and thrift and characterize themselves as 'the Yankees of South America." They are shrewd and aggressive individualists who are prouder of being Antioqueños than of being Colombians. Things were not always so.

The beautiful slopes of the ranges flanking the Cauca River had had a dense Indian population, but this largely disappeared under the impact of the European arms, diseases, and disruptions. Development of the land during the colonial period was based on gold mining, but the mines were poor, largely placer mines, and they were worked by individual laborers to a great extent. There were few Indians and few Negroes, so the Spaniards had to work. Thus a tradition of hard work was established. This is quite contrary to much of Latin America, where it is the tradition that men of Spanish descent do no manual labor. As was common in the Spanish colonial lands, there were few European women; the result here was a mixed race: a composite of Negro, Indian, and white.

The mines were poor, and by concentrating on them the people condemned themselves to poverty. Until the end of the colonial period they remained backward, illiterate peoples of the mountainous interior of Colombia. So far it would sound as if the interior location in high mountains was hampering development because of the remoteness of Colombia from the world centers of intellectual ferment and economic innovation and growth in the north Atlantic areas.

The royal inspector who was in the province from 1784 to 1787 initiated a series of changes that were to overcome the obstacles of mere physical environment. Juan Antonio Mon y Velarde instituted far-reaching social, economic, and judicial reforms. He encouraged the foundation of agricultural settlements in the cool, malaria-free uplands; he offered bounties for the introduction of new crops; vagrancy laws were drawn up to send idlers to the new agricultural settlements. Here we meet the role of the unusual individual and see that within a culture there is still room for the exercise of the individual will.

By the opening of the nineteenth century the highland population was growing and beginning to expand southward into the forest-covered mountain slopes to the south of Medellín. The main current of expansion was south and southwest, but it could equally well have been northward. The expansion was spurred on by a short-lived boom in rubber, rumors of gold, and dreams of vanilla, cinchona, and cochineal plantations. Dreams are not unimportant. Because men dreamed, they pushed into the forests, cleared land, and produced something. It turned out that for a time the fortunes to be made were in the more prosaic corn and hog industry so well known in Iowa. In Colombia, however, it was a shifting agriculture that produced corn and other crops for the hogs, followed by a 10-year fallow in forest. The hogs went to market on their own legs, driven over trails, sometimes in herds of hundreds. Later coffee became the important crop.

At first only gold and silver could stand the cost of export. Until late in the nineteenth century coffee had to be moved out over mountain trails by mule or ox train. The movement of some valuable goods encouraged the building of roads, and better roads encouraged more export and, of course, an increased return flow of goods and materials. This feedback system has functioned so well that the coffee production in the hands of the Antioqueños has risen to the level of 70 percent of Colombia's production. Small holdings in the hands of private owners produce this coffee as well as much of their foodstuffs. Cattle is another important element in the Antioqueño economy. They have their own special breed: a white animal with black ears. Speculation as to the origin of these peculiar animals includes the possibility that they are the descendants of a breed of cattle that the Romans introduced into Spain. These are thrifty animals able to eat tropical grasses and resist tropical diseases. The Colombians have recently been introducing grasses and managing pastures in order to improve their cattle production both for meat and for dairy because these products are among the critical shortages in many tropical regions. In the lower lands large landholders have managed pastures of 40 acres or more, fenced so that the cattle may be rotated from one pasture to the other, thus giving the grass a chance to recover between grazings.

In the search for better grasses suited to equatorial conditions *Melinis multiflora*, African molasses grass, was accidentally introduced. This grass provides nutritious grazing and a protective mantle for the land, is tolerant of poor and eroded land, and spreads itself rapidly. It has taken over large areas of the country, healing the abandoned lands and turning them into useful pastures. It is at home in the *tierra caliente* and reaches well up into the *tierra fría*. Since its accidental introduction in 1906, it has nearly taken over the temperate lands of Colombia. Far from the physical environment dominating man, here man is busily reordering the physical environment to meet his needs. There are further large potentials for even the adjacent tropical

lowlands, as is shown by the success of the King Ranch in spreading cattle keeping into the hot wet lowlands of northeast Australia, or as the IR rice and the potential for multiple cropping in the tropics is demonstrating in the Philippines.

Antioqueño expansion is continuing. The economic base is broadening, and the population is growing and continuing to spread into new territory. The population growth has been phenomenal; there has been a doubling every 28 years since 1828. They have grown from about 50,000 people in 1880 to nearly 100,000 people in 1900 to over three million people in 1950. This represents a shift in their percentage of the Colombian population from 10 percent in 1835 to over 25 percent in 1938. This population growth has been accomplished with increasing economic strength rather than declining status, for the increase of productive people has led to the growth of cities, the building of roads, and to general economic development.

All is not perfect by any means. The standard of living in comparison with the United States is low, and health standards are abysmally bad. And the population growth is in spite of heavy handicaps. Especially in the lands below the *tierra fría,* the incidence of hookworm, dysentery, malaria, typhoid, syphilis, and gonorrhea is high. Hookworm is the worst. Infection runs close to 90 percent. However, this is clearly the result of the habit of not wearing shoes and the failure to use latrines. Typhoid simply reflects the absence of modern water-supply measures, while malaria results from a combination of the absence of screening and the lack of mosquito control and antimalarial drugs. The description of malaria in Colombia applies equally well to the upper Mississippi Valley in America in the period when colonists first pushed into that area and lived in housing about comparable to Colombian housing.

A further health problem for the Antioqueños has increased with the growth of cities: prostitution has proliferated, and the results are very high rates of venereal disease.

These problems have varied climatic controls. Malaria is much more serious in the *tierra caliente* than in the *tierra templada* and is almost nonexistent in the *tierra fría*. Hookworm is limited by temperature but is even more limited if human patterns of hygiene are changed. The same is true of typhoid, and venereal disease is rightly called a social disease. It is not sensitive to physical climate but to moral climate.

The potential interoceanic canal and the immense hydroelectric works discussed earlier are in the lowlands to the west of the Antioqueño lands. If the dams are built to create the lakes envisioned, the immense power source that the Antioqueños need for the industrial development will become available. Industrialization is crucial if their spectacular success story is to be a continuing one. The theme that it is the marshalling of human energy that determines whether such developments occur, combined with the immense drive of the Antioqueños, suggests that Colombia is likely to develop its potential. Expectably the dams will be built, the power utilized, and the potential for the interoceanic canal will be capitalized upon.

The Antioqueños, then, supply a case history of a people at first seemingly handicapped by a mountainous interior location and then stimulated by a socio-economic program into a populational and territorial expansion along the adjacent mountain slopes. Their explosive population and economic growth has put them ahead of the racially and culturally similar countrymen who had the identical territory available for development. To the non-Antioqueños it was a remote, empty, mountainous, geographically handicapped area. To the Antioqueños it was a glorious opportunity.

The Mexican Highland

The other center of civilization in the highlands of America in pre-Columbian times was in Mexico. Here the Spanish found the brilliant Aztec empire flourishing when they entered America. The civilization had roots deep in the past. Cities with ceremonial centers were in existence on the plateau of Mexico at least as early as 1000 B.C. Cities always imply developed agriculture, political structures, trade, and transportation. Ceremonial structures of great size tell still more about the degree of political development, religious beliefs, and surpluses of labor that can be channeled into works of art and architecture. By all of these measures the Aztec developments were comparable to those of Egypt and Mesopotamia.

The center of the Mexican highland culture was the great city of Tenochtitlán, present-day Mexico City. The elevation here is 7347 feet. This height is not great enough to create the physiological stress that the heights of Peru and Bolivia create, but it is high enough to make the climate moderate. Typically the sun is warm, but the shade is cool. The Spanish found Mexico City standing in the marshes at the edge of a lake. It had causeways leading into the city, and aqueducts brought a supply of good water. The city had a circumference of about 10 miles and a population of about

Detail from the Temple of Quetzalcoatl, Teotihuacán, Mexico. *The name Quetzalcoatl combines the words for the Quetzal bird and for snake. It is often translated as Plumed Serpent and represents one of the great culture-bearing godlike beings who are so vivdly enshrined in Mexican Indian mythology.* (Mexican National Tourist Council)

300,000 people. It had immense pyramids with temples on their tops, huge palaces, and such things as zoos, great markets to which goods from over a wide area were brought, and so forth. It was a large, bustling center of empire.

In some ways the Aztec empire resembled the Inca empire: both were located on plateaus surrounded by high mountains, and both were preceded by earlier empires. The Incas were preceded by the Tiahuanaco, to name but one important one; the Aztecs had built upon the Toltec empire. In both areas there was a long history of cultural growth and development with civilization extending back at least into the first millennia B.C. Both of these empires had expanded by conquering the neighboring people, but the Incas had solidified their hold more firmly than the Aztecs had. The Aztec empire was composed of numerous city-states, and there was almost constant strife between the various groups which served the useful purpose of supplying thousands of captives who could be used for sacrifices demanded by their religious beliefs. It has been seriously argued that the expenditure of life in this way was great enough to hold the population down to the levels their way of life could support. This was certainly a grim way to control population growth.

The Mexican highland culture had reached down into the adjacent warm and humid lowlands and conquered Mayan people there. This provides sharp contrast with the Inca history, where the expansion was toward the arid lands and not into the hot wet lowlands. With most of the people living at an elevation of a mile to a mile and a half above sea level, the Mexicans must, to some extent, have been "highland" people with some, even if minor, adjustment to the conditions of cool dry air. (Annual range of temperature at Mexico City is from 20° F. to 85° F., and the rainfall is 26 inches.) If hot wet climates are difficult barriers to such people, why did they not expand to the north where there were more similar climates?

The reasons seem to have been that to the north of central Mexico there were only wild tribes, while in the adjacent lowlands there were great city-states. Thus the cultural situation was reversed: in Peru the city-states were in the arid lands, and the tribal peoples were in the wet lands, so the expansion was toward the city-states. In Mexico the city-states were in the hot wet lowlands, and tribal people were in the arid lands. Here the expansion still was toward the areas of dense, settled, civilized population.

This tendency continued when the Spanish entered Middle America. They rapidly surveyed the scene. There were numerous geographic areas, each with different climates and accessibility to Spain and varying plant and animal potential. The land most like Spain in soil and climate, and hence the land that the Spanish could most easily have developed in the ways they knew best, was in California and Chile, yet neither became a center of Spanish development. The lands nearest to Spain, the lands most easily reached by sea, were the Caribbean islands, yet the Spanish did little with these beautiful and fertile lands. They were left virtually empty to fall into the hands of the French and British freebooters. Instead the Spanish seized two of the three great centers of population and cultural development and made them the centers of the development of their New World empire. Both of these were highland empires, the Aztec and the Inca. They largely ignored, except to destroy, the lowland Mayan empire.

At first there seems to be a perfect correlation here between avoidance of the hot wet lowlands and preference for the high dry plateaus. This correlation loses almost all of its force when it is noted that it is also a perfect correlation with the distribution of major sources of precious metals. The limestone lowland of the Yucatán peninsula lacks metals, and the Spanish choice was for areas where there were settled peoples who had precious metals and where there were known mines. Remember that in California, where there were no settled peoples and no known mines, the Spanish did nothing with the land, despite its desirable climate and fine soils. They thereby missed the fabulous gold of California and the immense silver wealth of Nevada.

Modern Mexico: The highland heartland

There is not one Mexico but many. The north is dry, and the south is rainy. The Yucatán peninsula is low, as is the area along the United States border, while the peninsula of Baja California is mountainous. The heartland of Mexico is the high plateau, with Mexico City as its center. The dominant position that this city holds in the minds of Mexicans is measured by the fact that to them it is Mexico; they don't bother to add the noun "city." It is in this central highland that great cities flourished in pre-Columbian time, and it is here that Mexican population, industry, and wealth are centered today. In the early centuries of the Christian era Teotihuacán, just 34 miles from modern Mexico City, had a population of 25,000 people, possibly as large as Rome at that time and surely larger than Athens. Modern Mexico City has grown from 1.5 million in 1940 to 4.5 million in 1960 to 9 million in 1973. This is a metropolitan district figure: central city plus suburbs. It is one of the world's 10 largest cities and still growing phenomenally. Such growth of urban centers is typical of the developing nations today.

The area dominated by Mexico City has the greatest population concentration and is central to the remaining populations. This area absorbs large quantities of raw materials and foodstuffs from the rest of Mexico as well as most of the imported goods that enter the country. In contrast, it exports little: 80 percent by value of the exports of goods to foreign countries come from peripheral Mexico. Even in nuclear Mexico (Mexico City) one sees enormous contrasts. On modern superhighways the latest of modern diesel buses speed past men plowing fields with ox teams and tiny villages without a trace of modern amenities.

The dominance of Mexico City is now changing rapidly. Factories are spreading not only in Mexico City but into the surrounding cities. Mexico is already 50 percent urban dwelling. The guarantee for a continuing future is seemingly being underwritten by the extension of education throughout the country. The tiniest villages have a schoolhouse, and a literate population is being established. As the farm children leave the land, they will have some preparation to work in the cities.

All is not rosy, of course. Mexico's population is growing at an astonishing rate. The population was 15 million in 1910, 40 million in 1960, and 55.3 in 1973. This is one of the highest rates of increase in the world. Despite the fact, growing industry has been absorbing the increase in population, and the per capita income and standard of living have been rising. The biggest problem is in the rural areas where land has been promised to the people. Unfortunately, there is just not enough useful land to provide every family with some to make a good living. Continuing industrialization seems to be the answer to this problem.

Because Latin America has been so troubled with political instability, and because the fortunes of various countries have varied so widely, depending on the choice of political economy that it chose to follow, it is worthwhile to review briefly the Mexican course. Mexico has an authoritarian one-party government that emerged from a period of chaos and bloodshed that lasted from 1910 to 1929. During this period of turmoil the economy was wrecked, and the population declined by one million. Finally the generals who had been tearing the country apart with their armed rivalries formed a national coalition, the Revolutionary Party, and this party has ruled ever since.

At first the top men were generals, but by 1946 civilians began to be elected to the presidency—an interesting example of transformation of a military dictatorship into a civilian-dominated state with only a single party. In recent elections about 80 percent of the votes went to the party. The president has great power but uses it only in consultation with the country's major representatives of industry, labor, the military, and the Church. The selection of the candidate for a new president amounts to election. This selection is done in quiet, unofficial consultation, and the man selected is then announced. The form of a national election is then gone through with enormous energy, and it serves to unite the nation behind the new president. The system works and has given Mexicans nearly 50 years of stability. They have used this time to start their nation on the road to a modern economy. Though they have a long way to go, no one who has traveled widely in Mexico and seen how enterprising and hard working the Mexicans are can doubt that they will make it if they can maintain political stability. The major threat to such continued stability lies in the poverty-stricken and still too often landless peasant. The situation is a race between population growth, education, and industrialization. Up to now industrialization has been winning.

Perhaps the most interesting aspect of Mexico's position today is in the concentration of population and power on high plateau land. As we have seen, this stemmed in part from antiquity—but only in part, for in the past, centers of population were found in the well-watered lowlands such as the Yucatán peninsula. Attracted by the great capital city of Tenochtitlán, with its established position as the administrative center for central Mexico and with its location giving it access to the rich mining areas, the Spanish continued it as the seat of power, population, and education in Mexico. The role of continuity is thus well exemplified. With the collapse of the Spanish empire came the period of reorganization, the loss of Texas and the entire western United States, and increasing internal turmoil finally reaching such heights that the population began declining. Then came the political decision by the leaders of the nation to work together rather than against one another, and 50 years later the nation is the success story of Latin America, bursting with energy, bustling with new industry, and firmly in possession of a sense of its identity. Mexicans no longer identify themselves as Indians, Ladino or Spanish—they are Mexican. The center of this growth and development has been in the nuclear highland area, a plateau 7000 feet high and surrounded on three sides by towering mountain ranges, remote from the sea, and having most of the disadvantages usually used to explain lack of success of some highland area.

Summary

An overall view of the highlands of the tropical regions of the world shows a familiar pattern. Those areas that were in or near the centers of growth of civilization shared in that growth. The presence of ideas made it possible to develop mountain lands and high plateaus to the level of civilization and empires. Today there is no great world power located in the highlands of the tropical regions, but there is no apparent reason in physical geography that this should be so. The tropical plateaus and highlands, such as those of Ethiopia, East Africa, and Middle America, offer large enough bases with equable climates, abundance of strategic minerals, potential water power, and great variety of soil, vegetation, and animal life. Today the development of transportation makes their terrain and position less of a negative factor than they may have been in the past. Even in the past, however, these factors were not sufficient to prevent the development of civilization in such regions when men willed to do so.

MID-LATITUDE MOUNTAIN LANDS

For convenience, the mountains between the tropics and 60 degrees north and south will be considered the mid-latitude mountain lands. The American mountain masses of the western United States and Canada, the Andes of Chile in South America, and in Eurasia the immense belt of mountains reaching from Spain to China are all mid-latitude mountains. In Africa the area includes highlands at each end of the continent. The mountains in southeast Australia, Tasmania, New Zealand, and Japan are discussed elsewhere.

In the tropics the mountains gain temperature advantages by their elevation. In mid-latitudes the results of elevation are complex. On the west sides of continents in latitudes near 30 degrees there is aridity; there mountains have the advantage of greater moisture. Farther north, with steadily falling seasonal and annual average temperatures, mountain lands begin to suffer from too much cold. All the other features of mountains already discussed—the great variability of slope and exposure to sun and rain, accompanied by many different kinds of plants and animals within short distances—remain true of the mountains in mid-latitude lands.

Middle Eastern Developments

In the Old World, civilization began in the shadow, poetically speaking, of the mountain lands. Indeed, there is a possibility that the first important steps toward civilization were taken on the mountainous plateaus of Asia Minor. It is in this area that wheat and barley, our earliest domestic grains, originated. This is also the home of the stone fruits, cherry, peach, almond, and of such animals as the domestic sheep and goat—and quite probably the center of origin of domestication of animals. Whether civilization began in the highlands, the foothills, or the plains is something that we may never know. Unquestionably, however, the highland masses included in what are now Turkey, Iran, and Afghanistan early shared in the growing civilization of the Middle East.

Anatolia

The area of modern Turkey, somewhat larger than Texas, is to be thought of as like the western part of the United States. The coastal climate is like that of California, but there is a rapid ascent to a plateau that is much like the Great Basin country. There is much arid land, but there are also many ranges that stand high enough to catch moisture and send living streams down to the surface of the plateau, which averages 3000 feet in elevation, creating a rich irrigation potential. Some of the ranges were glaciated in the past and are now snow capped in winter today. During the glacial past this area was studded with lakes, just as in the American Great Basin.

As early as 2000 B.C. this area was the seat of the Hittites. Basing themselves on this highland peninsula, they were for centuries a dominant power in the Middle East. In later times Anatolia has been the seat of a mighty Turkish empire. It is one of the curiosities of history that we live in a time so shortly after the collapse of the Turkish empire, yet we tend to forget that from this highland a gigantic pincers movement was directed against Europe that nearly changed the course of history. The physical base that served for the development of civilization and empire is still there. The region is rich in its variety of climates, potential water power and irrigable areas, and such agricultural bases as good forests. It is not its highland geography but its present state of political and economic stagnation that handicaps this land.

Iran and Afghanistan

Much the same can be said of Iran and Afghanistan. They have been the scene of great civilizations. They could again be developed, productive countries, and indeed Iran is today developing rapidly. The turn of history, the direction of cultural growth, especially the effect of the Islamic conquest in shifting the focus of interest away from Iran, the changes of dynasties—all partially explain the lack of progress in these areas. But most of all, the total cultural pattern of the people has not been in tune with the economic revolution that began in northwestern Europe and has still not reached them in any effective way. These areas once were near the center of Eurasian growth and development, but they have been out of the mainstream of developments for the past 2000 years. This lag is not due to their position, for these highland countries lie in the crossroads of Eurasia and Africa. The location of political-economic growth, however, has ignored this crossroads area because of the culturally induced explosion on the "isolated" northwestern end of the European peninsula.

The spread of ideas, as we saw earlier, was from the eastern end of the Mediterranean westward. It went from the mountainous Middle East to mountainous Crete and from there to mountainous Italy. Civilization was later in time and lesser in development as distance to the west increased. Spain was late, and the cultural influence of the Middle East was markedly diminished. The beautiful Atlas mountain region, the exact climatic and physiographic equivalent of the southern California region, only belatedly entered the realm of the civilized world. That mountains of themselves are no barrier to development is well illustrated by Italy or Spain at various times.

Switzerland

If one were to pick a piece of land that was seemingly overwhelmingly handicapped by a mountainous terrain, Switzerland would make an excellent choice. Within the 16,000 square miles of Switzerland there is little flat ground and a great deal of very high mountainous land. Consider these figures: arable land, 12 percent; meadow and pasture, 41 percent; forest and woodland, 25 percent; other, 23 percent. In addition, the Swiss are divided in language, ethnic background, and religion. People of Italian, French, and Germanic ancestry and of Catholic and Protestant faith make up this nation. They have no outlet to the sea of their own, and they have few mineral resources.

The Swiss plateau, with an elevation of about 2500 feet, has the bulk of the population. The temperature is moderate, not hot in summer and not too cold in winter. Nevertheless, even at an elevation of 2500 feet, the decrease in temperature is about 10 degrees below what would be expected at sea level. Even the plateau of Switzerland, therefore, has summer and winter temperatures that would be expected at sea level in areas much farther to the north. On the high mountains, with temperature decreasing at the rate of 3 degrees with each 1000 feet of increased altitude, the limit of grain cultivation is found at about 4000 feet. Above 9000 feet snow stays the year around. Between the upper limit of grain growing and the zone of perpetual snow the land is used for forest and for pasture.

In all of the mountain lands of Europe there is a strongly developed system of vertical seasonal movements to take advantage of the summer grazing in the Alpine meadows, and this is particularly well known

View from the Säntis, Northeastern Switzerland. *The Swiss Alps are justly famous for their spectacular scenery. Their location in the heart of densely populated and affluent Europe aids in making them valuable for tourism.* (Swiss National Tourist Office)

in Switzerland. It should not be overlooked that it is also well developed in Spain, Italy, and generally throughout the mountains of Europe. However, such seasonal movements of cattle were rare or unknown elsewhere in the world before 1600. In Switzerland there is not much space for lateral movement. In Spain the movement of the herds to the mountains often involves long drives across the lowlands to reach the mountain masses. In Switzerland the movement is normally only one of a few miles up the local mountain side. The Swiss often have a major farmstead at low elevation, where grain is grown and fodder for the stock is stored for winter feeding. In spring as the snows begin to melt, the cattle are sent up the mountain to their summer pastures. Part of the farm family may go up with them to stay for the summer. At other times there are very complex movements of people and herds up and down the mountain according to the season, the work load, and the particular crop to be harvested.

There are also settlements in Switzerland well up in the mountains near the edge of the limit of grain growing. Villagers persist here with a way of life that was probably established during the Neolithic period. They live in snug houses whose back ends are built into the hill slope, with the only door and a single window facing downhill. The window is for light, not ventilation. The house has hay stored in its upper part, and cattle are kept in the back part of the one very large, ground-floor room. Cooking and eating are carried on at the front of the room, and sleeping is in curtained bedsteads along the sides of the room. The stove is for cooking only; heat is provided by the animals. This might be described as the original hay burning, heating, and air-conditioning system. The heat of the animal bodies keeps the room at a comfortable temperature; their respiration keeps the humidity high.

While the thought of living quite so closely with cattle is offensive to our modern sensibilities, especially to our noses, it was the way of life of our northern European ancestors for a very long time, indeed, until a relatively short time ago.

It often seems as if there must be an inherited tolerance to the smells of cows and horses. Most people find the smell of the stable tolerable, but this does not extend to all animals. While the stablelike smells in the Washington Zoo house for zebras and giraffes and their kin is pungent but pleasant, the house that holds the lions and tigers is rarely endurable. Most of mankind lived very closely with their animals throughout the Neolithic period, and many of the peasant people of Europe still do. It is not a mountaineer's trait; it is an old European trait that has been preserved in a few mountain villages in Switzerland. But even the perservation of old ways of doing things is by no means found only in mountains—old ways of doing things last today in such lowland spots as the Hebrides and other isolated areas.

The miracle of Switzerland is the building of a high level of productivity and hence a high standard of living on so small and mountainous a physical base. Many factors have entered into this. Switzerland occupies a position astride some of the principal passes that lead from Italy into central Europe. For many centuries trade has funneled through this highland area. Traders carry ideas as well as goods. Ideas are the key

to growth and the development of nations and areas. Switzerland therefore has had a good position, but this too can easily be overemphasized.

Why should the major flow of ideas go through the passes? Why not go by sea to the mouth of the Rhone River and thence up that easy valley to the heart of Europe? Or from the other side, why not follow the Danube to the head of the Rhine and thus again penetrate Europe without having the difficult task of crossing the Alpine passes? An explanation based on the mountain-pass theory is too simple. On close scrutiny we find that the principal pass, the Brenner, is not even in Swiss territory. Further, the use of passes through mountains when easier low routes exist immediately suggests that political considerations such as the blocking of other routes by unfriendly powers were important. A major cause for the use of the Alpine passes was, of course, the importance of northern Italy as a center of commercial and industrial activity in medieval times and later. However, this is not a geographical factor so much as it is a cultural-historical factor.

The Swiss Confederation came into being in 1291, but its present political unity dates to the nineteenth century. The agricultural-dairying way of life has long been well established, as have been such pursuits as skilled woodworking in the southern German and Swiss region. To these skills the Swiss have added watch making and have become the world's leaders in this field. They are also important manufacturers of textiles and machinery, and they have a notable chemical industry. All of this has been developed in a nation virtually lacking in what to our modern industrial age has seemed "required" raw materials: coal, iron, petroleum. Instead they have developed to the maximum the hydroelectric power that is potentially present in every mountain land within the humid climatic zones. As an example of Swiss enterprise, the Grande Dixence, with a height of 940 feet, is one of the world's highest dams. Within the past 150 years the Swiss have grown faster in industrial and commercial development than the majority of the European nations and now have one of the highest standards of living.

All this success has been in spite of their mountainous land with its "limitations." The energetic Swiss have made their mountains an asset by capitalizing on their scenic beauty and their opportunities for mountain climbing and skiing. What the Swiss have done is possible for most of the other mountain lands of the world, but it is particularly readily duplicated by the other mountainous areas of Europe, populated with similar people, located in the same highly developed part of the world, and possessing as many, and often more, basic resources. Yet some of the other mountainous areas of Europe are losing population—the Massif Central of France, for instance. The reason often advanced is that it is a mountainous region. The reason cannot be that simple.

Asiatic Areas

There are immense mountain lands in Asia. Japan is a mountainous land, and in some way its development is to be compared with Switzerland: this has already been discussed (see Chapter 6). Japan has made its solution one that in large part turns its back on its mountains. Its agriculture is largely based on rice (one-half the arable land being devoted to it), a flat-land crop. Even when the hill country is used, it is flattened at great cost by terracing. The Japanese solution has been to take the path of industrialization, as indeed every advanced nation today has done. It is not a matter of mountains or no mountains but of organization and motivation for industrial production.

China too can be discussed as a mid-latitude mountain land. China, as now developed, is not mountainous, but the China with the largest potential is. Most of China is mountainous to hilly land. This is well expressed in the figures for the distribution of its type of land: arable, 10 percent; meadow and pasture, 20 percent; forest and woodland, 8 percent; other, 62 percent. Most of the land in the ambiguous classification of "other" is rough, mountainous land. South China is largely mountainous, as is western China. Despite the immense hydroelectric potential in these regions and the immense mineral resources known to exist there, these are the least populated parts of China. Again it is not the mountains that prevent development; it is the total outlook of the agrarian-minded Chinese. Now that the Chinese are forging ahead with their version of the industrial revolution, these areas, now the emptiest of their land, will probably be the most important part of their territory.

Tibet

Tibet is the Old World equivalent of the Altiplano of Peru and Bolivia. It is even higher, for the average elevation is 16,000 feet. The mountains start from here and go up to Mt. Everest's world record, 29,000 feet. This land is so high and the 15,000- to 17,000-foot plateau is so surrounded by mountain ranges that very little moisture can reach it. The height

alone is sufficient to limit the precipitation to low amounts. Since the capacity of the air to hold moisture is so related to its temperature that the lower the temperature the lower the moisture capacity of the air, and since the air temperature decreases at the rate of 3 degrees per 1000 feet increase in elevation, it follows that the air on the Tibetan plateau must be about 50 degrees cooler than that at sea level in the same latitude. The consequences are cool air of low moisture-carrying capacity and a precipitation on the plateau of about 8 inches. In these altitudes precipitation does not increase as one goes on up the mountains; it tends to decrease. Therefore there is relatively little opportunity to use dependable water supplies coming down off the higher mountains for irrigation of the arid plateau. Unquestionably the same effects of high elevation on the physiology of man are effective in Tibet as they are on the Altiplano of Peru. However, the local variety of mankind must be assumed to be well adjusted to these conditions.

Tibet has an area of nearly 500,000 square miles and an estimated population of perhaps one and a quarter million prior to the Communist Chinese attack, which took the form of a veritable war of extermination. It is quite clear that the population is small and the density of population among the earth's lowest. Are the causes physical? Is it the high elevation, the mountains, or the aridity? Of these factors we have already seen that the mountains as such cannot stop developments. High elevation is not sufficient to prevent acclimatized races of mankind from flourishing even at such elevations as these; there is a town in Tibet, Gartok, at 15,000 feet elevation. Of the factors of the physical environment, the most limiting one seems to be the climatic combination of low temperatures and low precipitation. Are these sufficient to prevent man from developing a denser population on these high arid plateaus?

The present way of life of most of the Tibetans is that of herder; horses, sheep, goats, cattle, camels, and yaks are the principal herd animals. The yak is an oxlike animal with a heavy coat of hair and is capable of living and working at very high altitudes. It gives a milk that is rich in butter fat. The agriculture of Tibet is mostly carried on in the lower valleys; barley, wheat, turnips, and cabbages are typical Old World plants grown there. The Tibetans also grow the Andean potato, a plant capable of being grown in the great altitudes of the plateau. Actually there is a whole complex of domestic plants peculiar to the Andean plateau that could be grown on the Tibetan plateau. The following questions then arise: are the Tibetans limited in number by their preference for a pastoral way of life?

Monastery at Lhasa, Tibet, Built in the Seventh Century. *This magnificent building epitomizes the Tibet that was the high mountain land once a center of power and culture.* (Courtesy of The American Museum of Natural History)

Could their numbers be multiplied several times by shifting to a high-altitude type of agriculture? Could this larger population be profitably put to work mining some of the known, extensive, but largely untouched, mineral resources of this vast mountain area? Is it impossible to think of Tibet being developed as a Switzerland of the Orient? Certainly there are some limitations that Tibet has that Switzerland escapes. It is difficult to think of a tourist industry in a country where most of mankind would become violently ill from the effects of the altitude. Yet if tropical heat is oppressive, Tibet offers an immense recreation area that is so located as to be readily available to the greatest mass of mankind. It is not physical geography that decrees that the Asiatic masses cannot afford delightful trips to scenic Tibet.

Tibet is or was a Buddhist nation. It is more difficult to get reliable information about this always mysterious kingdom now that the Communist Chinese control it than it was when the Lamaists did. Lamaism, the form of Buddhism that developed in Tibet, centered around great monasteries that dominated the country. Prior to the tenth century their unity made Tibet a strong kingdom, but since then they have been divided. Prior to the Chinese Communist conquest of the country the number of Lamas was estimated as high as 500,-000. Since most of these were celibates, this created a strange population structure, and both polyandry and polygamy were practiced by the lay population. The main effect was limitation of the rate of popula-

tion growth. The Chinese domination of Tibet today is a recurrence of an old theme, for the Chinese have at various times controlled the area. This time the Chinese seem bent on destroying Lamaism and the great monasteries. Just what effect this will have on the course of development of Tibet remains to be seen. It is certain that this struggle, with destruction of old ways of life and the exile of many leaders, who, if they return will bring floods of new ideas, is bound to alter man's perception of the best use of this region.

While at first glance the great altitude, aridity, and cold of Tibet would seem to be sufficient causes for the low population of the region and the lack of development of industry, further consideration raises some doubts. If the people are maintaining themselves despite the adverse physical geography and social practices, there would seem to be little hindrance to man in the biological field. The presence of a good variety

Lake Louise, Western Canada. *In addition to their mineral, hydroelectric, and forestry potential, mountain lands often have breathtaking beauty, which makes them attractive to tourists. Mountain lands can be the seat of advanced nations, such as Switzerland, of great empires, such as the empire of the Incas, of stagnant cultures, as in Tibet, or they may be empty lands, as in much of North America. Their role can change even with the whim of fashion—skiing has already greatly changed our evaluation of the mountains.* (Information, Canada)

of domestic animals, some variety of domestic plants, and the certainty that there are still more useful high-altitude plants that could be introduced to make the land more productive of foodstuffs suggest that there are considerable unexploited possibilities for increasing the population of the land. Pastoral nomadism, Lamaism, and a position remote from the mainstream of the industrial revolution have combined to keep Tibet from achieving anything like the Inca empire.

Summary
The mid-latitude mountains, then, show much the same range of human achievement as the other climatic areas or the tropical mountains. Around the Mediterranean they are or have been at times centers of civilization, empire, and cultural growth. Mountainous lands in mid-latitudes may be nearly totally undeveloped as are the highlands of East Africa, which lay far beyond the range of penetration of ideas, or highly developed as in modern Switzerland. They may be largely ignored as in the Chinese highlands, despite the obvious (to Western eyes) advantages of the region, or alternately viewed as a center of wealth and learning and as a low point in achievement as in Bolivia. Such variability in human achievement in moderately similar environments cannot be a physical, environmentally determined result.

HIGH-LATITUDE MOUNTAIN LANDS

The high-latitude mountains pose one of the most formidable problems that man has faced. In Alaska and adjacent Canada there are high mountains, including the Brooks Range, running east and west to the north of the Yukon Valley; the Alaskan Range to the south; and the north-south trending ranges of the northern Rockies and coast ranges that join in this area. These are high mountains: Mount McKinley, 20,300 feet high, in the Alaskan Range is the highest peak in North America. While the Brooks Range is not now glaciated, this is not due to the lack of cold; it is the result of too little precipitation there to sustain a snow cover. The forest border is found along the southern foothills of these mountains. Their slopes are tundra, and their tops are barren rock. Much the same can be said of the Alaskan Range. These are not areas where any kind of agriculture can be carried on, and there is very little animal life. There is no attraction in such areas with the exception of possible mineral strikes or the development of a tourist industry based on the permanent snow supplies in such areas as the Alaskan Range. With

the shrinking of the world through the expansion of air transportation, it is becoming feasible to have ski resorts in Alaska to let those addicted to skiing develop it into a year-round sport. In the main, however, these Arctic mountain areas are the world's emptiest. Even the deserts have more people than the Arctic mountains. These are the extreme deserts of the northern lands.

In the Old World, the Scandinavian highland, the northern part of the Ural Mountains, and the northwestern part of Siberia are mountainous areas with latitudes north of 60 degrees. All of them are marked by the limitation of tree growth to low elevations and warm exposures. All of these mountain masses were glaciated in the past, and some of them retain minor glaciers today. As in America these are among the emptiest parts of our world. Man lives in the lowlands between these mountain masses, and he rarely finds any use for the highlands. They have little feed for his herds except on their lower slopes; their barren rocky upper slopes, having little or no vegetation, can support little or no animal life.

The physical environment of these Arctic mountains is as physically determined a barrier to man as we have on earth, but even this must be carefully stated. At our present state of development we see little of value in them; however, we can even now clearly foresee that any major find of mineral wealth in such areas would lead to the rapid development of transportation, settlement, and the exploitation of whatever this source of wealth might be. The parallel is the case of the development of the iron resources of the Quebec-Labrador area. Twenty-five years ago it would have been considered folly to predict that thousands of people would soon be settled in that seemingly forbidding region and that the probable trend would be toward still fuller development, with maintenance of fairly large populations in the region for an indefinite time.

As was suggested above, one of the uses that some of the Arctic mountain regions may have is for recreation. Rapid transportation places these areas within easy reach of the increasingly leisure-possessed working force of the industrial world. This same world is being pressed by an increasing population, which is crowding our traditional vacation areas. Who can say that the Arctic mountain areas may not have a future as the ultimate refuge for those who want to get away from the mass of mankind? The constant snow cover may attract the athletic; the stark beauty of the barren ranges may attract the esthetic. We have the technological ability to build and maintain cities in them if we wish. At the moment we do not wish to do so. This is different from saying that the physical environment denies the possibility of our doing so.

Summary

Mountains must be considered a special problem in cultural geography. They offer great variations in short distances, and for industrial man they are exceptionally favorable for discovering minerals and developing water power (in humid areas). In the deserts, mountains are centers of advantage because of their water catching. The steep slopes present difficulties for transportation, with water transportation the most hampered, overland transportation intermediately so, and air transport the least affected. Technological change clearly alters the potential of most mountain areas. Truly great elevations in mountains become virtually an absolute barrier to mankind, for most of mankind is barred from elevations above 11,000 feet, and only the races that have long lived in the highest mountains can reproduce above these levels. This becomes significant only in low-latitude mountains, for from mid-latitudes poleward, the lowered temperatures associated with increasing elevation place such elevations above the timber line. Only the obtaining of minerals provides any reason for going so high. At any rate, only a tiny fraction of the earth is actually above 17,500 feet, the approximate limit for man's permanent biological adaptation to residence in mountains.

In general man's accomplishments in mountains follow the familiar pattern of access to ideas. Those mountainous areas in or near great centers of political, social, and economic growth have shared in the advances. As the centers of cultural action have shifted, some mountain areas have alternately been advantageously and disadvantageously located, and their fortunes have varied accordingly. Once again it is demonstrated that the decisive factor is cultural rather than physical, with the limitation that in these lands we have met, through lowered oxygen content and air pressure changes with elevation, an absolute limit for man's expansion unless he is to be fed bottled oxygen or placed in pressurized housing. A less absolute limit is also set by the lowered temperatures even at moderate elevations in high altitudes. The decrease in vegetation and soil cover leads inevitably to a decrease in carrying capacity for animals, and soon there is little in the landscape except minerals. Only if there are concentrations of valuable minerals does such an environment hold any attraction for contemporary man. It seems excessively optimistic to think of very high elevations in the tropics or even moderate elevations in the high latitudes as ever becoming areas of great attraction for mankind.

Review Questions

1. If Great Britain had not seized South Yemen to protect Britain's lifeline to India, would South Yemen have become an important crossroads area? What factors must you consider?
2. What principle does Ethiopia's "northward reaction" illustrate? Give other examples of similar reactions.
3. Compare highland New Guinea with highland Peru and Tibet. Which has the physical advantage? Which has the cultural advantage? Explain.

4. The Incas knew how to use highland people and lowland people effectively. Why did the Spanish not know how to do so?
5. If mountains are barriers, what is to be said of the frequent invasion of Spain from France and of France from Spain?
6. Could Tibet be the Switzerland of Asia? What advantages and disadvantages would such a position bring to it?
7. The mountain lands in high latitudes are now the world's emptiest areas outside Antarctica. Why can we not take this as being physically-environmentally determined, in the past, in the present, and in the future?
8. At the opening of this century Aspen, Colorado, was a flourishing town. It then became virtually a ghost town, and it is now flourishing again. Why? Is this due to environmental change? Cultural change?
9. Why are there mountains? What does the idea that the surface of the earth is made of plates have to do with mountains?
10. If mountains are poor environments, why was civilization in tropical America so strongly associated with mountains? Compare other tropical mountain regions.
11. Why is the Mexican capital located in the high mountainous part of the country?
12. What determined the direction of expansion of the American highland Indian? Did similar causes operate in different climates and among other peoples? Can you derive a principle?

Suggested Readings

Atwood, Wallace W. *The Rocky Mountains.* American Mountain Series, vol. 3. New York: Vanguard Press, Inc., 1945.

Dundas, C. *Kilimanjaro and Its People.* London: H. F. & G. Witherby, Ltd., 1924.

Fifer, J. Valerie. *Bolivia: Land Location and Politics since 1825.* Cambridge, England: Cambridge University Press, 1972.

Ford, T. *Man and Land in Peru.* Gainesville: University of Florida Press, 1955.

Harrer, H. *Seven Years in Tibet.* New York: E. P. Dutton & Co., Inc., 1954.

Linke, Lilo. *Ecuador, Country of Contrasts.* 3d ed. New York: Oxford University Press, 1960.

Matthiesen, P. *Under the Mountain Wall, A Chronicle of Two Seasons in the Stone Age.* New York: The Viking Press, Inc., 1962. A description of life among primitive people in the mountains of New Guinea.

Monge, C. *Acclimatization in the Andes.* Baltimore: Johns Hopkins University Press, 1948.

Ogrizek, D., and Rufenacht, J. G., eds. *Switzerland.* New York: McGraw-Hill Book Company, Inc., 1949.

Osborne, H. *Bolivia: A Land Divided.* London: Royal Institute of International Affairs, 1954.

Parsons, James J. "Antioqueño Colonization in Western Colombia." Ibero Americana: no. 32. Berkeley, Calif.: University of California Press, 1949. This monograph was the major source for discussion of the Antioqueños. Interested students will find it valuable.

Peattie, R. *Mountain Geography.* Cambridge, Mass.: Harvard University Press, 1936.

Simoons, F. J. *Northwest Ethiopia: Economy and People.* Madison, Wis.: University of Wisconsin Press, 1960.

Thompson, C. H., and Woodruff, H. W. *Economic Development in Rhodesia and Nyasaland.* London: Dobson Books, Ltd., 1954.

Toniolo, A. R. "Studies of Depopulation in the Mountains of Italy." *The Geographical Review,* vol. 27 (1937), pp. 473–477.

FOCUS

Ethiopia, Kashmir, Nepal, Bolivia, Ecuador, Colombia, Peru

The Northern Forest Lands

CHAPTER NINE

THE northern forest lands are found only in the Northern Hemisphere. There are no lands in the latitudes of 60 to 70 degrees in the Southern Hemisphere, while in the Northern Hemisphere there are the vast areas of North America and of Eurasia in these positions. Instead of four or five comparable areas around the world, we have two for this climatic type.

The Physical Setting

In high latitudes the seasonal changes are extreme. In summer the sun may shine for 24 hours or disappear for only an hour or two. The mid-winter sun will rise above the horizon for only an hour or two. North of

67 degrees it will not rise at all. With no solar energy coming in, and with radiation from the earth escaping into outer space, the temperature of the land falls. In these areas there are great masses of land, and the interior of these lands is far from the sea, the only great reservoir of heat that could temper the extremes of the climate. Lacking such oceanic influences, both winters and summers can be extreme in their temperatures. Winter cold is at its greatest not along the continental margins in the north but in the deep continental interiors in high latitudes. An example of such a cold spot is found at Verkhoyansk at 67 degrees north in Siberia. Here the January average temperature falls to −60° F. The absolute cold record is about −90° F. This is truly great cold, yet note that Verkhoyansk is a town; people live under such conditions.

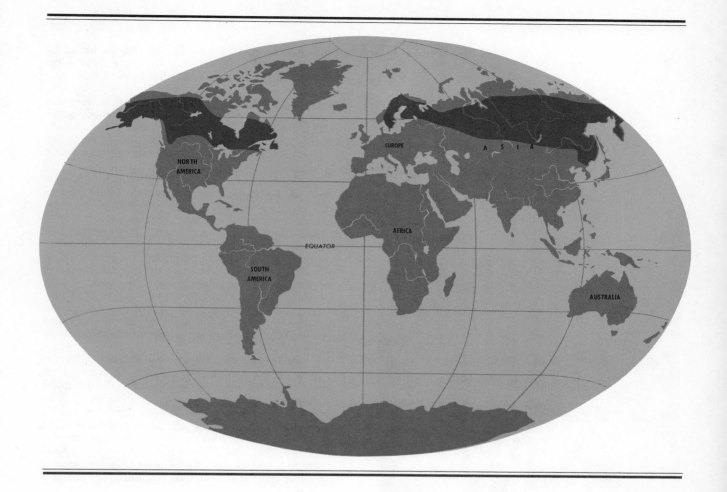

Surprisingly the interiors of continents in such latitudes can also be hot. Fairbanks, Alaska, at 64 degrees north and in the interior of Alaska, in most years has a summer day that reaches 90° F. At Fort Yukon the extreme temperatures recorded range from 100° F. to —78° F. In these latitudes there is considerable variation in temperature from the coast inland. On the coast, winters are milder, and summers are cooler than in the interiors. It is noticeable that this type of climate reaches the west side of continents at about 60 degrees but comes 10 to 20 degrees farther south on the east coasts of continents. This is because the average flow of the wind in these latitudes is from west to east, bringing oceanic-tempered weather to the west coasts and continental-interior extreme temperatures to the east coasts. Thus, although Seattle and Quebec have

about the same latitudes, Seattle has a mild marine west-coast type of climate, and Quebec has long, cold winters and short, hot summers. The range of temperature included in these lands is great. Along their southern boundaries the average for January is about 20° F., but along their cold boundaries it may be —20° F. to —60° F. Summer has much less variation: all these lands vary between 50° F. and 70° F. in July, comparatively mild temperatures. It is the winter cold that most markedly sets these areas apart from the rest of the world.

The natural vegetation is forest. This varies from good forests along their southern boundaries to an increasingly thin cover of smaller and smaller trees to the northern limit of tree growth. It is not the existence of extremely great cold but the lack of enough summer

warmth that limits tree growth. This limit is found to be approximately the line that separates lands with no month above 50° F. and those that have at least one month above that temperature. Trees will survive, though they can hardly be said to thrive, if at least one month goes beyond this temperature; for really good growth, trees need much more warm weather than that. Coniferous trees predominate, though there are also considerable birch and aspen. The northern forest is called *tiaga*.

Rainfall in the northern forest lands is moderate to low. In the continental interiors the totals are between 10 to 20 inches; on the continental margins, where the maritime air masses can penetrate, the rainfall may rise to 40 inches per year. Low temperature throughout most of the year allows little evaporation while the ground, frozen during half the year or more in some areas, prevents the draining away of the water underground. The vegetation has little solar energy with which to work and is virtually dormant during the cold period; hence it gives off relatively litttle water. The result is ground that is frequently soggy and seems like a land with much greater rainfall than there actually is.

The wetness of the landscape is increased by the presence of great numbers of lakes, a heritage of the glacial period. The glaciers so recently disrupted the previous drainage system by gouging holes here and dumping sands and gravels there that the streams have had little opportunity to cut through the dams and fill the hollows. Postglacial time in these regions varies from about 12,000 years ago along their southern borders to a still current glacial state in a few places. Parts of Alaska and Labrador, for example, are still

(Upper) **Olavinlinna Castle in the Eastern Lake District, Finland.** *Built in 1475, it stands as a monument to the inflow of ideas from the south. Note that there are no castles of such age in the American northern forest lands. The boats too reveal the inflow of ideas, for they are built of planks that overlap along the edges exactly the way the Viking ships were built. The clinker-built boat is a specifically Scandinavian invention. Its use here reveals directions of influence and also the tendency among men to continue the old ways of doing things.* (Finnish National Travel Office)

(Lower) **Logs at Varkaus, Finland.** *Out of the vast areas of coniferous trees in the northern forests comes an enormous amount of wood, much of which goes to meet the demand for pulp for paper. Although the rate of growth of the trees is slow, the extent of land where they grow is vast, and we have little other use for it at present.* (Finnish National Travel Office)

glaciated. These are not the coldest areas but the areas of higher precipitation, for glaciers were not the result of cold so much as they were of high amounts of precipitation in the form of snow falling in areas with summers too short to melt off all of the winter snow. Whatever the causes of the glaciers, the mark that they left on our northern lands is impressive. This is most easily seen from the air or on a map. The boundary between the land occupied by the last glacier and that land not recently glaciated can be drawn with considerable accuracy simply by marking off the areas with innumerable natural lakes from that part of the landscape with few or none.

The recency of the glacial past is of great importance also in determining the characteristics of the soils of these regions. Materials that have not been exposed to weathering for great lengths of time have more soluble minerals left in them. Therefore they are potentially rich. In most of these northern forest regions this is of little importance, for these are not agricultural lands except along their southern boundaries. The recent disappearance of the glaciers is also indicated by the exposure of much rock, for where the land stood relatively high, the ice stripped off much of the soil mantle; the time lapse has not been sufficient to produce from these exposed rocks another mantle of weathered materials. This condition has been a boon to prospecting in these areas, which have been the source of various metals and fossil fuels.

In discussing these lands it has been possible to speak of the climate and forests and past glacial history without differentiating the Eurasian lands from the American ones. In most ways this is one great area. The plant and animal distribution reflects the fact that the lands are grouped around the North Pole, and that the minor gap between Asia and America at the Bering Strait has frequently been nonexistent in the geological past. This has allowed plants and animals to move back and forth. Consequently the lands around the North Pole share similar species of trees and bushes, caribou and mice. That this exchange of plants and animals from Asia to America through this now Arctic gateway has gone on for a long time is illustrated by the redwoods. Asiatic in their origin, the redwood trees were carried millions of years ago through the Alaska region into America, and today they survive in only three areas: the coast of California, the Sierra Nevada of California, and in the inner part of China. Then much later in time elephants and buffalo migrated to America and horses out of America by this same route. It is through this ancient connection between the Old and the New Worlds that the first men probably made their way into the Americas.

The physical structure of the Old World lands and those of the New also have similarities. Both have great central plains that drain to the Arctic Sea, both have mountainous areas on their east and west sides—in both cases it is the Pacific edge that has the greater mountain mass—and both have great rivers. The Yukon in the American Arctic and the Ob, Yenisei, and Lena in Asia are listed among the 20 largest rivers in the world. In these vast areas, with their enormous exposures of rock, are found some of the world's great mineral reserves. Both areas face on the Arctic in the north, are bounded by the Atlantic and Pacific on their sides, and along their southern boundaries meet forest midlatitude lands, grasslands, and arid lands.

Given these physical similarities, what have been their human histories?

MAN IN THE NORTHERN FOREST LANDS

The human histories of the northern forest lands have to be shorter, or at least more interrupted, than the history of occupance in most of the world, because these lands have repeatedly been glaciated. The American lands were almost entirely ice covered, for only small areas of Alaska escaped glaciation. In Asia the ice cover was less extensive. The largest ice mass lay over Scandinavia and spread outward. In Siberia glaciation was largely limited to the mountain masses and their adjacent foothills. By the time of the second glacial about 400,000 years ago, Sinanthropus, the Chinese man, was already living on the edge of the northern forest lands. It is quite probable that by then man was occupying land along the southern edge of this climatic zone all the way to Europe, though our record in western Europe is clearest for man during interglacial periods; man apparently withdrew from Europe during glacial periods. Man finally stayed in Europe during the last glacial period, the Würm, that began about 70,000 years ago.

We know that in this immense expanse of time between 400,000 and 70,000 years ago man was developing physically to some extent, but even more importantly, he was developing culturally. He was making more and better tools from a greater variety of materials, making more adequate housing and clothing, and learning to control fire. Tailored skin clothing, the spear thrower, the oil lamp, and a wide assortment of stone tools evolved. As in the case of the industrial revolution—where centuries of hidden developments paved

Leon Lake, Nord Fjord, Norway. *The mountainous areas in the northern forest were carved by the glaciers and often left with lakes in the valleys. In Europe lands of this type are often used for agriculture, but elsewhere they are generally used only for extractive industries such as forestry and mining.* (Norwegian Information Office)

the way for the burst of the nineteenth-century industrial development—tens of millennia were going into the development of traits that were to make possible the use of the northern lands. This development finally gave man the equipment to live in cold climates and hunt the great game animals that lived on the forest-tundra border; men then followed their game wherever it led them.

Upper Paleolithic Hunters

One corridor led from southeastern Europe south of the Urals toward Lake Balkhash, on into the Selenga River drainage, and thence to Lake Baikal. During the last glaciation, probably as early as 30,000 years ago, men were living on the river terraces in this area; they were skilled hunters who killed mammoths, deer, horse, and reindeer. They built semisubterranean

houses, the roofs of which were formed of interlocking reindeer horns and whose side walls were shored up with mammoth bones. There are traces of this way of life roughly along the 50th parallel all the way from the Atlantic to the Pacific.

We do not know the origin of this new hunting skill. Because a great deal more archeological work has been done in western Europe than elsewhere, we know more about that region; since we know more, we tend to assume that this is the place of origin. It is much more likely that this way of life originated at some intermediate area, perhaps in the area of the Black, Caspian, and Aral seas, and then spread both northwest and northeast. That there was a common source is shown by the similarity of tools, the ways of life, and even art forms. In western Europe and Inner Asia men were not only hunting similar great animals in similar ways, they were also representing them in ivory carvings and making similar ceremonial staffs. While one may argue that life in a cold climate requires clothes and housing, clothes made to fit the limbs, and snug houses built underground, no such argument can account for the similarity of art forms and ceremonial forms. As a matter of fact, not even the form of houses and clothing is satisfactorily explained in this way, for the people living in the cold northern forest lands of Canada did not build similar underground houses or equally well-tailored clothing as those found in the Asiatic area.

The introduction of specialized hunting skills was to change much of the northern parts of the world. In a sense, the hunters were too successful, for they exterminated the largest animals. The mammoths were used up by the close of the ice age. As the ice withdrew, the hunters moved farther north. There the preferred prey was the reindeer, while in the more forested lands to the south horse and deer were important. The significance of the developing of the hunting way of life, with its abundance of meat and hides and the associated skills of making clothing and snug housing, is that these skills opened up to man the occupation of the cold continental interior lands on a scale not previously possible.

It is probably significant that at this time similar ways of life appear in the Americas. At present our most important information is coming from the unglaciated areas to the south and the west of the glaciated area. Skilled hunters, doing much the same things that the Eurasian skilled hunters were doing, appear widely spread over North and South America before the time of the last ice retreat. These people are known in America as the Folsom and Clovis people. They made distinctive, large spear points. They hunted mastodon

MAP 9-1   North Polar View of the World
Formerly glaciated areas are shown in dark shading. At Bering Strait the land is seen as it would be at a time of glacial lowering. The arrows suggest possible routes by which men may have reached North America. Probably all three of these routes were used at different times. Plants and animals as well as men frequently traveled back and forth through this northern gateway in late geologic time.

and mammoth, horse, camel, and ground sloth, as well as smaller game. As with the men of Eurasia, they were so successful that they exterminated most of these animals. We know that the Folsom spear points date to about 10,000 years ago, and the Clovis points are older. Sandia points are still older, for they have been found beneath a stalagmite layer in a cave, while the Folsom points were above the stalagmite layer. Some large interval of time associated with a climatic change is recorded by this placement.

It is in the zones of the world held by these specialized hunters that the greatest extinction of large game animals occurred. It was in northern Eurasia that the elephants, rhinoceros, and other large game animals disappeared. They survived in southern Asia and Africa, where specialization was in the direction of agriculture rather than hunting. In America the impact of the hunting cultures was especially great. Perhaps it was because the incoming hunting people spread more

suddenly through lands that had been previously held by people specialized in gathering, with hunting a minor activity. Whatever the cause, it is clear that after the hunting people appeared, many of the big game animals of the Americas disappeared. This is probably to be viewed as another of the human modifications of the landscape.

Reindeer Pastoral Nomadism in Eurasia

In Asia a complex series of changes took place that did not occur in America. The idea of domesticating animals instead of hunting them originated somewhere in southwestern Asia, and as we have seen, whole series of animals were brought under human control.

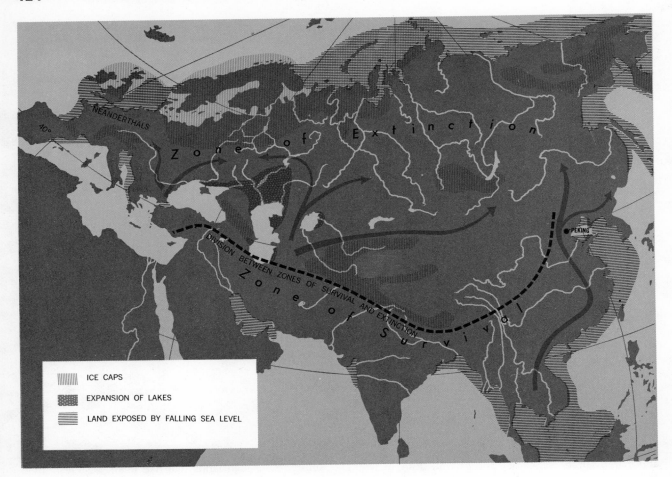

MAP 9-2   Early Man in the Old World Arctic

The principal glaciated areas of the north are indicated, as well as the larger areas of continental shelf exposed by the lowering of sea level. The arrows suggest major lines of the spread of men into the cold-winter zones of the Arctic forest. People with advanced hunting skills spread into Siberia during the last glacial epoch. There is an interesting correlation between the appearance of such "great hunters" adapted to Arctic life and the disappearance of big game, especially the elephants (mammoth and mastodon). Elephants survived into the Christian era in most areas south of the line drawn here but were exterminated quite early north of it. Wild elephants did survive in Northwest Africa into Roman times, in the Fertile Crescent into Old Testament times, in India, Burma, Thailand to the present, and in northern China to 1000 B.C. There surely has been continuous occupation by man of these areas of extreme winter cold for at least 20,000 to 30,000 years. It probably was representatives of these Arctic hunters who entered America during the last glaciation, displacing many of the earlier, simpler cultured people and exterminating much big game—the native American horses, camels (except in Andean South America), elephants, giant beavers, giant buffaloes, and the like. The arrow leading to Peking indicates the spread of men of Southeast Asian cultural pattern northward.

Reindeer in Valdres, Norway. *Reindeer and the closely related caribou are found throughout the Arctic forest region, but only in the Old World were they domesticated. If they were domesticated in imitation of the earlier domestication of the goat, horse, and cow, this is a remarkable transfer of ideas. As we have seen, such a transfer did not occur with animals in Africa or with plants in North America.* (Norwegian Information Office)

The idea of domestication spread to the northern forest people. Since their land was too cold and too poor in the fodder plants suited to horses and cows, they applied the idea to the available animal, the reindeer; an entirely new way of life using the northern forests was created. The reindeer can live on the Arctic and subarctic vegetation and withstand the great cold of the deep continental interiors without being sheltered in barns and without requiring any storage of hay or other foodstuffs to carry it through the winter. Its requirements are for suitable range.

A primary food for the reindeer is the so-called reindeer moss. This is a low growth, and a given area can sustain a herd for only a short time. The herds then must be moved frequently, and the reindeer people can have no permanent home. They must be free to move over broad areas in company with their herds. This way of life can never lead to dense settlement of the land, but it does offer a more secure way than does hunting in the same environment. Instead of being dependent on the presence or absence of the reindeer, with possible starvation occurring in the years that the wild herds bypass a territory in search of new pastures, the domestic herd assures a constant food supply kept at

hand and under control. The animals can also be milked, or ridden, or used to pull sleighs. They yield hides and furs, sinews, antler, and bone—useful raw materials for clothing, tents, thread, and tools.

The people of northern Eurasia have a fully developed way of life that is dependent on the reindeer. They have finely developed tents that enable them to move about freely and are snug even in the extreme cold of Inner Asia. The form of the tent varies greatly from area to area. In central Siberia it is made of felt and mounted over a portable wooden frame. There is an inner lining to the tent that increases the insulating quality and absorbs the moisture released by the humans. When the lining is saturated, it is unfastened and taken outdoors where the temperature may be −10° or −20° F. The moisture quickly freezes, and the lining is then beaten with sticks, and the ice crystals fly out. This is "dry cleaning" or, more accurately, cold drying. That these people use felt tents points the direction of the origin of the trait, for felt is made by the pastoral nomads of the Inner Asian grasslands.

The use of the reindeer varies. In some areas they are ridden, but in most areas they are not. Most people use them for pulling sleds. The sleds vary greatly in shape: from boat-shaped affairs to toboggans to sleds with runners. The reindeer may or may not be milked or ridden; this variation suggests that reindeer domestication was not a single invention that then spread as a developed cultural unit. Rather, the variability looks more like the result of a number of people from the grasslands to the south of the northern forest lands entering these cold lands and finding that their horses, cows, goats, and sheep could not stand the great cold; they then substituted the native reindeer. Alternatively, it may have been that the old residents of the northern forest, through long and intimate contact with the peoples along their southern border, adapted the idea of domestication to the reindeer. Those in the west would be in contact with people such as the Turks, who milked their domesticated animals. Those in the east would learn from such people as the Chinese, who did not milk them. The great horse nomads were in the east of Eurasia, and riding would be expected there more than in the west.

We do not yet know the time and sequence of events in man's occupation of Eurasia in any detail. After the specialized hunters who were there in the Upper Pleistocene, we know little until practically modern times. The ancient hunters of the elephants probably remained as forest hunters. When the idea of domestication spread through the grasslands of Central Asia and through the woodlands of western Europe, the hunters must have been displaced from the grasslands. It is tempting to account for the sudden expansion of some of the Mongoloid people by suggesting that they were the first to apply the idea of domestication in the Inner Asian lands. Domestication of sheep began about 10,000 years ago, and domestic cattle appeared 7000 years ago. Nomadism in Inner Asia may have begun about 3000 B.C. No one at the moment knows when reindeer nomadism began. Considering the slowness of man's learning, it would not be unexpected if it had taken another 2000 years or so for a beginning to be made.

Once the idea was born that the wild deerlike animals that ranged the treeless tundra in the summer and the wooded taiga in the winter could be treated like cows and horses, the northern forest man had a significantly surer source of subsistence. Reindeer herding is so recent that the animals are barely tame, and it is often a question of whether man is driving or following the herd. The animals are kept close, and they are attracted to the men in part by the salt that they supply. The herds are moved (or followed) from summer to winter ranges. Animals are killed for both meat and hides, which had many uses.

The thickly haired hides are the source of warm clothing. The clothing is tailored to fit, and double suits are worn to increase the protection against the cold. Tailored clothing is a widespread trait throughout Inner Asia. It is distinctly non-European and not at all typical of any of the tropical or subtropical people, all of whom use robes and similar flowing garments. The trousers and coat type of dress made possible the occupation of the high Arctic, but their development seems to have been in the high mid-latitudes.

The Absence of Reindeer Herding in North America

In the American forest lands there is a virtually identical physical environment, occupied by a closely related people and having in the landscape a wild animal, the caribou, that is very closely related to the reindeer. In America, however, this animal was never domesticated. Why? The opportunity, the need, even the access to the idea that it could be done were certainly all present.

Most of the land was held by two linguistic groups. Western Canada was held by the Athabascan-speaking peoples. Eastern Canada and the adjacent United States were held by Algonkian-speaking people. These people hunted in the forests, fished in the innumerable lakes and streams, and gathered berries in the short summer. They lived in tents and wore skin clothing, but their clothing was not nearly as warm as that of the Eskimos, who lived in areas subject to less extreme cold than the interior lands occupied by these forest Indians. Here again we meet the lack of compelling force in the environment. The Indians knew of the Eskimos, often fought with them, and had every opportunity to know about their superior clothing. The Indians certainly needed such clothing for greater comfort and efficiency, even if not for absolute survival. Like the Tierra del Fueguians, they simply did not choose to change.

It is sometimes stated that there are large numbers of caribou in our north, and this is true. But the area is immense, and the numbers are not great in proportion to the area. The large herds that are shown in pictures and described in stories appear only at certain times and places. They are the result of concentration of migrating animals coming from huge summer grazing areas and moving toward winter grazing grounds. Such herds frequently move through a given pass through mountains or between lakes. The Indian people used to get a large part of their winter food at such points by slaughtering at the time of the fall migration. All was fine if the herd appeared on schedule. However, the herd occasionally migrated by another route to an area not yet overgrazed. Famine then struck the Indian group that had staked its survival on the appearance of the herd at its usual passage.

Why did no genius arise to say, "Let us domesticate them!" Why did no word of what their neighbors in Siberia had done leak through to these people? It has been the theme of this book that geniuses with new ideas are the rarest thing on earth, and that even when a new idea is created and put to work, it may take an extremely long time to spread even to adjacent people in obvious need of the idea. We may well surmise that the idea of domestication reached at least some of these people, for many traits have a distribution right across the Bering Strait. The bow and arrow entered America via Alaska, and a long series of increasingly complex forms of bows suggest that there was a long-continued flow of ideas. The magic flight myth, a myth as recognizable as the George Washington and the cherry tree story, was common to northern Asia and northern North America. Ideas about how to make sleds, bark canoes, and snowshoes all crossed. Surely with all of these transfers of knowledge, someone must have mentioned the possibility of domesticating the reindeer-caribou.

The perfect example of how much opportunity there was for the transfer of ideas is found in the report of one of the Jesuit missionaries sent to Canada in the early colonial days and later transferred to Siberia. In Canada he worked among the northern forest people and converted many of them to Christianity. Years later, working among similar people in Siberia, he met a woman whom he had converted in eastern Canada. This woman had been captured in warfare, made a slave, and traded from tribe to tribe until she had reached Siberia. It is unlikely that this is the only instance of its kind, or that movement was only in one direction. Ideas clearly crossed from the Old World to the New, intermittently and over a long period of time. The idea of domestication of the American equivalent of the reindeer simply was not adopted.

The Eskimos had villages on both sides of the Bering Strait. Those on the Asiatic side were in contact with Asiatic reindeer herders, but the Eskimos never took up this way of life. In the opening decades of this century the United States government decided that what the Eskimos needed was a better source of income, and it seemed logical to get them to raise rein-

deer. Interestingly it was decided to get reindeer from Siberia. Why not start with the easily tamed caribou? Instead men were sent to Siberia to purchase a herd and drive it to Alaska, crossing the ice at the Bering Strait. The herd was managed for a time by the Eskimos in the Point Barrow region with the supervision of government employees. Shortly, however, the herd was turned over to the Eskimos. The herd prospered, and soon there were half a million reindeer in Eskimo hands. Such large numbers of animals cannot be kept near a permanent village but must be moved about over wide areas in order to have adequate grazing. For a time all went well, and the Eskimos had an abundance of meat and hides. They even had more animals than they could actually use, but there was no market for the surplus.

The tending of the herds was in conflict with the Eskimo way of life. The Eskimos are specialists in sea hunting. They are not fishermen, for their major prey are the sea mammals, and these are hunted, not fished. Not only have these animals been their major source of food, but their whole way of life—including their value system—is built around these animals. Of greatest prestige value is the killing of whales. Further, the Barrow Eskimos live in a large village. Social life is extensive and greatly appreciated. Reindeer herding interfered with all of this. Whaling time comes in the spring, and the men then form crews and take boats to the openings that are forming in the sea ice. There they wait for whales to come along; they stay there for weeks, and the entire whaling season lasts for a month or two. So great is the prestige attached to whaling that the men often waste the best of the sealing season waiting in vain for a whale. Eskimos were reluctant to forego the excitement and pleasure of either whaling or village life to follow herds of reindeer around.

The result of the cultural pressures was that the Eskimos took less and less care of their herds as they increased in size and had to be cared for at greater and greater distances from their villages. Presently the herd began to dwindle through loss to wolves, loss from wandering away and joining with the caribou, and just plain loss through neglect. Today there are no domestic caribou in the hands of the Point Barrow Eskimos. There was abundant need, there was a perfect physical setting, and there was abundant know-how supplied. A start was made, and success seemed assured—then the whole enterprise failed because herding and a nomadic way of life did not fit into the Eskimo's scheme of things. It is not easy to launch new ideas.

We have here the clues as to why herding did not take root among the Eskimos in the past. Since the Eskimos held both sides of the Bering Strait, this re-

An Eskimo Woman Cutting Up a White Whale. *The problems of people in transition are enormous. If the Eskimos use modern boats, will they overfish their waters? Do they want to cling to their old ways of life? How can they do so if each new device they use draws them deeper into modern commercial life? It is really desirable for people to live on the Arctic margins of the world and pursue an Upper Paleolithic way of life? (Information, Canada)*

duced to a low level the possibility of the forest Indians learning of reindeer herding. Since the Eskimos were on most unfriendly terms with the forest Indians, this still further reduced the learning possibility.

It was the forest Indians that might well have taken up the herding way of life; they were more accustomed to moving about following game and the seasonal abundance of fish and berries. Curiously there seems to have been almost no attempt in modern times to get this northern forest people of America interested in reindeer herding, although they were already more dependent on the caribou than the Eskimos were. Through the action of such forces as these the practice of herding caribou in America failed to develop. Not only did the idea not spread directly, it was not even hit upon independently. Nowhere in North America were herd animals handled as domesticates. Only in South America, where the llama was domesticated, was anything of this sort tried.

The reindeer was not the only useful animal on the scene. The musk ox was formerly widespread. This animal is a small, oxlike and sheeplike creature that can stand the greatest cold without shelter and will feed itself in the tundra or Arctic forest. It is now limited to a few spots in the remote Arctic. This is not because this is where it lives best but because this is the only area where the absence of man lets it live. Recent attempts at raising these animals in captivity in New England show that they are easily tamed. They yield excellent meat and very fine wool. They require no expensive barns, no cutting and storing of winter feed. Moreover, they are a dual purpose animal whose products would probably be more acceptable in the world market for meat and textiles than reindeer-caribou products. However, these animals were not domesticated in the past, nor are they being used today as a means of bringing the vast northern forest-tundra areas into our modern economy.

MODERN USE OF NORTHERN FOREST LANDS

Use by Native Peoples

In the Old World the reindeer-herding way of life persisted into the present from the Lapps in the north of Finland clear across the north of Siberia to the vicinity of the Bering Strait. It has come to dominate the whole area, even including the treeless border of the Arctic Ocean, which in America is held by the sea-ice hunters. This way of life has been left little disturbed by the modern world. Except for such extractive activities as mining, lumbering (on the southern margins), and obtaining furs, we have little use for these northern lands.

In the New World the picture has been of similarly little change. This is still Indian North America. The earliest modern manner of utilizing it was by drawing the Indians into the fur trade. This was a reasonable approach, for the Indians were skilled trappers dispersed over the country, and valuable fur-bearing animals were there. These Indians were producing nothing else of any value to modern Western society, and they wanted the goods that that economy was producing. The fur trade that grew up under the French and later the British neatly linked the wants of both.

It has sometimes been assumed that the northern forests are swarming with fur-bearing animals. After all, they have for centuries supplied large quantities of valuable pelts to the world. But this is another caribou-in-migration type of question. The area from which the pelts are assembled is vast. Each Indian trapper operates a trap line that covers many square miles of territory, but the yield per acre is small.

The ability of a land to support animals, in the last analysis, is related to the amount of plant food that the land provides. The amount of plant life is dependent on the length of the growing season, the distribution and amount of water, and the amount of solar energy available to the plants. In nearly all of these factors the northern forest lands fall into the low-productivity category. Water is present, but for most of the year it is locked up as ice. The part of the year that is above freezing is short and even then undependable, for frost may occur early or late on the southern border of these regions or in any month of the year along their northern borders. Solar energy varies from none north of 67 degrees in winter, to weak sun distributed over a long period of the day. Summer sun at 60 degrees north has a noon elevation equivalent to a winter noon sun in mid-latitudes. It only manages to warm the land because of the extra hours that it shines. Thus the total amount of insolation that the plants can utilize is limited to the brief period of long, low sun in the middle of the brief warm period. The result is a low amount of vegetative growth.

This condition shows up in many ways. It takes 3 to 10 times as long to produce a saw log in the northern forests as it does in our southern forests. The dream of vast herds of caribou grazing on the tundra during the summer and wintering in the shelter of the northern forest in the winter gets a particularly rude shock. The winter browse is the reindeer moss, which grows 6 to 19 inches high under the trees. However, experience shows that once it is grazed down it may take 10, 20, or 30 years to regrow to the level that it will again support a herd—which means that there must be approximately 20 times as much land in reserve as is grazed in any one winter. This land relationship reduces the efficiency of the grazing industry enormously. Under good management in the American South an acre of pasture can carry one cow per year. In the northern forest it would likely require several acres to supply grazing for one caribou for one winter, and 20 years of recovery time would have to be allowed. Similarly for the fur industry: a muskrat marsh in the Chesapeake Bay area probably yields tenfold the pelts per acre that a comparable area in the northern forest does; there are probably more deer per acre in Pennsylvania than caribou per square mile in the northern forest.

A Cree Brings Furs to a Hudson's Bay Company Post at Moose Factory Island, James Bay. *Indians still harvest furs in the Arctic forest of America. These vast nearly unoccupied forest lands continue to yield furs, but man has thought of no new use for them.* (Information, Canada)

The production of furs in the northern forest, then, is not the result of the area's being overcrowded with fur-bearing animals, but our having few other uses for these lands. This leaves the northern forest lands of both the Old World and the New among the least changed parts of the world.

Use by Industrial Society

For our modern industrial society, with its supporting mechanized agriculture, these lands supply needed pulp wood and exotic furs, but we have few uses for them that require our placing any great number of people in them. We can put people there, if we choose to do so. This is shown by the mining developments.

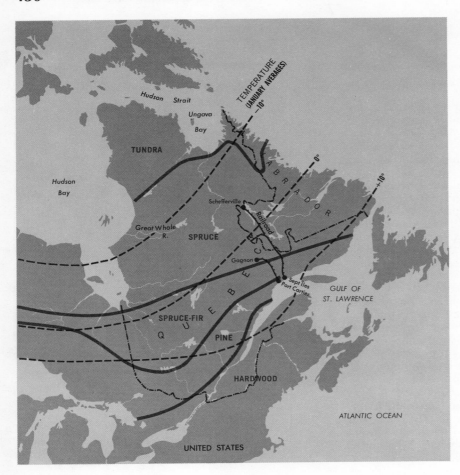

MAP 9-3 Vegetation and Temperature in Labrador

The vast empty wilderness of the Ungava peninsula is now being developed for its enormous iron resources. Development is in the provinces of Labrador and Quebec. The distribution of vegetation in the area as well as the January average temperatures indicate the increasingly Arctic conditions to the north. Canadian settlement has been in the hardwood zone with little occupation of the pine forest. The attraction of mineral wealth has now led to the extension of settlement into the spruce forest to well above the line for 0° in January. While this seems cold, many settled parts of the earth are colder in winter. The Labrador area, however, is unusually cool in summer for its latitude.

Mining

A spectacular example of the way in which man can move into and develop "impossible" areas is the iron rush into Canada. It has long been known that there were extensive iron-ore deposits in the Labrador-Quebec area. Few areas are as isolated, undeveloped and unpopulated with so rugged a climate. The great peninsula lying between Hudson Bay and the St. Lawrence has remained a virtual wasteland since the days of the European discovery. Almost the entire population of Quebec is concentrated along the St. Lawrence, and most of it along the Quebec to Ottawa sector. However, the view of this area has now begun to change.

Climatically this area is comparable to the Amur River region of the Soviet maritime provinces, where population densities are 2 to 25 per square mile and the sizable town of Komsomolsk is located north of 50 degrees. The Labrador-Quebec area is between 50 and 60 degrees north. The isotherm for 0° F. (the temperature line connecting points having 0° F.) in January runs through the center of the region, and most of the area is poleward of the 50° F. July isotherm. The isotherm for 0° F. in January also runs close to Winnipeg and Fairbanks, but Fairbanks is somewhat warmer, and Winnipeg is much warmer in summer. Only the northern part of this area is tundra; most of it is forest covered. Along the St. Lawrence the forest is made up of white, Norway, and jack pine. North of this is a mixed spruce and fir forest. Still farther north, spruce forest extends almost to Ungava Bay. The northern limits of the pine forest and a growing season of 90 days or more are approximately the same. These varied measures of climatic severity do not suggest that this is useless land, yet the French settlement in this area was limited to the strip of hardwoods that reached up the St. Lawrence. It is one of the oddities of the history of the spread of Western civilization that this land has been judged to be of such low utility, but our view is now changing.

The sources of iron ore in the United States have been dwindling. There is an oversupply in the world market, but no nation wants to be dependent on overseas movement of so critical a material as iron ore if it is possible to avoid it. Exploration has shown that the amount of Labrador-Quebec ores is enormous and the quality is very high. The main body of ore occupies a 50-mile-wide trough that extends north-south for 750 miles and reaches into Labrador. Once the decision was reached to develop these ores, the necessary steps were taken to raise funds and proceed with the required work.

For the original undertaking about 90 percent of the funds came from United States sources; eight United States steel companies pooled their capital to undertake what many people considered to be an impossible job. More than $300 million were spent in developing the first mines, and a 360-mile railroad was built from Sept-Îles to Schefferville. Until this project was begun, there was no such town as Schefferville: it was built to order to hold 5000 people, and it is complete with stores, churches, schools, hospitals, bowling alleys, and movies.

It had to be a whole town, for the kind of workers that are desired are permanent, settled, married family men. In isolated circumstances such as this, only this type of man can be expected to settle down and become part of a stable working force. Children, on the other hand, must have schools, and wives need some social life. The inclusion of churches completes the needs of any normal social group. The parallel is close to that of the Panama Canal. The difference is that the development in Labrador was designed to avoid the mistake of ever having a population of restless, quarrelsome, ever-changing single men for a working force.

In addition to this social problem, there were severe physical problems. What do you do about a water supply in a land where the average temperature for January is 0° F.? Either a deep source of water must be found, or surface water must be dammed up in reservoirs of such depth that they will be several times as deep as the thickest ice to be found. This is not as formidable as it sounds: 6 feet of ice is about the maximum expectation in severe continental climates. A reservoir with 60 feet of deep water is not difficult to

construct in any hilly terrain. In winter it is more of a problem to supply the water to buildings. Pipes must be deeply buried or heated, but in mid-winter the waste must eventually be dumped on some surface to freeze, for the great cold of the area stops all surface drainage. Freezing then stores up the enormous volume of wastes from the settlement until the summer sun melts the whole mass away.

There are also problems in house building. Where the ground freezes very deeply every winter and thaws every spring, buildings are heaved around by the expansion and contraction. Much of the northern forest area and most of the tundra area of the earth is perpetually frozen. This condition is called permafrost. Building on permafrost is particularly difficult; the building becomes a center of heat, and it then begins to melt its way into the frozen ground. This melting goes on unevenly, and presently the building has taken on a permanent tilt. But as for all such problems, there are possible solutions: the simplest is to insulate the ground from the effect of the heated building.

There are other problems. Automobiles are difficult to operate at very low temperatures, because oil turns into a solid below certain temperatures and metals become brittle. Ideally a car should not only have antifreeze but the motor should be kept warm even when the car is standing idle. Yet this is readily accomplished: electrical heating elements can be placed in the engine blocks and plugged into the electrical circuit at home or into a similar plug attached to each parking meter in town, with the cost of the electricity paid for as part of the parking fee. The point here is that although there are indeed many problems, there are also solutions. The particular solution used depends on the cost involved: the decision must be made in terms of whether the gain in occupying a given area is sufficient to justify the costs. So far it has been mining that has led to extensive modern settlement of the northern forest. Gold first, as in Alaska, but now a whole series of metals each of greater bulk and lower unit value than the preceding one have been developed as the costs of operating in this unusual environment have been reduced.

Environmental costs

Such costs are everywhere, not just in extreme environments. For instance, although cars in the far north may have to be heated during winter, we are finding that cars in most of our large urban centers are going to have to have costly fume controls attached to them. In the Los Angeles basin, where the smog situation has reached intolerable limits, the first attempt to

(Upper) **Wheat Growing in Arctic Forest Land at the Dominion Experimental Station, Canada.** *This area, at 58° N., located 100 miles north of the Peace River district, produces prize-winning grain. The past 50 years have been a period of increasing warmth in high latitudes, aiding in the growing of grain at such extreme latitudes. No one knows how long this climatic amelioration will continue.* (Information, Canada)

(Lower) **A Canadian Oil Well in the Midst of Wheat Lands, Edmonton.** *Discovery of extensive oil fields in the Canadian midwest has been an economic boon.* (Information, Canada)

control automobile exhaust fumes has been introduced. It is clear that most of our large urban centers are going to have to take similar measures. Because of the unusual structure of the lower atmosphere at Los Angeles, more costly measures are going to have to be taken there than elsewhere. Several million people believe that the costs are worthwhile to obtain the benefits of living in that area.

Similar analyses would show that each region bears its share of costs. Tropical regions bear the costs of heat, humidity, and associated accelerated rates of deterioration of most organic materials, and a higher cost of control of animal life, particularly those life forms that either attack man directly or indirectly. Mid-latitudes carry not a lesser burden but a double burden. Much of the eastern United States struggles with a super tropical summer with all of its costs and burdens, and then it has to shoulder the problems of living in a subarctic winter. Every fall when the last of the leaves has been raked up and dug out of the gutters on the roof, the storm windows put up, and the antifreeze put in the car, the Easterner must make the annual mental struggle of whether to bother about snow tires or just to depend on chains. This follows, of course, on a period when he wrestled with exhaust fans, discussed buying an air conditioner, and could not work comfortably at night because the light attracted bugs that somehow or another managed to get through any screening that man has yet devised that would not also cut off air circulation.

There are environmental costs everywhere. We tend to overlook those that we are familiar with and have come to take for granted. We tend to exaggerate those that are unfamiliar to us. While the costs of living in one area as compared with another are unequal, there seems to have been no concerted attempt to measure them realistically. It would not be easy to do so. Many of the costs are hidden, or indirect, or only partially environmental.

Future development in Canada

In the Labrador-Quebec wilderness today steam shovels dig the ore from the open pit mines in 8-ton bites, and ore trains run 24 hours a day eight months of the year. The entire operation is mechanized for the speedy handling of immense tonnages. Ore cars are dumped two at a time in 67 seconds. It takes less than three hours to load a 20,000-ton ore ship. Annual production from the Schefferville area has reached tens of millions of tons.

Another development at Gagnon, 200 miles to the south, represents an investment of $200 million. The ore at Gagnon is concentrated to get the higher grade produce demanded for modern smelters. Millions of tons of ore are now mined annually and reduced to highly concentrated iron ore at Gagnon. It is then shipped by rail 193 miles to Port Cartier on the Saint Lawrence River. Here a 50-foot deep harbor has been blasted out of the rocky shore line.

Changes in the techniques of steel making, together with a growing demand for steel and lessening demand for iron, make these expensive developments economically feasible. The concentrates are less bulky, and hence less costly to haul. When they are fed into furnaces, they increase the steel yield by 20 percent.

These two immense investments, which are changing the human geography of a piece of land twice as large as Texas, are apparently only a beginning. At Wabush Lake between Schefferville and Gagnon another giant project is underway. It repeats the previous ones: mines, concentrators, town site, railroad, and inevitably new ports and new ore carriers. A long chain of supporting industries throughout Canada and the United States from equipment makers to stockbrokers will be energized in the financing, building, maintaining, and profiting from this multi–hundred-million-dollar venture. Labor is pouring into the area because wages are high. The developments that are now all concentrated in the southern end of the iron trough will furnish the springboard for northward penetration of the area as these immense deposits are worked out. Geological exploration indicates that there are rich deposits along Ungava Bay and along Great Whale River on Hudson Bay. The deposits along Ungava Bay can be reached by short rail lines from ports along the Hudson Strait, but production will be hampered by a short shipping season due to ice conditions in the North Atlantic. The effect of the penetration of a former wilderness by three separate rail lines and the establishment of a permanent population that has been doubling and redoubling is difficult to forecast. Some of the effects, however, are already apparent.

Sept-Iles was an unimportant little town located on the shores of the Saint Lawrence 500 miles east of Montreal. In 1950 it had 1500 inhabitants; in 1960 it had 15,000—making it the fastest-growing city in the province of Quebec. All values go up with such a population growth: real estate values rose from $150,000 to over $56 million in this 10-year period. This region is now one of the major iron-producing and shipping areas of the world. Most of this will be shipped in giant ore carriers. It can move by sea to any world port or can go up the St. Lawrence Seaway to reach the immense inland market around the Great Lakes.

The rail lines could make possible the efficient tapping of the immense areas of forest for lumber and pulp. Since this area is almost entirely in the Canadian shield, a highly metamorphosed mass of ancient crystalline rock, further mineral concentrations are to be expected. The presence of a developed mining industry and a transportation network could make useful known deposits that until now have been too costly to work. The presence of a large industrial population should make it feasible in the southern part of this area to grow some farm produce, especially such crops as potatoes, for local consumption. A dairy industry to supply the local needs should also be profitable. The iron ore deposits may then serve as a means of opening up an area that has long been potentially useful but the development of which has lagged because it could not bear the cost of such developmental installations as railroads. The extensiveness of the ore deposits seems to assure at least a hundred years of mining activity in the area, and occupation and development of the area for that long are likely to lead to broader development. It seems unlikely that northern Quebec will ever revert to its former condition of virtually uninhabited wilderness.

An example of what may develop is found in the plans for the Churchill River in Labrador. The construction of dams and installation of generators there could make this one of the biggest hydroelectric complexes in North America, with power being sold as far away as New York. The Churchill River drops more than 1000 feet in 15 miles, and the sale of 2 million kilowatts of power to New York is being discussed. In northeastern Quebec near the Labrador border the Manikuagan River is also being developed. One of the highest concrete dams in the world, 700 feet high and 4000 feet long, is the key project in a five-dam project that will add 5.7 million kilowatts to the province's power supply. Man is expanding into an area long bypassed.

Transportation Problems

In the northern forest lands there are few roads or railroads; the cost of transportation is high. Where is cause and effect here? Is the northern forest a difficult area in which to build and maintain roads? Or is the absence of a good road network the result of the absence of population? Roads can be built: the Alcan Highway was pushed through to Alaska in amazingly short time. Despite a relatively low use, it is maintained through a vast and relatively empty territory which, because of the road, is gaining population. Much of the road follows well-drained divides, and the maintenance cost is relatively low; the major costs are met at river crossings where ice jams and floods make anything except a major bridge a temporary affair. The bridges can be built to stand, but the cost is high in proportion to the use now being made of the road. The use of the road will be held down as long as there are such weak links in the system.

In building the railroad to Fort Churchill on the Hudson Bay some years ago—in an attempt to provide a shorter route to the sea for central Canadian wheat—the problem of crossing swampy ground in summer was neatly solved. It was noted that there was no difficulty in winter, for the swampy ground was frozen like rock. The engineers therefore put a blanket of gravel over the muck sufficient to insulate it during the brief summer; this gave a ribbon of permafrost for the railroad to run on that was good throughout the year.

It has often been noted that the rivers in the northern forest regions run the wrong way: most of the large rivers run to the frozen sea. Taken with the innumerable lakes and ponds that interrupt the landscape, the waterways seem more of a nuisance to land travel than a help. This is a tropical or mid-latitude outlook on the landscape, for the time to travel in the northern forest is in the winter, not the summer.

The water areas in winter are frozen to great depth. They can then bear immense loads. The surrounding terrain is also covered with snow, and snow has the happy quality that it smoothes out the terrain somewhat and provides a cover offering slight resistance to friction—anyone who has been sledding knows these qualities of ice and snow. This then makes the northern forest easy terrain through which to transport goods. In actual practice tractor trains are made up to move goods through these territories; large caterpillar tractors pull long lines of cargo sleds. The frozen rivers, lakes, and streams function as highways; portages are easily made by taking to the snow. What to a tropical or mid-latitude people seems like a hopeless situation proves to be advantageous.

It is questionable whether a dense network of railroads will ever be built in the northern forest lands. If we can move heavy goods by tractor trailer in the winter, why go to the immense cost of building and maintaining railroads? Why build highways either? We have moved from a time dominated by the railroad to one dominated by the automobile and truck, and we seem quite surely to be headed for one where air travel will be dominant. Nowhere is this view given a more realistic flavor than in the northern forest areas.

At Fairbanks the outskirts of the city are ringed with airports, which are thickly parked with small airplanes. Highways are limited, but air travel is almost unlimited. In summer these light planes are put on pontoons, and they use the ever-present pond and stream as their landing fields. In winter they don skis

and can land wherever there is an open stretch of snow, thus opening up not only all lakes and streams as potential landing areas but any clearing. Since air travel is much faster and light planes can be bought for the price of an automobile and can operate on a mile for mile basis at lower rates, why bother with the huge costs of highway building and maintenance? Since air travel in the northern forest regions can take advantage of their much greater density of lakes to serve as natural landing fields in summer and winter, these regions actually have an advantage over other climatic areas.

Alaska as a Northern Forest Land

As with many political subdivisions Alaska does not fit neatly into the climatic divisions used in this book. The Panhandle is a typical marine climate and is in every way physically like the coast of Europe from southern Norway to Scotland. The mainland of Alaska is like the northern forested parts of Canada and the Soviet Union except for the polar fringe, which is a treeless tundra. The major center of development in the northern forested area is about Fairbanks in the Yukon Valley. For our purposes this part of Alaska can be discussed as a northern forest land. In some of the discussion, however, Alaska will be referred to as a whole, for its political unity means that developments in one area influence developments in other parts.

Alaska is as big as Finland, Norway, Sweden, Great Britain, Ireland, Denmark, and Switzerland combined, and these northern and mountainous lands have been chosen to increase the comparison. One could also use four Californias or two Texases to achieve an Alaska-size unit. Alaska divides into three major parts: the Arctic plain north of the Brooks Range, the Yukon Valley between the Brooks Range and the Alaskan Range, and the coastal areas of the Aleutians and the Panhandle. These three divisions are treeless polar land, northern forest land, and marine west-coast land, respectively.

The Yukon lowland between the two great mountain ranges is large and complex, and it has minor mountain ranges within it. The area is 250 miles wide on its north-south axis at its narrowest, 500 miles wide at the Bering Sea edge, and 750 miles along its northeast-southwest trending axis. The center of this vast region is at Fairbanks on the Tanana River, a tributary of the Yukon. Away from the Bering Sea one finds typical continental interior extremes of temperature ranging from −78° F. to 100° F. The summer heat that usually startles those unacquainted with northern interior lands is due to the long days of 20 or more hours of sunshine. This climate is like that of the strip of forested land reaching from the Urals to Lake Baikal or like Finland. In these regions it is the summer warmth that is important rather than the degree of winter cold, for if the summers are warm, trees grow and crops are possible if the frost-free season is long enough. In the Fairbanks area the frost-free season is 100 days, barely enough for some limited agriculture.

Alaska was occupied aboriginally by several separate cultures. The sea-oriented people of the Panhandle with cultural resemblances to other people around the Pacific have already been discussed. The Aleutian Islands were occupied by a people who are culturally related to the Eskimos and who occupied the continental margin from the Bering Sea along the Arctic Ocean to the Atlantic. In the interior lived northern forest-dwelling hunting people. In Alaska and adjacent Canada these were Athabascan-speaking people; some of their southern Canadian relatives broke off and wandered down through the Great Plains to form the Navaho and Apache of the American Southwest, where many took up agriculture, weaving, pottery making, and pastoralism.

In Canada and adjacent Alaska the way of life of these people was that of hunters and fishers of the forested regions. Deer, caribou, and moose were their principal game, with waterfowl and salmon in season and lake and stream fish throughout the year. Despite their living in a continental interior situation where they were exposed to maximum cold, they were not warmly dressed or housed—not nearly as well equipped as the Eskimos, for example. They had mocassins with leggings, leather shirts, and fur robes, but these were not as well made as the Eskimo garments. They lived in tents through the year rather than in snug underground houses or even comfortable snow houses. They did not see the caribou as a domesticable animal and so remained simple hunters in a landscape that could have supported reindeer nomadism. In modern times they have been drawn into harvesting the northern forest for furs. They are an excellent example of man's limited perception of the possibilities of an environment and the fact that sheer need often fails to spur inventions.

The Russian period in Alaska is also a source of insights into man's perception of his environment. The Russians have a centuries-long history of expansion across Siberia in the pursuit of furs. This eventually led them to the Pacific and the discovery of the rich sea-otter resources of the Aleutians, which in turn led to their establishment of bases along the coastal areas and to their extension on down the coast into California, where they provoked the Spanish settlement. Their perception of the environment was heavily concentrated on furs, especially the sea otter, so they were strongly maritime-oriented. The result was that they learned extremely little about the resources of the land, especially of the interior. Distance and transportation were important factors; they were more than halfway around the world from Moscow when they were in Alaska, and the connection with home across the North Pacific and then by land across the expanse of Siberia was a tenuous one. As California had been to the Spaniards, so Alaska was to the Russians: a land too far away, with too few visible resources for too few men to develop. So the United States bought this land, three times as big as Scandinavia, for 2 cents per acre over stormy protest that we were throwing good money away. As in California, gold was soon found. The rush of miners led to exploration and settlement, and the course of history was changed.

The pace of change has been leisurely up to now. Why indeed should men go to this faraway icebox when free land, growing industry, and vast opportunity were available closer to home in more familiar climates? Developments such as harvesting the salmon, with a great seasonal movement of men and equipment to do the work, and with a prompt outflow of the men and their wages and their product, brought little lasting benefit to Alaska. The Alaskans long demanded statehood and an end to this colonial exploitation. When they obtained statehood in 1959, they found that the problem of development was a huge one; a review of their problems will show that these are little different from those of any other underdeveloped country, whatever the climate.

Alaska has a tiny population. There are only about 337,000 people in this immense land. This is about 6 people per square mile; the United States has 57 per square mile, Great Britain has 598! Or compare Finland's 4.6 million—36 per square mile—in a smaller area of comparable climate. For so few people the burden of costs of development are huge. There is, to begin with, an inadequate transportation system: the Alcan highway through Canada is mostly unpaved; there are two short (500 miles total) railroad lines and a little over 6000 miles of roads in the highway system. This would describe an underdeveloped area anywhere.

Alaska is on the over-the-Pole route from Europe to the Orient, and Anchorage and Fairbanks are getting an increasing flow of transient passengers. Alaska would like to tap this flow to add to its already large tourist business. One does wonder about the ability of the tourist business to fuel growth and development in as many countries and regions as are utilizing it, but growing productiveness does make more people able to travel. Air transport makes distant parts of the world more accessible than ever before in human history, and mankind's curiosity about other lands and peoples is insatiable. Alaska has lots to offer: magnificent scenery, wonderful fishing, superlative hunting, great ski potential, and (while they remain unchanged) exotic people such as the Eskimos.

Costs are high, for nearly everything, including 90 percent of the food, is shipped in. Housing costs about twice as much as in Seattle. Using Seattle for a cost-of-living comparison, the effect of improving transportation is shown in a steady decline in the excess of cost in Alaska: 1948, 45 percent; 1958, 37 percent; 1962, 30 percent. The petroleum development in the 1970s, which increased the volume of goods moved, will surely continue this trend of decreasing the differential in costs. These costs are balanced by high wages, for the Alaskans have one of the highest per capita income in the United States.

These factors are worth studying to see how much is physical geography, how much is economics, and how much is a developmental stage. Is it pure physical-geographic distance that makes food so costly? Why should food be expensive in a land that has one of the world's richest fisheries, an immense wildlife reservoir (a potential food supply in the reindeer and moose and musk ox, if viewed similarly to that suggested for the elephant, giraffe, hippopotamus of Africa), and as good a dairy potential as such countries as Finland? A combination of factors is at work here. The fish, salmon especially, is largely exported. The wild meat animals are seen as game, not as cattle. Other ways of making a living are seen as more attractive than dairying. Yet dairying is more costly at Fairbanks, where cows must be housed and feed stockpiled in the brief growing season, than it is on the mild wet coast of Washington near Seattle, where there is a long growing season. Further, labor and capital are scarcer and more costly in Alaska than in Washington. Thus although dairy products must cost more in Alaska, the costs are set partly by labor and capital and partly by climate.

That the low population and developmental stage of Alaska is probably more important than the climatic factor can be seen by comparing it with the Scandinavian countries or Finland, where dairy products are not disproportionately expensive. Here too the cultural factor enters in, for the Scandinavian cultures share the Germanic small-grain and cattle tradition. In those lands dairying is a way of life. In Alaska there is no such tradition, and the flood of incoming people not only include few with such an outlook, but even fewer with the capital to buy and clear and fertilize land, build huge barns, accumulate a dairy herd, and find the farm labor to operate the enterprise. Because the construction and mining business pay laborers huge salaries, and labor is in short supply, there is little incentive to enter a business as demanding as dairying. But again the costs are only partly climatic, and even the labor costs are related more to stage of development, politics, and history than they are to pure physical geography. Had Alaska's history been like that of Finland, the labor supply would be different. Were Alaska still part of the Soviet Union, development would be in a quite different direction. As an indication of the effects of even minor changes in politics, the change in Alaska since statehood can be reviewed.

Perhaps the major change has been in the world of ideas. Many Americans now think of Alaska as offering great opportunities and go there fired with enthusiasm. Between 1950 and 1960 the population of Alaska grew 75 percent—the fastest of any state. With the oil boom, continued rapid growth seems assured. Granted, percentage figures are deceptive, but this is still a striking measure of change. The demand is for engineers, technicians, nurses, and people with business experience and capital. The enthusiasm of the Alaskans at having their affairs in their own hands instead of being run by the Federal government has been chilled by the realization that they have to assume costs of government. A small population in an immense area has a huge financial problem in supporting and expanding roads and public services. Schools, airports, and harbors are very costly; hence income taxes and gasoline taxes are quite high in Alaska. The state is aided by Federal programs and the military establishments in the state. Alaskans, however, look to natural resources, especially minerals, to supply the funds for development.

The Alaskans were certain that there was wealth in their land and set out to find it. The most successful finds have been minerals, and the other big development has been in forestry. Development of petroleum is providing a much needed source of revenue, and the Alaskan budget is being balanced by royalties and other fees from oil. The role of such resources for developing areas has been a repeating theme: compare oil in Venezuela or copper in Chile. The early developments were on the Kenai peninsula and Cook Inlet, but exploration is going on over large areas including the tundra areas east of Point Barrow on the Arctic coastal plain. A measure of the value of oil and gas to individual enterprises is the $3000 per year saving in power by one dairy in Anchorage.

Japan is investing broadly in Alaska—$70 million in one pulp and lumber company, $4 million in another—and is involved in joint financing of a project to supply the Tokyo Gas Company with up to 230,000 tons of liquified methane gas from Alaska. This will require an investment of several hundred million dollars in ships, machinery, and other development. The Alaskans comment that Japan needs everything Alaska has: fish, timber, pulp, and minerals. One oil company has discovered an estimated billion tons of iron ore 200 miles southeast of Anchorage. Copper development is under way at Kabuk, north of the Arctic circle but still in the Yukon basin. Such developments lead Alaskans to talk of a potential $5 billion per year annual mineral production.

In 1968 the decades-long exploration for oil on the North Slope of Alaska paid off with immense strikes. Initially there were estimates of 5 billion barrels, and these were soon increased to estimates of 10 billion barrels of reserves. This would make this new field twice as rich as the fabulous east Texas fields found in 1920. Alaskans were jubilant, for despite the location of this find on the remote, barren tundra of the North Slope, they knew that the oil supplies of the other 49 states were past their peak and there would be demand for this oil although it would be difficult to transport to market. To those who doubted that oil could be piped from this remote frozen area, over mountain ranges, to an ice-free seaport, it was immediately pointed out that the Soviets were already building a pipeline 3000 miles long to carry oil from their Arctic regions to their refineries. The proposals have sparked a flurry of actions.

A tanker was specially fitted out and sailed from the East coast of the United States through the Arctic ice north of Canada to the vicinity of Prudhoe Bay, the location of the first oil strike. The round trip was accomplished, but only with the aid of icebreakers and with damage to the ship. Future tankers for this service could be built to do the job, though it is questionable if this is the best solution. In addition there are political problems. The tanker must traverse Canadian territorial waters and the Canadians are not enthusiastic about huge tankers punching their way through ice floes with all the attendant risks of oil spill. Nor is a welcome mat out along the coast of Maine, where the deep-water ports to receive such cargoes invite the location of refineries from which to distribute the oil products to the eastern United States. We are at a moment of emotional fervor over ecological risks.

Alaska's stake in the development is enormous. It is counted on to bring in the revenue necessary to finance the initial development of the state and to lead to the growth of population and industries that will provide continuing growth. A petroleum development, now that the immense oil-potential area is known, can move rapidly while other developments may take years. The initial effect of the oil finds was to greatly enrich the state through leasing of lands for drilling. In the first burst of leasing the state received $900 million. This was budgeted for relatively rapid expenditure in the expectation that the production of oil would start relatively rapidly and that further income would be forthcoming. Instead the whole project was stalled by environmental questions to the dismay of the Alaskans.

The current developments at Prudhoe Bay where the enormous oil strikes have been made illustrate the problems of modern development in the Arctic. Costs of development are high. A well on the Alaskan North Slope costs from $1 to $4 million compared to $75,000 for a land well in the "Lower 48" or $500,000 for an offshore well in the Gulf of Mexico. The cost difference is largely due to transportation. It costs more than $500,000 simply to set up a drilling camp in the Arctic. Barge traffic must move through the Bering Sea, through Bering Strait, around Point Barrow, and into Prudhoe Bay through the Beaufort Sea. While this route is relatively cheap, it can only be used during a brief 6- to 8-week period when the shore ice moves out in summer. Some years the winds do not drive the ice offshore, and then this route becomes useless. Further, the shore is extremely shallow here—water depths of only 10 to 20 feet are found up to 3 miles offshore. Barges can move materials down the Mackenzie River from the Great Slave Lake, a 10-day 1800-mile trip. Materials can be flown in from Fairbanks at $160 per ton or brought in by caterpillar train or truck. The latter two are slow and more costly than air freight, but there are limits to what can be fitted into an airplane.

The truck and tractor traffic damages the tundra and starts gullies, so the oil companies are required to lay gravel strips up to 8 feet thick under roads, buildings, and landing strips. These insulate the frozen subsurface and prevent erosion. The difficulty with the permanently frozen ground is that it is often 60 to 80 percent ice, and melting leads to catastrophic sinking of the land surface. The proposed pipeline across this permafrost will not only be costly but poses myriad problems of expansion and contraction, thawing, and heaving with freezing. The oil is there, and it is needed by the rest of the United States. In one way or another, by one route or another, it will be brought out of the Arctic despite the difficulties and the great costs. Hordes of workmen will enter the Arctic in the boom period but not many will stay. The Eskimos will gain new skills and a taste for television and other exotic products.

Predictably some of them will find permanent work in the oil fields, but even more will use their new-found skills to satisfy their recently acquired tastes and then move out of their Arctic homeland.

The impact of a development of this kind is measured by the situation in Fairbanks, the center of the Alaskan settlement. New people flooded there in 1967 at such a rate that schools increased their enrollment by 20 percent, when 5 percent growth was expected. Housing shortages were immediate. In expectation of continuing development such towns as Anchorage saw construction of an inn 15 stories high with 200 rooms, a 22-story annex built on another hotel, and another 200-room inn. The Alaska Pipeline Service Company invested more than $50 million in pipe and equipment to start the pipeline from Prudhoe Bay to Valdez on the Pacific. This project would employ 7000 men for three years, an enormous boost to the Alaskan economy. The ecological controversy centers on the effect of a pipeline on wildlife and permafrost. Questions are asked about whether caribou will cross the pipeline or whether it will restrict their migrations. In regard to permafrost, the question is whether the pipeline can be insulated to keep it from melting into the permafrost. There is also worry about pipe ruptures and oil spills. The ecologists view Alaska as the one unspoiled area where there is a chance to avoid the mistakes made in the other states. The Alaskans are beginning to feel that the ecologists from the Lower 48 are stopping all development in their state, which desperately needs development, and that they themselves are taking sufficient measures to safeguard their environment. It is a classic confrontation of varied choices: development versus open space, pollution risks versus needed resources, local demands versus national needs, one vocal minority versus another.

Current power needs are being met by gas and oil discoveries, and further power is available from hydroelectric developments. The most ambitious project is the proposed Rampart Dam on the Yukon 100 miles northwest of Fairbanks. The proposed dam would cost $1.3 billion and create a lake 400 miles long and up to 80 miles wide. It would be larger than Lake Erie and the twelfth largest lake in the world. It is estimated that it might raise the temperature of the Fairbanks region by two to three degrees and add six frost-free days to the growing season. Climatic change would be a side effect, for the primary goal is cheap power, now about twice as expensive in Alaska as in the other states. Power at Rampart Dam would cost about 3 mills, one-twentieth the present cost. This would be very attractive to industry, and in a land where agriculture is a marginal activity, it is industry that is thought to be the

hope for the future. There are other power sites of less magnificence, but also less cost, that could produce power almost as cheaply. The problem is not in the potential but in the development. There are similar visionary plans for immense dams in South America. In all of these cases, the huge projects have enormous prospects for good but also each has immense ecological impact, and the debate today focuses on cost, benefit and impact balances.

This far-north agriculture, as now conceived, is marginal. Land is available for homesteading, virtually free for the development. There is no land rush, for it takes money and hard work to clear the land and bring it into production. Then there is the problem of a 100-day growing season compared to the United States–Canadian border area, which generally has 120 days, and the rest of the United States outside high mountain areas which has many more. This short season will probably make it uneconomic to grow anything there that can be shipped in reasonably. Yet what can be grown is epitomized by the passion in the Fairbanks area for growing tomatoes. These are started in boxes placed in sunny windows or in greenhouses, and set out only after the frost danger is past. People experiment with putting aluminum sheets on the north side of the plants to focus the solar energy. In the very long (20-hour) day of mid-summer, boosted by reflectors, a plant can do a lot of growing. Given a few 100-degree days, even tomatoes will flourish. However, tomatoes flourish more cheaply and easily in California.

Alaska, then, reveals again the multiple possibilities within an environment and some of the forces that determine men's perceptions of these forces. To the Indians the Yukon Valley was a fine hunting area for moose and caribou, with rivers good for salmon in season and trout the year round. The potential of the area for reindeer nomadism was not perceived. The Russians focused on furs, a high-value, low-bulk product that could be shipped large distances; they did little to explore other possibilities. The Americans found gold and peopled the land to a small degree, but they left it as a marginal colonial area until recently. Belatedly Alaska achieved statehood, and energetic exploration for minerals, utilization of forests, development of hydroelectric power, and discovery of oil suggest that this area may develop some industry. If it does, it will be reversing the worldwide trend of mankind to move out of these marginal areas. Perhaps the Alaskas of the world will be the refuge of those who consider the maximum density of mankind per square mile as the death of what is good in life. Such a choice would mean that some people would perceive these thinly occupied lands as offering a way of life that densely occupied lands do not. They could choose northern lands, mountain lands, desert lands, and many a space in the humid lands as an escape from the termitelike urban way of life.

The Soviet Union

The Soviet Union is a huge block of land that falls climatically into several categories, and to this extent it fails to fit the plan of this book, that is, comparing what a uniform man has done on uniform (especially in climate) parts of the earth. However, it is by no means the only country that does this, for, as we have seen, the United States, Chile, and Australia also occupy several climatic zones. We have varied our study a bit by including a topographic type, the mountain lands, and a vegetative type, the grasslands. The Soviet Union is a political type. Such an approach has been used before in this text for smaller units—for instance, Puerto Rico and Cuba—but in the Soviet Union we have the opportunity to examine the working of a political system on a grand scale and compare the developments in old Russia with those of the Soviet Union. With the country so prominent in political affairs, such comparisons are well worth making.

The Soviet Union covers a total area of 8,600,000 square miles—about one-sixth of the land area of the earth and almost two and a half times the size of the United States. The Soviet Union has long had this colossal amount of territory, though with some variation through time, and this vast area, centrally located as it is, has led to much theorizing about its potential as a world power. As was mentioned in the Introduction, Sir Halford Mackinder, in *The Pivot of History,* thought that control of the heartland of Eurasia, far and away the greatest land mass and also possessing the most people, was the single most important fact in political geography.

Mere area, however, is not enough to ensure greatness. The United States and Brazil are of about equal size, and Brazil has more useful land. Japan, a minuscule nation, is now number three in industrial output, and in times past such modest-sized nations as Great Britain or Spain have been among the great world powers.

While the land of the Soviet Union is vast, it is limited in its use, like that of the United States. The European part of the Soviet Union is a vast plain extending from the Baltic Sea to the Ural Mountains and from the Black Sea to the Barents Sea. In this huge area the major drainage is south by way of the Dnieper, Don, and Volga rivers. Beyond the Urals lies the Asiatic part of the Soviet Union, with plains extending from the Aral Sea in the south in a great northeastward sweep, and with mountains beginning near Lake Balkhash and extending eastward and northward. A huge amount of the Soviet Union, then, is low-lying relatively flat land. The great plains in Siberia are drained to the north by some of the major rivers of the earth: the Ob, Yenisei, and Lena. All three of these rivers are about the length of the Mississippi proper and the Ob-Irtysh combination is about the same length as the Mississippi-Missouri combination. We hear little about these rivers because they are located in a thinly populated part of the world and they drain into the Arctic Sea. Their role is much like that of the Mackenzie in Canada, one of the largest rivers in the world but of limited interest because it is remote and it empties into the frozen Arctic Ocean.

The Soviet Union lies far to the north; most of the land is between 50 and 70 degrees. Like Canada, a small part of the land lies well to the south, but this southward-projecting territory is in the arid region to the east of the Caspian Sea. It is so deep in the interior of the continent and so hemmed in by mountain barriers that it has both the great aridity and the seasonal extremes of temperature to be expected in such a location.

The vegetation, that handy summarizer of many climatic factors, characterizes the Soviet landscapes. In the far north along the Arctic Sea is a belt of tundra—treeless, long-wintered, almost summerless land. South of this is an immense belt of taiga, the northern coniferous forest of long, hard winters and brief summers. Leningrad lies on the southern edge of this forest in the west at 60 degrees north. The forest broadens by extending southward another 10 degrees on the eastern side of Siberia. To the south in the European region lies a belt of temperate mixed forest: conifers together with deciduous hardwood trees such as oak. Farther south there is a narrow strip of temperate deciduous forest somewhat similar to the forests of western Europe. These forests containing deciduous hardwoods measure the extent of moderate—for the Soviet Union—climate with adequate rainfall. They are limited to the southeast by decreasing rainfall, and the area becomes wedge shaped with its narrow tip reaching the Urals. South of the forests are the steppes and deserts of Inner Asia. In the maritime provinces of the easternmost part of the Soviet Union along the Pacific there are again modest-sized areas of temperate forest. Little of the Soviet Union can be said to have a temperate climate. It is a land of deep continental position in far northern latitudes, and this location gives rise to what

may better be called intemperate climates: ferocious winters followed by brief, hot summers.

The Soviet population is heavily concentrated in the west in the European region. On a map of Eurasia it is clear that there are two great clusters of population on the continent: one is in Europe, and the other is in the India-China area. The Soviet population is an eastward extension of the European population, and it forms a triangle pointing into central Asia along the zones of temperate forest lands. As far as the Urals the dense populations are continuous; beyond the Urals as far as Manchuria they are only spots of heavy occupation. In Manchuria, where the Chinese realm is met, large densities of men on the land are again found.

It is clear, then, that huge as the Soviet Union is, the land is not one of unlimited potential, especially for agriculture as we now know it. With Europe drawing its agriculture from the subtropical lands, and Asia drawing its agricultural stimulus in part from tropical areas, these far-northern lands were of limited use for agricultural men. Only the southern parts of the Soviet Union are good wheat lands, notably the Ukraine. Hardier grains such as rye do better to the north. The American potato has in more recent time become important in the cold and short-seasoned lands. The flow of ideas eastward from Europe expanded the cattle and small-grain keeping economy of northern Europe to the east of the familiar deciduous and mixed forest belts; the Soviet Union was the frontier, both in distance from centers of origins and for adaptation of plants, animals, and ways of life to northern forest lands. On the Asiatic side the situation was much the same. The Chinese expansion out of northern China was toward the warmer and better-watered lands of south China. Only in this century have the Chinese moved into Manchuria, displacing the pastoral economy that had dominated that area for millennia. For a long time the response of man to the problem of the extreme climates and dry lands of Inner Asia was pastoral nomadism—the moving of

herds and flocks across the face of the land following the seasons, the rainfall, and the natural harvest of grasses. To the north the people of the taiga mimicked this way of life but had to substitute the reindeer for the more familiar domestic animals. It was from this background of far-northern location and great cultural marginality that the Soviet Union of today came into being.

Just how great a handicap this marginal position was for the Russians is illustrated by the fact that in A.D. 862 they invited the Vikings to come be their kings and unify them. They were then primitive people, ignorant of ships and shipping, and knew nothing of such basic skills as iron casting and building in stone and brick.

The orientation of Russia toward the outer world has shifted through time. Early trade routes from the Baltic led south via the major rivers to the Black Sea where Greek and Byzantine merchants were met. Trade and access to ideas led to the development of a capital at Kiev on the Dnieper river at the steppe-forest margin, where the mingling of Nordic and Mediterranean cultures gave rise to a major cultural advance in the tenth and eleventh centuries. The Mongol invasions of the thirteenth century led to the sacking of Kiev and the driving of the Russians out of the steppes and back into the forests. The capital then shifted to Novgorod, and trade was oriented to the west. On the west, however, the expanding Germanic power under the Teutonic Order threatened Russians in the south, and the Swedish empire, then in a bellicose state, threatened them to the north; the Russians were forced back into their forested interior, where Moscow became the capital. The intertwining of location in relation to people and ideas, the impact of historic events, the rise and fall of people, and the changing attitudes of peoples are neatly illustrated in this thumbnail sketch of the shifting centers of Russian culture. The isolation of Russia that marks it at the dawn of modern history

Three Faces of the Soviet Union. *The north is a vast and largely empty land of long winters and intense cold. This is the Malozemelskaya tundra (opposite, above) in the Nenets National area.* (A. Kokhanov for Novesti Press Agency) *The central region is milder in climate, and to the south, verging on the arid regions, lies the wheat belt. Here we see the Kazakh region (opposite, lower left) east of the Urals and near Lake Balkhash. The huge fields and the gangs of heavy equipment are typical of Soviet collectivized farming.* (I. Budnevich for Novesti Press Agency) *The southern part of the country on the Black Sea is a sunbathing area (opposite, lower right), comparable to beaches on the Great Lakes in the United States.* (M. Alpert for Novesti Press Agency)

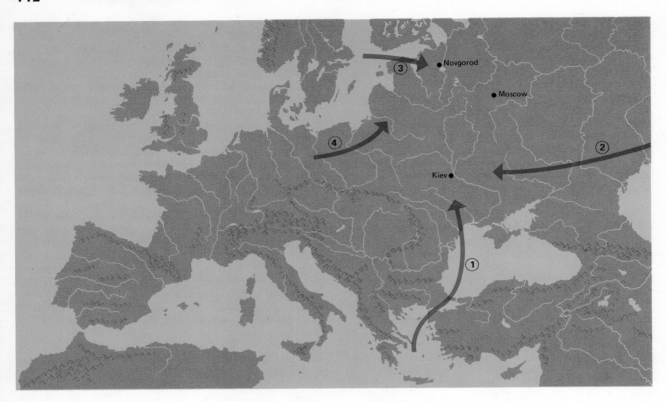

MAP 9-4   Cultural Thrusts and the Location of Russian Capitals
*Mediterranean influences on Russia led to the establishment of a capital at Kiev (1) by the
tenth century. In the thirteenth century Mongol invasions (2) drove the Russians back into the
forests. The capital was reestablished at Novgorod. During this period trade and cultural
influences came from the West. Swedish (3) and German (4) expansion (Teutonic Knights)
drove the Russians further east, and by the fifteenth century they had once again shifted their
capital, this time to Moscow.*

is something imposed upon the country by Mongol
horseman, Swedish armies, and Teutonic Knights.
Russians had to relocate, readjust, and start over again
to establish contact with the outer world and assimilate
once again ideas from the West.

There is nothing particularly new in this development. The Japanese at various times have imported
ideas wholesale from China and today from all over
the world. It is not too long ago that the United States
was borrowing ideas, especially from Great Britain;
some of the technology in the spinning and weaving
business was purloined and brought to the United
States. Further back in time the British were on the
fringes of the world of growing knowledge and absorbing the flow of ideas coming from the Mediterranean

powerhouse. And so it was with the Russians. A slow
development began after the thirteenth-century Mongol
invasions that nearly destroyed the earlier beginnings
and forced the Russians to pay tribute to the Mongols
for 240 years. In 1472 Ivan III ruled from Moscow, then
a wooden village in the forests; his court was virtually
illiterate. Ivan the Terrible began the opening of the
doors to the West, and at the end of the seventeenth
century Peter the Great swung them wide and brought
in as many ideas as he could. Under Catherine the
Great an Imperial Russian Academy of Science was
organized, and it contained no Russians. It was a neat
measure of the Russian determination to import brain
power, to westernize, and to modernize the country
as was seen fit.

Russians learned about trade and transportation from the West. Mining was a skill that they learned from Greeks, British, and Swedes. The inflow of technology begun under Ivan the Terrible has continued under the various Soviet governments through the twentieth century. The Czars struggled with the ideas of political change that came from the West. Although feudalism was long gone in the West, serfdom lingered on in Russia. Finally Alexander II freed the serfs with provisions for them to purchase land; the agreed price for the land was paid to the landed proprietor at once, and the peasant repaid the government on long-term loans at 5 percent. If the whole process sounds very much like that in the developing nations today, it is for the very good reason that Russia was then just that: a developing nation that was belatedly emerging from feudalism.

Again, as with so many of the developing nations today, the progress was dishearteningly slow. Some Czars made forward steps, but others were complete autocrats and stopped the reform movements. Autocratic rule slowed growth and development, and wars hindered progress. Against this difficult background came World War I, the Bolshevik revolution, and the Communist takeover. It is hard to state the case properly in a few pages. Russia had been struggling for centuries to emerge from its backward state, and it had made far more progress than it is often credited with. The revolution set off an immense and long-lasting internal struggle that was very destructive. The Communist regime, which finally took full control, now makes huge claims for its success. It tends to credit the Communists with miracles of growth, as if nothing had been accomplished by the pre-Communist regime. We need to assess the situation to determine what is actually the case here.

The Communist dream is a utopian world where workers will be rewarded according to their needs. Coercive, repressive government such as that in the Soviet Union today is expected to fade away as the system is perfected and each citizen is educated to serve the state willingly, not for his own aggrandizement but for the good of the state. Admittedly this is a goal to be reached; for the present the system is a socialistic one rather than a communistic one. Dictatorship is accepted, and is called the dictatorship of the proletariat (literally, of the working class). Actually, the proletariat is the Communist Party, which includes only about 10 percent of the population. It is ruled from the top with an iron hand as firm as that of any of the Czars. The future of the ideal communist state is continuously put off. In the 1960s it was supposed to arrive by 1980; in the mid-1970s, however, it seems no closer than it was decades ago.

In 1960 Soviet plans were programmed to exceed United States production by 150 percent by 1980; Soviet people were to have two and a half times as many consumer goods as United States citizens have. The record from 1950 to 1968 had been a two-thirds improvement—the Soviets were then getting one-third of the United States levels of consumer goods. There are many ways to measure such matters, and a few such indexes will be given here to show the relative positions of the United States and the Soviet Union in living standards.

Housing in the Soviet Union is far below American standards. In large cities people live three or four to a room. Bathrooms and kitchens are communal; several families share each one. The square footage allowed per person is 62, less than the United States government allowance for desk space in an office. In 1960 the Soviet goal was to increase the amount to 97 square feet in seven years. This would still be one-third of the United States average, and the actual level reached was only 70 square feet. In 1961 the occupancy per room was 2.72 people while in the United States the figure was 0.60. In Moscow 39 percent of the apartments have private baths, and one in 10 has hot water.

The daughter of an American steelworker, who had majored in Russian, went to the Soviet Union and was welcomed with open arms as a representative of the working class. The immediate bond turned into an immediate distrust when the Soviet people began asking her about her life in the United States. The American girl told them that her family owned their own home: a whole house, for just their family, with not just one but two bathrooms, and a whole kitchen for just their use. Only top men in the Communist Party or generals—certainly not a laborer in a steel mill—had anything like that. When the American girl added that they owned a car, a radio, a television set, a refrigerator, and a washer-dryer, her good standing as a member of the working class was totally destroyed. She had either to be a liar or a disguised representative of the capitalist class. The American girl for her part reported back home that for a woman to be just a homemaker was unthinkable in the Soviet Union; all women worked. Clothing was uniformly expensive and poor in quality and taste. Everywhere she saw what in America would be called vast poverty, no privacy for anyone, and deadening monotony for everyone. But observers generally report that the people are not discontented, for they are future oriented, believing that it is only proper to work hard now to bring the utopia that has been so long promised.

When the Soviet Union is looked at more closely to see just what it has in manpower, how it is using its resources of land and raw materials, and how its factories function, one begins to understand this picture of low standards of living. The country lost perhaps 20 million people in the war, which led to a real labor shortage. As a result most students must go to work at age 15 or 16. When they are working, however, they are only about one-half to one-third as efficient as American workers. This is largely because they have less capital input to aid them: less equipment, less electricity, and so forth.

Another measure of productivity is the use of auxiliary workers—maintenance men, loaders, and unloaders. The Soviet Union uses 85 such people for every basic production worker; the United States uses 38. Production workers in the Soviet Union are about 60 to 70 percent as efficient as in the United States, and their auxiliary workers only about 25 percent as efficient. To this must be added a high labor turnover of 20 to 30 percent per year; failure to use trained personnel (34 percent of all engineering positions in Soviet industry are staffed by university level engineers, but 22 percent of the trained engineers are in nontechnical employment); and an oversupply of trained personnel. For example, between 1965 and 1970, 3,500,000 people were trained as machine operators, but only 1 in 10 was so employed.

The distribution of the labor force is antique. Thirty-one percent of the labor force is in agriculture, while less than 5 percent of the United States labor force is so employed. In 1973 the average American farmer produced about $7,750 worth of foodstuffs per year, compared to $854 turned out by his Soviet counterpart. One American farmer's produce feeds 46 people, while the Soviet farmer feeds only 7. This again measures not only industrial development but the capital equipment and power available per worker. The American farmer has a whole array of machinery and tractors that enable one man to do the work of gangs of men working barehanded. The Soviet armed forces take up nearly 4 percent of the Soviet labor pool but less than 3 percent of the American pool. In the Soviet Union women make up 50 percent of the labor force; they do heavy labor such as ditch digging, paving streets, and shoveling snow. With the heavy drain on the available manpower by the agricultural sector of the economy, the Soviet Union, even with its greater population, ends up with less manpower available for industry, trade and service industries than the United States.

The ultimate test of economic systems is their productivity. For the Soviet system versus the United States it is socialism versus capitalism; bureaucratic planning and direction versus a free economy ruled by the economic forces of supply and demand regulated through a price system. In the Soviet Union there is no profit motive and no competition. The government decides what is to be produced, in what quantity, and in what quality. The order of priorities is military, industry, politics, consumer, and farmer, respectively. One might say that the peasants are still just peasants, the lowest on the line of national priorities, and that the workers are included with them.

The advantages of a planned economy can be seen in the kind of growth that the Soviets have been able to achieve. In the 1950s, starting from an abnormally low level, a growth rate of 9 percent per year was achieved in industry. Between 1960 and 1972 the Soviet Union increased its gross national product by 80 percent (the United States by 62 percent), yet the Soviet economy was still less than half as productive as that of the United States. Further, during 1966–1970 the Soviet growth rate was down to 5.5 percent, dropping to 3.5 percent in 1971, and settling at about 5 percent annual growth thereafter. The Soviets tried to promote economic growth by expanding the labor force. Between 1950 and 1970 this was done by drawing on womanpower: 60 percent of the additional workers were women. By 1970, 92 percent of the able-bodied population was either employed or studying. The corresponding United States figure was 76 percent. With a 20 percent larger population the Soviet Union had a 40 percent larger working force, but it lagged in production and growth. The Soviets are now importing labor: Bulgarians, North Koreans, and Finns. Growth rate for the gross national product was below 3.5 percent, even though consumer goods and the service industries were being starved to feed industrial growth. The Soviet Union has used its absolute power resolutely to build the military and industrial base, while holding consumers down to barely tolerable living conditions and promising them that things will be better in the future. The Soviet Union has maximized the power of the state to build heavy industry and the military at the expense of the consumer. The United States and the West, on the other hand, have used their power to also multiply individual choices among goods, occupations, and patterns of living.

A long-term student of the Soviet Union has assessed the situation as one of impressive growth in industrial output but at a cost of forcing students into the labor pool, employing women in huge numbers, and holding consumer goods down to low limits. One-third of the gross national product is ploughed back into industry versus less than 20 percent in the United States. The result is a rapidly growing industrial and

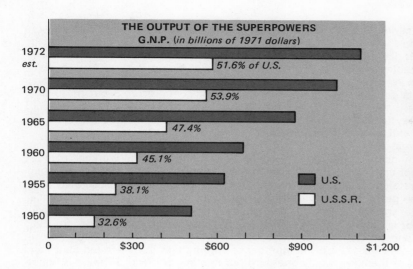

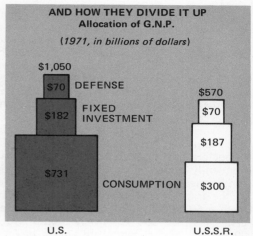

FIGURE 9-1   Productivity and Consumption: United States versus Soviet Union
*In terms of total output, the Soviet economy grew much faster than ours in the 1950s,
but since then the Soviet growth rate has dwindled. The absolute gap between the world's two
largest economies kept widening even in the 1950s. In 1972, partly because of the disastrous
wheat harvest, the Soviet economy grew only marginally—about 1.5 percent. With an economy
roughly half as big as that of the U.S., the U.S.S.R. spends about as much as the U.S. for
defense and for fixed investment (plant and equipment). The pinch, inescapably, comes in
consumption. Aggregate consumption in the Soviet Union amounts to only about 40 percent
of the U.S. level, or about one-third on a per capita basis. [Courtesy of Fortune Magazine
(February 1973)/Tom Cardamone]*

space technology alongside a standard of living for the people that is one-third that of the United States. One could well say that there are two economies in the Soviet Union: a strong and growing industrial economy zooming toward the twenty-first century, and a housing and consumer industry still struggling along in the nineteenth century. While growth figures for GNP in the United States, western Europe, and the Soviet Union are all about the same—just under 24 percent over a five-year period 1958–1963—the gross national product for the United States was $1.2 trillion, that for the Soviet Union well under $600 billion. As this review shows, there are huge discrepancies, and the figures show different things when analyzed in different ways. *U.S. News and World Report* (April 2, 1973), commenting on a Soviet claim of an overall growth of 4 percent in 1972, stated that by United States measures the figure would be closer to 1.5 percent.

The Soviets themselves tell jokes about how their system works. In one story about a nail factory and

its quotas, the planners chided the factory for not producing enough units, and they set a quota in numbers of nails to be produced. That year the factory turned out tacks in huge quantities. The planners from Moscow pointed out that they had managed to produce a 50-year supply of tacks, which had more than met their quota for quantity of units, but they had not produced the bulk of products so greatly needed; consequently the next year's quota was to be in tons. The manager then shifted to producing nothing but railroad spikes and turned out a huge quota that again far exceeded his requirements. Annoyed, the planners sent down a directive that quality was to be considered, so that year the factory turned out hand-wrought nails suitable for ornamenting doors for fancy homes. It produced enough to supply the country for several centuries, and in despair the planners sent the manager to Siberia. If there is a lesson, it may be to minimize the government controls. As this example shows, such controls tend to do more harm than good.

In the Soviet Union the quota is the all-important thing, for that is the basis on which industry and management are measured and the economy is planned and run. Every effort is made to get a soft quota, and all the kinds of lobbying and bribery known to capitalism are unleashed to influence the quota setters. Huge efforts are expended to obtain and hoard scarce materials—to the contrary of a cost-conscious system—for success is not measured in price or cost efficiency but in meeting the quota. Meeting the quota, then, leads to producing units at any cost and of any quality. Inspectors, appointed to see that quality is kept up, are frequently bribed. How efficient the whole system is can be measured by the following figures. Ten percent of the tractors coming off the assembly line are defective; 50 percent of all fuel jets are rejected; 60 percent of all TV sets fail in six months. At one plant 90 percent of the dental drills were below specifications though the inspectors passed 98 percent of them. This situation is forcing the Soviets toward capitalistic measures: cost accounting, limited competition, rewards tied to productivity. On the other hand, the United States seems to be moving toward centralized planning and a managed economy—the very things that the Soviets are demonstrating do not produce a worker's paradise.

Leonard Kantorovich, a brilliant Soviet mathematician, has estimated that the distortions and inefficiencies stemming from the Soviet Union's centralized planning, which replaces a normal price system, reduced total output by perhaps one-third. A fair example, which also illustrates a tendency toward large, showy projects, is the huge hydroelectric plant being built on the Angara River at Ust-Ilimsik in Siberia. There is no industry there, and no plants were under construction in 1973 in anticipation of 1974 plant completion. A huge investment was made, but no provision for utilization of this capital outlay accompanied it.

In 1971 the Soviets launched a new five-year plan. It still projected little increase in consumer goods. Andrei Sakharov, a well-known Soviet physicist, commented that they were catching up with the West in some of the old traditional industries such as steel. However, he found that they were falling behind in newer industries such as automation, computers, petrochemicals, and industrial research and development. Perhaps this realization is what has prompted the Soviets to turn to the West again. Fiat of Italy is building a huge auto plant, and Great Britain is building chemical plants for them. Japan is busy aiding them in harvesting timber. And the Soviets are borrowing capital from the capitalists. By 1970 they had obtained $1.5 billion and were asking for $2 to $3 billion more in credit. And still they lagged. In Great Britain or the United States it takes one to two years to build a chemical plant; in the Soviet Union it takes five to seven years. On a given day in Russia 6 out of 10 of the scarce automobiles are out of commission, 3 out of 10 farm tractors are down, 1 out of 4 construction machines are out of order, and 1 out of 3 machine tools are inoperable.

The Soviets have endured low standards of living for 50 years on the promise that the workers' paradise is just around the corner. Yet today they have only one-third of the United States railroad trackage to serve their much larger country. They have only 6 percent of our highways, and most of them are primitive dirt roads. They are not greatly inconvenienced, for even with their greater population they have only 22 percent of the number of automobiles in the United States and only one-third our number of trucks. While it is tempting to point to World War II as a primary cause, one has only to review the spectacular recovery of Japan or West Germany, countries that were equally devastated, to see that this is an insufficient answer. The Soviet system of a centrally directed planned economy simply is not competitive in world markets, and it cannot even supply its people a decent standard of living at home. The figures in Table 9-1 show how great these discrepancies are. Note that instead of dollars or rubles, time required to earn the item is used as the measure since this gives a more meaningful comparison.

TABLE 9-1   Production: U.S. vs U.S.S.R.

Consumer Goods	Time to Earn (hr.)	
	U.S.S.R.	U.S.
TV set	695	57
Small car	5716	720
Bar of soap	25 (min)	2 (min)
Refrigerator	343	32
Man's shoes	49	45
Diet (per capita consumption per year)		
Meat and fat	106	219
Eggs	159	309
Fish	34	14
Grain	326	142
Potatoes	280	106

U.S. diet is high in proteins; U.S.S.R. diet is high in starches.

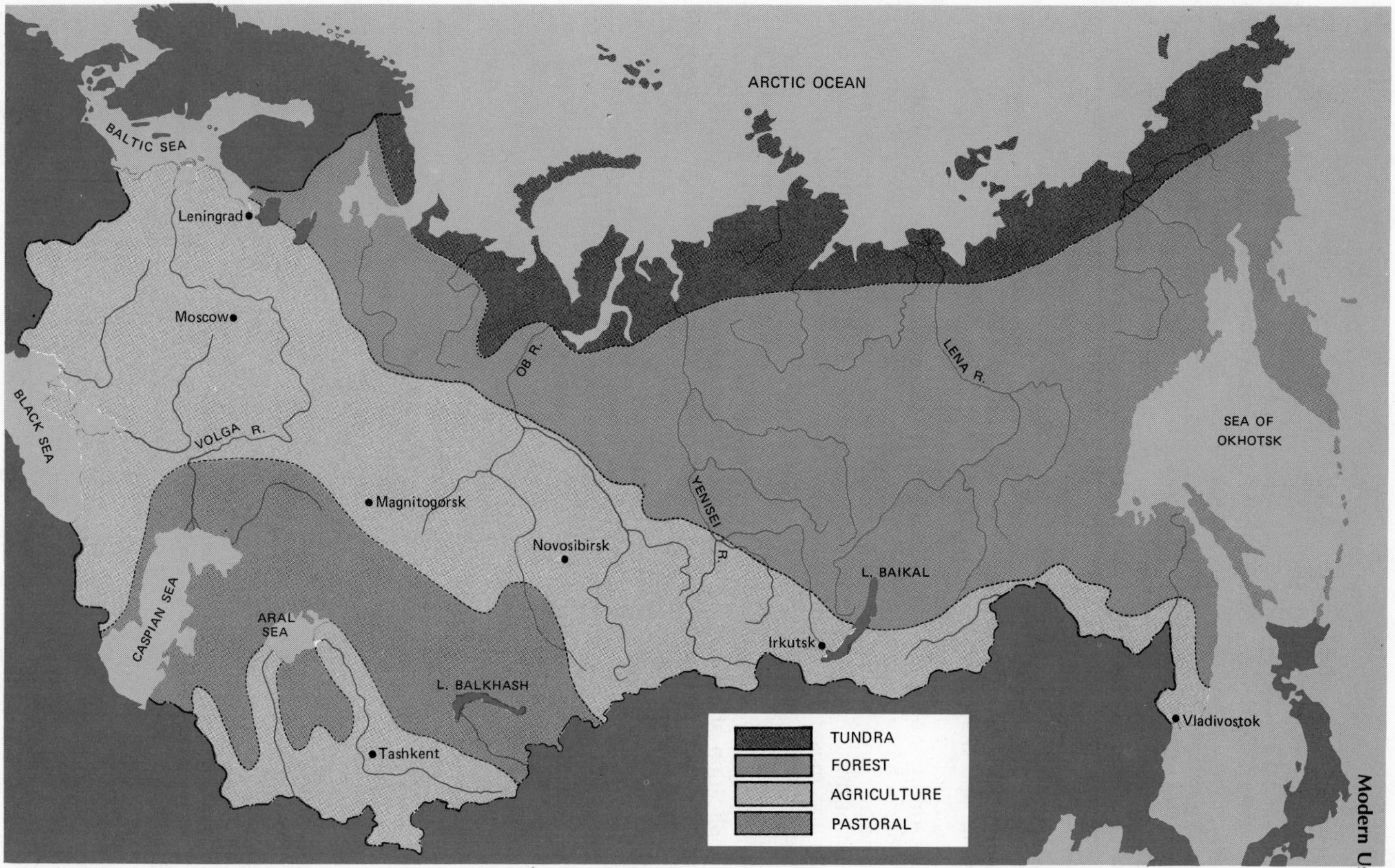

MAP 9-5   Soviet Union: Land Use
Land use is described here by means of four general categories: tundra, forest, agriculture, and pastoral. The tundra is very little used except for summer reindeer herding. The forest zone is used for timber, fur, and reindeer herding. The agricultural zone is marked by rye to the north and wheat to the south. Cotton is grown along the rivers feeding the Black and Caspian seas from the south. The arid regions around the Aral Sea and Lake Balkhash are used by horse-camel-sheep pastoralists.

The agricultural picture in the Soviet Union is equally disadvantageous. The Soviets employ nearly one-third of their people on farms while we employ less than 5 percent. They have 500 million acres in crop land and run a deficit in food. We have 300 million acres of crop land and not only feed our own people but export huge amounts as well. The oddity of this picture is that we make our knowledge freely available to all, including the Soviets. We have had missions from the Soviets here repeatedly, have toured them through our corn, hog, and wheat belts, and have given them our agricultural publications. The knowledge is all theirs if they want to take it. They take the facts, but they do not take the system, and they have not succeeded as a result.

As the review of the Soviet Union's physical geography indicated, the land is laid out in broad zones with treeless tundra along the Arctic Sea and to the south a huge belt of taiga; only south of this area does one enter the agricultural realm. In the mixed forest belt the soils are podzols—an acid, often poorly drained, and often nutrient-poor soil type. The needs are obviously drainage, such as is extensively provided for fields in our corn belt, and the application of nutrients. Southward in the Ukraine and its vicinity lie the most fertile lands. The latitude is still northern, and this means short seasons; the location is on the dry side of the climatic zones, so there is an increasing risk of drought. Most recently the Soviets have been trying to open up grasslands that until now have been considered too risky to farm because of the threat of drought and wind. These so-called virgin lands are very much like Montana and Saskatchewan in climate and risk. In 1954 Nikita Khrushchev gambled by ploughing up 100,000,000 acres, equal to one-third of the cultivated acreage in the United States. The first attempts to open them up led to disastrous losses— in labor, energy, and seed, and in land due to the blowing away of topsoil in the drought period. The situation was much like that in our Great Plains in the years of drought, earning the area the title of the Dust Bowl.

The Soviets have tried to rationalize farming, and they have overdone large-scale operations hoping to do better than the individual could hope to do by means of economy of scale and superior planning. They have consolidated the small peasant holdings into collective farms. One field measures 20,000 acres, or 30 square miles. It is a totally unmanageable unit, for within such an extent, one would expect to have a variety of soil types, drainage situations, and micro-climates. Not all of a piece of land such as this would be suited to one crop nor would it be ready to be planted or ploughed all at one time. According to Khrushchev the Soviet state farms used 80 percent more man-hours of labor than comparable units in the United States; on the cooperatively run collective farms the figure was 630 percent more. Some of this inefficiency can be charged to the system. If 10 percent of the tractors coming off the assembly line are defective and 3 out of 10 tractors are out of commission on any given day, the farm program that is geared to mammoth operations is already in trouble. If one adds the fact that the farmers have no incentive to produce and the management is engaged in hoarding, bribing, and quota meeting, the low productivity and inefficiency begin to be understandable.

With 70 percent more land in cultivation than the United States, the Soviets have only 40 percent as many trucks and tractors and 75 percent as many combines. Weather is usually blamed for frequent crop failures. However, the Canadians grow wheat under equal climatic risks; they salvage more even in bad years with better equipment and greater freedom to act. In the Soviet Union excessive planning leaves the farmer little room for decision. He is told what to grow, when to sow, when to fertilize, and exactly when to harvest each field. Only a farmer can fully comprehend the practice of farming by a computer printout prepared six months in advance by technicians in the national capital. One would do better following *The Old Farmer's Almanac,* for at least one is free to disregard it if it calls for snow on the Fourth of July. Young people are now leaving the land, and only mechanization can stave off an agricultural decline in the Soviet Union. Already a need is seen to expend up to 35 percent of their total investment in support of farm production. Much will have to go to trucks, roads, and granaries to support wheat production.

The ultimate test of the system would be a private sector within the agricultural economy with the same people in the same areas competing with the state system. Incredibly this actually exists, and it gives us a yardstick that measures productivity when everything is held constant except the economic system. The peasants on the collective farms are allowed small private plots, usually half an acre. In 1972 only 3 percent of the area sown was in such plots. On these tiny bits of land making up but a small fraction of the agricultural land of the nation, the following production figures were achieved: 51 percent of the nation's meat, 35 percent of the milk, 20 percent of the wool, 38 percent of the green vegetables, and 63 percent of the potatoes. Overall this amounted to one-half of the total agricultural production of the Soviet Union. It is a classic case of the desire of men to produce if they get a return on their labor and of the near total failure of the socialist system to change men.

Siberia looms large in any thinking about the Soviet Union. It is a huge area of 5 million square miles, far larger than the whole United States. It makes up one-half of the land area of the Soviet Union, but it has only one-tenth of the population—about 25 million people. A large part of this population is concentrated in a few cities. Novosibirsk, for instance, is a city of more than one million. Most of the population is along the trans-Siberian railroad, which runs through the southern part of the country, and there are few people indeed in the north. It is a land of cruel winters with temperatures dipping at times as low as −90° F. There is a great deal of permafrost, land that is frozen to great depths and never thaws except briefly at the surface. Frozen ground does not allow for underground drainage, and this creates swampy conditions. The great rivers flow to the north, thaw first in their headwaters, and flood over immense areas as they pour their water down into the still-frozen lower reaches of the system. To the east there are high mountains with all of the associated rough and broken terrain.

So vast a territory is bound to contain wealth, and Siberia more than lives up to expectations. Iron, coal, oil, tin, lead, copper, nickel, tungsten, and timber are a sampling of the resources found there in major quantities. Ninety percent of Soviet coal, 50 percent of its iron, 33 percent of its natural gas, and 75 percent of its timber is in Siberia. The Soviets plan to extract these valuables resources, having budgeted $50 billion for development. These plans include new rail lines: a northern and a mid-Siberian line. The Czars used to exile troublesome politicians by the hundreds to Siberia, and Stalin sent millions of people there under any pretext, using them as slave labor. Today incentives are used to try to move up to 1.5 million settlers into Siberia, but the area is like the Amazon in that its history makes it most unattractive to settlers. Further, it is truly a land of isolation and hardship; it is so unattractive that in the period from 1957 to 1963 there was a net loss of population of 400,000 people. The question arises whether the Soviets can make Siberia work in a socialist system or whether they will have to go back to the old days of simply seizing the requisite number of people under whatever pretext and sending them to work in the Siberian labor camps.

The Soviets now have a huge, complex, agroindustrial economy. It is subject to great stress, for they have squeezed the surplus out of the system and have

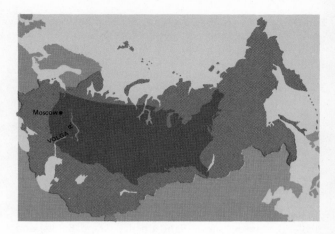

MAP 9-6   Soviet Union and United States: Areas Compared
Note that the latitudes are dissimilar. Moscow is at 55°50′ N. This is the latitude of Peace River in Canada, the northern limit of agriculture. The Soviet Union has two and one-half times the area of the United States, but 70 percent of the population lives west of the Volga river on 13 percent of the area, and 30 percent lives east of the Volga on 18 percent of the area.

exhausted their labor pool. In general, planning has failed them, but after a hesitant move toward reform programs in the 1960s, they reverted to more planning. For a brief period the Liberman reforms, which gave managers more leeway to use judgment and hold them responsible for good quality salable goods, were tried. However, neither enough responsibility was delegated nor were there enough adventurous managers. An attempt to accelerate the economy by giving the people more money also failed. There was little to buy, and the little available was of such poor quality that the people simply saved their money.

For the short run the Soviets can make do with exports of raw materials to pay for imports from the West. Oil, and especially natural gas, scarce metals such as chromium, vanadium, and gold, and timber can pay for a lot of imports. However, the problems of the Soviet Union are too large to be settled externally. Eventually they have to come to grips with the problem of making their industry productive. After 50 years of socialism under the dictatorship of the proletariat, this vast land with its inherent wealth of all major resources is short of food and consumer goods, with its people poorly housed and inadequately fed and clothed. After uneven economic growth with advances balanced and paid for by stagnation of consumer goods and services, the whole economy seems now to be bogging down. Neither socialistic centralized planning nor education of the people to true communism has worked. Only the peasants acting as individualistic entrepreneurs on their tiny plots of land are highly productive.

The Soviets have also gone to the sea to meet their protein needs. They have found that it is easier to expend capital on ships and harvest fish for protein than it is to change their agricultural system to obtain the needed protein. The expansion at sea is an interesting one. After exhausting their local fisheries, such as the Barents sea fishery, the Soviets expanded into the North Sea fishery, an area already heavily used by the nations around the North Sea. Presently they were off Iceland, then Newfoundland on the Grand Banks, and thence they have shifted to worldwide fishing. This has involved the development of huge freezer and factory ships, which process fish in every way—canning, extracting oil, making fish meal, and so forth. They have developed fleet techniques with groups of catch boats supplying a factory ship, which in turn unloads onto a freighter that carries the product back to the Soviet Union. Ships stay at sea for long periods, and the crews are rotated. The principal port for receiving fish is Murmansk, and the fish is distributed from there by rail to much of the country.

The oceans are common property as far as fishing is concerned, except for territorial waters. Traditionally this was set by the 3-mile limit, but as we have seen in the case of Peru, the tendency is for nations to extend this limit to protect their interests in offshore minerals and fisheries. The United States, after watching the Soviet factory ships operating within sight of American shores and using their fleets to sweep clean the fish stocks that we have carefully husbanded, has reluctantly extended its territorial waters, as far as fishing is concerned, to a 12-mile limit. It is estimated that 90 percent of the fish of the sea are to be found on the continental shelves and in a few other spots, such as the zones of upwelling water off the coast of Peru. The great expanses of the world oceans are virtually deserts in terms of fish life. The good spots are now rapidly being depleted by Soviet, Japanese, Polish, and other worldwide fishers equipped with ships, nets, and radarlike devices for locating schools of fish, and often they use light planes launched from the factory ship to locate the schools. The efficiency of harvesting combined with the number of nations with ships at sea has already exceeded the ability of the sea to provide fish, and the world fisheries face certain decline. This will pose problems for the Soviets, for they have come to depend on the protein from the sea, and they have a huge investment in ships to harvest the sea. The capital investment and the need are among the reasons for their reluctance to stop whaling despite the clear danger of extinction for these unique creatures of the sea.

We return finally to the problem of men and their lands. Is land the problem in the Soviet Union? The review of the physical setting made clear that the far-northern location does indeed pose problems. However, Finland, Scandinavia, and Canada all have similar physical-geographical problems related to their far-northern position, yet all these countries have far higher standards of living; it is clear that physical geography is an insufficient answer. History helps explain the Soviet Union's belated beginning, its numerous setbacks by Mongols and others, and yet this is not a wholly satisfying answer. Canada starts later, but with a British-based socioeconomic system in a precisely similar environment, it provides a far better standard of living. If the Canadian and Soviet starts were measured from the time when Catherine the Great began importing western scientists and Canada was being settled by Europeans, it is clear that Canada has bettered the Soviet Union in most measures of development. It lags, of course, in armies, navies, and imperialist adventures, for Canada is as notable for not interfering in the internal affairs of other nations as the Soviet Union is notorious for just this expensive activity. The emphasis given to the social-economic-political system of the Soviet Union was deliberate, for this is the real key to the Soviet successes and failures with its land and its people.

Potential Uses

The northern forest lands *can* be used: their potential usefulness has steadily been increasing as we have mastered the problems of transportation, housing, water supply, and heating. For a long time their utilization remained at the level of man's first penetration. For early men they were a source of game; for European man they were long a source of furs. Later we used their timber and minerals. The possibility of making far greater use exists, for we could use the vast forests for reindeer-caribou meat-and-hide production or for musk ox production. They could also be summer resort areas. Because of the increasing crowding of other parts of the world and the ease of air travel, this northern lake land is becoming more attractive today.

These lands cannot be expected to produce either plant or animal materials as cheaply as those of lands in lower latitudes, for such production is dependent on solar energy. These are lands of most unequal seasonal distribution of solar energy, with an annual total that is far lower than the mid-latitude and tropical areas. Nonetheless, there is considerable potential in these lands, more than is now being utilized. Will we then begin to use them? If so, for what purposes? And who will benefit from such exploitation?

There are two conflicting tendencies. The world population is growing rapidly. This would seem to suggest that we should be putting more and more people into these lands. In practice, however, we find that potentially more useful lands (nearer to centers of civilization, with better developed transportation networks and so forth) are faced with decreasing population. In 18 farm states in mid-America 61 percent of the counties lost population between 1950 and 1960, and this trend continued in the 1960–1970 census decade. Further, this is a worldwide trend, as has been mentioned in several previous sections. Even in the Soviet Union where people are moved about under greater governmental control than elsewhere, population in Siberia declined as soon as governmental pressure was relaxed. The population explosion that we hear so much about would seem to be better described as an implosion than an explosion. Mankind is concentrating in a few, largely urban areas. The empty parts of the world are becoming emptier, not fuller. In general the northern forest lands seem more likely to continue to empty out than to fill up. This will not be due to the land as much as it will reflect the decisions of man concerning the kind of life he chooses to lead. Should populations of sufficient size develop in the now relatively empty northern forest regions to support the amenities that people find today in urban centers, people might come and stay. Long-term mineral developments such those in the Ungava peninsula of Canada could lead to such settlement, and we may see the equivalent of Finland appearing in Alaska, western Canada, and the Labrador area.

Summary

The two northern forest areas of the world—areas in the hands of the Soviet Union and in American, mostly Canadian, hands—are strikingly similar in physical makeup. They have similar climates, landforms, vegetation, and animal life, and both areas have mineralized zones. Men entered both areas at least as early as 20,000 years ago as specialized hunters of such great mammals of the Arctic as the mammoth. In both areas their hunting efficiency was great enough to exterminate the biggest of the big game, and the men then came into an ecologic balance with the more numerous, more rapid-breeding, smaller game including the reindeer, caribou, moose, deer, and rabbit. Relatively recently the uses of these lands diverged. The introduction of the idea of domestication of animals and their utilization for transportation, milking, and nomadism in the Old World gave rise, by the transfer of these ideas to the hunting people of the northern forest, to a new way of life in the northern forest that was also extensible into the tundra. The lesser time, the limited contact, and the orientation of the peoples living at the Bering Strait unfortunately blocked the spread of these ideas to the New World.

In the Old World the exploitation of the northern forest lands for timber began early but was limited by transportation problems. The extraction of products of high value in proportion to their weight, such as furs, very early led the Europeans, especially the Russians, to spread through the northern forest areas. The fur trade led the Russians clear across Siberia, into the Pacific, and down the Pacific coast to pose a threat to Spain's colonial claims. The French and later the British development of fur trade throughout Canada was a later duplication

of this type of utilization of northern forest lands. Currently both the Soviets and the Americans are pushing agricultural developments northward into these areas, extracting forest products from them in increasing quantity and exploiting the mineral resources. The similarities are partly geographic, partly economic (the comparative advantage of these lands at our present stage of development leaves them in the extractive stage), and partly cultural. Striking dissimilarities arise now because of the differing socioeconomic philosophies of the Soviet Union and of the West.

Although it is easy to overstress the similarities, there are differences. The development in America is largely done by individual corporations in a free market economy. Development is heavily mechanized to reduce the need for labor. Labor in the northern forest lands is scarce; it must largely be imported and supported at great cost. In addition, above average pay must be given to attract and retain a labor force. Soviet development contrasts sharply with the development of the Canadian-Alaskan region, the physical-geographical parallel to the Soviet Union. As a labor force, political exiles have played a large role in Siberian development, but they have had no role in the West. Despite the fact that fewer political exiles are now used and the government has provided incentives to encourage people to migrate, Siberia remains unattractive to the Soviet people. Canada, on the other hand, is developing with the increase in world demand for raw materials. Alaska has lagged because the other states offer more advantages, but it will also develop with the increase in demand of raw materials, particularly minerals. Canada and Alaska have a far better standard of living than the Soviet Union. Soviets feel most strongly challenged by the United States and boast of rapidly surpassing the Americans. By retarding the growth of consumer goods and pouring capital into armaments and heavy industry, they have managed to catch up with the West in certain industries and in defense, but this has been achieved at the expense of the people.

Occupation and development are proceeding in various parts of the northern forest lands according to various plans, for different purposes, at different rates, and at different human costs. Although the North American lands are little different from those of Eurasia, the patterns of development are vastly different. Ancient development as in Finland is absent from North America and Siberia. Developments in North America are similar despite the political boundaries between the Canadian and United States Arctic because of the basic economic and political systems of the two countries. The Soviet development diverges not only from the American but also from the Swedish and Finnish because of ideological differences. All of this development takes place in a uniform physical environment.

Southeastern Alaska. *The northern forest lands are covered with conifers. Though these grow slowly, the vastness of the areas where they are found makes the timber resource abundant. The northern forest lands were glaciated until recently, and the glaciers have left innumerable lakes and frequent rapids and falls on the water courses. Industrial society uses these lands largely for obtaining materials such as timber, pulp, and metals. They are nearly empty, and despite growth at certain centers of mineral activity, the trend is to move out of them still further.* (Alaska Travel Division Photo)

Review Questions

1. What is the evidence for ancient and long-continued connections between northeast Asia and northwest America? Why did civilization not grow in that part of America as a result?
2. Why did the natives of northern Asia domesticate the reindeer?
3. What are the principles illustrated by the Eskimo-caribou story?
4. What is the physical basis for the low number of fur-bearing animals in the Arctic forest lands? Why, then, are these major fur-producing areas?
5. Since the physical base is so similar in the northern forest lands, why are there such contrasts in development and productivity as in Finland, Siberia, Alaska, and Canada?
6. How good is the correlation between climatic need and human response in housing and clothing in northern forest lands?
7. Since the demand for meat is so great, why isn't reindeer herding a big business in America?
8. Discuss the pros and cons of an Alaskan resource problem in terms of the oil pipeline or the dam on the Yukon.
9. The Russians moved their capital from one city to another. What cities have been the capital? Why? What do these moves show?
10. If the Soviet economy is as inefficient as portrayed here, how can the country be so advanced in military and space developments?
11. With such vast land area, why can't the Soviets feed themselves?

Suggested Readings

Bergson, Abram and Kuznets, Simon, eds. *Economic Trends in the Soviet Union.* Cambridge, Mass.: Harvard University Press, 1963.

Canada, An Introduction to the Geography of the Canadian Arctic. Canadian Geography Information Series, no. 2. Ottawa, 1951.

Cooley, Richard A. *Alaska: a Challenge in Conservation.* Madison, Wis.: University of Wisconsin Press, 1966.

Gjessing, G. *Changing Lapps, A Study in Culture Relations in Northernmost Norway.* London: University of London, 1954.

Hare, F. K. "Climate and Zonal Divisions of the Boreal Forest Formation in Eastern Canada." *The Geographical Review,* vol. 40 (1950), pp. 615–636. Focuses on the Labrador-Ungava region where a great mining development is now proceeding. Also valuable for its description of northern forest conditions.

Humphreys, Graham. "Schefferville, Quebec: A New Pioneering Town." *The Geographical Review,* vol. 48 (1958), pp. 151–166. The first of the towns built for the exploitation of eastern Canadian iron ore.

Kimble, G. H. T., and Good, Dorothy. *Geography of the Northlands.* American Geographical Society Special Publication, no. 32. New York: John Wiley & Sons, Inc., 1955.

Mead, W. R. *An Economic Geography of the Scandinavian States and Finland.* London: University of London Press, 1961.

Naysmith, John K. *Canada North—Man and the Land.* Ottawa: Canadian Department of Indian Affairs and Northern Development, 1971.

Nutter, Gilbert Warren. *The Growth of Industrial Production in the Soviet Union: A Study.* National Bureau of Economic Research, General Series, no. 75. Princeton, N.J.: Princeton University Press, 1962.

Parker, William A. *An Historical Geography of Russia.* Chicago: Aldine-Atherton, Inc., 1968.

Phillips, Robert A. J. *Canada's North.* New York: St. Martin's Press, 1967.

Putnam, D. F., and Kerr, D. P. *A Regional Geography of Canada.* Don Mills, Ontario: J. M. Dent & Sons (Canada), Ltd., 1961.

Stefansson, V. "The Colonization of Northern Lands." In *Climate and Man: Yearbook of Agriculture, 1941.* Washington, D.C.: U.S. Department of Agriculture.

Stone, K. H. "Populating Alaska: The United States Phase." *The Geographical Review,* vol. 42 (1942), pp. 384–404.

Taylor, G. *Canada, A Study of Cool Continental Environments and Their Effects on British and French Settlement.* London: Methuen & Co., Ltd., 1947.

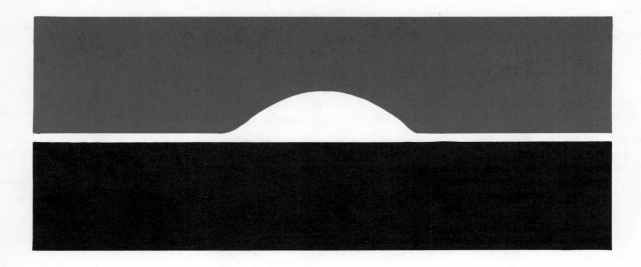

The Polar Lands

CHAPTER TEN

THERE are two polar areas. One, Antarctica, is an ice-capped continent which, until the present, has had no human occupation. The other is an ice-covered ocean with human occupation on its shores and the edges of its ice floes. Man learned to live on the edge of the treeless zones near perpetual ice when the ice masses covered most of northwestern Europe and all of northern North America. When the last great ice retreat began about 10,000 years ago, men had amassed enough environmental control through housing and clothing to be able to follow their familiar environments northward to the shore of the Arctic Ocean. It was probably at this time that one of the groups of men moving into the Arctic found its way from Siberia into Alaska. Man stayed in the Arctic thereafter.

The Physical Setting

The treeless Arctic areas, north of the isotherm for 50° F. for the warmest month, resemble the grasslands. They are covered with vegetation, but it is a uniformly low-growing plant cover. There are mosses, lichens, sedges, grasses, and herbaceous plants. In the brief summer there are bright flowers and much green. The plants are adapted to withstand great cold, a short growing season, and a perpetually frozen subsoil. In the tundra areas of the world the ground never thaws for more than a few inches. Beneath that the ground is always frozen and often to very great depths: as we have mentioned earlier, such deeply frozen ground is

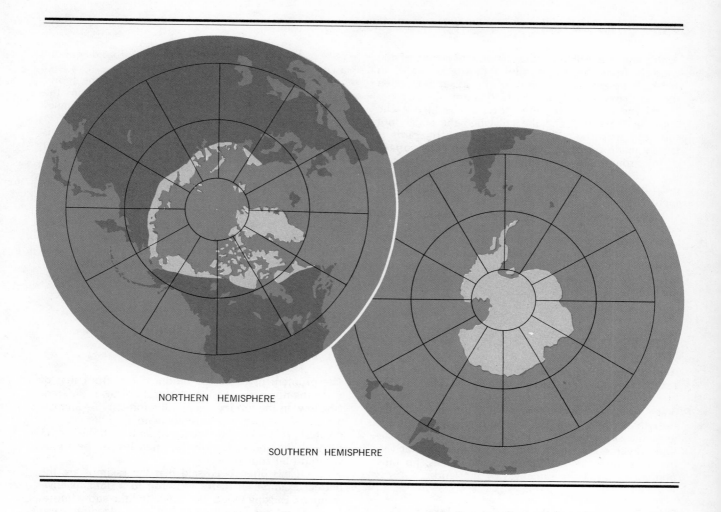

NORTHERN HEMISPHERE

SOUTHERN HEMISPHERE

called permafrost. The soils are often classified as muck soils, though on well-drained ridges there are more normal soils. With the ground frozen, water can drain away only with great difficulty. The result is that when the summer thaw comes, much of the land is a great, wet, spongy surface.

This wetness persists in spite of the fact that the total rainfall is low. Tundra areas generally receive less than 10 inches of rainfall per year, and this comes mostly in spring and fall. Winter is particularly dry, and the total depth of snow cover is shallow. The major centers of the glaciers were not in the high Arctic but lay in lower latitudes: over Scandinavia, northern Quebec, and the Rockies along the Canadian border.

There was so little snowfall in Alaska that the Yukon Valley was never glaciated, nor was much of the Arctic coastal plain of Siberia and Alaska.

The climatology of this situation is quite simple. In winter these areas are intensely cold; the capacity of air masses to hold moisture is related to their temperature in such a way that the lower the temperature the less the moisture capacity. Consequently the very cold winter air masses contain very little moisture. Unless oceanic air masses can be brought into these areas, there can be little winter precipitation, which would, of course, come in the form of snow. The Arctic Ocean does not supply much moisture to the Arctic regions since it is frozen over most of the time. In both the

Arctic and the Antarctic the major movement of the wind is from the frozen, dry interiors outward. In Alaska there are local reversals of this in summer when the sea thaws in part and the land warms greatly.

The result is that the snows of the high Arctic come in the spring and the fall when the air is "not too cold to snow." The snow of winter blizzards is largely made up of snow whipped up off the surface of the ground. However, there are not many gales in the Arctic in winter, for in winter it is the center of the polar high, and in the center of a high pressure system there is little wind. Strong winds are related to large changes of air pressure over short distances. In the Arctic in winter the land is cold, and the ice-covered sea is cold: there are no great contrasts. Therefore the wind is moderate in strength and largely from the northeast. Much the same is true of the Antarctic: the violent winds are on the continental margins where the land-sea contrasts are great.

THE ESKIMOS

The dominant winds are the polar easterlies. Their reality is brought out vividly by the fact that the Eskimos at Point Barrow, the northernmost point of the United States, use them to find their way. Out on the tundra everything looks alike. On an overcast day the snow is gray, the sky is gray, and there may be no clearly defined skyline: gray melts into gray. The sun is not visible, and there are few landmarks: no trees, no hills, nothing to break the sameness of the land.

Eskimo Pottery Lamp from the Yukon Tundra, Alaska. *The bowl is filled with oil, and dry moss is used as a wick. Such lamps as these first appeared about 20,000 years ago in the archeology of southern France.* (Smithsonian Institution)

"How can you find your way around without a compass?" you may ask an Eskimo.

"Easy," he says. "See those ripples in the snow? Up here in winter the wind always blows from the northeast, and those ripples always form at right angles to the wind. So they run northwest and southeast."

The snow in the Arctic is not deep. There is usually less than 2 feet, but it covers everything. At Point Barrow, Alaska, heavy equipment is moved across the land on broad-runnered sleds pulled by tractors. Engineering crews go out in houses mounted on such sleds, so they can sleep in warm comfort wherever the day's work ends. The Eskimos move about freely with their light dog-drawn sleds. Where contractors, exploring for oil in the area, dump trash, the Eskimos come, load their sleds, and haul off ponderous loads for such light equipment as a sled drawn by half-a-dozen dogs. The sliding power of the snow is a real virtue.

Origin of Some Eskimo Cultural Traits

The problem of using the tundra differs from that of using the northern forest lands, for the snow is different (shallow on the tundra, deep in the forests), the animals are different, and exposure to wind is different. The greatest problem is the absence of timber on the tundra for building and fuel. How can men live in the Arctic without fuel and shelter?

It has often been said that the Eskimos are the perfect example of man's response to a difficult environment. Lacking wood, they invented the snow house. Lacking fuel, they invented the oil lamp. Needing warm clothing, they invented the finest tailored skin clothing and the bone needle with which to sew the skins. They did this in the Arctic, we are given to understand, in response to the needs of life there. Only one thing is wrong with this argument: no one, not even a Tierra del Fueguian, could live long enough in the Arctic to make this series of inventions. The inventions, with the possible exception of the snow house, had to be made elsewhere and then utilized to make living possible in the Arctic. The situation is exactly like the American agricultural expansion into the grasslands.

Some Eskimo traits can be traced to quite different climates from those where the Eskimos now use them. The oil lamp is an example. In the Arctic there is little or no wood; the Eskimo heats the interior of his house by burning oil in an open dish with a wick of moss. In principle this is exactly like the kerosene lamps that were in common use in Western society

An Eskimo in a Kayak, White Bay, Canada. *The harpoon and harpoon line are laid out on the deck. Eskimo-type harpoons appear as early as the Upper Paleolithic period in western Europe, where skin boats also survived into modern times in Ireland.* (Information, Canada)

prior to the development of gas lights and electricity and that are still widely used in nonelectrified rural areas. It is also identical in principle to the oil lamps that were used in Greece, Egypt, and Rome. We can trace these lamps far back in time. They were used in western Europe 20,000 years ago by the Upper Paleolithic people. Hence the Eskimo probably did not invent this useful item but got it from some other people. Just who did invent it is not known. It may well have come out of the Southwest Asian center of inventiveness and then spread west into Europe and finally east into Siberia.

Eskimo clothing most probably was not invented in the Arctic but in lands where it was much less required for survival than desired for comfort. We also find that bone needles appear in the Upper Paleolithic cultures in Europe. So far as our present evidence goes, it would appear that tailored skin clothing was developed somewhere 20 degrees or more to the south of the Arctic lands of the Eskimos.

The Eskimo weapons are highly specialized items. Bone and horn are very widely used to make harpoon points. These points are slotted, barbed, and toggled, and they have inset stone blades in their tips and on their sides in a way that duplicates the work of the reindeer hunters of the Upper Paleolithic period of Europe. The spear thrower, an important weapon

for the Eskimo when hunting sea mammals in his kayak, is also an ancient weapon, appearing at least as early as the Upper Paleolithic. The Eskimo underground house is also at least of this same age, and it was in use over wide areas far to the south of the Arctic. In short, all of the technical inventions that the Eskimos have that are essential to life in the high Arctic were in use 20,000 years or more earlier in more southern latitudes. The accumulation of these inventions in other areas made possible man's invasion of the treeless Arctic.

It has often been noticed that many of the Eskimo cultural traits are like those of the Upper Paleolithic people of Europe. How this is to be interpreted has varied through time. The stone lamp, tailored skin clothing, harpoons, spear throwers, semisubterranean houses, engraving on bone and ivory, and carving figures in the round are only a partial list of the traits that the Eskimos share with the Upper Paleolithic people of Europe of 20,000 years ago. Further, these and other facets of culture such as magic flight myths and bear ceremonialism have a nearly continuous distribution from western Europe into North America. Recent Soviet archeological discoveries and analyses indicate that the Upper Paleolithic people near Lake Baikal are Europoid men who followed the ice border to the east, and they suggest that these similarities are actually due to the spread of men and ideas. They certainly support a view that ways of life are enormously persistent, inventions are uncommon, and the spread of ideas is dominant over independent invention.

Eskimo culture in the form that we know it today is often said to be only about 2000 years old. That Eskimo speech is relatively uniform from one end of the Arctic to the other suggests that their spread across the American north is not very ancient. On the other hand, this tells us nothing about how ancient their culture may be in Siberia. Further, archeological work in Alaska and Canada adds continuously to the list of pre-Eskimo cultures that occupied that area. Man surely has lived there for 10,000 years, and Eskimolike equipment is required for life there.

Eskimo Solutions to Arctic Housing

The underground house

The Eskimos are often thought of as living in snow houses. However, only about 25 percent of them have ever seen one, for only the Central Eskimos build them, and they are the least numerous of the Eskimo groups. The largest population of Eskimos is in Alaska. There the ancient house was the semisubterranean house, almost exactly like those built in Siberia 20,000 years ago. The house is dug down into the ground. A roof is formed of timbers, bone, antler, or a combination of all of these materials. In the past mammoth bone was commonly used in Siberia, and in Alaska whalebone was similarly used. The roof was covered with sticks, grass, and finally sod, making a thick, well-insulated covering. Entrance was either through a hatch in the roof or through a tunnellike entrance to the side.

Inside the house a relatively large number of people occupied a small space. Their body heat alone was sufficient to raise the temperature, and further heat was added by cooking over oil lamps. Even when not cooking, the oil lamp was kept going to provide light. Temperature in these houses was often in the 80s and sometimes higher, and the humidity was very high. Under these conditions there was no need for clothing. In addition, if clothing was worn indoors, it would become saturated with moisture; when worn outdoors, its insulating qualities would fail, and the wearer would freeze. It was not only comfortable but intensely practical to shed the clothing in the entry and live indoors in a state of nearly complete undress. The Upper Paleolithic artists, who lived in similar houses from Siberia

to Europe, portrayed people as unclothed, suggesting a similar way of life at that time.

The snow house

The snow house is a remarkable invention. It involves the principle of the arch, rotated to form a dome. The Eskimo was not aware of this as an abstract principle. An arch of stone must be supported until the keystone is put in place. The arch will then bear many

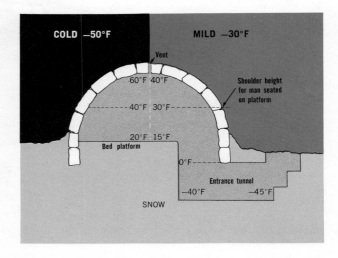

FIGURE 10-1   Cross Section of an Eskimo Snow House
The method of constructing an Eskimo house from blocks of snow is suggested in the drawing. The house is normally banked up with snow about the base, and a tunnel is dug for an entrance. It is important to have the platform inside at least 18 inches higher than the level of the entrance tunnel. The heat produced by the occupants will then make the interior moderately warm. If an oil lamp is used, the house can become comfortably warm.

It is an oddity of such a structure that the colder it is outside, the warmer it can be inside. If the temperature outside is extremely cold, the air inside can get quite warm before melting of the roof begins. In this diagram, at −50° F. outside, a temperature of about 60° is indicated just under the roof and a temperature of 40° at shoulder height for a man sitting on the bed platform. Somewhat colder inside temperatures are suggested for outside temperatures in the −30° to −40° range. With outside temperatures warmer than 20° below zero, the snow house is not useful, and tents must be used instead.

times its own weight. In the snow house the Eskimo is able to build his arch without supporting scaffolding because snow has adhesive qualities. The right kind of snow is required: it must be of adequate thickness, and it must have been packed by the wind to the proper consistency. Wind-packed snow can be cut into large blocks, lifted, and stood on edge. Building a snow house consists of finding this kind of snow, cutting large blocks, setting them on edge in a circular plan, and spiraling the blocks upward and inward as the walls are raised. It is a neat trick requiring some skill, acquired mostly by directed practice.

The snow house has enormous advantages. Once completed, it is strong: wind cannot blow it away or make it shake and flap in the annoying manner that a tent does. It is completely windproof and practically soundproof. Once fully stabilized, in part by the occupants' breath being absorbed into the snow and thus freezing the blocks more tightly together, it is strong enough that a polar bear can stand on top of it. It makes carrying a tent unnecessary, and it makes it possible to build a snug shelter where there is no wood and where the ground is too frozen to allow one to dig an underground house. A house can be set up in about an hour's time; it can then be abandoned without a qualm after a day or a week if the conditions of the hunt suggest that it would be better to move on.

It is usually thought that the interior of a snow house must be cold. This is not true, for they can be kept very warm. The floor is dug out, leaving a bench along the sides. Furs are placed on this bench to provide insulation so that sitting there will not melt the snow beneath. The body heat of several people in a snow house will raise the inside temperature to 20° to 30° F. when the outside temperature is −20° to −30° F. If an oil lamp is lit for cooking or for light, the temperature will then go still higher.

Whether the snow roof will melt is determined by the temperature inside the igloo, the thickness of the roof, and the temperature outside the igloo. If the heat flow through the roof keeps in balance so that the inner wall of the roof stays at 32° F. or less, all is well. If the heat is not flowing through the roof fast enough, the inside of the roof will begin to melt, and drops of water will form. To an Eskimo this is notice that action must be taken. He can throw on his parka, duck outside, and scrape some snow off the roof. There will then be a more rapid heat loss through the thinner insulating barrier, and the inside temperature will balance. Or he can poke a piece of stove pipe up through the roof, providing a ventilator through which some of the interior heat can escape. This also can be regulated; if too much heat escapes and it begins to be cold, one can stuff a glove in the ventilator to cut the heat loss down.

The interior of a snow house is light in daylight, and even moonlight will give sufficient light for dressing and cooking. In bright sunshine men have been

Eskimo Igloo under Construction, Northwest Territories, Canada. *This large triple-domed igloo is for the ceremonial drum dance.* (Information, Canada)

known to go snow blind inside a snow house. If artificial light is needed, a candle gives ample, soft, diffuse light because of the reflecting ice crystals in the snow dome.

Entrance to an igloo is by a sunken tunnel. This requires locating the house in an area having about 3 feet of snow cover. The tunnel leads into the sunken floor to take advantage of the properties of air: warm air rises and is trapped near the ceiling; the cold air lies in the tunnel and in the well-like central area. It is essential that the sleeping bench be at least 18 inches above the level of the tunnel entrance. The air temperature in the floor of the tunnel may be −40° F.; the bench level where people sit, 20° F.; and the apex of the ceiling, 60° F. When the igloo temperature rises, some clothing must be shed lest it become damp. The clothing is then hung in the tunnel where it will be kept dry by the cold and will be handy when the Eskimo is ready to go out.

The temperature contrasts in a tent and a snow house are striking. With no heat and with an outside temperature of −50° F., the interior of the tent will reach −10° to −20° F. Under the same conditions in a snow house the temperature near the roof will remain above freezing and at the sleeping platform level will not drop even close to 0° F.

Life in such a house can be exciting at times. The dome shape catches the sounds coming through the snow crust and focuses them on the interior. When the sled dogs are bedded down in the snow outside and the family is snug inside, there is little for the Eskimo to fear, except the polar bear. If a bear should come into camp and attack the sled dogs, the Eskimo must get out quickly. Dogs are precious because the Eskimo depends on them for his ability to move rapidly through his snow-covered land. Fortunately, the approach of a polar bear is signaled by the crunching noise of his feet in the snow, which is picked up by the snow house and amplified to warn the Eskimo.

Americans who have used the snow house in their explorations in the Arctic have evolved a system for such emergencies. Each night one man has the duty in case of such attack to spring out of his sleeping bag, grab his gun, and rush out of the igloo to shoot at the bear before he can destroy any of the dogs. The rest of the crew stop to put on clothes before they rush out. A nearly naked man can stand a few minutes of −20° or −30° F. if he has his boots on. By that time his companions are on the scene, and he can duck back into shelter and warmth.

In one recorded experience men on a mountain-climbing expedition built a snow house. On entering it they noted that the temperature was −27° F. After 20 minutes their body heat had raised the temperature 60 degrees to 33° F. When they lit a single burner gasoline stove, the temperature became very warm, and 20 minutes after turning the stove off, the temperature of their igloo was 65° F.

The snow house is a tremendously useful invention that every people in extremely high mountains or extremely cold snowy conditions "need," but it has remained the possession of the Central Eskimos. It did not spread into Siberia, nor was it reinvented there, in Scandinavia, the Alps, or the Himalayas. Even more surprising, it has very reluctantly been adopted by polar explorers, especially in Antarctica where it would be at maximum usefulness. Instead, if one reads the accounts of the early expeditions there, men struggled with heavy tents, which gave them little relief from the cold while the wind made the canvas crack and bang like cannonading.

The failure to reinvent when spurred by need is a recurrent theme. The repeated failure to adopt useful inventions is even more startling. It is salutary to see that even modern, rational Western man is just as subject to these failures in perception as most other men at most other times. Note also the stubborn American refusal to adopt the obviously superior and well-known metric system. Indeed, it makes one wonder even more at the apparent transfers of knowledge. How and why did the Siberians adopt and apply the idea of domestication to reindeer when Africans did not make the transfer to useful native animals and the American Indian overlooked the native wild rice? The ability to transfer knowledge from one setting to another requires a deeper perception of the nature of things than we often think, and it is this difficulty that makes the clichés of "environmental determinism" and "necessity as the mother of invention" such dubious guides to understanding man's relationship to his environment.

Eskimo Clothing

A man fully clothed in Eskimo fashion worries as much about overheating while working as he does about freezing. Eskimo clothing utilizes the insulation value of fur, that is, the trapping of air in the fur to form an insulating layer. The effect of this insulation is spoiled by dampness; hence the clothing must be kept dry both in the house and in use. It is imperative that the Eskimo not overexert himself so as to perspire actively. When heavy work is necessary, the Eskimos may be seen loosening their parkas at the throat and at the waist in order to let body heat escape or even shed-

Eskimo with Sled. *The Eskimos are famed for their ingenuity in making implements and adapting to their environment. Here an Eskimo has made a sled using frozen fish wrapped in sealskin for runners. These are long hard girders when frozen, and the sealskin provides some resistance to fracturing. Water added to the bottom of the runners will sheet them with ice, serving as a wearing surface. The crosspieces are of reindeer antler. Notice the Eskimo's tailored skin clothing and that he is doing this work with heavy gloves on.* (Information, Canada)

with him into the Arctic world, and he is normally snug and warm even under conditions of extreme external cold.

An anecdote is told of the comfort that the Eskimo experiences at home in his own clothes and environment. An Arctic explorer brought an Eskimo couple to visit New York. He outfitted them in standard Western clothing and showed them the town. When it was time for the Eskimos to return, he saw them off on the ship that was to take them to northernmost Greenland. The Eskimo woman burst into tears at the departure. The explorer was touched at such a display of emotion and asked why the tears, thinking, of course, that she would reply that she was going to miss him. Not so. She explained that she felt sorry for him, being left behind to live in uncomfortable clothes in this awfully cold place.

It is not true, of course, that the Eskimo is never cold. Some of his work requires long periods of sitting still, and the cold can then be a problem. The Eskimo's sleeves are so loose fitting that under these circumstances he can withdraw his arms and fold them over his chest. This reduces his surface of radiation still further and aids him in keeping warm. Further, both the Eskimo's face and hands are noticeably warmer than those of other people. In the Eskimo skull the openings for the blood vessels that supply the face region are large, indicating that an increased-blood-flow type of physical adaptation has occurred. A similar adaptation making possible increased blood flow into the hands so that they remain warm and capable of doing dexterous work under extremely cold conditions is further indication that the Eskimo is physiologically adapted to his extremely cold environment. Whatever the ultimate source of the classic Eskimo culture, its bearers have surely been living in cold environments for a very long time.

Sea-Mammal Hunting

The maritime Eskimos obtain their food largely from sea mammals. In the middle of the winter seals are caught when they come up to breathe through holes in the ice that they maintain as the sea ice thickens. The Eskimos may sit for long hours at these holes. In late spring seals are harpooned on the ice. All of the hunting techniques are changing now that rifles have been introduced, and this description is more concerned with the traditional way of life. Walrus are harpooned; this is big game, for a walrus weighs a ton or more and can deal destructive blows with his tusks.

ding one layer of clothing. Clothing is worn in two layers: one set, fur side in; another set, fur side out. Boots and gloves are worn. The clothing is loose fitting for active work but tight at the top to prevent the loss of body heat, which means a tight belt and a tight throat. The hood is important for cutting down loss of heat from the head and neck areas. One of the few Upper Paleolithic statues with indications of clothing shows a woman wearing a parkalike hood. It has been said that the Eskimo does not live physiologically in an Arctic climate but in a tropical climate: in his house he lives in high heat and humidity; in his double-fur suit, boots, and gloves he carries this warm climate

Umiak and Crew, 1869. *This type of large skin boat is capable of extended voyages, as the Irish demonstrated with their very similar craft called curaghs. Eskimos used such a boat for whale hunting as well as general transportation.* (The American Geographical Society)

Yet this sport pales when compared to the Eskimo's hunting of the whale.

The most important whale hunting in the north is done in the spring when the whales begin moving northward along the cracks that are opening in the ice. Eskimo crews station themselves along these "leads." Each crew has a large open boat, the umiak, made of skin stretched over a light wooden frame. The boat is mounted on blocks at the edge of the open water, and the crew settles down to wait. When a whale comes along the edge of the ice, the harpooner gets in the bow of the boat, and the crew lines the sides, set to hurl the boat, harpooner and all, at the whale. The harpooner plunges the harpoon into the great beast; the whale dives, and the harpoon line is let run out. With it go floats of inflated seal skins. The whale, wounded and hampered by the pull of the floats, sooner or later must surface. The original boat and any other boats in the vicinity are waiting for him when he comes up in order to drive more harpoons home. This process goes on until the exhausted whale lies weakened on the surface. The Eskimos then come alongside the whale and drive lances into vital areas. It is a hazardous and thrilling business with an enormous yield of meat as the goal. Whales run a ton to the foot, and 20- and 30-foot whales are the usual catch.

The whale is now awash in the sea. "Land" is a shelf of ice standing 3 to 5 feet out of the sea. How can they lift this tonnage onto that shelf? When it is on the shelf, the distance to the village may be one or more miles away over sea ice. Our estimate of primitive man's ability gains enormously from watching the Eskimos handle this problem.

First a ramp is chiseled out of the sea ice. The Eskimo knows no formal physics, but here he is employing the principle of the inclined plane. The whale's head is towed to this ramp, and slits are cut in the side of the head. Long ropes are then led through the slits, up onto the ice, around blocks of ice, and back again. Poles are then put through the lengths of rope, and by twisting, a powerful purchase is created that drags the massive head up onto the ice a ways. Cutting tools are then brought into play, and as much of the flesh as can be reached is cut away. The lines are then rerigged, and the carcass is hauled out another short distance.

The scene is one of wild, colorful activity. The meat is hauled back onto the ice in chunks. The whole village descends on the scene with all the carrying equipment obtainable. Sleds, dogs, women, and children all come and work. The ice is soon a great mass of pink to red from the blood of the blocks of meat that are being dragged back from the butchery. Joy is intense, for this mass of meat is insurance against hunger for a long time to come. It will be eaten greedily on the spot and in the village. Much of it will be buried in the frozen ground to be preserved for later need. The whole carcass will be shared with the whole village ultimately. The original crew will get certain privileged pieces, but the rule is that all share in the kill. By this means, as long as there is meat coming into the village, all are kept alive. The fortunate hunter simply cannot keep his family fat while others starve. When he runs into bad luck, he can in turn claim meat from those who have had the good fortune to get a seal, a bear, or a caribou. This is a normal outlook on food among primitive peoples, and it offers the kinds of security our insurance policies accomplish today.

The Eskimos illustrate the fact that men can live in almost any part of the earth. They have a highly specialized and ingenious culture that makes of the treeless Arctic sea edge a useful environment. Far from being unhappy with their lot, the Eskimos are always described as among the happiest of people on the earth. They are not greatly depressed by the long night of winter, even though they prefer the light, even the very cold light of spring. They are comfortable in the winter, and summer has its disadvantages. Summer is a time of slush and wet, travel overland is difficult, and the myriads of mosquitoes are a terrific nuisance. Tents are poor substitutes for the snug snow houses of the Central Eskimos, but the thaw often makes the underground houses of the western Eskimos damp, so then they move out into tents. The return of the good, cold, dry winter with its crisp snow brings ease of travel by dog sled swiftly gliding across the land. Darkness is not so bad: the moonlight shining on the snow makes things almost as light as day.

The Impact of Industrial Society

At Point Barrow the Navy exploration for oil during the late 1940s and early 1950s brought the Eskimos high wages and contact with the benefits of civilization. The Eskimo men were employed as equipment operators as well as in the shops of the Arctic contractors. They proved to be quick to learn to handle modern machinery and became not just proficient but expert. Since they were paid the same as imported non-native labor, they found themselves with an abundance of money. This was spent in interesting ways. They built frame houses, which look much like Midwestern frame cottages, and they heated these with stoves using coal mined many miles away and hauled to the village. Warm houses demanded refrigeration, refrigerators came into use in the Arctic, and skin boats had outboard motors attached to them.

When the search for oil ceased, the high income was cut off. The large number of Eskimos who had gathered at Point Barrow for this work then had a number of choices. They could go to Fairbanks or farther south and get employment, for they were now skilled tractor drivers, machinists, and carpenters. They could not stay where they were and use these skills except for sporadic periods of defense projects or mining developments. They could revert to their old hunting way of life. Or they could just sit there, using up the money that they had and hoping that something would happen. Actually all three courses were taken.

Some of the Eskimos went to Fairbanks for employment. For a number of them the way of life was too strange, and they tended to return to the Arctic. Some stayed at Point Barrow and tried to get by on hunting and trapping and occasional employment op-

An Eskimo Learning Tailoring at the Eskimo Rehabilitation Center, Frobisher Bay, Northwest Territories, Canada. *The Eskimo is quick to learn new skills and can then move easily into a modern economy. As he does so, the Arctic Ocean fringe land will be emptied of inhabitants, and a way of life that has roots going back 30,000 or more years in Eurasia will probably end.* (Information, Canada)

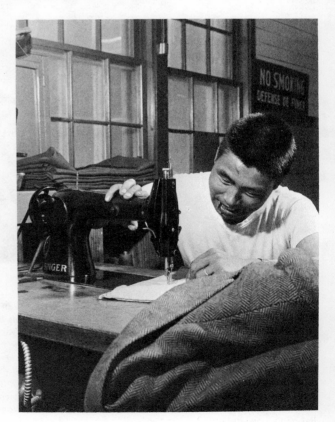

portunities. They continued to operate the coal mine in order to heat their frame houses, for oil lamps will not heat the airy space of a frame house, but now oil and gas from local fields are available to them. A few went back to moving along the edge of the Arctic ice seeking seals, for trapping fits better into the modern economy since the furs have value in our society. The overall effect, however, was to move some of the Eskimos out of the Arctic-ice edge way of life.

The impact of industrial society on similarly situ-

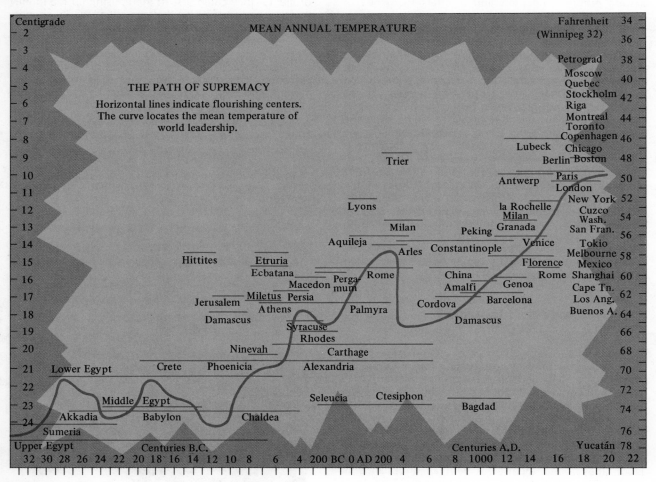

FIGURE 10-2   The Northward Course of Empire

This famous diagram prepared by S. C. Gilfillan was seized upon by Ellsworth Huntington and Vilhjalmur Stefansson. For Stefansson it became the title of his book, The Northward Course of Empire, *and a major prop in his thesis that the Arctic was a land with a great future. Note that Yucatán is in the twentieth century, placed at the bottom of the cultural scale. From 1000 B.C. to A.D. 1500 it would merit a horizontal line in keeping with its role as a flourishing center. South China would also qualify as would India. Notice also that Egypt at 3400 B.C. is the earliest civilization noted; today Mesopotamian civilizations would be listed much earlier. The figure is historically inaccurate, it focuses entirely on European history in the Old World, and it totally ignores the New World. Nonetheless, this diagram has had a powerful influence on geographic thought.* [*Reprinted with permission from the* Political Science Quarterly, *35 (September 1920), 395*]

ated peoples will probably have much the same result. New ways of life will be introduced, and new wants will arise that can only be met by continuing to live within the industrial society or at least producing something that the industrial society wants. The overall result is likely to be the abandonment of the simpler economies that have survived in isolated regions such as the Arctic and the withdrawal of the people who have lived in these areas to centers where employment is more readily available. This will be a continuation of the general withdrawal of mankind from areas that for tens of millennia have been considered useful lands, increasing the concentration of mankind in areas now considered attractive. This is an uneven matter for Alaska, as we have seen elsewhere in this text, is growing, and when special situations arise, mineral finds for instance, some areas formerly of little value and thinly held may become attractive and grow in population.

THE FRIENDLY ARCTIC: A MATTER OF PERCEPTION

Just after the turn of the century Vilhjalmur Stefansson traveled widely in the Arctic and became the advocate of high latitudes as the lands of the future. He made many pertinent observations that are useful even today for understanding man's perception of his environment in general. He cited an inquiry by Huntington of experts' views of the civilization of Iceland. On a scale of 10, with 10 for high civilization and 1 for savagery, the results were as follows:

Experts Questioned	Score for Iceland
Asiatics	3
Latin Americans	4
Americans	5
British	6
Scandinavians	8

At least two things are being measured: real knowledge, and tropical versus mid-latitude attitudes toward the Arctic. Around 1900 it seems that Americans, whether Latin or Anglo, neither knew much about Iceland nor could conceive of civilization in near polar areas. To this Stefansson opposed Gilfillan's famous chart on which Huntington and others based their dictum: northward the course of empire makes its way.

The measure of even northern European underestimation of the Arctic and the blanketing of all northern underdeveloped areas as having little value is also found in Great Britain's and France's haggling over whether France would surrender Canada or the sugar-producing island of Guadeloupe. Benjamin Franklin finally convinced the British that although Canada

(Upper) **The Stone-and-Sod House outside Reykjavík.** *Rarely used in Iceland today, it was the prevalent house type in Viking days. House ruins such as these have been excavated in Greenland and in the Ungava region of Canada.* (Hugh Gillespie for World Bank)

(Lower) **Modern Farm Houses in Iceland.** *Imported materials are used, and the houses are built to plans usually found in European-dominated areas. The bleak background is typical of Arctic regions.* (World Bank)

might not be worth much, politically it would be wise to remove the French from our borders. And so vast, rich Canada was traded for 600 square miles of a Caribbean island. It was comparable to Seward's purchasing all of Alaska at 2 cents per acre and being roundly criticized for it, and a perfect measure of Europeans, even northern Europeans, giving a low rating even to the subarctic.

Stefansson has listed the common misunderstandings of the Arctic as follows:

1. The Arctic is unendurably cold. Actually many areas in Montana and Alberta get 10 or more degrees colder during most winters than the coldest ever recorded on the Arctic Sea coast.
2. The Arctic is never warm. There frequently are 100-degree maximums in interior locations, and that is hot anywhere. In reality, under humid Arctic summer conditions one feels that heat more than one would in the desert.
3. Summer is so short that it limits the usefulness of the land. Stefansson replied that the summers were short but intense, and in a classic overstatement he said—"The North is the greatest potential grazing area of the world."
4. The Arctic is snow covered all year round. This is true only of the mountains. The lowlands are great prairies "green with grass and golden with flowers" in the summertime, and Stefansson said that Alaska alone has such areas equal in extent to all of England and Scotland.
5. The total amount of snow is large. Actually, away from the coast, the amount of snow is small, and the grass on the plains usually rises through the snow all winter.

Despite Stefansson's persuasive writing and lecturing, America's perception of the Arctic is still that it is a frozen wasteland from which we propose to obtain some oil but otherwise a most unfriendly and unproductive environment. It is not unusual that the man who had lived there found it friendly and could see economic potential, while those who had no experience in the area could see no good in it. The importance of perception of the environment as being more important than reality could scarcely be better illustrated.

ANTARCTICA: UNINHABITED OR UNINHABITABLE?

The Antarctic has had no human population until the present period of exploration. There are now so-called permanent bases on the edge of the ice cap, and in this period of exploration we have learned a number of things about this vast area. There are some ice-free spots, and they have been in existence for quite some time, for they contain the carcasses of seals that have been dated by carbon-14 as about 2000 years old. But these are mere specks of bare rock in an immensity of ice and snow.

Ice is now melting off of Antarctica. Should it all melt off, the sea level would rise at least 100 feet and probably 250 feet, and the habitable plains of the world would be inundated. Florida, Denmark, the Netherlands, and similar areas would be lost to mankind. Gaining Antarctica would probably be viewed as small recompense by people not accustomed to polar life.

Antarctica can be approached only with difficulty. It is surrounded by the world's stormiest seas. In most of the area a floating shelf of ice extends out to sea from the land, although this is hardly discernible since the land is buried under one to two miles of ice. Only on the Palmer Peninsula does the south polar land approach another land. Here the South Shetland Islands, off the end of the Palmer Peninsula, lie in nearly the same latitudes as their namesakes to the north of Scotland. They are separated from the islands of Tierra del Fuego by the Drake Passage. This is a 350-mile-wide waterway, swept by frequent wild storms and often containing floating ice. This was an absolute barrier to man.

Is this correctly stated? Would it not be more correct to say that this was an absolute barrier to the kind of men who occupied the southern tip of South America in prehistoric times? Certainly we can and do cross this water gap today. Captain Edward Palmer sailed to Antarctica in a small sloop, and a British man in 1967 singlehandedly sailed a small yacht through these stormy waters. The Norse crossed greater distances in more northern waters that also were storm swept and ice filled and settled in Iceland and Greenland. The failure of man to colonize the unglaciated tip of the Palmer Peninsula is much more realistically described as being due to the absence of the required means of navigation in the hands of the men in the adjacent land.

But could men, any men, live on a continent that, with insignificant exceptions, is covered a mile or so thick with ice? The answer is that they surely could. Had the Eskimos reached Antarctica, they would have found it a good land for their way of life. After all, they were not dependent on the land for any of their requirements. They could build a highly satisfactory house out of snow, make thoroughly satisfactory weapons and utensils out of bone, and fine clothing out of hides and furs. In Antarctica they would have found a sea filled with seals and whales. The seal population on the islands off the Palmer Peninsula is estimated to have been in the millions before our fur hunting virtually exterminated them. There are still great numbers of seals in the area. These include such huge forms as the leopard seal. The penguins also would be a valuable food source. The abundance of whales in the Antarctic Ocean makes this the major whale fishery of the world today, and as we have seen, the Eskimos knew how to catch these animals. The oil from whale and seal would supply the needed fuel and light that the Eskimos knew how to use in oil lamps.

It is one of the accidents of history that the parts of the world closest to Antarctica were not held by people with boating skills to reach it. However, even if the Tasmanians or the Tierra del Fueguians with their rafts and their bark canoes managed to reach the frozen continent, they lacked the clothing, housing, and economic skills to have survived. As with the Eskimo, it is too late to be stimulated to invent clothing, housing, oil lamps, and all the rest of the needs of survival in

the high latitudes when you arrive there. It is necessary to arrive in such difficult conditions with the equipment already in existence.

The earliest boat people to approach these frozen southern lands were probably the Polynesians. The people of New Zealand mentioned in their accounts great voyages so far south that they saw ice fields in the sea. But these were a tropical maritime people, and ice floating in the sea was sufficient reason for them to turn back. They had the boats but none of the other cultural equipment or the psychological orientation to take up life in such a land, even if they had reached it. Antarctica is empty, then, not because no men could have lived there but because no men with adequate equipment for living there ever came within striking distance of it. The solution to life in high latitudes was reached in the continental mid-latitudes, probably in proximity to the glacial ice masses. But the Southern Hemisphere does not have an environment to encourage the solution of polar-life problems, so a cold-adapted way of life never developed there. Antarctica is not uninhabitable; it is simply uninhabited.

The U.S.S. Glacier in the Ice Floes of the Antarctic. *An American discovered Antarctica, and the United States has led in its exploration. Although we now have few ideas on a future use for this ice-buried south polar continent, we also had little to suggest for Alaska at the time the United States purchased it.* (Official U.S. Navy Photograph)

Iceland. *Iceland has a great glacial cap, volcanoes that supply heat for some of the towns, and rich fisheries surrounding it. It is a treeless land and beyond the limit of grain agriculture. Yet it has been occupied by man for at least 1000 years and has maintained a high level of civilization. Men in seemingly better lands have often accomplished less.* (Haraldur-Olafsson for World Bank)

Summary

It is not likely that men are now going to move into the Arctic or the Antarctic in any appreciable numbers. The tendency for mankind to pile up in great urban centers is the rule of the day. We are more likely to continue this concentration of people than revert to the ancient pattern of a much more uniform distribution of mankind over the entire earth. The Arctic is neither so forbidding nor so attractive as has been claimed. It can be used. It cannot be made to yield vegetable and animal life in proportion to those parts of the world that receive greater sunlight and moisture. Sunlight is the energy source that underlies the entire food chain of plant production and animal feeding. The Arctic receives a minimum of solar energy; therefore it can produce only a minimal amount of the biological base for animal existence. The point, however, is that we are not using that minimum; and use of it is a human, not environmental, decision.

Review Questions

1. How does the case of the Eskimo living in the Arctic test the adage "necessity is the mother of invention"?
2. Why was much of the polar area in Siberia and part of Alaska unglaciated?
3. In what way does the Eskimo occupation of the Arctic resemble the American occupation of the grasslands?
4. Food sharing among the Eskimos is similar to that among the Yuma. Why should this be?
5. Discuss the reindeer-to-the-Eskimo case with special attention to the principles involved.
6. Since men lived at the southern tip of South America for at least 10,000 years, why did they not enter Antarctica just as men of the northern hemisphere invaded the Arctic?

Suggested Readings

Baird, Patrick D. *The Polar World*. London: Longmans, Green & Co., Ltd.; New York: John Wiley & Sons, Inc., 1964.

Birket-Smith, Kaj. *The Eskimos*. Translated from Danish by W. E. Calvert. London: Methuen & Co., Ltd., 1959.

Freuchen, Peter. *Book of the Eskimos*. Cleveland: The World Publishing Company, 1961.

Kimble, G. H. T., and Good, Dorothy, eds. *Geography of the Northlands*. American Geographical Society Special Publication, no. 32. New York: John Wiley & Sons, Inc., 1955.

Lee, C. *Snow, Ice, and Penguins*. New York: Dodd, Mead & Company, Inc., 1950.

Siple, Paul Allman. *90° South, The Story of the American South Pole Conquest*. New York: G. P. Putnam's Sons, 1959.

Sonnenfeld, J. "An Arctic Reindeer Industry: Growth and Decline." *The Geographical Review*, vol. 49 (1959), pp. 76–94.

Stefansson, V. *The Friendly Arctic*. New York: The Macmillan Company, 1943.

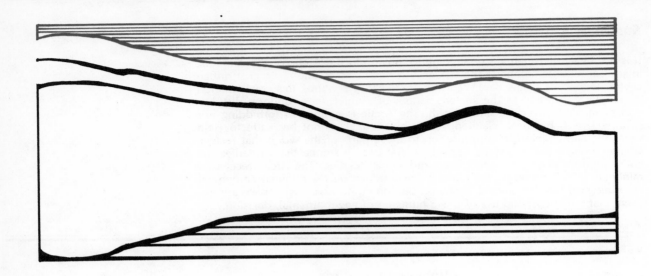

The Role of Physical Environment and Culture

THE problem with which we began is the nature of the relationship between man and his environment. The ancient and perennial answer has been that the physical environment makes man do what he does. Although this is sometimes softened to "influences him in what he does," to the physical-environmental determinist the physical environment is of primary importance. At times this book may seem to have said that the physical environment has very little to do with man's achievements, but obviously the environment exists and must be dealt with. The real need is for a balanced view. Fundamentally the answer our examination of man on the earth has given is that man's culture is the *primary* determining factor. The environment is *secondary* in its role in almost all cases. The same physical environment plays an entirely dif-

ferent role for the farmer, the manufacturer, the hunter, the pastoralist; therefore it cannot be the determining factor.

Emphasis on simple physical-environmental control over human achievement has been challenged on several counts. It has been pointed out by one group that man often alters his environment, and numerous examples of this have been given in the preceding chapters. We still have to ask just why and in what way man alters his environment, and the answer to these questions has been seen to vary from group to group, place to place, and time to time, even when the physical environment is held reasonably constant. A second approach notes that, except in very rare cases, man has choices open to him, and out of the many possibilities he chooses one or another. This is the position of the

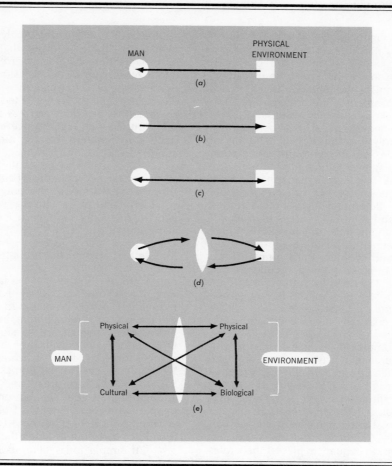

FIGURE 11-1 Man-Land Relationships

The nature of the relationship of man to his environment has been variously described: (a) The physical environment acts directly on man. (b) Man acts directly on the physical environment. (c) Each acts directly on the other. (d) Reciprocal action is indirect because it is conditioned by culture. (e) In reality the situation is complicated because of the great number of factors and the variety of their relationships.

possibilist school of thought, associated with the French geographer Vidal de la Blache. Like the statement that the environment is changed by man, the possibilist position leaves the reason for changes to be explained. A third, still later school of thought claims that man's perception of his environment is the principal determinant in how he lives in any particular time and place. Again one must ask where men get their perceptions. It is quite unfair to imply that those advocating any of these approaches were, or are, unaware of another deeper level of analysis: that they still had to pursue the "why" of differing impacts of man on the land, choices, and perceptions. The perceptions a people bring to the landscape determine their choices among possible ways of life and establish the use man makes of the physical environment, thus explaining the impact of man on the environment. The three approaches to solution of the man-land problem must ultimately go back to study of the origins of the particular people and the sets of ideas that derived from their past. Ideas govern perceptions, choices, and actions in any natural setting and result in the complex of man and nature that can be called the culturally determined landscape. That has been the theme of this book.

Culture—the complex of learned behavior, attitudes, language, preference for type of housing and dress, for milking or bleeding an animal, and so forth—is the dominant factor in human society. It is virtually impossible to define culture in simple terms meaningful to the nonspecialist. It is best understood through example, and we have studied many examples of culture at work. Thus in California, Indians, Mexicans, Spanish, and Americans, in sequence, played their roles in the identical environment, each making their choice of way of life from the numerous possibilities that existed in terms of their culturally determined perceptions. Their perceptions, the result of a long history peculiar to each group, determined, in the given circumstances of their moment in history, how they would utilize this one physical environment. The choices of the Indians, the Mexicans, and the Spanish have been examined. In this example and others the differences clearly were not physical-environmental, for the environment stayed the same. The possibilities that different men perceived changed again and again. Were the differences due to the variable races?

Race

One of the variables first considered in Chapter 1 was race, and the discussion showed that we cannot expect to find mental differences among the races comparable to the bodily differences. Analysis of the functions of minds and bodies showed that while human bodies have had different kinds of problems to solve, human minds have been faced with similar problems. If selection has operated on man, mental similarity and bodily divergence are to be expected. There is widespread agreement today that man has been subject to strong selection for the ability to live in human societies, and since this required certain mental attributes, men's minds should be much alike.

This seems to be contradicted by accomplishment and race correlations, for the claim is often made that the Negro race is the only one not associated at some time and place with a great civilization. The danger of deducing biological potential from cultural achievement has been a persistent theme of this book. The early center of growth and development in the Middle East favored the brunette Europeans and made cultural laggards of the lighter-skinned northwest Europeans. No one argues that the latter lagged because of race, for we now see clearly that their tardiness was a momentary thing. We tend to overlook the fact that civilization is a late happening in human history and the processes of change are so slow that the amount of time during which civilization has existed on the earth is too short for it to have achieved a universal spread. So simple an idea as the domestication of animals has had 10,000 years to spread, and all races are intelligent enough to grasp the idea, but only a limited number of people (with representatives from all races rejecting or accepting) took it up before A.D. 1500, simply because it had not had time to reach them effectively.

If simple ideas take so long, then more complex ideas may be expected to take correspondingly longer. One frequently hears of the cultural lag of the dark-skinned, curly-haired races as evidence of a racial inability. Examination of the accomplishment of these races shows that their advance has been nearly in proportion to their geographic proximity to the center of early growth of ideas: the people of India most advanced, Africa next, Melanesia next, Australia and Tasmania least. This is almost a simple distance relationship. It is reinforced by recognition of the role of the Sahara in reducing contact with Negro Africa and of the Indian Ocean monsoon in facilitating contact with India. Lest the physical environment be given more than its due weight, it should be noted that the Indian Ocean monsoon can facilitate contact only for seaborne cultures and the unchanging Sahara is less of a barrier now than it was 5000 years ago. The environmental restraint is always culturally conditioned.

One can put a mischievous twist to the race and civilization argument by presenting the hypothesis that there has never been a civilization without a Negroid element. Certainly all the circum-Mediterranean peoples have Negroid genes that date back at least as far as the Upper Paleolithic, when Negroid art appears in Spain alongside the European cave art and Negroid skeletons appear in northern Italy at Grimaldi. Egypt, Mesopotamia, India, and China were all once Negroid, and their people carry Negroid genes today. It is amusing to note that a racist-type explanation of the quick-subtle-shrewd characteristics of the south Chinese would be to point out that these are the people who inherited from the racial mixing that created the south Chinese subrace more Negroid genes than the purer Mongoloid northerners. That American Indians had Negroid traits among the earliest settlers has been a persistent suggestion by leading anthropologists such as Earnest A. Hooton and Roland B. Dixon, both of Harvard. One may then hypothesize Negroid genes in the American Indian civilizations, and an antiracist could then counter the stock argument, "No Negro group has ever produced a civilization" with "No civilization has ever arisen without Negroid genes." Actually neither argument has much validity, and one is as racist as the other. Neither really explains the localization of civilization in the way that the cultural-historical-geographical analysis does. Further, both overemphasize the Negro race. Examples of cultural lag characterize all the races at one time or another.

Similar situations of distant position, long lag, and sudden spurt of progress have been discussed for other races and places. For example, the American Southwest lagged in a hunting and gathering stage for millennia after agriculture began in Mexico. Although agriculture reached the edge of the Colorado plateau region about 3000 B.C., the plateau people did not take it up until about the time of Christ. Up to that time one might well have argued that they were too stupid to shift to agriculture. Thereafter, however, they rapidly changed and established the Pueblo culture, one of the cultural high spots north of Mexico in pre-Columbian time. Meanwhile, their kinsmen of the Great Basin continued to lag—not from racial causes or even from environmental causes other than distance from the source of ideas, with a consequent dilution of impact. For a relatively sudden change such as is observable in the Colorado plateau region, we postulated an intense contact situation, and the existence of such a situation in the plateau region is strongly supported in recent archeological analyses. In some as yet undiscovered way, but quite possibly due to immigrant groups, the plateau people were jolted out of their old ways and started on new paths. Neither man nor land changed, but the culture did, and man then perceived new possibilities in the environment. As a result, man used the environment differently thereafter.

Equally revealing are cases where with no racial change culture has declined. The Incas, Mesopotamians, Greeks, Italians, Spaniards, British, Egyptians, Chinese, Indians (of Asia) are a small sampling, including representatives of all the major races of mankind, of those who have scaled the heights and fallen in some degree. Europoids, Mongoloids, and Negroids at some time or place have been cultural tarriers. Neither accomplishment nor failure nor inventiveness seems to be racially linked.

No people at any time anywhere have invented more than a minute part of their culture, and people who were so located that they did not have access to the mainstreams of ideas stagnated. This too was regardless of race, for the isolation of the Ainus, the Bushmen, the Tierra del Fueguians, and the Veddas spelled stagnation. Isolation is not necessarily a matter of distance. The boatmen of the Pacific were less isolated on their specks of land lost in the immensity of the Pacific than the people of the Great Basin with advanced ideas nearby in coastal Oregon and southern Arizona. The Polynesians were mobile, and their fast-sailing craft compensated for the vast distances of the Pacific. The Great Basin people were sedentary, place-bound footsloggers, culturally limited to their local environment.

Often isolation of an area led to so great a cultural lag that advanced people bypassed it, causing the cultural gap to widen further. This was conspicuously so with aboriginal California and Australia. In both cases unsophisticated observers concluded that the cultural laggards must be genetically inferior. Such a view suffers from a shallow time perspective and lack of comprehension of cultural processes. Melanesia is another example of an area where cultural lag is probably the cause of cultural bypassing persisting into the twentieth century.

At other times race is invoked as the cause when it is seen that biologically distinct groups living side by side persist in culturally different ways of life, often associated with different economic levels, for long periods of time. Most cases are readily explained as examples of cultural persistence of value systems linked with educational and economic opportunities or liabilities. Gypsies, for instance, are a biological unit which, as their blood groups show, has maintained racial purity through intermarriage over a period of centuries—since they left their homeland in northwest India. People from this area would be Europoid in race. Their culture makes of them a parasitic group, for they refuse to do what all others call productive work, living instead by fortune-telling, confidence games, and theft. This they do regardless of the country, climate, or opportunities they have for education and thus entry into the normal culture that surrounds them. No one argues seriously that this is a genetic pattern. Why then argue genes when some of the American Indians have failed to make the change to twentieth-century technology? Or, for the reverse case, why argue climatic stimulus or genes when descendants of the Pilgrims continue to display a penchant for hard work and success in business and professions even after several generations spent in climates far different from stormbound, long-winter New England? The American South, another example, was an area of leadership in the colonial period and then stagnated; this was "explained" early in this century as deterioration due to climate. Today the area is showing greater economic vitality than New England, and no one now argues for climatic stimulation. As for the race argument, the same races have been there throughout. With the burden of management resting on the white race, accomplishment or lack of it is a simple measure of changing economic, political, and cultural forces.

The situation of the Negro race is equally interesting, showing little accomplishment in the recent past, with brilliant exceptions such as George W. Carver, but rising accomplishment now, with cabinet members, mayors of large cities, and millionaires as measures. Predictably the change among the mass of Negroes will be slow. Detailed studies would show it was slow in the beginning for others: for the British before the time of Christ, for the Japanese shortly thereafter, and for the dwellers of the Colorado plateau who were exposed to agricultural beginnings by 3000 B.C. but failed to take up the opportunity until about A.D. 1. Slow cultural change has characterized all the races; hence cultural lag does not prove genetic inferiority. We must add that it does not prove genetic equality either. We conclude that of the major factors—race, physical environment, culture—race is probably the least important variant.

Physical Environment and Culture

Much of the problem can be put into simple diagrams, as in Figure 11-1. At *a*, the arrow suggests that the physical environment acts directly on man and that this is the only action occurring. This is clearly false. At *b*, the diagram shows the opposite: man acts on his environment, and there is no contrary action. This too is false. At *c*, the double-pointed arrow suggests action and reaction: the environment acts on man and man on the environment. This is better, but it still is misleading, for the environment rarely acts directly on man,

and the way in which man reacts on the environment is not described when it is simply attributed to man in general. Men with different cultures react differently to similar environments. Partly this is due to their cultural interpositions—clothing, housing, heating, and so forth—and partly it is due simply to their outlook. Similarly the direct action of men on the environment is culturally formed. Farmers may cut and burn that part of the forest which they need for agriculture, but pastoralists are likely to burn down the whole forest.

Cases exemplifying most of these relationships have been presented in the preceding chapters. Striking examples of the direct action of the environment on man come from the time when the races were formed. Thus dark skins are a direct response to high intensities of sunlight and light skins a response to low intensities. Blocky bodies and small extremities are a response to great cold, and their opposite to heat load (either high heat or moderate heat with high humidity). Shifts in basal metabolism (Tierra del Fuegians) or in circulation (Australians) are examples of direct environmental (climatic) impact on physiology. But such reactions are rare and associated with low cultural levels.

The opposite set of reactions (man on land) are probably always culturally moderated. When man uses fire, grazing animals, plows, hoes, or any other tools, he is operating on the environment through a cultural intermediary. More subtle differences are evident where the culturally determined psychology of the individual determines his feeling toward his environment, and his perception of the nature of his environment determines his response. One classic case showed the effect of the grasslands on people accustomed to forest land. One group trekked across the richest soils in America, seeing no promise of fertility in a land that did not grow trees. Among the women of the first generation that settled on the treeless grassy plains, there was a high frequency of mental breakdown, partly induced by the barren, treeless, flat, monotonous, strange land. Their daughters had little of this feeling. Today we still see the southerner overreacting to mild cold and the New Englander overreacting to summer heat in the South. Individual reactions determining physiological changes were given earlier. While there are physiological shifts to be made in adjusting to heat and cold, these normally occur within a month, but psychological (cultural) adjustment often takes years. For more concrete cases we can cite the stubborn pressing of agriculture beyond a reasonable climatic limit in the American Southwest by people such as the Hopi, the excessive use of good farming land for pastoral land in much of Argentina, Uruguay, and Chile, and the near mania for terracing land in some parts of the world when men in other similar parts of the world see no need for it.

The diagram at *d* in Figure 11–1 shows a lense-like figure between man and the environment. This lense can be thought of as the filter through which the physical-environmental influences must pass before they can act on man, and through which man will view the environment and thus have his action on the environment determined. The diagrams could be complicated indefinitely; the final diagram *(e)* only begins to indicate the complexity of the problem. Man is physiological, psychological, economical, political, spiritual, and sociological. The physical environment is composed of weather and climate, fresh- and salt-water bodies, flora and fauna, both macro- and microgeology, soils, and landforms. Each of these items could be subdivided. There is potential action and reaction between every element of the cultural and the physical environments.

The environment may act directly on man. No one is comfortable at 120° F. Few men can live above 15,000 feet, and most men are not comfortable at 11,000. Although some men do live in places of such heat (the Sahara) and great altitudes (Peru, Bolivia, Tibet), areas of such extremes are relatively scarce on our earth. In contrast, over most of the earth the environment is acting on man less than man on the environment. When man burns the forest and changes an area to a grassland, he has obviously acted directly and decisively on the land. He makes equally great changes when he clears land for farming or builds cities, and he may make very extensive changes with grazing. Even the slash-and-burn migratory forest farming changes the nature of the forest that is so used. Thus for most of the earth—the farmed, grazed, lumbered, occupied part—the impact of man on the land is much more evident than is any effect of the land (climate, soil, resources) on man. As we have seen, the impact of the environment on man has become more remote, indirect, and limited as man's culture has grown, and his impact on the land has become ever more direct and extensive. The impact of the physical environment on unclothed and unhoused man was quite different from what it became once he could defend himself against climatic forces, and it is increasingly lessened today with heating and air conditioning, disease control, and so forth. The impact of man on the land increased when he got fire, when he began farming, when he substituted bulldozers for horses; with nuclear power he now proposes even greater modifications of the earth. The man-land relationship over at least the past million years is culturally conditioned.

Theoretically there should be an enormous number of combinations of man and land. In one sense there are: no two humanly occupied parts of the earth are exactly alike. Nevertheless, there are broad similarities among them. The order that emerges from a study of the potential chaos is primarily a cultural one. There are cultural worlds, and their similarities are greater than the innumerable potential combinations would lead one to expect. Further, these cultural worlds cut across physical-geographical boundaries. The European cultural world is a good example. It not only covers a variety of climates and landforms in Europe but is spreading rapidly into tropical rainy climates, where great cities such as Rio de Janeiro resemble their European counterparts in city design, architecture, dress, and automobiles. In many ways these and similar items would be better in one climate than another and hence should not be alike in different climates. Northwest European male clothing, for example, is an adaptation splendidly suited to Arctic conditions and excessively ill-suited for heat and humidity. But Europeans wear their jackets and Hindus their turbans with little regard for heat or cold. China's cultural pattern is an amalgam of ideas assembled from all the adjacent land areas, including some from the arid northwest, a climate somewhat like that of South Dakota. The accumulation of ways of doing things, their extension into other areas, and their maintenance over long periods of time are the forces that create the cultural worlds we see. These are remarkably free of climatic bounding and are historically derived.

To understand man's use of the land, his effect on the land, and the land's effect on human cultures, we must then focus on the cultural processes. The most important of these for our problem are invention and diffusion. These two processes act through time and space causing cultural change, and it is cultural change that leads to the variety of human landscapes of the earth. Not every cultural change appears in the landscape, and not every change in the landscape is due to man. The leaves on trees in mid-latitudes change color and fall without any reference to man's activity, but the amount and kind of tree left in the landscape may be largely determined by man. Changes in jazz styles probably do not affect the landscape, but changes in art styles may, as seen in the modernistic buildings that add their sometimes odd accents to the landscape.

Invention and Diffusion

Two processes have dominated this discussion: invention and diffusion. An invention always stems from an unusual individual. He does not operate in a vacuum, however, for he is enmeshed in a network of physical and cultural forces. Rare indeed are the cases where the environment created a need that led or compelled man to make an invention, and inventions that we can prove to have been deliberate to meet a felt need occur late in human history. The Americans on entering the grasslands assembled inventions from over half the earth and adapted them to their use. The Eskimos entering the Arctic did much the same thing. The human response to environmental challenges is more accurately described as adaptive than inventive. Still, invention had to be made by someone, somewhere, for some reason. Histories of inventions that we know well suggest that play, accident, and antecedents are the major causes.

Bicycles, airplanes, and automobiles are inventions which perhaps were playthings that became useful. Their usefulness was a result of their invention; they did not appear in answer to a need. The bow and arrow has been hypothesized as the by-product of the musical bow or some game, for no one has been able to imagine how the completed device could have sprung into being. How does one simultaneously link up the springiness of wood with the storage and release of directed power, via shaping of the wood and tensing it with a string, and combine this with a projectile notched to take the string, pointed and balanced for accurate flight, and propelled with sufficient power to be a lethal weapon? The device is not functional until all the elements are combined, and it is most reasonable to look for intermediate steps. Musical bows are ancient and important in Africa, the probable place of the relatively recent invention of the bow and arrow as weapon. Perhaps this invention derives from a play impulse and the accidental propulsion of a stick normally used to strike the string (instead of plucking it). This may have led to a game of flip-the-stick, and eventually to shoot-the-stick and thus the arrow. Rather than a deliberate invention based on need, the weapon seems more likely to have been the result of play. The musical bow may be the antecedent to it, and the steps may have included accident and play.

Similarly the boat and the steam engine are the antecedents of the steamboat, which Fulton did not invent but only made a paying proposition, by marrying the governor's daughter and getting an exclusive steamboat franchise on the Hudson River. A long line of French and British inventors preceded Fulton, and propellers and jet propulsion were tried and rejected in favor of the paddle wheel, whose name tells the story of its origin.

Planned inventions in response to felt needs are very late phenomena, and they are poor guides to past conditions. Sugar from beets was a direct result of the British blockade of Europe during the Napoleonic Wars and a resultant conscious search for sugar. An accurate clock was the result of a felt need for determining longitude, and this required, at that time though not now, a means of comparing time at a fixed point on the earth with local sun time. Today we have vast research institutes seeking to make new combinations of old ideas or to find new ideas to produce solutions to problems. As a result we have produced utilitarian items such as refrigerators and what began as nonutilitarian items such as radios. In the past the stock of ideas was small, the number of men free to indulge in combining them was minute, and there was no systematic seeking of new ideas and combinations. At rare intervals some genius made the mental or physical connection, and a new tool (idea or item) was born. It is exceedingly difficult to find one that is probably a response to a physical-environmental need. Accident plays a role: a man turning a crystal in a band of sunlight accidentally discovered how to break light into its component colors; electrical impulse and nerve action resulted from observation of freshly dissected frogs legs hung out where a breeze bumped them against an iron railing where an electrical charge caused the legs to twitch; the discovery of penicillin resulted from accidental contamination of a dish in which deadly germs were being grown in a laboratory. In all these cases antecedents were also present, and unusual men perceived the accident as significant.

However it arises, an idea has a life history. It may die young. The idea of the wheel certainly existed in Mexico in pre-Columbian time, but it seemingly died, having achieved only a slight spread as a novelty —wheels on toy animals. The true arch was known in Mexico, for we know of at least three examples, but it too failed to spread, and the idea seemingly died out. We do not know whether the wheel and the true arch were ideas unsuccessfully introduced from the outside or ideas created independently in America. Since independent inventions have been shown to be rare, it seems more probable that they are introductions that failed to receive acceptance. Whether the idea occurred locally or was introduced, once the Mexicans had rejected the trait, those who were learning from them were deprived of the opportunity to adopt it. Thus, though the wheel might have had little utility in forested and mountainous middle America, it could have been easily used on the plains of North America had the Mexicans only adopted it and spread the idea.

Failure to adopt inventions is naturally very hard to trace in prehistory, for the very rejection prevents a record from being made. In modern time there are numerous examples such as the temporary rejection of the Colt pistol. The Browning automatic gun, the submarine, and many other inventions were at first re-

A. UNIFORM EARTH

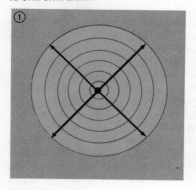

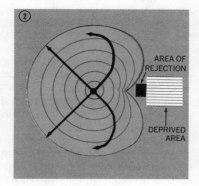

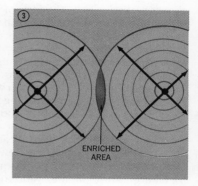

B. NONUNIFORM EARTH

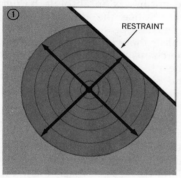

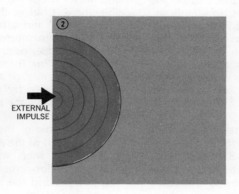

FIGURE 11-2   The Flow of Ideas on a Uniform Earth and on a Nonuniform Earth
*On a uniform earth an idea could flow out from its center of origin in concentric circles
(A1). It could meet with rejection on a purely cultural, nonracial basis, and the further flow
of the idea would then be deformed, with some parts of the uniform earth thereafter deprived of
access to the idea (A2). Ideas arising at separate points could overlap in the spread, and the
area with the double set of ideas would be enriched (A3). If one idea were "boats" and the
other "steam engines," the enriched area could leap to "steam boats."*

*On a nonuniform earth ideas introduced from outside might spread from a continental
margin (B2). An example would be the British colonization of North America. There could, of
course, be many restraints, either political-cultural or physical (B1). In the settlement of
North America the British were constrained by the French to the north and the Spanish to the
south. The spread of agriculture was constrained in North America by cold in the north, but
this was a secondary-level decision, for the prior decision to live by agriculture created this
constraint. For the fur trader, there was no such constraint.*

jected, and the inventors had to wander about, often
from country to country, to find a receptive group. In
the past the opportunity to travel was more restricted,
so more ideas probably died untried.

We have innumerable cases of the spread of
ideas. The idea of domesticating plants is most useful
because the plants add certainty to our judgment that
we are looking at the same idea in two areas. It is

harder to be sure of a concept such as a king, but if there is transfer of a complex of ideas such as divine-king-priest combined with royal incest, we are more certain that it is the same concept in two locations. Similarly the alphabet is a very complex idea, and its travels can be charted quite accurately. From studies of known and probable diffusions, as far back as our knowledge reaches, we see men at rare intervals creating new ideas and then spreading them to others. Stone tools of a kind called pebble tools first appear in East Africa about two million years ago and spread over nearly all of the Old World outside the Arctic regions. They are also present in the New World, but whether by diffusion or independent invention is hotly debated. Metallurgy arose in the Middle East and slowly and steadily spread by conquest, invasion, trade, and simple learning. The so-called Australian ballot, so simple an idea as privacy in voting, is a recent invention, made at one point in time and space, which has spread widely through the world. Rarity of invention and the frequency of spread (diffusion) of ideas dominate the growth and development of men's way of life on this earth. Therefore those who will not, or are so located that they cannot, share in the flow of available ideas fall behind in cultural growth.

There are geographical constraints: distant peoples, especially in the past, fell behind because of their isolation. Tasmania, Tierra del Fuego, Baja California, and the Great Basin are classic cases that have been discussed. However, physical geography is not now, and for some millennia has not been, an absolute determinant. Ideas crossed the vast oceans to reach America, bypassing in the Pacific the closer lands of New Guinea, Australia, and Tasmania. Distant New Zealand was more wtihin the flow of ideas into the Pacific from the mainland than was nearby New Guinea. This pattern continues today as New Zealand, despite its position at the antipodes of the land hemisphere, shares fully in the modern world's growth and development while Africa, even Mediterranean North Africa, lags. Human, political, economic, and historical forces can overcome mere geographical distance.

A series of situations of inventions and diffusions is presented in diagram in Figure 11-2. If an idea was born near the center of a uniform earth similar to that discussed in Chapter 1, a simple radial spread might be expected (A1). However, simple individualistic acceptance and rejection could make a more complex pattern (A2). Recall that even with uniform man (no racial differences), individual reactions or situations of cultural lag can create just such patterns. The existence of two centers of origin could create a pattern such as is shown in A3. An approximation of A1 is the spread of agriculture in the Old World; of A2, the flow of ideas past Australia to the New World; of A3, the meeting of Chinese and Indian influences in Indonesia. Ideas coming from elsewhere might have a pattern such as is diagramed in B2. The British settlement of North America approximates this. In many situations the spread of ideas or ways of life may be subject to restraints. These might be nonracial, nonphysical-environmental as in A2, or they might be cultural or primarily cultural and secondarily physical constraints: the French and later the British to the north, the strong Iroquois Confederation in what is now New York, the French and Spanish claims along and beyond the Mississippi, and the Spanish claims in the south and west. There were also physical restraints of variable force. Cold in the north was an absolute barrier to European agriculture. Notice the double cultural valuation—agriculture and European type—and recall that for reindeer pastoralists there would have been no barrier, and for fur merchants there actually was no barrier. The role of the physical-environment barrier was conditioned by the culture in question. Accordingly culture was the primary cause, and the role of the physical environment was secondary.

The role of the Appalachian barrier in controlling the spread of the American colonists has often been exaggerated. Men with similar equipment (wagons, axes, and rifles) crossed the vastly greater barriers of the Rockies and the Sierra Nevadas to reach the West although they were constrained by the physiography and their use of wagons to certain passes. In broad interpretations of American history little importance has been placed upon these physical barriers as compared with that attached to the friendly little Appalachian hills. Of equal significance one wave of settlers found the grasslands a physical-geographical restraint to be overcome in their search for good forest lands, while later settlers found them to be enormously useful. Similarly the high mountains that were avoided in favor of the low passes in the days of covered wagons are now so attractive to ski enthusiasts that some of the most expensive real estate in the world is in mountain resorts such as Aspen, Colorado.

The problem of overcoming human resistance to ideas is more significant than the problem of overcoming mere distance. Actual cases illustrate this most clearly. Southern Italy held the cultural lead when influences from the eastern Mediterranean focused on that region in pre-Roman times. When in post-Roman times warfare and piracy made it an exposed area, especially during the wars with the Muslims, the standard of civilization declined. In more modern times the growth of northern Europe placed northern Italy in an advantageous position, and the cultural positions of northern and southern Italy are now reversed. Energetic redevelopment plans suggest a reemergence for southern Italy. The stage setting, the physical environment, has changed little while the various acts of the play were performed.

Africa may be treated diagramatically, ignoring the physiography and climate, to obtain a pattern approximating the reality of the cultural scene. Given the general directions, times, and intensities of cultural impacts, a pattern emerges that approximates the facts, and this tends to validate the predominance of the cultural-historical factors. Two constraints are added: the Sahara, a climatic limit, and the tsetse fly, a biological factor. In this case they probably reinforced the cultural factors.

The spread of ideas is dependent upon the generation somewhere of new ideas to spread. Perhaps the outstanding feature of any study of this situation is the scarcity of new ideas. The obtuseness of the human mind in inventive situations is remarkable. Australian women dug tubers for millennia, carefully replacing the vines, knowing that a new tuber would then grow. Good growing areas are described as looking almost like plowed fields. Surely here mankind trembled on the verge of the invention of agriculture, but the final step—the deliberate multiplication under human care (weeding, defending against animals and insects)—was not made. It is typical of ideas that even simple ones are simple only in retrospect.

This becomes even clearer when the cases of failure to transfer an idea are examined. While the idea of domesticating a plant or an animal is actually an extremely difficult concept, once the idea that man can take other organisms and manage their propagation is perceived, it should be easy to transfer it. If dogs, why not pigs? If cows, why not buffalo? If corn, why not

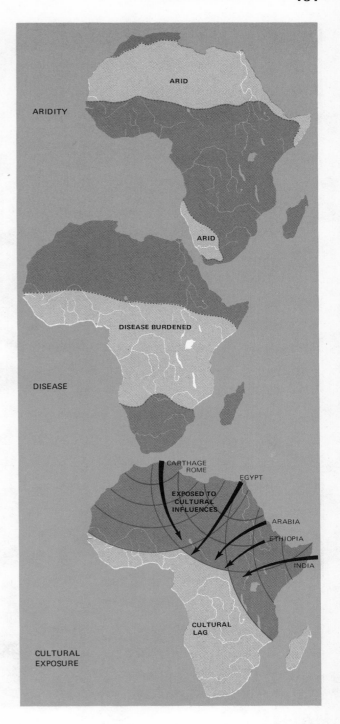

MAP 11-1 Africa: Cultural Influences and Environmental Restraints

Cultural growth in Africa was most conspicuous in the north and east, the regions most exposed to ideas from other peoples, despite physical restraints. North Africa is arid, and central Africa is disease burdened, notably by sleeping sickness spread by the tsetse fly. The best land is Africa is probably in Southeast Africa, where neither aridity nor disease is so prominent. However, this area was distant from the world of ideas and was less advanced than areas that were poorer in physical geography.

Threshing Grain: Diffusion or Invention?
The Near Eastern pattern of threshing grain (upper) includes driving cattle around on the grain, often dragging a sled of some kind. This has been practiced for many thousands of years. (World Bank) In Asia men thresh rice by beating the heads over a tub (lower). Which is the better way? Since both had tubs and both had cattle, why did one choose one way while the other chose another? Since each knew of the alternative method for centuries, why did one or the other not change? Could you predict what pattern the American Indians adopted in terms of wheat growing? (Mary M. Hill for World Bank)

wild rice? If corn, beans, and squash, why not potatoes or tobacco? But we have reviewed a whole series of failures of this type. It was not obvious to such dedicated agriculturists as the Hopi that the wild potatoes they gathered could be cultivated. In part they lacked a close parallel, for corn, beans, and squash are fruit and seed crops, and the potato is a root crop. (As an example of the difficulty in transferring, we eat the Eastern tree squirrel and consider it a delicacy, but in the treeless West the ground-dwelling squirrels are never eaten!) More important, corn, beans, and squash came to the Hopi as agricultural plants. The wild potato was a plant they had probably gathered for millennia, and it was indelibly implanted in their culture as a wild plant to be gathered; planting it was unthinkable to the masses. It was not that the Hopi were averse to accepting new food sources. They accepted peaches, watermelons, and wheat from the Spanish, but these were introduced as agricultural plants. The solitary genius who might have said, "If we can plant and manage corn, beans, and squash, we can manage any useful plant, especially one that we know well such as the wild potato," either did not exist or was not listened to. One might ask if they needed the potato. The answer must be that they did, for despite careful storage of food, the Hopi occasionally ran into drought years that caused famine so acute that in one drought they threw children off the cliffs rather than watch them starve to death. Their maize, beans, and squash were being grown far from their native habitats, were vulnerable to minor weather changes, and can better be described as growing in the Hopi country in spite of the environment rather than because of it. The native potato would have been at home in the climate and probably would have been easier to grow and more resistant to the vagaries of the local weather.

Even more startling was the Hopi attitude toward tobacco, which they smoked as avidly as our chain smokers of today. If tobacco came up in their fields, they carefully protected it and removed a leaf at a time for curing. When the plant matured seed, they might even shake the stock gently to spread the seed slightly. But why did they not gather the seed and plant it and have an abundant supply? As we have seen, such action, or lack of it, is commonplace and cuts across race, place, and time. Negro Africa strove to use Middle Eastern cattle in the zone of nutrient-poor grasses and tsetse fly infection when there was an abundance of animals present that were adapted to the tropical grasses and resistant to the tsetse fly. That this lack of perception was neither a racial fault nor a result of the "deadening tropical climate" can be seen by considering other cases. The Americans ignored the wild cow of America, the buffalo, which was highly adapted to the heat and cold of the American grasslands. European cattle must be housed and fed in northern plains winters, and southern plains cattle may be killed by cold waves. The acclimated, toothsome buffalo would have been far better, but the nearest approach to attempting to use it was to cross buffalo with cattle to produce a hybrid called cattalo. Why not just buffalo?

We have reviewed the failure of the Maori to perceive that the flightless wild birds of New Zealand, which they called chickens but treated as wild game, could have been superchickens. Nearly all of mankind had comparable opportunities and did not see them. Nearly all of mankind had domestic dogs, and most of mankind until recently had no other domestic animals. Further, much of mankind ate dogs, kept pets of many kinds, and frequently went hungry. People did not make the seemingly easy transfer of ideas even when spurred by need. Thus in North America alone closely related forms of pigs, goats, sheep, cows, chickens (the turkey outside Mexico) were all common as pets but never utilized as domestics.

Let us extend the idea of domestication. What is gathered, hunted, or perhaps expensively eradicated today that is potentially useful? There are showers of acorns to a nuisance level in many American yards, yet acorns are more nutritious than maize. Moreover, oak trees require no annual planting, cultivating, and harvesting, and they expose the soil to no erosion risk. Parts of America are plagued with locusts; these are nutritious and delicious. Starlings by the millions foul our cities. Why not harvest and eat them as the Spanish do? Or for slightly more adventurous application of the idea of domestication in this day of talk of harvesting the sea, why do we not milk whales or seals or porpoises? Or at least herd them with submarines, keeping track of them by means of electronic equipment (replacing the old cowbell), and so use whales to graze the seas as our cattle do the grasslands? Some of these thoughts are bound to seem farfetched or impractical, perhaps even repulsive, but this is the normal human reaction to a new idea. The list could be extended immeasurably, though the average human being is not aware of this. Through most of human time man has not paused to philosophize, that is, try to generalize and thus see the possibilities about him.

This leads inevitably to the problem of multiple and plural inventions. Both anesthesia and calculus were invented simultaneously in recent time. Does this mean that independent inventions probably occurred in the past? The answer seems to be no. Modern simultaneous inventions arise from a stock of ideas held in common by the equivalent of a tribal group of the past. In the eighteenth and nineteenth centuries this tribe could be called the Learned Ones. As with political units there has been much fragmentation, and we must now deal with various tribelets such as chemists, physicists, and mathematicians. An odd characteristic of these groups is that they transcend place boundaries and with modern means of communication know more of what members of their group on the other side of the world are doing, thinking, and saying than earlier men knew about brethren on the other side of the mountain. That they say, think, and do similar things and make similar discoveries is due to their participating in a closely knit intellectual society. We confuse the geographical separation of such societies with mental separation. Formerly men in local areas were isolated in their thought patterns; now ideas travel widely throughout the world.

There are numerous examples. In colonial days Benjamin Franklin was one of the world's leading experimenters with electricity. We remember the kite experiment and overlook numerous others, including static generators capable of working machinery. His work was stimulated by European experiments, reports of which he awaited eagerly. By return ship went reports of his work. The European and the American work built on a common foundation in a specific area of interest shared by a few men at a particular time and place. These electrical inventions were parallel but hardly independent discoveries, and they were localized—northwest Europe and northeast America. They did not pop up in China, South Africa, and Alaska, yet the phenomenon of electricity, especially in lightning and static electricity, is everywhere. Men wearing furs in dry Arctic air must often have got a rude jolt from built-up static charges. The cold dry air of the physical environment set the stage; the use of furs ensured that static electricity would be generated. The sparks must often have been seen and felt, but only when men pooled their knowledge in a shared interest was the physics of the generation and application of electrical power solved.

The invention of anesthesia was similar. In one part of the world chemistry produced nitrous oxide. The glue sniffers of the day found that they could get a kick out of breathing it, and the practice became moderately widespread in some circles. It was noted that people under the influence of nitrous oxide often injured themselves but felt no pain. Two or more doctors and dentists seized on it as a useful means of controlling pain during operations. This could only occur where a peculiar set of circumstances existed:

chemistry, a group of gas sniffers, a medical profession sufficiently developed that it conducted painful operations and hence could see in the gas a means of controlling pain. It is hardly surprising that two men made the connection more or less simultaneously and even less surprising that this happened where and when it did. It could only have happened at that time and at that place.

All inventions whose history we know well show localized origin or, as the cases above show, origin in one cultural area where men shared a set of antecedent ideas, and then spread. With a system such as the alphabet no one disputes this, although it is admitted that if we did not have the record, we would be hard pressed to prove the relationship of some of the extreme ends of the series. With the growth of knowledge in this century it has come to be recognized that strong links existed between what were formerly considered to be separate centers of growth and development. Egypt, Mesopotamia, India, and China were once thought of as autonomous. Now we know that they were linked together and many if not most of the independent inventions credited to these centers were the result of spreads from one to another.

Shifting to systems that are more closely tied to the environment changes little. As we have seen, there is little uniformity in house types in the arid lands, and the one zone of uniformity (the Old World dry lands north of the equator) is composed entirely of contiguous areas where we have abundant evidence for the spread of ideas. Neither are grasslands marked by one type of dwelling, nor are forested regions. The log cabin, so firmly fixed in American minds as a natural response to a forested environment, was brought to America by the Danes and spread from Delaware. The Pilgrims of Plymouth and the Cavaliers of Virginia did not know it. Even need for housing is ineffective. Natives of South Australia, Tasmania, and Tierra del Fuego, for example, all needed houses by any usual measure of need but lacked them. The problem of invention is far too complex to be explained by the environmental determinist's saying that "necessity is the mother of invention." Necessity is determined by perception. Hopis needed potatoes; Tierra del Fueguians needed houses, good boats, and adequate clothing. The Eskimos and their Athabascan neighbors needed and still need domestic reindeer. Either the perceptive and creative minds had not occurred, in those societies and concerned themselves with the problems, or they had failed in the task of selling their idea. Clearly then, the perception and adoption of new ideas are rare happenings among mankind.

Population and Knowledge

Once the creative mind has produced an idea that increases man's control over his environment, there is a potential and usually a real population growth. This means that there will now be an increased chance for the occurrence of great minds. The assumption here is that great minds occur as a proportion of a population; they are simply the very high end of the normal distribution curve of ability. If their rate of occurrence is 1 per 100,000, small populations must have very few of them, and to the extent that advancement is dependent on the production of superior minds, small populations must lag. The creation of ideas by increasing the number of potential creators must, then, all other things being equal, provide more opportunities for further ideas that should in turn further increase the population; hence more creative people and still more ideas. If this progression is true, the growth of knowledge and of population over the long run of human history should follow a curve of acceleration. This is a commonplace in diagramming the growth of knowledge. This also is the probable shape of the human population growth curve. The very long Lower Paleolithic period was one when little happened over great lengths of time. It was also a time of extremely low human population. For example, a population figure for the British Isles during a Lower Paleolithic period as low as 1000 people has been suggested. While this may be inaccurate by a factor of 10 or more, it is true that the population was small compared with today. If genius occurs at a 1 per 100,000 rate, there would be 1 per 100 generations, and he would have no like mind with which to communicate.

As inventions were made and population began to grow, there were more men in contact with other men. More ideas were exchanged and the brighter men received greater stimulation from current ideas. We realized in the discussion of Ethiopia (Chapter 8) that men with advanced knowledge communicate more fully with men of like advancement. The feedback system tends to reinforce the advantage of the region with the head start. Finally in some areas the stage was reached where every man need no longer be a jack-of-all-trades; some men could specialize in certain pursuits and not only become more skillful but could create better ways of doing things. Ultimately growth of knowledge led to increases of productivity so that some men could be spared for such seemingly unproductive pursuits as astronomy, mathematics, map making, and the like. These pursuits laid the foundation for even greater growth of knowledge.

Diffusion Today. *On Hokkaido, Japan's northern island, dairy farms have been established (upper). The scene could easily take place in Wisconsin, which has a similar climate. (United Nations for World Bank) Note the dairy in Kenya (lower). Not only are the cows similar but modern milking machines are in use (Per Gunvall for World Bank) Clearly we are looking at the spread of a whole complex of ideas. This is diffusion, one of the most constant and characteristic cultural traits of mankind. This case is easy to decide, but what about inventions or ideas that seem natural for man to develop? Consider terracing. For how many lands is it illustrated and discussed? Is this more or less natural to do? How would we decide whether it is diffusion or invention or at times both?*

Today in our alarm over the population explosion we tend to overlook a number of things. To begin with, we are not experiencing a population explosion but an *implosion*. Everyone is trying to move into the large urban centers; this is true even in relatively empty lands. For example, in Argentina, Brazil, and Australia, all lands with enormous areas and relatively small populations, about half of the total population lives in the three major cities of each country. In the United States half of all the rural counties lost population in the census decade 1960–1970, while the cities and urban counties grew. Settled areas are returning to wilderness. Bears are so common in the southern Appalachians as to be a menace to campers, deer are increasingly troublesome in many orchard areas, and beaver are coming back so strongly that they are a nuisance because of their tendency to build dams in drains and culverts. For those who yearn for the wide, open spaces, there are more available each year as those who have lived in them choose the amenities of schools, hospitals, symphonies, art galleries, and the economic opportunities that are concentrated in urban clusters.

There is almost no limit to this movement. All the people in the world could stand up in Baltimore County, Maryland, if one allows as space 2 feet by 3 feet for each person. All the people in the world can be housed with great ease in 100-story apartment houses on half of Long Island. The usual first reaction to these figures is utter disbelief, and this is a measure of the degree to which the population alarmists have created a false image of the mass of existing mankind. A second reaction is to state that such a concentration of people would be both undesirable and impossible to feed. The point of the illustration is to argue neither for the desirability of such a concentration nor for the possibility of supplying it but only to show the real magnitude of the mass of mankind. It should make clear that whatever else we are about to run out of, space is not included.

Whether the drive toward urbanization is desirable is not only beside the point but quite simply a matter of taste. Many a Manhattan native would consider a camp in a wilderness area a close approximation of a hell, while Manhattan would be sourly viewed by most natives of open lands. Nevertheless, the tendency today is toward urbanization to the extent that open areas, even wilderness areas, are increasing. Upper Maine, for instance, has thousands of square miles without a permanent habitation or even a road. The coastal plain of the American South has thousands of square miles of wilderness, much of which was once settled land.

The efficiency of transportation now makes it possible to supply urban centers with the enormous tonnages of food and material they require. Even water, one of the commodities needed in vast amounts, can be provided by pipeline, and at seaboard locations it will probably soon be available by desalinization since the achievement of economic desalinization of sea water is likely in the near future.

The problem of food supply is also more cultural than technological. We know how to multiply food productivity far better than we know how to get people to take the steps necessary to do so. The King Ranch has demonstrated how to convert tropical forest lands into productive cattle land, but little of the tropical forest is being so converted. The IR rice work and its extension to wheat greatly increases the food-producing potential of the world, but only small parts of the world are as yet affected. The world's grasses are still being explored for better pasture grasses, and the potential increases in productivity are factorial, at times as much as tenfold increases. We have noted the effect of accidental introductions of grasses into Colombia and the southeastern United States. Still, wide areas of our own country are covered with grasses of such low quality that a cow would starve to death eating them. We can make rice, wheat, or corn yield several times the "normal"' crop. We can plant grasses that have several times the nutritive values of the native grasses. We can fertilize and lime soils to make poor land rich. What is startling is how little of the world has been active to maximize potential yields. We have all the knowledge needed to do these things, and our knowledge continues to grow at an explosive rate. Since we are using increasingly less of the world's surface for actual living, and in areas such as the United States increasingly less for agricultural production because of the increased productivity per acre, since it has already been demonstrated that similar results can be obtained in other lands, and since we are adding further knowledge useful for multiplying productivity at an accelerating rate, the population-explosion problem changes into a problem of diffusion of knowledge. The population problem remains real, grimly real, in areas like India, but it is important to keep in mind that we are not running out of space, nor are we really running out of potential productivity. The gap between actual and potential is cultural, not technological and not physical-geographical. We are looking again at the problem of the spread of ideas.

Men produce ideas. Ideas make it possible for there to be more men to produce more ideas. This feedback system may be unbelievably slow to start, but once it gains momentum its acceleration is exponential; the long view shows just such a growth. We are currently experiencing a normal situation, a population explosion accompanying our exploding growth of knowledge.

Location and Ideas

The growth of knowledge made a geographically localized start, and this has had great consequences, for the spread of knowledge (diffusion) requires contact and, all other things being equal, proximity promotes contact. There are other variables, not the least of which is cultural receptivity. The Yuma were an example of nonreceptiveness. Time is also a factor, and it is probable that given more time the Yuma would have become more receptive, for the ability to take up ideas is related to cultural level. Diffusion then proceeds in time and space; it is culturally conditioned in both these dimensions and meets some geographical conditioning in its space dimension.

Time and cultural change are diagramatically shown in Figure 11-3. The duration of the differing cultural periods is strikingly uneven. Depending on what dates are accepted, the beginning of stone tool-making by man was somewhere between two and three million years ago. In the enormously long early period that is known as the Lower Paleolithic, hundreds of thousands of years passed with almost no change visible in the archeological remains. The chart here suggests something like three major increments of knowledge in perhaps 900,000 years. This would include such items as learning to flake stone in two fundamentally different patterns, to make simple chopping tools, and use fire. In the Middle Paleolithic at least an equal number of inventions seem to have been made in 150,000 years. New tools appear, and man becomes a skillful hunter of big game and able to live in very cold climates. This suggests fairly efficient clothing. In the Upper Paleolithic perhaps twice as much advancement was accomplished in about 20,000 years. Then about 10,000 years ago agriculture appears, so far as our present knowledge goes, in the Middle East. Thereafter the growth of knowledge and population become explosive.

Notice the change of scale that is necessary if this graph is to be kept within limits: hundreds of thousands, then tens of thousands, then thousands. The duration of the periods is shown on an unchanging scale in the upper part of Figure 11-3. Notice that on this scale it is not feasible to show any time division since the appearance of metals about 5000 years ago.

It is worth mentioning again that the curve of growth of knowledge is of the same shape as the curve of growth of population. In the overall picture it is clear, first, that the growth of knowledge makes possible the growth of population and, second, that population follows with some lag behind the potential that exists at any moment. At the present time of incredibly rapid pyramiding of knowledge, it seems clear that the actual population is lagging very far behind the potential.

The spread of ideas from a center can be illustrated diagramatically, mathematically, or verbally, on a small or a large scale. A diagramatic presentation on a very large scale is used here as well as verbal description of the processes at work. The role of the individual, momentarily, is omitted. We will use the device of a uniform world again in order to simplify the problem and will assume a uniform mankind in a uniform primitive cultural stage. Since inventiveness creates cultural differences, we would expect a network of centers of difference to arise, but we could assume some of these differences to be vastly more important than others. Specialization in hunting or in pastoralism leads to limited ends, while the food gatherers who became agriculturists started a line of development with seemingly no limit. We will first consider a diagramatic model (Figure 11-4) of the invention of agriculture.

A uniform world is assumed with mankind in various Upper Paleolithic stages of existence. The beginning of agriculture is assumed to occur at X. The circles A, B, C, and D mark equal distances from the center X. The assumption is made that ideas will flow radially (right side of figure) and laterally (left side of figure). In the radial flow the ideas will reach the closest people A during the first time interval (in practice some thousands of years). After some time they too will enter the agricultural revolution. Their numbers will grow, and their cultural growth will begin to accelerate. In the next time interval the ideas that have reached A will be passed on to the next closest people B.

Other things will also be happening. Cultural growth will be proceeding at an exponential rate at X, and there will be an increasing flow of ideas toward A. This is indicated by the wider arrow from X to A that is labeled 2. Having made a start, A will be increasingly receptive to ideas and will also begin to have a weak influence on X, thereby giving X a still greater advantage in cultural accumulation. The beginning of a flow of ideas toward X in stage 2 is indicated by the arrow from A to X that is labeled 2. The successive flows of ideas to A are shown by arrows with thickness indicating greater flow. Thus the flow from X to A is assumed to increase greatly through time due to the

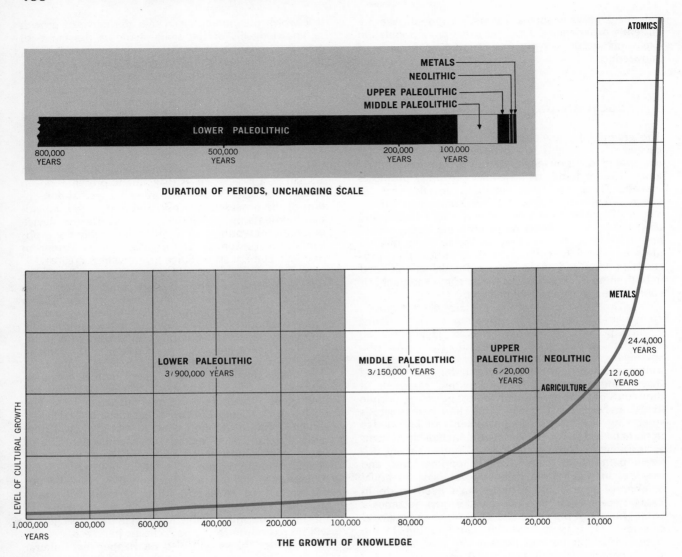

DURATION OF PERIODS, UNCHANGING SCALE

FIGURE 11-3   Cultural Growth Acceleration
*Rates of cultural growth are suggested for the major cultural periods. The curve is exponential.
Major or revolutionary increments of knowledge are shown as fractions: 3/900,000 years,
3/150,000 years. The extremely slow growth of knowledge during the time of small populations
and the immensely accelerated growth of knowledge during periods of increasing population
are such that on this level of generalization the curve of population and the curve of
knowledge are identical. In time it is clear that population growth tends to increase knowledge
and grow further due to the increased well-being created by the new knowledge. Such a
feedback system will create a curve like this. For purposes of comparison the duration of
periods is shown on an unchanging scale.*

FIGURE 11-4 The Flow of Ideas from a Center of Origin

The flow of ideas from a center of origin of a massive breakthrough such as the invention of agriculture might be so diagramed. The assumptions are a uniform earth and uniform man. If a people X achieves this breakthrough, its population and learning will begin to grow at an accelerated rate. The flow of ideas to the people at A is shown in four time intervals. The outflow from A is assumed to increase in time, and this growth is suggested by progressively thicker arrows. The reach of ideas is considered to be progressively farther through the time intervals and to be progressively diminished by distance. A cultural effect is also implied. At stage 4 a people located at D would be unprepared to accept the full flood of ideas that a people at A, having experienced three preparatory stages, could accept. The lower right quadrant diagrams the return flow of ideas. These must start later and be less under the conditions hypothesized here. The left half of the diagram adds the lateral-flow picture, here portrayed as it might be at stage 4. Outflow and inflow, as well as lateral or peripheral flow, are shown by proportionally thick arrows.

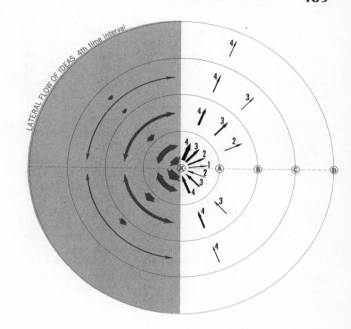

increasing advancement of X as the major producer of cultural growth, and because of the parallel growth of A, constantly increasing its receptivity to ideas.

The return flow of ideas from A to X will increase steadily with cultural growth at A. For the more distant areas, the same assumptions are made. The great advantage of the area of origination is clear: given a cultural breakthrough at any point on the surface of a uniform earth, a radial pattern of spread with retention of dominance by the center and with declining cultural level accompanying increasing distance is to be expected.

On the left side of the figure, arrows are entered to suggest the nature and extent of lateral flow of ideas that would become important as the field of cultural advancement widened. They are shown here as they might be in the fourth time interval. The inner areas, being more advanced, are both more productive and more receptive to ideas. Consequently the greatest lateral flow of ideas is nearer the center. Hence both radial and lateral flow tend to reinforce the initial lead of the cultural areas, as shown by the weighted arrows.

Let us select from history one of the great turning points, such as the invention of agriculture, and put it into our model to see what would happen. The initial effect of an agricultural invention would be to slightly increase but greatly stabilize food production in the area. The people living there would not have so great

a famine risk, for they would have begun to control production. The effect of agriculture would approximate Justus von Liebig's law of the minimum, which states that the factor in short supply controls production. For a primitive population the limiting factor may be the food supply at one season of the year. The total number of persons might be limited to the survivors of the winter starving time. This might limit a band of seed gatherers to 100, even though in spring, summer, and fall their territory could well support 500. It is too easy to assume that men would naturally store food during the good periods. The Yuma, for instance, did not do so sufficiently to avoid an annual starving time. The introduction of agriculture gives a potential for controlling this type of situation, but again the Yuma illustrate the slowness with which application is made.

Thus the first effect of the agricultural revolution would probably be small, but it should result in some increase in population. Later, fully developed agriculture would lead to a tenfold increase in population. Usually it is argued that there would also be an increase in leisure. Perhaps we have overstressed this, for when man undertook to control nature he had to work more, not less. In the early city-state that grew out of the agricultural village stage leisure was for the elite only. Primitive people have lots of leisure, but they usually have high risks as well. The game may move away or take a different route, and drought or fire may destroy the wild-grass fields. Indians of the northern forests have been known to wait in leisure for arrival of the reindeer herds that normally migrate along a certain route. When the herds switched elsewhere, many of the tribe died of starvation. Reindeer herders would have had no such leisurely waiting period; they would have had to keep more continuously busy with their herding, but there would have been less likelihood of their starving. Similarly a chance grass fire could destroy fields of wild grasses for which the gatherers were lazily awaiting the ripening, but the corn fields of the planters would be less exposed to fire because of the cleared and weeded land that steady labor had maintained. It probably was not leisure but numbers, and later specialization, that accelerated cultural growth.

Numbers would increase slowly at first, and specialization would come only later. There was little specialization even on the level of simple village agriculturists, and the millennia spent in their way of life before towns and then cities began to grow show how slowly these initial steps were taken. Eventually numbers increased and with them the frequency of occurrence of unusual minds capable of making significant advances in thought. Clearly those in the area that took the first steps would be more likely to lead, and once they were well underway, they would hold a long lead over other potential starters.

Their neighbors would be influenced by them. Either they would learn to plant and herd and multiply their numbers, or they would presently be overwhelmed by the increasing numbers of the farmers. If we assume that ideas flowed and men were receptive to new ideas (a large assumption), and if we divide our space into units, we can discuss the spread in terms of increments—additions to knowledge. The people of the beginning area would, in the first unit of time, slowly advance in numbers, stability, and social complexity and then start to generate technological skills. Notice that in the case of agricultural origins in the Middle East this period lasted a few thousand years. Although the process was a continuous one, we may break it into units. In the first unit of time, perhaps 3000 years, the development of simple agricultural villages proceeded slowly in the area of origin. In the second interval further growth occurred in the central region but at a quickening tempo. Meanwhile, the influence of the new ideas would have reached surrounding people. We may conceive of these people as forming a belt around the original center. They would be at the beginning steps of entry into a settled agricultural life. Judging by people that we have actually seen in such circumstance, some of the Apache or the Yuma, for instance, they would only slowly settle down to being full-time farmers. Only very slowly would they

abandon the leisurely ways of the hunters and gatherers for the orderly toil of the food producers. For a while they would try to have the best of both worlds, and during this time they would have little to contribute to the world of ideas. They would be taking, not giving.

In another increment of time, a shorter one now, for things would be accelerating at the center, more new ideas would begin to radiate out from the center where more people would mean more potential geniuses, and more ideas would mean more people and more production. The desultory farmers next to this center of growth and development would conceivably see the advantage of settling down to full-scale production, and they too would begin, though only slowly as yet, to increase in numbers and try their hand at developing an idea or two. Already they could be expected to have slightly influenced the people in the ring beyond them, and these people twice removed from the center could now be expected to enter the stage of part-time farming. The sequence so far reads: in interval 1 farming is invented at the center. In interval 2 farming develops at the center, and a beginning in farming is made in the circle closest to it. In interval 3 farming becomes advanced at the center. It becomes regular in the circle closest to the center, and desultory farming begins in the next circle. The center by this analysis always retains its lead.

Actually its lead should increase for two reasons. The growth of culture tends to be exponential. Further, as the surrounding areas reach the level where they begin to produce ideas of their own, it is the central region that is both located best to obtain them and is, due to its level of development, best able to utilize them. Primitive cultures have difficulty accepting and using advanced ideas. For example, the North American Indians could use iron but did not take up smelting; they accepted firearms but could not make gunpowder. Given sufficient contact time, they could, of course, have advanced the next step.

An example of an adjacent people creating something to spread back to their benefactors in the world of ideas would be the Palestinian people who took the pictographic-ideographic writing system of the Egyptians and simplified it into the alphabet. A further step was the Greek modification of this alphabet by adding vowels. These innovations flowed back into the central region and were reradiated. There are exceptions: the Semitic people did not take up the use of vowels, for instance, but most people who adopted the alphabetical system did. The slowness of these processes is well exemplified by writing. Notice that in the mid-twentieth century large parts of the world are not yet literate and in A.D. 1500 writing was unknown in enormous areas of the world. The center that invented writing sent it out into the surrounding belt and received back the improved idea of the alphabet. It then sent out this new system and received back the addition of vowels. Finally it sent out this still better system. The advance and feedback of ideas and the associated accelerated growth of ideas occurred in the central areas, while the distant areas such as South Africa, Australia, and New Guinea lagged further and further behind.

In America, with Middle America as the center of our model, Tierra del Fuego, California, and Canada were the distant lands. As in the Old World the centers of growth and development influenced each other markedly. Thus the Toltec-Maya civilizations were mutually stimulating centers, and the Mexican civilizations both stimulated and were stimulated by the South American civilizations. As far as we can see, however, the flow between Mexico and the eastern United States went one way only. Given time, perhaps the pre-Columbian North Americans would have produced something of value to the Mexicans, but the two thousand years or so of available time in a beginning agricultural village stage seem not to have been enough for the production of genius and ideas of sufficient importance to produce anything interesting for the civilized Mexicans.

Time and Ideas

In Figure 11-5 cultural growth is shown in exponential curves in terms of the four time stages indicated in Figure 11-4. Here the people A are assumed to start with a level of 1 at stage 1, by which time the people X have doubled their cultural wealth and reached level 2. One time interval (stage 2) would see a doubling of cultural content for A and sufficient influence from them on the adjacent people B to give these a cultural level of 1. Meanwhile, X has again doubled its cultural wealth and is at level 4. A will have started B on its road of development. All areas beyond the reach of the new ideas will have remained at or near their original cultural levels. Their numbers are too small to allow for much self-starting.

In the next time interval (by stage 3) A will have advanced to level 4, and B can be assumed to have reached the cultural level of 2 and influenced the adjacent people C to give them a cultural level of 1. By this time X has again doubled its cultural wealth to level 8 on self-generation of ideas, and it would not be unreasonable to credit this people with a pickup of one level from the surrounding peoples who have made a start in cultural growth beyond the Paleolithic levels.

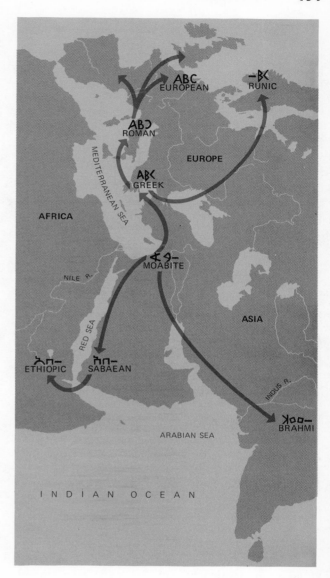

MAP 11-2 Spread of the ABC's
The alphabet is conceded to have been invented but once—in the Middle East. From the Middle East the idea reached the Greeks, who turned the letter A 90 degrees to the right, modified the B, and added a C. These letters spread north to the Scandinavians. With modifications, the letters became runes. Another line of diffusion via Rome gave us our ABC's. Other lines reached south to Ethiopia via Saba and also to India. India's letter A was turned 180 degrees to the right. Where a letter was lacking in an alphabet, a dash is shown on the map.

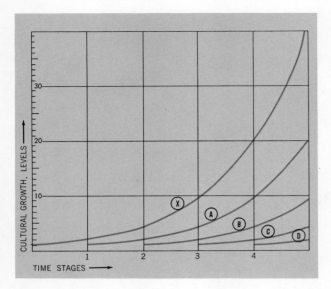

FIGURE 11-5   The Rate of Growth of Population and Ideas
This diagram shows the same phenomena as Figure 11-4. X, having made a start, begins accelerating growth. Even assuming perfect transmission of ideas, people at succeedingly greater distances will begin their cultural growth at a later date. All other things being equal, the other cultures will never catch up with X, B will never catch up with A, and so forth. Historically the influence of an individual (Perón in Argentina) or a politico-economic system (communism in Russia) can and normally does eventually slow the leader's (X's) pace and allows a late starter to exceed in development. It is usually a nearby people who overtakes when the leader falters, not a distant group. The sequence in history has been Middle East, Greece, Rome—not Middle East, Australia, Tierra del Fuego.

The continuation of this process gives the initial group an ever-widening lead. Groups at even modest distances from the center will rapidly be left so far behind that the situation we described for the California Indians or the Australian natives comes into play. When the cultural gap becomes too wide, cultural transmission is inhibited, and the tendency is either for one people to replace the other or for the advanced people to bypass the cultural laggards. In a diagramatic way this is what should have happened on the earth. As we have seen in many case histories, this is in fact somewhat the way things did occur. Of course, abstract treatments such as this leave out the human factors. Collapse of an empire or freezing of a bureaucratic caste system or an Alexander the Great can break up the beautiful symmetry of the curves.

Although the curves indicate that the people making the original breakthrough in knowledge should never lose their lead, the record shows that they always do. The people of East Africa made one of the earliest radical advances when about two million years ago they invented stone tools. They apparently held the lead a long time, for they seem also to have been the inventors of the bifacially flaked hand axe and possibly the bola. Perhaps a million years passed before other centers of innovation surpassed them. The agricultural revolution about 10,000 years ago gave the Middle East a lead it held for about 8000 years. The industrial revolution gave northwest Europe a lead it is now rapidly losing to such outposts as the United States and Japan. On a smaller scale, nations rise and fall; we have reviewed a number of cases. Rome became increasingly bureaucratic, despotic, and centralized, and its tax system combined with an excessive tax burden drove land out of production and men off the land. When the barbarians pushed, the walls came down as if termites had eaten away the timbers—in a sense, they had.

In Greece a lack of centralized organization, reflecting almost a mania for the independence of individuals and cities, usually prevented a united effort in the face of aggression. When empires became powerful units, the Greek city-states fell prey to conquerors, and in a long succession of destructive wars Greece's greatness became only a memory. In Spain the view that business was a somewhat disreputable occupation as compared with gentlemanly land ownership—a prevalent Mediterranean outlook—combined with a thirst for empire, paved the way for dissipation of the enormous treasure the country derived from the New World. Torn by internal and external warfare, Spain lost its commanding lead and sank into centuries of decadence, from which it is just beginning to emerge. These examples and others we have reviewed are important for the insight they provide into the enigma of loss of leadership—something that should not happen according to the theoretical diagram but is a cultural regularity.

The causes of breakdowns among leaders are extremely complex. Overcentralization as under the dictators in Rome or overfragmentation as in Greece, overvaluation of the possession of land in the Mediterranean generally and especially in Spain, and wars everywhere are just the beginning of the story. At one time the Muslims were rulers of a vast empire stretching from Gibraltar to India, but they later divided into segments and are now one of the lagging groups in the world. The Muslim religion, with its long-maintained ban on such things as printing and insurance and its view in the past that Muslims are rulers and warriors but not workers (Christians and Jews are laborers), has been a deterrent to development in Muslim lands. The Muslims spread their faith widely on an east-west axis, largely through the arid lands with which they were familiar but also to wet lands such as present Bangladesh and Indonesia. The lack of advancement that marks all these lands, some of which were once in positions of world leadership, is in part due to a marked tendency of the Muslim world to exclude the outsider, to look inward, and to be intensely conservative. That these ideas are still powerful deterrents to development is widely apparent in North Africa today.

Agriculture

Since agriculture is a specific invention and is tied to the physical environment through the plant's requirements, it provides an excellent test of the role of cultural and environmental restraints in the spread of ideas. We can start with the invention of Middle Eastern agriculture (wheat and barley) and compare the spread through time with the theoretical diagram.

The spread of agriculture is shown on Map 11-3. Concentric circles are drawn around the area in the northern end of Mesopotamia that at the moment seems to be the best candidate for the place of origin of agriculture in the Old World. The first circle is at a distance of 500 miles, the second at 1000, the third at 2000, and the fourth at 3000. The spread of agriculture by 5000 B.C. was largely to contiguous, similar regions, with the exception of its early appearance in the dissimilar region of the Nile Valley. By 4000 B.C. the spread elongated along a few natural routes: down the Mediterranean to northwest Africa, up the Nile, eastward toward India. The Mesopotamia plain was also occupied by agriculturists in this time interval. The next thousand years saw the shores of the Mediterranean widely settled by agriculturists, and agriculture began to spread up the major river valleys into Europe, moved into the edge of the desert in northwest Africa, and leaped to China and to India. By 2000 B.C. agriculture had been pushed north nearly to its northern climatic limits and had spanned the continent of Eurasia. The areas of notable lag are Africa south of the Sahara and Oceania beyond the nearer parts of Malaysia. The general conformity to a spreading out from a single center is moderately good.

The distortions from a radial spread are largely the effect of the physical geography. The major distortion parallels the very strong east-west axis of the physical features of Eurasia and North Africa. The limit of agriculture to the north was determined by cold, to the south by aridity. The major mountain chains run east-west. The Mediterranean Sea–Persian Gulf axis is also an east-west one. The importance of the sea axis was probably accentuated by the cultural effect of sea travel. Thus the Mediterranean seems very early to have been an axis for the long-distance movement of ideas. There is a strong suggestion of a very early movement of agricultural ideas coastwise from the Persian Gulf toward India and from that area toward the mouth of the Red Sea. In slightly later time the India to Red Sea via southern Arabia route was to become so important that it is known as the Sabaean Lane. The distributions here suggest that it may have been assuming importance as early as 2000 B.C. The vast desert area of the Sahara seems to have been an important barrier to the spread of agriculture. It was penetrated by the Nile corridor and later was flanked by the sea routes, but the total effect was a great diminution of the contact.

The physical environment clearly played a role in shaping the early spread of agriculture. Notice, however, that use or nonuse of sea routes is culturally determined. It is also true that the northern limit is culturally determined; it is the decision to practice agriculture that is primary. The climatic limit relates to a cultural way of life and is actually associated with particular plants whose use is culturally determined.

A difference in agriculture would have changed the model. Suppose agriculture had begun with rye instead of wheat. The expected elongation of early spread would still have been east-west, but now belong the axis of the continental climate stretching from the north German plain to China via Lake Baikal. One famous model for diffusion dealt with the spread of ideas along a lake shore. The spread would be utterly different if the people had boats or lacked them or if they had political and economic ties across the lake or down the shore. The environment is there and is not to be ignored, but the exact nature of its role is determined by the culture.

Any analysis such as this is at the mercy of new information. The beginning of agriculture in the Middle East, formerly estimated at 5000 years ago, is now dated at about 10,000 years ago. The dates for the arrival of agriculture in distant parts of the world are imperfectly known. There is a question as to whether there were separate agricultural origins in the Far East and whether there was even the possibility that the origin of agriculture was in the Southeast Asian area. The analysis

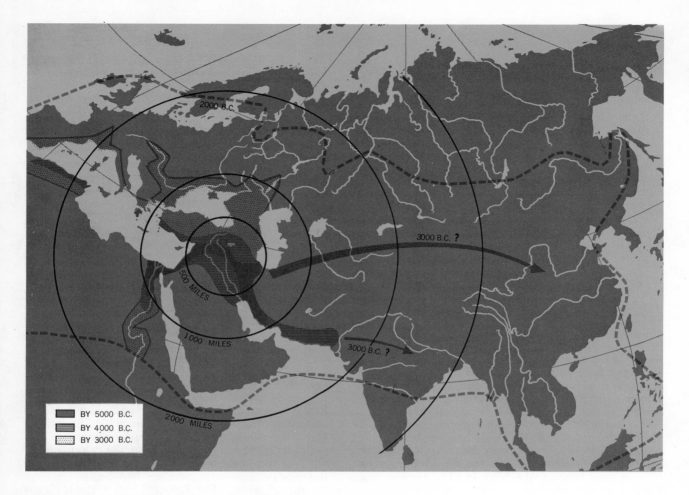

MAP 11-3   The Spread of Agriculture over the Old World

*Circles 500, 1000, 2000, and 3000 miles in radius are centered on what is presently
believed to have been the area of the origin of agriculture in the Middle East. The areas
actually occupied are designated by shading, advancing from 5000 B.C. to 4000 B.C.
to 3000 B.C. The limits of agriculture by 2000 B.C. are suggested by the dashed line. The
interaction of cultural and physical geography is apparent in the east-west distortion
of the distributions. This is in part purely physical; the lay of the land and the climatic
zones run that way. In part it is also cultural; the possession of shipping made the
Mediterranean, the Persian Gulf, the Indian Ocean, and the island world of Southeast Asia
highways of diffusion. Without boats, they would have been barriers.*

presented here tends to weaken the thesis of a Southeast Asian origin for agriculture, since it was not Southeast Asia that experienced the earliest population and
cultural growth, as might have been expected if it had
priority in agricultural origins. On the other hand, since
Southeast Asia and especially India are extremely poorly
known archeologically, it would be premature to be-

come dogmatic about it. Nonetheless, the total picture
as we know it today suggests agricultural beginnings
in the Middle East with a consequent initial start that
gave the area cultural primacy for at least 5000 years.

According to the pattern in Figure 11-5, an area
once starting off with such a lead should increase its
advantage through time at an accelerating rate and

never be surpassed. Actually, as we have reviewed in endless cases, social disorganization, bureaucratic over-organization, wars, and similar conditions seem inevitably to topple such leaders. There seems to be a social entropy for all leading nations, and even regions have eventually lost their lead. The Middle East, the earliest starter, held the lead longest. Perhaps leadership from 8000 B.C. to the time of Christ is sufficient evidence of the importance of a head start.

It should be abundantly clear by now that the patterning of ways of life on this earth is the result of a few fundamental processes that function on this most *un*uniform earth. The results are more understandable in terms of these processes than through a view that shifts the primary shaping of the cultural world to the physical environment. As we have repeatedly seen, similar physical environments do not cause similar cultural environments. On the other hand, similar cultural backgrounds will repeatedly convert otherwise dissimilar physical environments into reasonable facsimiles of the homeland where the culture originated. The overriding forces in human geography, then, are cultural forces. Neither geographical determinism nor racial determinism is nearly as productive of insights into the growth and development of culture as is a cultural-historical study.

Other Models

Further examples of the dominance of cultural attraction over physical-environmental restraints are diagramed in Map 11-4, with the aim of showing the predictive value of the cultural-historical approach. The Spanish occupation was early concentrated in the areas of American Indian civilizations. Within these civilized areas choices were made by the Spanish predominantly in terms of the availability of gold and silver, and this evaluation was strong enough at times to offset the tendency to go to the civilized center. For example, the Spanish avoided the nonmetalliferous but civilized Maya country and moved to the northwest in Mexico to silver-rich areas outside the civilized zones. We cannot escape the cultural determinism here by pointing to the physical-environmental contrast between limestone in the Maya country and ore-rich igneous rocks in the Mexican highlands and in the Andes. If the Spaniards had not been gold-and-silver mad, they would have evaluated the scene differently. Had they been Portuguese sugar producers, for example, they would have approached the low wet peninsula of the Maya with a different perspective.

Map 11-5 shows the reverse of this analysis, for in this case civilized areas were sought out. We are just beginning to learn about pre-Columbian agricultural developments in America, and we are only a little more advanced in our knowledge of Asiatic and Mediterranean impacts on America. The Asiatics clearly were more advanced than the American Indians, and we can use the Spanish situation as our model. The Asiatics went to Mexico and Peru much as the Spaniards did. This suggests that these areas already had taken the lead in pre-Columbian America, and the possibility that there were agricultural beginnings and hence settled peoples in this zone would seem a reasonable hypothesis. Recent work in Mexico tends to place agricultural beginnings there at least as early as 7000 B.C., and our earliest evidence of Asiatic and European contacts is several millennia after that, following a period long enough to suggest considerable population growth, settlement, and a resultant potential for conquest or trade or colonization by a more advanced people. The tropical focus of Asiatic influence in America seems more logically explained in cultural terms, as in the Spanish model, than in physical-environmental terms. How otherwise can we explain the fact that the Asiatics ignored the beauties of California and Chile and chose instead Ecuador and southern Mexico? "Beauty is in the eye of the beholder," and settled people look very handsome indeed to those who seek manpower for development, productive people for taxation, and at least minimal resources to aid any settlement.

That there is nothing unusual about this can be further illustrated by recalling the direction of imperial expansion in pre-Columbian Peru and Mexico. The highland empires of Tiahuanaco (centered in Bolivia) and of the Incas of Peru expanded into the adjacent arid regions and avoided the adjacent humid tropical lands. The highland empire of Mexico expanded into the humid lowlands and not into the adjacent arid lands. The contrast was culturally motivated, for the invaded areas were culturally advanced and the ignored areas were culturally undeveloped.

Still another cultural-physical contrast is shown in Map 11-8, on which the major ocean currents are indicated. If contact was controlled primarily by natural currents of air and water—and these two circulations are generally in parallel—once men took to the sea, Asiatic influence should appear in mid-latitude America on the Northwest coast, and American influence should appear on Europe's northwest coast. In equatorial regions there should be Eurafrican influence on South America's east coast and South American influence in Polynesia–New Guinea and Southeast Asia. These expectations are only partially fulfilled and more

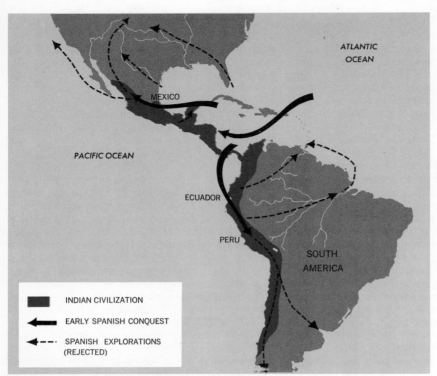

MAP 11-4   Spain in the New World
Penetration of the mainland by the Spanish hardly began before 1500. By 1540 they had explored most of the major areas of North and South America except northern interior North America. Their choice of civilized and populated areas and their rejection, at least initially, of areas now thought to be most desirable (California and Argentina, for instance) are conspicuous.

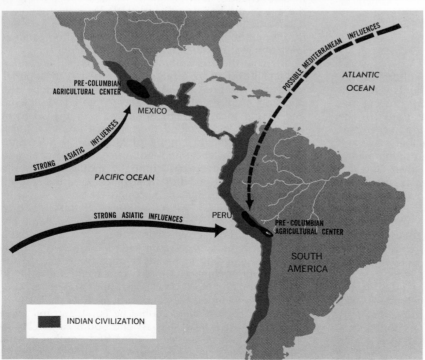

MAP 11-5   Asiatic and Mediterranean Influences in the New World
The focusing of Asiatic and Mediterranean influences in the zones of agricultural origins is indicated here. The spread of civilization was into the adjacent areas. The situation seemingly was parallel to the Spanish impact. The existence of settled people makes an area attractive to incoming people of higher culture. Advancement attracts advanced people, and further growth and development is fostered; adjacent regions benefit first. The text discusses environmental strictures and choices, including the contrast between the Aztec-Toltec expansion into the hot lowlands and the Inca expansion into the dry coastal lands.

often reversed. Asiatic influence does appear on the northwest coast of America in many details such as art forms, use of aconite to poison whales, and Japanese-type body armor. A far greater Asiatic impact is evident, however, on the Pacific coast of Mexico, Ecuador, and Peru. This is seemingly against wind and current. One could retain a physical-geographical control by postulating a great circle sailing around the Pacific (Japan, Aleutians, down the North American coast), but there is remarkably little evidence along the route. Alternatively one could credit the Asiatics with cultural means to sail across, or even against; the winds and currents of the tropical Pacific. The Asiatics were in fact so equipped (centerboards, rudders, and fore-and-aft rig sails) and could have overcome the physical-environmental restraint. The cultural evidence suggests that they did. Theoretically the Mexican and South American civilized areas should have been able easily to reach Polynesia, Melanesia, and Southeast Asia. Although there is indeed evidence of American influences in these areas, it is slight compared to the evidence for Asiatic influence on the Americas.

On the Atlantic side the winds and currents lead to an expectation of northern North American influence on northwest Europe and a Eurafrican influence on Middle America with a focus in the Caribbean to the hump of Brazil. In the north there is only slight evidence of stray arrivals of Eskimos and rare arrivals of birchbark-canoe people in Europe, and we know of no tangible cultural impacts. A weak cultural current carried by Norse and Irish probably flowed the other way, from Europe to America, against the prevailing winds and currents. The cultural differential was slight, for until later historical times the northwest Europeans were only slightly more advanced than the northeastern Indians. Although there probably was some flow of knowledge, it seemingly was slight.

In the tropical regions the winds and currents would lead to a prediction of influences from the vicinity of the Strait of Gibraltar reaching the Venezuela-Caribbean coast. The cultural gradient would, after about 4000 B.C., be in the same direction. In terms of the models discussed here, then, we would expect to see some Mediterranean influences in parts of America. The Caribbean pineapple portrayed at Pompeii, the portraits of Mediterranean race types in Mexico, along with the appearance of wheeled toys, true arches, and J. H. Rowe's list of 60 specifically Mediterranean traits in the Andean region of America are some of the evidence that this route, supported by winds, ocean currents, and cultural gradients, was an important one. Our model suggests that evidence for influences along this route is likely to increase as we focus attention on

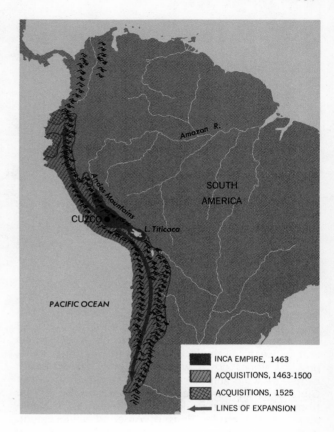

MAP 11-6   The Inca Empire
The pronounced lineal expansion of the Inca empire along the Andes and the adjacent coast is evident here. While a predilection for the highlands is understandable in a highland people, expansion into the arid coastal lowlands clearly reflects the attraction of the irrigated valley civilizations there. Avoidance of the wet tropical areas to the east seemingly reflects both the highlanders' avoidance of the lowlands and the lack of any attractive developed peoples there who could be profitably subjugated.

the problem. The small distance from Africa to America and the persistent winds and currents lead to an expectation of strong African influences on northeastern South America, but few signs are known.

Here we meet again another principle that we have dealt with constantly. Ideas tend to flow from advanced centers of growth and development to lower areas, and only weak return flow is expected. Stated

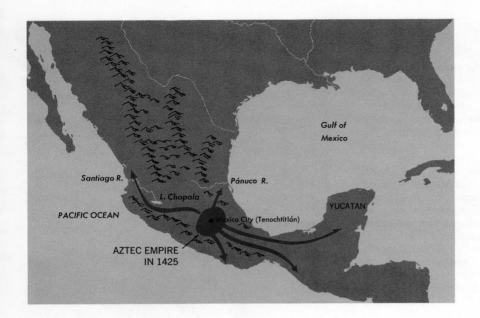

MAP 11-7 The Aztec Empire
In Mexico one of the centers of civilization, the Aztec empire, was in the highlands centering on the valley of Mexico. There was an adjacent civilization in the tropical forested lowland of Yucatán. Expansion was toward the hot wet but civilized lowlands, just as a thesis of the attracting power of civilization would lead one to expect. Notice the weakness of the flow of civilization to the north into the land of uncivilized people, as compared with the spread to the south, areas generally civilized or semicivilized.

another way, ideas tend to flow from high to low. We have seen this in the case of the Middle Eastern and the Middle American centers. It is to be expected also in transoceanic examples. We might think of there being physical slopes down which one may sail with ease carried by wind and ocean currents—from the Strait of Gibraltar to the Caribbean. There might also be a slope from the towering peaks of Mediterranean achievements to the modest hills of Amerind achievement. With the two forces acting in unison, a strong flow might be expected *if* the Europeans were adequate mariners. On the Pacific side the cultural slope was from Asia to America, and the physical slope (winds and currents in equatorial regions) was opposed. It would take a high cultural slope to overcome this, but the one feature that would do it would be advanced technology in sailing, which the Asiatics actually possessed.

After 1500 the northern part of this tropical Atlantic "gradient" became the Spanish route to America. The southern branch took the Portuguese to Brazil. An interesting aside that emphasizes the cultural forces again is found in the fact that early contact of the Portuguese with the hump of Brazil failed to stimulate them to make an early settlement. The causes were cultural because they were seeking India, civilization, and spices. The hump of Brazil offered only poor forest-dwelling villagers who possessed little of value. It was not that Brazil was tropical and that the Portuguese were mid-latitude Mediterranean land people, for they were on their way to a tropical land, seeking exotic

goods of great value. It was not the tropics that they were rejecting but the absence of settled people with valuable products—a cultural evaluation again.

Our models show then that the force of cultural factors is sufficient to offset physical factors. Further, they allow us to project known cases back in time to hypothesize past conditions and permit the statement that since both physical and cultural gradients slope from the Strait of Gibraltar to America, a strong Mediterranean influence on the Caribbean-Venezuela area is to be expected.

Population, Space, Resources, Food, and Energy

There is much agitation in some circles today about the whole population and resource complex, and some people even worry about man's running out of space. There can be no disagreement that there is a limit somewhere on how many people can live on this earth, but there is room for disagreement on what that limit is. At the ultimate end of the scale there is just so much space, air, water, and sunlight. Estimates vary from our now being at that limit, to a 10- to 20-fold increase being the ultimate possible, though clearly not desirable. The alarmists see man as about to destroy himself through overpopulation, famine, and exhaustion of resources. To a surprising degree, agricultural experts,

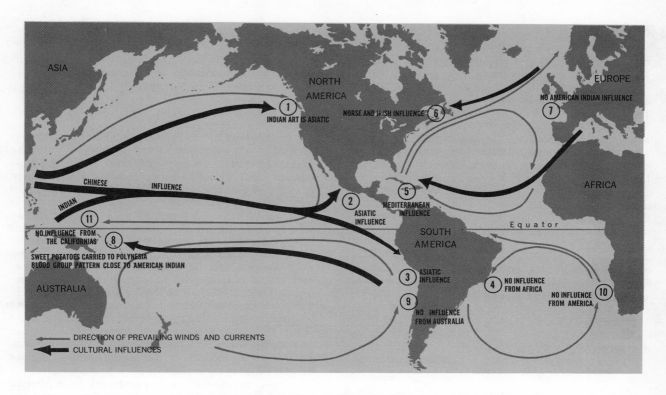

MAP 11-8 Winds, Currents, and Cultural Influences across the World's Oceans

When the prevailing winds and currents on a world map are viewed as possible highways for the spread of ideas, one has a test of physical-environmental as against cultural determinism. The first step in the spreading of ideas is cultural. Without watercraft the great wind and water currents are useless to man. With watercraft he can follow the winds and currents, and if he is sufficiently skilled he can sail across the currents or against them. This map presents eleven cases in each of which the currents, favorable or unfavorable, and cultural influences, or their absence, are indicated. All conditions are found: favorable conditions and cultural transfers, favorable conditions but no cultural transfer, unfavorable conditions but cultural transfers nonetheless.

Currents favorable and cultural flow evident (positive cases):
> Asian influences on northwest coast of America (1)
> Mediterranean influences in the Caribbean and on the adjacent mainland (5)
> American Indian influences in the Pacific (8)

Currents favorable but no influence (negative cases):
> Equatorial Africa to South America (4)
> Northeast North America to Europe (7)
> Australia to South America (9)
> South America to South Africa (10)
> California and Baja California to Polynesia and Southeast Asia (11)

Currents unfavorable but cultural evidence present (negative cases):
> Asiatic Indian influences in Mexico (2)
> Chinese influences in Mexico and Peru (2, 3)
> Norse and Irish influences in northeast North America (6)

resource authorities, and atomic energy specialists tend to take a more hopeful view.

My grandfather's generation had large families, and it was the boast of my mother's side of the family that they had had a baker's dozen children. The reaction was remarkable, for the next generation "nearly exterminated themselves," as one man put it, having had only one or two children. My parents had three children, a large family to some. Then came a wave of large families and now a wave of reduced families. The current reduction is unquestionably related to the alarm expressed over future overcrowding. Yet similar ups and downs in family size occurred when there was no such alarm being broadcast.

TABLE 11-1   Population and Area

Land area of the world excluding Antarctica	52,180,000 sq. mi.
Population of the earth (1972 estimate)	4,000,000,000 people
Average population density (round numbers)	80 people per sq. mi.

Area	People per Square Mile
Asia (50 per cent of all mankind)	117
Europe	161
North America	33
South America	26
Africa	28
Australia	4
New Zealand	27

If a tolerable limit of world population lies between 10 and 20 times the present situation, something like a world average of 1000 people per square mile is suggested. Presumably much of this population would be urban, and a Netherlands type of landscape could be envisioned as the optimum. The figures above show that there is a maximum population density in Europe while the continents of North and South America, Africa, and Australia are relatively empty.

One of the points seemingly being overlooked at present is that a large part of the increasing population of the world is due to increased longevity. This must inevitably lead to an aging population, and this will mean that a larger part of the population will be living for longer periods of time after their reproductive years have passed. This aging population will lead to higher death rates, for lives saved in infancy are simply postponed deaths. It is notable that this trend is already appearing in the United States where an increase in the death rate was noted in 1972. An increasing population based on prolongation of life must then become self-limiting as the population ages, provided that longevity markedly passes the fertility span.

Space

We have dealt with the problem of space at various points, especially when pointing out that many of the areas that we discussed were either relatively or even really virtually empty. Tasmania is an excellent example of a nearly empty land with abundant opportunity for development and occupation. Yet today it is not unusual to find a seemingly well-informed person who believes that we are on the verge of running out of standing room on the earth. I once calculated that if all of mankind were put in the Grand Canyon, the mass would barely fill the inner gorge at the bottom and would not be noticeable from the canyon rim, or, as stated earlier, all of mankind could stand up in Baltimore County, Maryland. This exercise is not meant to point to a desirable practice but to show that there is no immediate or foreseeable danger of running out of standing room. The problem of what is a desirable density of occupation of the earth is quite another matter. What is crowding?

Most of mankind, as has frequently been pointed out in this text, is now moving into the major cities of the world. The land is emptying out. Mankind is choosing crowding over isolation, economic and social advantage in cities over rural poverty. and deprivation. Where the choice need not be that, as with the choice between living in the largest American cities or in the surrounding suburbs, the movement is from the city into the suburbs. Americans feel the great city to be crowded in relation to the opportunities, and our huge cities are now losing population. It is commonly overlooked that the "overcrowded" slums of New York are less crowded now than they were at the turn of the century.

Crowding is a relative thing. Some of the most desirable places in the world are the most crowded. Monaco, that delectable spot on the Mediterranean, has a population density of 40,000 people per square mile. People pay handsomely for the privilege of staying there despite what seems to be an incredible crowding. Admittedly that is a rather special situation, for the Principality of Monaco is really a city, but there are whole countries such as the Netherlands with population densities approaching 1000 per square mile. As we have seen in the Netherlands, only about 10 percent of the land is occupied by houses, streets, factories, and the like, and a very high percentage of the land is in pastures. Further, the Dutch standard of living is high, and their country one of the cleanest and most attractive in the world. A traveler in the European countryside has to remind himself constantly that he is on a minor peninsula of the Eurasian land mass and that one-fourth of all mankind lives here, for the rural landscape is open and uncrowded, and wild deer are commonly seen grazing in the Bavarian countryside. The United States with 54 people per square mile, Brazil with 26, and a whole continent such as Africa with 28 suggest that the earth is far from crowded yet. One must make endless disclaimers. Africa has the huge Sahara and Kalahari-Namib dry areas where sizable population is not to be expected unless there are breakthroughs in technology. Nevertheless, except for spots here and there, much of wet tropical Africa is thinly populated. This area of Africa has the tsetse fly, but here we can expect modern technology to solve the problem. The screw-worm fly has been suppressed in the United States without the use of the hard pesticides such as DDT but by biological attack: the release of sterile male flies. The so-called third generation insecticides are already under development. These are hormones that interrupt the complicated growth cycles of the insect (egg, larva, pupa, adult), each stage of which requires a hormone to trigger it but which must not be present in the next stage. These substances are very specific, affecting only one type of insect. They are very powerful, requiring only minute amounts per acre. For example, one kind of European beetle cannot be raised in American laboratories on American-made paper towels, for a chemical in American fir trees inhibits the growth. Even the minute traces that survive the paper-making process are enough to stop growth. The chemical clearly is a natural defense of this species of tree against this kind of beetle, of the family *Pyrrhocoridae*, which includes some of the most destructive pests of the cotton plant. We have in these hormones a whole new avenue of attack on the insect world, one that is highly specific rather than deadly to a wide range of insects and one that the insects cannot easily develop an immunity to. Presumably this line of attack will allow Africa to eliminate the tsetse fly. This discussion is introduced here only as an example of what can be done in the world of technology and to indicate that the usefulness of land is not a stable factor. Instead it is a constantly shifting one, which is subject to technological and psychological changes. New controls over insects may change the utility of the tropics. Shifts in outlook may change man's view of desirable developments; for example, the American reversal of the desirability of the deserts.

The gist of this is that the world is far from full; neither good areas such as Tasmania, nor difficult but useful areas such as Brazil, nor potential areas such as insect-burdened Africa or drought-burdened desert shores. The earth is not now fully occupied in terms of our present technology, and our rapidly expanding knowledge promises to make more useful. Despite this, mankind is not spreading out into the thinly occupied areas but is imploding into the densely settled spots on the earth. Man seems to have an enormous ability to withstand crowding.

Food

The problems of basic resources and food are twin specters for the population alarmists. The great famine in India, predicted for the early 1970s, has not yet materialized. However, changing national priorities have once again caused a severe shortage of food supplies. The threat of famine has not yet disappeared, and mismanagement can bring it back. On the other hand, management of development as skillful as Japan's or Taiwan's could banish it seemingly forever. There are more cattle in India than in the United States, but the Indians won't eat them. India has less population density and more basic resources than Great Britain, yet it languishes agriculturally and industrially. India needs to develop its economic infrastructure (transportation, education, power resources), but it indulges in atomic armaments and builds up its heavy industry instead. Famine may well result, but it will be due less to overpopulation than to improper use of the land.

Some decades ago one of the great limitations on the agriculture of the world was the availability of nitrogen. True, the atmosphere is 80 percent nitrogen, but we were dependent on a few plants for the fixation of this invaluable material. Then technology made it possible to use electricity to fix nitrogen, and now we have it available to pour on the land—to an alarming degree, according to some. How many more of these situations exist?

The IR rice type of development has changed the world grain situation. There is an oversupply in countries that have been traditional importers of grain. The food technologists are busily working on milk substitutes made from vegetable products such as soybeans. Cow's milk is relatively indigestible for most of mankind, and we may be able to design a fluid that is better suited to human consumption, less perishable, and many times cheaper. When feed is given a cow to obtain milk, the conversion is extremely inefficient: about 1 to 10 in terms of energy. We should be able to manufacture a milk substitute for a fraction of that price and simultaneously free thousands of acres of land from the support of cow feed to the more efficient production of soybeans. The soybean is now being used to produce meat substitutes such as "beef steaks" that some epicures cannot distinguish from the real thing. Again the potential gain in productivity of the land is immense. The list of such food technological revolutions is already large, and the potential is almost unlimited.

In a survey of the world food crisis presented at the American Association for the Advancement of Science meetings in 1968 and published by the Association in 1970, the opening statement on world food supplies and population growth stated that the food gap can be closed and the population stabilized. The two problems are related, and we are doing better at increasing food than stabilizing population. A survey of African problems dealt with the great diversity of that continent and concluded that the greatest block to progress in food supply was social and political. For Europe the progress was shown to be 1 hectare of land to feed one adult in medieval time, one-half a hectare in the mid-nineteenth century, and one-tenth a hectare in 1967. The survey of South and Southeast Asia includes in its final statement: "If there is any need for family planning and population control, it should not be out of fear that man will not be able to produce food for himself, but as a measure to produce a surplus and improve his living standard." In the same AAAS report, reviewing United States agriculture, one finds such facts as an increase of corn yields from about 25 bushels per acre at the beginning of this century to over 40 bushels by 1950, to about 80 bushels by 1970, and a theoretical future yield of 400 bushels per acre. The timing is very significant: the greatest increases came with the burst of science and technology in the twentieth century. Wheat, rice, soybeans, chickens, and cows are actually being redesigned to provide similar huge increases in productivity. In a sweeping review of the United States achievement in relation to the world situation, this AAAS report states that the rate of increase in food supply is far outrunning the rate of increase in population. In the United States alone, if restrictions on agricultural production were lifted, the productivity of the land would double or triple in three to five years. These international agricultural experts simply scoff at the population explosion—food crisis envisioned by people whom they frankly categorize as alarmists. In the United States the Federal government has paid our farmers not to produce. The Soviets insist on maintaining an economic system that stifles food production; in contrast, recall the productivity of the peasants' privately owned plots. China and India are trying to feed nearly half of mankind, but their agricultural methods virtually date back to the agriculture of Neolithic times. Agriculture in Africa is primitive, as is that in Latin America. If famine threatens, it is partially the result of failing to use land properly.

Mineral resources

What about basic mineral resources? Despite claims to the contrary there seems to be little reason for concern, with the possible exception of the fossil fuels. Some of us can remember that all our lives we have been told that we were about to run out of resource X in Y years. At the coming of year Y the date of exhaustion was found to be somewhat farther down the road. Perhaps it is a bit like crying wolf, and perhaps there is a real wolf on the horizon, but he is remarkably hard to find. The exceptions, of course, are oil today and coal tomorrow. If atomic energy is advanced for future use, then even these problems are minimized.

The supply of iron ore is virtually limitless, and our problem really is that iron is so cheap that instead of reusing it when it is in diffuse form such as old auto bodies, we simply litter the landscape with them. Further, within this century iron has been to a large extent superseded by aluminum and magnesium. The supply of magnesium from sea water is unlimited, and aluminum is one of the commonest materials in the earth's crust. The cost of both is fixed more by the energy required to refine the metal than by its scarcity. Since the 1920s aluminum has changed from a rare material, a small sample of which we viewed wonderingly in class, to a competitive mineral that displaces iron and wood and other materials. We are dependent on bauxite as our source for this useful mineral, but the supply is limited. Should we learn an economical way to break the aluminum-silicon bond, we would have a virtually endless supply of this mineral.

It is only a few years ago that we were told that burning petroleum for fuel was criminal, for it was the only known source of reasonably priced lubricants, and without lubricants our economy grinds to a friction-generated, overheated halt. But today we have various coatings that provide oil-free lubrication. Copper, while more critical, is available in immense amounts in the deposits on the ocean floor, an area that is already becoming economic to mine. The mineral now sought on the ocean floor is manganese, which is found in potato-sized nodules that also contain a wide range of other minerals including copper, cobalt, and silver. Three dredge ships at sea can supply the world demand for manganese and drive most continental mines out of business. As a by-product, other minerals will be produced. One, cobalt, will be in such supply in relation to world demand as to virtually destroy the present market. Metals, along with other minerals, can readily be substituted one for another. Consider a building. It can be all wood, brick, stone, concrete, iron, glass, and so on. What material is used is determined by style, availability, and cost. Because of the high cost of traditional buildings, America is momentarily being flooded with mobile homes made predominantly of metal. On a grander scale we have moved through the copper, bronze, iron, and steel ages and may now be entering a magnesium-aluminum age. In fuels we have gone from wood to coal to petroleum and seem headed for nearly total replacement of these fuels by nuclear energy. One simply cannot predict future uses of any resource by a simple projection of present levels of use, although this is regularly done.

The fossil fuels are, of course, in a different category. They are finite in supply and not being renewed at a rapid rate. Despite seemingly endless new finds of petroleum there is in reality an end to this somewhere. We do not yet know what this end may be. The Arctic Basin may prove to be a huge reservoir, as the early finds in Alaska, Canada, and the Soviet Union suggest. We have only begun to drill the continental shelves, and they are proving to be highly productive. Startlingly we have now found salt domes on the bottom of the deep ocean, and these suggest that oil-bearing structures may be widely found over the ocean floor. We are also developing technology to the extent that we will be able to drill the ocean floor at any point. Should the entire ocean floor prove to be an oil source, the potential supply situation would be revised upward by many orders of magnitude. This is particularly important in light of the world oil crisis, with the Arab nations threatening the international economy by their manipulation of supplies and prices. In the short run this is an acute difficulty that will be resolved by the development of oil in other areas and the shift to other energy sources (coal, geothermal, wind, tide, and ultimately atomic). The use of the blockade, boycott, and similar actions in economic history has usually ended in substitutions for the seemingly irreplaceable commodity. Beet sugar resulted from blockades during the Napoleonic wars; synthetic rubber came into use because of restricted supplies during World War II. The Arab economic oil war expectably will lead to similar shifts. Coal, which is in larger supply, seems not to have the possible expansion of sources that petroleum does. For both of these resources, however, the heavy de-

mand has been for their use as fuel. If nuclear power develops, as now seems highly likely, this gross demand will diminish, and they will be used as bases for chemical industries, at least until a newer technology diminishes or eliminates their value for this purpose.

The point of this discussion is that we too easily view the current situation as static, which is, of course, not the case. We are in a world of such revolutionary change that the use of any material cannot be guaranteed beyond tomorrow. Coal and oil loom so large on our horizon today that this is hard to envision, but not too very long ago so did special woods for making wagons, English oak for making battleships, cinchona trees for making quinine, and rubber trees for natural rubber. Technological change is not slowing but accelerating, and no resource needs and limits are estimatable beyond a few decades at most.

Air and water

Water and air are the most basic resources of all. The water problem, outside the arid regions, is more one of misuse than of shortage. We invented a sewage system that produces vast amounts of slightly polluted water that we then find difficult to dispose of. Technology can be expected to tackle the problem from one of two avenues: either devise a nonwater-based sewage system or devise a way to reuse the water. Either or both are possible, and a solution will be found once we decide that the matter is pressing enough. Vast amounts of water are now emptied into the Arctic Sea from North America and Eurasia, and it is entirely within technologic feasibility to deliver this water to southern areas that could well use it. Mexico looks longingly at Canadian Arctic rivers and points to the feasibility of delivering some of this water to the arid northern part of Mexico. The difficulty is that Canada does not want anyone else to use its water. Neither, for that matter, does the state of Washington view with pleasure California's designs on the water of the Columbia River. Not daunted, the technicians have pointed out that when the Columbia River reaches the sea, at which point the state of Washington no longer controls the water, it is quite feasible to pipe the water beneath the sea along the continental shelf down to the California coast where the major cities would be delighted to use it. On similar grand scales such vast estuaries as Long Island Sound could easily be dammed up and thereby converted into vast fresh-water reservoirs to supply the adjacent cities as well as serve as recharge areas for the aquifers that are now being invaded by salt water. Similar possibilities exist for

Delaware Bay, Chesapeake Bay, San Francisco Bay, and so on around the coast of America. For some of these projects, drafts of the costs and benefits have been worked out. A survey of the world would surely show that there are many more such areas. The La Plata estuary in Argentina comes readily to mind, and there are many more. The Dutch combine water use with storm and flood control to make immense gains in useful fresh water both on the surface and in the recharging of aquifers otherwise liable to invasion by salt water. For the arid regions the possibility of desalinizing sea water has already moved from the theoretical stage to the practical though still expensive stage. Should nuclear energy developments be as radical as now seems possible, fresh water from the sea will revolutionize the water problem for all regions bordering seas.

Air pollution control is seen as a great problem, but this is a much simpler problem than any of the others discussed so far. There is no problem with supplying pollution; we can produce vast quantities. On the other hand, there is no question about how to reduce it. We can devise methods whenever we are willing to pay the cost and to exert the political force to compel the polluters to cease and desist. The situation is complex, for we have inherited a situation that came into being when pollution was less of a problem and waste disposal methods that have turned out to be very destructive were allowed; at the time they were not viewed as destructive nor were their long-range effects considered. We now have to undo the problems created at that time, change the outlook of all concerned at this time, work out the technological problems, and pay the bill. The cost, as always, will be passed on to the ultimate consumer. If cleaning up the effluent from a steel plant makes steel more costly, you will have to pay more for your automobile. We are already paying for the reduction of smog through expensive devices on automobiles. Yet it is not all a losing proposition. Some of the pollution controls actually pay for themselves through the recapture of useful materials; the sulfur recaptured in the stacks of oil refineries, for instance. Moreover, clean air and water are important in terms of health, esthetics, savings on deterioration of paint and clothing, and so forth. Pollution control is probably an all-round paying proposition if all the costs are added in. That we do not move instantly to do what obviously needs to be done is an excellent measure of the inertia of social systems.

The complexity of these situations is epitomized today under the term "ecology." If we change the

major estuaries of the world into fresh-water lakes, we will also change the fisheries of the sea, for many fish are dependent for part of their life cycle on the availability of the brackish waters of the estuaries for their nurseries. This is particularly true of the shrimp of the Gulf coast of the United States. The Netherlands shrinks from reclaiming the Walden Zee, simply because it is a major wildlife center. The plain truth of the matter is that it is impossible to use the earth without changing its ecology. It is, then, less a matter of changing the ecology here and there than of being aware of what the changes are likely to be and deciding whether we are willing to make these changes to attain the end we have in view. We have become aware that our every act has ecological consequences, and we are, to a far greater degree than in the past, weighing the consequences of our actions. We must not, as some seem to do, drift into the position that all change is to be resisted lest there be some ecologic impact, for that viewpoint could paralyze growth and development.

The nuclear energy field is in many ways the test area for studies of population growth, economic development, pollution, ecologic change, water supply, and so forth. Our population demands more and more energy, far out of proportion to the increase in numbers of people. This puts pressure on the energy supplies of the earth, and we would quickly exhaust the oil and coal if nuclear energy were not on the horizon. But many fear the pollution aspects of such energy: excess heat to be dissipated and long-lived radioactive by-products to be disposed of.

The role of energy is enormous, and the greatest source of energy is the sun. Through photosynthesis plants transform the energy and produce the food and raw materials that are basic to our society and economy. The efficiency of our use of this energy is enormously increased through the application of auxiliary energy. By manufacturing fertilizers, pesticides, and farm machinery, all of which require energy in one form or another, we immensely increase the yields of the land. The people freed from labor on the land then are employed in industry, education, and research, which in turn feed back into the production line. There is great concern now that the developing nations are using a disproportionate share of the energy and raw materials of the earth—so much so that if the resources of the earth were utilized by all people at the rate the United States is using them, there would very quickly be nearly total exhaustion of the world supplies. The alternatives are simple. The advanced nations can go on at their present rates of consumption and keep the rest of the world in an underdeveloped state, or they can

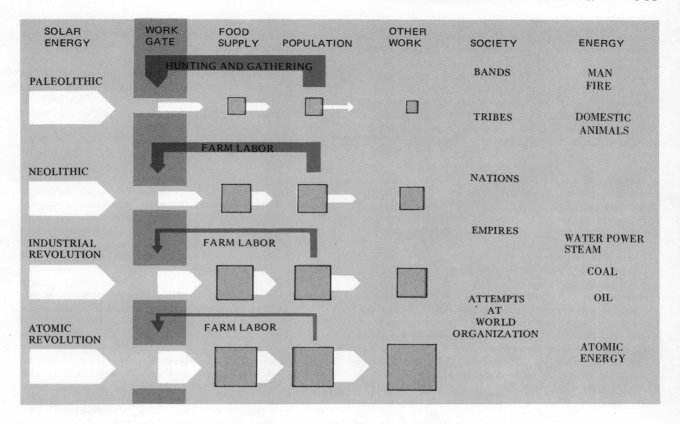

FIGURE 11-6   Energy in Relation to Food, Population, and Society

Solar energy is shown as a constant quantity. The amount man is able to use is shown as if it passed through a gate or window whose opening depended on man's work. The food and population relationship is used to illustrate the role of energy in production of not only food but "other work" as well. In the Paleolithic period man utilized a very modest amount of plant and animal life by expending most of his time and energy in hunting and gathering. Food supply and population were both small. Only a tiny amount of man's time and energy was free for other activities. In succeeding economic revolutions the proportion of work expended in relation to the solar energy harvested decreased for the work done by man and led to greater efficiency in food production. At each step food supply and population increased, and the excess of energy free for work other than primary food production increased. This can be expressed in many ways, but here the changes in society from bands of a few families to tribes, to nations, and to our present striving for a unified world is used. As man has been able to substitute energy, at first animal and then inanimate, for human labor in agriculture, he has been able to divert more and more time to "other work," including research and development. This has formed a feedback system whereby more research makes possible more efficient use of solar energy (opens the gate) and provides both for more food for more people and more people with more time for more research. The atomic revolution suggests a still greater acceleration of this process.

cut back their own use of materials with the accompanying lowering of standards of living. Either way the resources will eventually run out. These are not particularly attractive alternatives, and fortunately the nuclear energy field offers a better one. We can move from a unidirectional flow of resource use to a looped or circular use. We currently use materials and discard them with very little recycling. This leads both to increasing pollution and a rapid rate of exhaustion of basic resources. We are so accustomed to this that we even tend to measure our economic status or gross national product in part in terms of items consumed. Our need is to move to a closed system where materials do not move through and out of our economy but simply circulate within the economy. To do this we will need a new technology with a virtually unlimited supply of energy, and this is what the newest developments in the nuclear field promise.

The first atomic energy sources were dependent on nuclear fuels such as uranium 235, of which there is a limited supply. However, new systems such as the fusion process can draw on unlimited sources of supply such as tritium which are estimated as sufficient for more than 100 million years. It hardly seems necessary to plan farther ahead than that. Nuclear fusion is the basic energy process of the sun and the stars. It was first reproduced in the laboratory in 1932, first used in a thermonuclear test explosion in 1952, and in 1972 moved from theoretical discussion to technological, economic, and sociological discussion of the application of the developing process. No one at the moment can say for sure that the technological problems can be solved, but optimists believe that with proper funding they can be solved in 20 years.

One of the important concepts related to the energy source is the fusion torch. The ultrahigh-density plasmas of the fusion reactor could be used to vaporize and ionize any material. The atom could then be reassembled in any desired arrangement. While this is futuristic at the moment, that moment may be of only 20 to 50 years duration, depending upon the financial support for development. Should the fusion process produce unlimited power, this power combined with the technological ability to rearrange atoms would leave us with no resource problems. The fusion process will not use oxygen or hydrocarbons; hence it will not add carbon dioxide or other combustion products to the atmosphere. Nor will there be radioactive wastes except tritium. Tritium, however, is a fusion fuel and can be returned to the system for burning. The fusion system does not produce dangerous radioactive wastes and is incapable of a "runaway" accident. The biologi-

cal hazard of the fusion system is estimated at about one-millionth that of the fission reactor. Possibilities exist for fusion plants with no thermal pollution.

This has not been an attempt to explain the fusion process but only to call attention to the existence of such a process. Its importance is that it could almost totally change our view of energy availability and resource use and exhaustion. Most of the predictions of the impossibility of the rest of the world to rise to United States standards because there is not enough energy (coal, oil, hydropower, geothermal power) or basic resources (copper, wood, tin) are correct for the situation in 1975. They may not be applicable to the year 2000, however, for technology is advancing so rapidly that projections from the immediate past may have little validity in the immediate future. We are in a period of such rapid and fundamental change that in retrospect we may realize that this was a period comparable to the agricultural revolution. Indeed, should the fusion process allow us to disassociate materials and then reassemble the atoms to suit our needs, we may be on the verge of an atomic revolution more far-reaching than we have dreamed. Should it become possible to use a fusion torch to disassemble chemical compounds and reassemble them into other compounds, we could, among other things, simply do away with agriculture. It surely is an old-fashioned, inefficient, and highly risky business. If we could simply manufacture food from carbon, hydrogen, and other atoms, we would have made a quantum jump greater than that from the Paleolithic to the Neolithic. The world divides between those who believe we can and those who believe we can't. Jules Verne correctly foresaw atomic power, and the visionary's view of the world, characterized as a Buck Rogers' world, has come to pass within the past few decades. Space ships, atomic energy—these were parts of a dream world in the 1920s. Fifty years later they are a reality in a world of accelerating scientific discovery and technological application.

Human Will, Cultural Variety, and the Nature of Knowledge

The lesson that emerges is the simple one of human control over human actions. It is human will that is decisive, not the physical environment. The human will is channeled in its action by a fabric of social customs, attitudes, and laws that is tough, resistant to change, and persistent through time. Yet culture is delicate, like a fabric that can under some circumstances rip or become unraveled. Culture has enormous power to shape man's actions and color his view of the actual world. It has the ability to persist through great lengths of time with only slow change as new ideas are tried or not tried, adopted or rejected. Though man may adjust to his environment, we must always remember that he adjusts to it as he sees it and his perspective is culturally determined. This human image of the world is the ultimate reason why various cultural worlds exist and why their resemblances cut across environmental differences and similarities.

Perception of these complex relationships is easy once they are pointed out. The presentation here is an elementary, introductory one, but it should make clear how valuable it is to be able to deal with wide ranges of knowledge. How can we make judgments about man unless we know something about him? The need for such knowledge takes us to anthropology. It is also necessary to know what happened when, and this means that history must be studied. In the study of where things happened, we learn what places are like with regard to landform, climate, and resources, and this leads to the study of geography, geology, and other earth sciences. Ideally we should have deep knowledge of all subjects to understand man's relationship to the land he lives on. Few people today try for the breadth of knowledge that was once the mark of the educated man and is exemplified in such men as Alexander von Humboldt and Thomas Jefferson in past generations and the American geographer Carl Sauer, the anthropologist A. L. Kroeber, and the historian James Malin in our day. Geography is one of the few scholarly disciplines in which men still try to learn about many things and thus gain both greater depth of perception and guard against drifting into one-sided interpretations. This book should be taken as an introduction, an invitation to go on to studies of earth sciences, the humanities, and social sciences so that the problems and questions raised can be better understood and more easily answered. If there are unusual insights, if the material stimulates inquiry leading to reading and course work, this probably is because the combination of humanities, social sciences, and physical sciences yields greater depth of perception than any narrowly specialized approach. This is not meant to attack the specialists, for the value of the detailed studies that mark the present day is incalculable. We have tended, however, to lose sight of the wholeness of things, the unity of reality, the fact that the world is an enormously complex, interacting whole.

Man is the life form with the most complex mental development on this earth, and he has used this mental development to create culture—the pattern of learned behavior. This pattern of learned behavior is now so complex and powerful that it colors his views, reactions, and actions in and on the physical environment. Only in extreme and rare cases does the physical environment act directly on man without some cultural modulation. Reaction to the highest mountains on earth is the outstanding example. More commonly one finds the type of reaction reported by Douglas Lee for the Colorado Desert, where attitude modified physiological reaction to high heat.

It has often been said that geography and history cannot be separated without irreparable damage to understanding, and the truth of this should be abundantly clear by now. We must aim at a broad, philosophical, cultural history and geography. Mere accumulation of dates and places do not have meaning. Meaning springs from inquiry, from eternally asking why this instead of that.

Education comes in the same way. No one can educate anyone else, for education is the original do-it-yourself project. The aim of this book has been to provide a stimulus to education by laying out the biggest and most exciting questions and suggesting ways of answering them. By no means are all the answers here. We know remarkably little about physical-environmental impacts as modulated by cultural psychology. What we know about the details of rates of spread of ideas, the processes that control acceptance and rejection, and the role of cultural patterns and individuals within cultural patterns is hardly out of the infantile stage. Of the origins (or is it origin?) of agriculture, that great fundamental revolution in economy, and its spread over the world, we know little. The influences of the great early growth centers (the Middle East, Egypt, India, China, Peru, Mexico) are better known now than they were 40 years ago, but only beginnings have been made, and the oceanic doorways are barely ajar at this point in time.

The aim here has been to provide a start toward a liberal education in the Aristotelian sense—that a person becomes competent to judge correctly which parts of an exposition are satisfactory and which are not. Presumably anyone who has completed this book is now better able to deal with biological man, physical environment, and cultural man and should be better able to judge other expositions in these areas. If some of the explanations in this exposition are less than satisfactory, let that be an invitation to inquiry and the doorway to a liberal education.

Appendix

MAP PROJECTIONS

The fact that the earth is round and a map is flat presents insoluble problems, for we can no more represent the surface of the globe on a flat piece of paper without disruptions than we can take the skin off an orange and press it out without cutting, tearing, and stretching it.

Ideally a map of the world would show all parts of the earth on the same scale (true scale), in their real shapes (conformal), with true area (equal area), and with correct directional relationships (azimuthal). In reality we can never have all these qualities on one map. Some maps have none of these qualities; some have one or another. For small areas this is unimpor-

tant. A city can be mapped without worrying about such distortions, and the United States can be mapped with very small error in shape, scale, and area relationships, but no projection can show the whole earth without great distortion. Only a globe can do that, and globes are expensive to make in sufficiently large size to show details and are far too cumbersome even in small sizes to carry around. However, since they show true shapes, true areas, true directional relationships, and true scale, they are indispensable in the classroom, for reference to them can always correct false impressions gained from flat maps of the world.

If the world map presents the poles as lines, with Siberia and Alaska far apart, the globe shows these to be neighbors on the shores of the Arctic Ocean. If a map based on a Mercator projection shows Greenland as larger than South America, the globe corrects that impression and portrays it as only one-ninth the size of South America. Maps are useful, but they must be understood, or they can be misleading. The basis for understanding maps is consideration of the problem posed by representing a round earth on a flat map and solution of this problem by geometric arrangement of reference lines.

The arrangement of lines on which a map is drawn is called a projection. The lines are the frame of reference for locating things on the earth, just as the street system of our cities is the framework that allows us to describe a store as being at 5th Street and Broadway. Fixed points on our sphere are established by the rotation of the earth, for the axis of the earth establishes the North and South Poles. Lines from pole to pole are called meridians. We can then establish an east-west line by drawing a line around the earth equidistant from the poles. This is the equator. Lines drawn parallel to this line, called parallels, also run east and west and complete a network of intersecting lines quite comparable to the street layout in a grid-pattern town. By reference to the meridians and parallels we can locate any place on earth. New York City is at 41 degrees north latitude and 74 degrees west longitude.

The reference lines are divided into degrees, and locations are given in degrees with distance north and south called latitude and distance east and west called longitude. It is important to keep parallels and meridians in mind as lines, and latitudes and longitudes as distance if confusion is to be avoided. Latitude is counted from the equator north and south to the poles. This gives four quadrants of 90 degrees each. Longitude is counted from the meridian, called the prime meridian, that passes through Greenwich, England, and intersects the equator off the West African coast. From this zero point longitude is counted east and west to the 180th meridian. Distance on the earth is measured in degrees of angle north or south and east or west from these starting points.

To make a map of the earth we must locate a large number of places in relation to this grid. For instance, to map the coastline of North America we must first lay out an arrangement of meridians and parallels. Then we must locate a great number of coastal places in terms of this grid, putting a dot for instance at 41 degrees north and 74 degrees west to indicate the location of New York City. Once the dots are correctly placed, a line connecting all the dots will represent the outline of North America.

Our ability to construct maps developed late, for it required the accumulation of an enormous amount of place-location data, some of which were very diffi-

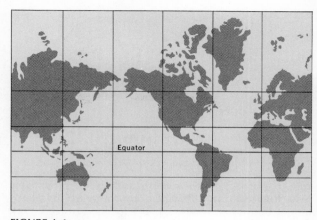

FIGURE A-1

Note the polar regions as they appear on cylindrical projections (Mercator's projection).

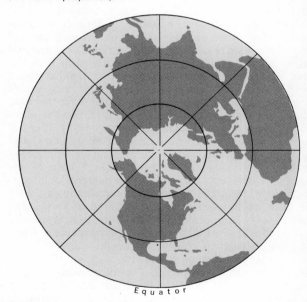

FIGURE A-2

The continents grouped around the pole as shown on a globe.

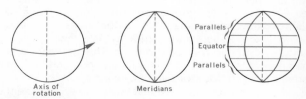

FIGURE A-3

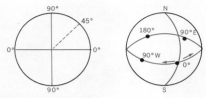

FIGURE A-4
Latitude is counted from zero at the equator to 90 degrees at the poles. Longitude is counted from the prime meridian east and west to 180 degrees.

cult to obtain. Distance north and south is relatively easy, for it can be obtained from observing the elevation of the sun and the stars. The location of New York City can be determined by measuring the height of the North Star. It is 41 degrees above the horizon, and that is the latitude. In more complicated fashion the sun and other stars can also be used to determine latitude. Distance east and west, however, is difficult to establish, and distance at sea was not determined beyond estimates of a day's sailing until accurate clocks allowed us to compare time at one part of the earth with the sun time where we were. Everyone is familiar today with the difference in time on east-west trips across any large area such as the United States or the time change in flights to Europe or Hawaii. Since the earth rotates on its axis through 360 degrees in 24 hours, it turns 15 degrees each hour, and morning, noon, and night move at this same rate. Today we adjust our watches after long east-west flights. Each hour of adjustment represents a change of 15 degrees longitude; thus our watches can be used to compute the east-west distance we traveled. By using this kind of knowledge and accurate long-running clocks in the past, we finally got good east-west distance measurement of the earth and could begin to construct accurate maps.

The problem of portraying a spherical surface on a plane surface remained. The solutions are largely mathematical, but some can be visualized. Two curved surfaces are easily converted into plane surfaces. A cone or a sphere can be cut and rolled out flat. Let us assume that we have a hollow glass sphere with meridians and parallels painted on it. By means of a light we project these lines onto a cone, or cylinder, made of light-sensitive paper placed over the globe. This will record the parallels and meridians. We then cut the cone, or cylinder, and roll it out flat. The meridians and parallels of the earth are now on the paper, and we can proceed to plot locations and make a map of the earth.

It is obvious at once that since the arrangement of parallels and meridians is different on the two projections, places plotted in relation to these lines will be

arranged differently. The results are not exactly like the real earth. Shapes change, or the areas may not be comparable from one part of the map to the other (South America and Greenland), or the scale may vary from place to place, or directions may have been radically changed. For instance, in the cylindrical projection everything around the North and South Poles has been stretched out into a line instead of grouped around a center. There is no way to portray a spherical surface on a plane surface without some distortion.

The map makers have invented many ways of portraying the earth. The parallels and meridians can be thought of as projected as shadows thrown against a plane, and this plane can be thought of as resting on either pole, or at any point on the equator, or at any other desired point on the earth's surface. The light source can be thought of as being at the center of the globe, the edge of the globe, at an infinite distance so that the light rays are traveling in parallel instead of diverging. In each case the qualities of the map change. Some of the projections against a plane have desirable qualities, but they can show only half the earth. This can be overcome by adding one-quarter of a diameter to each side of the hemisphere and extending the parallels and constructing meridians. The meridians can be constructed to give good shape in mid-latitudes (see sinusoidal) or in high latitudes (see Mollweide), and ingenious cartographers have even cut the top off one and married it to the middle of the other to get the best of each (homolosine).

On such world maps the shapes are best along the meridian running through the center of the map and are poor along the margins. This can be overcome by interrupting the map and having two or three or more central meridians. Too many interruptions destroy the continuity of the map, but a few of them greatly improve shapes, and an equal-area map of quite good shape can be so made.

Although the beginning student is easily lost in the complexities of map constructions, there are only a few points that must be remembered. No flat map of the whole earth or any major part of it can be true in all respects. Something must be distorted; some quality must be sacrificed. The important thing is to be aware of this and to view maps accordingly. If comparisons are to be made of areas, such as wheat growing, then select an equal-area map, and especially avoid a cylindrical, or Mercator, projection. If directions are important, select a map with good directional relationships, such as Mercators. Notice that there is good and bad usage for one projection. If interest is in the polar regions, use a projection that is centered on the poles. Always refer to a globe as the antidote to the distortions that are found in all world maps.

In defense of projections and cartographers, it should be added that they are seldom at fault, but unskilled users of maps often are. There is a projection for nearly every conceivable purpose, and when it is correctly used, it is a good projection. When used for some purpose other than that for which it was invented, it is not a bad projection, it is simply a misused projection.

Bonne's projection: A conic projection

This modification of a conic projection has one standard parallel, the one where the cones touches the earth. Areas are comparable all over the map. Distances are comparable along all the parallels and the standard meridian. But scale (distance) and shapes change toward the margins of the map.

Conic projection with two standard parallels

If the cone is thought of as entering the globe at one parallel and emerging at another parallel, scale is true along both parallels and nearly true between them. The parallels can be selected to give good map qualities for areas the size of the United States.

Sinusoidal projection

This is an equal-area projection, and scale is true along all parallels and the central meridian. Shape is poor in high latitudes. Notice Alaska and the lower tip of South America.

The classic example is Mercator's projection. This is an excellent projection for showing mariners winds and currents because directions on it are true. Use of this projection to portray any world areal distribution is a bad misapplication, for area and scale change rapidly from north to south. Some common and useful projections are illustrated here and their qualities described.

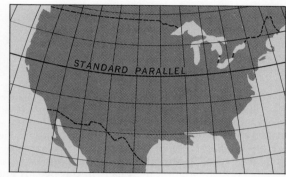

FIGURE A-5   Bonne's Projection

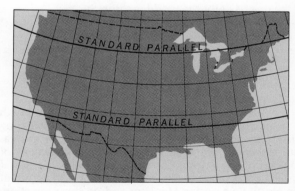

FIGURE A-6   Conic Projection with Two Standard Parallels

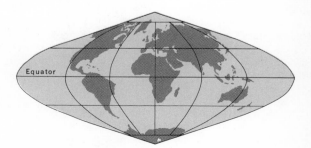

FIGURE A-7   Sinusoidal Projection

Mollweide's homolographic projection

Although scale changes both north and south and east and west on this map, areas are comparable, and shapes are relatively good in high latitudes. Compare North America on this map and on the preceding one.

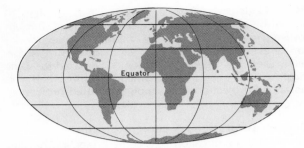

FIGURE A-8   Mollweide's Homolographic Projection

Interrupted homolographic projection

Since shape is best along the central meridian, the homalographic projection has been interrupted so that there is a central meridian for each continent. The principle is that of tearing the skin of the orange a little. Such "torn," or interrupted, projections can also be made with the central meridians in the oceans and the interruptions in the continents. This equal-area projection with its good shape for most of the world is the one used for the world maps in this text.

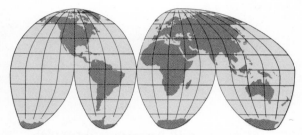

FIGURE A-9   Interrupted Homolographic Projection

Lambert's azimuthal equal-area projection, polar projection

The relationships of the lands about the North Pole are correctly shown here, but notice the increasing distortion away from the pole. However, since areas and directions are both true and the arrangement of land about the poles is correctly shown, this is a useful projection. Compare the arrangement of lands about the North Pole on the Mercator and on this projection. This projection is useful also for other than polar areas.

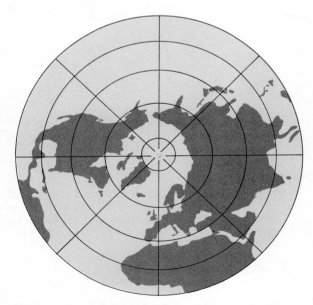

FIGURE A-10   Lambert's Azimuthal Equal-Area Projection

Mercator's projection

This modification of a cylindrical projection has good shapes over small areas and retains good directional relationships. This last feature makes it useful to mariners and desirable for plotting data such as winds and currents where directions are important. East-west and north-south distances enlarge greatly away from the equator, and areas in high latitudes are enormously exaggerated. This is a good map to use when direction is important, but its usefulness in the classroom is limited by its distortion of sizes and relationships of polar lands.

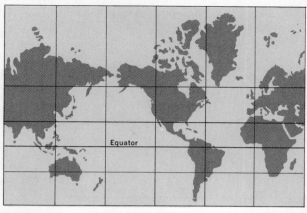

FIGURE A-11   Mercator's Projection

Land Form

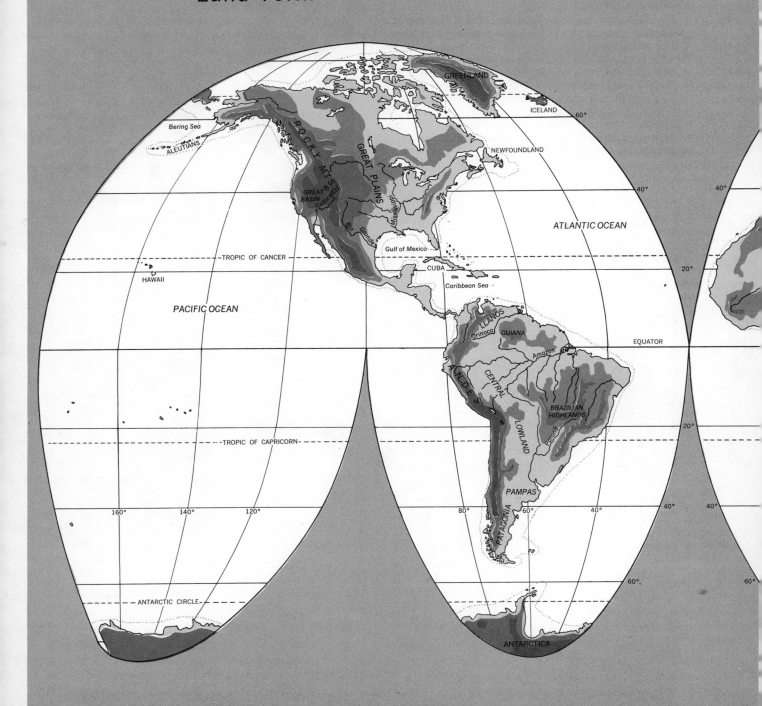

GREENLAND

ICELAND

60°

Bering Sea

ALEUTIANS

NEWFOUNDLAND

ROCKY MTS.

GREAT PLAINS

40°

40°

GREAT BASIN

Colorado

ATLANTIC OCEAN

Missouri

TROPIC OF CANCER

Rio Grande

Gulf of Mexico

20°

HAWAII

CUBA

CUBA

Caribbean Sea

PACIFIC OCEAN

LLANOS

Orinoco

GUIANA

EQUATOR

Amazon

ANDES

CENTRAL

BRAZILIAN
HIGHLANDS

20°

LOWLAND

Paraná

TROPIC OF CAPRICORN

PAMPAS

160°   140°   120°

PATAGONIA

80°   60°   40°   40°   40°

60°

60°

ANTARCTIC CIRCLE

ANTARCTICA

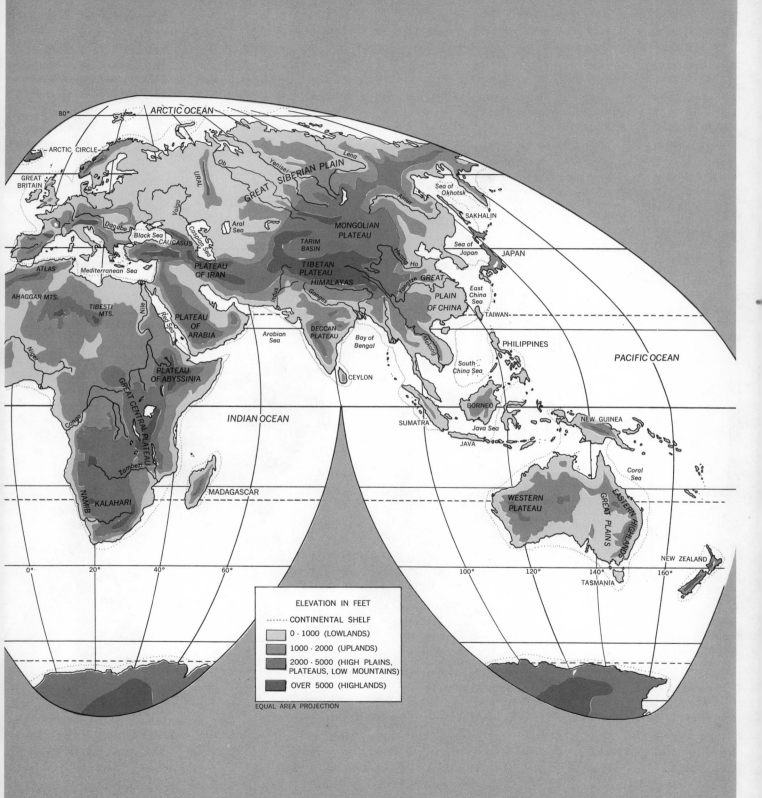

ARCTIC OCEAN

80°

ARCTIC CIRCLE

GREAT
BRITAIN

URAL

Ob

Yenisei

GREAT SIBERIAN PLAIN

Lena

Amur

Sea of
Okhotsk

SAKHALIN

Volga

Danube

Aral
Sea

MONGOLIAN
PLATEAU

Sea of
Japan

JAPAN

Black Sea

CAUCASUS

Caspian Sea

TARIM
BASIN

Hwang Ho

ATLAS

Mediterranean Sea

PLATEAU
OF IRAN

TIBETAN
PLATEAU

Yangtze

GREAT

East
China
Sea

AHAGGAR MTS.

HIMALAYAS

PLAIN

TAIWAN

TIBESTI
MTS.

Nile

Red Sea

PLATEAU
OF
ARABIA

Indus

Ganges

Arabian
Sea

DECCAN
PLATEAU

Bay of
Bengal

Mekong

OF CHINA

PHILIPPINES

PACIFIC OCEAN

Niger

CEYLON

South
China Sea

PLATEAU
OF ABYSSINIA

GREAT CENTRAL PLATEAU

Congo

INDIAN OCEAN

SUMATRA

BORNEO

Java Sea

JAVA

NEW GUINEA

Coral
Sea

Zambezi

MADAGASCAR

NAMIB

KALAHARI

WESTERN
PLATEAU

GREAT PLAINS

EASTERN HIGHLANDS

NEW ZEALAND

0°   20°   40°   60°

100°   120°   140°   160°

TASMANIA

ELEVATION IN FEET

CONTINENTAL SHELF

0 - 1000 (LOWLANDS)

1000 - 2000 (UPLANDS)

2000 - 5000 (HIGH PLAINS,
PLATEAUS, LOW MOUNTAINS)

OVER 5000 (HIGHLANDS)

EQUAL AREA PROJECTION

Types of Climate

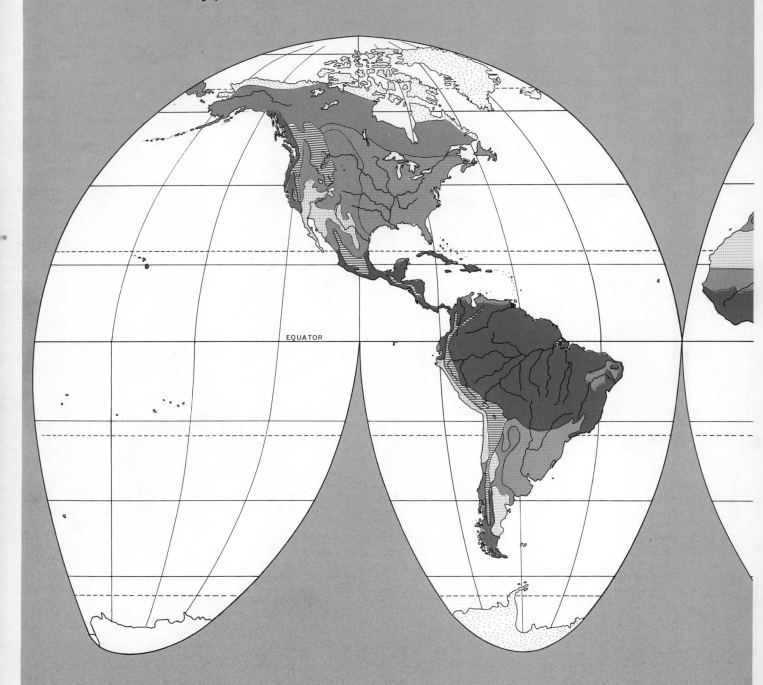

EQUATOR

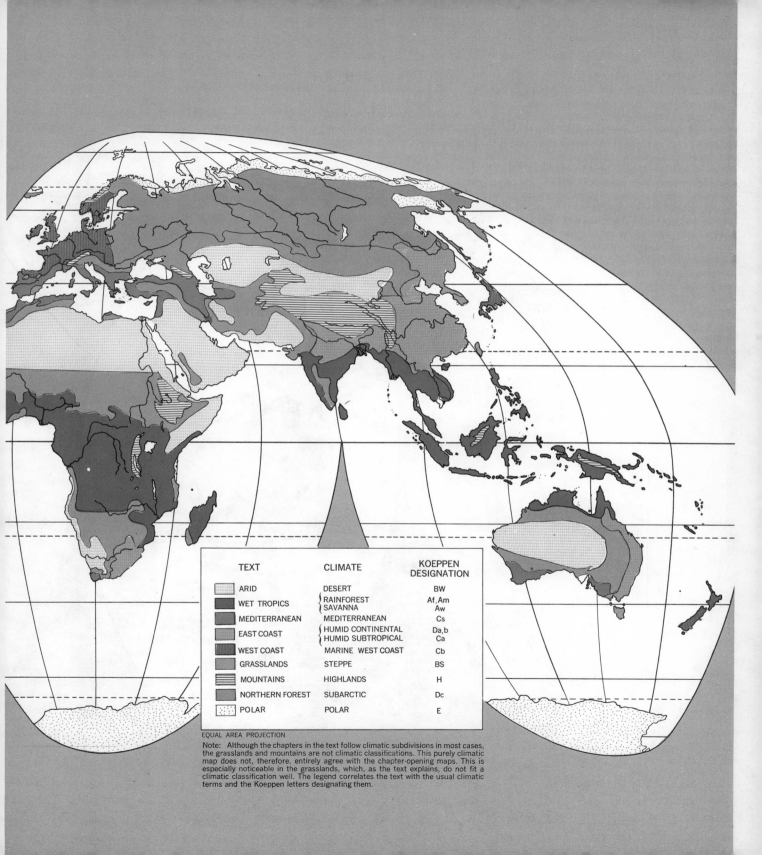

TEXT	CLIMATE	KOEPPEN DESIGNATION
ARID	DESERT	BW
WET TROPICS	RAINFOREST	Af, Am
	SAVANNA	Aw
MEDITERRANEAN	MEDITERRANEAN	Cs
EAST COAST	HUMID CONTINENTAL	Da,b
	HUMID SUBTROPICAL	Ca
WEST COAST	MARINE WEST COAST	Cb
GRASSLANDS	STEPPE	BS
MOUNTAINS	HIGHLANDS	H
NORTHERN FOREST	SUBARCTIC	Dc
POLAR	POLAR	E

EQUAL AREA PROJECTION

Note: Although the chapters in the text follow climatic subdivisions in most cases, the grasslands and mountains are not climatic classifications. This purely climatic map does not, therefore, entirely agree with the chapter-opening maps. This is especially noticeable in the grasslands, which, as the text explains, do not fit a climatic classification well. The legend correlates the text with the usual climatic terms and the Koeppen letters designating them.

Population

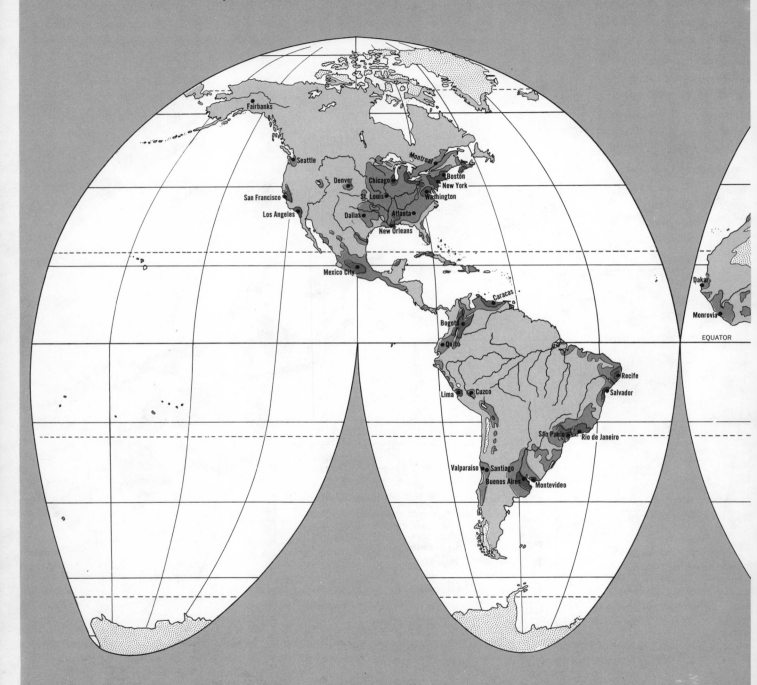

Fairbanks

Seattle
Montreal
Denver
Chicago
Boston
San Francisco
St. Louis
New York
Washington
Los Angeles
Dallas
Atlanta
New Orleans

Mexico City

Caracas

Bogotá

Quito

Lima
Cuzco

Recife

Salvador

São Paulo
Rio de Janeiro

Valparaiso
Santiago

Buenos Aires
Montevideo

Dakar

Monrovia

EQUATOR

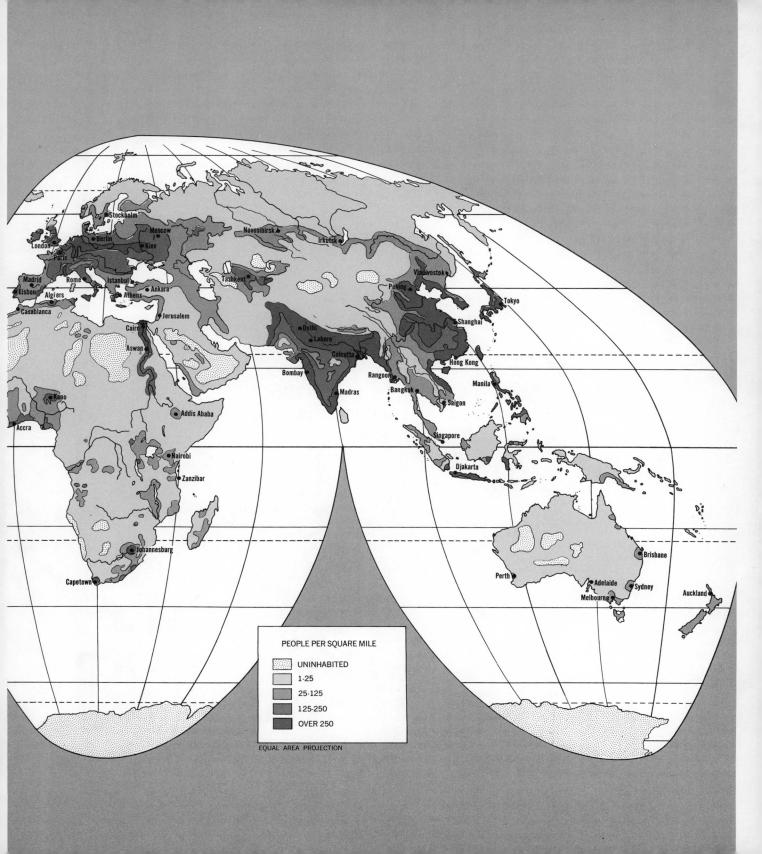

PEOPLE PER SQUARE MILE

UNINHABITED
1-25
25-125
125-250
OVER 250

EQUAL AREA PROJECTION

Index